METALLOTHERAPEUTIC DRUGS AND METAL-BASED DIAGNOSTIC AGENTS

METALLOTHERAPEUTIC DRUGS AND METAL-BASED DIAGNOSTIC AGENTS

THE USE OF METALS IN MEDICINE

Marcel Gielen
Edward R.T. Tiekink

John Wiley & Sons, Ltd

Telephone (+44) 1243 779777

Email (for orders and customer service enquiries): cs-books@wiley.co.uk
Visit our Home Page on www.wiley.com

Other Wiley Editorial Offices

John Wiley & Sons Inc., 111 River Street, Hoboken, NJ 07030, USA

Jossey-Bass, 989 Market Street, San Francisco, CA 94103-1741, USA

Wiley-VCH Verlag GmbH, Boschstr. 12, D-69469 Weinheim, Germany

John Wiley & Sons Australia Ltd, 33 Park Road, Milton, Queensland 4064, Australia

John Wiley & Sons (Asia) Pte Ltd, 2 Clementi Loop #02-01, Jin Xing Distripark, Singapore 129809

John Wiley & Sons Canada Ltd, 22 Worcester Road, Etobicoke, Ontario, Canada M9W 1L1

Wiley also publishes its books in a variety of electronic formats. Some content that appears in print may not be available in electronic books.

Library of Congress Cataloging in Publication Data

Tiekink, Edward R.
Metallotherapeutic drugs and metal-based diagnostic agents – the use of metals in medicine / Edward R. Tiekink & Marcel Gielen.
p. cm.
Includes index.
ISBN 0-470-86403-6
1. Metals—Therapetuic use. 2. Metals—Diagnostic use. I. Gielen, M. (Marcel), 1938– II. Title.
RM666.M513T546 2005
615′.23—dc22

2004022079

British Library Cataloguing in Publication Data

A catalogue record for this book is available from the British Library

ISBN 0-470-86403-6 (HB)

Typeset in 10/12pt Times by Integra Software Services Pvt. Ltd, Pondicherry, India
Printed and bound in Great Britain by TJ International, Padstow, Cornwall
This book is printed on acid-free paper responsibly manufactured from sustainable forestry in which at least two trees are planted for each one used for paper production.

Contents

Contributors

Silvio Aime
Universita degli Studi di Torino
Dipartimento di Chimica IFM e Centro per il Molecular Imaging
7 Via Pietro Giuria
I-10125 Torino
Italy

Mario Barbagallo
Department of Internal Medicine and Geriatrics
University of Palermo
via F. Scaduto 6/c
I-90144 Palermo
Italy

Alessandro Barge
Dipartimento di Chimica IFM e Centro per il Molecular Imaging
7 Via Pietro Giuria
I-10125 Torino
Italy

Lawrence R. Bernstein
285 Willow Road
Menlo Park
CA94025
USA

Tanushree Bose
University of Isfahan
Department of Biology
Isfahan
Iran

Viktor Brabec
Institute of Biophysics
Academy of Sciences of the Czech Republic
CZ-61265 Brno
Czech Republic

Neil Burford
Department of Chemistry
Dalhousie University
Halifax
Nova Scotia
B3H 4J3
Canada

Michael Carland
Chemistry Department
University of Melbourne
Parkville
Victoria 3052
Australia

Hui Chao
Department of Chemistry
The Key Laboratory of Gene Engineering of Ministry of Education
State Key Laboratory of Optoelectronic Materials and Technologies
Sun Yat-Sen University
Guangzhou 510275
P.R. China

Simonetta Geninatti Crich
Dipartimento di Chimica IFM e Centro per il Molecular Imaging
7 Via Pietro Giuria
I-10125 Torino
Italy

Walter Dastrù
Dipartimento di Chimica IFM e Centro per il Molecular Imaging
7 Via Pietro Giuria
I-10125 Torino
Italy

Pietro Delva
Department of Biomedical and Surgical Sciences
University of Verona
Policlinico G.B. Rossi
I-37134 Verona
Italy

Jonathan R. Dilworth
Inorganic Chemistry Laboratory
University of Oxford
South Parks Road
Oxford
OX1 3QR
UK

Ligia J. Dominguez
Department of Internal Medicine and Geriatrics
University of Palermo
via F. Scaduto 6/c
I-90144 Palermo
Italy

Paul S. Donnelly
Inorganic Chemistry Laboratory
University of Oxford
South Parks Road
Oxford
OX1 3QR
UK

Erich Dubler
Institute of Inorganic Chemistry
University of Zurich
Winterthurerstrasse 190
CH-8057 Zurich
Switzerland

Melanie D. Eelman
Department of Chemistry
Dalhousie University
Halifax
Nova Scotia B3H 4J3
Canada

Tahli Fenner
Chemistry Department
University of Melbourne
Parkville
Victoria 3052
Australia

Jeanne H. Freeland-Graves
College of Natural Sciences
The University of Texas at Austin
Human Ecology
1 University Station Stop A2700
Austin
TX 78712
USA

Achilleas Garoufis
Section of Inorganic and Analytical Chemistry
Department of Chemistry
University of Ioannina
45110 Ioannina
Greece

Eliana Gianolio
Dipartimento di Chimica IFM e Centro per il Molecular Imaging
7 Via Pietro Giuria
I-10125 Torino
Italy

Marcel Gielen
HNMR Department
Vrije Universiteit Brussel
2 Pleinlaan
B-1050 Brussels
Belgium

Mei Lin Go
Department of Pharmacy
National University of Singapore
Singapore 117543

Francisco González-Vílchez
Department of Inorganic Chemistry
Chemistry Faculty

University of Sevilla
41071 Sevilla
Spain

Sotiris K. Hadjikakou
Section of Inorganic and Analytical Chemistry
Department of Chemistry
University of Ioannina
45110 Ioannina
Greece

Nick Hadjiliadis
University of Ioannina
Department of Chemistry
Laboratory of Inorganic and General Chemistry
Ioannina 45110
Greece

Adrian J. Harwood
Wolfson Institute for Biomedical Research
Cruciform Building
University College London
Gower St.
London
WC1 E6BT
UK

Jane V. Higdon
Linus Pauling Institute
Oregon State University
Corvallis OR 97331
USA

Emily Ho
Linus Pauling Institute
Department of Nutrition and Food Management
Oregon State University
Corvallis OR 97331
USA

Paul C. Ho
Department of Pharmacy
National University of Singapore
Singapore 117543

Soo Yei Ho
Department of Chemistry
National University of Singapore
Singapore 117543

Narayan S. Hosmane
Department of Chemistry and Biochemistry
Northern Illinois University
DeKalb
Illinois 60115-2862
USA

Luba Ignatovich
Latvian Institute of Organic Synthesis
21 Aizkraukles Str.
Riga
LV-1006 Latvia

Tamás Jakusch
Bioinorganic Research Group of the Hungarian Academy of Sciences
Department of Chemistry
University of Szeged
Hungary

Liang-Nian Ji
Department of Chemistry
Zhongshan (Sun Yat-Sen) University
Guangzhou 510275
P.R. China

Lan Jin
Department of Chemistry and Open Laboratory of Chemical Biology
The University of Hong Kong
Hong Kong SAR

Abbass Karbassian
College of Natural Sciences
The University of Texas at Austin
Human Ecology
1 University Station Stop A2700
Austin
TX 78712
USA

Jana Kasparkova
Institute of Biophysics
Academy of Sciences of the Czech Republic
CZ-61265 Brno
Czech Republic

Tamás Kiss
Department of Inorganic and Analytical Chemistry
Joszef Attila University
PO Box 440
H-6701 Szeged
Hungary

Olivier Lentzen
Université Libre de Bruxelles
Organic Chemistry and Photochemistry
CP 160/08
50 Avenue F.D. Roosevelt
B-1050 Brussels
Belgium

Edmunds Lukevics
Institute of Organic Synthesis
Latvian Academy of Sciences
21 Aizkraukles
LV-1006 Riga
Latvia

John A. Maguire
Department of Chemistry
Southern Methodist University
Dallas
Texas 75275-0314
USA

Andrée Kirsch-De Mesmaeker
Universite Libre de Bruxelles
Faculté des Sciences
Chimie Organique Physique
CP 160/08
50 Avenue F.D. Roosevelt
B-1050 Bruxelles
Belgium

Cécile Moucheron
Université Libre de Bruxelles
Organic Chemistry and Photochemistry
CP 160/08
50 Avenue F.D. Roosevelt
B-1050 Brussels
Belgium

Florian P. Pruchnik
Faculty of Chemistry University of Wrocław
Joliot-Curie 14
50-383 Wrocław
Poland

Geeta Rana
Department of Chemistry and Biochemistry
Northern Illinois University
DeKalb
Illinois 60115-2862
USA

Thanos Salifoglou
Department of Chemistry
University of Crete
Leoforos Knossos 300
Heraklion 71409
Greece

Oyebola O. Sogbein
Department of Chemistry and The McMaster Institute of Applied Radiation Sciences
McMaster University
Hamilton
Ontario
Canada
L8S 4M1

Hongzhe Sun
Department of Chemistry
University of Hong Kong
Pokfulam Road
Hong Kong SAR

Edward R.T. Tiekink
Department of Chemistry
3 Science Drive
National University of Singapore
Singapore 117543

Fulvio Uggeri
Bracco Imaging S.p.A.
50 Via Folli
I-20135 Milano
Italy

John F. Valliant
Department of Chemistry
McMaster University
Hamilton L8S 4M1
Ontario
Canada

Rosario Vilaplana
Inorganic Chemistry Department
Bioinorganic Chemistry Laboratory
Chemistry Faculty
University of Seville
41071 Seville
Spain

Kamesh Vyakaranam
Department of Chemistry and Biochemistry
Northern Illinois University
DeKalb
Illinois 60115-2862
USA

Robin S.B. Williams
Wolfson Institute for Biomedical Research
Cruciform Building
University College London
Gower St.
London
WC1 E6BT
UK

Xiang Wu
Department of Pharmacy
National University of Singapore
18 Science Drive 4
Singapore 117543

Siucheong Yan
Department of Chemistry and Open Laboratory of
Chemical Biology
The University of Hong Kong
Hong Kong SAR

Preface

It takes but just a cursory glance at the number of chapters and their titles in this book, *Metallotherapeutic Drugs and Metal-Based Diagnostic Agents: The Use of Metals in Medicine*, to gain an appreciation of the astonishing number and variety of roles that metals play in contemporary medicine. From the outset, the editors acknowledge that they have adopted a very liberal interpretation of the term 'metal'. Thus, elements that would not typically be classified as metals are nevertheless integrated in this book because they are not normally regarded as constituents of 'organic' drugs. With this justification, the reader will find chapters on the medicinal uses of inorganics, that is, of elements such as boron, lithium, selenium, to name a few, as well as metal-containing species. In the same way, the book is not restricted to coordination complexes as organometallic species are also well represented, for example those of germanium, iron and titanium. In the same way, 'minerals', such as arsenic oxide, are included. In keeping with the notion that healthy mammals rely on (bio-essential) metals for the normal functioning of approximately a third of their proteins and enzymes, a large number of drugs are metal-based and considerable effort is being devoted to developing both second- and third-generation drugs as well as generating novel metal-based drugs. While there is no doubt that there is an emphasis on 'metallotherapeutics' throughout the book, the use of metals in medicine is not restricted to metal-based drugs. Contemporary medicine relies heavily on metal-based diagnostic agents. Magnetic resonance imaging (MRI) exploits the unique property of cationic forms of some metals, that is, their high paramagnetism. For example, gadolinium(III) has seven unpaired electrons making it a most useful contrast agent in this context; manganese(II) is also utilized. Similarly, non-invasive radiopharmaceuticals, for example complexes containing various isotopes of cobalt, gallium, rhenium, rhodium, ruthenium and technetium-99m, are increasingly being used in medicine, as they can be exploited to deliver sterilizing radiation to targeted sites. Boron in boron nuclear capture therapy (BNCT) is still at the forefront of the treatment of malignant brain tumours. The above notwithstanding, the majority of chapters in this book are devoted to describing the use of 'inorganics' as drugs and the development of new metal-based drugs.

The use of metals in medicine traces back to antiquity with various elements, such as arsenic, gold and iron, being used to treat different ailments. In more modern times, perhaps the first metal-based drug to find wide use was an arsenical, Salvarsan®, developed by Paul Ehrlich, used in the treatment of syphilis; the later availability of penicillin made this drug redundant. The use of metals such as arsenic in medicine raises a very important issue, that is, the issue of toxicity. Most people would associate an element such as arsenic with a poison and yet, arsenic oxide has been developed as a drug for the treatment of promyelocytic leukaemia. Clearly there is a subtle balance between therapeutic benefit on the one hand and deleterious side effects on the other. As mentioned earlier, there are a large number of bio-essential trace metals required for healthy mammals, and deviations from normal concentrations of these are causes of disease. Chelation therapy can be employed, for example, in the treatment of Wilson's disease (copper poisoning) as can treatment with zinc supplements. Indeed, over the counter (OTC) dietary supplements of calcium, manganese and selenium are commonly available to redress imbalances. Other OTCs include antacids, containing aluminium, and zinc-containing lozenges for treating the common cold. In terms of contemporary therapeutic medicine, a large number of metal-based drugs may be prescribed to patients.

Diseases such as anaemia (iron), asthma (gold and magnesium), bipolar disorder (lithium), diabetes (vanadium), rheumatoid arthritis (gold), stroke (magnesium), tropical diseases (antimony and rhodium) and ulcers (bismuth) can all be treated by metal-based drugs. Arguably, the most prominent drug containing a metal centre is cisplatin, the most widely used anti-cancer drug. The success of cisplatin in chemotherapy has motivated a large number of studies into the potential of other platinum complexes as well as other metal (like palladium) complexes as anti-tumour drugs. Indeed, in most chapters there will be at least a short section devoted to describing research into the anti-cancer potential of compounds containing the element under focus. The role of the metal in the agents/drugs also attracts discussion in many chapters. Thus, the 'metal' may simply be a substitute for an atom in an 'organic' drug, for example boron, selenium and silicon, giving rise to a more potent drug. Alternatively, the metal may serve as a carrier of a pharmaceutically active molecule, that is, the metal is exploited as a platform to transport the organic drug by altering the normal metabolic pathway of the drug, introducing different reaction kinetics, and so on. Finally, the metal may play a very real role in terms of being the effective therapeutic agent. An example of this is the platinum centre in cisplatin which is known to coordinate DNA, thereby disrupting DNA replication.

From the foregoing and, more importantly, from the contents of the various chapters, the reader will gain an appreciation of the very real role metal-based drugs play in modern medicine and of the considerable effort being devoted to the development of novel complexes with greater efficacy and reduced toxic side effects (an inevitable consequence of all chemotherapy). Emphasizing this very point, at the time of writing this Preface (June, 2004), a tin porphyrin complex, tin ethyl etiopurpurin, exploiting the principles of Photodynamic Therapy, was given approval by the United States FDA for use in the treatment of age-related macular degeneration (AMD), a common cause of vision impairment in the aged.

Marcel Gielen
Edward R.T. Tiekink

1 $_3$Li Lithium Metallotherapeutics

Robin S.B. Williams and Adrian J. Harwood

1.1 Introduction

Lithium, from the Greek word 'lithos' or stone, is the smallest and lightest solid element – a monovalent cation appearing third on the periodic table. It is well known for its use in batteries, metal alloys, glass manufacture and for its clinical use in the treatment of manic depression or bipolar disorder (BD), a chronic disorder which affects between 1 and 2% of the population and is characterized by episodic periods of elevated and/or depressed mood.[1] This disorder severely reduces patients' quality of life and dramatically increases the likelihood of these patients committing suicide.[2] Lithium's clinical role has given rise to extensive amounts of scientific research; however, despite this, there is still considerable debate regarding its medicinal targets. It is also unclear how such a structurally simple ion can have such a profound effect on specific and complex medical disorders such as BD.

For many years lithium's structural simplicity may have discouraged the development and patenting of alternative anti-manic agents – how can one modify and patent a monovalent ion? The fact that it has largely been ignored by pharmaceutical companies can give the impression that it is an 'old drug', with little new therapeutic potential. However, in truth, it is still highly successful in the treatment for BD. In addition, it is now increasingly being used as a tool by developmental and cell biologists, and also shows great promise in the treatment of other medical disorders.

Metallotherapeutic Drugs and Metal-Based Diagnostic Agents: The Use of Metals in Medicine
Edited by Gielen and Tiekink

In this chapter we will introduce lithium and its medicinal history, describe its biological actions, its known targets in biological systems and its therapeutic applications.

1.2 The Inorganic Chemistry of Lithium

Lithium is a member of group 1 (1A) alkali metals, has an atomic number of 3, an atomic weight of 6.941 and contains a single valence electron. In its pure form it is a soft silvery white metal, turning grey on exposure to air. Because of its reactivity in the metallic state, it is only found in nature as an ore or at low concentrations as a soluble salt, e.g. it is present in river waters at ~0.002 mg/l. But it has also been found at higher levels in naturally occurring springs which have previously been associated with health improvements (e.g. Lithia Springs Mineral Water, Georgia, USA).

The most important commercial ore containing lithium is spodumene, $LiAl(SiO_3)_2$. The extraction of lithium from this ore occurs by converting it to the soluble chloride salt. This process starts by heating spodumene to around 1100 °C. This is then mixed with sulphuric acid and water to form lithium sulphate (Li_2SO_4). This is washed with sodium carbonate (Na_2CO_3) to form a precipitate of the lithium carbonate (Li_2CO_3) (Equation 1.1). Reaction of lithium carbonate with hydrochloric acid gives rise to lithium chloride (Equation 1.2).

$$Li_2SO_4 + Na_2CO_3 \longrightarrow Na_2SO_4 + Li_2CO_3 \tag{1.1}$$

$$Li_2CO_3 + 2HCl \longrightarrow 2LiCl + CO_2 + H_2O \tag{1.2}$$

As a salt, lithium has the smallest ionic radius, the largest field density at its surface and is the least reactive of the alkali metals. The very small diameter of the lithium ion in relation to water means that molecular packing gives it an anomalously large hydration shell in proportion to other group 1 elements. Thus in experiments examining the effect of lithium salts on osmolarity, the corresponding sodium salt is often used as a control, as this provides the closest approximation to lithium's hydration shell size.

Although the hydration shell of lithium is similar in size to sodium, its ionic radius is much closer to that of the magnesium ion (Table 1.1). This similarity allows lithium ions to compete with magnesium ions for binding sites in proteins.[3–5] As magnesium ions interact with hundreds of proteins, this suggests that lithium may inhibit numerous enzymes. However, the relative specificity of the lithium effect suggests that only proteins with very low affinities for magnesium are targeted by therapeutic concentrations of the drug. The total magnesium concentration of the cell is between 10 and 15 mM, but the free concentration of magnesium is between 0.5 and 1.0 mM. This is the same concentration of lithium found in plasma from patients undergoing lithium therapy.[6]

Table 1.1 The chemical symbol, atomic charge and atomic number of lithium and its neighbouring elements on the periodic and their ionic radii table in different coordination states. Units are given in pico moles

	Lithium	Beryllium	Boron	Sodium	Magnesium
Symbol	Li	Be	B	Na	Mg
Charge	I	II	III	I	II
Atomic no.	3	4	5	11	12
4 coordinate	73	41	25	113	71
6 coordinate	90	59	41	116	86

1.3 Biology of Lithium

1.3.1 The history of lithium therapeutics

Garrod described the medical use of lithium in 1859 for the treatment of rheumatic conditions and gout – and in particular 'brain gout'. The hypothesis behind the use of lithium at this point was based on its ability to dissolve nitrogen-containing compounds, called urates or uric acid, which were thought to build up in the body giving rise to many illnesses. This is certainly true for gout, but the idea was extended to include many other human physical disorders. By the 1880s, Carl Lange and others[7] were using lithium for the treatment of BD, and lithium carbonate and citrate were described in the British Pharmacopoeia of 1885.

In addition to the treatment of patients, the urea hypothesis spurred the development of the lithium tonic and its increasing use in common food stuffs. The most notable today is 7 UP, originally introduced in 1929 as 'Bib-Label Lithiated Lemon-Lime Soda'. The lithium was removed in 1950. The decline of the urea hypothesis and the lack of a credible therapeutic mechanism for lithium lead to its disuse as a therapeutic agent until it was re-discovered in 1949 by John Cade.[8] He was continuing the search for a toxin which may cause BD, and was working on the premise that these toxins may be excreted in the urine of BD patients. He thus injected concentrated urine samples from these patients and control subjects into the abdomen of Guinea pigs, and found that bipolar patient urine had greater toxicity than that of the control subject. To identify what this toxic compound was, he then tested different forms of nitrogen-containing compounds found in urine, and identified urea as being the most toxic. His line of reasoning continued, that dissolving this urea may alleviate the toxicity, so he tried co-injecting urea with alkali metals, only to find that lithium causes sedation and stupor in the animals. This suggested that lithium may have sedative or mood-control properties. Following testing of lithium toxicity on himself, he then administered it to BD, depressed and schizophrenic patients, and found huge improvements in the BD patient group. As no effective drug treatments existed at the time for any major psychiatric disease, this was an extremely exciting event and effectively places lithium as the first modern psychopharmacological agent.

Cade's work, however, rapidly became overshadowed by two events. First, on the same year as Cade's discovery, the Food and Drug Administration (FDA) banned lithium following the death of four US patients due to lithium overdose. It was thought at the time that lithium chloride could be used to reduce sodium chloride intake, and would therefore be useful in the treatment of hypertension. Doses of up to 14 g per day were used in these cases, which were extremely high compared with modern lithium use. The second event to overshadow Cade's re-discovery of lithium was the discovery of chlorpromazine, the first anti-psychotic. This molecule heralded a new era in the development of psychotropic drugs for the treatment of schizophrenia and major depression.

Fortunately, throughout the 1950s Schou[9] and others continued to investigate the use of lithium and demonstrated that it could be used safely in the treatment of BD. It was re-approved by the FDA in 1972 and is now prescribed in over 50% of BD cases,[10] although it is very common that it is used in combination with other drugs. It is marketed under a variety of names including Lithicarb®, Quilonum SR®, Lithotabs® and Librium®. It has also clearly been shown to reduce the risk of suicide in mood disorder patients,[11] and it has considerable socioeconomic impact, e.g. it has been estimated that it has saved around $8 billion in the US alone in 1991.[12]

1.3.2 Lithium and the body

Lithium therapy is always taken orally, usually as lithium carbonate as this causes least irritation to the stomach, to a total dose of up to 30 mmol (<2 g) per day. Treatment is monitored by blood concentration 12 h after administration. The therapeutic index for lithium, which represents the concentration window from efficacy to toxicity, is very narrow, occurring between 0.4 and 0.8 mM. Interestingly, lithium treatment alters magnesium balance, and concentrations of magnesium ions are altered in both the blood and the urine following lithium treatment. This is consistent with displacement of magnesium by lithium from cellular binding sites.

In general, lithium distributes uniformly among the body tissues, and most cells of the body experience an external lithium concentration of less than 2 mM, even at the highest clinical doses. However, local accumulation occurs in some tissues, but not those of the brain. In particular, lithium accumulates to a moderate degree in endocrine glands and to a high concentration at the tip of the papilla in the kidney. At a blood concentration of 1 mM, the renal papillary can reach 60–65 mM.

Lithium is not soluble in lipids and hence will not cross the plasma membrane. There are a number of routes for lithium entry into the cell – the most prevalent being lithium–sodium counter-transport,[13] anion exchange[14] and through other unrelated transport molecules. It is also not bound to proteins in plasma or tissue, unlike other BD treatments.

In addition to its therapeutic uses, embryologists have long known lithium to be a teratogen – a drug that affects patterning and proportion of cell types in the developing embryo. For example, in sea urchins, lithium causes vegetalization of the animal blastomeres,[15] and in vertebrates such as *Xenopus*[16] and zebrafish[17] it causes expansion of the dorsal mesoderm and duplication of the dorsal axis. These effects are not restricted to animals, as lithium treatment of *Dictyostelium*, a non-metazoan eukaryote, leads to mis-specification of spore and basal disc cell fate.[18] It was therefore assumed that lithium treatment will be teratogenic in humans, as seen for the other mood stabilizers valproic acid (VPA) and carbamazepine (CBZ).[19] However, despite the assumed clinical risk, there are surprisingly few reports of teratogenic effects of lithium on human development, although some reports have suggested an increased risk of congenital heart defects.[19] The difference in these teratogenic rates may reflect dosage, as embryological experiments carried out in the laboratory generally use higher than therapeutic lithium concentrations.

1.4 Targets of Lithium

It is well accepted that there are two general targets of lithium in the cell – the enzymes glycogen synthase kinase-3 (GSK-3) and the phosphomonoesterase family (PMEs).[20] No other direct targets of lithium have been found over the last 30 years, despite considerable amounts of research, although many other targets have been suggested. These include inhibiting adenyl cyclase[21] and G proteins,[22] and interacting with cell surface receptors to affect guanidine trisphosphate (GTP) binding.[23] These lithium targets are more likely to be affected indirectly through altered GSK-3 or PMEs' activities, or were found to be lithium targets at artificially high lithium concentrations that do not occur in the body during lithium therapy.

1.4.1 Glycogen synthase kinase-3

Lithium inhibits GSK-3 kinases, a family of highly conserved serine/threonine protein kinases that have been identified in all eukaryotes examined to date. In vertebrates, GSK-3 is found as two isoforms: GSK-3α and GSK-3β. Lithium has been demonstrated to inhibit both these isoforms *in vitro* and *in vivo* through competition for magnesium ion binding (see section 1.2).[24,25] GSK-3 phosphorylates its amino acid target at an unusual substrate recognition site which contains phosphorylated serine or threonine four residues towards the

Figure 1.1 Single-letter amino acid sequence for the GSK-3 phosphorylation site on glycogen synthase, defined by single-letter abbreviations. Prior phosphorylation the residue four amino acids towards the 'C'-terminal of the target serine/threonine (S/T) allow the enzyme to recognize the site. The subsequent phosphorylated amino acid may then function to enable further phosphorylation

'C' terminus of the phospho-donor site (Figure 1.1). This acts as a docking site to anchor GSK-3 on its substrate. The resulting phosphorylation can then be used to prime further neighbouring residues. Loss of this priming phosphorylation decreases affinity with the substrate by 100–1000-fold.[26] This also means that in most cases GSK-3 works in concert with other kinases.

Glycogen synthase kinase-3 was first identified as an enzyme that phosphorylates and inactivates glycogen synthase[27] in the insulin response pathway (Figure 1.2). In this pathway, insulin turns on GSK-3 activity to inhibit glycogen synthesis. Consequently, in diabetic rat hepatocytes, lithium restores glycogen synthesis in the presence of an overactive insulin pathway. However, GSK-3 has many other targets within the cell including a number of important transcription factors and cytoskeletal proteins.

A significant GSK-3 substrate is the protein β-catenin, which regulates gene expression by binding to the LEF-1/TCF-3 family of transcription factors.[28] GSK-3 phosphorylation of β-catenin leads to its degradation and hence prevents its accumulation in the nucleus where it binds T-cell factor transcription factors (TCFs) (Figure 1.2). This phosphorylation event is regulated by binding to axin, which brings GSK-3 and a priming kinase, casein kinase 1 (CK 1), in contact with β-catenin.[29] Formation of the axin complex is regulated by the binding of the extra-cellular glycoprotein Wnt to its cell surface receptor Frizzled. In this way Wnt signalling can regulate gene expression through control of β-catenin degradation. The number of cellular events found to be regulated by Wnt signalling is increasing and includes the control of cell fate during development and cell proliferation. Loss of the GSK-3 phosphorylation sites from β-catenin, deletions of adenomatous polyposis coli protein (APC), another component of the axin–β-catenin protein complex, and mutations of axin itself are all associated with various forms of human cancer.[30] β-Catenin is also a component of adherens junctions which anchor the actin cytoskeleton to points of cell–cell contact via cadherin proteins on the cell surface. GSK-3-mediated changes in β-catenin stability do not affect adherens

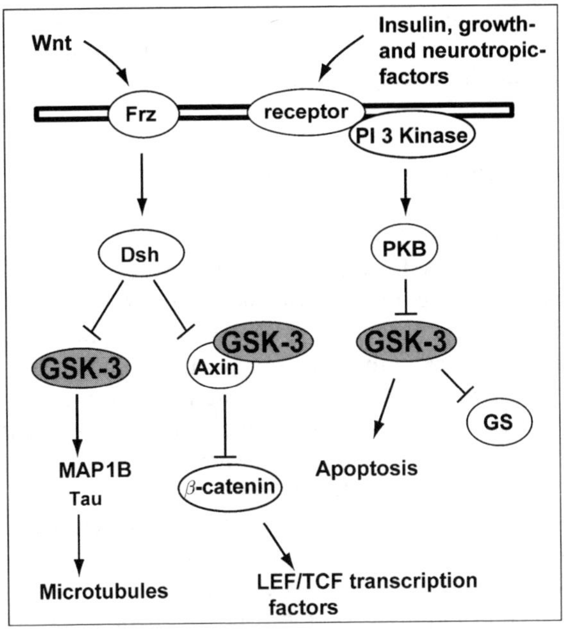

Figure 1.2 Glycogen synthase kinase-3 functions in a variety of signalling pathways. Wnt signalling via Frizzled (Frz) and Dishevelled (Dsh) functions to inhibit GSK-3 activity. This effects the phosphorylation and thus stability of the cell cytoskeleton via cell structural components such as microtubule-associated protein MAP 1B and Tau, and the degradation of β-catenin which effects transcriptional events in the cell. GSK-3 also functions to modulate apoptotic events within the cell via protein kinase B (PKB) and PI3 kinase signalling, and inhibits glycogen synthase (GS) activity

junction-mediated cell contact, and cell contact does not alter regulation of β-catenin-mediated gene expression.

In addition to its role in cell signalling, GSK-3 is also involved in the regulation of the cell cytoskeleton. Inhibition of GSK-3 by lithium, or other GSK-3 inhibitors, affects both microtubule dynamics[31] and microtubule polarity.[32] These effects are clearly seen in neurons which rely heavily on the microtubule cytoskeleton to grow and maintain their morphology. Lithium treatment causes changes in axonal branching and behaviour of the developing growth cone and thus, the structure of the synapse. Inhibition of GSK-3 in dividing cells also leads to misalignment of the mitotic spindle, affecting chromosome segregation[33] and the plane of cell division during development.[34] GSK-3 substrates involved in these structural changes include microtubule-associated protein 1B (MAP1B), Tau and APC. Both MAP1B and Tau are microtubule-binding proteins that regulate microtubule dynamics. Tau is hyper-phosphorylated by GSK-3, leading to the generation of the paired helical filaments (PHF) seen in Alzheimer's disease.[35] In addition to its association with the axin complex, APC also binds microtubules and when lost leads to chromosomal instability during mitosis.

1.4.2 Inositol phosphate signalling

Inositol monophosphatase (IMP) is the best known member of a super-family of structurally related PMEs that can hydrolyze all substrates with sugar phosphate backbones (Figure 1.3). Biochemical and structural studies of a number of family members, including IMP, inositol polyphosphatase (IPP), fructose 1,6-bisphosphatase and the rat 3′-phosphoadenosine 5′-phosphate PAP phosphatase (RnPIP), have found that this enzyme family contains a common amino acid sequence motif which constitutes a lithium-sensitive magnesium ion binding site.[20] The magnesium ion is not required for direct hydrolysis of the phosphodiester bond, but is required for binding the cleaved product and phosphate. Lithium functions to trap the cleaved product in the active site, resulting in the uncompetitive inhibition of enzyme activity.

Inhibition of IMP reduces the cellular concentration of myo-inositol (Figures 1.4 and 1.5), which is required for the production of a spectrum of inositol phosphate-based soluble compounds and membrane lipids. The most studied of these is Ins(1,4,5)P_3 or InsP_3, an intracellular signal molecule that binds to receptors on the endoplasmic reticulum to release calcium, which in turn elicits a range of cell responses, including activation of protein kinases such as PKC. Lithium treatment would also be expected to affect other compounds, such as Ins(1,3,4,5)P_4 and InsP_6, which have signalling functions; PIP_2, which is hydrolyzed by phospholipase C to produce InsP_3 and diacyl glycerol (DAG), and a variety of phosphatidylinositides phosphorylated at the 3′ carbon, which form membrane-binding sites for proteins such as PKB.

Valuable insights regarding the mechanism of action of lithium have been gained using a single-celled eukaryotic amoeba, *Dictyostelium discoideum*.[18,36]

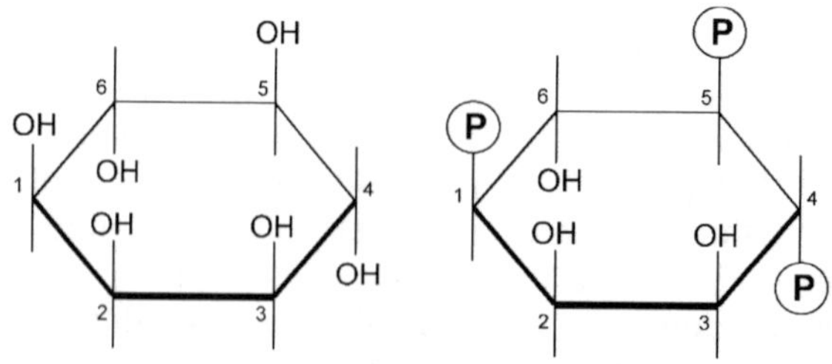

Figure 1.3 Structure of the sugar myo-inositol and its most well-known signalling derivative, inositol 1,4,5-trisphosphate (Ins(1,4,5)P_3 or InsP_3). Carbon atoms are numbered as indicated, showing the phosphorylation ('P') at carbons 1,4 and 5

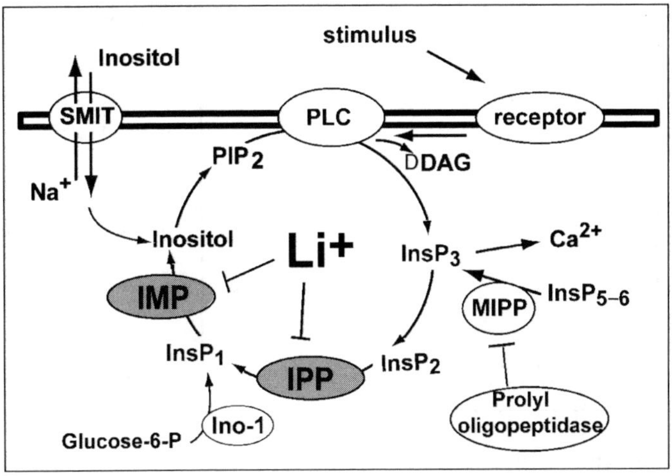

Figure 1.4 The role of lithium in the inhibition of inositol phosphate signalling. Lithium inhibits two enzymes in the phosphomonoesterase family – IMP and IPP. These enzymes enable the recycling of (myo-) inositol to produce phosphatidylinositol bisphosphate (PIP_2). PIP_2 is hydrolyzed on the cell membrane by phospholipase C (PLC) to produce DAG and inositol 1,4,5-trisphosphate ($InsP_3$). $InsP_3$ functions to release calcium from the endoplasmic reticulum. The signalling properties of $InsP_3$ are terminated by its hydrolysis to $InsP_2$. The 'inositol depletion' theory of BD treatment proposes that the therapeutic effect of lithium is to block inositol recycling via IMP/IPP inhibition, and thus reduce inositol levels in the cell. This would result in the attenuation of an over-stimulated $InsP_3$ signalling pathway. Prolyl oligopeptidase regulates the breakdown of higher-order inositol phosphates (inositol pentakisphosphate and hexakisphosphate, $InsP_5$ and $InsP_6$) to $InsP_3$ via modulating the activity of multiple inositol polyphosphate phosphatase (MIPP). The *de novo* biosynthesis of inositol, from glucose-6-phosphate, catalyzed by Inositol synthase (Ino-1), is also blocked by lithium through IMP inhibition. Inositol uptake may also provide a source of inositol via a sodium myo-inositol transporter (SMIT)

This research has identified mutants resistant to the effect of lithium – and one of these mutants, *lisA*, has an elevated $InsP_3$ concentration due to the up-regulation of multiple inositol polyphosphate phosphatase (MIPP) activity (Figure 1.4). This enzyme generates $InsP_3$ from an unconventional route, by the dephosphorylation of $InsP_{5-6}$. The *lisA* gene encodes prolyl oligopeptidase (PO), a cytosolic enzyme characterized by its ability to cleave peptides of less than 3 kDa at a proline residue. PO activity is important for brain function as its inhibitors enhance memory; lithium has the opposite effect on memory. This inverse relationship between PO activity and $InsP_3$ concentration has been demonstrated in mammalian neurons and cell lines derived from astrocytes, ancillary cells associated with brain neurons. Of particular interest is the association of abnormal PO activity with mood disorders and a number of other mental illnesses.[37–39]

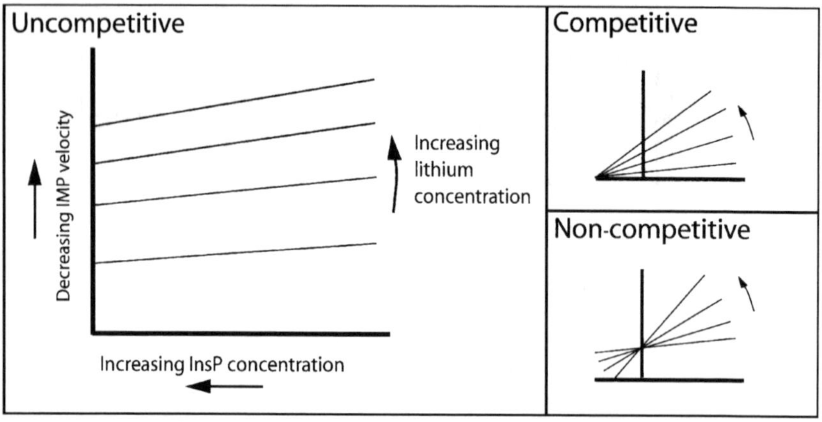

Figure 1.5 Schematic graphs showing different modes of enzyme inhibition. The uncompetitive nature of lithium's inhibition of IMP means that enzyme activity decreases with increasing lithium concentration, but at high substrate concentrations (low x-axis values) the inhibitory effect of lithium is stronger than that found in other forms of enzyme inhibition (competitive and non-competitive)[42]

In addition to lithium's effect on inositol phosphate metabolism, some PMEs, such as RnPIP and the *Saccharomyces cerevisiae* protein Hal2, can also efficiently dephosphorylate di-phosphonucleotide substrates, such as 3′-phosphoadenosine 5′-phosphate (PAP) and inositol 1,4-bisphosphate. In many cases the PAP phosphatase activity is the dominant enzyme activity. PAP phosphatase catalyzes the hydrolysis of PAP to form adenosine monophosphate (AMP). Lithium inhibition, therefore, may lead to increased PAP levels, which is a potent inhibitor of enzymes that utilize PAPs, such as the sulphotransferases. Loss of PAP phosphatase genes in yeasts causes dependence on exogenous sulphur and methioinine and inhibits RNA processing. This pathway is involved in the salt tolerance response in plants and fungi. It is not yet clear whether blocking PAP phosphatase contributes to the therapeutic action of lithium in the treatment of BD.

1.5 Lithium Therapeutics

1.5.1 Bipolar disorder and schizophrenia

As described above, lithium inhibits GSK-3 and inositol phosphate signalling and both pathways have therefore been suggested to be involved in BD. The best current hypothesis describing how lithium functions in the treatment of BD is the 'inositol depletion' hypothesis of Berridge *et al.*[40] This theory suggests that the cause of BD may be overactive InsP signalling in the brain of these patients, and this may be reduced by the inhibitory effect of lithium on InsP signalling (Figure 1.4). In principle, therefore, lithium-based inositol-depletion effects should be reversed by addition of inositol to the test system, and this has been shown in many experiments. *In vivo*, it has been shown that lithium causes the suppression of rearing in rats, and this effect is overcome by intra-cerebroventricular injection of inositol. A more defined effect is seen with limbic seizures induced by pilocarpine, a compound that induces epileptic fits, which can be blocked by injection of lithium. Again, this effect can be suppressed by inositol. Using this 'inositol rescue' criterion, a number of brain physiological and behavioural effects of lithium have been ascribed to changes in InsP signalling. In reality, experiments using this 'inositol rescue' process are complicated by the fact that inositol uptake can be inhibited at high exogenous inositol levels,[41] therefore causing similar effects to lithium treatment. Caution must therefore be taken when inositol addition fails to reverse lithium effects.

One way that scientists are trying to identify which of lithium's targets is responsible for mood control is by examining other BD drugs to see if they also affect these targets. This has proved to be a contentious area of research. In support of the involvement of InsP signalling as the therapeutic target of lithium, researchers have found that all three commonly used BD drugs (lithium, VPA and CBZ) function to increase the size of neuronal growth cones, the developing tip of a mammalian neuron.[36] This increase was shown to be overcome by the addition of inositol to these cells, thus implicating inositol depletion as the common mechanism of action of BD drugs. Furthermore, both lithium and VPA decrease the amount of inositol in the rat brain[43] and change membrane lipid concentration, a process directly linked to inositol signalling.[44] Indeed, a recent study in BD patients suggests altered InsP signalling may be corrected by both VPA and lithium,[45] and these patients show altered activity of PO (see section 4.2),[37,39] an enzyme which controls InsP signalling.[18,36,46] The common effect of structurally diverse BD drugs on InsP signalling therefore strongly suggests this is the therapeutic target of these drugs.

The case for lithium acting on GSK-3 in the treatment of BD is weaker and centred around two sets of observations. First, inhibition of GSK-3 alters the structure of cerebella neurons. If a similar process occurred in other parts of the brain, e.g. the hippocampus and the frontal cortex, this could change brain function. Secondly, lithium is a neuro-protective agent, and can reduce the hypersensitivity to toxic insults seen in cells that overexpress GSK-3. It is therefore possible that lithium could protect against disease-induced cell death. Studies have shown that both bipolar and unipolar depressed patients show a decreased volume of prefrontal cortex grey matter,[47] and this may be associated with a reduction in the number of glial cells,[48] and reduced numbers of both glial and neuronal cells have been found in postmortem studies of these patients.[49] Lithium treatment has also been shown to increase the volume of grey matter in the human brain.[50]

Interestingly, GSK-3 has been implicated in the origins of schizophrenia. Both the Wnt receptor Frizzled-3 (Fz3) and PKB protein have been associated with schizophrenia[51–53] (Figure 1.2). Although the Fz3 association is only genetic, PKB polymorphisms have been found to directly reduce PKB expression. Treatment with the anti-psychotic haloperidol increases activity of the kinase PDK-1, an upstream activator of PKB. As PKB inhibits GSK-3 activity, these observations suggest that GSK-3 may be elevated in schizophrenic patients. Early reports on the use of lithium suggest that it may be active against schizophrenic patients, but with the availability of many anti-psychotic drugs it is not in common usage.

1.5.2 Alzheimer's disease

Alzheimer's disease is a neurodegenerative brain disorder causing neuronal dysfunction and ultimately cell death, giving rise to dementia. Its prevalence is increasing, as it was estimated to affect more than 10% of people over 65, or 48% of people over 85 in the US.[54] The onset of Alzheimer's disease occurs with the accumulation of extra-cellular senile plaques composed of amyloid-β (Aβ) peptides and with the accumulation of intercellular neurofibrillary tangles. Aβ peptides are composed of a mixture of mainly 40 and 42 amino acid peptides[55] and are generated from the transmembrane glycoprotein amyloid precursor protein (APP) by sequential proteolysis catalyzed by aspartyl protease β-secretase (BACE)[56] and presenilin-dependent γ-secretase.[57]

There are several direct roles of lithium in treatment of Alzheimer's disease. First, GSK-3 functions to phosphorylate Tau (section 1.4.1) – a process which is necessary for the accumulation of microfibrile tangles. Secondly, inhibition of GSK-α causes a reduced production of Aβ peptides.[58] These processes are affected at therapeutic levels of lithium. A reduction of Aβ peptides is caused by the reduced processing of APP by γ-secretase, although it does not affect Notch processing, another signalling pathway modulated by this enzyme. Instead, lithium may function to modulate the access of specific substrates or the activity of these substrates towards the γ-secretase complex, a process linked to GSK-α signalling.[58] Thirdly, some mutations of PS-1 are also associated with familial Alzheimer's disease, and loss of PS-1 in mice leads to increased β-catenin protein and cell division; these results implicate PS-1 and GSK-3 in this disorder.

In addition to the direct involvement of GSK-3 in Alzheimer's disease, the neuro-protective or anti-apoptotic effects of lithium may also play a role in the treatment of Alzheimer's disease. The neuro-protective role of lithium can be clearly seen with regard to glutamate toxicity. Pretreatment of primary cerebella granule cells with lithium gives a dose-dependent protection against this toxicity, in a concentration range that is found in patients undergoing lithium treatment. This toxicity occurs via *N*-methyl-D-aspartate (NMDA) receptors, as antagonists to these receptors also block toxicity. It is interesting to note that this effect is unlikely to be a direct action of lithium, as pretreatment is necessary for protection[59] and this effect occurs at even lower concentrations in cortically derived neurons.[60]

In addition to glutamate, lithium has been shown to protect against induced apoptosis in neuronally derived cultures by a number of agents including the seizure-inducing compounds kainite,[60] α-amino-3-hydroxy-5-methylisoxozole-4-propionic acid (AMPA),[60] the anti-convulsants Phenytoin and CBZ[61] and deprivation of high potassium and high serum.[62] The broad range of compounds that lead to cell death suggests that a central signalling process is being activated in this apoptotic process and it is this common feature which is targeted by lithium. It is of considerable interest that VPA also functions to protect cerebella granule cells from glutamate exocitoxicity.[60] This suggests that inositol trisphosphate depletion, which is the common action of the mood stabilizers, lithium and VPA,[36] may function to prevent cell death through excitotoxicity. In support of this, pretreatment of both cerebella granule cells and cortical neurons with lithium for 7 seven days reduces the NMDA-receptor-mediated Ca^{2+} entry into cells, but does not change subunit levels.

Glycogen synthase kinase-3 may act as a modulator of apoptosis. Treatment of rat hippocampal neurons with the Aβ peptide, which builds up in patients with Alzheimer's disease, both increases GSK-3β expression and induces apoptosis.[63] The apoptotic effect is blocked by antisense oligonucleotides directed at GSK-3. Wnt stimulation also protects against apoptosis, although this may be due to indirect induction of insulin-like growth factor (IGF) proteins. Insulin, IGFs, nerve growth factor (NGF), and brain-derived neurotrophic factor (BDNF) can all inhibit GSK 3 through activation of PI3 kinase (Figure 1.2). Inhibition of PI3 kinase, by use of chemical inhibitors or serum withdrawal, leads to increased GSK-3 activity, and this correlates with apoptosis. In addition, full GSK-3 activity requires phosphorylation at an internal tyrosine (Tyr216).[64] Interestingly, it has been shown that several apoptotic stimuli induce an increase in Tyr216 phosphorylation and increase GSK-3 activity. Consistent with these observations, overexpression of GSK-3 also correlates with neuronal degeneration.[65] Moderate increases in GSK-3 activity in human neuroblastoma SH-SY5Y cells did not increase the basal rate of apoptosis or caspase-3, which sits within the apoptotic signal transduction pathway, but they are associated with increased sensitivity to apoptotic stimuli. These effects are blocked by expression of the anti-apoptotic protein Bcl-2 and expression of a dominant negative form of a tumor suppressor p53. Consistent with a role of GSK-3, many of these effects can be reduced by lithium treatment.

Lithium also functions as a neuro-protective agent by altering gene transcription. P53 promotes the expression of Bax, which is a pro-apoptotic gene whose product may cause the release of cytochrome c from the mitochondria and induce subsequent caspase activation and protein degradation. Long-term lithium treatment increases p53 expression in cerebellar granule cells, as are Bax mRNA and protein levels.[66] Similarly, the product of the Bcl-2 gene is a cytoprotective protein, which interacts with the mitochondrial membrane to prevent Bax-induced cytochrome c release and the induction of this apoptotic pathway. Long-term lithium treatment increases both Bcl-2 mRNA and protein levels.[66] This has also been reported in rat brains following treatment with either lithium or VPA.[67]

1.5.3 Ischemia (stroke)

Animal models of stroke have been extensively used to examine possible therapeutic effects of lithium. Pretreatment of mice for 16 days prior to stroke induction greatly reduces the damage caused by middle cerebral artery occlusion (MCAO) which is the cause of brain damage.[68] This neuro-protective effect is also seen in a model of stroke which has been suggested to more closely resemble acute stroke cases, that is, transient ischemia models of stroke. In this system, animals are exposed to MCAO for 1 h, with reperfusion immediately afterwards. Administration of a single dose of lithium up to 3 h after this transient exposure significantly reduces the damage caused by this process.[68] It is interesting that this protection does not require extended pretreatment, which may suggest a separate mechanism of protection to the neuro-protective effects discussed earlier. In addition, the apoptotic effects of lithium (p. 13) also may protect brain regions post stroke, as lithium decreases Bax protein levels in this transient model by around 70%.[68]

1.5.4 Adverse effects

The narrow therapeutic window of lithium makes monitoring blood levels essential during treatment. Increasing lithium blood levels over 1.2 mM gives rise to tremors – usually noticed in the fingers – and dosage reduction can usually reduce this effect to mild tremors. Some patients complain of slowed mental agility and forgetfulness. Memory problems are one of the leading causes of non-compliance and the third most common side effect. Extreme lithium doses produce chronic nausea and diarrhea; thus lithium poisoning is rare, but is sometimes seen after poor monitoring of clinical application or an unsuccessful suicide attempt. In a study of 15 deliberate self-poisonings, no deaths were seen, although as discussed in section 1.3.1, the lithium-induced deaths of the 1950s clearly show that some individuals are at risk. Episodic nausea can usually be relieved by taking lithium with food. Some patients gain weight progressively on lithium and it is the second most common reason patients stop taking it. Weight gain is greater in patients who are overweight to begin with. Some patients show decreased thyroid levels and rarely goiter. About 5% develop hypothyroidism and 30% have elevated thyroid-stimulating hormone levels. Polyuria (passing an excessive quantity of urine) or polydipsia (excessive thirst) occurs in one out of five patients. Aggravation of psoriasis and alopecia can occur but hair usually re-grows with or without the lithium.

Based on reports from the 1970s, kidney damage has been a concern for patients taking long-term lithium treatment. However, this worry has not been borne out in further study of correctly treated patients and in practice there appears to be a low incidence of renal damage, although kidney examination is recommended every 6–12 months. Harm can occur if patients are stabilized on lithium and given drugs which may affect renal balance. For instance, non-steroidal anti-inflammatory drugs can cause a 60% increase in blood lithium concentration because of their effect of reducing lithium clearance through the kidneys. These drugs can cause a doubling of the blood lithium levels and the patient may therefore show lithium toxicity. This can also occur with anti-inflammatory drugs like Ibuprofen®, Naproxen® and Indomethacin®.

Acknowledgements

R.S.B. Williams and A.J. Harwood are supported by a Wellcome Trust Career Development Fellowship and a Wellcome Trust Senior Fellowship, respectively.

References

1. ten Have M, Vollebergh W, Bijl R, Nolen WA. *J. Affect. Disord.* 2002; 68: 203.
2. Muller-Oerlinghausen B. *Eur. Arch. Psychiatry Clin. Neurosci.* 2001; 251: II72.
3. Frausto da Silva JJ, Williams RJ. *Nature.* 1976; 263: 237.
4. Raju B, Murphy E, Levy LA *et al. Am. J. Physiol.* 1989; 256: C540.
5. Ryves WJ, Dajani R, Pearl L, Harwood AJ. *Biochem. Biophys. Res. Commun.* 2002; 290: 967.
6. Kerwin R (ed.). *The Bethlem and Maudsley NHS Trust Prescribing Guidelines,* Martin Duntiz 5th Edn. London, 1999.
7. Hammond WA. *Insanity in a Treatise on Diseases of the Nervous System,* D. Appleton, New York, 1873.
8. Cade JF. *Med. J. Aust.* 1949; 36: 349.
9. Schou M. *Arch. Gen. Psychiatry.* 1997; 54: 9.
10. Levine J, Chengappa KN, Brar JS *et al. Bipolar Disorder.* 2000; 2: 120.
11. Goodwin FK, Fireman B, Simon GE *et al. JAMA.* 2003; 290: 1467.
12. Wyatt RJ, Henter ID, Jamison JC. *Psychiatry Q.* 2001; 72: 149.
13. Duhm J, Becker BF. *Prog. Clin. Biol. Res.* 1978; 21: 551.
14. Becker BF, Duhm J. *J. Physiol.* 1978; 282: 149.

15. Livingston BT, Wilt FH. *Proc. Natl. Acad. Sci. USA*. 1989; 86: 3669.
16. Emily-Fenouil F, Ghiglione C, Lhomond G *et al. Development*. 1998; 125: 2489.
17. D'Amico LA, Cooper MS. *Biochem. Cell Biol.* 1997; 75: 563.
18. Williams RS, Eames M, Ryves WJ *et al. EMBO J.* 1999; 18: 2734.
19. Gagliardi JP, Krishnan KR. *Psychopharmacol. Bull.* 2003; 37: 59.
20. York JD, Ponder JW, Majerus PW. *Proc. Natl. Acad. Sci. USA*. 1995; 92: 5149.
21. Thams P, Geisler A. *Acta Pharmacol. Toxicol.* 1979; 45: 329.
22. Avissar S, Schreiber G, Danon A, Belmaker RH. *Nature*. 1988; 331: 440.
23. Mota dF, Amari L, Srinivasan C *et al. Biochemistry*. 1994; 33: 4101.
24. Klein PS, Melton DA. *Proc. Natl. Acad. Sci. USA*. 1996; 93: 8455.
25. Stambolic V, Ruel L, Woodgett JR. *Curr. Biol.* 1996; 6: 1664.
26. Thomas GM, Frame S, Goedert M *et al. FEBS Lett.* 1999; 458: 247.
27. Embi N, Rylatt DB, Cohen P. *Eur. J. Biochem.* 1980; 107: 519.
28. van Noort M, Meeldijk J, van der ZR *et al. J. Biol. Chem.* 2002; 277: 17901.
29. Levina E, Oren M, Ben Ze'ev A. *Oncogene*. 2004; 1.
30. Polakis P. *Genes Dev.* 2000; 14: 1837.
31. Goold RG, Owen R, Gordon-Weeks PR. *J. Cell Sci.* 1999; 112: 3373.
32. Etienne-Manneville S, Hall A. *Nature*. 2003; 421: 753.
33. Wakefield JG, Stephens DJ, Tavare JM. *J. Cell Sci.* 2003; 116: 637.
34. Schlesinger A, Shelton CA, Maloof JN *et al. Genes Dev.* 1999; 13: 2028.
35. Sperber BR, Leight S, Goedert M, Lee VM. *Neurosci. Lett.* 1995; 197: 149.
36. Williams RS, Cheng L, Mudge AW, Harwood AJ. *Nature*. 2002; 417: 292.
37. Breen G, Harwood AJ, Gregory K *et al. Bipolar Disorder*. 2004; 6: 156.
38. Maes M, Goossens F, Scharpe S *et al. Biol. Psychiatry*. 1994; 35: 545.
39. Maes M, Goossens F, Scharpe S *et al. Psychiatry Res.* 1995; 58: 217.
40. Berridge MJ, Downes CP, Hanley MR. *Cell*. 1989; 59: 411.
41. Wolfson M, Bersudsky Y, Zinger E *et al. Brain Res.* 2000; 855: 158.
42. Cleland WW. *Adv. Enz.* 1970; 29: 1.
43. O'Donnell T, Rotzinger S, Nakashima TT *et al. Eur. Neuropsychopharmacol.* 2003; 13: 199.
44. Ding D, Greenberg ML. *Mol. Microbiol.* 2003; 47: 373.
45. Silverstone PH, Wu RH, O'Donnell T *et al. Hum. Psychopharmacol.* 2002; 17: 321.
46. Schulz I, Gerhartz B, Neubauer A *et al. Eur. J. Biochem.* 2002; 269: 5813.
47. Drevets WC, Price JL, Simpson JR Jr *et al. Nature*. 1997; 386: 824.
48. Ongur D, Drevets WC, Price JL. *Proc. Natl. Acad. Sci. USA*. 1998; 95: 13290.
49. Rajkowska G, Miguel-Hidalgo JJ, Wei J *et al. Biol. Psychiatry*. 1999; 45: 1085.
50. Moore GJ, Bebchuk JM, Hasanat K *et al. Biol. Psychiatry*. 2000; 48: 1.
51. Emamian ES, Hall D, Birnbaum MJ *et al. Nat. Genet.* 2004; 36: 131.
52. Katsu T, Ujike H, Nakano T *et al. Neurosci. Lett.* 2003; 353: 53.
53. Yang J, Si T, Ling Y *et al. Biol. Psychiatry*. 2003; 54: 1298.
54. Lacy C. *Pharma. Guide Hos. Med.* 1998; 10: 1.
55. Selkoe DJ. *Proc. Natl. Acad. Sci. USA*. 2001; 98: 11039.
56. Vassar R, Bennett BD, Babu-Khan S *et al. Science*. 1999; 286: 735.
57. De Strooper B, Saftig P, Craessaerts K *et al. Nature*. 1998; 391: 387.
58. Phiel CJ, Wilson CA, Lee VM, Klein PS. *Nature*. 2003; 423: 435.
59. Nonaka S, Hough CJ, Chuang DM. *Proc. Natl. Acad. Sci. USA*. 1998; 95: 2642.
60. Hashimoto R, Hough C, Nakazawa T *et al. J. Neurochem.* 2002; 80: 589.

61. Nonaka S, Katsube N, Chuang DM. *J. Pharmacol. Exp. Ther.* 1998; 286: 539.
62. D'Mello SR, Anelli R, Calissano P. *Exp. Cell Res.* 1994; 211: 332.
63. Cedazo-Minguez A, Popescu BO, Blanco-Millan JM *et al. J. Neurochem.* 2003; 87: 1152.
64. Fan G, Ballou LM, Lin RZ. *J. Biol. Chem.* 2003; 278: 52432.
65. Lucas JJ, Hernandez F, Gomez-Ramos P *et al. EMBO J.* 2001; 20: 27.
66. Chen RW, Chuang DM. *J. Biol. Chem.* 1999; 274: 6039.
67. Chen G, Zeng WZ, Yuan PX *et al. J. Neurochem.* 1999; 72: 879.
68. Nonaka S, Chuang DM. *Neuroreport.* 1998; 9: 2081.

2 $_5$B Boron Compounds as Therapeutic Drugs

Geeta Rana, Kamesh Vyakaranam, John A. Maguire and Narayan S. Hosmane

Boron analogues of a variety of biologically important molecules have been found to play an increasingly important role in the areas of pharmacology, medicine, biochemistry and nutrition. Bearing close structural resemblance to their natural organic counterparts, the boron analogues exhibit potent biological activities. Compounds containing one or more boron atoms have been prepared for a variety of applications. It has been shown in model studies that boron analogues of nucleosides,[1] nucleotides,[2–7] amino acids[8–10] and nucleic acids possess potent anti-cancer,[11–13] anti-inflammatory,[14] hypolipidaemic and anti-osteoporotic[15] properties. The primary focus of this chapter is on the applications of boronated compounds for boron neutron capture therapy (BNCT).

2.1 Boron Neutron Capture Therapy

Boron neutron capture therapy is a binary form of treatment of cancer in which a compound containing ^{10}B is selectively delivered to tumor tissues prior to irradiation by neutrons.[16] Upon irradiation with thermal neutrons, interaction of a ^{10}B atom with a thermal neutron produces an α-particle, a high-energy Li-7 ion and a low-energy γ-ray Figure 2.1. The linear energy transfer (LET) of these heavily charged particles have a range of about one-cell diameter, which confines

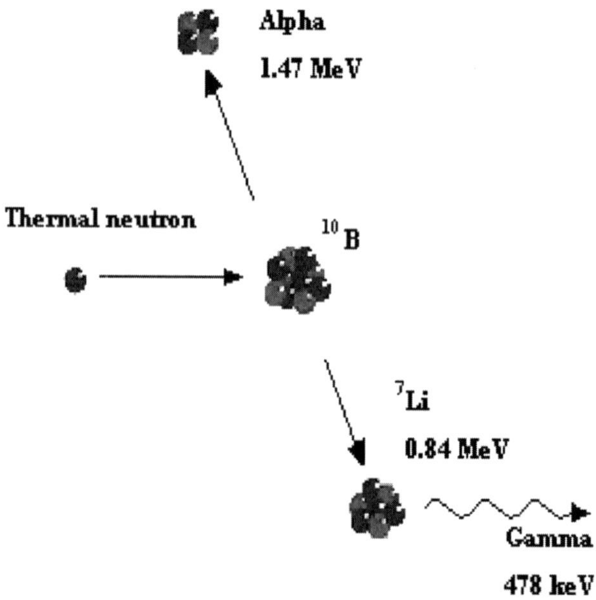

Figure 2.1 The mechanism of boron neutron capture therapy (BNCT) (see Plate 1)

the dissipated energy to the cell from which they arise, hence minimizing the effect on surrounding tissue. The capture cross section of boron for neutrons is more than three orders of magnitude higher than for other nuclei common in living tissue, so the target region can be dosed with neutrons at a sizeable flux and still have only minimal effect on boron-free regions in the beam path.[17] Additional advantages of using ^{10}B compared to other nuclides (e.g. ^{235}U, ^{7}Li) are that the fission products have high LET and the boron compounds can be synthesized having hydrolytically stable linkages between boron and elements such as C, O and N.[18]

If boron can be transported to the target tissue with sufficient specificity and quantity using tumor-selective agents, tumor cell death may be accomplished. The required boron concentration is generally estimated at 10^9 atoms of ^{10}B (natural abundance 19.9%) per cell, which translates to approximately 35 μg of ^{10}B per gram of tissue. To prevent damage to healthy tissue in the path of the neutron beam, the surrounding tissue should contain no more than 5 μg of ^{10}B per gram of tissue.[19] A variety of carrier molecules have been investigated as boron delivery systems to tumor cells.[19c,20] These include carbohydrates,[21] polyamines,[22] nucleosides,[23] antibodies,[24] porphyrins,[25] liposomes[26] and amino acids.[27] In order to obtain more favorable tumor to normal brain (T/N) and tumor to blood (T/B) ratios, and to significantly increase the amount of boron that can be incorporated into the tumor, a variety of structural modifications, using borane and carborane compounds, have been investigated.

Fast neutron therapy has long been used to treat cancer, including brain tumors.[28] However, for certain brain tumors, e.g. glioblastoma multiforme, the prognosis is poor.[29] After diagnosis, the expected survival time of 8 months can be extended by another 4 months by treatment using fast neutron therapy. There is essentially no difference in survival time between fast neutron and photon therapy for patients but the cause of death does depend upon the treatment. With fast neutron therapy, death results from significant damage to normal tissue. This is because of the close dose–effect relationship between tumor and healthy tissue. In photon therapy, regrowth of the tumor is the primary cause of death. A selective dose enhancement in tumor tissue using BNC (from the thermal and epithermal components of the fast neutron beam) would permit a reduction in overall dose thereby lowering the dose to normal tissue. Also, at depth, fast neutrons are thermalized allowing capture by ^{10}B, thus enhancing the applied dose. Even a relatively small percentage increase could have a significant clinical effect. The concentration of boron in tumors to bring about the BNC-enhanced fast neutron effect is thought to be 100 ppm. The application of current BNCT clinical trial agents, $Na_2B_{12}H_{11}SH$ (BSH) and *p*-boronophenylalanine (BPA) to assist in fast neutron therapy, is currently under consideration. However, such an application might not be practical because BSH, despite its high boron content, may actually have radio-protectant characteristics because of its thiol group.[30]

Several requirements must be met in order for this therapy to be effective: (i) a concentration of 25–30 μg of ^{10}B atoms per gram of tumor must be achieved; (ii) a tumor to normal tissue (T:N) ratio of the boron delivery agent greater than 1 is necessary; (iii) the agent should be of low toxicity;[19] and (iv) balance between the compound's lipophilic and hydrophilic components must be achieved. A series of compounds have been evaluated since the 1950s that include compounds containing one boron atom (aromatic boronic acids and related analogues) or many boron atoms (polyhedral borane anions, $B_{10}H_{10}^{2-}$ and $B_{12}H_{12}^{2-}$, and carborane cages, $C_2B_{10}H_{10}$ and $C_2B_9H_{10}$).[19c]

The primary focus of BNCT has been in the treatment of malignant brain tumors because these tumors have little history of metastasizing to other organ sites. Hence, if tumor eradication can be achieved in the brain, then there should be a significant increase in life expectancy. Only two compounds, BSH and BPA, are currently available for clinical trials.[19c] The results with these compounds are not promising; hence, there is an immediate need for potential precursor molecules and more sophisticated technologies that can selectively destroy tumor cells and lead to the expansion of BNCT. While the BSH is an anionic polyhedral borane icosahedron in which a thiol group replaces an *exo*-polyhedral hydrogen atom, the BPA is a boronophenylalanine. The structures of these two boron compounds are shown in Scheme 2.1.

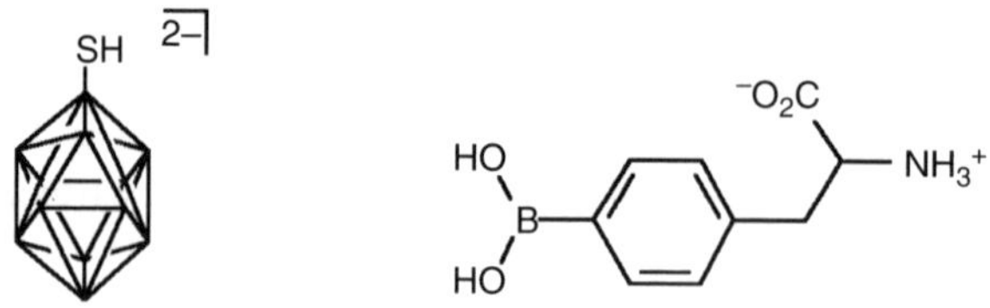

Scheme 2.1 Boron compounds used in clinical trials, BSH (left) and BPA (right)

2.2 Classes of Boron Compounds for Potential Use in BNCT

2.2.1 DNA binders

There has been a special interest in the development of boron compounds that could be concentrated at or near the cell's DNA and a number of boron-containing DNA intercalators have been evaluated as BNCT agents. The development of boron-containing pyrimidines, purines, nucleosides and nucleotides is predicated on the basis that the effectiveness of a boron species incorporated into the tumor's DNA would enhance the effectiveness of the capture reaction by a factor of 2–5 over what would be if the boron compounds were evenly distributed in the cytoplasm.[31] Neidle and coworkers[32] have recently shown that substituted fluorenones and anthraquinones act as telomerase inhibitors. Since the telomerase enzyme is a reverse transcriptase which elongates the 3′ ends of the telomerase DNA,[33a] a number of carborane-substituted fluorenone, anthracene-9,10-dione, and sulfone compounds have been synthesized (Figure 2.2).[33b] In fact, the use of large-cage polyhedral boranes, such as $B_{10}H_{10}^{2-}$ and $B_{12}H_{12}^{2-}$, and carboranes, such as the *o*-carborane, as efficient boron transfer agents in BNCT has been the subject of a number of studies.[34,35] In Figure 2.2, and subsequent figures, the boranes are normally depicted as idealized polyhedra with B—H units understood to be at the unlabeled vertices. Only when a non-boron atom replaces a boron vertex, its symbol is shown. Thus the carborane cages shown in Figure 2.2 represent the $C_2B_9H_9$ cage. They have taken advantage of the DNA affinity of the fluorenones and anthraquinones and the high boron content of the polyhedral boron cages in the synthesis of a series of structurally related anthracene-9,10-diones, fluorenones, phenylsulfones, along with the corresponding open cage *nido* derivatives as possible precursors to BNCT agents that could bind to DNA.

It should be noted that in the carborane cage all the atoms are boron except the two carbons drawn. Chemotherapeutic alkylating agents are electrophiles that can be covalently incorporated into tumor DNA and, as a result, interfere with tumor replication. Therefore, a number of boron analogues have been synthesized; some examples are shown in Figure 2.3.[36]

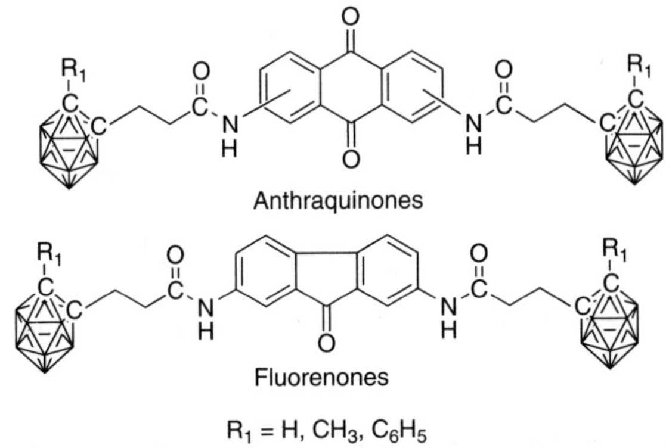

Figure 2.2 Boronated anthraquinones and fluorenones

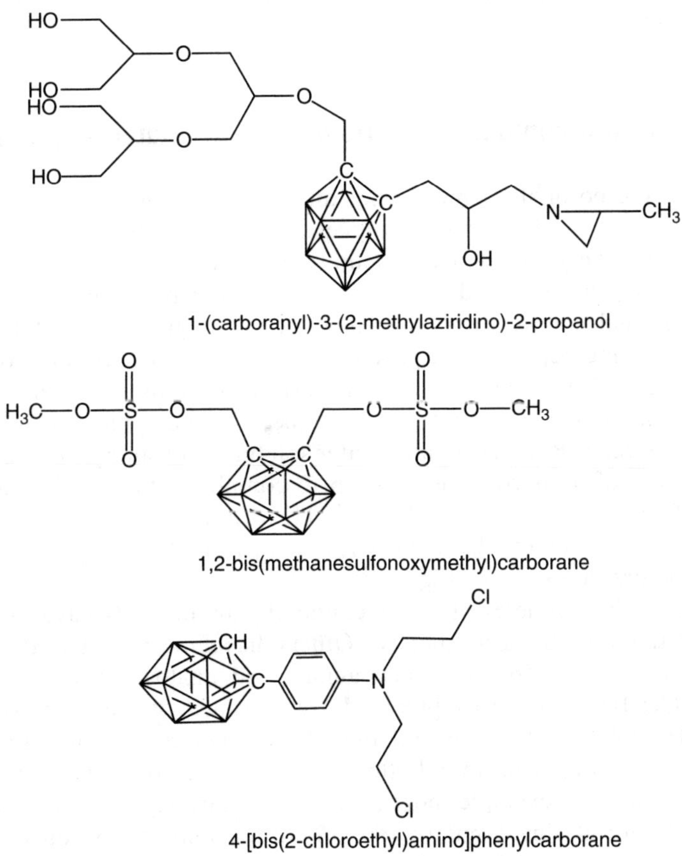

Figure 2.3 Alkylating agents

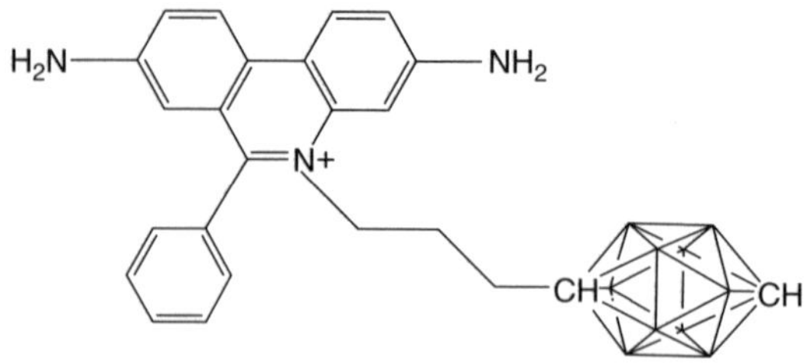

Figure 2.4 Boron containing the phenanthradinium moiety

More recent work has focused on the development of boron-containing phenanthridinium analogues.[37] The basic ring structure in these tricyclic planar compounds intercalates readily into the DNA molecule; this has been the basis for the synthesis of the boronated analogues – a representative structure is given in Figure 2.4.

2.2.2 Boron-containing amino acids and related peptides

Interest in the boronated amino acids and peptides stems from the fact that *p*-BPA is one of only two clinically used BNCT agents. The strategy for concentrating the boronated compounds in tumor cells is that the cellular building blocks are required in greater number in rapidly dividing cells than in their normal counterparts; this may lead to a greater concentration of boron-containing peptides and proteins in the tumor cell. In addition to the *para*-isomer of BPA, the *ortho*- and *meta*-isomers have also been evaluated (Figure 2.5).[38] However, only the *para*-isomer has undergone extensive biological testing. Since BPA has limited water solubility, advantage has been taken of the ability of its boronic acid group to complex with carbohydrates as a means of increasing water solubility.[39] There have been other attempts to increase water solubility by incorporating hydrophilic groups into BPA.[40] Some structures are given in Figure 2.6.

Not all of the syntheses of boron-containing amino acids have focused on those of the boronic acid moiety. Others have incorporated the borane zwitterionic groups. The first such amino acid analogue was boroglycine, H_3NBH_2COOH, which was prepared from trimethylamine carboxyborane (Me_3NBH_2COOH) by amine exchange.[41] The replacement of carbon by a boron atom showed a profound effect on the pK_as of both the carboxyl and amine protons. For example, boroglycine has a carboxyl pK_a of 8.3 as compared to 2.4 for glycine. Similarly, the pK_a of the ammonia nitrogen deprotonation in H_3NBH_2COOH is >11.[42] The synthesis of boron analogues of other common amino acids such as alanine, leucine, etc. have also been reported.[41]

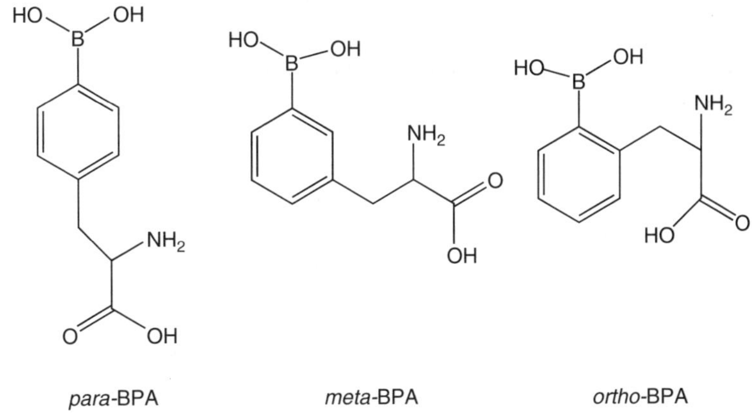

Figure 2.5 Structure of BPA isomers

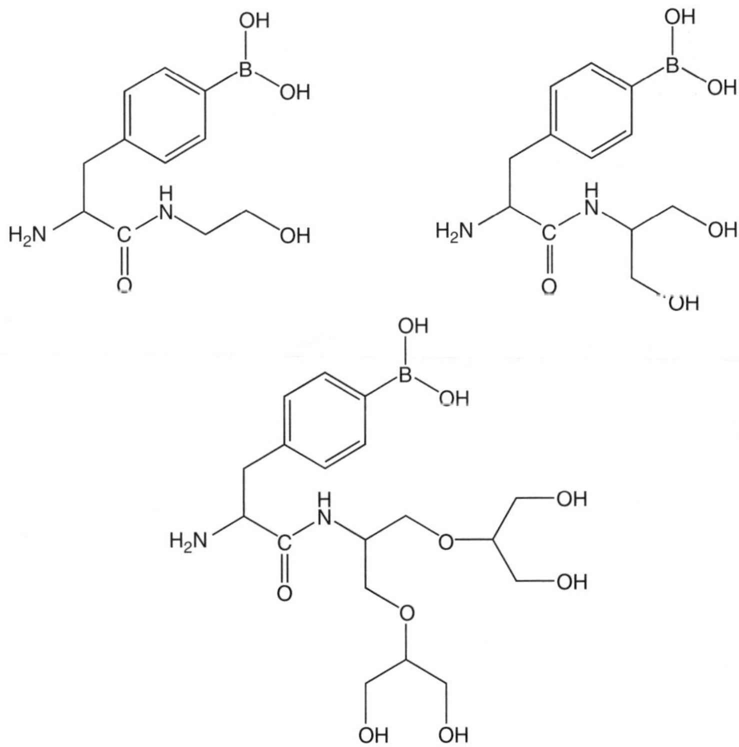

Figure 2.6 Derivatives of BPA

Several of these borane-containing amino acids are shown in Figure 2.7. There has also been interest in making analogues that contain clusters of boron atoms, with special interest centered on the carboranes. Figure 2.8 shows the isomers of *o*-carboranylalanine.[43,19c]

Such analogues are biologically active and demonstrate the advantages of incorporating functionalities that will balance the carboranyl's lipophilic character. A number of different carborane-containing amino acids have been synthesized and evaluated,[44] some of which are shown in Figure 2.9.

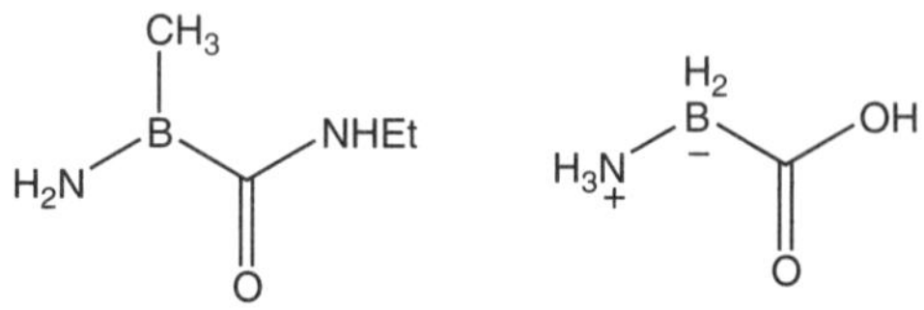

Figure 2.7 Boranoamino acid derivatives

(R)-*o*-carboranylalanine

(S)-*o*-carboranylalanine

Figure 2.8 Structures of carboranylalanine

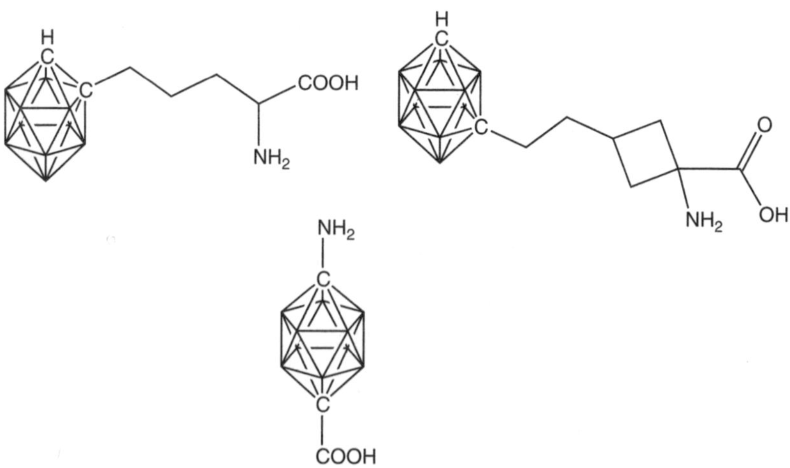

Figure 2.9 Carborane-containing amino acids

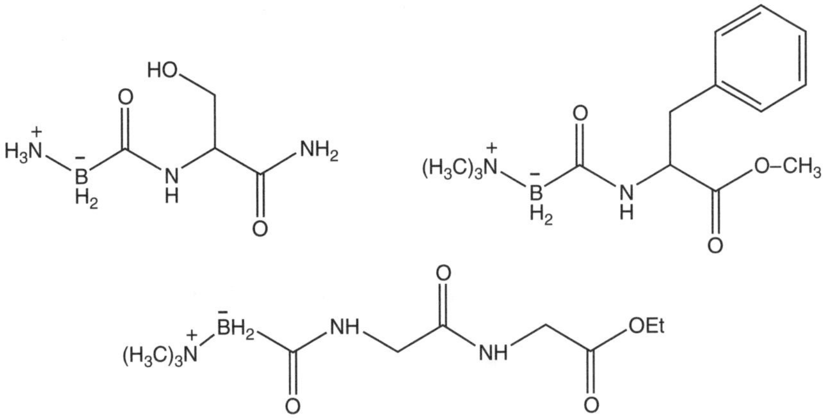

Figure 2.10 Some boron-containing peptides

The synthesis of these boranoamino acids leads to the development of boron-containing peptides. Since tumor cells require increased amounts of proteins, small peptides that may cross cell membranes can be utilized by tumor cells. Examples of these different types of peptides are shown in Figure 2.10.

Soloway suggested that boronated amino acids could also produce a possible twofold attack on tumors, they could serve as anti-metabolites and inhibit tumor growth as well as being selectively incorporated into the proteins of the neoplasm.[21] The preparation and evaluation of small boron-containing peptides are only in their early stages of development.

2.2.3 Boron-containing nucleic acid precursors

Since the presumed target of the high LET radiation generated by a boron capture reaction is the cell nucleus,[43b] a great deal of effort has been devoted to the design and synthesis of potential precursors of the nucleic acids that could be incorporated into cellular DNA.

Initial studies in this effort were focused on the nucleic acid bases, the purines and the pyrimidines. Some structures of these boronated compounds are given in Figure 2.11.

Many of these compounds are of low hydrolytic stability. When stability was enhanced by introducing aromatic group on the borons, the compounds ceased to emulate biochemical processes. However, it was found that when the hydroxyboryl group was bonded directly to a nitrogen atom, as shown in Figure 2.12, an increased hydrolytic stability was achieved. However, it remains to be seen how such compounds will behave under physiological conditions.

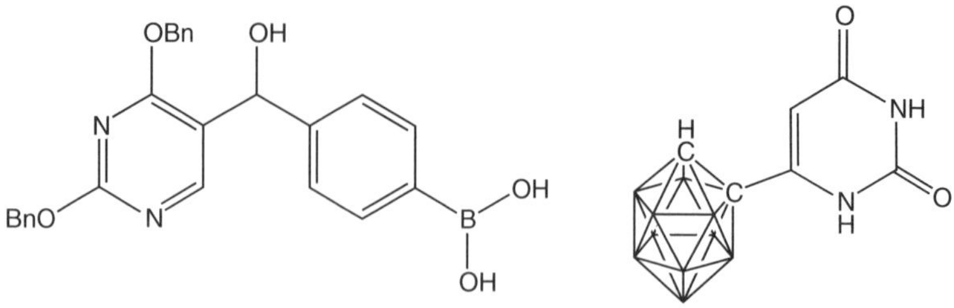

Figure 2.11 Boron-containing nucleic acid bases

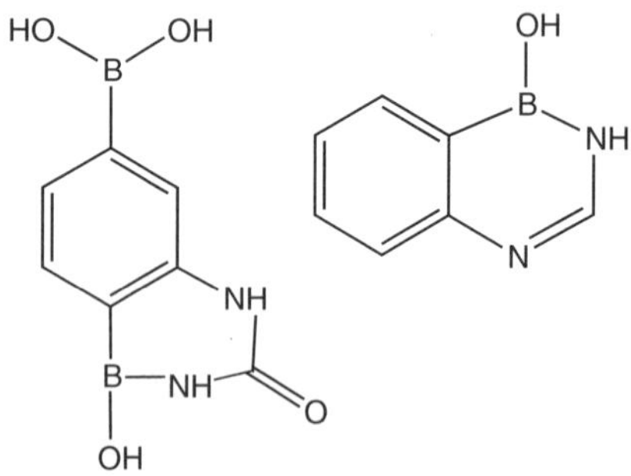

Figure 2.12 Hydroxoboryl isoesters of benzopyrimidines

Another approach is to attach the boron moiety directly on a pyrimidine nucleus, giving boron-containing nucleosides; the first such compound that was synthesized is shown in Figure 2.13.[23a] Other boronated nucleosides, such as those shown in Figure 2.14, have been synthesized starting with the respective 5-halogenated nucleoside derivatives.[40,19c]

A series of single-boron-atom nucleosides, such as those shown in Figure 2.15, have been reported by Spielvogel and coworkers.[4] One of the limitations of these

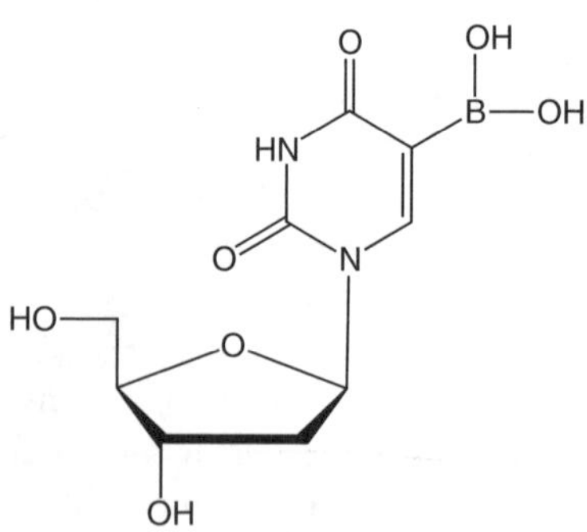

Figure 2.13 Structure of 5-(dihydroxylboryl)-2′-deoxyuridine

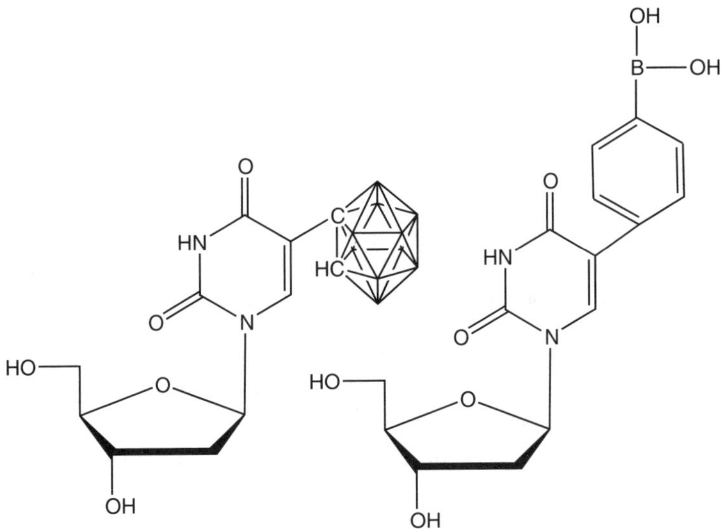

Figure 2.14 Boron-containing nucleosides

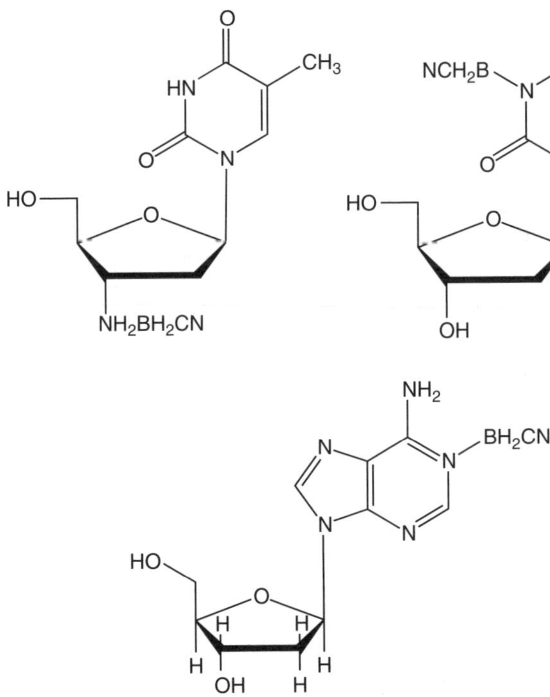

Figure 2.15 Single-boron-atom nucleosides

compounds is that they have a low percentage of boron. Therefore, work has been directed toward incorporating stable boron clusters into nucleosides.[45] Examples are shown in Figures 2.14 and 2.16. Because of their chemical inertness, derivatives of *closo*-$C_2B_{10}H_{12}$ clusters are most often incorporated (see Figures 2.16 and 2.17) into nucleosides. These carboranyl nucleosides suffer from their significant lipophilicities that compromise any assessment of their enzymatic reactivity or *in vitro* cellular evaluation.

Polyhydroxyl groups have been introduced in order to achieve suitable hydrophilic/lipophilic balance (Figure 2.18).

In addition to nucleosides, several boronated nucleotides have been prepared and evaluated (Figure 2.19).[10,46] Although such ionic structures at present lack an ability to cross the cell membrane, their penetrability will be based ultimately upon the overall lipophilic/hydrophilic balance, which could be adjusted by chemical manipulation. Moreover, the enzymatic cleavage of such phosphates under *in vivo* conditions may make these compounds a basis for prototype drugs.

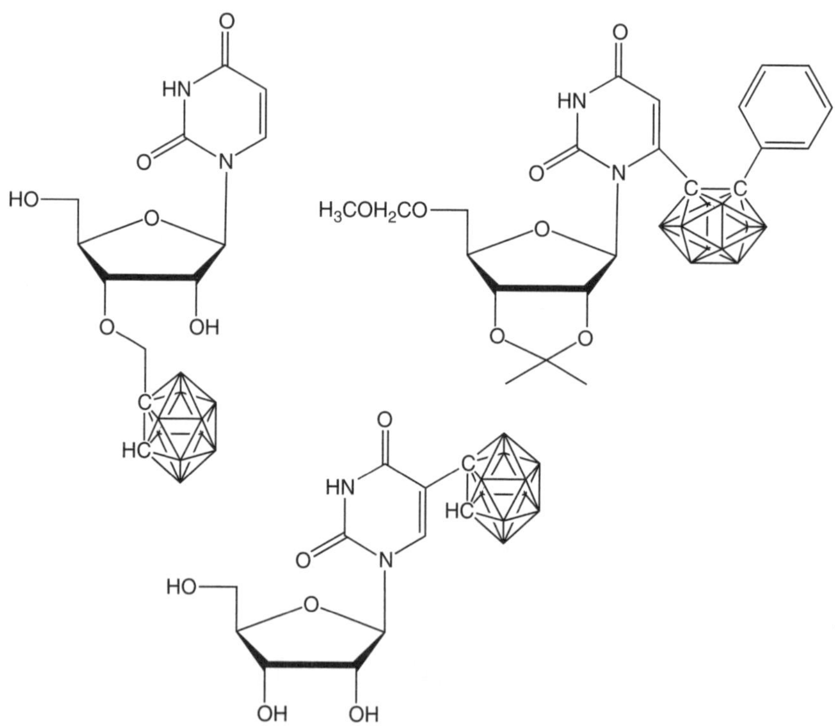

Figure 2.16 Carborane-containing nucleoside derivatives

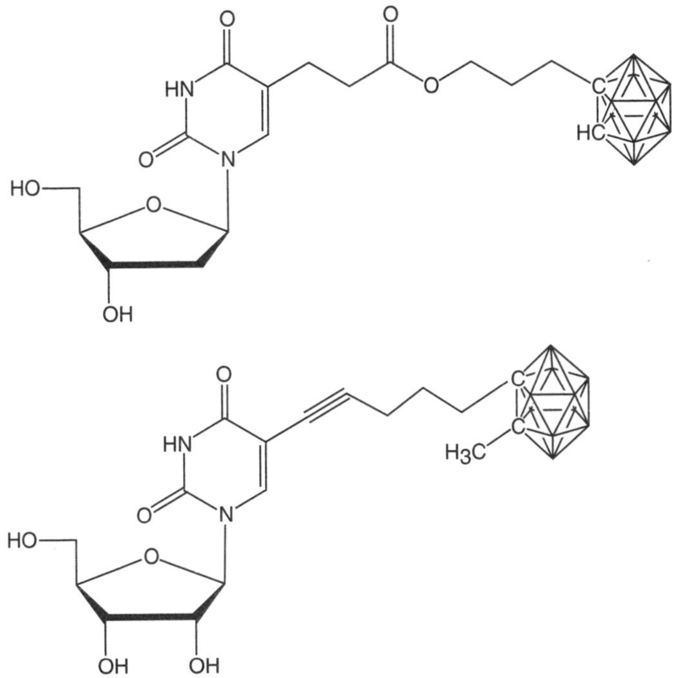

Figure 2.17 Tethered carborane nucleoside pyrimidines

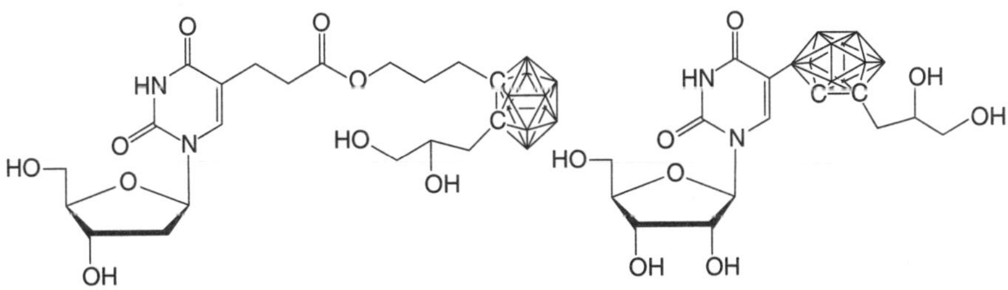

Figure 2.18 Hydrophilically enhanced carboranyl pyrimidine nucleosides

2.3 Phosphates, Phosphonates and Phosphoramidates

It has been observed that ^{32}P was concentrated in and persisted to an extraordinary high degree in a variety of different malignant tumors.[47] This led to the syntheses of phosphorus-containing boron compounds. Early work was by

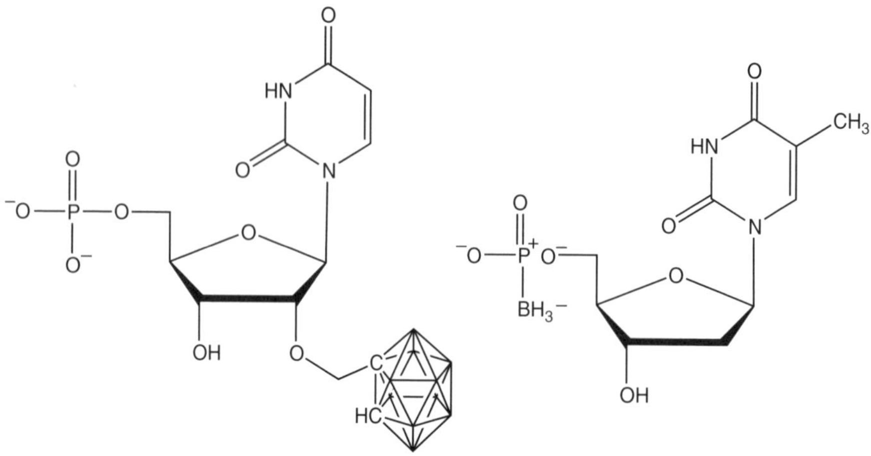

Figure 2.19 Boron-containing nucleotides

Kaczmarczyk on the syntheses of some phosphate esters having attached boron cages, as shown in Figure 2.20.[48] Other, neutral, phosphorus-containing carboranes have also been reported, several of which are shown in Figure 2.21.

Preliminary studies show that many of these compounds are highly toxic. However, more complete biological assessments of boron-containing phosphates and phosphonates, as potential tumor-targeting BNCT agents, remain to be carried out.

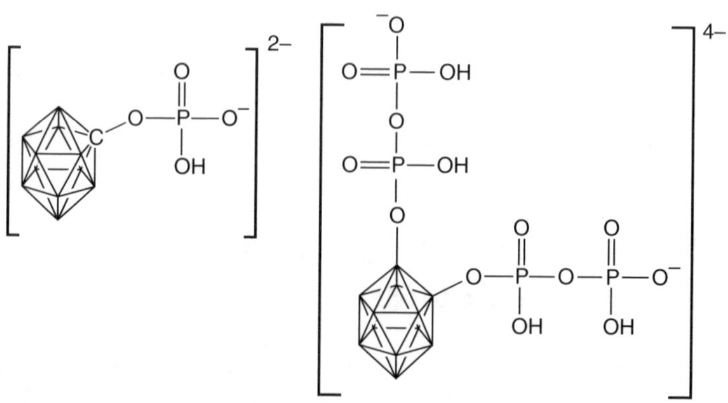

Figure 2.20 Phosphate esters of polyhedral boranes

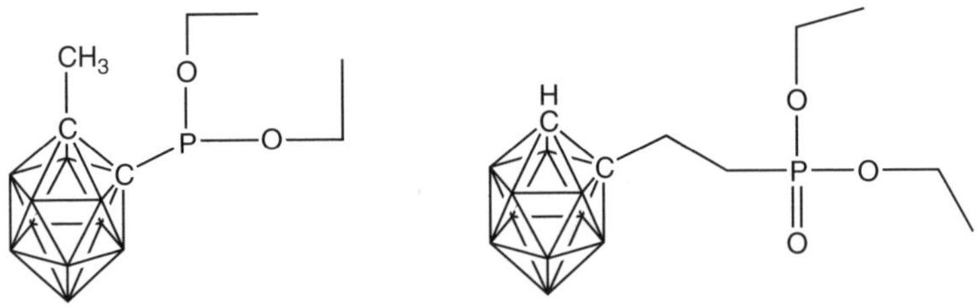

Figure 2.21 Phosphorus-containing carboranes

2.4 Amines

Although boron-containing amines themselves do not act as tumor-targeting agents, such compounds can serve as intermediates in the development of such delivery agents. The syntheses of boronated nitrogen-containing compounds are well established. While the synthesis of boron-substituted pyrimidines from the reaction of iodomethaneboronic esters with mercaptopyrimidines was reported by Matteson and Cheng in 1968,[49] more recently several carborane amides have been synthesized; some representative structures of boron-containing amines are shown in Figure 2.22.[50]

2.5 Boron Analogues of Pyrophosphates

Pyrophosphates with (P—O—P) linkage and methylene bisphosphonates (P—CH_2—P) have been shown to possess potent anti-viral activity against

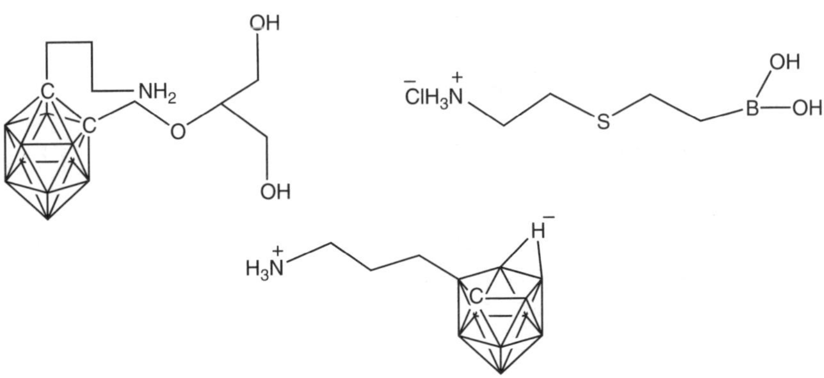

Figure 2.22 Boronated amines

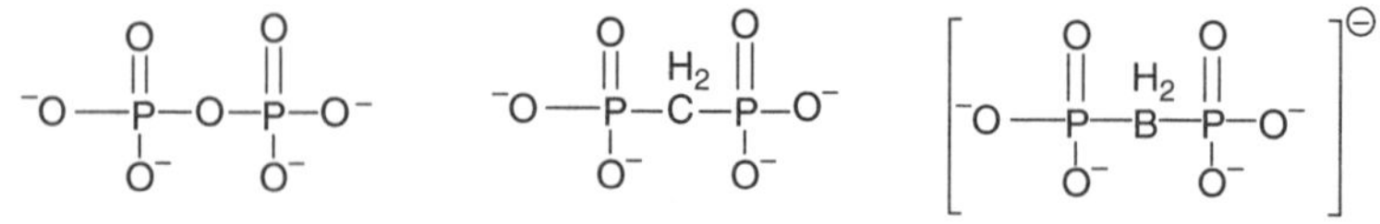

Figure 2.23 Analogues of pyrophosphates

the growth of Herpes viruses as well as other viruses. Phosphonoacetic acid and phosphonoformic acid derivatives have been clinically proven to be effective against a host of viral diseases.[51] Pyrophosphates also play a significant role in the regulation of the calcification and decalcification process *in vivo* and are, therefore, important in calcium metabolism and prevention of bone disorders.[51] Unfortunately, due to low hydrolytic stability, *in vivo* therapeutic use of pyrophosphates is limited. Bisphosphonates, on the other hand, have widespread use for treating bone disorders, such as Paget's disease.[51] A dihydroboride group (BH_2^-) is isoelectronic with atomic oxygen and the methylene group. Thus, the boron analogues of pyrophosphates and geminal bisphosphonates in which the bridging oxygen or methylene group is replaced with a dihydroboron group have been synthesized (Figure 2.23). A significant difference between the boron and non-boron analogues is the additional negative charge accompanying boron substitution.

The synthetic scheme to prepare boron analogues of pyrophosphates is given in Scheme 2.2. These boron analogues, which have an additional positive charge, are being evaluated for their anti-viral activities.

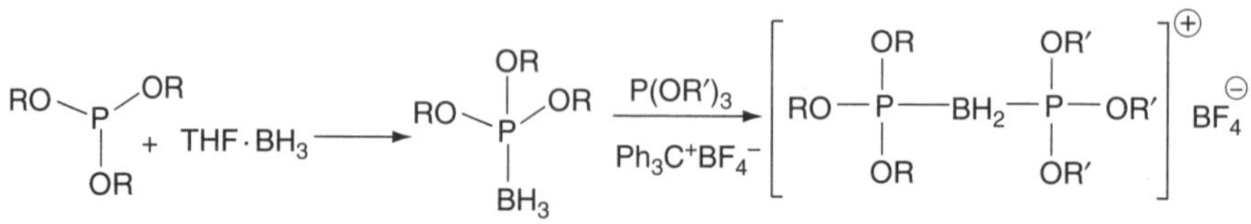

Scheme 2.2 Synthesis of boronated pyrophosphates

2.6 Boronated Polyamines

Polyamines, such as spermidine, spermine, and putrescine, are essential for the growth and proliferation of all the cells (Figure 2.24).[52] The presence of quickly dividing cells, such as cancer cells, is closely correlated with an

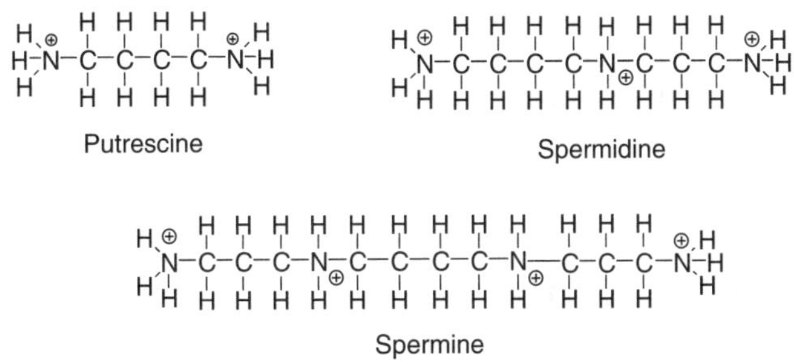

Figure 2.24 Naturally occurring polyamines

increase in the cellular polyamines.[53] Because of their tumor specificity, the polyamines have been suggested as potential targeting agents for delivering toxic moieties preferentially to cancer cells.[54] Other advantages of using the polyamine system are that the transporter is very aggressive, increasing the cellular polyamine concentration to 6 mM within a few hours,[55] can tolerate structural restrictions,[56] and accumulates in close proximity to nuclear and mitochondrial DNA, thereby increasing the potential for lethal damage by BNCT.

Strategies could consist of replacing a carbon sequence, such as the four carbons in putrescine, with a carborane moiety, such as a 1,12-$C_2B_{10}H_{10}$ (*p*-carboranyl) group. Other approaches would include using carboranes as pendant groups on the polyamine. The polyamines having carborane cages would combine the advantages of tumor specificity and increased boron concentration in tumor cells, which is ideal for BNCT. Representative compounds related to spermine and spermidine are shown in Figure 2.25.

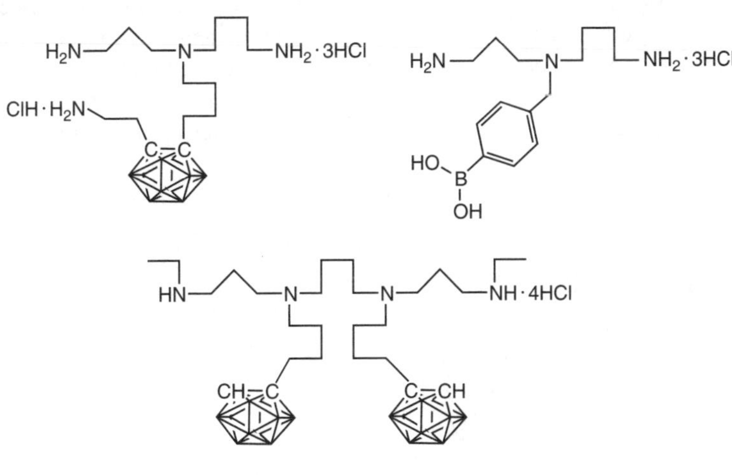

Figure 2.25 Polyamines

DNA-binding studies demonstrated that these structures simulated and competed with the naturally occurring polyamines. At present, the major limitations observed with these boron-containing polyamines is both their cellular and *in vivo* toxicity. Less toxic analogues have been prepared,[57] but it remains to be determined what clinical utility they can offer.

2.7 Carbohydrates

It is known that there are significant differences in the carbohydrate composition of the cell membrane surfaces of malignant cells when compared to normal cells.[58] It is possible that boron carbohydrate compounds may achieve elevated concentrations in tumor cells through the action of the glucose transport system.[59] The initial boron cage compounds containing a carbohydrate moiety were those that were developed for binding to antibodies.[60] The purpose of the carbohydrate functionality was solely to increase the compound's aqueous solubility. Subsequently, carborane-containing analogues of glucose, and gulose have been prepared (Figure 2.26).

More recently, a series of carbohydrates have been synthesized in which the BSH moiety has been attached to the carbohydrate group (Figure 2.27). However, the evaluation of their bio-distribution is limited and much more remains to be accomplished.

2.8 Lipoproteins

The rate of metabolism of low density lipoproteins (LDLs) is higher in tumor cells than in the normal cells.[61] This difference is due to the increased need that tumor cells possess for cholesterol in order to facilitate new membrane formation. The overexpression of the LDL receptor on the tumor cell membrane is responsible for its LDL accretion. This observation led Kahl to first propose the use of such receptor mediation as the basis for selectively delivering boron to tumor

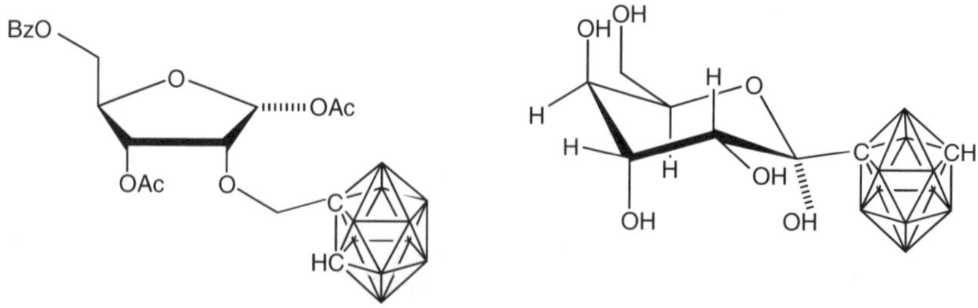

Figure 2.26 Carborane-containing analogues of carbohydrates

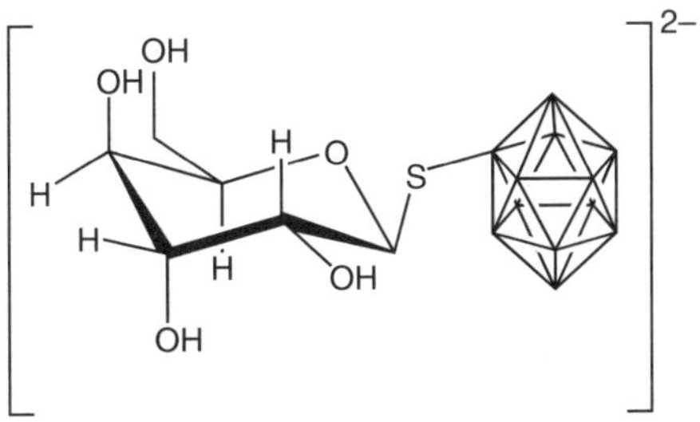

Figure 2.27 Thiododecarborate-containing sugars

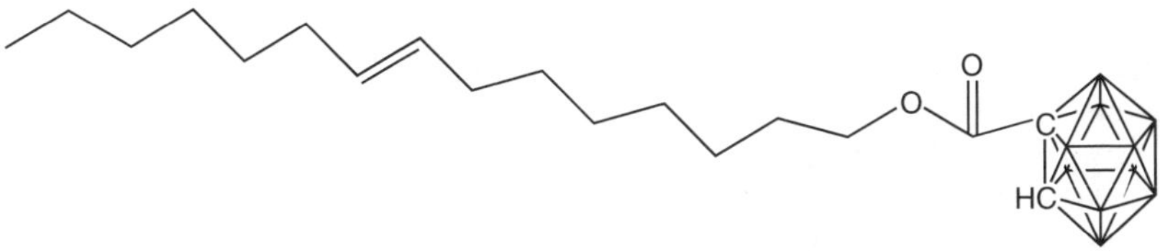

Figure 2.28 A boronated component for reconstitution with LDLs

cells.[62] The initial compounds synthesized were esters of carborane carboxylic acid with various fatty acid alcohols; an example is shown in Figure 2.28.

Since this initial work, Kallio and coworkers have additionally explored the potential use of boron-containing LDLs as tumor delivery agents, but further studies need to be done in this area.[63]

2.9 Lipids and Phospholipids

It has been found that various natural and synthetic ether lipids possess a proclivity for and persistence in a variety of spontaneous and chemically induced tumors. This is due, in part, to the absence of the enzyme O-alkylglycerol mono-oxidase in their neoplasms. Accordingly, Lemmen and Werner have synthesized the carborane-containing phospholipid shown in Figure 2.29.[64] However, the results both from cell culture studies and in tumor-bearing

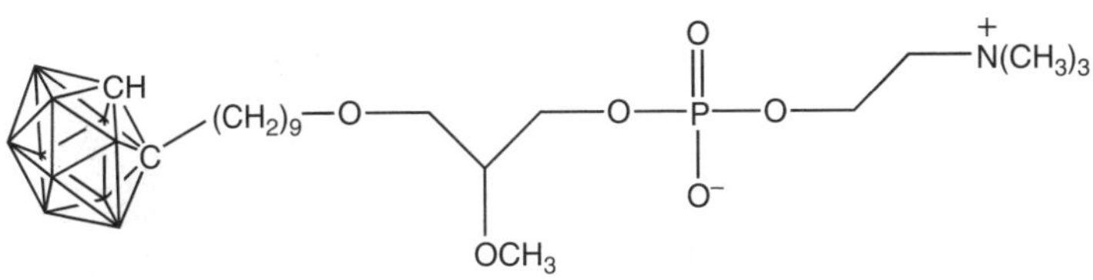

Figure 2.29 A carborane-containing phospholipid

animals were not particularly encouraging. Since then, very little additional work has appeared in the literature.

2.10 Radiation Sensitizers

The development of radiation sensitizers arose from the observation that various nitroimidazoles appear to be taken up selectively by poorly vascularized areas of tumors and are retained therein by their reductive alkylation through the metabolic formation of electrophiles.[65] On this basis, Scobie and Threadgill first proposed the investigation of boronated analogues of nitroimidazoles.[66] Such polyether-linked carbonylnitroimidazoles possessed improved aqueous solubility, and preliminary biological studies in tumor-bearing mice demonstrated both tumor uptake and retention.[67] Another approach in the development of boron-containing radiation sensitizers was to couple BSH to 1-(2-bromoethyl)-2-methyl-5-nitroimidazole to give the boronated imidocaptate, as described by Swenson *et al.*[68] Structures of this and other boronated nitroimidazoles are shown in Figure 2.30.

2.11 Cyclic Thiourea Derivatives

The rationale behind the development of boronated cyclic thiourea derivatives is that their boron-free analogues have the capacity for becoming incorporated into the polymeric structure of melanin through covalent bonding into melanin precursors, based on the enol form of the thiourea moiety.[69] Tjarks and Gabel synthesized carborane-containing cyclic thioureas as well as the first dihydroxyl-

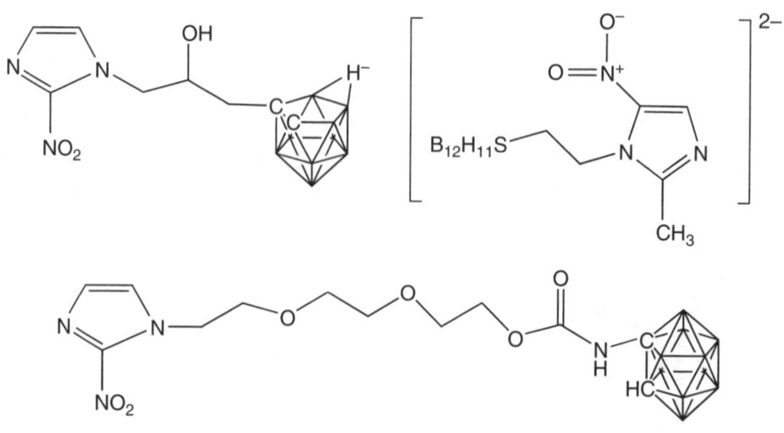

Figure 2.30 Boronated nitroimidazoles

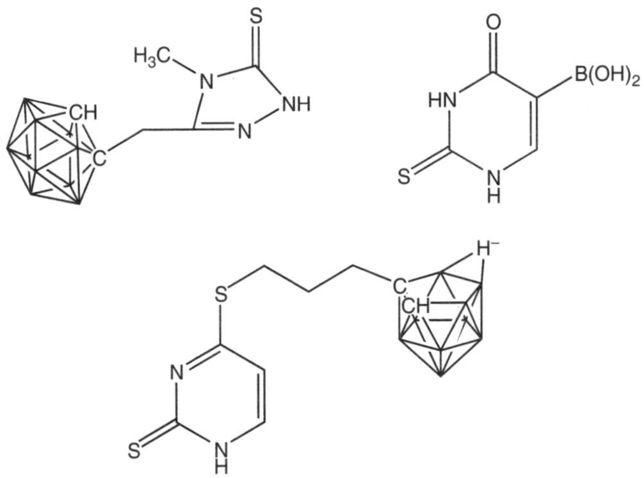

Figure 2.31 Boronated thioureas

boryl-containing thiouracils.[70] This compound showed greater tumor persistence than did BPA.[70] Substantial synthetic work has been focused on the synthesis of *closo*-carboranylthiouracils.[71] Representative structures are depicted in Figure 2.31.

2.12 Central Nervous System (CNS) Depressants: Promazines, Hydantoins and Barbiturates

The basis for synthesizing boronated promazines arose from localization studies of chloropromazines (CPZ) in melanoma-bearing cells.[72,19c] The first such boron-containing compounds were those in which the boron moieties were attached to the aliphatic nitrogen of CPZ.[73] Unfortunately, the results with these compounds were not promising. After these initial syntheses, carborane-containing promazines were described in which the boron entity was incorporated into the aromatic rings of the phenothiazine nucleus. Figure 2.32 shows some of the examples of these compounds.

2.13 Hydantoins and Barbiturates

Certain depressants have been shown to enter the brain rapidly and also achieve concentration ratios of the order of 4:1 between primary brain tumors

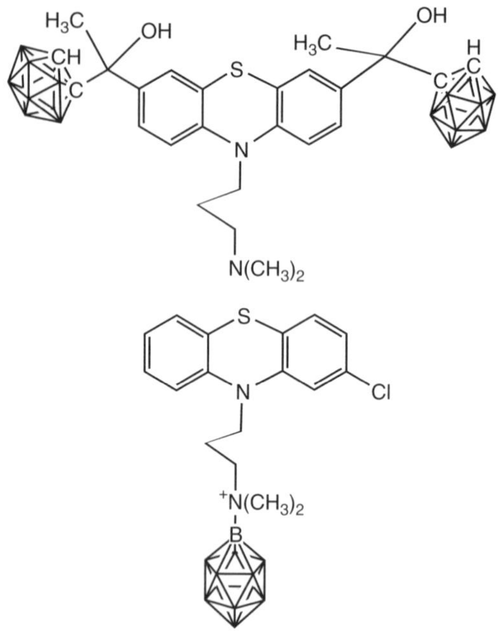

Figure 2.32 Boron-containing promazines

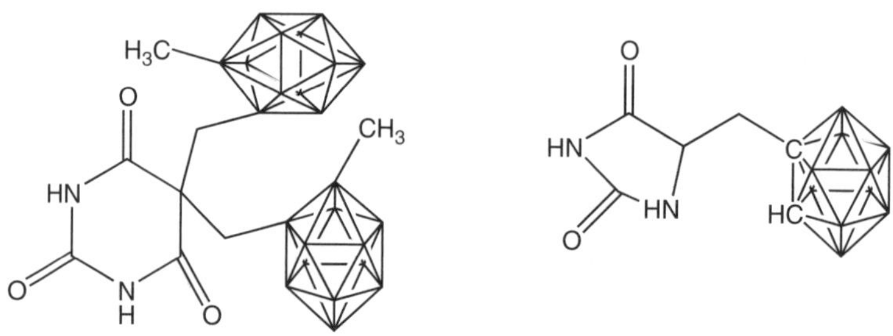

Figure 2.33 Boronated barbiturate and hydantoin

and normal brain.[74] Therefore, some efforts have been directed toward attaching boron cages on barbiturate and hydantoin molecules (Figure 2.33). Their bio-distribution studies need to be investigated further.

2.14 Oligonucleotide Antisense Agents

Over the past decade, there has been extensive work in the design, synthesis and application of nuclease-resistant oligonucleotide analogues for therapeutics, research and diagnostic use.[75] The research in this area arises from the fact that most forms of diseases are the result of a function of protease enzymes, and most therapeutic drugs are directed toward these proteins. Antisense technology deals with the destruction of disease-causing proteins. This is done by inhibiting the particular protein's production through hydrogen bonding of nuclease-resistant oligonucleotides. More specifically, this therapy relies upon the hybridization of sequence-specific hydrogen bonding of oligonucleotides to RNA and single-stranded DNA so as to interfere in the translation process (protein synthesis).[76] In this regard, research has been carried out in modification or replacement of the basic structure of the nucleic acids, e.g. phosphorus backbone, sugar or base moieties.

Oligonucleotides, in which non-bridging oxygen atoms are replaced by borane (BH_3) groups, constitute a very important class of modified nucleic acids.[4] These compounds seek to take advantage of the fact that the BH_3 group is isoelectronic with atomic oxygen in natural oligonucleotides, and isoelectronic and isostructural with the oligonucleotide methyl phosphonates, which are nuclease-resistant. The structures of several boron-containing dinucleotide analogues are shown in Figure 2.34.[77]

On the other hand, the α-borano triphosphates are good substrates for DNA polymerases and incorporation of boranophosphates into DNA causes an increase in the resistance to exo- and endo-nucleases as compared to non-modified DNA.[3a] In addition, important new diagnostic applications have been reported in the areas of polymerase chain reaction (PCR) sequencing[76] and DNA diagnostics using boronated DNA. Figure 2.35 summarizes some of the applications of boronated nucleosides, nucleotides and oligonucleotides.[78]

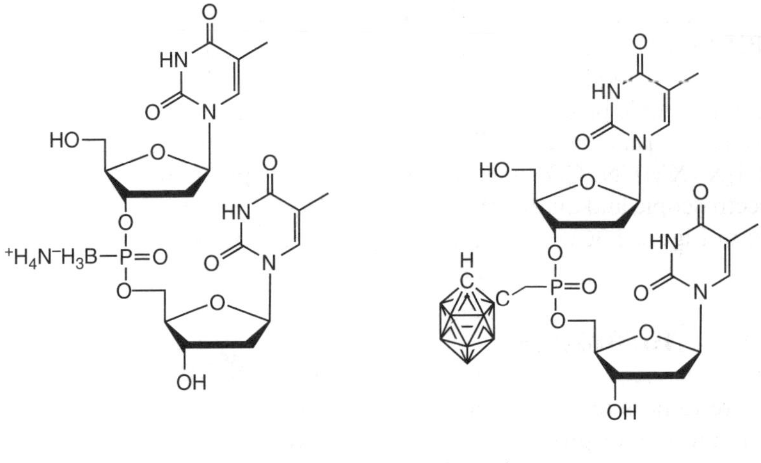

Figure 2.34 Dinucleotide analogues

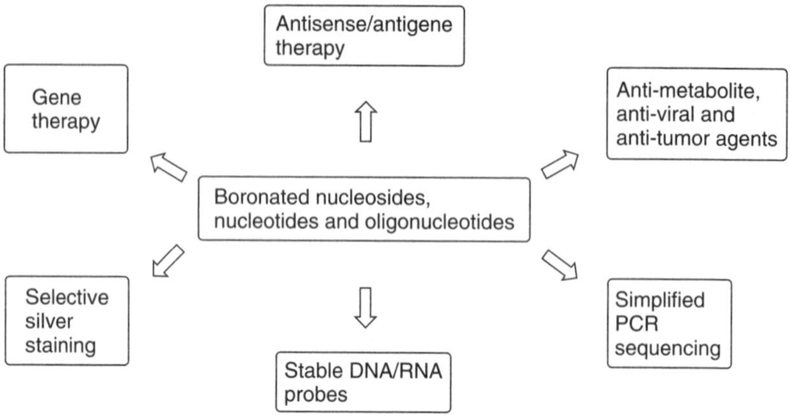

Figure 2.35 Applications of boronated nucleosides, nucleotides and oligonucleotides

Although considerable efforts have been devoted to taking advantage of the boranophosphate linkage, numerous limitations are inherent in the use of the BH_3 moiety, especially in chemical syntheses. The highly reducing nature of this group can cause base degradation. It has been reported that the borane group is incompatible with some commonly used protecting groups in modified oligonucleotide synthesis.[29] Likewise, the BH_3 moiety has severe toxicity implications in that borohydrides, boranocarbonates and amineboranes typically have LD_{50} values in the tens of mg/kg by ip injection in mice. The syntheses and properties of biologically important molecules that contain substituted boranes of the form BH_2X (X = COOR, C(O)NHR, CN, etc.) have been investigated extensively.[1,79] The use of these substituted boranes in oligonucleotide modification may ameliorate some of the problems encountered with the unsubstituted boranonucleic acids. Since the boron analogue of glycine, H_3NBH_2COOH, was found to have a very low toxicity (LD_{50} > 2000 mg/kg),[30] there is a good chance that the use of substituted boranonucleosides will result in less toxic products.

Hosmane *et al.* have synthesized modified oligonucleotides containing P—BH_2X (X = CN, COOMe and CONHEt) linkages and characterized them by spectroscopic and analytical techniques.[80] The corresponding structures are shown in Figure 2.36.

2.15 Hormones

There are tumors that are formed in various tissues and organs whose development and rate of proliferation are strongly hormone-dependent. For example, the steroid hormones have receptors that are localized in the cell's nucleus,

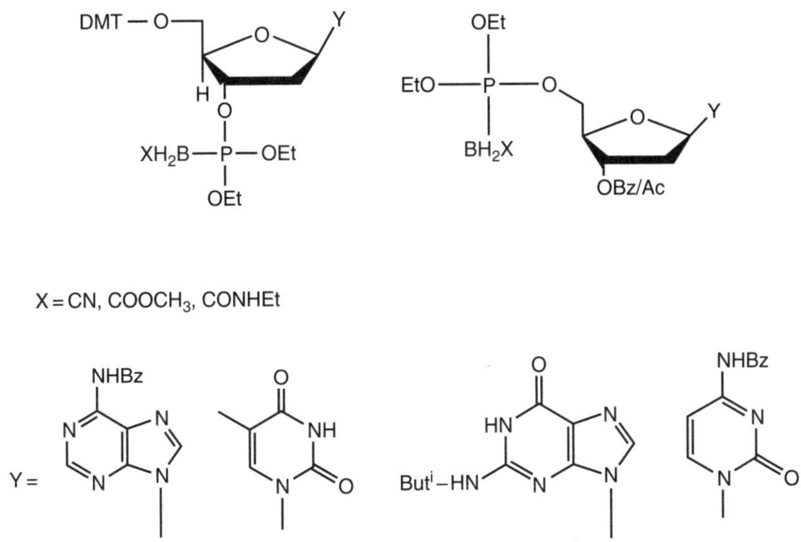

Figure 2.36 Some nucleoside derivatives

the key target for high LET radiation. It was this observation that prompted the exploration of hormones as BNCT agents. Sweet synthesized the first carboranyl derivative of estradiol (Figure 2.37).[81]

Subsequently, Wellmann and Gabel have synthesized boron analogues of hormonal antagonists and other estrogenic and androgenic compounds.[82] One of the advantages of these boronated steroid hormones is their relatively low

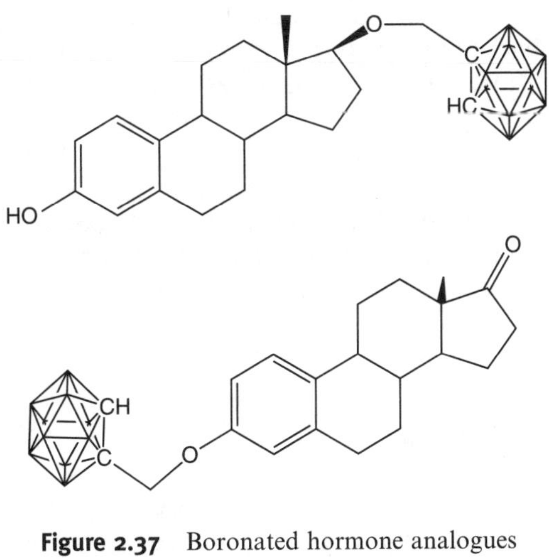

Figure 2.37 Boronated hormone analogues

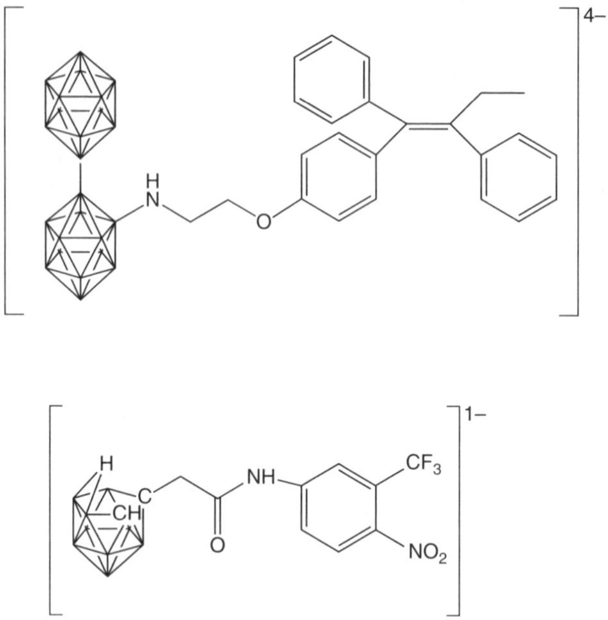

Figure 2.38 Boronated hormonal antagonists

molecular weights, offering the potential for the rapid targeting of tumor cells in contrast to other growth factors. Key requirements for such compounds are that they be chemically stable under *in vivo* conditions, be non-toxic and retain their hormonal activities. Several of these esterogen-related compounds have demonstrated the anticipated specificity for malignant cells; however, the maximum number of ^{10}B atoms that could be achieved with these carboranyl compounds was in the order of 10^5–10^6 atoms of ^{10}B atoms per cell.

There has been renewed interest in the synthesis of boron-containing steroids.[83] Also, a recent work in this area has exemplified the coupling of a boron-containing moiety to a ligand possessing binding specificity for intracellular hormone receptors (Figure 2.38).[84]

Though these compounds have intrinsic advantages, especially with regard to the rapidity of tumor targeting, the fact that adequate receptor site density is a *sine qua non* in the development of useful BNCT agents has not been adequately addressed.

2.16 Liposomes

While LDLs are natural lipoproteins with a proclivity for the tumor cells in which the receptor for the vesicle is overexpressed, liposomes can be viewed as

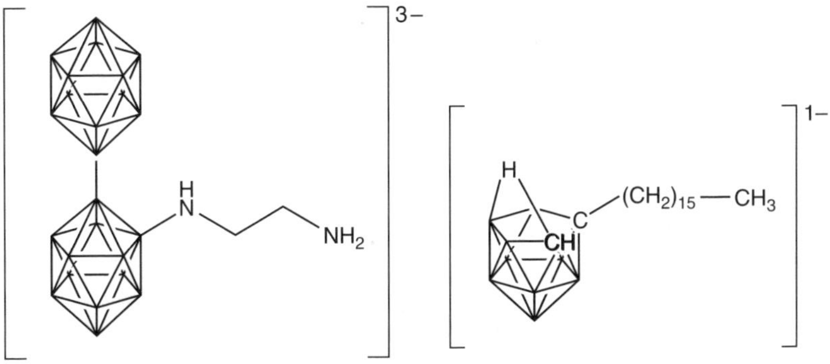

Figure 2.39 Boron compounds for incorporation into liposomes

related synthetic vesicles. Hawthorne and his associates have been involved in a major effort to develop liposomes as boron-delivery agents to tumors. Small unilamellar liposomes have been shown to penetrate the tumor cell membrane and localize intracellularly. Therefore, incorporating boron compounds within such vesicles should provide a method for achieving selectivity between tumor and normal cells. Several of the compounds developed for this purpose are shown in Figure 2.39.[19c,26,84]

A key question is whether the liposome can pass through the blood/brain barrier (BBB). Unilamellar liposomes with diameters as small as 60 nm are incapable of crossing the BBB. Liposomes have also been used by others to deliver boronated thiouracils to melanoma cells and as a targeting vehicle for BSH through intracerebral administration. An advantage of using liposomes is that the boron species itself does not need to possess tumor-targeting properties. Once the compound is deposited either intracellularly within the tumor cell or interstitially within the tumor by the liposome and has properties that result in cell penetration and binding to subcellular organelles, persistence would be achieved. With respect to brain tumors, the unique restrictive nature of the normal BBB limits penetrability of the CNS, especially for liposomes greater than 60 nm.[19c] However, in the total scheme of things for all tumors, liposomes may well be an important delivery system for BNCT agents.[84]

2.17 Conclusions

Boronated compounds have found their usefulness in all areas of medicine, pharmacology and science. Though there has been a number of extensive studies done on the use of a variety of boron compounds as BNCT agents, other areas should be explored. In this regard, it is essential that any new compound must be compared biologically against the two agents, BSH and

BPA. The main thrust of the research to date has been in the syntheses of boronated analogues of compounds that are known to concentrate preferentially in cancer cells. The list is extensive and the search for more efficient boron-delivery agents is an ongoing endeavor.

Acknowledgements

This work was supported by grants from the National Science Foundation (CHE-0241319), the donors of the Petroleum Research Fund, administered by the American Chemical Society, The Robert A. Welch Foundation (N-1322 to J.A.M.), and Northern Illinois University through a Presidential Research Professorship. The Forschungspreis der Alexander von Humboldt-Stiftung (to N.S.H.) is also hereby gratefully acknowledged.

References

1. (a) K. Vyakaranam; G. Rana; N.S. Hosmane and B.F. Spielvogel, *Metal-Based Drugs*, **3**, 145 (2001). (b) K. Vyakaranam; G. Rana; B.F. Spielvogel and N.S. Hosmane, *Inorg. Chem. Commun.*, **4**, 629 (2001).
2. A. Hasan; H. Li; J. Tomasz and B.R. Shaw, *Nucl. Acid Res.*, **24**, 2150 (1996).
3. (a) H. Li; K. Porter; F. Huang and B.R. Shaw, *Nucl. Acid Res.*, **23**, 4495 (1995). (b) A. Sood; B.R. Shaw and B.F. Spielvogel, *J. Am. Chem. Soc.*, **112**, 9000 (1990). (c) A. Sood; B.F. Spielvogel and B.R. Shaw, *J. Am. Chem. Soc.*, **111**, 9234 (1989).
4. B.F. Spielvogel; A. Sood; B.R. Shaw *et al.*, *Progress in Neutron Capture Therapy for Cancer*, Proceedings of International Symposium, Sydney, Australia, **4** (1992), pp. 211–213.
5. B.R. Shaw; J. Madison; A. Sood and B.F. Spielvogel, *Methods Mol. Biol.*, **20**, 225 (1993).
6. R.R. Kane; K. Drechsel and M.F. Hawthorne, *J. Am. Chem. Soc.*, **115**, 8853 (1993).
7. B.F. Spielvogel; A. Sood; B.R. Shaw and I.H. Hall, *Pure Appl. Chem.*, **63**, 415 (1991).
8. (a) B.F. Spielvogel; A. Sood; J. Tomasz *et al.*, In *Advances in Neutron Capture Therapy*; Soloway, A.H., Barth, R.F. and Carpenter, D.E., Eds; Plenum Press, New York, Vol. 5, 1993, pp. 361–365. (b) M. Takagaki; W. Powell; A. Sood *et al.*, *Radiat. Res.*, **156**, 118 (2001).
9. R.R. Kane; R.H. Pack and M.F. Hawthorne, *J. Org. Chem.*, **58**, 991 (1993).
10. (a) I.H. Hall; E.S. Hall; L. Chi *et al.*, *Anticancer Res.*, **12**, 1091 (1992). (b) G. Rana; K. Vyakaranam; B.F. Spielvogel *et al.*, *Inorg. Chim. Acta*, **342**, 255 (2003).
11. A. Sood; B.F. Spielvogel; B.R. Shaw *et al.*, *Anticancer Res.*, **12**, 335 (1992).
12. A. Sood; B.R. Shaw; B.F. Spielvogel *et al.*, *Pharmazie*, **47**, 833 (1992).
13. B.F. Spielvogel; A. Sood; B.R. Shaw and I.H. Hall, *Curr. Top. Chem. Boron*, Kabalka, G.W., Ed.; Royal Society of Chemistry: Cambridge, United Kingdom, 1994 (Sp. publ. 143), pp. 193–198.
14. K.G. Rajendran; B.S. Burnham; S.Y. Chen *et al.*, *J. Pharm. Sci.*, **83**, 1391 (1994).
15. J. Tomasz; B.R. Shaw; K.W. Porter *et al.*, *Angew. Chem., Int. Ed. Engl.*, **31**, 1373 (1992).

16. G.L. Locker, *Am. J. Roentgenol. Radiat. Ther.*, **36**, 1 (1936).
17. M. Javid; G.L. Brownell and W.H. Sweet, *J. Clin. Invest.*, **31**, 604 (1952).
18. G. Friedlander; J.W. Kennedy; E.S. Macias and J.M. Miller, *Nucl. Radiochem.*, 3rd edn, John Wiley & Sons, New York, 1981.
19. (a) R.F. Barth; A.H. Soloway and R.M. Brugger, *Cancer Invest.*, **14**, 534 (1996). (b) R.G. Zamenhof; A.M. Kalend and W.D. Bloomer, *J. Natl. Cancer Inst.*, **84**, 1290 (1992). (c) A.H. Soloway; W. Tjarks; B.A. Barnum *et al.*, *Chem. Rev.*, **98**, 1515 (1998).
20. See http://www.cnea.gov.ar/xxi/bnct/eng/. Last date visited. April 14, 2003.
21. W. Tjarks; A.K.M. Anisuzzaman; L. Liu *et al.*, *J. Med. Chem.*, **35**, 1628 (1992).
22. H. Ghaneolhosseini; W. Tjarks and S. Sjoberg, *Tetrahedron*, **54**, 3877 (1998).
23. (a) Z.J. Lesnikowski and R.F. Schinazi, *Pol. J. Chem.*, **69**, 827 (1995). (b) Y. Yamamoto and K.I. Imamura, *Bioorg. Med. Chem. Lett.*, **6**, 1855 (1996). (c) F.G. Rong; A.H. Soloway; S. Ikeda and D.H. Ives, *Nucleosides Nucleotides Nucleic Acids*, **16**, 379 (1997).
24. (a) M.F. Hawthorne; R.J. Wiersema and M. Takasugi, *J. Med. Chem.*, **15**, 449 (1972). (b) E.A. Mizusawa; H.L. Dahlman; S.J. Bennett and M.F. Hawthorne, *Proc. Natl. Acad. Sci. USA*, **79**, 3011 (1982).
25. S.B. Kahl and M.S. Koo, *J. Chem. Soc., Chem. Commun.*, 1769 (1990).
26. K. Shelly; D.A. Feakes; M.F. Hawthorne *et al.*, *Proc. Natl. Acad. Sci. USA*, **89**, 9039 (1992).
27. (a) J.L. Fauchere; O. Leukart; A. Eberle and R. Schwyzer, *Helv. Chim. Acta*, **62**, 1382 (1979). (b) J. Malmqist and S. Sjoberg, *Tetrahedron*, **52**, 9207 (1996).
28. W.H. Sweet; A.H. Soloway and G.L. Brownell, *Acta Union Int. Contre Cancer*, **16**, 1216 (1960).
29. D.A. Allen and T.D. Beynon, *Phys. Med. Biol.*, **40**, 807 (1995).
30. J.A. Coderre; M.S. Makar; P.L. Micca *et al.*, *Abstracts of the Fifth International Symposium on Neutron Capture Therapy for Cancer*; Columbus, OH, 1992.
31. T. Hartman and J. Carllson, *Radiother. Oncol.*, **31**, 61 (1994).
32. P.J. Perry; M.A. Read; R.T. Davies *et al.*, *J. Med. Chem.*, **42**, 2679 (1999).
33. (a) P.J. Perry; A.P. Reszka; A.A. Wood *et al.*, *J. Med. Chem.*, **41**, 4873 (1998). (b) K. Vyakaranam; G. Rana; J.A. Maguire and N.S. Hosmane, *Organometallics*, **21**, 3905 (2002).
34. (a) F. Wellmann; R. Abraham; R. Muller and D. Gabel, *Z. Naturforsch. C*, **46**, 252 (1991). (b) M. Miura; D. Gabel; G. Oenbrink and R.G. Fairchild, *Tetrahedron Lett.*, **31**, 2247 (1990).
35. F. Li; K. Shelly; R.R. Kane *et al.*, *J. Am. Chem. Soc.*, **118**, 6506 (1996). (b) R.A. Watson-Clark; C.B. Knobler and M.F. Hawthorne, *Inorg. Chem.*, **35**, 2963 (1996).
36. F. Haslinger and A.H. Soloway, *J. Med. Chem.*, **9**, 792 (1966).
37. W. Tjarks; H. Ghaneolhosseini; C.L.A. Henssen *et al.*, *Tetrahedron Lett.*, **37**, 6905 (1996).
38. K. Yoshino; T. Maruyama; H. Takahashi *et al.*, *Advances in Neutron Capture Therapy*; Plenum Press, New York, 1993.
39. Y. Mori; A. Suzuki; K. Yoshino and H. Kakihana, *Pigment Cell Res.*, **2**, 273 (1989).
40. H. Nemoto; S. Iwamoto; H. Nakamura and Y. Yamamoto, *Chem. Lett.*, **22**, 465 (1993).
41. B.F. Spielvogel; A.T. McPhail; M.K. Das and I.H. Hall, *J. Am. Chem. Soc.*, **102**, 6343 (1980).
42. K.H. Scheller; R.B. Martin; B.F. Spielvogel and A.T. McPhail, *Inorg. Chim. Acta*, **57**, 227 (1982).

43. (a) W. Karnbrock; H.J. Musiol and L. Moroder, *Tetrahedron*, **51**, 1187 (1995). (b) S. Sjoberg; M.F. Hawthorne; P. Lindstrom *et al.*, In *Advances in Neutron Capture Therapy*; Soloway, A.H., Barth, R.F. and Carpenter, D.E., Eds.; Plenum Press, New York, Vol. 5, 1993, pp. 269–272.
44. (a) I.M. Wyzlic and A.H. Soloway, *Tetrahedron Lett.*, **33**, 7489 (1992). (b) P.A. Radel and S.B. Kahl, *J. Org. Chem.*, **61**, 4582 (1996). (c) J.K. Prashar and D.E. Moore, *J. Chem. Soc., Perkin. Trans.*, 1051 (1993).
45. (a) Y. Yamamoto, *Pure Appl. Chem.*, **63**, 423 (1991). (b) G. Palmisano and M. Santagostino, *Tetrahedron*, **49**, 2533 (1993).
46. A.H. Soloway; A.K.M. Anisuzzaman; F. Alam *et al.*, *Pure Appl. Chem.*, **63**, 411 (1991).
47. B. Selverstone; W.H. Sweet and C. Robinson, *Ann. Surg.*, **130**, 643 (1949).
48. K.C. John; A. Kaczmarczyk and A.H. Soloway, *J. Med. Chem.*, **12**, 54 (1969).
49. D.S. Matteson and T.C. Cheng, *J. Org. Chem.*, **33**, 3055 (1968).
50. (a) K. Shelly; M.F. Hawthorne and C.B. Knobler, *Inorg. Chem.*, **31**, 2889 (1992). (b) G. Rana; K. Vyakaranam; S.C. Ledger *et al.*, *Appl. Organomet. Chem.*, **17**, 361 (2003).
51. (a) H. Fleish, *Drugs.*, **42**, 919 (1991). (b) G. Rana; K. Vyakaranam; B.F. Spielvogel *et al.*, *Inorg. Chim. Acta*, **344C**, 249 (2003).
52. (a) A.E. Pegg and P.P. McCann, *Am. J. Physiol.*, **C212** (1982). (b) C.W. Tabor and H. Tabor, *Annu. Rev. Biochem.*, **53**, 749 (1984).
53. B.K. Chang; P.R. Libby; R.J. Bergeron and C.W. Porter, *Biochem. Biophys. Res. Commun.*, **157**, 264 (1988).
54. P.A. Stark; B.D. Thrall; G.G. Meadows and M.M. Abdel-Monem, *J. Med. Chem.*, **35**, 4264 (1992).
55. J.L.A. Mitchell; R.R. Diveley and A. Bareyal-Leyser, *Biochem. Biophys. Res. Commun.*, **186**, 81 (1992).
56. N.A. Khan; V. Quemener and J.-P. Moulinoux, *Cell Biol. Int. Rep.*, **15**, 9 (1991).
57. (a) J.C. Zhuo; J. Cai; A.H. Soloway *et al.*, *J. Med. Chem.*, **42**, 1282 (1999). (b) J. Cai; A.H. Soloway; R.F. Barth *et al.*, *J. Med. Chem.*, **40**, 3887 (1997).
58. J.R. Mangiardi, In *Brain Tumors: An Encyclopedic Approach*. Kaye, A.H. and Laws, E.R. Jr, Eds; Churchill Livingstone, Edinburgh, p. 99, 1995.
59. T. Peymann, *Sauerstoff und Schwefel am Undecahydro-closo-dodecaborate (2-) als Nucleophile*; Ph.D. Thesis, University of Bremen, Germany, 1995.
60. M.F. Hawthorne, *Pure Appl. Chem.*, **63**, 327 (1991).
61. Y.K. Ho; R.G. Smith; M.S. Brown and J.L. Goldstein, *Blood*, **52**, 1099 (1978).
62. S.B. Kahl, *Proceedings of the Workshop of the Radiation Research Program of NCI*; Annapolis, Maryland and Zink, S., Eds; National Cancer Institute, Bethesda, MD; March 3–4, 1988.
63. M. Kallio; J. Callaway; E. Saario and S. Kahl, *Cancer Neutron Capture Therapy*, Mishima, Y., Ed.; Plenum Press, New York, 1996.
64. P. Lemmen and B. Werner, *Chem. Phys. Lipids*, **62**, 185 (1992).
65. K.A. Kennedy, *Anticancer Drug Des.*, **2**, 181 (1987).
66. M. Scobie and M.D. Threadgill, *J. Chem. Soc., Perkin. Trans.*, 203 (1994).
67. P.J. Wood; M. Scobie and M.D. Threadgill, *Int. J. Radiat. Biol.*, **70**, 587 (1996).
68. D.H. Swenson; B.H. Laster and R.L. Metzger, *J. Med. Chem.*, **39**, 1540 (1996).
69. J.R. Whittaker, *J. Biol. Chem.*, **111**, 6217 (1971).
70. W. Tjarks, *Borhaltige Thioharnstoffe in der Neutroneneinfangtherapie*; Ph.D. Thesis, University of Bremen, Germany, 1989.

71. B.S. Larsson; K. Olander; L. Dencker and L. Br. Holmquist, *J. Cancer*, **46**, 538 (1982).
72. R.G. Fairchild; D. Greenberg; K.P. Watts *et al.*, *Cancer Res.*, **42**, 556 (1982).
73. T. Nakagawa and K. Aono, *Chem. Pharm. Bull.*, **24**, 778 (1976).
74. T. Rosenblum and A. Stein, *Biochem. Biopharmacol.*, **12**, 1453 (1963).
75. (a) M.I. Phillips, Ed., *Antisense Technology Part A: General Methods, Methods of Delivery and RNA Studies; Methods in Enzymology* 313; Academic Press, New York, 226, 2000. (b) M.I. Phillips, *Antisense Technology Part B: Applications; Methods in Enzymology* 314; Academic Press, New York, 237, 1999.
76. (a) E. Uhlman and A. Peyman, *Chem. Rev.*, **90**, 543–548 (1990). (b) R.W. Wagner, *Nature*, **372**, 333–335 (1994). (c) J.F. Milligan; M.D. Matteucci and J.C. Martin, *J. Med. Chem.*, **36**, 1923–1937 (1993).
77. Z.J. Lesnikowski and R.F. Schinazi, *J. Org. Chem.*, **58**, 6531 (1993).
78. B.F. Spielvogel; W. Powell and A. Sood, *Main Group Met. Chem.*, **19**, 699 (1996).
79. B.F. Spielvogel; G. Rana; K. Vyakaranam *et al.*, *Collect. Czech. Chem. Commun.*, **67**, 1095 (2002).
80. (a) K. Vyakaranam; G. Rana; B.F. Spielvogel *et al.*, *Nucleosides Nucleotides Nucleic Acids*, **21**, 581 (2002). (b) K. Vyakaranam and N.S. Hosmane, *Bioinorg. Chem. Appl.*, **2**, 31 (2004).
81. F. Sweet, *Steroids*, **37**, 223 (1981).
82. F. Wellmann and D. Gabel, *Proc. First Int. Symp. Neutron Capture Ther*; Brookhaven Natl. Lab. BNL, *BNL 51730*, 1983.
83. L. Schneiderova; O. Strouf; B. Gruner *et al.*, *Collect. Czech. Chem. Commun.*, **57**, 463 (1992).
84. M.F. Hawthorne, *Cancer Neutron Capture Therapy*, Mishima, Y., Ed.; Plenum Press, New York, 1996.

3 $_{12}$Mg The Role of Magnesium as a Metallotherapeutic Drug

Pietro Delva

3.1 Introduction

Magnesium is an element with an atomic number of 12 and a mass of 24.32 Da. It is the fourth most abundant cation in the body and the second most abundant cation in intracellular fluid. Nevertheless, the mechanisms involved in its regulation are still unknown. Magnesium is an essential cofactor for more than 300 cellular enzymes located in many intracellular compartments including nucleus, mitochondria, cytosol and microsomes. Magnesium is mainly involved in the activity of enzymes deputated to energy transfer through the action of adenosine nucleotides and also plays a role in protein and nucleic acid synthesis within the cell.

The knowledge of the detailed mechanisms regulating magnesium cellular homeostasis has been hampered by technical and methodological difficulties. In fact, until quite recently only the total magnesium concentration was measurable. Magnesium is present in three different states in most biological systems: freely coordinated to water, associated with anions and bound to protein. Because protein-bound and complexed magnesium are unavailable for biochemical processes, only ionized magnesium has biological activity. The synthesis of fluorescent probes quite specific for magnesium has greatly improved the knowledge about intracellular magnesium homeostasis.

Metallotherapeutic Drugs and Metal-Based Diagnostic Agents: The Use of Metals in Medicine
Edited by Gielen and Tiekink

Cytosolic magnesium concentration is far from its electrochemical equilibrium. If cytosolic magnesium was at electrochemical equilibrium, assuming a membrane potential of −70 mV and an extracellular magnesium concentration of about 0.8 mmol/l, its concentration would be approximately 190 mmol/l. The total intracellular magnesium concentration is about 12 mmol/l, thus it is evident that magnesium must be pumped out of the cell.[1–4] The great majority of cellular magnesium is bound to intracellular constituents such as nucleic acids, phospholipids ATP and proteins, and ionized magnesium concentration is in the 0.3–1 mmol/l range.[5,6]

On average, the human body contains approximately 1 mol (24 g) of magnesium.[7] About half the magnesium is present in the bone, and the other half is intracellular in soft tissue and muscle. Less than 1% of total body magnesium is present in plasma. This is particularly important to remember when interpreting laboratory data about magnesium. For example, plasma levels of magnesium reflect a very small fraction of total body magnesium and may not provide information on its status.

Recent studies indicate that the majority of magnesium is absorbed in the ilium and colon,[8] and that there is an inverse curvilinear relationship between intake of magnesium and fractional absorption which ranges from 65 to 10% absorption at low and high intake, respectively.[9] This suggests that for treating a magnesium deficiency a long oral supplementation period is required. The kidney is the major excretory organ for magnesium and is responsible for controlling plasma magnesium levels. The majority (70–80%) of plasma magnesium is filtered through the glomerular membrane. About 10–15% of the filtered magnesium is reabsorbed along the proximal tubule, about 60–70% is reabsorbed along the thick ascending limb of the Henle's loop and another 10–15% in the distal convoluted tubule.[10]

The majority of body magnesium is located in the bone and part of this magnesium is adsorbed to the surface and is in equilibrium with extracellular magnesium. At reduced plasma concentrations, magnesium can rapidly be released from the bone. Bone magnesium therefore represents a rapidly exchangeable pool that buffers magnesium extracellular concentration.[11]

The usual diet provides 2–7.5 mg of Mg^{2+} per kg of body weight. For adults the recommended intakes of magnesium are between 300 and 420 mg/day.[12] In animal models it has been shown that a reduction of total intracellular magnesium concentration can only be achieved by feeding animals with a severely deficient diet.[13] Only if plasma magnesium is reduced below 0.2 mmol/l, a small decrease in intracellular magnesium concentration can be measured, indicating that the effects of magnesium deficiency are mainly restricted to extracellular magnesium.

3.2 Magnesium as a Drug

Few other drugs have been touted for as many roles as magnesium-based drugs have. Potential uses of magnesium include the treatment of:

- eclampsia
- asthma
- ischemic stroke
- myocardial infarction
- arrhythmias.

3.2.1 Gestational hypertension, preeclampsia and eclampsia

Gestational hypertension is the most frequent cause of hypertension during pregnancy and the rate varies between 6 and 17%.[14,15] Some women with gestational hypertension will subsequently progress to preeclampsia and the rate of progression depends on gestational age at the time of diagnosis. The rate reaches 50% when gestational hypertension develops before 30 weeks gestation. Preeclampsia is primarily defined as gestational hypertension plus proteinuria. Preeclampsia is considered severe if there is severe gestational hypertension in association with proteinuria (in the normal kidney, during the filtration process that takes place in the glomerulus, proteins do not enter the urinary space and the presence of proteins in urine is always a sign of glomerular disease) or in the presence of multiorgan involvement such as pulmonary edema (i.e. the passage of fluids from the pulmonar interstitial compartment to the alveolar lumen causing a severe obstacle to the diffusion of oxygen from the alveolar space to the blood), seizures, oliguria (i.e. a reduction in the daily urine volume that is associated with a reduction in glomerular filtration rate), thrombocytopenia (i.e. a decrease in platelet count in the blood), abnormal liver enzymes or persistent severe central nervous system symptoms. During preeclampsia both cardiac output and plasma volume are reduced, whereas systemic vascular resistances are increased. These variations cause a reduced placenta, kidney and brain perfusion, resulting in increased foetal and maternal morbidity (i.e. an increase in the incidence of any non-mortal disease) and mortality. The etiology of preeclampsia is still unknown but an endothelial dysfunction (i.e. a decrease in the endothelial function that leads to a decrease in the production of endothelium-derived substances, mainly vasodilators and inhibitors of platelet aggregation) may play a crucial role. Magnesium has been shown to ameliorate endothelial dysfunction in preeclampsia and this may be due either to its direct vasodilatatory properties or to the ability to stimulate endothelial release of the endothelial vasodilator prostacyclin.

Eclampsia, the occurrence of a seizure in association with preeclampsia, remains a rare but serious complication of pregnancy and is estimated to complicate around 1 in 2000 deliveries in Europe and in other developed countries.[16]

There are few randomized trials evaluating magnesium to prevent preeclampsia. All these trials have limited sample size but reveal minimal to no benefit;[17] thus magnesium should not be routinely used for this purpose. The results of these studies were the subject of a review.[18]

As far as the use of magnesium for eclampsia is concerned, a recent review involving 897 women has summarized the evidence about the differential effects of magnesium sulphate when compared with phenytoin for the care of women with eclampsia.[19] The comparison was in terms of maternal mortality, recurrence of convulsions and other serious morbidity that could lead to death. Magnesium sulphate reduces the risk of further fits compared with phenytoin and the trend in maternal mortality is also in favour of magnesium sulphate. The review should be viewed in conjunction with those comparing magnesium sulphate with diazepam[20] and with lytic cocktail.[21] Magnesium sulphate had a 52% lower risk of recurrent convulsions versus diazepam and 67% lower risk of recurrent convulsions versus phenytoin. Overall, there is now compelling evidence in favour of magnesium sulphate rather than phenytoin, diazepam or lytic cocktail for the treatment of eclampsia.

Several protocols for treatment have been used. The most common is a loading dose of 4–6 g of $Mg_2SO_4{\cdot}7H_2O$ infused intravenously over 15 to 20 min, then a continuous intravenous infusion at 1 g/h for around 24 hours. Because magnesium is excreted in the urine, woman with oliguria or elevated creatinine should be monitored carefully and dosed accordingly. Magnesium sulphate is believed to exert its effects through a non-specific mechanism of action by competing with calcium for entry into myometrial (i.e. muscular cells in the uterus) cells through voltage-operated Ca^{2+} channels. Because its effect occurs at other sites, magnesium sulphate is associated with a wide range of side effects. Common side effects include flushing, nausea, chest tightness and lethargy, which are directly related to high serum magnesium concentrations. Fetal adverse events have been reported with magnesium sulphate, with a decrease in fetal heart rate and heart rate variability.

3.2.2 Asthma

The prevalence of asthma has been increasing worldwide despite aggressive efforts to treat and prevent it.[22] Asthma is a chronic disease affecting millions around the world. In the United States alone there are approximately 15 million patients with asthma. The airways of a person with asthma are hyper-responsive to a variety of allergic and nonallergic triggers. Alone or in various combinations,[23] these triggers can initiate inflammatory cascades activating resident cells (i.e. mast cells and airway epithelium) and recruiting a variety of immune cells including eosinophils and neutrophils. Histamines, leukotrienes, prostaglandins, quinines and cytokines are among a myriad of activated substances released by these cells leading to epithelial damage, increased mucus production, mucosal edema, smooth muscle contraction and bronchial hyper-responsiveness.[24] The outcomes are narrowed airways, airflow obstruction and elevated airways resistance.

Use of magnesium therapy in asthma was first described in 1936 by two Uruguayan physicians.[25] Despite this, it is not a standard recommendation of the National Heart, Lung and Blood Institute (NHLBI) guidelines[26] for diagnosis and management of asthma because of conflicting reports regarding its efficacy.

The mechanism by which magnesium works in asthma is still unclear but it is likely that it is related to relaxation of bronchial smooth muscle cells.[27,28] It has been demonstrated that when magnesium is administered to animals, relaxation of bronchial smooth muscle cells ensues.[29] Magnesium seems to inhibit the release of histamine from mast cells which recruits inflammatory mediators.[30] Adverse events associated with magnesium administration include facial flushing, tachycardia (i.e. a high frequency of ventricular contraction per minute, at least 90/min), muscle weakness, nausea and vomiting.[31] Most literature suggests that magnesium provides no benefit in the treatment of asthma in adults.[32,33] Rowe *et al.* performed a meta-analysis (i.e. a statistical procedure in which a number of human-based studies comparing the effects of a new drug treatment with a conventional therapy are pooled together to increase the statistical power of the findings) identifying seven randomized, controlled trials which evaluated intravenous magnesium sulphate in acute asthma for a total of 668 patients.[33] Overall, the outcome measures were not significantly different when compared to standard care.[34] Data in paediatric patients tend to be slightly more favourable.[35,36]

In general, the drug is well tolerated and inexpensive at the usual clinical dose of 1.2–2.0 g administered intravenously over 20 min. In spite of this the overall data present in the literature do not support routine use of magnesium in patients with acute asthma. Furthermore, inhaled preparations of magnesium sulphate have been reported to provide no additional benefit in addition to therapy with albuterol, which is a drug with β-adrenergic properties, i.e. acting as a β-adrenergic agonist, in adult patients with mild or moderate asthma exacerbations.[37]

3.2.3 Stroke

For 15 years many neuroprotective agents have been described as effective in the treatment of ischaemic stroke, i.e. an infarct of brain tissue which is different from haemorrhagic stroke, i.e. an haemorrhagia in the context of brain tissue. Magnesium sulphate is currently undergoing clinical trials in the treatment of acute stroke. It is well known how Mg^{2+} plays a physiological role in processes pertinent to ischaemia. In the brain it is complexed with ATP and is an important cofactor in cellular energy metabolism and protein synthesis.[38] Brain magnesium concentrations are regulated by active blood–brain barrier transport that maintains cerebrospinal fluid (CSF) concentrations higher than those in serum, typically 1.1 mmol/l compared to 0.8 mmol/l. Magnesium concentration increases in situations of focal ischaemia and seizures in animal models.[39,40] Intracellular ionized magnesium concentrations increase in ischaemic stroke presumably because of a decrease in cellular ATP, the main intracellular buffer of magnesium.[41]

Magnesium is an attractive therapeutic agent because it is inexpensive, widely available and intravenous, and intramuscular administration yields predictable serum concentrations. Magnesium is normally excreted by the kidneys with a half life of 4 hours or less and adverse effects generally are restricted to patients with advanced renal failure. Overdose can be detected clinically by loss of deep tendon reflexes and administration of calcium gluconate generally avoids further problems. The wide therapeutic index of magnesium contrasts with the majority of neuroprotective agents which have been associated with important side effects. There are a number of possible mechanisms of action of magnesium in the protection of neurones and glia (i.e. the connective tissue in the central nervous system) from ischaemic damage. Reduction of infarct size by magnesium may be the consequence of an effect on blood flow or a direct neuroprotective effect or a combination of both. After cerebral artery occlusion, a core region rapidly necroses. A surrounding region of ischaemia is known as ischaemic penumbra which may also proceed towards necroses following a complex range of metabolic reactions. These processes include excessive release of neurotransmitters such as glutamate, excessive activation of glutamate receptors, excessive sodium and calcium entry into the cell via receptor-operated and voltage-dependent channels, and activation of calcium-dependent enzymes that lead to free-radical production. This latter phenomenon leads to membrane lipid breakdown, proteolysis and initiation of apoptosis and inflammatory response (for review see Ref. 42).

Magnesium may act at different levels in this complex chain of reactions. It may inhibit ischaemia-induced glutamate release, an excitatory amino acid neurotransmitter,[43] and has antagonist properties at the *N*-methyl-D-aspartate (NMDA) receptor ion channel level[44,45] leading to increased Ca^{2+} intracellular concentrations. Magnesium antagonizes calcium entry via voltage-gated channels, antagonizes mitochondrial calcium overload and may prevent cellular ATP depletion.[46] Despite doubts about brain penetration by systemically administered magnesium, preclinical and clinical studies show slow entrance of magnesium into the CSF and brain tissue.[47] The beneficial effects of magnesium administration in animal models of ischaemia are numerous. Magnesium ameliorates brain injury induced by injection of NMDA,[48] prevents NMDA-induced seizures,[49] increases ionized magnesium after head trauma[50] and induces modifications of NMDA binding properties.[51] In addition to neuronal effects, magnesium also has a number of vascular effects that may be important, particularly increased cerebral blood supply[52] and antagonism of vasoconstrictor substances.[41]

A series of six randomized controlled trials on magnesium in stroke were recently performed but only four are suitable for meta-analysis.[53] All trials have reported a reduction in the end-point of mortality but the small number of patients has hampered the true significance of these studies.[54,55] Most clinical trials have administered magnesium as an intravenous loading infusion usually over 15 min followed by a maintenance infusion of about 24 hours. Overall the results of these small trials are encouraging and have constituted the basis for a large ongoing trial, the intravenous magnesium efficacy in stroke (IMAGES) trial. IMAGES is a multicentric study involving over 130 centres and designed to detect a 5.5% absolute difference in death and disability 3 months after stroke.[56] A substudy of IMAGES, MR-IMAGES, utilizes magnetic resonance imaging to compare lesion size at acute and later time points.

Clinical data are presently too limited to permit definite conclusions on the effects of magnesium in patients with stroke. Magnesium has a profile comparable with or superior to synthetic neuroprotective drugs. Nevertheless, no neuroprotective agent has so far been successful in clinical trials and there have been a large number of failures with promising compounds.

3.2.4 Acute myocardial infarction

Interest in the therapeutic use of magnesium in patients with acute myocardial infarction comes from non-controlled studies carried out in Australia, Asia and Europe concerning a supposed decrease in arrhythmic events and an increase in survival in patients treated with magnesium.[57,58] Epidemiological studies have also reported that the incidence of sudden death is greater in geographical areas characterized by soft water, which are relatively poor in calcium and magnesium.[59] An epidemiological perspective study has been carried out into the relation between plasma levels of magnesium and the incidence of coronary artery disease,[60] and another study investigated the relationship between dietary magnesium and coronary artery disease.[61] In the latter study, 13 922 middle-aged subjects were studied for a follow-up period of from 4 to 7 years. The risk of the onset of coronary artery disease was significantly higher in subjects with low plasma levels of magnesium and there was a significant inverse correlation between the two parameters. Furthermore, patients with plurimetabolic syndrome, that is, a syndrome characterized by obesity, dyslipidemia, reduced peripheral glucose tolerance in the presence or absence of arterial hypertension, and patients who are characterized by increased risk of coronary artery disease show a decrease in ionized intralymphocyte magnesium and the level of magnesium is inversely related to peripheral insulin resistance.[62–65]

A weak inverse correlation was also found between dietary intake of magnesium and the incidence of coronary artery disease. It is well known that a key element in the physiopathology of acute myocardial infarction is platelet activation. Under normal conditions, platelets circulate in a 'resting' condition without interacting with each other or with other cells such as leucocytes or endothelial cells. After contact with agonists such as thrombin, collagen and ADP, the platelets become active and exhibit fibrinogen binding sites on the IIb-IIIa glycoprotein. At this point plasma fibrinogen may bind to the platelets thus allowing them to form microaggregates. This is the first, still reversible, step in platelet aggregation. After this, the platelets release into circulation a multitude of molecules which were stored in the granules.

Some studies have shown that magnesium is able to inhibit platelet activation both by inhibiting certain factors which stimulate the platelets such as thromboxanes A2 and by inhibiting the production of platelet-inhibiting factors such as prostacyclin.[66,67]

Many years ago it was reported that magnesium was able to prolong coagulation time if added in small quantities to human plasma.[68] Adams and Mitchel found that magnesium inhibited thrombus formation and increased the minimum concentration of ADP necessary to start the formation of a thrombus.[69] In an experimental *in vitro* model of platelet-dependent thrombosis, Ravn and collaborators recently described how the *in vitro* administration of magnesium was able to inhibit the formation of arterial thrombosis.[70] Hypomagnesaemia (low magnesium plasmatic levels) seems to selectively inhibit the release of nitric oxide by the endothelium.[71] Since nitric oxide is a powerful inhibitor of platelet aggregation and adhesion, hypomagnesaemia could be accompanied by a predisposition to thrombosis.

A global analysis by Horner[72] of magnesium therapy in acute myocardial infarction found that this treatment is safe and useful. However, two double-blind controlled mega-trials have cast doubt on this point of view. The first is the Leicester intravenous magnesium interventional trial (LIMIT-2) which recruited 2316 patients.[73] Magnesium was infused either at the moment of thrombolitic therapy or before spontaneous riperfusion (when an acute myocardial infarction is caused by thrombosis on an atherosclerotic plaque in a coronary artery leading to complete vessel occlusion) a spontaneous riperfusion may occur. Total mortality measured at day 28 was 7.8% in the group randomized to magnesium and 10.3% in the group randomized to placebo with a relative reduction of 24%. Moreover, in the group randomized to magnesium there was a 25% decrease in episodes of left cardiac failure. In the group treated with magnesium, no improvement was recorded in terms of episodes of hypotension or of arrhythmic events and, in fact, in this group there was no reduction in the use of anti-arrhythmic drugs or cardiac pacing. The effect of the reduction of mortality obtained with magnesium was long lasting, as shown by a reduction in total mortality of 16% with a mean follow-up of 2.7 years. Side-effects of the treatment included an increase in bradiarrhythmia (i.e. a series of arrhythmia characterized by either decrease in frequency of ventricular contractions or an increased duration of atrio-ventricular conduction, i.e. a slower propagation of excitation along the specialized tissue of conduction) and skin flushing on administration.

The second mega-trial to deal with the question of treatment with infusions of magnesium in patients with acute myocardial infarction was called the Fourth International Study of Infarction Survival (ISIS-4).[74] In addition to the infusion of magnesium, patients were allocated randomly to thrombolitic, nitroderivative or ACE (angiotensin-converting enzyme)-inhibitor therapy. The patients were recruited with a mean interval of 8 hours (maximum 24 hours) from the onset of symptoms and treatment with magnesium was interrupted during riperfusion. The death rate measured after 5 weeks in patients randomized to treatment with magnesium did not show any improvement compared to placebo, and there was even a trend towards an increase in the death rate, with an increase in the incidence of cardiogenic shock and heart failure in the group treated with magnesium (7.6% vs 7.2%), although there was a significant reduction in the incidence of ventricular fibrillation. Since this study, the use of magnesium in the treatment of myocardial infarction has been significantly scaled down.

In any case, the two studies described above, which show clearly contrasting results, are characterized by differences in their design. In LIMIT-2, the dose of magnesium was a bolus of 8 mmol followed by 65 mmol infused over 24 hours, while in ISIS-4 the dose was of 8 mmol in bolus followed by 72 mmol over 24 hours. The timing of the infusion of magnesium was different since in LIMIT-2 it was earlier than in ISIS-4. Some authors hypothesized that the delay in starting the infusion of magnesium may have limited the ability of the magnesium to inhibit the damage from riperfusion, which is one of its main beneficial actions. During the first minutes of riperfusion, the calcium accumulates in the myocardium, particularly in the mitochondria, and there is depletion of high-energy phosphates and a contractile dysfunction. If magnesium is to carry out in full its protective effect, the concentration of magnesium in the plasma must be increased before riperfusion. The timing of the infusion of magnesium is therefore critical in relation to both spontaneous riperfusion and after thrombolysis. The ideal therapeutic window for the infusion of magnesium is probably limited to the first 1–2 min after riperfusion and it has been shown that the protective effect of magnesium is reduced when it is administered after an hour from the beginning of riperfusion.[75] Some authors have shown in an experimental model that magnesium is able to reduce the influx of calcium in myocardial cells only when it is administered within 15 min from the onset of riperfusion. The delicate question of the exact timing of the infusion of magnesium has been reconsidered in the MAGIC (magnesium in coronaries) trial.

Unfortunately, the recent results of the MAGIC study are not encouraging.[76] MAGIC is a vast-scale trial designed to show the potential beneficial effect of the infusion of magnesium. The experiment involves the enrolment of 10 400 patients with a high risk of acute myocardial infarction and elevation of the ST-segment, and their random assignment to early treatment with infusion of magnesium or placebo. The primary end-point was the mortality at 30 days.[76] At 30 days, 15.3% patients in the magnesium group and 15.2% in the placebo group had died. No benefit or harm due to magnesium was observed. Extensive univariate analysis did not identify any subgroup that benefited from magnesium nor were there beneficial effects from magnesium in a multivariate model that adjusted for factors shown to affect mortality risk.

From all these data taken together, in current coronary care practice, there is no indication for the routine administration of intravenous magnesium in patients with acute myocardial infarction.

3.2.5 Arrhythmias

Arrhythmias is any cardiac rhythm different from the so-called sinusal rhythm, i.e. generated from the sinus node, a group of cells situated in the right atrium characterized by pacemaker function for their capacity to depolarize spontaneously. Numerous studies have been carried out in an attempt to define the role of both intracellular and extracellular magnesium in the physiopathology and treatment of arrhythmias.

Variations in both intracellular and extracellular magnesium can influence directly the activity of numerous ionic membrane channels, particularly calcium and potassium channels, as described in recent reviews.[77,78]

It is very difficult to establish the role of a lack of magnesium *in vivo* and at the same time the possible arrhythmogenic risk. This depends on various factors. In the first place, it is very difficult to know whether a patient is lacking in magnesium when we consider its prevalent location inside cells. The Framingham Heart Study fails to show any correlation between hypomagnesaemia and the incidence of ventricular premature complexes with a rate of more than 10 per hour.[79] However, they did show an increased risk of frequent or complex ventricular premature complexes in the case of serious reductions in magnesaemia. Although both supraventricular and ventricular arrhythmias have been associated with hypomagnesaemia,[80] the interpretation of this relationship is made less clear by the limited correlation between the intracellular concentration of magnesium in myocites (i.e. myocardial contractile cells) and in plasma.[81]

'Torsade de pointes' or polymorphic ventricular tachycardia is the form of arrhythmia for which treatment with magnesium seems most appropriate. The American Heart Association's 1992 Guidelines for Cardiopulmonary Resuscitation and Emergency Cardiac Care included the infusion of magnesium sulphate in the treatment of torsade de pointes. In the past, the treatment aimed at reducing the (QT) interval (i.e. a portion of the electrocardiogram, in particular, an interval which comprises the electrical activity corresponding to both ventricular contraction and diastole or ventricular depolarization and repolarization) through sympathicomimetics or cardiac pacing. In 1984 the use of magnesium sulphate as a treatment for torsade de pointes was described for the first time.[82] The treatment consisted in the infusion of an 8 mmol IV bolus of magnesium sulphate and was followed by the resolution of the arrhythmia without relapse in all three patients treated. Since then there have been many reports of the efficacy of magnesium in the treatment of this kind of arrhythmia,[83] with the exception of one report in which neither the infusion of magnesium alone, nor cardiac pacing was sufficient to restore stability to the sinusal rhythm but the two associated methods were successful.[84] Magnesium therapy is not useful in the long-term management of long QT syndrome, a disorder of cardiac repolarization that may precipitate torsade de pointes. For other forms of arrhythmias such as atrial fibrillation, ventricular tachycardia, re-entrant supraventricular tachycardia there are only anecdotical reports. Thus, the role of magnesium appears to be secondary and limited to those patients for whom other drugs are contraindicated or have been shown to be ineffective (for review see Ref. 78).

3.2.6 Miscellaneous

The use of magnesium has been reported in many pathological conditions. In kidney diseases it has been evaluated for stone prevention and for treatment of hyperphospataemia. The role of magnesium in stone prevention is poorly defined. Although *in vitro* studies show that magnesium inhibits calcium oxalate and calcium phosphate crystal growth and aggregation, no beneficial effect of magnesium therapy has been demonstrated in a population of recurrent calcium stone formers.[85] More recent data show that in patients who had experienced stone formation, oral magnesium supplements are able to produce a significant reduction in stone recurrences.[86]

In chronic kidney disease, magnesium has been used experimentally in recent decades as an alternative to calcium-based phosphate binders. O'Donovan *et al.* used magnesium carbonate in 28 patients for 2 years as a substitute for aluminium hydroxide and found a significant drop in predialysis aluminium levels and controlled serum phosphate levels.[87] Other investigators observed the same efficacy but also significant side effects. Thus, though magnesium carbonate is an alternative to calcium-containing binders,[88] it is not widely used.

The effect of oral magnesium supplementation has been studied in osteoporosis. In fact, significant reductions in serum and bone magnesium have been described in several studies on post-menopausal women with osteoporosis. Recently, dietary magnesium has been observed to be positively associated with bone mineral density (BMD) in a community-based study in elderly men and women.[89] Despite this, epidemiological studies relating magnesium intake to bone mass or rate of bone mass loss have been conflicting. There are few studies on the effect of magnesium supplementation on bone loss. Two recent studies on this subject report conflicting results[90,91] thus illustrating the ongoing controversy on the role of magnesium in bone turnover.

One of the pathogenetic mechanisms responsible for sickling in patients with sickle cell anaemia (haemoglobin SS) is the decreased hydration status of the cells. One of the main determinants of cell dehydration is the membrane K^{+}-Cl^{-} cotransport system, and the magnesium-induced inhibition of this transport system may be beneficial in this disease. A small unblinded and uncontrolled study in patients with haemoglobin SS has shown that oral magnesium supplementation for 6 months can induce a significant reduction in the number of painful crises due to vaso-occlusion.[92] A double-blind, placebo-controlled study is in progress to evaluate the effect of magnesium pidolate on adult and paediatric cases of haemoglobin SS disease.

Magnesium is also supposed to have a role in affective disorders as it seems to influence several systems related to the pathophysiology of depression (see Ref. 93 for review). The action of magnesium to reduce hypothalamus–pituitary–adrenocortical system activity is suggestive of its possible use as a mood stabilizer and controlled studies are warranted to test its clinical efficacy.

Magnesium also has a possible role in the treatment of migraine. A recent small controlled trial showed no beneficial effect in the magnesium group[94] and even if it is suggested that the effectiveness of magnesium is related to plasma magnesium levels, there is still insufficient evidence to support its use in migraine.

References

1. Flatman PW. *Annu Rev Physiol* 1991; **53**: 259–271.
2. Murphy E, Freudenrich CC, Lieberman M. *Annu Rev Physiol* 1991; **53**: 273–287.
3. Romani A, Marfella C, Scarpa A. *Miner Electrolyte Metab* 1993; **19**: 282–289.
4. Romani A, Scarpa A. *Nature* 1990; **346**: 841–844.
5. Delva P, Pastori C, Degan M *et al. Hypertension* 2000; **35**: 113–117.
6. Suguyama T, Goldman WF. *Am J Physiol* 1995; **269**: C698–C705.
7. Lentner C. *Geigy Scientific Tables*. Ciba-Geigy Basel, Switzerland 1981, Vol. 1, pp. 217–220.
8. Hardwick LL, Jones MR, Brautbar N, Lee DBN. *Miner Electrolyte Metab* 1990; **16**: 174–180.
9. Fine KD, Santa Ana CA, Porter JL, Fordtran JS. *J Clin Invest* 1991; **88**: 396–402.
10. Quamme GA, de Rouffignac C. *Front Biosci* 2000; **5**: D694–D711.
11. Vormann J, *Mol Asp Med* 2004; **24**: 27–37.
12. Vormann J, Anke M. *J Clin Basic Cardiol* 2002; **5**: 49–53.
13. Vormann J, Gunther T, Hollriegl V, Schumann K. *Eur J Nutr* 1998; **37** (Suppl. 1): 92–97.
14. Hauth JC, Ewell MG, Levine RL *et al. Obstet Gynecol* 2000; **95**: 24–28.
15. Knuist M, Bonsel GJ, Zondervan HA, Treffers PE. *Gynecol Obstet* 1998; **61**: 127.

16. Douglas K, Redman C. *Br Med J* 1994; **309**: 1395–1400.
17. Ehrenberg A. *Acta Obstet Gynecol Scand* 1997; **164**: 108–110.
18. Sibai BM. *Am J Obstet Gynecol* 1998; **179**: 1275–1278.
19. Duley L, Henderson-Smart D. In: *The Cochrane Library*, Issue 1, John Wiley & Sons Ltd, Chichester, UK, 2004; 1–16.
20. Duley L, Henderson-Smart D. In: *The Cochrane Library*, Issue 1, Oxford: Update Software. CD000127 (2003).
21. Duley L, Gulmezoglu AM. In: *The Cochrane Library*, Issue 1, Oxford: Update Software. CD002960 (2003).
22. Mannino DM, Homa DM, Pertowski CA *et al. MMWR Morb Mortal Wkly Rep* 1998; 1–28.
23. Green RM, Custovic A, Sanderson G *et al. BMJ* 2002; **324**: 1–5.
24. Sitesh RR, Milgrom H. *Curr. Allergy Asthma Rep* 2003; **3**: 179–189.
25. Resello JC, Pla JC. *Prens Med Argent* 1936; **23**: 1677–1680.
26. National Asthma Education and Prevention Program Expert Panel Report 2. *NIH-publication* no. 97–4051. Bethesda MD Department of Health and Human Services (1997).
27. Okoyama H, Takashi A, Okayama M *et al. JAMA* 1987; **257**: 1076–1078.
28. McLean RM. *Am J Med* 1994; **96**: 63–76.
29. Spivey WH, Skobeloff EM, Levin RM. *Ann Emerg Med* 1990; **19**: 1107–1112.
30. Scarfone RJ, Loiselle JM, Joffe MD *et al. Ann Emerg Med* 2000; **36**: 572–578.
31. Pabon H, Monem G, Kissoon N. *Pediatr Emerg Care* 1994; **10**: 200–203.
32. Green SM, Rothrock SG. *Ann Emerg Med* 1992; **21**: 260–265.
33. Tiffany BR, Berk WA, Todd IK *et al. Ann Emerg Med* 2000; **36**: 181–190.
34. Ciarallo L, Sauer AH, Shannon MW. *L Pediatr* 1996; **129**: 809–814.
35. Darcie DS, Varsha BM, Cary EJ. *Ann Pharmacother* 2002; **36**: 1249–1260.
36. Bessmertny O, DiGregorio RV, Cohen H *et al. Ann Emerg Med* 2002; **39**: 585–591.
37. Ebel H, Gunther T. *J Clin Chem Clin Biochem* 1980; **18**: 257–270.
38. Sjostrom LG, Wester P. *Cerebrovasc Dis* 1995; **5**: 241.
39. Hallak M, Berman RF, Irlenkauf SM *et al. Am J Obstet Gynecol* 1992; **167**: 1605–1610.
40. Helpern JA, Vande Linde AM, Welch KM *et al. Neurology* 1993; **43**: 1577–1581.
41. Muir KW. *Postgrad Med J* 2002; **78**: 641–645.
42. Smith DAS, Connick JH, Stone TW. *Br J Pharmacol* 1989; **97**: 475–482.
43. Jhonson JW, Ascher P. *Biophys J* 1990; **57**: 1085–1090.
44. Brocard JB, Rajdev S, Reynolds U. *Neuron* 1993; **11**: 751–757.
45. McIntosh TK, Faden AL, Yamakami I *et al. J Neurotrauma* 1988; **5**: 17–31.
46. Fuchs-Buder T, Tramer MR, Tassonyl E. *J Neurosurg Anesthesiol* 1997; **9**: 324–328.
47. McDonald JW, Silverstein FS, Johnston MV. *Neurosci Lett* 1990; **109**: 234–238.
48. Mason BA, Standley CA, Irtenkauf SM *et al. Am J Obstet Gynecol* 1994; **171**: 999–1002.
49. Heath DL, Vink R. *Neurosci Lett* 1997; **228**: 175–178.
50. Hallack M. *Magnes Res* 1998; **11**: 117–131.
51. Chi OZ, Pollak P, Weiss HR. *Arch Int Pharmacodyn Ther* 1990; **304**: 196–205.
52. IMAGES Study Group, Bradford APJ, Muir KW, Lees KR. *Cerebrovasc Dis* 1998; **8** (Suppl. 4): 86.
53. Wester PO, Asplund K, Eriksson S *et al. Acta Neurol Scand* 1984; **70**: 143.

54. Muir KW, Lees KP. *Stroke* 2002; **33**: 386.
55. Bradford APJ, Lees KR. *Curr Controlled Trials Cardiovasc Med* 2000; **1**: 184–190.
57. Schroeder HA. *JAMA* 1960; **172**: 1902–1908.
58. Marier J, Neri LC. *Magnesium* 1985; **4**: 53–59.
59. Anderson TW, Leriche WH, Mackay JS. *N Engl J Med* 1969; **280**: 805–807.
60. Gartside PS, Glueck CL. *J Am Coll Nutr* 1995; **14**: 71–79.
61. Liao F, Folsom AR, Brancati F. *Am Heart J* 1998; **136**: 480–490.
62. Delva P, Pastori C, Degan M *et al. Hypertension* 1996; **28**: 433–439.
63. Delva P, Pastori C, Montesi G *et al. Life Sci* 1998; **63**: 1405–1415.
64. Delva P, Pastori C, Degan M *et al. Life Sci* 1998; **62**(24): 2231–2240.
65. Delva P. *Mol Asp Med* 2003; **24**(1–3): 63–78.
66. Gawaz M, Ott I, Reininger AJ, Neumann FJ. *Thromb Haemost* 1994; **72**: 912–918.
67. Nadler JL, Goodson S, Rude RK. *Hypertension* 1987; **9**: 379–383.
68. Greville GD, Lehmann H. *J Physiol* 1943; **103**: 175–184.
69. Adams JH, Mitchel JR. *Thromb Haemost* 1979; **42**: 603–610.
70. Ravn HB, Kristensen SD, Hjoridal VE *et al. Arterioscler Thromb Vasc Biol* 1997; **17**: 3620–3625.
71. Gold ME, Buga GM, Wood KS *et al. Circ Res* 1990; **66**: 355–366.
72. Horner SM. *Circulation* 1992 Sep; **86**(3): 774–779.
73. Woods KL, Fletcher S, Roffe C, Haider Y. *Lancet* 1992; **339**: 1553–1558.
74. ISIS-4 Collaborative Group, ISIS-4. *Lancet* 1995; **345**: 669–685.
75. Baxter G, Sumeray M, Walker J. *Lancet* 1996; **348**: 1424–1426.
76. The MAGIC trial investigators. *Lancet* 2002; **360**: 1189–1196.
77. Agus MSD, Agus ZS. *Crit Care Clinics* 2001; **17**: 175–186.
78. Delva P. *Mol Asp Med* 2003; **24**: 53–62.
79. Tsuji H, Venditti FJ Jr, Evans JC *et al. Am J Cardiol* 1994; **74**: 232.
80. Dyckner T. *Acta Med Scand* 1980; **207**: 59–66.
81. Keller P, Aronson R. *Prog Cardiovasc Dis* 1990; **32**: 433–448.
82. Tzivoni D, Keren A, Cohen AM, Stern S. *Am J Cardiol* 1984; **53**: 528.
83. Perticone F, Adinolfi L, Bonaduce D. *Am Heart J* 1986; **112**: 847–849.
84. Toivonen LK, Leinonen H. *Int J Cardiol* 1986; **12**: 260–262.
85. Ettinger B, Citron JT, Livermore B, Dolman LI. *J Urol* 1988; **139**: 679–684.
86. Schwille PO, Schmiedl A, Herrmann U *et al. Urol Res* 2000; **28**: 167–177.
87. O'Donovan R, Baldwin D, Hammer M *et al. Lancet* 1986; **I**: 880–881.
88. Parsons V, Baldwin D, Moniz C *et al. Nephron* 1993; **63**: 79–83.
89. Tucker KL, Hannan MT, Chen H *et al. Am J Clin Nutr* 1999; **69**: 727–736.
90. Diami HP, Porta S, Wirnsberger G *et al. J Clin Endocrinol Metab* 1998; **83**: 2742–2748.
91. Doyle L, Flynn A, Cashman K. *Eur J Clin Nutr* 1999; **53**(4): 255–261.
92. De Franceschi L, Bachir D, Galacteros F *et al. J Clin Invest* 1997; **100**: 1847–1852.
93. Murck H. *Nutritional Neurosci* 2002; **5**: 375–389.
94. Corbo J. *Ann Emerg Med* 2001; **38**: 621–627.

4 $_{13}$Al Aluminum Metallotherapeutics

Thanos Salifoglou

4.1 Introduction

Metallopharmaceuticals have been at the forefront of research and development for some decades. Their use in preventing disease, ameliorating clinical pathological manifestations and curing human ailments has propelled research in a variety of directions, covering a diverse spectrum of chemical elements across the periodic table. To this end, various metals have been used in the preparation of complexes with appropriate organic ligands, reflecting biological activity arising from metal–biomolecule interactions in human pathophysiological states. Aluminum is one of those metals, present in pharmaceuticals used for the treatment of a number of diseases.

The variety of reasons for which aluminum compounds find applications in preventive and therapeutic medicine necessitated pertinent research into their nature and structure–function relationships in human biology. Hence, the justified quest for improved aluminum-containing drugs, which are characterized by enhanced preventive and healing action, safety and avid biological activity.

In this chapter, a treatise of aluminum-containing therapeutics is presented with the purpose of (a) outlining the families of aluminum-containing pharmaceuticals currently available for medicinal use, (b) delving into their potential mechanistic

Metallotherapeutic Drugs and Metal-Based Diagnostic Agents: The Use of Metals in Medicine
Edited by Gielen and Tiekink

involvement of biologically relevant pharmacological action(s), (c) reflecting their potential in the future development of next-generation pharmaceuticals targeting human ailments and (d) projecting their medically supervised use, in order to avoid toxic manifestations, potentially associating aluminum with human diseases.

4.2 Adjuvants

An adjuvant is an agent or a mixture of agents which possesses the ability to bind to a specific antigen. Upon administration to a host, it enhances cell-mediated immunity (CMI) triggered by the presence of the antigen.[1] Immunoadjuvant effects have been associated with the induction of T helper cell type 2 responses, while the specific underlying mechanisms are still under investigation. T helper cell type 2 responses are intimately associated with the human immune system's production of anti-inflammatory cytokines (primarily IL-4, IL-10 and IL-13) and stimulation of β-cells to proliferate, differentiate and produce antibodies. Literature work has focused on immunoadjuvants describing microscopic as well as macroscopic phenomena associated with their incipient chemistry at the biological level. One of the remarkable properties of adjuvants is their capacity to bind or adsorb a large number of antigenic molecules over a wide surface area, thus enhancing the interaction of immune cells with the presenting antigens and stimulating[2] immune response. Among the oldest adjuvant species known to date are alum and metal oxides, aluminum–oxygen-containing compounds like aluminum hydroxide, aluminum oxide as well as others (ZrO_2, silica, hematite and magnetite, etc.). The aluminum–oxygen compounds are the oldest adjuvants experimented with and constitute a very specific family of metallopharmaceuticals approved for safe use at the clinical level.

Of the aluminum-containing adjuvants used today, three appear to be the most prevalent: aluminum hydroxide, aluminum phosphate and alum ($KAl(SO_4)_2{\cdot}12H_2O$). These three agents are included in many vaccines and have been licensed for use by the US Food and Drug Administration.[3] The total aluminum currently licensed to be present in a vaccine is in the range 0.85–0.125 mg/dose. Following is an account of these adjuvants widely used in metallopharmaceuticals.

4.2.1 Alum

The chemical composition of alum is $KAl(SO_4)_2{\cdot}12H_2O$. It is a widely used adjuvant in a number of pharmaceutical products linked to immunization.[4] The pharmaceutical itself is a mineral gel adjuvant, prepared by two different approaches. The first one is based on the alum–antigen co-precipitation method, involving treatment of alum with NaOH to a final pH 7, leading to the precipitation of $Al(OH)_3$. The method tends to yield heterogeneous precipitates. The advantages of this method are quite important in that (a) it induces a granulomatous reaction at the injection site and (b) it promotes antigen presentation by macrophages and activation of the ensuing immunocascade. The second method employs initial adsorption of antigen onto preformed $Al(OH)_3$ gels. It has the advantage that it facilitates freezing or spray-freeze-drying (SFD) of antigen–adjuvant complexes. SFD is a novel and effective drying process for the production of stable adjuvanted vaccine formulations.[5] It has been shown to (a) minimize adjuvant coagulation and (b) preserve immunogenicity of the vaccine antigen.

The mechanisms by which immune adjuvants function are still a matter of scientific conjecture. Key to delineating those mechanisms is the role of aluminum or its derivative $Al(OH)_3$, derived from alum through the treatment process. It is logical to deduce that interactions arise between Al(III) and cellular targets such as proteins, as a result of which biological activity develops or is enhanced through mediated action(s) of the associated presenting antigen in the mineral gel adjuvant. One such role that aluminum might play originates in its ability as a trivalent cationic metal ion to neutralize negatively charged molecular targets (e.g. plasma protein fibrinogen), raising hydrophobic effects that potentially lead to (a) protein precipitation on hydrophilic surfaces (e.g. $Al(OH)_3$) and (b) changes in the tertiary structure of interacting proteins adsorbed onto hydrophilic/hydrophobic surfaces. In this respect, critical is (a) the link between hydrophilicity/hydrophobicity of aluminum–oxygen species and the adjuvant cytochemistry and (b) the correlation of the magnitude of the adjuvancy with the magnitude of the inflammatory response (granulomatous reaction), produced by the metal-oxygen species and the surface area for antigen presentation.

4.2.2 $Al(OH)_3$

Aluminum hydroxide has proven to be a safe adjuvant in a number of vaccines destined for humans and animals. It can act alone or synergistically with other adjuvants (co-adjuvants). Characteristic is the case of ginseng (the dry extract prepared from the *Panax ginseng C.A. Meyer-root*) known to contain ginsenocides, which demonstrably exhibit adjuvant properties. In this sense, it has been shown that T-ginseng and $Al(OH)_3$ act synergistically and improve potently antibody response to various immune antigens (e.g. porcine parvovirus-antigen).[6] The mechanisms of interactions between $Al(OH)_3$ and ginseng are not known. In general, it is thought that the existence of a co-adjuvant in an $Al(OH)_3$-adjuvanted vaccine may imply interactions with $Al(OH)_3$ at the site of injection and concomitant stimulation of antigen-presenting cells in a more efficient manner as a whole. Alternatively, the two adjuvants may stimulate a broader spectrum of immune cells than each adjuvant alone, thus providing for an enhanced antibody titer of the vaccine. It appears that co-adjuvants such as ginseng may (a) enhance the potency of the aluminum-adjuvanted vaccine in a simple, safe and cheap manner, (b) offer stability in the antibody titer induced by the vaccines, and (c) help reduce the total amount of aluminum for the establishment of specific bioactivity in the pharmaceutical preparation of the employed vaccines.

It is worth noting that $Al(OH)_3$-adjuvanted vaccines induce higher levels of IgG1 antibodies than IgG2. Interestingly, the presence of ginseng as a co-adjuvant results in a shift of the IgG profile, favoring high levels of IgG2 over IgG1. Thus, a preferential induction of IgG factors through the concerted action of $Al(OH)_3$ and ginseng over that observed for the $Al(OH)_3$ alone indicates the importance of the co-adjuvant in the clinical profile of the vaccine. Active ingredients in ginseng that might contribute to the enhanced immunogenic activity of the employed vaccines with two co-adjuvants include the potent saponins, shown to (a) promote killer-cell activity and interferon production and (b) stimulate the phagocytic activity of macrophages and polymorphonuclear cells.

Significant insight into the interactions between the major molecular components of human immune vaccines and aluminum-containing adjuvants was provided in the case of the recombinant anthrax vaccine containing $Al(OH)_3$ as an adjuvant.[7] The study focused on the major component of the second-generation vaccine, the recombinant protective antigen (rPA) protein, and compared the effect of the presence of $Al(OH)_3$ (zero charge point (zcp) = 11.5) versus that of $AlPO_4$ (zcp = 5.4). The rPA binds rapidly and efficiently to $Al(OH)_3$ under physiological conditions, in contrast to poor binding of the same molecule to $AlPO_4$. Taking into consideration the pI of rPA (5.6 at pH = 7.4) and the zcp of the employed adjuvants, it was shown how important the surface charge is in the adsorption of rPA to aluminum-containing adjuvants. It is noted that the effectiveness of the adjuvant in a vaccine, preparation is dependent on the complete adsorption of the antigen on the adjuvant surface. To that end, the type of interactions between the primary molecular target and the aluminum adjuvant is a determining factor of the overall biological activity of the vaccine. These interactions lead to the formation of complexes, with electrostatics often dictating the nature of the molecular assembly in such complexes. The significance of electrostatic interactions has previously been studied[3,8] and reported using model proteins, thus lending credence to the observations made on the recombinant vaccine protein. It appears that as the pI/zcp of the aluminum adjuvant approaches the corresponding pI of the protein (rPA), the decrease of the charge difference between the two species in the assembled complex is reduced with a concomitant release of the antigen in the vaccine preparation. In this sense, phosphate ions contribute to the desorption of the antigen from the aluminum adjuvant, resulting in an antigen comparable in biological activity to unadsorbed control material.

Past studies have shown that addition of phosphate anion to an antigen–aluminum-hydroxide adjuvant complex lowers significantly the zcp of the adjuvant.

4.2.3 Aluminum phosphate

Aluminum phosphate has been used extensively as an adjuvant in vaccines. A characteristic example is the Chiron (*Haemophilus influenza* type b) Hib-CRM_{197} conjugate vaccine, a liquid formulation containing $AlPO_4$ as an adjuvant. It is a stabler preparation than the adjuvanted vaccine containing $Al(OH)_3$. Model studies of the test vaccine $AlPO_4$/Hib-CRM_{197} in mice and guinea pigs demonstrated an increased immunogenicity, despite the fact that the conjugate remains largely unabsorbed in this particular formulation.

Undoubtedly, the presence of $AlPO_4$ enhances significantly the immunogenicity of Hib-CRM_{197} conjugate vaccine in comparison with the unadjuvanted vaccine. A pilot study demonstrated that the new formulation was safe and immunogenic to infants,[9] resulting in a significant increase of the anti-PRP (anti-Polyribosyl Ribitol Phosphate) antibody titer. In light of these advances, the presence of $AlPO_4$ opens up the area of development of adjuvanted conjugate vaccines in combination with other vaccines, like Meningococci, DTwP, DtaP, etc.

4.3 Antacids

Antacids are substances that exhibit cytoprotective effects toward attacks against the gastric mucosa. They are thought not only to protect but also to restore and heal ulcerations through actions, which have been and still are under research scrutiny. The antacids are used either alone or in synergistic mixtures with other compounds. The combination formulations in the various antacids target wider goals through their neutralizing activity and reduction in potential side effects arising from single components. To this end, antacids primarily contain aluminum either alone or in combination with calcium or magnesium. The first formulation (Al–Ca) is set to combat diarrhea, whereas the second formulation (Al–Mg) reduces constipation. Below is a comprehensive account of aluminum-containing antacids, their use and potential mode of action as the latter emerges from ongoing research studies.[10]

4.3.1 Aluminum hydroxide

Aluminum hydroxide is a very well-known antacid, present in a number of over the counter (OTC) pharmaceutical preparations targeting heartburn, reflux oesophagitis, gastritis, peptic ulcer and hyperphosphatemia. Representative

$Al(OH)_3$-containing antacid preparations include Maalox®, Di-Gel®, Mylanta®, Riopan®, Rolaids®, Aludrox®, etc. Antacids in combination with other active components constitute the core of pharmaceuticals used throughout the world for combating human disorders of the peptic and other systems.

4.3.2 Aluminum glycinate

The molecular composition $Al(NH_2CH_2COO)(OH)_2$, with two hydroxide moieties, reflects the full formulation of dihydroxy aluminum glycinate. The species has been tested experimentally[11] and clinically as a new intragastric buffer solution for achievement of prolonged physiological pH value.[12] The same species has been tested as a gastric acid neutralizer in the case of peptic ulcer. In contemporary pharmacological preparations, it is being administered in the formulation marketed as Gastralgine®. This pharmaceutical was tested as an antacid and mucosal coating agent.[13] Gastralgine® contains $Al(OH)_3$ and aluminum glycinate, magnesium trisilicate and simethicone (a nonprescription drug, which is used for short-term relief of excess gas in the gastrointestinal tract; it is also used to relieve symptoms of infant colic) and works by promoting protective effects from ulcers.

4.3.3 Peptic ulcer disease

Peptic ulcers are thought to arise as a result of excessive amounts of gastric acid and pepsin or negligible mucosal resistance.[14] It is logical, then, that any medication to combat the problem relates to the aforementioned factors and enhances the corresponding defense mechanism(s). Among the various categories of anti-ulcer drugs in the market, specific families contain aluminum. These include (a) anti-secretory medication and antacids and (b) mucosal protection agents.

A significant discovery in the treatment of peptic ulcers has emerged from advancements in the research of histamine H_2-receptors and their antagonists. Akin to the H_2-antagonists are antacids, which are well known to the majority of people suffering from ulcers and ulcer-related ailments. Antacids are recognized as agents capable of (a) managing acute and chronic gastroduodenal ulcerations[15] and (b) contributing to ulcer healing.[16]

Mucosal protection agents act by increasing mucosal resistance or coating ulcer craters. Dominant among such agents is Sucralfate®. It is a complex of sulfated sucrose and $Al(OH)_3$. The complex inhibits peptic hydrolysis. It also exhibits high affinity for ulcerated mucosal loci. Sucralfate® does not possess anti-secretory or acid-neutralizing activity. It acts by forming a larger complex with albumin, fibrinogen as well as other proteins. The arising species act as a barrier to acid, pepsin or bile acid intrusion, more like a molecular ulcer sealant (bandage). To this end, pharmacologically, Sucralfate® has similar gastric and duodenal ulcer healing rates with the H_2-antagonist Cimetidine®.[17] The major advantage of Sucralfate®, however, over other anti-ulcer drugs is that it does not adsorb strongly to the ulcerated areas, thus avoiding potentially ensuing systemic side effects. The disadvantages of using Sucralfate® include constipation and a delay in voiding.

Do antacids possess ulcer-healing properties? What is the mechanism of their healing action? In fact, the actual mechanism of action is not known. The currently held view rides on the premise that the ulcer-healing action of antacids is primarily due to their neutralizing capacity of the luminal gastric acid. The case, though, may be more complex than that. To answer such a question one has to clarify the process of ulcer healing, which involves reconstruction of the mucosal architecture. The latter implies that the existing mucosal defect has to be restored through filling in the gap with proliferating epithelial and connective tissue cells. A number of growth factors participating actively in this process include the epidermal growth factor (EGF), its receptor EGF-R, Transforming growth factor (TGFβ), basic fibroblast growth factor (bFGF) and others. EGF is an important growth factor, which (a) accelerates cell migration to achieve re-epithelialization of the ulcer base and (b) triggers cell proliferation and divisions, essential for filling ulcer craters, leading to reconstruction of epithelial structures within the ulcer scar.[18] TGFβ is a growth factor, involved in signaling pathways supporting endothelial cell proliferation leading to ulcer healing. bFGF is a growth factor stimulating granulation tissue development and formation of microvessels through angiogenesis, thus contributing to ulcer healing and mucosal regeneration.[19] Studies suggest that the pharmacological action of the medication emanates from activation of gastric mucosa genes encoding for EGF and EGF-R expression.[20] EGF binds to EGF-R on the epithelial cell surface and activates a number of intracellular signaling pathways eventually leading to cell migration and proliferation. The latter process is crucial for filling in the injured ulcer locus (the ulcer crater). Concomitantly, the process of healing the ulcer scar is activated.

Consistent with the results of the aforementioned studies are studies conducted with antacids, like Talcid®, containing aluminum.[21] Talcid® bears the active ingredient hydrotalcite, with the molecular formulation $Mg_6Al_2(OH)_{10}CO_3{\cdot}4H_2O$. Hydrotalcite is characterized by a layered structure with the layers consisting of $[Al(OH)_2]^+$ and $Mg(OH)_2$. In this layered structure (Figure 4.1), the aluminum ions bear positive charge. Water-bearing interlayers contain carbonate anions (CO_3^{2-}), which provide the counteracting negative charge.

Collectively, cell migration, re-epithelialization and ultimate architectural restoration of the ulcer scar are the events contributing to ulcer healing. In this sense, the ulcer-healing process described above provides new insight into the molecular mechanism of action of pharmacological preparations in antacids, thus challenging the so far conventional belief on antacid healing action. Undoubtedly, the interaction of aluminum-containing antacids with the gastric

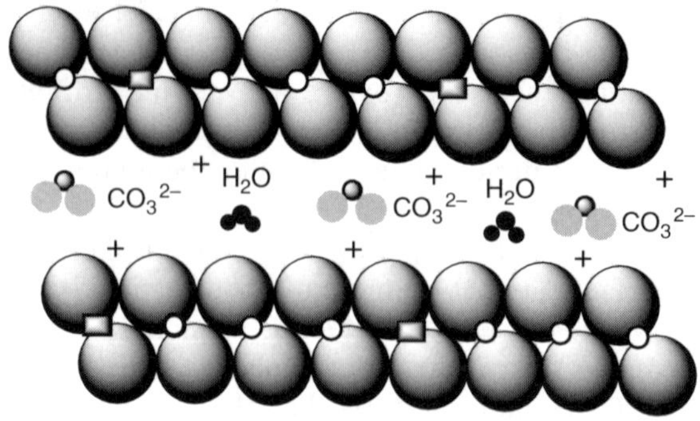

Figure 4.1 Molecular rendition of the hydrotalcite-layered structure, with positive charged layers composed of $[Al(OH)_2]^+$ (□) and $Mg(OH)_2$ (○), with a ratio of 1:3. The water-containing interlayers enclave carbonate anions neutralizing the positive charge

mucus is responsible for the activation of a cascade of (bio)chemical events, which most likely contribute to the protective function of antacids.

Equally appealing is the case of *Helicobacter pylori*, the bacterium causing chronic active gastritis and gastroduodenal ulcers.[22] *H. pylori* can induce cellular damage to the gastric mucosa through cytotoxins (cytotoxin VacA) and enzymes[23] that it excretes. This is accomplished by adhering to the surface epithelial cells, thus enabling the delivery of cytotoxins. The first such identified virulence factor, a cytotoxin (VacA), was termed 'vacuolating cytotoxin', due to its ability to generate vacuoles in eukaryotic cells and induce inflammatory response. The enzyme urease was also found to play an important role in the survival of the bacterium by degrading urea to ammonia, thus protecting the organism under extreme gastric pH conditions. Proteins (larger than 10 kDa) have been found to interfere with EGF binding to its receptor and EGF's stimulation of gastric cell proliferative responses. Other enzymes like lipase and proteases have also been found to be involved in the injury and creation of ulcerations, while factors like GroEL- and GroES-like heat-shock proteins (HSP) were found[24] to help the bacterium tolerate extreme conditions (pH change). GroEL- and GroES-like HSP factors are chaperone proteins, belonging to the HSP family and contributing to the assembly or correct folding of target proteins by causing the target proteins to acquire or stabilize a specific conformation.

Employment of the $Al(OH)_3$-containing hydrotalcite antacid (oral administration) was found to promote adsorption of all proteins secreted by *H. pylori*, including HSP, cytotoxin VacA and urease, and to counteract inhibition of EGF-stimulated cell proliferation. The latter is a very important process in ulcer healing. Moreover, the antacid Talcid® has been noted to induce mucosal activation of genes encoding for the EGF factor and its receptor EGF-R.[25,26] Consequently, the aluminum-containing antacid removes factors which cause damage to the gastric epithelial cells (urease which generates ammonia and cytotoxins). Ultimately, the inhibition reversal of EGF-stimulated cell proliferation offers a new perspective in the pharmacological and mechanistic[27] action of the drug and its clinical healing process.

On the basis of the aforementioned data, it appears that hydrotalcite can work efficiently toward healing of ulcerations due to bacterial as well as non-bacterial causes. Analogous studies conducted with other antacids, like Maalox® ($Al(OH)_3$ and $Mg(OH)_2$), emphasized the importance of the biochemical processes in the cytoprotection of gastric mucosa from *H. pylori* and its pathogenicity on gastroduodenal ulcers. The results indicate that the antacid decreases *H. pylori* HSP60 expression, thereby decreasing the inflammatory response in the gastric mucosa.[28]

The concurrent administration of aluminum pharmaceuticals with antibiotics is an issue of contemporary significance, as the interaction of both categories of agents results in variable activities.[29] The underlying aluminum–antibiotic aqueous chemistry emerges as an important factor in the biological activity of the administered antacid and antibiotic drugs, often limiting the bioavailability of the antibiotic.[30] Hence, concurrent use of both pharmaceuticals should proceed under strict medical supervision.

4.3.4 Bismuth aluminum carbonate

Bismuth aluminum carbonate has been used in clinical studies investigating the treatment of gastroduodenal disorders.[31] The studies project the remarkable action of bismuth aluminum carbonate on ulcers, emphasizing its drug efficacy on patients accepting the pharmaceutical satisfactorily (with no side effects, like nausea and constipation). The pharmaceutical, when administered alone or in combination with atropins, constitutes an excellent therapeutic for the conservative treatment of gastroduodenal ulcers.

4.3.5 Bismuth–magnesium–sodium alumino-silicate

This polymetallic silicic derivative contains a number of metals known to act as antacids, under specific conditions. In this regard, the bismuth–magnesium–sodium alumino-silicate pharmacological preparation has been employed in the treatment of common intestinal dyspepsia of the newborn.[32]

4.4 Phosphate Binders

In the case of renal disease, the use of aluminum-containing drugs is often suggested. The conservative management of intrinsic renal failure requires careful attention to fluid balance, maintenance of adequate nutrition and prevention or correction of disorders like hyperphosphatemia. Hyperphosphatemia is a disorder responsible for osteodystrophy commonly seen in patients with end-stage renal disease.[33] It requires medication, commonest of which are pharmaceutical preparations of aluminum, promoting efficient phosphate binding. Use, however, of such medication contradicts efforts to exclude aluminum from such patients, with all the dire consequences arising from aluminum absorption on one hand and the detrimental repercussions from a protracted hyperphosphatemia on the other. This dilemma has propelled research into developing secure aluminum-containing phosphate binders. The most widely used aluminum drugs in this case are antacids, containing $Al(OH)_3$. Normo-phosphatemia can be promoted by using $Al(OH)_3$ alone or by administering in combination with other drugs. Characteristic in this sense is the case of chronic renal insufficiency patients to whom antibiotics (e.g. ofloxacin) are administered in conjunction with the $Al(OH)_3$ phosphate-binder therapy.[34]

Studies[35,36] aimed at understanding the nature and reactivity of aluminum species, which promote reduction of hyperphosphatemia through the use of aluminum-containing antacids,[37] delve into issues relating to the availability of Al(III) ions and the concomitant phosphate-binding forms that Al(III) generates under the pH-specific conditions prevalent in the alimentary tract. The studies suggest that aluminum-containing antacids (of variable composition and formulations like Alucap®, Aludrox® and others) provide the necessary pool of Al(III) ions when dissociated in the acidic environment of the stomach following oral ingestion. Some of the Al(III) will be absorbed there, with the majority of the metal ion being absorbed through the gut. There, the pH of the lumen is high, ~6, and remains between 6 and 8.5 in the distal intestines. In this fairly high pH range, Al(III) is freshly precipitated yielding colloidal amorphous $Al(OH)_3$. In the presence of a large surface of such a preformed $Al(OH)_3$ precipitate, phosphate is adsorbed[38] (through chemistry involving the OH^- groups of $Al(OH)_3$[39] and species like the predominantly present HPO_4^{2-} ions) and will pass intact through the intestine without any decomposition because of the high pH. Any aluminum phosphate complex that might form under acidic pH conditions is also driven through due to insolubility under such conditions.[40] Finally, the phosphate–aluminum product will be excreted through the feces. Therefore, as far as phosphate binding is concerned, it is important from the pharmacological point of view to deliver $Al(OH)_3$ in its most amorphous form into the proximal intestine, concurrently trying to avoid its release in the gastric fluids at low pH. Low pH would clearly generate soluble forms[41,42] of Al(III) capable of promoting long-term toxic effects.[43] Due to risks of aluminum toxicity, calcium-based lanthanum carbonate and other agents have also been introduced in phosphate-binding therapy.

4.4.1 Basic aluminum carbonate

Basic aluminum carbonate has been employed in studies on gender differences in antacid-induced phosphate deprivation. In this regard, an animal model was used to study the susceptibility of women to antacid-induced phosphate deficiency. In the study, the male and female rats used were given basic aluminum carbonate gel (Basaljel®) as the antacid.[44] The study shows that female patients are more susceptible to phosphate-depleting effects of aluminum hydroxide antacids than are males.

4.4.2 Alumino-silicates

Alumino-silicate formulations (e.g. Malinal®) require prior exposure to acid to produce free Al(III) before they can act as phosphate binders, whereas amorphous $Al(OH)_3$ acts as an efficient phosphate binder without prior exposure to acid.

4.5 Alginate Raft Formulations

Heartburn is one of the commonly experienced gastrointestinal symptoms. In fact it is the primary symptom of gastro-oesophageal reflux disease (GERD).[45] The symptoms emerge, most often, following oesophageal influx of gastric acid. Thus, heartburn is strongly correlated with acidification of the oesophagus. The heartburn symptoms usually emerge in individuals after meals or nocturnally, often followed by burning substernal pain. The phenomena of gastrointestinal reflux and regurgitation are also clinical disorders frequently observed in infants and children.[46] High incidence heartburn has also been observed in asthmatic patients[47] and pregnant women on a daily basis.[48]

Alleviation of symptoms in these cases comes with alginate-based medication. Traditional medication includes antacids, containing $Al(OH)_3$, $AlPO_4$ or combinations thereof with $Mg(OH)_2$ and magnesium trisilicate,[49] characterized by a rapid onset in their action. Antacids raise the intragastric pH within minutes and offer relief of the heartburn symptoms. H_2-receptor antagonists can also be used as nonprescription medication (with characteristic examples of H_2-antagonists being Cimetidine®, Ranitidine®, Famotidine®, etc.), effectively controlling heartburn. These require systemic absorption for their action. Thus, they have a longer delay for the onset of their acid neutralization action, a disadvantage compared to antacids.[50]

Effective pharmaceuticals, combining the advantages of both, are alginate-based raft formulations. Alginates are natural polysaccharide polymers isolated from brown seaweed (Phaeophyceae) (Figure 4.2).

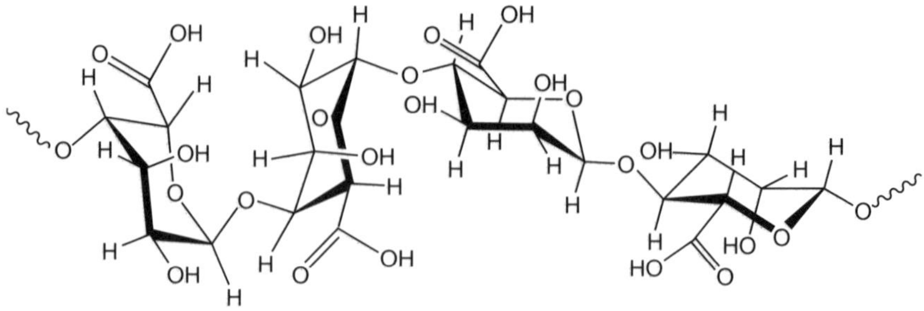

Figure 4.2 Alginic acids are linear unbranched polymers containing β-(1–4)-linked D-mannuronic acid (**M**) and α-(1–4)-linked L-guluronic acid (**G**) monomers

In the acidic environment of the stomach, alginate salts and alginic acids precipitate to form a low-density viscous gel. The gel forms immediately upon contact with gastric acid. The marketed formulations often vary in composition (Gaviscon®), with the commonest form containing alginic acid, sodium bicarbonate, magnesium trisilicate and $Al(OH)_3$. The latter agent is a well-known constituent of antacid preparations frequently used to counteract gastric acid secretion. This component appears to play a significant role in the action of the alginate-based drug(s).[51]

Upon contact of the medication with gastric acid, formation of a gel matrix takes place. Bicarbonate included in the formulation is trapped in the assembled 'eggbox' structure and leads to the generation of CO_2 gas. The gas bubbles entrapped in the gel matrix convert it into foam and provide buoyancy, which allows the gel to float on the surface of the gastric contents, much like a raft on water, while concurrently providing neutralizing capacity. The raft physically acts as a barrier to gastric acid reflux and moves into the oesophagus during reflux. It essentially acts as a mobile neutralizing sealant in the oesophageal space when the gastric pressure is high. When the pressure drops, the raft is lowered back into the stomach area.

Studies conducted *in vitro* have looked into the interplay of visco-elastic raft strength and neutralizing capacity of raft formulations in the case of aluminum antacids. It was found that $Al(OH)_3$ increases the neutralization profile of the raft, supporting the idea that the antacid-associated activity can be trapped into the raft of the gel matrix. That effect, however, results in a parallel reduction of the raft strength.[52] Alginate-based antacids combine the advantages of rapid onset of relief, brought about by antacids, while concurrently offering significantly longer duration of action compared to conventional antacids. Finally, studies on variably formulated alginate raft preparations suggest that the generated CO_2 from the bicarbonate along with antacid components, like $Al(OH)_3$, within a specific formulation[53] stay trapped in the assembled foaming raft, effectively providing it with lasting neutralization capacity.

4.6 Blistering Diseases in the Elderly

Blistering diseases in the elderly are a rare group of diseases mediated through the immune system, induced by drugs or arising from other primary systemic physiological aberrations. Pathogenetically, they are autoimmune diseases which affect the cutaneous and mucosal tissues. Treatment requires an arsenal of approaches, including localized and systemic therapies. Localized treatments involve care of the local area of the blister (topical care) through diverse means. Of those, crucial is soaking of the blister in an antiseptic solution. For that, the two major chemicals used are potassium permanganate and aluminum subacetate.[54] Aluminum subacetate (a basic aluminum acetate solution) is used locally on the skin and mucous membranes as an astringent and plays the role of the topical antiseptic. Solutions of the same chemical are also used as a wet dressing in a number of other skin diseases.

4.7 Metabolic Diseases and Aluminum

Aberrations in lipid metabolism lead to a plethora of clinical symptomatologies often with dire consequences for humans. Therefore, regulating lipid metabolism constitutes a significant step toward prevention, alleviation of clinical symptoms and potential cure for some of the diseases arising from aberrant lipid metabolism.

Alufibrate® is a common drug employed in the regulation of lipid metabolism. Its action concentrates on the decrease of the concentrations of cholesterol and triglycerides in blood.[55] As such, it is used for combating atherosclerosis and hypertension. To this end, characteristic diseases toward which Alufibrate® acts include hyperliproteinemias (e.g. type IIa, type IIb and type IV hyperlipidemias).[56]

The drug bears the formulation bis[2-(*p*-chlorophenoxy)-2-methylpropionato]-hydroxyaluminum, essentially the basic aluminum salt of clofibric acid (Figure 4.3).

The clinical efficacy of Alufibrate® in hyperlipidemic patients has been well established.[57] The drug is known to lower serum levels of triglycerides more so than cholesterol. A number of clinical investigations have attempted to delineate the mechanism of action of the drug, yet its mode of action is not fully understood. According to those investigations, inhibition of acetyl-CoA carboxylase by hypolipidemic medication, like Alufibrate®, may be linked to less acetyl-CoA

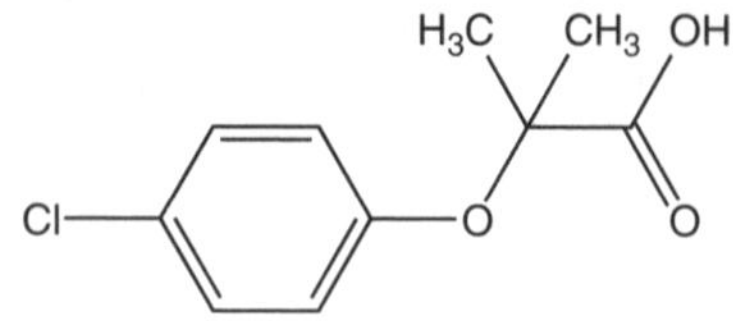

Figure 4.3 Structure of clofibric acid

being used for fatty acid synthesis, thus providing more of that substrate for oxidation.[58] Alufibrate® was also observed to reduce the concentration of very low density lipoproteins (VLDL), which may be attributed to diminished synthesis of VLDL lipids.

4.8 Anti-malarial Substances

The malaria organism *Plasmodium falciparum* detoxifies heme, which is released upon degradation of host erythrocyte hemoglobin, by sequestering it within the parasite digestive vacuole in the form of a polymer called hemozoin. The antimalarial agent Chloroquine® acts by interrupting heme polymerization, yet its efficacy is impaired by drug-resistant organisms. The synthetic complex ethylenediamine-*N*,*N*′-bis[propyl(2-hydroxy-(R)-benzylimino)]aluminum(III) (Figure 4.4) kills intraerythrocytic malaria parasites in a stage-specific manner, thus reflecting a new class of probes for Chloroquine®-resistant mechanisms and a fresh prospect in antimalarial drug development.[59]

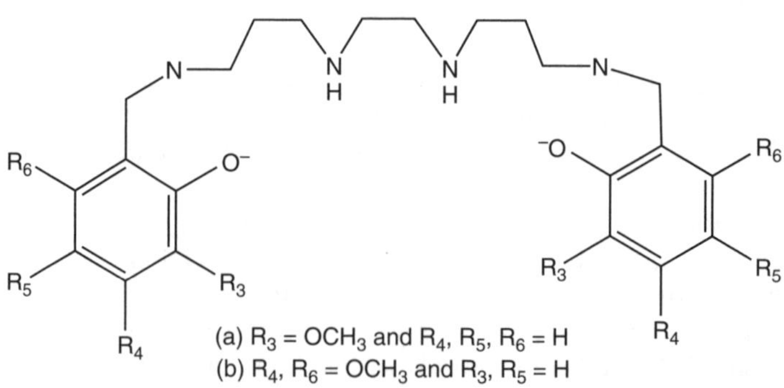

Figure 4.4 The Schiff-base reduced amino derivative ligand used for the synthesis of the proposed octahedral (*R*)-ENBPA-Al(III) complex

4.9 Potential Aluminum Toxicity

Aside from the positive influence of aluminum pharmaceuticals in human health, excessive and unattended use of such preparations may have adverse effects in physiology. To this end, generally:

1 Excessive use of aluminum preparations influences negatively the human health.[60] Specifically, among other clinical manifestations, aberrations are observed in calcium and phosphate metabolism resulting in damage of the skeletal bone system. Excessive intake of aluminum (through the food chain, drug ingestion, water supply, etc.) has been noted to accumulate in sensitive loci, resulting in pathological aberrations such as amnesia in young people. Prominent, also, is the presence of aluminum in the brain of patients with Alzheimer's disease,[61] promoting an association of that element with the disease and its neuropathological hallmarks.

2 In a number of OTC drugs, aluminum is a major component. In fact, excessive use of aluminum-containing antacids, available as OTC medication, could contribute quantitatively to aluminum exposure of humans. Hence, a warning on potential dangers from aluminum-containing drugs serves to raise consciousness against their unsupervised use. Potential links of aluminum with neurodiseases, like idiopathic Parkinson's disease, have also been reported.[62]

Given the fact that aluminum has been studied as a potential inducer and/or contributor to a number of exposure-related pathologies, like embryonic and fetal toxic effects in humans and animals, osteomalacias, etc. it is quite critical that (a) the (bio)chemistry of that element in a biological environment be studied and (b) the associated speciation under conditions of ingestion and metabolic processing be investigated for potential toxicity effects. Clinical trials suggest that drug toxicity emerges through increased absorption of aluminum upon concurrent administration of aluminum-containing medication and citrate or citrate-containing beverages,[63] use of antacids during pregnancies, etc. Crucial, therefore, in this respect appears to be in-depth research, addressing toxicity issues and providing answers that could formulate the conditions under which well-defined non-toxic aluminum drugs should be administered to humans. Through such research, safer formulations of aluminum-containing medication could emerge that enable biologically effective treatment of human pathologies. To date, warnings of aluminum overexposure from such medications vary. It is advisable, therefore, that patients be fully informed on the aluminum-containing drug that they use,[64] consistent with their specific physiology or pathological conditions, to the point that medical administration minimizes potential risks. Patient-tailored administration of well-defined aluminum drugs appears to be the safest way of providing medical assistance to human pathologies increasingly burdening modern societies.

4.10 Conclusions

Clinical use of aluminum in medications has been around for long. Aluminum compounds

- have been primary components of vaccines in the form of immunoadjuvants – specific forms of aluminum adjuvants have been shown to enhance antigenicity in a number of vaccines (e.g. diphtheria and tetanus toxoid);
- are extensively used in antacids, for the treatment of gastroenterological conditions, including secretory disturbances of the alimentary tract;
- exhibit anti-inflammatory activity;
- act as antiseptic agents;
- regulate lipid metabolism, combating atherosclerosis and hypertension;
- are used in phosphate binder therapy on end-stage renal failure patients.

Evidently, aluminum has found numerous applications in modern pharmacology. Its inclusion in prescription and nonprescription drugs reflects its usefulness and concurrently necessitates perusal of mechanistic action(s) in medicinal bioactivity, the enhancement of which remains a driving force for the development of a new generation of aluminum-containing drugs targeting human diseases. To this end, more potent and effective drugs require higher safety standards, addressing unknown risks often lurking in as yet unexplored cytochemistries of aberrant human pathophysiologies. It logically ensues, therefore, that physicochemical, biological, clinical and toxicological research of aluminum compounds toward structure–function relationships and strong safeguards emerges as a prominent contemporary metallopharmaceutical field, full of challenges and scientifically sound medical advances.

References

1. J.M. Brewer, M. Conacher, C.A. Hunter *et al. J. Immunol.* **163**, 6448–6454 (1999).
2. M. Ulanova, A. Tarkowski, M. Hahn-Zoric and L.Å. Hanson *Infect. Immun.* **69**, 1151–1159 (2001).
3. N. Baylor, W. Egan and P. Richman *Vaccine* **20**, S18–S23 (2002).
4. J.O. Naim, C.J. van Oss, W. Wu *et al. Vaccine* **15**, 1183–1193 (1997).
5. Y.-F. Maa, L. Zhao, L.G. Payne and D. Chen *J. Pharm. Sci.* **92**, 319–332 (2003).
6. E. Rivera, A. Daggfeldt and S. Hu *Vet. Immun. Immunopath.* **91**, 19–27 (2003).
7. S. Jendrek, S.F. Little, S. Hem *et al. Vaccine* **21**, 3011–3018 (2003).
8. P.M. Callahan, A.L. Shorter and S.L. Hem *Pharm. Res.* **8**, 851–857 (1991).

9. G. Kanra, S. Viviani, K. Yurdakök *et al. Ped. Int.* **45**, 314–318 (2003).
10. A. Tarnawski, D. Hollander, D. Cummings *et al. Gastroenterology* **86**, 1276–1281 (1984).
11. T. Clark and J.N. Hunt *Practitioner* **180**(1077), 334–339 (1958).
12. B.E. Moret and W. Kloppe *Arztl. Wochensch.* **10**(34–35), 788–790 (1955).
13. F. Ramaholimihaso, A. Duchateau, F. Scheck and G. Thiefin *Gastroenterol. Clin. Biol.* **21**, 370–376 (1997).
14. G. Bertaccini and G. Corruzi *Dig. Dis. Sci.* **30**(Nov. 1985 Suppl.), 43S–51S (1985).
15. W.L. Peterson, R.A. Sturdevant, H.D. Frankl *et al. N. Eng. J. Med.* **297**, 341–345 (1977).
16. A. Tarnawski, J. Stachura, W.J. Krause *et al. J. Clin. Gastroenterol.* **13**(Suppl. 1), S42–S47 (1991).
17. I.M. Marks, I.M. Samloff, M. Aarimaa and M. Siurala in 'Proceedings of the 2nd International Sucralfate Symposium' *Scand. J. Gastroenterol.* **18**(Suppl. 83), 82 (1983).
18. A. Tarnawski, I.L. Szabo, S.S. Husain and B. Soreghan *J. Physiol. (Paris)* **95**, 337–344 (2001).
19. N. Ferrara *Kidney Int.* **56**, 794–814 (1999).
20. A. Tarnawski, M. Tomikawa, M. Ohta and I.J. Sarfeh *J. Physiol. (Paris)* **94**, 93–98 (2000).
21. P. Grübel, K.R. Bhaskar, D.R. Cave *et al. Aliment. Pharmacol. Ther.* **11**, 139–145 (1997).
22. T.L. Cover and M.J. Blaser *Adv. Intern. Med.* **41**, 85–117 (1996).
23. D.T. Smoot *Gastroenterology* **113**(Suppl.), 31–34 (1997).
24. H.L.T. Mobley *Gastroenterology* **113**(Suppl.), S21–S28 (1997).
25. A. Tarnawski, K.J. Whalstrom, M. Ohta and I.J. Sarfeh *Gastroenterology* **109**, 109–276 (1996).
26. R. Pai and A. Tarnawski *Dig. Dis. Sci.* **43**(Suppl. 9), 14S–22S (1998).
27. A. Tarnawski, R. Pai, R. Itani and F.A. Wyle *Digest.* **60**, 449–455 (1999).
28. S. Kamiya, H. Yamaguchi, T. Osaki *et al. Scand. J. Gastroenterol.* **7**, 663–670 (1999).
29. M. Lazzaroni, B.P. Imbimbo, S. Bargiggia *et al. Antimicrob. Agents Chemother.* **37**, 2212–2216 (1993).
30. M. Tanaka, T. Kurata, V. Fujisawa, *et al. Antimicrob. Agents Chemother.* **37**, 2173–2178 (1993).
31. P. Uhry and M. Duizend *Progr. Med. (Paris)* **90**, 530–534 (1962).
32. A. Combes-Hamelle and R. Frot *Ann. Pediatr. (Paris)* **14**, 2482–2483 (1967).
33. A.H. Lau, J.M. Kuk and K.L. Franson *Int. J. Artif. Organs* **21**, 19–22 (1998).
34. W. Schulz and A. Dorfler *Infection* **14**(Suppl. 1), S97–S101 (1986).
35. E.A. Larson, S.R. Ash, J.L. White and S.L. Hem *Kidney Int.* **29**, 1131–1135 (1986).
36. R.W. Cargill, M. Dutkowskij, A. Prescott *et al. J. Pharm. Pharmacol.* **41**, 11–16 (1989).
37. A.C. Alfrey *Kidney Int.* **29**(Suppl. 18), S8–S11 (1986).
38. K.F.N.K. Kwong and P.M. Huang *Nature* **271**, 336–337 (1978).
39. N. Mikami, M. Sasaki, K. Hachiya *et al. J. Phys. Chem.* **87**, 1454–1458 (1983).
40. J.W. Akitt, N.N. Greenwood and G.D. Lester *J. Chem. Soc. A*, Part II 2450–2457 (1971).
41. R.B. Martin *Clin. Chem.* **32**, 1797–1806 (1986).
42. A. Salifoglou *Coord. Chem. Rev.* **228**, 297–317 (2002).

43. D.P. Perl and A.R. Brody *Science* **208**, 297–299 (1980).
44. K.B. Schwarz, D.C. Zimmerman, D.H. Alpers and L.V. Avioli *Gastroenterology* **89**, 313–320 (1985).
45. K. Devault and D.O. Castell *Arch. Intern. Med.* **155**, 2165–2172 (1995).
46. E.W. Fonkalsrud and M.E. Ament *Curr. Probl. Surg.* **33**, 1–70 (1996).
47. S.J. Sontag *Am. J. Med.* **103**, 84S–90S (1997).
48. H.I. Jacoby and D.A. Brodie in *Emerging Gastrointestinal Therapy Markets*, Decision Resources, Waltham, MA, 1997.
49. W.A. Ritschel in *Antacids and Other Drugs in GI Diseases,* Drug Intelligence Publications, Inc., Hamilton, IL, USA, 1984.
50. P.J. Kahrilas in Gastroesophageal reflux disease and its complications in *Gastrointestinal and Liver Disease*, 6th Edition, M. Feldman, B.F. Scharschmidt, M. Sleisenger and R. Zorab (Eds). W.B. Saunders, Co., New York, USA, 1998, pp. 498–517.
51. K.G. Mandel, B.P. Daggy, D.A. Brodie and H.I. Jacoby *Aliment. Pharmacol. Ther.* **14**, 669–690 (2000).
52. N. Washington, C. Washington, C.G. Wilson and S.S. Davis *Int. J. Pharm.* **28**, 139–143 (1986).
53. N. Washington, C.G. Wilson and S.S. Davis *Int. J. Pharm.* **27**, 279–286 (1985).
54. N. Sami, S.W. Yeh and A.R. Ahmed *Dermatol. Clin.* **22**, 73–86 (2004).
55. J. Dalmau, B. Vicens, J. Zapatero and L. Bruseghini *Arzneimittelforschung* **31**, 1840–1844 (1981).
56. T. Tomita, R. Takeshita, T. Nagakura and E. Hayashi *Jpn. J. Pharmacol.* **23**, 281–287 (1973).
57. H.B. Stähelin, J.T. Locher and R. Maier *Clin. Chim. Acta* **54**, 115–122 (1974).
58. M.E. Maragoudakis, H. Hankin and J.M. Wasvary *J. Biol. Chem.* **247**, 342–347 (1972).
59. D.E. Goldberg, V. Sharma, A. Oksman *et al. J. Biol. Chem.* **272**, 6567–6572 (1997).
60. P. Altmann, J. Cunningham, U. Dhanesha *et al. Brit. Med. J.* **319**, 807–811 (1999).
61. C. Exley and J.D. Birchall *J. Theor. Biol.* **159**, 83–98 (1992).
62. E. Altschuler *Med. Hypotheses* **53**, 22–23 (1999).
63. S. Fairweather-Tait, K. Hickson, B. McGaw and M. Reid *Eur. J. Clin. Nutr.* **48**, 71–73 (1994).
64. C.M. Reinke, J. Breitkreutz and H. Leuenberger *Drug Saf.* **26**, 1011–1025 (2003).

5 $_{14}$Si Biological Activity of Organosilicon Compounds

Edmunds Lukevics and Luba Ignatovich

5.1 Introduction

Two main approaches aimed at the preparation of new pharmaceuticals using organosilicon compounds are now in vogue:

1. application of organosilicon compounds to prepare biologically active non-silicon-containing compounds;[1] and
2. synthesis of organosilicon compounds possessing intrinsic biological activity.[1–29]

In the first case organosilicon compounds are used for the temporary protection of a functional group,[1,17,27,30] as reagents mainly for reduction and cyanation,[27] or as reaction intermediates facilitating the synthesis of the target nucleosides[31] or β-lactam antibiotics.[32]

The anti-cancer drug Tegafur (Ftorafur®, Futraful), 5-fluoro-1-(2-tetrahydrofuryl)-2,4(1H,3H)-pyrimidinedione, was one of the first nucleoside analogues industrially prepared by the silyl method.[31] The silylation converts the starting 5-fluorouracil into the more reactive liquid bis-silyl derivative. The reaction of this organosilicon intermediate with 2-chlorotetrahydrofuran already starts at −50 °C and proceeds smoothly at room temperature. The remaining protective group is easily removed by crystallization from alcohols.

Metallotherapeutic Drugs and Metal-Based Diagnostic Agents: The Use of Metals in Medicine
Edited by Gielen and Tiekink

In the second case the biological properties of organosilicon compounds themselves are exploited in drug design. There are three ways for obtaining biologically active derivatives of organosilicon compounds:

1 modification of biologically active organic compounds by introducing silicon-containing substituents;[6,7,9,18,22,26]

2 preparation of silicon analogues of known drugs by substitution of one or more carbon atoms for silicon atoms;[6,7,10,14,15,19,22,25,28,29] and

3 synthesis and study of the biological activity of structurally specific organosilicon compounds lacking organic analogues.[6,8,11,13,22,28,29]

Using these methods several thousands of biologically active organosilicon compounds have been prepared; nine of them have entered human clinical trials, another four are used as agrochemicals.

5.2 Organosilicon Modification

One of the most convenient ways of obtaining biologically active organosilicon compounds is the modification of bioactive organic compounds by silylation. In the course of this process organosilicon groups are used to substitute one or several hydrogen atoms linked either with heteroatom (oxygen, sulphur, nitrogen) or carbon atoms:

X = O, S, N, C

5.2.1 O-, S- and N-Silylation

O-Silylation of hydroxyl-group-containing drugs leads to prodrugs with an enhanced lipophilicity.[18,26,33–36] They are more rapidly transported across the lipid barrier followed by hydrolysis liberating the parent drug. The rate of the hydrolysis can be regulated by the substituents at the silicon atom. Under acidic conditions the hydrolysis rate diminishes in the order:[34,37,38]

$$Me_3Si > Me_2EtSi > Et_3Si > t\text{-}BuMe_2Si > TxMe_2Si > i\text{-}Pr_3Si > t\text{-}BuPh_2Si$$

If the silylating agent $R_{4-n}SiX_n$ contains more than one functional group at the silicon atom ($n > 1$), two to four residues of the biologically active substance can be introduced to one silicon atom.[7,9,33,39–43] In the case of polyhydroxy compounds heterocyclic siloxa derivatives have been obtained.[44–49] Biologically active siloxa heterocycles have been prepared also by intramolecular cyclization.[50–52]

O-Silylation can be used not only for the modification of hydroxyl group containing biologically active organic compounds but also for the modification of organic derivatives of aluminium,[53] germanium,[54,55] titanium[56] and phosphorus.[57]

In many cases such a modification increases the activity of the parent drug[7,9,22,26,36,46,48,58–68] or prolongs the duration of its action.[7,9,18,22,26,69–71] Reduction of toxicity has also been observed.[7,9,63,71,72]

As a result, the anabolic agent Silabolin® (**1**), an *O*-trimethylsilyl derivative of 19-nortestosterone, has been developed and registered in Russia.[72] It exhibits high anabolic potency, long duration of action, lower toxicity than other anabolic steroids and good tolerability.[72]

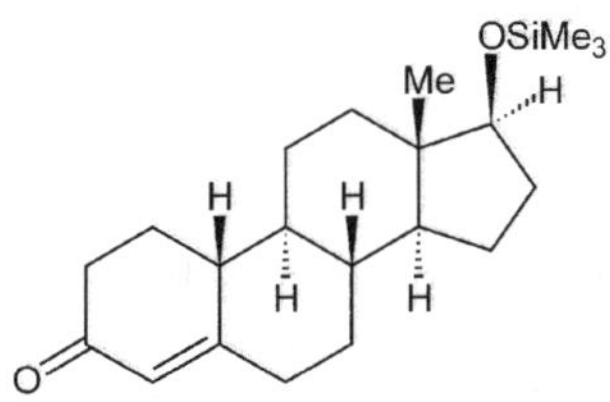

1 Silabolin®

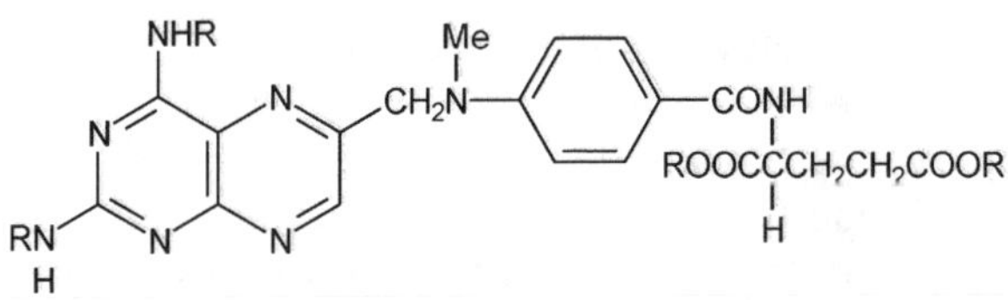

2a R = H, **2b** R = *t*-BuMe$_2$Si

The introduction of an organosilicon substituent into the molecular framework of a drug can generate some biological properties not inherent in the initial drug. Thus, for example, the introduction of dimethyl(*tert*-butyl)silyl groups into the anti-tumour drug methotrexate (**2a**) yields silylated methotrexate (**2b**) which exhibits an activity against experimental allergic encephalomyelitis and, in contrast to **2a**, prevents serious paralysis and mortality in animals.[73] It can be assumed that the activity of **2b** is due to the increase of lipophilicity and to the ability to pass across the blood–brain barrier.

It has been found that, in contrast to uridine (**3a**), 5-*O*-*tert*-butyldimethylsilyluridine (**3b**) exhibits anti-tumour activity suppressing the growth of human lung fibrosarcoma cells, HT-1080.[63] Some growth inhibition of human breast

(MCF-7), central nervous system (SF-268) and NCI-H40 lung cancer has been observed for 3′-*O*-silatranylthymidine (**4**).[74]

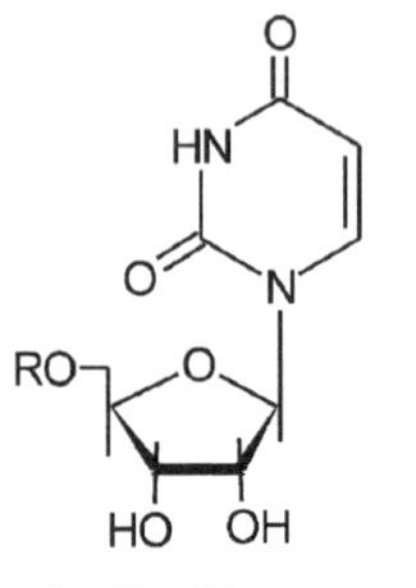

3a R = H
3b R = *t*-BuMe$_2$Si

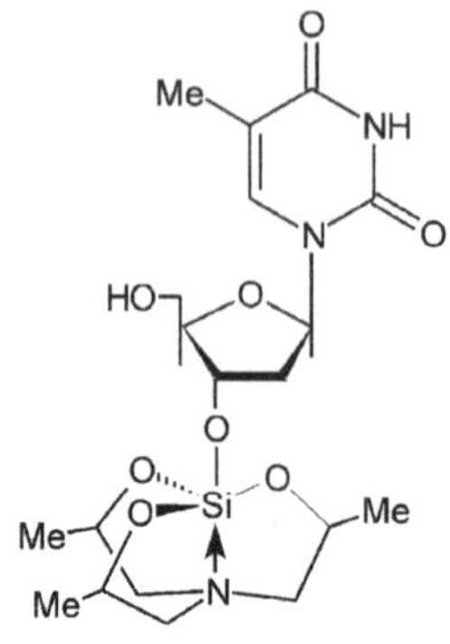

4

Silylation of hydroxyl groups of (β-D-ribofuranosyl)-3′-spiro-5″-(4″-amino-1″,2″-oxathiole-2″,2″-dioxide) nucleosides led to a new family of highly specific non-nucleoside reverse transcriptase inhibitors designated as TSAO.[75–98] Several hundreds of silyl derivatives of these nucleoside analogues have been synthesized, their human immunodeficiency virus type 1 (HIV-1)-inhibiting activity has been tested and structure–activity relationships have been established.

The introduction of bulky *tert*-butyldimethylsilyl group at the C-2′ position of the ribose moiety led to the appearance of a slight anti-HIV-1 activity. The C-5′ silyl group increased the activity but the presence of *tert*-butyldimethylsilyl group at both C-2′ and C-5′ positions is necessary for high anti-HIV-1 activity.[76,78,83,94,98] A larger thexyldimethylsilyl substituent is also accepted at the 5′ position with only a marginal loss of activity.[83] The presence of the unique 3-spiro group having a D-ribo configuration is also a prerequisite for anti-viral activity.[77,84,85]

The prototype compound is [2′,5′-bis-*O*-(*tert*-butyldimethylsilyl)-β-D-ribofuranosyl]-3′-spiro-5″-(4″-amino-1″,2″-oxathiole-2″,2″-dioxide)thymine (TSAO-T, **5a**). The thymine moiety of TSAO-T can be replaced by a number of other pyrimidines, purines and 1,2,3-triazoles without a marked decrease of anti-viral activity.[76,78–81,87] Urea derivatives, which can mimic to a large extent both the shape and the electrostatic potential of a thymine ring, can effectively replace this nucleic base.[88] Also, the dihydrouracil TSAO derivatives retained pronounced anti-HIV-1 activity.[88] The oxalyl group in the position 4″ of the 3′-spiro moiety was important for the anti-HIV-1 activity against both wild type and TSAO-resistant strains.[95] The most selective compound is the 3-*N*-methyl derivative **5b** but the introduction of a 3-hydroxypropyl group at the N-3 (compound **5c**) led to a 2–6-fold improvement in anti-viral potency.[93] Several TSAO derivatives bearing, at the N-3 position of the thymine, an L-amino acid,[90] (*E*)-hydroxypropenyl or vinyl group (**5d**) showed some activity against HIV-2 in addition to their inhibitory effect on HIV-1.[97]

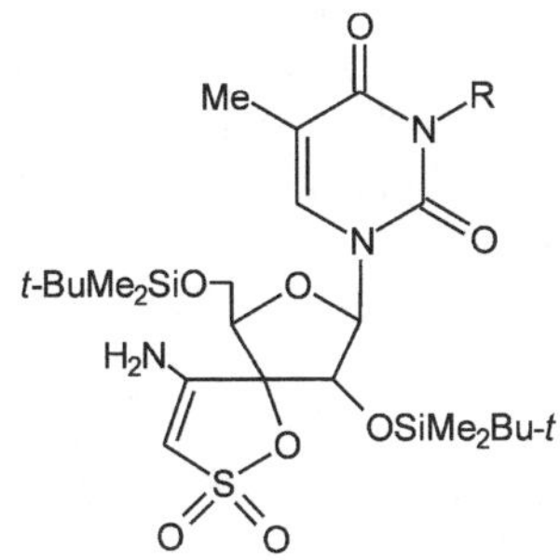

TSAO
5a R = H; **5c** R = $(CH_2)_3OH$;
5b R = Me; **5d** R = CH = CH_2

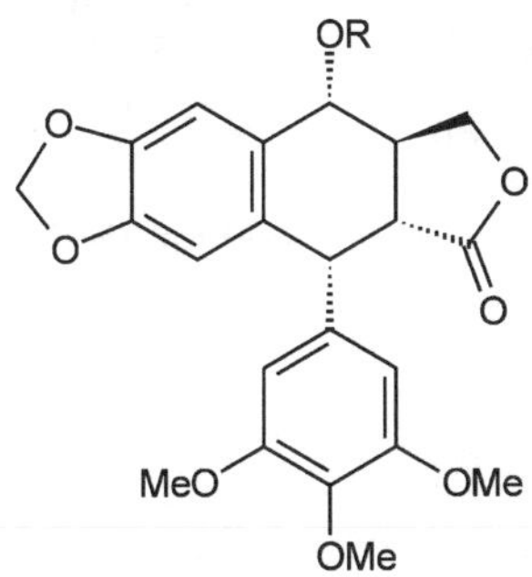

6a R = H
6b R = *t*-$BuMe_2Si$

(*S*)-4-*tert*-Butyldimethylsiloxy-2-cyclopenten-1-one, containing the cyclopentenone ring of prostaglandine A1, inhibited herpes simplex virus (HSV-1) replication in cell cultures at sub-toxic concentrations. It was also active against acyclovir-resistant strain of HSV-1 but did not affect two laboratory strains of HSV-2.[99] At the same time, the *O*-silyl derivative **6b** of podophyllotoxin (**6a**) showed an unusual 10-fold selectivity for HSV-2 as compared to HSV-1.[100]

S-Silylation, similarly to O-silylation, can increase the activity[101–107] and lower the toxicity.[107–112] N-silylation lowers the toxicity of aziridines[113] and nystatine.[114] Its *N*-silyl (Me_3Si, Et_3Si) derivatives possess a higher fungicidal activity than nystatine itself but the introduction of bulky triisopropylsilyl group at the pyrrole nitrogen atom in 3-pyrrolyl-2,5-dihydro-1H-2,5-pyrrole-dione decreased the anti-bacterial activity against methicillin- and ciprofloxacin-resistant bacterium *Staphylococcus aureus 134/94*.[115]

The most investigated field of medical applications of compounds containing Si—N bonds is cancer photodynamic therapy. It is a method for treatment of tumours involving the systematic administration of tumour-localizing photosensitizers and their subsequent excitation with visible light (600–800 nm) to produce tumour necrosis.[116,117] Silicon phthalocyanines (**7**),[53,118–130] 2,3-naphthalocyanines (**8**)[131–139] and tetradibenzobarrelenophthalocyanines[140] are suitable for this purpose.

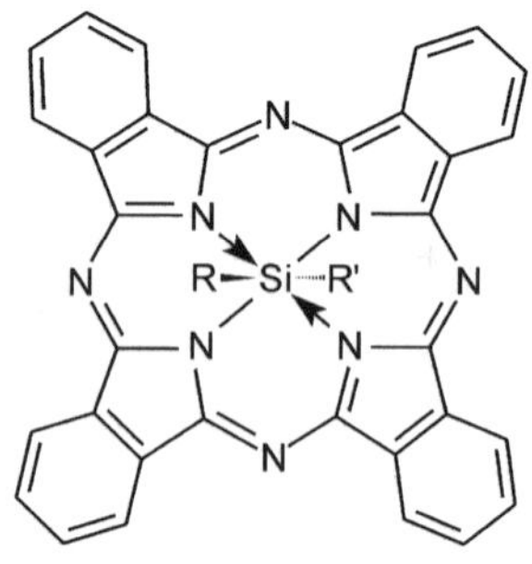

7a R = R' = Cl;
7b R = R' = R''_3SiO;
7c R = OH, R' = $OSiMe_2(CH_2)_3NMe_2$

Pc-4

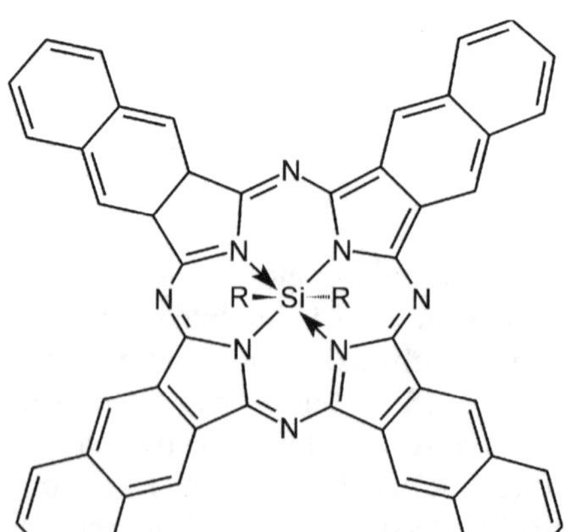

8 R = R' R''_2SiO

They have a high efficiency of tumour targeting, a large selectivity of accumulation in the tumour, a strong absorption in the red visible region and a high photoactivity.

The photoinduced photokilling of achromic M6 melanoma cells is correlated with an increase in cell-lipid peroxidation and superoxide dismutase activities, a decrease in the concentrations of glutathione and its oxidized form and a reduction of glutathione peroxidase and catalase activities.[124]

The nature and shape of two axial ligands on the central silicon can modify the lipophilic character of a substance and impede the aggregation in solution.[53,119–122,130,135–139] The substituents can be equal (**7a,b, 8**) or different (**7c**). The silicon phthalocyanine (**7c**)[53,120,123,125,126–128] has been chosen for clinical trials in cancer photodynamic therapy. This compound combined with red light irradiation also inactivated pathogenic viruses in red blood cells.[127]

5.2.2 C-Silylation

C-Silylation of drugs has been widely used to design new bioactive agents.[6,7,22,29] It can increase the activity[141–149] or significantly change the properties of the parent compound.[150–155]

For example, 17-(triethylsilyl)ethynylestradiol[143] and its derivatives[144] possess extremely high anti-fertility properties and, at the same time, reduced estrogenic activity. Odour tests of the trialkylsilylated benzyl mercaptans and benzenethiols have revealed that the trimethylsilyl substituent at the benzene ring has a remarkable effect in reducing the foul smell of the parent benzyl mercaptan and benzenethiol.[154]

The advantage of a C-modification by organosilyl groups lies in the possibility of wide variations of substituents at the silicon atom thus providing the opportunity to obtain libraries of organosilicon compounds representing a broad spectrum of pharmacological activities. Organosilicon amines are one of such class of compounds exhibiting anti-bacterial,[6,7,9,21,156] anti-fungal,[6,7,9,21] anti-cancer,[21,149,156 170] central nervous system,[6,7,166,171–176] cardiovascular[7,171,177,178] and other activities.[6,7]

Four organosilicon compounds designed by this approach have entered clinical studies: the muscle relaxants Sandoz 58–112 (**9**)[179] and Silperisone (**10**),[180–183] the long-lasting acetylcholinesterase inhibitor Zifrosilone (**11**) proposed for the treatment of patients with Alzheimer's disease[184–188] and the anti-cancer drug Karenitecin (**12b**).[149,189]

Silylcamptothecins (silatecans), possessing a high anti-cancer activity, obviously demonstrate the importance of C-silylation for the enhancement of the biological activity of known drugs. Camptothecins are potent cytotoxic agents that inhibit DNA topoisomerase I but their clinical utility is limited by the instability of the δ-lactone ring present in the structure, some side effects and a narrow spectrum of activity.

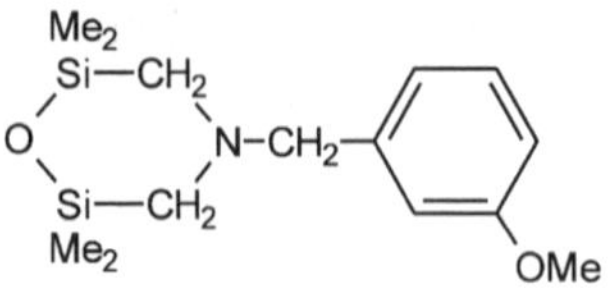

9 Sandoz 58–112

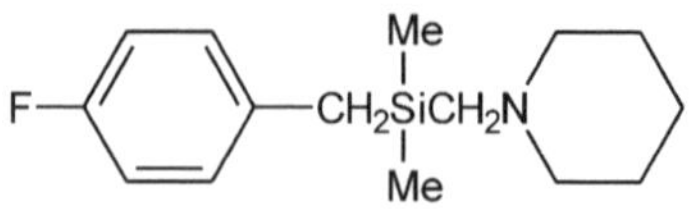

10 Silperisone

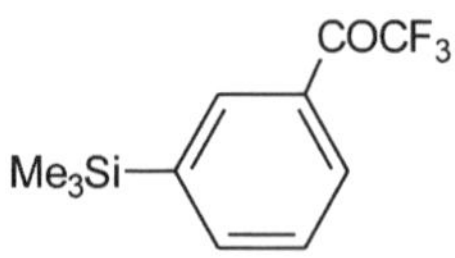

11 Zifrosilone

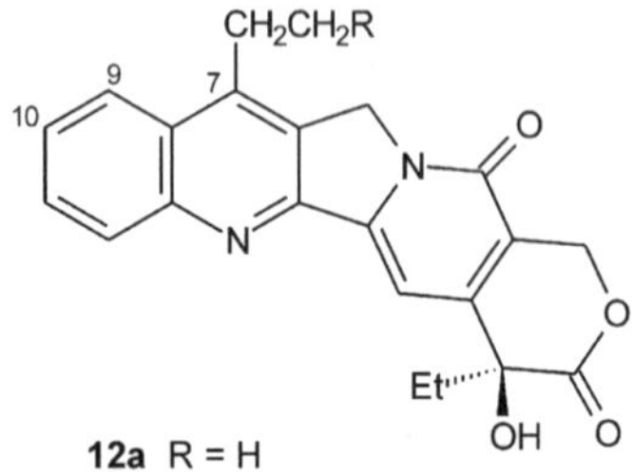

12a R = H
12b R = Me_3Si, Karenitecin

The incorporation of the lipophilic organosilicon substituents into the camptothecin structure provides enhanced blood stability, increased cell penetration and improved pharmacokinetics.[149,190–199] A series of silatecans were prepared with different triorganylsilyl groups attached to the 7-position. 7-Trimethylsilyl camptothecin was about two times more potent than camptothecin in inhibiting HL-60 human promyelocytic leukaemic cells, 833K teratocarcinoma solid tumour cells and DC-3F hamster lung cells.[190] Combination of a 7-trimethylsilyl group with an 11-fluorine provides a compound that is about five times more potent. Combination with a 10-amino group increases the activity 10 times but introduction of all three substituents gives a compound that is more than 20 times more potent than camptothecin.[149,190] The influence of the 7-triorganosilyl group decreases in the following order:

$$t\text{-BuMe}_2\text{Si} > \text{Me}_3\text{Si} > t\text{-BuPh}_2\text{Si} > i\text{-Pr}_3\text{Si}$$

7-*tert*-Butyldimethylsilyl-10-hydroxycamptothecin displays superior stability in human blood when compared with clinically used camptothecins. It is 25 times more lipophilic than camptothecin, and it is readily incorporated into cellular and liposomal bilayers. This silatecan exhibits high cytotoxicity and very high intrinsic potency against topoisomerase I.[194] It can effectively cure mice with intracranially implanted human gliomas.[192]

7-[2-(Trimethylsilyl)ethyl]-20(*S*)-camptothecin (**12b**) (Karenitecin) is significantly more potent than other camptothecins. It has a broad anti-tumour activity, superior oral bioavailability, increased lactone stability and lack of multi-drug resistance susceptibility.[191,195,197] In a wide spectrum of human malignant cell lines the Karenitecin exhibits a nanomolar to picomolar activity.[196] It is more potent than the parent 7-ethylcamptothecin (**12a**) and topotecan in experimental human ovarian cancer.[198]

Silatecans containing a 7-membered lactone ring have been also prepared and shown to be potent topoisomerase I inhibitors that are stable in human blood.[193,199]

C-Silylation of triazole has led to a series of potent fungicides.[200–211] Bis(4-fluorophenyl)methyl(1*H*-1,2,4-triazol-1-ylmethyl)silane (**13**) (Flusilazole) is a highly potent ergosterol biosynthesis inhibiting fungicide which has found acceptance world-wide for controlling diseases of cereals, fruits, vegetables and other agricultural products.[200–204] It inhibits also the growth of the human pathogenic fungi *Trichophyton mentagrophytes* (minimal inhibiting concentration 1 μg ml^{-1}) and *Candida albicans* (8 μg ml^{-1}).[203]

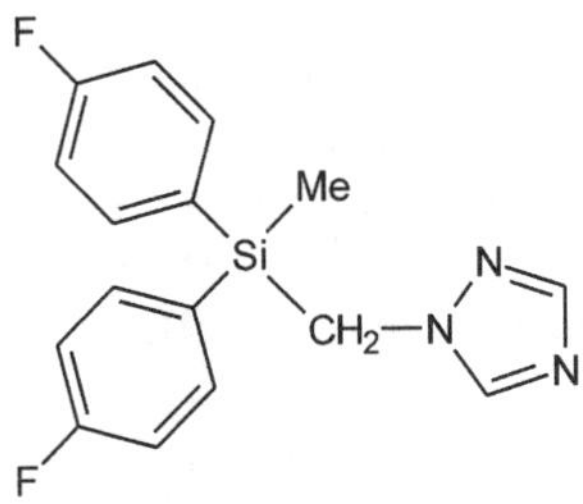

13 Flusilazole

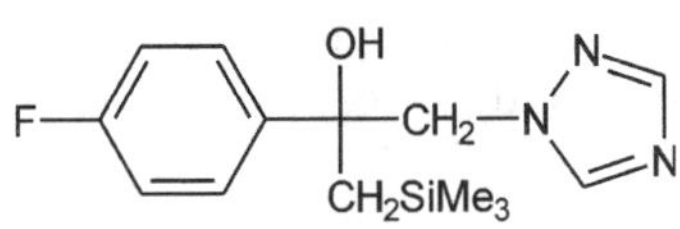

14 Simeconazole

2-(4-Fluorophenyl)-1-(1*H*-1,2,4-triazol-1-yl)-3-trimethylsilylpropan-2-ol (**14**) (Simeconazole) combines a prominent systemic fungicidal activity with a good crop safety.[208–210] It is also effective against rice sheath blight by submerged application, rice blast, sharp eyespot and powdery mildew.[210]

Another potent fungicide for the control of take-all disease of cereal crops caused by the soil-borne fungus *Gaeumannomyces graminis* has been found through a screening of compounds prepared by C-silylation of aromatic and heterocyclic amides.[212,213] 4,5-Dimethyl-*N*-(2-propenyl)-2-trimethylsilyl-3-thiophenecarboxamide (**15**) (Silthiopham) had outstanding protectant activity

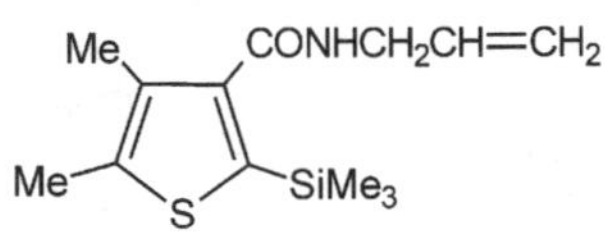

15 Silthiopham

providing long duration of control of root infection,[213–216] and it was selected for marketing in formulation under the trade name Latitude®.

5.3 Sila Analogues

Being in the same group as carbon in the periodic table of elements, silicon differs from the latter in its physicochemical properties (considerably larger covalent radii; lower electronegativity; different electronic energies for their outer electrons; longer bond lengths with hydrogen, carbon and oxygen; and lower bond energies with hydrogen and carbon and higher bond energy with oxygen[25]). It can be expected that these physicochemical effects may, in principle, result in different biological effects.

In order to compare the biological activity of carbon and silicon analogues having exactly the same chemical structure, with a silicon atom replacing one of the carbon atoms, a large number of sila-substituted drugs have been synthesized and biologically tested.[6,7,10,14,15,19,22,29] Two groups of compounds have been found. The main group embraces analogous derivatives of carbon and silicon with close biological activity (bioisosters);[7,22,29,217] the second incorporates carbon and silicon structural analogues exhibiting different biological properties.[7,22,148,210,218–225]

In the course of these investigations a variety of sila analogues of the muscarinic antagonists have been prepared.[22,29] Two of them, hexahydrosiladifenidol (**16a**) (HHSiD)[226–231] and *p*-fluoro-hexahydrosiladifenidol (**16b**) (*p*-F-HHSiD)[226,228,231–234] are used worldwide as selective tools for the classification of muscarinic receptor subtypes.[22,29]

Optically active silanols and hydroxymethylsilanes of this series of muscarinic antagonists have been prepared. Their (*R*)-enantiomers showed a significantly higher affinity for muscarinic M_2 and M_3 receptors than the corresponding (S)-antipodes.[235]

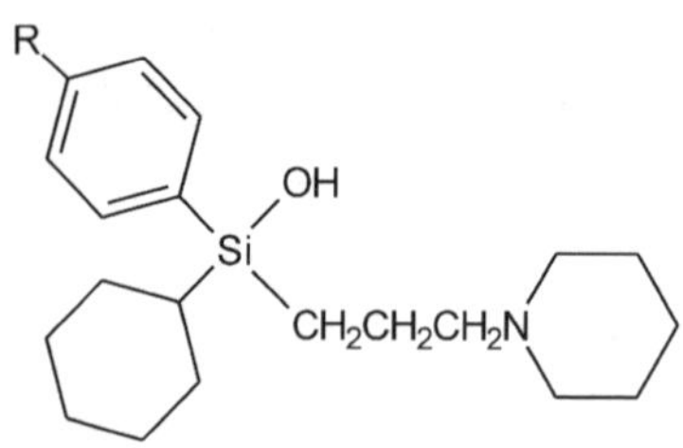

16a R = H (HHSiD)
16b R = F (*p*-F-HHSiD)

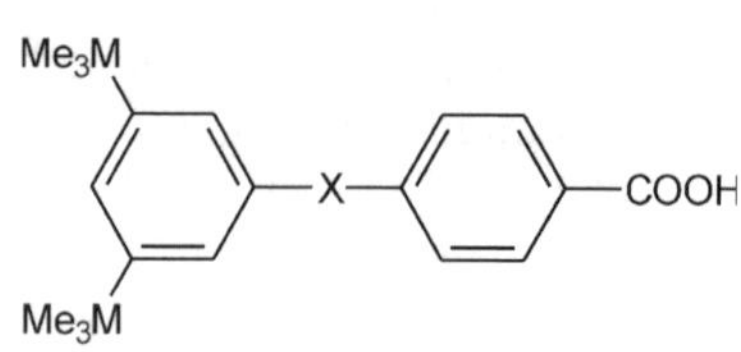

a M = C; **b** M = Si

17 X = COCH=CH
18 X = NHCO
19 X = CONH (**b** TAC-101)

Some trimethylsilyl derivatives of retinobenzoic acids show high retinoidal activity in human promyelotic leukaemia cells HL-60.[224,236] Thus, (E)-4-{3-[3,5-bis(trimethylsilyl)phenyl]-3-oxo-1-propenyl}benzoic acid (**17b**) is one order of magnitude more potent than retinoic acid. It shows an activity similar to that of the carbon analogue **17a**.[224] If only one R_3M group exists in the *meta*-position, the silicon derivative is more active than the carbon analogue but inferior to retinoic acid in activity, whereas in the case of *para*-substituted chalcones the substitution of a *t*-butyl group for trimethylsilyl leads to complete loss of activity. Me_3M-substituted retinobenzoic acids **18** (M = C, Si), with an amide group between the aromatic rings, possess similar retinoidal activities. At the same time, in the pair of isomeric amides **19**, the trimethylsilyl derivative **19b** was five times more active than the carbon analogue **19a**.[224]

4-[3,5-Bis(trimethylsilyl)benzamido]benzoic acid (TAC-101) has significant anti-metastatic and life-prolonging effect in a model of experimental liver metastasis of human gastric cancer cell line AZ-521.[237,238] It also shows some efficacy in the prevention of experimental liver metastasis of colon cancer in mice, which was associated with the induction of apoptopic cell death and suppression of cancer cell invasion.[239] TAC-101 was found to be an orally active anti-angiogenic agent and it has been suggested that this anti-angiogenic property may contribute to its efficacy against liver metastasis of human stomach cancer cells.[240,241] Based on high efficacy in several animal models,[240–248] a clinical trial of TAC-101 was started in patients with cancer.[241]

Additionally, TAC-101 suppressed the onset and development of collagen-induced arthritis in mice and experimental autoimmune encephalomyelitis.[249]

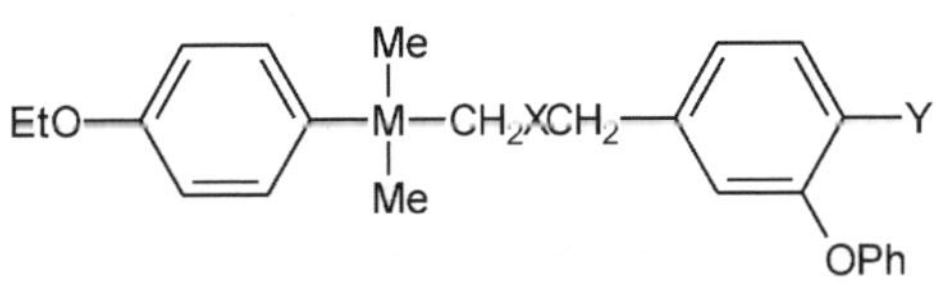

a M = C; **b** M = Si

20 X = O, Y = H
21 X = CH_2, Y = F (**b** Silafluofen)

The silicon analogues of the insecticides etofenprox (**20a**) and MTI-800 (**21a**) were slightly less toxic to insects (Mexican bean beetle, cabbage looper, pea aphid, American cockroach) than the carbon analogues.[250–253] The compounds **21** with a methylene group instead of oxygen and fluorine at the 4-position of phenoxyphenyl were more active than compounds **20**, but the most important property of 4-ethoxyphenyl-[3-(4-fluoro-3-phenoxyphenyl)propyl]dimethylsilane (insecticide Silafluofen **21b**) in contrast to pyrethroids is its negligible fish toxicity.[250]

5.4 Specific Organosilicon Compounds

There are several classes of organosilicon compounds lacking organic analogues and groups of compounds with specific properties based on the distinguished characteristics of silicon.

Silicon can form stable penta-[8,12] and hexa-coordinated[129,135] organic derivatives which is possible for carbon only in some special cases. One of such class of compounds with penta-coordinated silicon is silatranes (**22**) having a wide variety of biological effects.[6–9,12,13]

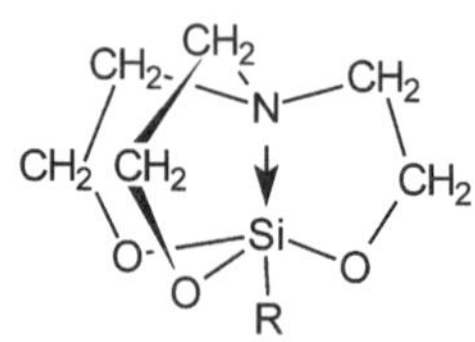

22 Silatranes

Silatranes bearing aryl[7,8] and 2-thienyl[254] groups at the silicon atom have been shown to be very toxic compounds (LD_{50} 0.3–1.0 mg kg^{-1}). 1-(4-Chlorophenyl)-silatrane[255] has been proposed as a single-dose rodenticide.[256–259] It has several distinct features: self-detoxification, which minimizes or eliminates the hazard of persistent or long-lasting residues, no secondary poisoning hazard and low dermal toxicity.[260]

Alkyl, alkenyl and alkoxysilatranes are compounds exhibiting a low toxicity (LD_{50} 2000–3000 mg kg^{-1}).[6–8] Methyl,[9] chloromethyl[9,261] and ethoxysilatrane[9] have been found to stimulate collagen biosynthesis in cartilaginous tissues of chick embryos *in vitro* (increasing the formation of peptide-bound ^{14}C-hydroxyproline from the ^{14}C-proline of the incubation medium). These compounds did not change the activity of collagen prolylhydrolase but stimulated protein biosynthesis (the incorporation of ^{14}C-proline increased 1.5-fold).[9]

Some other biological properties for these low toxic silatranes have been described: anti-tumour,[8,12,74,262] hepatoprotective,[263] hypocholesterolemic,[264] intestine peristalsis stimulating,[265] wound and bone fracture healing[8,266] and plant growth stimulating.[8]

Silicones (oligo and polysiloxanes) are the most important organosilicon compounds used in medical and medicinal application, for which there are no carbon analogues. They are widely used in plastic and reconstructive surgery, as implanted prosthetic devices, shunts and catheters. Silicone fluids for their anti-coagulant and anti-foaming properties (simethicone) have been used as lubricants in medical devices and anti-flatulents (also in veterinary medicine). They are also frequently used as depot forms for pharmaceuticals, in cosmetic and other formulations for topical administration (e.g. anti-perspirants).[6,7,267]

For a long time, these silicones without functional groups have been assumed to be biologically inert.[6,7,27]

Unexpectedly, some cyclosiloxanes were found to be active estrogens.[11,268] The most active compound was 2,6-*cis*-diphenylhexamethylcyclotetrasiloxane (**23**).[269–272] It entered clinical studies for the treatment of prostatic carcinoma.[272–275]

There are substantial differences in the chemical properties of hydroxyl derivatives of carbon (carbinols) and silicon (silanols). Silanols are stronger acids than the corresponding carbinols, and silanediols are more stable than geminal organic diols that readily eliminate water to give the corresponding ketones. These differences have been used to design biologically active silanols.

Diphenylsilanediol $Ph_2Si(OH)_2$ has been found to exhibit anti-convulsant properties.[11,276,277] Unfortunately its use as an anti-epileptic agent was prevented by toxic side effects.[278]

Such simple silanediols did not inhibit hydrolases. Therefore dipeptides and tripeptides containing a more complex silanediol group that mimic angiotensin-converting enzyme (ACE), HIV protease and thermolysin inhibitors were synthesized.[279–282] One of them, **24**, inhibited ACE with an IC_{50} value as low as 14 nM presumably acting as an active site zinc chelator.[282]

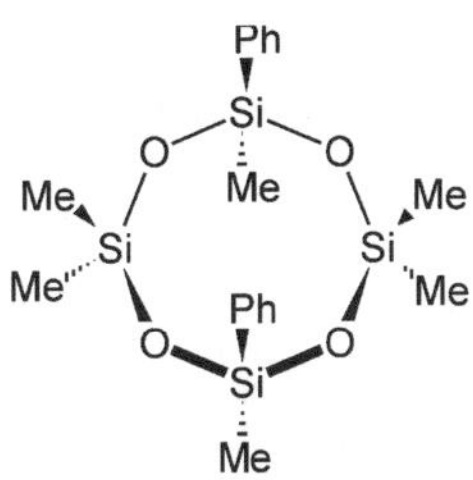

23 Cisobitan®

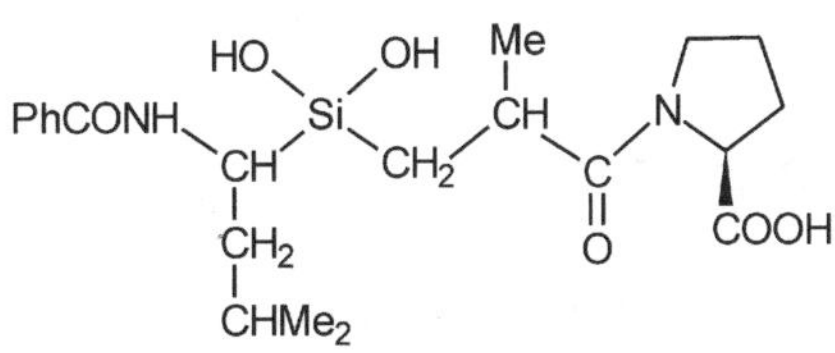

24

$MeSi(OH)_2OK \cdot nRCOOH \cdot mH_2O$

possibly

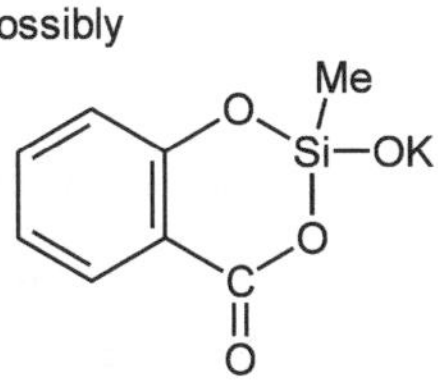

25 Conjonctyl®

Methylsilanetriol complexes with sodium and potassium salts of various hydroxycarboxylic and amino acids (salicylic, citric, ascorbic, nicotinic, benzylpenicillin) have been reported as useful for the treatment of certain cardiovascular diseases,[6,7,283–286] cancers[287] and virus infections.[288,289] The compound formed by the reaction of methylsilanetriol with potassium salicylate (**25**) (Conjonctyl®) has been proposed for the treatment of painful polycystic mastopathies.[290]

A specific property of trimethoxysilanes $RSi(OMe)_3$ is their ready hydrolysis to silanetriols $RSi(OH)_3$, which bind to a wide range of surfaces such as glass, ceramic, cotton, wool, leather, polyester, rubber and wood. If the substituent R contains a quaternary ammonium group, the compound exhibits anti-microbial activity in solution. It binds strongly to the surface of materials and imparts an anti-microbial activity to the treated surface. The practically used compound is dimethyloctadecyl[3-(trimethoxysilyl)propyl]ammonium chloride (**26**).[291–295] It has a broad spectrum of activity against Gram positive and Gram negative bacteria, fungi and algae.[291] Silicon rubber coated with **26** showed anti-microbial activity against adhering bacteria both *in vitro* and *in vivo*.[296]

$$(MeO)_3SiCH_2CH_2CH_2\overset{+}{N}Me_2C_{18}H_{37}\ Cl^- \qquad (PhCH_2O)_2SiMeCH_2CH_2Cl$$

26 Dow Corning 5700 **27** CGA 15281

Another specific reaction of organosilicon compounds unknown for the carbon analogues is the β-elimination of 2-halogenoethylsilanes yielding ethylene. Based on this property, (2-chloroethyl)*bis*(benzyloxy)methylsilane (**27**) has been proposed as an ethylene-releasing plant growth regulator.[7,22,297–300]

References

1. A. Kleeman, J. Engel, B. Kutscher and D. Reichert, *Pharmaceutical Substances*, Thieme, Stuttgart, New York, 1999.
2. R.J. Fessenden and J.S. Fessenden, *Adv. Drug Res.*, **4**, 95–132 (1967).
3. M.G. Voronkov and E. Lukevics, *Russ. Chem. Rev.*, **38**, 975–986 (1969).
4. L.R. Garson and L.K. Kirchner, *J. Pharm. Sci.*, **60**, 1113–1127 (1971).
5. M.G. Voronkov, *Ann. Rep. Med. Chem.*, **10**, 265–273 (1975).
6. M.G. Voronkov, G.I. Zelchan and E. Lukevics, *Silizium und Leben*, Akademie-Verlag, Berlin, 1975.
7. M.G. Voronkov, G.I. Zelchan and E. Lukevics, *Silicon and Life*, Zinatne, Riga, 1978 (in Russian).
8. M.G. Voronkov, in *Biochemistry of Silicon and Related Problems*, G. Bendz and I. Lindquist (Eds), Plenum Press, New York, London, 1978, 395–433.
9. E. Lukevics, in *Biochemistry of Silicon and Related Problems*, G. Bendz and I. Lindquist (Eds), Plenum Press, New York, London, 1978, 435–445.
10. U. Wannagat, in *Biochemistry of Silicon and Related Problems*, G. Bendz and I. Lindquist (Eds), Plenum Press, New York, London, 1978, 447–472.
11. R.R. LeVier, M.L. Chandler and S.R. Wendel, in *Biochemistry of Silicon and Related Problems*, G. Bendz and I. Lindquist (Eds), Plenum Press, New York, London, 1978, 473–514.

12. M.G. Voronkov and V.M. Dyakov, *Silatranes*, Nauka, Novosibirsk, 1978 (in Russian).
13. M.G. Voronkov, *Top. Curr. Chem.*, **84**, 77–135 (1979).
14. R. Tacke and U. Wannagat, *Top. Curr. Chem.*, **84**, 1–75 (1979).
15. R.J. Fessenden and J.S. Fessenden, *Adv. Organomet. Chem.*, **18**, 275–299 (1980).
16. R. Tacke, *Chemie in unserer Zeit*, **14**, 197–207 (1980).
17. R.C. Treadgold, *Process Biochem.*, **18**, 30–33 (1983).
18. J.S. Millership and M.L. Shanks, *Int. J. Pharm.*, **28**, 1–9 (1986).
19. R. Tacke and H. Zilch, *Endeavour*, **10**, 191–197 (1986).
20. R. Tacke and B. Becker, *Main Group Met. Chem.*, **10**, 169–197 (1987).
21. E. Lukevics, T. Lapina and I. Segal, *Latv. PSR Zinat. Akad. Vestis*, N 11, 111–124 (1987).
22. R. Tacke and H. Linch, in *The Chemistry of Organic Silicon Compunds*, S. Patai and Z. Rappoport (Eds), John Wiley & Sons, Chichester, UK, 1989, 1143–1206.
23. A. Ricci, G. Seroni and M. Toddei, *Chimicaoggi*, 1989 (Sept.), 15.
24. P. Hencsei and E. Szalay, in *A kémia újabb eredményei*, Akadémiai Kiadó, Budapest, 1989, 125–181 (in Hungarian).
25. E. Lukevics and L. Ignatovich, *Appl. Organomet. Chem.*, **6**, 113–126 (1992).
26. E. Lukevics and A. Zablotskaya, *Metallorg. Khim.*, **6**, 263–284 (1993).
27. M.A. Brook, *Silicon in Organic, Organometallic, and Polymer Chemistry*, John Wiley & Sons, New York, Chichester, 2000.
28. G.A. Showell and J. Mills, *Drug Discov. Today*, **8**, 551–556 (2003).
29. W. Bains and R. Tacke, *Cur. Opin. Drug Discov. Dev.*, **6**, 526–543 (2003).
30. P. Kocienski, *Protecting Groups*, Thieme, Stuttgart, 2003.
31. E. Lukevics and A. Zablocka, *Nucleoside Synthesis Organosilicon Methods*, E. Horwood, Chichester, 1991.
32. G.A. Veinberg and E. Lukevics, *Heterocycles*, **38**, 2309–2342 (1994).
33. E. Chang and V.K. Jain, *J. Med. Chem.* **9**, 433–435 (1966).
34. J.S. Davies, C.L. Higginbotham, E.J. Tremeer *et al.*, *J. Chem. Soc., Perkin Trans. I*, 1992, 3043–3048.
35. S.J. Tremont, P.W. Collins, W.E. Perkins *et al.*, *J. Med. Chem.* **36**, 3087–3097 (1993).
36. A. Zablotskaya, I. Segal, A. Kemme *et al.*, *Chem. Heterocycl. Comp.*, **38**, 477–489 (2002).
37. A.H. Beckett, D.C. Taylor and J.W. Gorrod, *J. Pharm. Pharmacol.*, **27**, 588–593 (1975).
38. J.R. Hwu and N. Wang, *Chem. Rev.*, **89**, 1599–1615 (1989).
39. C.B. Warren, J.F. Butler, R.A. Wilson *et al.*, US Pat. 5633236 (1997).
40. J.J. Bonire, P.F. Olurinola, J.O. Ehinmidu *et al.*, *Nigerian J. Chem. Res.*, **5**, 21–24 (2000).
41. B. Keppler, Ger. DE 10 114 106 (2002), *Chem. Abstr.*, **136**, 183932z (2002).
42. B. Keppler, Ger. DE 10 114 222 (2002), *Chem. Abstr.*, **136**, 369838d (2002).
43. B. Keppler, Ger. DE 10 218 319 (2003), *Chem. Abstr.*, **139**, 323655a (2003).
44. G.B. Sterling and C.E. Pawloski, US Pat. 3256308 (1966), *Chem. Abstr.*, **65**, 7217 (1966).
45. C.L. Frye, US Pat. 3465020 (1969), *Chem. Abstr.*, **71**, 124636x (1969).

46. J.S. Davies, E.J. Tremeer and R.C. Treadgold, *J. Chem. Soc., Perkin Trans. I*, 1987, 1107–1115.
47. M.-C. Seguin, J. Gueyne, J.-F. Nicolay and A. Franco, PCT WO 96/10574 (1996).
48. S. Manfredini, D. Simoni, R. Feroni *et al.*, *J. Med. Chem.*, **40**, 3851–3857 (1997).
49. A. Hosoda, M. Nakauchi, S. Ikemoto *et al.*, *Jap. Kokai Tokkyo Koho*, JP 2002 155094 (2002); *Chem. Abstr.*, **136**, 401975j (2002).
50. H.J. Koetzsch, US Pat. 4924021 (1990); *Chem. Abstr.*, **114**, 43165x (1991).
51. E.A. Mamaeva, O.V. Agafonova, V.N. Negrebetsky *et al.*, *Khim.-Farm. Zh.*, **28**, N 6, 26–29 (1994).
52. A.V. Kurochka, O.V. Agafonova, A.S. Losev *et al.*, *Metal-Based Drugs*, **5**, 25–33 (1998).
53. N.L. Oleinick, A.R. Antunez, M.E. Clay *et al.*, *Photochem. Photobiol.*, **57**, 242–247 (1993).
54. E. Lukevics, L. Ignatovich, N. Shilina and S. Germane, *Appl. Organomet. Chem.*, **6**, 261–266 (1992).
55. M. Soncin, L. Polo, E. Reddi *et al.*, *Cancer Lett.*, **89**, 101–106 (1995).
56. H.J. Cohen, US Pat. 3334119 (1967); *Chem. Abstr.*, **69**, 717511z (1968).
57. K.D. Thiel, P. Wutzler, K. Issleib *et al.*, *Pharmazie*, **41**, 295–296 (1986).
58. P.J. Stang and W.L. Treptow, *J. Med. Chem.*, **24**, 468–472 (1981).
59. M. Bols, L. Binderup, J. Hansen and P. Rasmussen, *J. Med. Chem.*, **35**, 2768–2771 (1992).
60. A. Zablotskaya, S. Germane, I. Segal and E. Lukevics, *Latv. Kim. Z.*, 79–82 (1993).
61. E. Lukevics, A. Zablotskaya, S. Germane and I. Segal, *Latv. Kim. Z.*, 472–480 (1994).
62. T. Hayashi, T. Sugimoto, N. Takewaki *et al.*, *Bioorg. Med. Chem. Lett.*, **5**, 2943–2946 (1995).
63. E. Lukevics, I. Segal, I. Birgele and A. Zablotskaya, *Chem. Heterocycl. Comp.*, **34**, 1076–1080 (1999).
64. M. Itoh, J. Kuwahara, K. Itoh *et al.*, *Bioorg. Med. Chem. Lett.*, **12**, 2069–2072 (2002).
65. A. Zablotskaya, I. Segal, S. Germane *et al.*, *Chem. Heterocycl. Comp.*, **38**, 859–866 (2002).
66. E.E. Shul'ts, I.V. Nechepurenko, G.A. Tolstikov *et al.*, RU Pat. 2210561 (2003).
67. V. Casaña-Giner, J.E. Oliver, E. Jang and L. Carvalho, *J. Entomol. Sci.*, **38**, 111–119 (2003).
68. G. Cainelli, P. Galletti, S. Garbisa *et al.*, *Bioorg. Med. Chem. Lett.*, **11**, 5391–5399 (2003).
69. F.J. Saunders, *Proc. Soc. Exp. Biol. Med.*, **123**, 303–304 (1966).
70. C.R. Creveling, J.W. Daly, T. Tokuyama and B. Witkop, *Experientia*, **25**, 26–27 (1969).
71. M. Demina, A. Velikanov, A. Medvedeva *et al.*, *J. Organomet. Chem.*, **553**, 129–133 (1998).
72. A.A. Shishkina, T.I. Ivanenko, N.A. Zarubina *et al.*, *Khim.-Farm. Zh.*, **20**, 232–237 (1986).

73. H. Przuntek, M.E. Westarp, M.L. Vohl *et al.*, *Neuropharmacology*, **26**, 255–260 (1987).
74. C.A. Black, J.W. Ucci, J.S. Vorpagel *et al.*, *Bioorg. Med. Chem. Lett.*, **12**, 3521–3523 (2002).
75. J. Balzarini, M.-J. Pérez-Pérez, A. San-Félix *et al.*, *Proc. Natl. Acad. Sci. USA*, **89**, 4392–4396 (1992).
76. J. Balzarini, M.-J. Pérez-Pérez, A. San-Félix *et al.*, *Antimicrob. Agents Chemother.*, **36**, 1073–1080 (1992).
77. M.-J. Camarasa, M.-J. Pérez-Pérez, A. San-Félix *et al.*, *J. Med. Chem.*, **35**, 2721–2727 (1992).
78. M.-J. Pérez-Pérez, A. San-Félix, J. Balzarini *et al.*, *J. Med. Chem.*, **35**, 2988–2995 (1992).
79. S. Velázquez, A. San-Félix, M.-J. Pérez-Pérez *et al.*, *J. Med. Chem.*, **36**, 3230–3239 (1993).
80. A. San-Félix, S. Velázquez, M.-J. Pérez-Pérez *et al.*, *J. Med. Chem.*, **37**, 453–460 (1994).
81. R. Alvarez, S. Velázquez, A. San-Félix *et al.*, *J. Med. Chem.*, **37**, 4185–4194 (1994).
82. E. De Clerq, *Expert Opin. Invest. Drugs*, **3**, 253–271 (1994).
83. S. Ingate, M.-J. Pérez-Pérez, E. De Clerq *et al.*, *Antivir. Res.*, **27**, 281–299 (1995).
84. S.T. Ingate, M.-J. Camarasa, E. De Clerq and J. Balzarini, *Antivir. Chem. Chemother.*, **6**, 365–370 (1995).
85. S. Ingate, A. San-Félix, E. De Clerq *et al.*, *Nucleosides Nucleotides*, **14**, 299–301 (1995).
86. M.-J. Camarasa, M.-J. Pérez-Pérez, S. Velázquez *et al.*, in *Anti-infectives: Recent Advances in Chemistry and Structure–Activity Relationships*. P.H. Bentley, P.J. O'Hanlon (Eds), *The Royal Society of Chem.*, Cambridge, 1997, pp. 259–268.
87. S. Velázquez, R. Alvarez, C. Pérez *et al.*, *Antivir. Chem. Chemother.*, **9**, 481–489 (1998).
88. S. Velázquez, C. Chamorro, M.-J. Pérez-Pérez *et al.*, *J. Med. Chem.*, **41**, 4636–4647 (1998).
89. R. Garg, S.P. Gupta, H. Gao *et al.*, *Chem. Rev.*, **99**, 3525–3601 (1999).
90. C. Chamorro, E. De Clerq, J. Balzarini *et al.*, *Antivir. Chem. Chemother.*, **11**, 61–69 (2000).
91. K. Van Laethem, J.-C. Schmit, H. Pelemans *et al.*, *AIDS Res. Hum. Retroviruses*, **16**, 825–833 (2000).
92. M.-J. Camarasa, A. San-Félix, M.-J. Pérez-Pérez *et al.*, *J. Carbohydr. Chem.*, **19**, 451–469 (2000).
93. F. Rodriguez-Barrios, C. Pérez, E. Lobatón *et al.*, *J. Med. Chem.*, **44**, 1853–1865 (2001).
94. C. Chamorro, M.-J. Pérez-Pérez, F. Rodriguez-Barrios *et al.*, *Antivir. Res.*, **50**, 207–222 (2001).
95. E. Lobatón, S. Velázquez, M.-J. Pérez-Pérez *et al.*, *Nucleosides Nucleotides Nucleic Acids*, **20**, 707–710 (2002).
96. E. Lobatón, S. Velázquez, A. San-Félix *et al.*, *Nucleosides Nucleotides Nucleic Acids*, **20**, 711–714 (2001).

97. E. Lobatón, F. Rodrigues-Barrios, F. Gago *et al.*, *J. Med. Chem.*, **45**, 3934–3945 (2002).
98. M.-C. Bonache, C. Chamorro, E. Lobatón *et al.*, *Antivir. Chem. Chemoth.*, **14**, 249–262 (2003).
99. T. Fitzmaurice, D.R. Harper and H.J. Field, *Antivir. Chem. Chemoth.*, **14**, 217–223 (2003).
100. D.S. VanVliet, Y. Tachibana, K.F. Bastow *et al.*, *J. Med. Chem.*, **44**, 1422–1428 (2001).
101. J. Satgé, A. Gazes, M. Bouchat *et al.*, *Eur. J. Med., Chem.-Chim. Ther.*, **17**, 433–436 (1982).
102. M.L. Vohl, M.E. Westarp, H. Wekerle *et al.*, *Arzneim.-Forsch./Drug Res.*, **40**, 80–84 (1990).
103. D. Singh, A. Kumari, R.V. Singh *et al.*, *Appl. Organomet. Chem.*, **7**, 289–292 (1993).
104. D. Singh and R.V. Singh, *Main Group Met. Chem.*, **16**, 209–212 (1993).
105. A. Phor, A. Chaudhary, M. Jain *et al.*, *Main Group Met. Chem.*, **24**, 439–444 (2001).
106. R.V. Singh, S.C. Joshi, A. Gajraj and P. Nagpal, *Appl. Organomet. Chem.*, **16**, 713–720 (2002).
107. B. Célaris, C. Amourette, C. Lion and G. Rima, *Appl. Organomet. Chem.*, **17**, 561–569 (2003).
108. J. Satgé, G. Rima, M. Fatome *et al.*, *Eur. J. Med. Chem.*, **24**, 48–54 (1989).
109. G. Rima, J. Satgé, H. Sentenac-Roumanou *et al.*, *Appl. Organomet. Chem.*, **8**, 481–490 (1994).
110. G. Rima, J. Satgé, H. Sentenac-Roumanou *et al.*, *Main Group Metal Chem.*, **20**, 255–267 (1997).
111. G. Rima, J. Satgé, R. Dagiral *et al.*, *Metal-Based Drugs*, **5**, 139–146 (1998).
112. G. Rima, J. Satgé, R. Dagiral *et al.*, *Appl. Organomet. Chem.*, **13**, 583–594 (1999).
113. E.V. Bakhareva, M.G. Voronkov, P.V. Arbuzov *et al.*, *Khim.-Farm. Zh.*, **29**, N 3, 31–32 (1995).
114. V. Belakhov, A.A. Levina, Yu.D. Shenin and B.I. Ionin, *Khim.-Farm. Zh.*, **25**, N 3, 86–87 (1991).
115. S. Mahboobi, A. Popp and W.F. Fleck, *Pharmazie*, **54**, 730–733 (1999).
116. I. Diamond, S.G. Granelli, A.F. McDonagh *et al.*, *Lancet*, **2**, 1175–1177 (1972).
117. J. Moan, *Photochem. Photobiol.*, **43**, 681–690 (1986).
118. S.I. Zaidi, R. Agarwal, G. Eichler *et al.*, *Photochem. Photobiol.*, **58**, 204–210 (1993).
119. J.-P. Daziano, S. Steenken, C. Chabannon *et al.*, *Photochem. Photobiol.*, **64**, 712–719 (1996).
120. J. He, H.E. Larkin, Y.S. Li *et al.*, *Photochem. Photobiol.*, **65**, 581–586 (1997).
121. R. Decreau, A. Viola, C. Hadjur *et al.*, *C. R. Soc. Biol.*, **191**, 639–657 (1997).
122. R. Decreau, A. Viola, M.-J. Richard *et al.*, *J. Porphyrins Phthalocyanines*, **2**, 405–414 (1998).
123. C.Y. Anderson, K. Freye, K.A. Tubesing *et al.*, *Porphyrins Phthalocyanines*, **67**, 332–336 (1998).

124. R. Decreau, M.J. Richard, P. Verrando *et al.*, *J. Photochem. Photobiol.*, **48B**, 48–56 (1999).
125. M.J. Egorin, E.G. Zuhowski, D.L. Sentz *et al.*, *Cancer Chemother. Pharmacol.*, **44**, 283–294 (1999).
126. N.S. Trivedi, H.W. Wang, A.L. Nieminen *et al.*, *Photochem. Photobiol.*, **71**, 634–639 (2000).
127. E. Ben-Hur, W.-S. Chan, Z. Yim *et al.*, *Proc. SPIE-Int. Soc. Opt. Eng.*, **3909**, 34–39 (2000).
128. L.Y. Xue, S.M. Chiu and N.L. Oleinick, *Oncogene*, **20**, 3420–3427 (2001).
129. R.L. Morris, M.E. Varnes, M.E. Kenney *et al.*, *Photochem. Photobiol.*, **75**, 652–661 (2002).
130. J.-D. Huang, W.-P. Fong, E.Y.M. Chan *et al.*, Ng, *Tetrahedron Lett.*, **44**, 8029–8032 (2003).
131. P.A. Firey and M.A.J. Rodgers, *Photochem. Photobiol.*, **45**, 535–538 (1987).
132. V. Cuomo, G. Jori, B. Rihter *et al.*, *Brit. J. Cancer*, **62**, 966–970 (1990).
133. C. Bellemo, G. Jori, B.D. Rihter *et al.*, *Cancer Lett.*, **65**, 145–150 (1992).
134. N. Brasseur, T.-L. Nguyen, R. Langlois *et al.*, *J. Chim. Phys.*, **91**, 1011–1017 (1994).
135. N. Brasseur, T.-L. Nguyen, R. Langlois *et al.*, *J. Med. Chem.*, **37**, 415–420 (1994).
136. N. Brasseur, R. Ouellet, K. Lewis *et al.*, *Photochem. Photobiol.*, **62**, 1058–1065 (1995).
137. M. Kreimer-Birnbaum, M.M. Zuk, B. Rihter *et al.*, *Proc. SPIE-Int. Soc. Opt. Eng.*, **2625**, 105–116 (1996).
138. N. Brasseur, R. Ouellet, S. Marengo *et al.*, *J. Chim. Phys.*, **93**, 29–33 (1996).
139. M.M. Zuk, B.D. Rihter, M.E. Kenney *et al.*, *Photochem. Photobiol.*, **63**, 132–140 (1996).
140. M. Soncin, A. Busetti, E. Reddi *et al.*, *J. Photochem. Photobiol.*, **40B**, 163–167 (1997).
141. I. Belsky, D. Gertner and A. Zilkha, *J. Med. Chem.*, **11**, 448–450 (1968).
142. I. Belsky, D. Gertner and A. Zilkha, *J. Med. Chem.*, **11**, 451–453 (1968).
143. R.H. Peters, D.F. Crowe, M. Tanabe *et al.*, *J. Med. Chem.*, **30**, 646–652 (1987).
144. R.H. Peters, D.F. Crowe, M.A. Avery *et al.*, *J. Med. Chem.*, **31**, 572–576 (1988).
145. E. Lukevics, L. Demicheva, N. Erchak and S. Germane, *Appl. Organomet. Chem.*, **7**, 543–551 (1993).
146. B. Delord, M.C. Guillorit, J. Lafay *et al.*, *Eur J. Med. Chem.*, **31**, 111–112 (1996).
147. H. Ishikawa, H. Yamanaka, T. Kawamoto and A. Tanaka, *Appl. Microbiol. Biotechnol.*, **53**, 19–22 (1999).
148. E. Lukevics, L. Ignatovich and I. Shestakova, *Appl. Organomet. Chem.*, **17**, 898–905 (2003).
149. R.W. Versace, *Expert Opin. Ther. Patents*, **13**, 751–760 (2003).
150. M. Frankel, M. Broze, D. Gertner *et al.*, *J. Med. Chem.*, **11**, 857–860 (1968).
151. R. Tacke, A. Bentlage, R. Towart and E. Moller, *Eur. J. Med. Chem.*, **18**, 155–161 (1983).
152. F. Cavalier, R. Jacquier, J.-L. Mercadier *et al.*, *J. Pept. Res.*, **50**, 94–101 (1997).

153. M. Krause, X. Ligneau, H. Stark *et al.*, *J. Med. Chem.*, **41**, 4171–4176 (1998).
154. K. Nishide, T. Miyamoto, K. Kumar *et al.*, *Tetrahedron Lett.*, **43**, 8569–8573 (2002).
155. I. Hutchinson, T.D. Bradshaw, C.S. Matthews *et al.*, *Bioorg. Med.Chem. Lett.*, **13**, 471–474 (2003).
156. R. Sturkovich, Y. Goldberg, V.E. Verovskii *et al.*, *Appl. Organomet. Chem.*, **3**, 393–399 (1989).
157. R.D. Haugwitz, W.K. Anderson, J. Plowman *et al.*, *Appl. Organomet. Chem.*, **4**, 375–378 (1990).
158. W.K. Anderson and R.D. Haugwitz, US Pat. 4994591 (1991); *Chem. Abstr.*, **115**, 92600q (1991).
159. W.K. Anderson, R. Kasliwal, D.M. Houston *et al.*, *J. Med. Chem.*, **38**, 3789–3797 (1995).
160. N. Seiler, J.-G. Delcros, M. Vaultier *et al.*, *Cancer Res.*, **56**, 5624–5630 (1996).
161. M. Kato, K. Kataoka, Y. Nagasaki and T. Takezawa, PCT Int. Appl. WO 97 27859 (1997); *Chem. Abstr.*, **127**, 200037v (1997).
162. N. Seiler, F. Douaud, R. Havouis *et al.*, *Int. J. Oncol.*, **11**, 835–841 (1997).
163. J.-G. Delcros, M. Vaultier, N. Le Roch *et al.*, *Anticancer Drug Des.*, **12**, 35–48 (1997).
164. F. Douaud, N. Le Roch, J. Renault *et al.*, *Anticancer Drug Des.*, **12**, 621–633 (1997).
165. M. Kato, K. Kataoka, Y. Nagasaki and T. Takezawa, US Pat. 5910485 (1999).
166. E. Lukevics, P. Arsenyan, S. Germane and I. Shestakova, *Appl. Organomet. Chem.*, **13**, 795–798 (1999).
167. E. Lukevics, P. Arsenyan, I. Shestakova *et al.*, *Eur. J. Med. Chem.*, **36**, 507–515 (2001).
168. P. Boudjouk, T.E. Ready, S. Stafslien and B.P.S. Chauhan, PCT Int. Appl. WO 0260972 (2002); *Chem. Abstr.*, **137**, 150221j (2002).
169. F. Gaboriau, M. Saint-Pierre-Chazalet, J. Homand *et al.*, *Coll. Surf. B: Biointerfaces*, **24**, 133–143 (2002).
170. N. Le Roch, F. Douaud, R. Havoulis *et al.*, *Anticancer Res.*, **22**, 3765–3776 (2002).
171. S. Barcza, US Pat. 3853994 (1974); *Chem. Abstr.*, **82**, 98127n (1975).
172. E. Lukevics, I. Segal, S. Germane and M. Veveris, *Latv. Kim. Z.*, 106–112 (1991).
173. F. Wang, B. Venkataraman, M.E. Klein and L.M. Savre, *J. Org. Chem.*, **57**, 6687–6689 (1992).
174. C. Danzin, M. Zreika, P. Marchal *et al.*, *Biochem. Soc. Trans.*, **22**, 767–773 (1994).
175. J.-M. Kim, S.E. Hoegy and P.S. Mariano, *J. Am. Chem. Soc.*, **117**, 100–105 (1995).
176. C. Danzin, J.-N. Collard, P. Marchal and D. Shirlin, *Bioorg. Med. Chem. Lett.*, **5**, 2363–2366 (1995).
177. E. Lukevics, M. Veveris and V. Dirnens, *Appl. Organomet. Chem.*, **11**, 805–811 (1997).
178. E. Lukevics, P. Arsenyan and M. Veveris, *Metal-Based Drugs*, **5**, 251–257 (1998).
179. F.L. Tse, J.M. Jaffe and J.G. Dain, *J. Clin. Pharmacol.*, **24**, 47–57 (1984).
180. S. Farkas, S. Földeak, E. Kárpáti *et al.*, Eur. Pat. 468825 (1992); *Chem. Abstr.*, **116**, 214697v (1992).

181. S. Farkas, S. Földeak, E. Kárpáti *et al.*, Eur. Pat. 472304 (1992); *Chem. Abstr.*, **116**, 241938f (1992).
182. S. Farkas and E. Kárpáti, *Pharmacol. Res.*, **25**, Suppl. 2, 25–26 (1992).
183. T. During and E. Koppenhofer, *General Physiol. Biophys.*, **20**, 157–173 (2001).
184. J. Prous, A. Graul and J. Castañer, *Drugs of the Future*, **19**, 854–855 (1994).
185. J.-M. Hornsperger, J.-N. Collard, J.-G. Heydt *et al.*, *Biochem. Soc. Trans.*, **22**, 758–763 (1994).
186. X.D. Zhu, E. Giacobini and J.-M. Hornsperger, *Eur. J. Pharmacol.*, **276**, 93–99 (1995).
187. N.R. Cutler, R.D. Seifert, M.M. Schleman *et al.*, *Clin. Pharmacol. Therap.*, **58**, 54–61 (1995).
188. J. Dow, B.D. Dulery, J.-M. Hornsperger *et al.*, *Arzn.-Forsch. Drug Res.*, **45**, 1245–1252 (1995).
189. D.F.S. Kehrer, O. Soepenberg, W.J. Loos *et al.*, *Anticancer Drugs*, **12**, 89–105 (2001).
190. H. Josien, D. Bom, D.P. Curran *et al.*, *Bioorg. Med. Chem. Lett.*, **7**, 3189–3194 (1997).
191. F. Hausheer, K. Haridas, M. Zhao *et al.*, *Proc. Amer. Assoc. Cancer Res.*, **39**, 420–421 (1998).
192. I.F. Pollack, M. Erff, D. Bom *et al.*, *Cancer Res.*, **59**, 4898–4905 (1999).
193. D. Bom, D.P. Curran, A.J. Chavan *et al.*, *J. Med. Chem.*, **42**, 3018–3022 (1999).
194. D. Bom, D.P. Curran, S. Kruszewski *et al.*, *J. Med. Chem.*, **43**, 3970–3980 (2000).
195. A.H. Van Hattum, H.M. Pinedo, H.M.M. Schlüper *et al.*, *Int. J. Cancer*, **88**, 260–266 (2000).
196. F.H. Hausheer, M. Zhao, H. Kochat *et al.*, *Proc. Amer. Assoc. Cancer Res.*, **41**, 213 (2000).
197. S.T. Keir, F. Hausheer, A.A. Lawless *et al.*, *Cancer Chemother. Pharmacol.*, **48**, 83–87 (2001).
198. A.H. Van Hattum, H.M.M. Schlüpper, F.H. Hausheer *et al.*, *Int. J. Cancer*, **100**, 22–29 (2002).
199. A.E. Gabarda and D.P. Curran, *J. Comb. Chem.*, **5**, 617–624 (2003).
200. W.K. Moberg, Eur. Pat. Appl. EP 68813 (1983); *Chem. Abstr.*, **98**, 19430e (1983).
201. W.K. Moberg, US Pat. 4510136 (1985); *Chem. Abstr.*, **104**, 207438k (1985).
202. W.K. Moberg, G.S. Basarab, J. Cuomo and P.H. Liang, *ACS Symp. Ser.*, **355**, 288–301 (1987).
203. R. Tacke, B. Becker, D. Berg *et al.*, *J. Organomet. Chem.*, **438**, 45–55 (1992).
204. J.J. Anderson, L.M. Shalaby and D.S. Berg. *J. Agricult. Food Chem.*, **47**, 2439–2446 (1999).
205. B.R. Yoo, M.Y. Suk, Y.-M. Yu *et al.*, *Bull. Korean Chem. Soc.*, **19**, 358–362 (1998).
206. B.R. Yoo, M.Y. Suk, J.S. Han *et al.*, *Pesticide Sci.*, **52**, 138–144 (1998).
207. J.F. Chollet, J.L. Bonnemain, L. Miginiac and O. Rohr, *Pesticide Sci.*, **29**, 427–435 (1990).
208. H. Itoh, R. Yoneda, J. Tobitsuka *et al.*, *Chem. Pharm. Bull.*, **48**, 1148–1153 (2000).

209. H. Itoh, H. Tanaka, H. Ohta and H. Takeshiba, *Chem. Pharm. Bull.*, **49**, 909–911 (2001).
210. H. Itoh, H. Kajino, T. Tsukiyama *et al.*, *Chem. Pharm. Bull.*, **10**, 4029–4034 (2002).
211. H. Itoh, R. Yoneda, J. Tobitsuka *et al.*, *Chem. Pharm. Bull.*, **51**, 1113–1116 (2003).
212. D.P. Phillion, D.S. Bracolino, M.J. Graneto *et al.*, Eur. Pat. 538231 (1992).
213. D. Phillion, S.C. Wong and B. Short, US Pat. 5486621 (1996).
214. G. Mindt, *Gesunde Pflanzen*, **51**, 240–247 (1999).
215. T. Joseph-Horne, C. Heppner, J. Headrick and D.W. Hollomon, *Pesticide Biochem. Physiol.*, **67**, 168–186 (2000).
216. G. Philips, T.L. Fevig, P.H. Lau *et al.*, *Org. Process Res. Dev.*, **6**, 357–366 (2002).
217. T. Heinrich, C. Burschka, J. Warneck and T. Tacke, *Organometallics*, **23**, 361–366 (2004).
218. C.J. Palmer, L.M. Cole, I.H. Smith *et al.*, *J. Agric. Food Chem.*, **39**, 1335–1341 (1991).
219. P. Heinonen, H. Sipilä, K. Neuvonen *et al.*, *Eur. J. Med. Chem.*, **31**, 725–729 (1996).
220. E. Lukevics, P. Arsenyan, I. Shestakova *et al.*, *Metal-Based Drugs*, **7**, 63–66 (2000).
221. L. Ignatovich, D. Zarina, I. Shestakova *et al.*, *Metal-Based Drugs*, **8**, 211–214 (2001).
222. R.W. Jackson, R. Gelinas, T.A. Baughman *et al.*, *Bioorg. Med. Chem. Lett.*, **12**, 1093–1097 (2002).
223. I. Iovel, L. Golomba, L. Zvejniece *et al.*, *Chem. Heterocycl. Comp.*, **39**, 449–454 (2003).
224. T. Yamakawa, H. Kagechika, E. Kawachi *et al.*, *J. Med. Chem.*, **33**, 1430–1437 (1990).
225. D. Wrobel and U. Wannagat, *J. Organomet. Chem.*, **225**, 203–210 (1982).
226. G. Lambrecht, R. Feifel, B. Forth *et al.*, *Eur. J. Pharmacol.*, **152**, 193–194 (1988).
227. M. Eltze, G. Gmelin, J. Wess *et al.*, *Eur. J. Pharmacol.*, **158**, 233–242 (1988).
228. G. Lambrecht, R. Feifel, M. Wagner-Röder *et al.*, *Eur. J. Pharmacol.*, **168**, 71–80 (1989).
229. M. Waelbroeck, M. Tastenoy, J. Camus *et al.*, *Br. J. Pharmacol.*, **98**, 197–205 (1988).
230. N.J. Bucley, T.I. Bonner, C.M. Buckley and M.R. Brann, *Mol. Pharmacol.*, **35**, 469–476 (1989).
231. M. Waelbroeck, J. Camus, M. Tastenoy *et al.*, *Eur. J. Pharmacol., Mol. Pharmacol. Sect.*, **206**, 95–103 (1991).
232. R.M. Eglen, C.M. Cornett and R.L. Whiting, *Naunyn-Schmiedeberg's Arch. Pharmacol.*, **342**, 394–399 (1990).
233. R. Tacke, K. Mahner, C. Strohmann *et al.*, *J. Organomet. Chem.*, **417**, 339–353 (1991).
234. J.Y. An, H.S. Yun, Y.P. Lee *et al.*, *Br. J. Pharmacol.*, **137**, 1001–1010 (2002).
235. R. Tacke and S.A. Wagner, in *The Chemistry of Organic Silicon Compounds*, Z. Rappoport, Y. Apeloig (Eds), John Wiley & Sons, Chichester, vol. 2, pp. 2363–2400 (1998).
236. Y. Hashimoto, H. Kagechika, E. Kawachi *et al.*, *Biol. Pharm. Bull.*, **19**, 1322–1328 (1996).

237. J. Shibata, K. Wierzba, K. Murakami *et al.*, PCT Int. Appl. WO 96 32 101 (1996); *Chem. Abstr.*, **125**, 339081g (1996).
238. K. Murakami, K. Wierzba, M. Sano *et al.*, *Clin. Exp. Metastasis*, **16**, 323–331 (1998).
239. R. Sakukawa, K. Murakami, T. Ikeda *et al.*, *Oncol. Res.*, **10**, 287–293 (1998).
240. K. Murakami, R. Sakukawa, M. Sano *et al.*, *Clin. Cancer Res.*, **5**, 2304–2310 (1999).
241. T. Oikawa, K. Murakami, M.Sano *et al.*, *Jap. J. Cancer Res.*, **92**, 1225–1234 (2001).
242. K. Murakami, T. Matsuura, M. Sano *et al.*, *Clin. Exp. Metastasis*, **16**, 633–643 (1998).
243. K. Fujimoto, R. Hosotani, R. Doi *et al.*, *Int. J. Cancer*, **81**, 637–644 (1999).
244. K. Murakami, T. Yamaura, K. Suda *et al.*, *Jap. J. Cancer Res.*, **90**, 1254–1261 (1999).
245. J. Shibata, K. Murakami, K. Wierzba *et al.*, *Anticancer Res.*, **20**, 3169–3176 (2000).
246. J. Shibata, K. Murakami, Y. Aoyagi *et al., Anticancer Res.*, **20**, 3583–3590 (2000).
247. T. Miyaguchi, K. Nomata, M. Noguchi *et al.*, *Anticancer Res.*, **21**, 4025–4030 (2001).
248. S.J. Lee, Y. Ohashi, H. Sakura and I. Saiki, *Cancer Lett.*, **198**, 169–177 (2003).
249. N. Miyagawa, T. Homma, H. Kagechika *et al.*, *Pharmacology*, **67**, 21–31 (2003).
250. S. McN. Sieburth, C.J. Manly and D.W. Gammon, *Pesticide Sci.*, **28**, 289–307 (1990).
251. S. McN. Sieburth, S.Y. Lin and T.G. Cullen, *Pesticide Sci.*, **29**, 215–225 (1990).
252. H. Okimoto, K. Nishimura, K. Matsuda *et al.*, *Pestic. Biochem. Physiol.*, **49**, 83–93 (1994).
253. K. Nishimura and H. Okimoto, *Pest. Manag. Sci.*, **57**, 509–513 (2001).
254. E. Lukevics, S. Germane, O.A. Pudova and N.P. Erchak, *Khim.-Farm. Zh.*, **13**. N 10, 52–57 (1979).
255. M.K. Ticku and R.W. Olsen, *Neuropharmacology*, **18**, 315–318 (1979).
256. C.B. Beiter, M. Schwarcz and G. Crabtree, *Soap Chem. Spec.*, **46**, 38–46 (1970).
257. J.H. Greaves, R. Redfern and H. Tinworth, *J. Hyg.*, **73**, 39–43 (1974).
258. B.D. Rennison, *J. Hyg.*, **73**, 45–48 (1974).
259. F.P. Rowe, T. Swinney and A. Bradfield, *J. Hyg.*, **73**, 49–52 (1974).
260. C.B. Beiter, M. Schwarcz and G. Crabtree, *5-p-Chlorophenyl Silatrane: A New Single-Dose Rodenticide*, M & T Chemicals Inc., Rahway, NJ.
261. V.B. Kazimirovskaya, V.M. D'yakov and M.G. Voronkov, *Khim.-Farm. Zh.*, **35**, N 9, 3–5 (2001).
262. A. Grna, P.H. Koo and J. Hogan, *Anticancer Res.*, **12**, 565–569 (1992).
263. M.G. Voronkov, M.K. Nurbekov and M.M. Rasulov, *Byul. Exp. Biol. Med.*, **134**, 397–400 (2002).
264. P.P. Mehta, T. Ramasarma and C.K. Kurup, *Biochim. Biophys. Acta*, **920**, 102–104 (1987).
265. G.V. Porjadin, V.G. Pukhal'skaja, Y.I. Baukov *et al.*, RU Pat. 2190617 (2002); *Chem. Abstr.*, **138**, 281125z (2003).
266. L.A. Mansurova, A.B. Skomyakova, V.B. Kazimirovskaya *et al.*, *Dokl. Akad. Nauk*, **346**, 129–131 (1996).

267. E. Lukevics and Z. Zelmene, *Biological Activity of Silicon Compounds*, Institute of Organic Synthesis, Riga, 1984 (in Russian).
268. J.F. Hayden and S.A. Barlow, *Toxicol. Appl. Pharmacol.*, **21**, 68–79 (1972).
269. D.R. Bennett, S.J. Gorzinski and J.E. Le Beau, *Toxicol. Appl. Pharmacol.*, **21**, 55–67 (1972).
270. R.R. Le Vier and W.F. Boley, *Acta Pharmacol. Toxicol.*, **36**, Suppl. 3, 55–67 (1975).
271. R.R. Le Vier, D.R. Bennet and M.J. Hunter, *Acta Pharmacol. Toxicol.*, **36**, Suppl. 3, 68–80 (1975).
272. R.R. Le Vier and M.E. Jankowiak, *Acta Pharmacol. Toxicol.*, **36**, Suppl. 3, 81–92 (1975).
273. F. Edsmyr, P.L. Esposti, B. Johansson and B. Strindberg, *J. Urol.*, **120**, 705–707 (1978).
274. L. Collste, T. Berlin, B. von Garreits *et al.*, *Eur. Urol.*, **7**, 85–88 (1981).
275. O. Alfthan, L. Anderson, P.L. Esposti *et al.*, *Scand. J. Urol. Nephrol.*, **17**, 37–43 (1983).
276. V.D. Hulce and R.H. Rech, *Pharmacologists*, **16**, 228 (1974).
277. V.D. Hulce and R.H. Rech, *Pharmacologists*, **17**, 178 (1975).
278. J.G. Cunningham, R.B. Ford, J.A. Gifford *et al.*, *Am. J. Vet. Res.*, **42**, 2178–2181 (1981).
279. S.McN. Sieburth, T. Nittoli, A.M. Mutahi and L. Guo, *Angew. Chem. Int. Ed.*, **37**, 812–814 (1998).
280. C.A. Chen, S.McN. Sieburth, A. Glekas *et al.*, *Chem. Biol.*, **8**, 1161–1166 (2001).
281. J. Kim, A. Glekas and S.McN. Sieburth, *Bioorg. Med. Chem. Lett.*, **12**, 3625–3627 (2002).
282. M. wa Mutahi, T. Nittoli, L. Guo and S.McN. Sieburth, *J. Am. Chem. Soc.*, **124**, 7363–7375 (2002).
283. J. Gueyene and M.I. Duffaut, Fr. Pat. 1069M (1962); *Chem. Abstr.*, **59**, 12911 (1963).
284. P. Gendre, *C. R. Soc. Biol.*, **161**, 2177–2180 (1967).
285. J. Loeper and J. Loeper, *C. R. Soc. Biol.*, **161**, 589–591 (1967).
286. P. Gendre, *C. R. Acad. Sci.*, **270D**, 1356–1358 (1970).
287. J. Janet, *Proc.-verb. Soc. Sci. Phys. Nat. Bordeaux*, **1967**, 143–148.
288. J. Gueyne, N. Duffaut and R. Quilichini, *Therapie*, **17**, 549–557 (1962).
289. J. Janet, *Proc.-verb. Soc. Sci. Phys. Nat. Bordeaux*, 163–168 (1964).
290. M. Levrier, J. Reboul, N. Duffaut and Dilhuydy, *Senologia*, N 4, 3–9 (1977).
291. A.J. Isquith, E.A. Abbott and P.A. Walters, *Appl. Microbiol.*, **24**, 859–863 (1972).
292. E.A. Abbott and A.J. Isquith, Ger. Offen. 2 226 823 (1973); *Chem. Abstr.*, **79**, 28392j (1973).
293. E.A. Abbott, A.J. Isquith and P.A. Walters, Ger. Offen. 2 229 580 (1973); *Chem. Abstr.*, **79**, 39346y (1973).
294. E.A. Abbott and A.J. Isquith, Ger. Offen. 2 246 493 (1973); *Chem. Abstr.*, **79**, 62219t (1973).
295. P.R. Murray, A.C. Niles and R.L. Heeren, *J. Clin. Microbiol.*, **26**, 1884–1886 (1988).
296. B. Gottenbos, H.C. van der Mei, F. Klatter *et al.*, *Biomaterials*, **23**, 1417–1423 (2002).

297. W. Foery and H.P. Fisher, Ger. Offen. 2 149 680 (1972); *Chem. Abstr.*, **77**, 5605a (1972).
298. W. Foery, H.P. Fisher, D. Lohmann and G. Greber, Ger. Offen. 2 239 412 (1973); *Chem. Abstr.*, **78**, 159842h (1973).
299. W. Foery, H. Fisher, D. Lohmann and G. Greber, Ger. Offen. 2 309 762 (1973); *Chem. Abstr.*, **81**, 56472z (1974).
300. S.D. Seeley, G.A. Couvillon and S.J. Kays, *J. Amer. Soc. Hort. Sci.*, **107**, 682–687 (1982).

6 $_{20}$Ca The Role of Calcium as a Metallotherapeutic Drug

Mario Barbagallo and Ligia J. Dominguez

6.1 Introduction

Calcium is the most abundant inorganic element in the human body and is an essential key for many physiological processes.[1]

At present, numerous intracellular and extracellular physiological roles are recognized for this ion (Table 6.1). The universal role of calcium as a cellular messenger is well recognized. Unique among all the mineral ion species because of its large intra- versus extra-cellular gradient (1:10 000), calcium is critical in regulating cellular responsiveness to a wide variety of physiological signals, including cAMP and phosphatidylinositol cell signaling systems. The critical rôle of calcium in muscle contraction has been appreciated for over 100 years, since Ringer[2] first described the heart's contraction requirement for extracellular calcium. It is now understood that calcium acts as part of a final common pathway mediating stimulus-contraction coupling in cardiac, skeletal and smooth muscle. The intracellular, cytosolic soluble calcium, which amounts to only about 0.2 mg, is not only involved in regulating cardiac and smooth muscle contractile function, but is a critical factor in several life-defining

Metallotherapeutic Drugs and Metal-Based Diagnostic Agents: The Use of Metals in Medicine
Edited by Gielen and Tiekink

Table 6.1 Physiological roles of calcium

Intracellular level
Regulation of cAMP-dependent reactions
Regulation of ATP hydrolysis
Regulation of phosphorylation reactions
Coupling factor in muscular contraction
Release of hormones and neurotransmitters
Cellular enzymatic activation
Regulation of the cellular response to hormones
Stability of cellular membrane
Stability of tight junction and desmosomes
Regulation of permeability
Electrical properties (depolarization = excitability)
Elasticity
Extracellular level
Enzymatic activation
Calcification (bone and other tissues)
Coagulation cascade
Fertility (sperm activity)

biochemical processes, such as the cellular responses to a wide variety of external stimuli, including stimulus-secretion coupling in endocrine cells, neurotransmitter release from nerve terminals both centrally and peripherally, the coagulation cascade and the permeability of plasma membranes to sodium ions and the renal excretory function.[3] As such, endocrine, neural and renal aspects of blood pressure homeostasis as well as cardiac and vascular smooth muscle function are all determined, at least in part, by the steady-state intracellular calcium content of body tissues. The role of cellular calcium (Ca^{2+}) in stimulus-contraction coupling and relaxation is itself dependent on intracellular steady-state activities of other mineral ions (Mg^{2+}, K^{+}), and pH.[4,5] One of the factors that influences cellular calcium levels is dietary calcium, which functions through the action of calcitropic hormones (Figure 6.1).

6.2 Calcium Homeostasis

A normal adult body contains approximately 1000 g of calcium, of which around 99% is extracellular, most of which is stored in bones and teeth (around 990 g). In bone, calcium is important for preserving the structural integrity of the skeleton, as well as a reservoir of calcium ion for other physiological functions. This compartment represents a dynamic storage of calcium, since the ion is continually and freely exchanged under the influence of several calciotropic hormones (Figure 6.1). The remaining 1% of total body calcium (900 mg/g) is

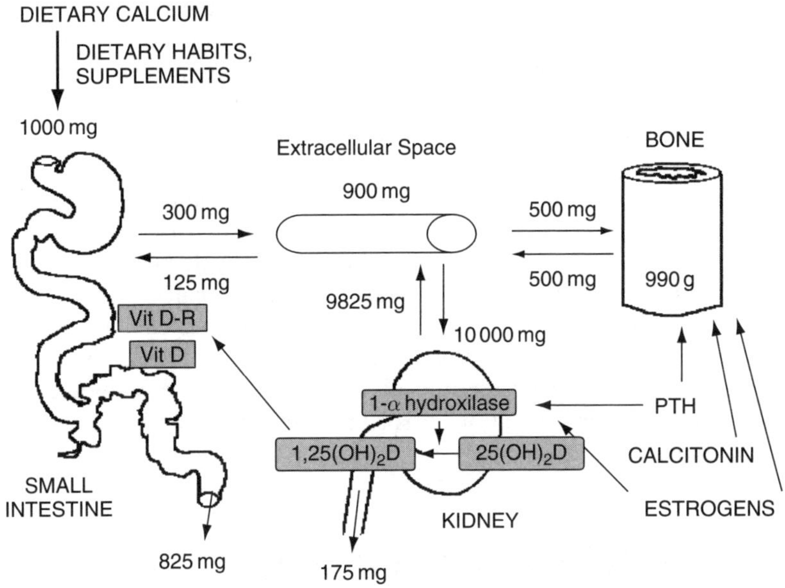

Figure 6.1 Calcium homeostasis

in the extracellular space including plasma, lymph, extracellular water and other tissues (Figure 6.1). Because calcium is so important to so many physiological processes, its concentration in the extracellular and intracellular milieu is rigorously controlled,[1,4] although intracellular calcium content may tend to increase with aging and certain diseases.[5] The elevated gradient between intra- and extracellular compartments (approximately 100 nmol/l and 1 mmol/l respectively) is necessary to maintain the cellular responsiveness to diverse extracellular stimuli. This difference in calcium concentration is maintained by the concerted effects of a variety of transport mechanisms operating at various subcellular levels.[4,5] Calcium enters the cell via voltage-dependent and hormone-dependent channels; and leaves the cell through the operation of a magnesium-dependent calcium-ATPase and a sodium–calcium exchanger. Uptake of calcium by intracellular organelles such as the mitochondria and the endoplasmic reticulum helps to maintain a low cytoplasmatic calcium level.[4,6]

6.3 Hormonal Regulation of Calcium Metabolism

Calcium homeostasis t is regulated by complex and integrated mechanisms that modify calcium balance and bone mass. Calcium balance is attained by concerted calcium fluxes through the gut, skeleton and kidney, the three major organs involved in normal serum calcium regulation. Calcium homeostasis is normally in equilibrium (i.e. the same quantity of calcium that enters the body by dietary consumption is excreted via the kidney and the gut) (Figure 6.1). These fluxes are continually regulated by diverse hormonal stimuli, some directly related to calcium and phosphorus metabolism, such as parathyroid hormone (PTH), vitamin D and calcitonin, secretion of which is regulated by ionized seric calcium, therefore called calciotropic hormones (Figure 6.1). Other hormones, not regulated by calcium levels, hence not considered calciotropic, have also a considerable influence on calcium balance. These non-calciotropic hormones include sexual hormones, thyroid hormone, adrenal steroids and growth factors. Other factors that may influence calcium balance include medicaments and physical activity.[1] Considering all these important physiological roles, calcium metabolism modifications associated with aging and/or diseases may affect the entire economy.[1,5]

6.4 Optimal Amount of Dietary Calcium Intake and Benefits of Calcium Supplementation

In the Stone Age the diet of human adults is estimated to have contained from 50 to 75 mmol of Ca (2000–3000 mg)/day, three to five times the median calcium intake of the present day in Western countries.[7] As the human genetic make-up has not modified, our body needs remain suited to a much higher calcium intake than we currently consume. Although mammalian physiology contains mechanisms by which organisms can adjust to temporary environmental shortages, chronic calcium retention has a number of health consequences. A number of chronic diseases have been linked to a low dietary calcium intake, including bone fragility, high blood pressure and colon cancer in susceptible individuals.[8] Recent studies have also linked weight control and regulation of body fat to an adequate dietary calcium intake. We present here evidence indicating that an optimal dietary calcium intake may prevent to some extent each of these multifactorial problems.

Calcium is an essential nutrient and the natural diets of all mammals are rich in calcium. Calcium requirements vary throughout an individual's lifetime, with greater needs in three critical periods of life: (a) during the times of childhood and adolescence, (b) in women during pregnancy and lactation and (c) in later adult life (Table 6.1). From birth until about age 18, bones are forming and growing, and during late adolescence, through young adulthood, adult bone is formed and reaches its maximum strength and density. A sufficient calcium intake is essential to those processes. Pregnancy represents a significant physiological stress on calcium homeostasis. A full-term infant accumulates approximately 30 g of calcium during gestation, most of which is assimilated into the fetal skeleton during the third trimester. During lactation, 160–300 mg/day of maternal calcium is lost through production of breast milk. Later in life, aging is physiologically associated with several modifications of calcium metabolism (Table 6.2). It has been shown that an altered cellular ion metabolism is associated with aging,[5,9] with a continuous age-dependent elevation of intracellular calcium and a reciprocal fall in serum ionized free calcium levels[10] (Figure 6.2). Calcium intake, intestinal absorption and renal conservation all tend to decrease with age[1,11] (Table 6.2), and may cause a state of negative calcium balance and of a progressive calcium deficiency in elderly subjects.

The Consensus Conference from the NIH has stated the recommended calcium intake throughout the different life periods (Table 6.3).[12] On the basis of the most current information available, optimal calcium intake has been estimated to be 400 mg/day (birth to 6 months) to 600 mg/day (6–12 months) in

Table 6.2 Alterations of calcium metabolism associated with aging

Decreased calcium intake
Decreased calcium absorption
Increased intracellular calcium
Decreased serum ionized calcium
Increased serum levels of PTH
Altered vitamin D status
Decreased renal conservation
Increased bone turnover
Decreased physical activity

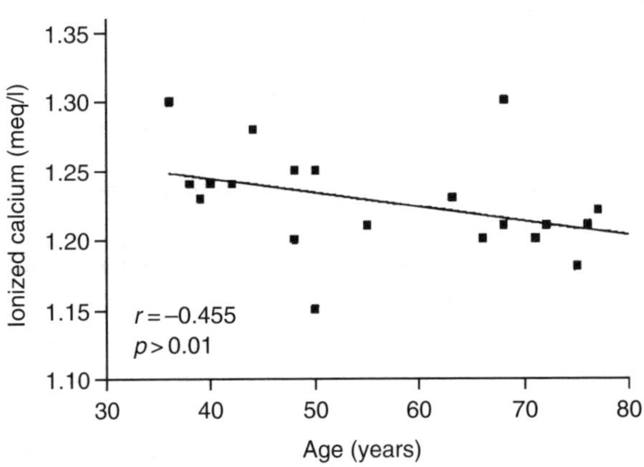

Figure 6.2 Effect of aging on serum ionized calcium in normal subjects[10]

Table 6.3 Optimal daily calcium intake according to NIH consensus conference[12]

Age	mg/day
Neonates	
0–6 months	400
6 months to 1 year	600
Children	
1–5 years	800
6–10 years	800–1200
Adolescents	
11–24 years	1200–1500
Male adults	
25–65 years	1000
Elderly (>65 years)	1500
Female adults	
20–25 years	1000
Pregnant and nursing	1200–1500
Postmenopausal (>50 years)	1500
Elderly (>65 years)	1500

infants; and 800 mg/day in young children (1–5 years) and 800–1200 mg/day for older children (6–10 years); 1200–1500 mg/day for adolescents and young adults (11–24 years). In adulthood the recommended intake is of around 1000 mg/day for women between 25 and 50 years and in adult men between 25 and 65 years. These suggested intakes increase in special situations to 1200–1500 mg/day for pregnant or lactating women, and to 1500 mg/day for post-menopausal women not on estrogen therapy. The recommended calcium needs also tend to increase in the elderly and has been estimated to be, for all women and men over 65, around 1500 mg/day.

Other factors play a role in calcium intake, including vitamin D, which is needed for adequate calcium absorption. In elderly men and women, an adequate calcium intake is of extreme importance. Indeed, in the older subjects the daily requirement is higher (1500 mg/day) than in the young individual to counteract the age-related alterations of calcium balance regulatory mechanisms mentioned earlier (Table 6.2).

In contrast to the NIH suggestions to increase the calcium content in the diet, several studies have demonstrated that dietary calcium content tends to be reduced in men as well as in women with advancing age.[13,14] In men and women 65 years of age and older, calcium intake of less than 600 mg/day (about one-third of the NIH recommendation) is common.[12] Optimal calcium intake may be achieved through diet, calcium-fortified foods, calcium supplements or various combinations of these. Many factors can negatively influence calcium availability, such as certain medications or food components.

Several studies have addressed the possible role of calcium supplementation in the prevention and treatment of a number of chronic disorders. This includes in the first place the prevention and treatment of osteoporosis, and other bone diseases, but the role of calcium has been suggested also in the prevention and treatment of salt-sensitive and pregnancy-associated hypertension, and to reduce the development of precancerous changes in colon cancer. The purpose of this chapter is to review the possible role of dietary calcium supplementation in the therapy and/or in the prevention of calcium-deficiency-related diseases.

6.5 Osteoporosis

Because 99% of total body calcium is found in the bone, the need for calcium is largely determined by skeletal requirements. Calcium is a major component of mineralized tissues and is required for normal growth and development of the skeleton and teeth.

Osteoporosis is the major underlying cause of bone fractures in postmenopausal women and in the elderly. It is clear and not disputed that calcium deficiency caused by very low intake is the cause of bone loss and osteoporosis. Since maintenance of serum calcium is of vital importance for the body homeostasis, when the dietary intake is reduced, the calcium needed is taken from the bone stores.

However, epidemiological data are conflicting in finding a relationship between calcium intake and bone health, although racial and other confounding factors may complicate the picture. A weak but positive relation between calcium intake and bone mass was found in a study conducted in the USA.[15] There is strong evidence for the importance of an adequate calcium intake in the prevention of bone loss associated with aging and/or menopause, in the reduction of risk fracture[16] and of a close positive relation between calcium intake and bone mass.[17] Absorption efficiency for calcium at the recommended dietary allowance can vary from 10 to almost 60% in postmenopausal women of the same age[18] and there is convincing evidence of a decreased absorption efficiency for calcium with age.[19]

Thus, wide variations in intake are required to obtain the same calcium balance. In the elderly, administration of vitamin D should be considered in order to correct the deficit of this hormone, and to facilitate the calcium absorption. The results of balance studies suggest a threshold effect for calcium intake: body retention of calcium increases with increasing calcium intake up to a threshold, beyond which further calcium intake causes no additional increment in calcium retention.

Two important factors that influence the occurrence of osteoporosis are optimal peak bone mass attained in the first two to three decades of life and the rate at which bone is lost in later years. Adequate calcium intake is critical to achieving optimal peak bone mass and modifies the rate of bone loss associated with aging.

Calcium insufficiency due to low calcium intake and reduced absorption can translate into an accelerated rate of age-related bone loss in older individuals. Among the homebound elderly and persons residing in long-term care facilities, vitamin D deficiency has been detected and may contribute to reduced calcium absorption. Numerous studies have shown that calcium supplementation in the range of 1200–1500 mg/day might help to prevent osteoporosis and has been shown to reduce the rates of bone loss in selected sites of the skeleton such as the femoral neck, and to decrease fracture incidence.[20,21] Thus, there is convincing evidence that a supplementation with calcium and vitamin D in women later in the menopause is able to reduce the risk of hip fractures and other non-vertebral fractures. Chapuy *et al.* reported a 40% reduction in hip and other extremity fractures within 18 months amongst elderly women living in retirement homes.[20] Dawson-Hughes *et al.* found a 55% reduction in non-vertebral fractures by 3 years of supplementation with calcium and vitamin D amongst men and women 65 years of age or older living in the community.[22] It has been suggested that the effect of calcium supplementation is particularly evident in studies in older osteoporotic women where subjects in the first 5 years after menopause were excluded, in whom calcium deficit and elevated PTH level are less important factors in the accelerated bone turnover than the fall in estrogen levels.[23]

In the evaluation of the role of calcium supplementation, it should be emphasized that an adequate calcium intake should not be considered just a therapy for bone mass loss but an essential strategy for the maintenance of health in the aging organism. In conditions of estrogen deficit, such as menopause and, further, in conditions in which bone mass is already reduced such as postmenopausal and senile osteoporosis, an adequate calcium intake should be associated with any medications that reduce fracture risk.[21,24]

6.6 Hypertension

The regulation of cellular calcium metabolism is central to blood pressure homeostasis; the higher the level of cytosolic free calcium, the greater the smooth muscle vasoconstrictor tone, the catecholamine secretion, the sympathetic nervous system activity, and thus blood pressure. An increase of cellular calcium, as part of the final common pathway leading to vascular smooth muscle contraction, may directly explain the increased peripheral resistance characteristic of essential hypertension. Animal, clinical and some epidemiological evidence suggests that high blood pressure is associated with abnormalities of calcium metabolism and secondary activation of the parathyroid gland.[4,25–30]

The effect of calcium dietary modifications on blood pressure control has generated great controversy. There are considerable epidemiological and clinical trial data on the relationship between blood pressure levels and calcium intake.[31] Epidemiologically, a consistent inverse relationship has been observed between dietary calcium intake and blood pressure. The epidemiological relationship of dietary calcium to blood pressure was definitely assessed by McCarron, who reanalyzed the data of the National Health and Nutrition Examination Survey (NHANES), observing an inverse relationship of calcium intake to blood pressure,[32,33] suggesting that a dietary calcium 'deficiency' may contribute to elevation of blood pressure in humans. The reanalysis of the NHANES data by McCarron indicated that a dietary calcium intake of at least 1000 mg was associated with a 40–50% reduction in hypertension prevalence.[33] Following this observation, a number of experimental and observational studies have supported the hypothesis that calcium supplementation can reduce blood pressure.[34–39]

However, it should be emphasized that almost 80 years before the NHANES results, W. Addison first suggested a link of dietary calcium to clinical hypertension. He reported that oral calcium supplementation could lower blood pressure in hypertensive individuals.[34] The findings of Addison were ignored for more than 50 years until similar results were reported in rats by Ayachi,[35] in normotensive humans by Belizan *et al.*[36] and in human hypertensives by others.[31,37–39] The results of randomized controlled trials of calcium supplementation on blood pressure have not been uniform. Pooled analyses showed a small but statistically significant reduction in systolic blood pressure and a non-significant trend toward reduction in diastolic blood pressure.[40–42]

The non-uniform results of the calcium supplementation trials performed has led some investigators to hypothesize that only a subgroup of individuals respond to calcium supplementation. Resnick *et al.* have proposed that the calcium supplementation was more beneficial among low-renin, salt-dependent forms of hypertension (around 30% of all hypertensive individuals). Those subjects exhibiting a more hypotensive response to calcium supplementation were also those that were more responsive to salt restriction (salt-volume-dependent hypertension).[43] Thus, the effect of calcium in lowering blood pressure is related and synergistic to the restriction of salt, in the subgroup of responsive subjects (salt-sensitive hypertension). A role for calcium-regulating hormones, 1,25 vitamin D and PTH in mediating both the hypertensive effect of salt and the hypotensive action of increased calcium intake was also suggested.[43,44] In particular, the beneficial effect of calcium supplementation may be mediated by a suppression of PTH secretion and reduction of circulating levels. St John *et al.*[45] have showed a weak, but significant, relationship between blood pressure and calcitropic hormones in a group of elderly people, suggesting that bigger changes in systolic blood pressure, may be observed in older subjects with an underlying hypovitaminosis D associated with elevated PTH levels.

The recently published Dietary Approaches to Stop Hypertension (DASH) trial has provided outcome data for hypertension control and dietary patterns rich in dairy products, fruits, vegetables, grains and lean meat. The dietary pattern reached in the DASH diet reflects the intake of these foods at the levels considered appropriate, but rarely achieved in the Western diet.[46]

Additional information is needed to identify sub-populations in a state of calcium deficiency that may benefit more from calcium supplementation. In particular, as previously discussed, elderly people may require greater calcium intake due to calcium malabsorption and vitamin D deficiency,[1,11,12] and old age is also a period associated with an increased risk of elevated blood pressure.[47]

Although the modest response in systolic and diastolic blood pressure reduction found in the trials does not justify the use of calcium supplementation as a sole treatment for patients with mild hypertension, a recommendation for increased calcium intake or calcium supplementation may be a useful advisory to be added to the other non-pharmacological suggestions for prevention and treatment of hypertension, at least in subjects with low-renin, salt-sensitive hypertension, and in special population more at risk of calcium deficiency, such as elderly hypertensives and women with gestational hypertension.

6.7 Hypertension in Pregnancy and Preeclampsia

Preeclampsia is a pregnancy-induced condition, affecting at least 5–8% of all pregnancies, occurring during the second half of gestation, and rapidly progressive. The main signs of preeclampsia are high blood pressure and protein in the urine. Preeclampsia causes 50–70% of cases of hypertension in pregnancy. Women with preeclampsia may also have swelling (edema) of the hands and feet, sudden weight gain, blurred vision, severe headaches and dizziness. Rarely, preeclampsia may progress to a life-threatening condition called eclampsia, which includes convulsions and sometimes leads to coma and death of the mother and baby. The cause of preeclampsia remains unknown. Low calcium intake has been implicated as a determinant of preeclampsia. Data regarding the role of supplemental calcium in reducing preeclampsia are conflicting. The 'calcium deficiency' hypothesis of gestational hypertension was first proposed by Belizan *et al.* to explain the surprisingly low incidence of preeclampsia and other pregnancy-specific forms of hypertensive disease in poor countries with a high dietary calcium intake, compared to the uniformly high incidence of gestational hypertension in equally poor countries with little access to calcium in the diet.[48,49]

Published meta-analysis (combination of data from several studies) of calcium supplementation in pregnancy has consistently found an important reduction in systolic and diastolic blood pressure and reduced incidence of preeclampsia.[50,51] Calcium supplementation might be expected to be of greater benefit to women with lower intakes of dietary calcium. Belizan showed that a low ratio of urinary calcium to creatinine at the time of enrolment, which could reflect a low dietary intake, was a predictor of benefit.[49] It has been suggested that calcium supplementation might reduce the incidence of preterm deliveries, caesarean deliveries, births of infants small for their gestational ages and delivery-related death resulting from direct effects on uterine smooth muscle to reduce contractility and prevent preterm labor;[50,51] but these data have not been confirmed.[52] A more recent large trial conducted on 4589 healthy women who were 13–21 weeks pregnant receiving daily treatment with either 2 g of elemental calcium or placebo for the remainder of their pregnancies did not show any significant reduction of the incidence or severity of preeclampsia or the delay of its onset after calcium supplementation.[52]

6.8 Colon Cancer

Laboratory, clinical and epidemiological evidence suggests that calcium may help prevent colorectal adenomas. In some epidemiological studies, higher calcium intake has been associated with a lower risk for the development of colon cancer. The Calcium Polyp Prevention Study, a randomized, double-blind trial of the effect of supplementation with calcium carbonate on the recurrence of colorectal adenomas, showed that in subjects with a recent history of colorectal adenomas receiving calcium carbonate (3 g, corresponding to 1200 mg of elemental calcium daily), there was a significant although moderate reduction in the risk of recurrent colorectal adenomas. The effect of calcium was independent of initial dietary fat and calcium intake.[53] Recent observational studies have reported significant inverse associations between calcium intake, including dairy products, and colon cancer risk.[54–55] The Health Professionals Follow Up Study has suggested that calcium may have a threshold effect on the risk of colon cancer: very low levels appear to increase risk, but once a certain level is reached (about 700 mg/day), there is no significant benefit with increased intake, either through diet or supplements.[54] Lupton has suggested that dairy sources of calcium may be more effective in combating colon cancer promotion than are equivalent intakes of calcium carbonate.[56]

6.9 Weight Control and Regulation of Body Fat

A link between dietary calcium and weight management in humans has been hypothesized. Data from the third National Health and Nutrition Examination Survey (NHANES III), which is representative of the US civilian non-institutionalized population, have been recently re-evaluated. Data have suggested that after controlling for energy intake and physical activity, contrary to common thought, an adequate calcium intake may actually be associated with a reduced likelihood of being in the highest quartile for adiposity.[57] Zemel *et al.* reported an almost 80% difference in obesity rates between those Americans in the lowest quartile of dietary calcium intake versus those in the highest quartile.[57] Several studies have reported an impact of supplemental dietary calcium on weight loss and have suggested that diets low in calories but high in calcium may provide some better protection against obesity. Experimental data in animals and in humans have also suggested a possible role of dietary calcium in regulating body fat, and an increase in lipolysis with a high-calcium/low-fat diet, consistent with an increase in energy metabolism.[58–61] Data from six observational studies and three controlled trials in which calcium intake was the independent variable (and either bone mass or blood pressure the original outcome variable) have been reanalyzed to evaluate the effect of calcium intake on body weight and body fat. Heaney *et al.* suggested that dietary calcium intake of at least 1200 mg/day was associated with a nearly fivefold reduction in the proportion of individuals who were overweight, compared to those consuming less than 500 mg of dietary calcium/day, and that an increased calcium intake across the population could reduce overweight and obesity prevalence by 60–80%, respectively.[62] The latter part of previous sentence is ambiguous: Are both overweight and obesity reduced by 60–80% or otherwise? Analysis reveals a consistent effect of higher calcium intakes, expressed as lower body fat and/or body weight, and reduced weight gain at midlife.[62,63] The Coronary Artery Risk Development in Young Adults (CARDIA) documented 30 and 65% reductions in the development of obesity and hypertension, respectively, in young overweight adults from increased dairy food intake over a 10-year period, but noted that as dairy intake increased, intake of fruit, vegetables and grains also tended to rise.[64]

Thus, the available evidence indicates that increasing calcium intake may substantially reduce the risk of being overweight, although long-term, large-scale prospective clinical trials need to be conducted to confirm and better clarify this association.

6.10 Periodontal Disease

Dietary calcium has long been a candidate to modulate periodontal disease. A study has reanalyzed the role of dietary calcium intake as a contributing risk factor for periodontal disease utilizing the data from the NHANES III, in nearly 13 000 people representing US adults.[65] They found that men and women who had calcium intakes of less than 500 mg/day, or about half the recommended dietary allowance, were almost twice as likely to have periodontal disease, as measured by the loss of attachment of the gums from the teeth.

These results suggest that low dietary intake of calcium results in more severe periodontal disease. Further studies will be needed to better define the role of calcium in periodontal disease and to determine the extent to which calcium supplementation will modulate periodontal disease and tooth loss.

6.11 Kidney Stones

Since 20–40% of recurrent stones are associated with elevated urinary calcium, it has been thought that consumption of high levels of calcium might cause or contribute to stone formation, and for many years it was suggested that a low-calcium diet was the best way to prevent kidney stones. But recent research has reversed that thinking. There is evidence that calcium restriction may actually increase the risk of kidney stones under certain conditions, and that additional calcium intake may actually lower the risk for kidney stones. A large prospective trial conducted to study the relation between dietary calcium and the risk of symptomatic kidney stones in a cohort of 45 619 men, with a follow-up period of 4 years,[66] concluded that high calcium intake (above 850 mg/day) decreased the risk of symptomatic kidney stones.

Borghi *et al.* conducted a five-year randomized trial comparing the effect of two diets in 120 men with recurrent calcium oxalate stones and hypercalciuria – a low-calcium diet and a diet relatively rich in calcium but low in animal protein and salt – for the prevention of recurrent stones in men with a history of kidney stones. The amount of oxalate was similar in the two diets, since most calcium stones are calcium oxalate stones and it was therefore important to equalize the oxalate intake. Over a follow-up period of 5 years, they found that those consuming the higher calcium diet (1200 mg/day) compared to those on low calcium (400 mg/day) had a 50% reduction in recurrence risk, providing evidence that this approach offers greater protection than the traditional low-calcium diet,[67] and suggesting that calcium restriction may actually lead to an increase in absorption and excretion of oxalate in the urine in both normal subjects and patients with kidney stones.

At present it is possible to conclude that, in normal individuals, calcium supplementation is not associated with stones formation, while further studies are needed to confirm that calcium supplementation may reduce stone formation in stones-forming subjects.

6.12 Calcium Supplements: Side Effects

Calcium supplementation has been extensively tested. Several large and long-term studies, using varying amounts of supplemental calcium, typically between 1000 and 2500 mg/day, have provided meaningful evidence that this amount of calcium intake is safe. Gastrointestinal side effects of calcium supplements have been observed, usually at relatively high dosages.[68,69] Some individuals may experience bloating or constipation when beginning calcium supplementation, and it has been reported in several clinical trials. In most cases bloating and constipation are temporary, as the body adjusts to the effects of the daily calcium supplement. Although calcium is better absorbed in an acid environment, only little acidity is necessary for calcium absorption.[70] However, subjects with hypochloridia (lack of adequate HCl production in the stomach), including the very elderly, those on high doses of antiulcer medications, and individuals who have fasted, should take calcium carbonate with a meal since that is when it is most effectively absorbed.[71]

6.13 Conclusions

During the past two to three decades, an impressive amount of data in the medical literature have strongly suggested a role for an adequate intake of calcium and dairy foods in the reduction of the disease burden of several common medical conditions. In Western countries, diets are generally poor in calcium and the mean dietary calcium consumption tends to be much lower than those needed to maintain an equilibrated calcium balance, as suggested in the NIH recommendation, especially in older individuals.[12]

Considerable epidemiological and clinical trial data have confirmed the beneficial effect of a dietary approach with calcium supplementation in the prevention and treatment of a number of chronic disorders, such as osteoporosis and periodontal disease, hypertension and preeclampsia, obesity and insulin resistance and some cancers.

Optimal calcium intake may be achieved through diet, calcium-fortified foods or calcium supplements. An adequate dietary intake of calcium by both nutritional and supplemental means is a relatively inexpensive approach that may ensure several health beneficial effects. McCarron and Heaney have recently reviewed the evidence of the net benefits of an increased dairy food intake on these chronic medical conditions and the estimated improvements in outcomes were combined with available data on annual costs of the respective disorders. They have calculated that, using conservative estimates of potential benefit, the projected savings was (only in the US) of approximately $26 billion in the first year and of approximately $200 billion after 5 years.[72] These data confirm the need for public health strategies that would help to ensure optimal calcium intake in the Western populations, and to develop health-promoting programs to change Western population behavior with respect to calcium intakes.

References

1. Dominguez LJ, Ruggero R, Bruno D *et al.* (1998) *Recent Advances in Geriatrics*, Barbagallo M, Licata G, Sowers JR (eds), Plenum Press, New York and London, 239–250.
2. Ringer S. (1883) *J Physiol*, 22, 2–5.
3. Rubin RP. (1970) *Pharmacol Rev*, 22, 389–427.
4. Barbagallo M, Resnick LM. (1996) *Endocrinology of the Vasculature, Contemporary Endocrinology* series. Sowers J (ed.), The Humana Press Inc., Totowa NJ, USA, 283–300.
5. Barbagallo M, Gupta R, Dominguez LJ, Resnick LM. (2000) *J Am Soc Geriatr*, 48, 1111–1116.
6. Borle AB. (1981) *Rev Physiol Biochem Pharmacol*, 90, 13–153.
7. Eaton SB, Nelson DA. (1991) *Am J Clin Nutr*, 54: 281S–287S.
8. Barger-Lux MJ, Heaney RP. (1994) *J Nutr*, 124 (8 Suppl.): 1406S–1411S.
9. Barbagallo M, Resnick LM, Dominguez LJ, Licata G. (1997) *Diabetes Metab*, 23, 281–294.
10. Barbagallo M, Licata G, Dominguez LJ, Resnick LM. (1999) *Hypertension*, 34, 902–906.
11. Heaney RP, Recker RR, Stegman MR, Moy AJ. (1989) *J Bone Miner Res*, 4, 469–475.
12. NIH Consensus Development Panel on Optimal Calcium Intake. (1994) *JAMA*, 272, 1942–1948.
13. Nordin BEC. (1962) *Am J Clin Nutr*, 10, 384–390.
14. Riggs BL, Wahner HW, Melton LJ III *et al.* (1987) *J Clin Invest*, 80, 979–982.
15. Garn SM, Solomon MA, Friendl J. (1981) *Ecol Food Nutr*, 10, 131–133.
16. Heaney RP. (1991) *Am J Med*, 91 (Suppl. 5B), 23S–27S.
17. Cummings RG. (1990) *Calc Tissue Int*, 47, 194–201.
18. Heaney RP, Recker RR. (1986) *Am J Clin Nutr*, 43, 229–305.
19. Bullamore JR, Wilkinson R, Gallagher JC, Nordin BEC. (1970) *Lancet*, 2, 535–537.
20. Chapuy MC, Arlot ME, Dubouef F *et al.* (1992) *New Engl J Med*, 327, 1637–1642.
21. Reid IR, Ames RW, Evans MC *et al.* (1993) *N Engl J Med*, 328, 460–464.
22. Dawson-Hughes B, Harris SS, Krall EA, Dallal GE. (1997) *N Engl J Med*, 337, 670–676.
23. Delmas PD. (1993) *Am J Med*, 95 (Suppl. 5A): 11S–16S.
24. Riggs BL, Melton LJ. (1992) *N Engl J Med*, 327, 620–627.
25. McCarron DA, Pingree PA, Rubin RJ *et al.* (1980) *Hypertension*, 2, 162–168.
26. Hvarfner A, Bergstrom R, Morlin C *et al.* (1987) *J Hypertens*, 5, 451–456.
27. Young EW, Morris CD, McCarron DA. (1992) *J Lab Clin Med*, 120, 624–632.
28. Barbagallo M, Resnick LM, Sosa RE *et al.* (1992) *Clin Sci*, 83, 561–565.
29. Resnick LM. (1990) *Am J Hypertens*, 3, 171S–178S.
30. Resnick LM, Gupta RK, Bhargava KK *et al.* (1991) *Hypertension*, 17, 951–957.
31. Resnick LM. (1999) *Am J Hypertens*, 12, 99–112.
32. McCarron D, Morris C, Cile C. (1982) *Science*, 217, 267–269.
33. McCarron D, Morris C, Henry J, Stanton J. (1984) *Science*, 224, 1392–1398.
34. Addison W. (1924) *Can Med Assoc J*, 14, 1059–1061.
35. Ayachi S. (1979) *Metabolism*, 28, 1234–1238.

36. Belizan JM, Villar J, Pineda O *et al.* (1983) *JAMA*, 249, 1161–1165.
37. Resnick LM, Nicholson J, Laragh JH. (1986) *Fed Proc*, 45, 2739–2745.
38. McCarron D, Morris C. (1985) *Ann Int Med*, 103, 825–831.
39. Grobbee, D, Hofman A. (1986) *Lancet*, ii, 703–707.
40. Bucher HC, Cook RJ, Guyatt GH *et al.* (1996) *JAMA*, 275, 1016–1022.
41. Cappuccio FP. (1996) *JAMA*, 276, 1385–1386.
42. Griffith LE, Guyatt GH, Cook RJ *et al.* (1999) *Am J Hypertens*, 12, 84–92.
43. Resnick LM. (1989) Laragh LM, Brenner BM (eds). *Hypertension: Pathophysiology, Diagnosis, and Management*, 2nd edn Raven Press, NY, 2037–2059.
44. Resnick LM. (1987) *Am J Med*, 82 (Suppl. 1B), 16–26.
45. St John A, Dick I, Hoad K *et al.* (1994) *Eur J Endocrinol*, 130, 446–450.
46. Appel LJ, Moore TJ, Obarzanek E *et al.* (1997) *N Engl J Med*, 336, 1117–1124.
47. The Seventh Report of The Joint National Committee on Prevention, Detection, Evaluation, and Treatment of High Blood Pressure. (2003) *Hypertension*, 42, 1206–1252.
48. Belizan JM, Villar J. (1980) *Am J Clin Nutr*, 33, 2202–2210.
49. Belizan JM, Villar J, Gonzalez L *et al.* (1991) *N Engl J Med*, 325, 1399–1405.
50. Carroli G, Duley L, Belizan JM, Villar J. (1994) *Br J Obstet Gynaecol*, 101, 753–758.
51. Bucher HC, Guyatt GH, Cook RJ *et al.* (1996) *JAMA*, 275, 1113–1117.
52. Levine RJ, Hauth JC, Curet LB *et al.* (1997) *N Engl J Med*, 337, 69–76.
53. Baron A, Beach M, Mandel JS *et al.*, For The Calcium Polyp Prevention Study Group. (1999) *N Engl J Med*, 340, 101–107.
54. Wu K, Willett WC, Fuchs CS *et al.* (2002) *J Natl Cancer Inst*, 94, 437–446.
55. Terry P, Baron JA, Bergkvist L *et al.* (2002) *Nutr Cancer*, 43, 39–46.
56. Lupton JR. (1997) *Am J Clin Nutr*, 66, 1065–1066.
57. Zemel MB, Shi H, Greer B *et al.* (2000) *FASEB J*, 14, 1132–1138.
58. Zemel MB. (2003) *Lipids*, 38, 139–146.
59. Heaney RP. (2003) *J Nutr*, 133, 268S–270S.
60. Parikh SJ, Yanovski JA. (2003) *Am J Clin Nutr*, 77, 281–287.
61. Shi H, Dirienzo D, Zemel MB. (2001) *FASEB J*, 15, 291–293.
62. Heaney RP, Davies KM, Barger-Lux MJ. (2002) *J Am Coll Nutr*, 21, 152S–155S.
63. Davies KM, Heaney RP, Recker RR *et al.* (2000) *J Clin Endocrinol Metab*, 85, 4635–4638.
64. Pereira MA, Jacobs DR, Van horn L *et al.* (2002) *JAMA*, 287, 2081–2089.
65. Nishida M, Grossi SG, Dunford RG *et al.* (2000) *J Periodontol*, 71, 1057–1066.
66. Curhan GC, Willett WC, Rimm EB, Stampfer MJ. (1993) *New Engl J Med*, 328, 833–838.
67. Borghi L, Schianchi T, Meschi T *et al.* (2002) *N Engl J Med*, 346, 77–84.
68. Caspary WF. (1996) *Eur J Gastroenterol Hepatol*, 8, 545–547.
69. O'Connell MA, Lindberg JS, Peller TP *et al.* (1989) *Clin Pharmacol*, 8, 425–427.
70. Bo-Linn GW, Davis GR, Buddrus DJ *et al.* (1984) *J Clin Invest*, 73, 640–647.
71. Recker RR. (1985) *N Engl J Med*, 313, 70–73.
72. McCarron DA, Heaney RP. (2004) *Am J Hypertens*, 17, 88–97.

7 $_{22}$Ti Anti-tumor Titanium Drugs

Erich Dubler

7.1 Introduction

This review will highlight studies on the biological and medical aspects of titanium and titanium complexes. In this context, relevant chemical properties of titanium may be summarized as follows.

Titanium is the seventh most abundant metal in the earth's crust, after aluminum, iron, calcium, sodium, magnesium and potassium. The most stable and common oxidation state is Ti(IV). Compounds in lower oxidation states −I, 0, II and III are quite readily oxidized to Ti(IV) by air, water or other reagents. Despite this fact, the energy for removal of four electrons is high so that the Ti^{4+} ion (estimated ionic radius 0.69 Å) may not exist. Further, the extremely polarizing nature of the small Ti^{4+} ion results in considerable covalency of metal–ligand bonds in Ti(IV) complexes.[1]

Many complexes of central atoms with d^0 configuration and higher oxidation states such as Ti(IV), Cr(VI) and Mn(VII) adopt tetrahedral structures, both because of the small size of the metal ion and because of the electrostatic repulsion of the ligands, often involving oxygen atoms with a considerable negative charge.[2] It is a characteristic of Ti(IV) compounds that they undergo hydrolysis to species with Ti—O bonds, forming (polymeric) octahedral structures. Due to the inherent lability of Ti—N bonds toward hydrolysis and the

Metallotherapeutic Drugs and Metal-Based Diagnostic Agents: The Use of Metals in Medicine
Edited by Gielen and Tiekink

greater thermodynamic stability of Ti—O bonds, only relatively few titanium complexes with amine ligands have been characterized.[3] It is generally accepted that titanium complexes with anti-cancer activity contain Ti(IV), although Ti(III) is also used in a recent patent.[4]

7.2 The Biochemistry of Titanium

Despite being one of the most abundant metals, titanium, together with aluminum, is apparently not used by organisms. Both elements do not satisfy the rigorous requirements for essentiality in studies to date. Titanium inhibits numerous hydrolytic enzymes, and as titanate can be stimulatory, but there is no evidence that it is bioessential.[5] At least two explanations can be suggested to account for these findings.[2]

One suggestion is that despite its abundance, titanium may not be sufficiently available for use by organisms, because it is extensively hydrolyzed in its usual oxidation state Ti(IV) to form insoluble hydroxides in neutral aqueous media. This argument would be consistent with the general observation that the elements' abundance in living systems today generally reflects their relative amounts in the sea, where life possibly started. However, the same argument could be made with regard to Fe(III), and iron is essential. Therefore, if organisms had found titanium to be useful, they could well have developed mechanisms to extract it, just as they did with iron. The second possibility therefore is that titanium simply has no useful role in organisms and hence is not utilized.

Because titanium metal has uniquely useful properties in modern technology, about 5×10^6 tons of titanium ores are manufactured worldwide per year.[6] The question arises as to whether this fact gives rise to toxicity problems due to the accumulation of titanium in the biosphere, for example by blocking essential functional groups of biomolecules or by replacing essential metal ions in biomolecules by titanium. The answer is most probably no.

Titanium is generally considered as one of the most biocompatible and corrosion-resistant metals available for clinical applications. Since the first publication on the orthopedic application of titanium metal in 1951, research activity and clinical experience resulted in new developments in the manufacturing and use of this metal and a variety of its alloys.[7] Medical application of titanium and its alloys include hip or knee endoprotheses, heart pacemakers, central venous ports connected to a catheter for deliverance of long-term chemotherapy, and calcium phosphate-coated titanium dental implants. Three classes of composites based on titanium materials offer interesting properties:[8] (i) Ti/porous composites have special mechanical properties for isoelastic implants, (ii) Ti/ceramic composites show improved osseointegration of the implants and (iii) the special physical properties of Ti/ceramic composites make them suitable for heart pacemaker leads. In addition to the use of titanium in medical devices, titanium dioxide is a UV-reflecting component of many sunscreen agents, has replaced lead in paints, and is present in many foods as a whitening pigment. Two recent papers deal with the potential mutagenic risk of the exposure to titanium dioxide as well as to wear particles of joint implants. No significant trends for exposure risk associations for cancer were observed.[9,10]

Not only bio-compatibility of titanium compounds, but even bio-activity of titanium-containing composites has been claimed. For example, a $K_2Ti_6O_{13}$ bio-ceramic coating on the surface of a Ti/Al/Zr/Sn/Nb alloy has recently been tested in a simulated body liquid.[11] Rough surfaces and small holes on the Ti-alloy coating provided sites for stimulated bone attachment by encouraging the active formation of a Ca—P layer with a ratio of calcium and phosphorus appropriate to human bone.

In discussing the potential toxicity of titanium anti-cancer drugs described in the following sections, it must be taken into account that, in contrast to TiO_2 and titanium-alloys, these titanium drugs probably contain easily exchangeable ligands. In addition, the drugs usually are applied parenterally, e.g. iv injection, and therefore are immediately available for biochemical reactions.

7.3 Titanium Anti-cancer Drugs

After the discovery of the tumor-inhibiting properties of *cis*-diammine dichloro-platinum(II) (cisplatin) in 1969[12] and its routine use as a leading cytostatic drug since 1979,[13] there has been increasing interest in the development of new anti-tumor metal agents including non-platinum metal complexes. Among these, two monomeric titanium(IV) complexes have qualified for clinical trials (Figure 7.1): $[Ti^{IV}(bzac)_2(OEt)_2]$ (budotitane) and $[Cp_2Ti^{IV}Cl_2]$ (titanocene dichloride).

Relevant reviews on titanium complexes in cancer treatment have been given by Köpf-Maier and Köpf,[14–17] Keppler *et al.*,[18–21] Harding and Mokdsi,[22] Caruso, Rossi and Pettinari [4] and Meléndez.[23]

Most anti-tumor agents act by interfering with molecular processes in the cell replication cycle.[4,13] Cisplatin is well known as a DNA-damaging agent. It is generally believed that cisplatin arrives at the tumor cell intact, crosses the plasma cell membrane to areas of low concentration of Cl^- ions (extracellular $[Cl^-]$ is 100 mM and intracellular $[Cl^-]$ is 3 mM). It then undergoes hydrolysis with H_2O and/or OH^- groups replacing one or both of the Cl^- anions. These new complexes become active, and subsequent intra- or inter-strand Pt–DNA bonds, mainly involving guanine N(7) sites, are formed. Cisplatin then causes cells to arrest at either the G1-, S- or G2-phase of the cell cycle. Failing adequate repair, the cells eventually undergo an aberrant mitosis followed by

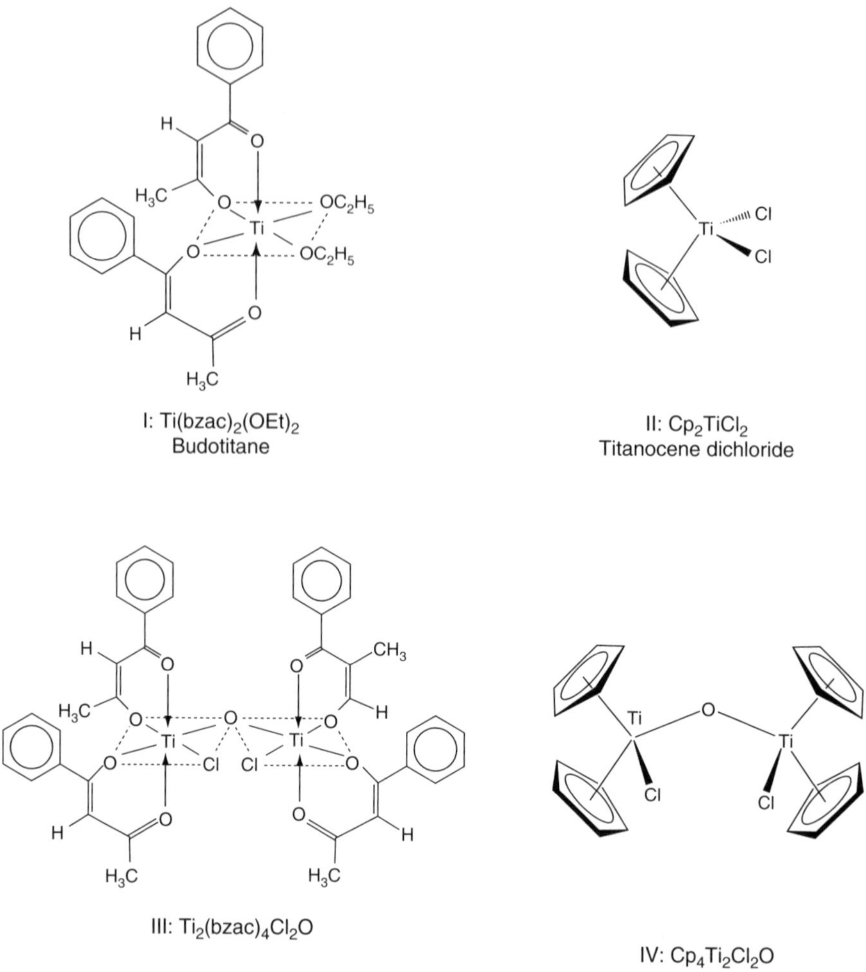

Figure 7.1 Structural formula of the anti-tumor titanium complexes $[Ti^{IV}(bzac)_2(OEt)_2]$ (budotitane, I, *cis-cis-cis* configuration shown) and $[Cp_2Ti^{IV}Cl_2]$ (titanocene dichloride, II) and of their possible dimeric hydrolysis products $[Ti^{IV}{}_2(bzac)_4Cl_2O]$ (III) and $[Cp_4Ti^{IV}{}_2Cl_2O]$ (IV)

apoptosis. A key question in view of such a mechanism is, why does cisplatin reach guanine-N(7) with competing S-donor ligands as glutathione or methionine available in the blood and in the cell?[24]

A reaction mechanism analogous to that of cisplatin may be suggested for the anti-cancer compounds budotitane and titanocene dichloride, since interaction with DNA has been described for both drugs.[16,17] In a study of the intracellular localization of titanium in xenografted human adenocarcinomas of the colon after application of a single therapeutic dose (80 mg/m^2) of titanocene dichloride in athymic mice,[25] titanium was first detected in the nuclear chromatin 12 h after application. Titanium location was near phosphorus-rich areas, suggesting titanium–DNA interactions. In addition, cisplatin and the two titanium drugs have two labile groups in the *cis* position (chloride or ethoxy) as a common structural feature. This gives rise to initial speculation that the three drugs might have similar biological mechanisms. However, their spectrum of action is very different. Cisplatin is active against fast-growing tumors and can be used to cure testicular cancer. In addition, it is successfully used in ovarian, neck, lung and other cancers. The main targets of budotitane, in contrast, are gastrointestinal tumors, indicating a use against slow-growing cancers, which pose the biggest problem in cancer therapy today. Titanocene dichloride also has a spectrum of activity which is different from that of cisplatin. It showed activity against human adenocarcinomas of the stomach and colon which are insensitive to common cytostatic agents. In addition, several crystallographic studies indicate that in $Cp_2Ti^{IV}Cl_2$ the Cp groups sterically prevent cross-linking DNA, and ^{31}P NMR studies indicate a prevalence of Ti(IV) binding to phosphate oxygen.[26] Therefore, titanocene dichloride seems not to function by analogy to cisplatin.

In contrast to the well-characterized platinum anti-cancer drugs, the active species for the anti-cancer activity of the titanium complexes *in vivo* has not yet been identified, and the mechanism including irreparable DNA damage and/or structural modification of DNA or other cellular targets is poorly understood.[22]

The actual ambiguity concerning the mechanism of action of titanium anti-cancer drugs has been summarized by Guo and Sadler.[27] Metal anti-cancer complexes are often electrophilic and can react with many biomolecules, including amino acids, polyphosphates, proteins and nucleic acids. However, these biomolecules may be located in different extra- or intra-cellular compartments. Therefore, carrier molecules (e.g. proteins such as albumin and transferrin, or small molecules such as ATP, glutathione or citrate) may be used to communicate between the compartments. Indeed, a detailed recent study has confirmed the uptake of Ti(IV) from $Cp_2Ti^{IV}Cl_2$ by human transferrin at blood plasma pH values, release of bound Ti(IV) to ATP at cellular endosomal pH values, and replacement of Ti(IV) by Fe(III), as well as uptake of a Ti_2–human-transferrin complex into placental cancer cells.[28] The substrate binding properties are controlled by natural gradients of e.g. pH, ATP or by ionic gradients. The gradients could also alter the affinity of drug molecules for different cellular components and facilitate drug binding to its target. The hard Lewis acid Ti(IV) finally may be transported into the cell by transferrins, iron-binding single-chain glycoproteins containing approximately 700 amino acids with molecular masses of about 80 kDa, which offer distorted octahedral coordination sites for metal ions involving oxygen and nitrogen donors.[29] Titanium(IV) then may bind to DNA at both negatively charged phosphates on the backbone and base nitrogen donors.[30] The high DNA concentration in the cell nucleus and potentially low pH close to the surface of DNA may favor DNA as a target for Ti(IV) binding under these conditions, since Ti(IV) does not strongly bind to DNA bases at physiological pH, but forms strong complexes with nucleotides only at pH values below 5.[27]

Discussing possible reaction mechanisms of the two clinically tested titanium drugs, an interesting aspect has to be taken into account. Given the hydrolytic instability of budotitane and of titanocene dichloride at physiological conditions, hydrolytic decomposition products should also be considered as possible active species. For example, the dimeric complex $[Ti^{IV}_2(bzac)_4Cl_2O]$, μ-oxo-dichlorotetrakis(1-phenylbutane-1,3-dionato)dititanium(IV) (Figure 7.1, III) may be formed already during the storage of budotitane and shows good anti-tumor activity.[19] Polynuclear species such as $[Cp_4Ti^{IV}_2Cl_2O]$ (Figure 7.1, IV), which are typical hydrolysis products of titanocene dichloride and precursors to the insoluble polymer $[(CpTi^{IV}O)_4O_2]n$, have also shown reduced or sporadic anti-tumor activity in animal studies.[15,22]

In addition to the two Ti(IV) anti-tumor drugs described above, a third Ti(IV) compound deserves a special comment – TiO_2 exhibits anti-tumor activity when finely dispersed and activated photochemically.[4] The effect may be described as photokilling of malignant cells by TiO_2 powder (photodynamic cancer therapy). For example, in a study using TiO_2 particles with UV irradiation, the photoexcited TiO_2 particles significantly suppressed the growth of HeLa cells that were implanted in nude mice.[31] The killing effect of photoexcited ultrafine TiO_2 particles has also been studied using U937 cells.[32] The electrophorogram shows that reactive oxygen species produced by photoexcited TiO_2 can damage DNA, which results in cell death. In this context, it is interesting to note that TiO_2 is also the final stage of budotitane hydrolysis.[18] Therefore, a new strategy could imply first bringing a titanium drug to the tumor (Ti(IV) tends to concentrate in DNA regions of the cell) and inducing titanium hydrolysis to generate TiO_2. Then, by using tumor imaging, a well-focused beam of light could be applied. Since the radiation needed for TiO_2 activation is much less energetic than X-rays, less damage to the surrounding normal tissue would be expected.[4]

7.4 Budotitane

7.4.1 Chemistry and anti-cancer activity

The $[Ti^{IV}(bzac)_2(OEt)_2]$ complex (budotitane, bzac = 1-phenylbutane-1,3-dionate = benzoylacetonate) was the first non-platinum complex tested in clinical trials.[18] Budotitane and related bis(β-diketonato) metal complexes can be synthesized from the corresponding metal tetrahalogenides and the diketonates in an anhydrous organic solvent. Budotitane was selected from about 300 related derivatives for further development. The structure–activity relation of complexes of the type M(β-diketonate)$_2$X$_2$ has been extensively tested by variation of the β-diketonate ligand, of the leaving group X and by variation of the central metal atom M, using the animal sarcoma 180 ascitic tumor model. The main results of these tests may be summarized as follows.[19]

The [Ti(acac)$_2$(OEt)$_2$] complex, which bears a symmetric acetylacetonate ligand, does not produce any anti-tumor activity, whereas the introduction of a phenyl group into the diketonate ligand may turn an inactive complex into a highly active substance. Unsubstituted aromatic ring systems in the periphery of the molecule have significant positive effects on the anti-tumor activity of such complexes, while substitutions at the aromatic ring system may be considered to be detrimental to the activity. It appears that a planar aromatic group in the β-diketonate ligand is responsible for the anti-tumor activity of the titanium(IV) complexes of the type M(β-diketonate)$_2$X$_2$. This situation leads to the suggestion that the activity of the complexes is determined by a DNA intercalating mechanism. It is interesting to note that the organometallic compound titanocene dichloride shows similar aspects exhibiting two π-bonded aromatic cyclopentadienyl rings, which also could be involved in a DNA intercalating mechanism of action.

The nature of the *cis*-configurated hydrolizable group X does not seem to contribute much to the anti-tumor activity of the diketonate complexes. However, the galenic behavior is considerably influenced by X because stability in water clearly increases in the order $I^- < Br^- < Cl^- < F^- < OR^-$. The bromine and fluorine complexes in addition have disadvantages over ethoxide as to the way the hydrolizable group is physiologically tolerated.

The activities of the titanium and zirconium complexes with the benzoylacetonate ligand are relatively similar, but they then decrease markedly in the order Hf > Mo > Sn > Ge, the germanium compound being virtually inactive even at higher doses.

7.4.2 Isomer abundance of budotitane

In any case, the spatial arrangement of the monomeric budotitane and/or the resulting hydrolyzed species seems to be of importance for the understanding of its anti-cancer activity. Complexes of the general formula [Ti(β-diketonato)$_2$X$_2$] can occur with a *cis*- or a *trans*- arrangement of the X ligands, the *cis* configuration being more stable. According to NMR data and the results of force field calculations, only those benzoyl–acetonato complexes with the extremely bulky substituents iodide or *p*-dimethylaminophenoxy as the hydrolyzable group X have the *trans* form.[18] Steric interactions are thought to be a key feature for the relative stabilities of the β-diketonato complexes, but electronic effects may also be important.[33]

As the β-diketonate ligand in $[Ti(bzac)_2X_2]$ complexes is unsymmetrical, there are eight possible isomers of the six-coordinate octahedrally configured complexes (Figure 7.2). Six of these have the X ligands *cis*, forming three enantiomeric pairs, which are diastereomers to each other, whereas the two *trans*-isomers are achiral. The isomeric configuration has been defined by three *cis* or *trans* prefixes which specify first the relative position of the X ligands, then the relative orientation of the benzoyl groups and finally the relative orientation of the acetyl ends of the benzoylacetonate ligands.[34] It is not known which one of these isomers exhibit anti-cancer activity. Isomer distributions have so far only been determined in $CDCl_3$ solutions by interpretation of NMR data,[33] but not in the galenic CremophorEL/propyleneglycol formulations used for the *in vivo* tests.

Force field calculations lead to calculated isomer distributions that are in satisfactory agreement with the experimental data derived from

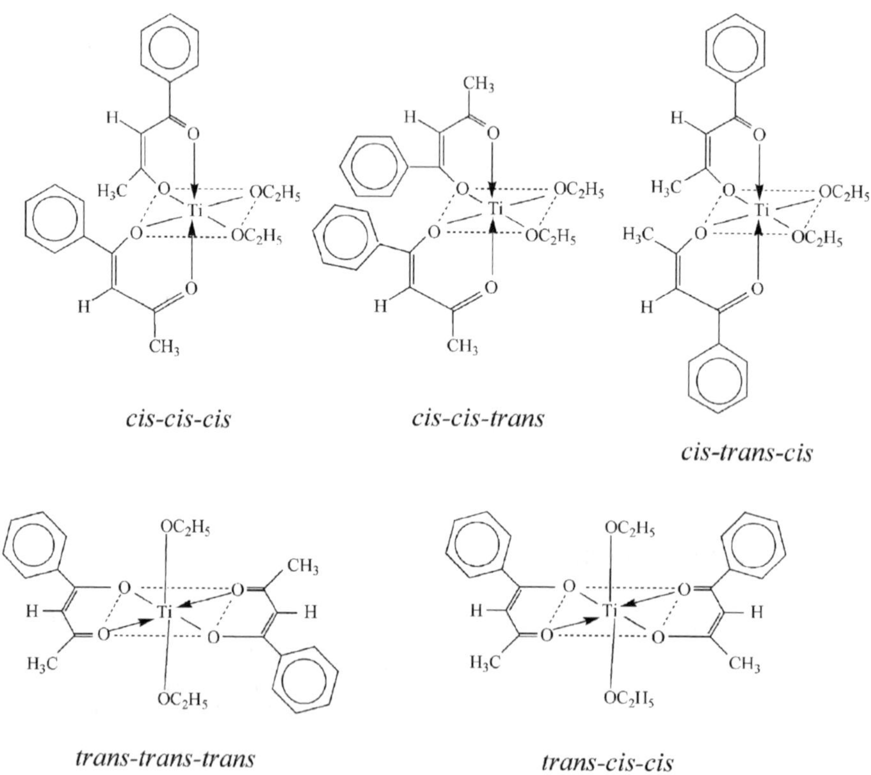

Figure 7.2 Isomers of budotitane $[Ti^{IV}(bzac)_2(OEt)_2]$. The three *cis*-isomers form three enantiomeric pairs (Δ and Λ configuration) which are diastereomers to each other. The two *trans*-isomers are achiral. In the crystal structure of budotitane, all molecules exhibit the *cis-cis-trans* configuration

temperature-dependent ^1H-NMR spectroscopy in $CDCl_3$ solution.[18,33] In budotitane $[Ti^{IV}(bzac)_2(OEt)_2]$, the observed/calculated isomer distributions (Figure 7.2) are *cis-cis-cis* 60%/57%, *cis-cis-trans* 19%/17% and *cis-trans-cis* 21%/26%, respectively.

Probably due to the fact that no isomerically pure fractions may be isolated from the isomeric mixture that exists in solution, budotitane usually does not give diffraction-quality crystals. Nevertheless, we have recently been able to crystallize budotitane $[Ti^{IV}(bzac)_2(OEt)_2]$ and its dichloro-derivative $[Ti^{IV}(bzac)_2Cl_2]$ from ethanol and toluene solutions, respectively.[35] Budotitane exhibits two crystallographically independent molecules involving Ti(1) and Ti(2) centers, both of which are octahedrally coordinated by six oxygen atoms with the two ethoxy ligands *cis*. Both molecules adopt the *cis-cis-trans* configuration with the acetyl ends of the benzoylacetonate ligands *trans* (Figure 7.2). The molecule involving Ti(1) exists in the Δ configuration, whereas the molecule involving Ti(2) shows the enantiomeric Λ configuration. An analysis of the crystal packing of budotitane shows that there are no intra- or inter-molecular π-π-stackings of the phenyl moieties of the benzoylacetonate ligands.[35] According to the spectroscopic data mentioned above, the *cis-cis-trans* configuration uniquely observed in the solid state occurs at only 19% in $CDCl_3$ solutions.

In the dichloro-derivative of budotitane, $[Ti^{IV}(bzac)_2Cl_2]$, the titanium atom is octahedrally coordinated within a Cl_2O_4 donor set with the two chloride atoms *cis*. In contrast to budotitane, the corresponding dichloro-derivative shows a *cis-trans-cis* arrangement with the benzoyl groups *trans*. The centrosymmetric space group accomodates equal numbers of Δ and Λ enantiomers within the unit cell. In contrast to budotitane, its dichloro-derivative clearly exhibits inter-molecular π-π-stackings of neighboring phenyl moieties.[35]

The phenyl groups in budotitane and in its dichloro-derivative are in approximately coplanar conjugation to the metal enolate rings. This is evidence of a certain aromaticity of the system which could support DNA intercalating mechanisms of the compounds.

7.4.3 Reaction with biomolecules

Because of the hardness of Ti(IV) and its affinity toward oxygen, many biomolecules offering oxygen donors are of interest in the context of possible metabolic pathways. For this reason, numerous steroid, sugar and nucleoside derivatives of budotitane have been synthesized and characterized mainly by spectroscopic methods, and in each case coordination by oxygen has been found.[20] Synthesis was possible with steroids like cholesterol or stigmasterol, with the bile acid dehydrocholic acid, as well as with steroid hormones like testosterone. Stable adducts of the nucleosides cytidine, uridine and thymidine were also synthesized. In these adducts, relatively fast binding to the hydroxy functions of the sugar moieties of the nucleosides took place, whereas coordination to the basic nitrogen of the nucleobases was not observed. These findings support the assumption that titanium–oxygen bonding involving the sugar or the phosphate moieties of DNA has to be considered in discussing the mode of action of budotitane.[20]

7.4.4 Animal studies

A preclinical study of budotitane in human tumor xenografts in nude mice has been reported in 1984.[36] In this investigation, the anti-tumor activities of budotitane and of the established anti-tumor drug cisplatin have been shown to be comparable.

Anti-tumor activity in transplantable animal tumor models in mice has been extensively studied by Keppler *et al.*[19] The results are presented in terms of T/C values, T/C (%) = (median survival time or medium tumor weight of treated animals vs median survival time or medium tumor weight of control animals) × 100. T/C values > 300%, indicating that a high percentage of animals are cured, have been found as a result of the budotitane therapy in the Stockholm ascitic tumor, the Ehrlich ascitic tumor and in the MAC 15A colon tumor, a transplantable colon adenocarcinoma. In contrast to cisplatin, activity against the quick-growing P338 and L1210 leukemias was marginal.

These studies were performed using transplantable tumors, which can be transplanted from one animal to another. An important disadvantage of such models is that they cannot ultimately define the organ tumors against which a new compound will be active in humans. Therefore, the evaluation of anti-tumor activity has also been performed using autochthonous tumors, which are mainly induced by a carcinogen, and which mimic the human situation fairly closely.[37] When comparing the activity of the established anti-tumor drugs 5-fluorouracil and cisplatin with that of budotitane, using AMMN (acetoxy-methyl-methyl-nitrosamine)-induced colorectal tumors, it was observed that budotitane is the most active drug. It reduces tumor volume to about 20% of the initial value. 5-Fluorouracil effects a tumor remission to about 40% of tumor volume, whereas cisplatin, with a value of about 120%, stimulates tumor growth a little. In addition, budotitane is the only compound to increase the lifespan of adenotumor-bearing animals, from 25 weeks in the control group to 36 weeks in the budotitane-treated group.[19]

7.4.5 Clinical investigations

For budotitane, a galenic formulation[19] had to be found which guarantees some water solubility and insensitivity to hydrolysis. The use of CremophorEL, a glycerine-polyethylene-glycolericin-oleate, was successful. Addition of 1,2-propyleneglycol resulted in a co-precipitate of the drug. Budotitane, CremophorEL and propyleneglycol are dissolved separately in water-free ethanol in the weight ratio of 1:9:1. The solutions are then mixed and evaporated at 30–40 °C. Within the resulting co-precipitate, budotitane is enveloped between layers of the solubilizer and is protected against hydrolysis. This co-precipitate can be dissolved in water under the formation of micelles. The co-precipitate solutions used in the clinic have favorable concentrations of 100–200 mg of budotitane per 100 ml. While such formulated solutions are stable for hours, the components possibly cannot be sufficiently well characterized to advance to further clinical trials.[38]

Budotitane has been extensively investigated in numerous preclinical studies.[20] It entered phase I clinical trials in 1986.[39] In 1996, a clinical phase I study and pharmacokinetic trial with budotitane administered as iv infusion twice weekly was reported.[40] The trial was performed on 18 patients refractory to all other known treatment. No objective tumor response was observed. The maximum tolerated dose of budotitane administered twice weekly was 230 mg/m^2, with cardiac arrhythmia as dose-limiting toxicity. Fifteen patients reported loss of taste at the day of infusion, reversible within 24 h. Budotitane, unlike cisplatin, did not cause vomiting. From these studies, the recommended dose for further clinical trials was 180 mg/m^2. However, subsequent clinical evaluations of budotitane have been stopped because its galenic formulation did not meet modern standards, although new formulations are being developed.[41]

The use of isomerically pure crystals of budotitane in the galenic formulation of the drug could eventually help to overcome these difficulties and may also increase the activity and/or bioavailability of budotitane.

7.5 Titanocene Dichloride

7.5.1 Chemistry and anti-cancer activity

In 1979, Köpf and Köpf-Maier reported[42] on the first metallocene with anti-tumor activity, the complex [$Cp_2Ti^{IV}Cl_2$] (bis(π-cyclopentadienyl)titanium(IV) dichloride, titanocene dichloride). The red crystalline titanium(IV) complex was first prepared by Wilkinson and Birmingham in 1954.[43] The two planar $C_5H_5^-$ cyclopentadienyl rings each have a delocalized negative charge and are coordinated to the metal center via pentahapto metal–carbon coordination in a 'bent sandwich' configuration. X-ray data show that the coordination about the titanium atom formed by two chlorine atoms and the two centroids of the cyclopentadienyl rings is that of a distorted tetrahedron. The average Ti—Cl bond distance is 2.364(3) Å, and the average Ti–Cp-(centroid) distance is 2.06 Å.[44]

Again, a number of structure–activity relation studies of the complexes $Cp_2M^{IV}X_2$ have been carried out[15] using, for example, the Ehrlich ascites tumor in mice. The main results of these tests may be summarized as follows.[22]

Titanocene derivatives with alkyl-substituted Cp rings, bridged Cp rings or only one Cp ring showed significantly reduced anti-tumor activity compared to the parent compound $Cp_2Ti^{IV}Cl_2$. Modification of the cyclopentadienyl ligand, for instance by replacement of H by R (CH_3, C_2H_5, $Si(CH_3)_3$ and $N(CH_3)_2$) groups ranging from mono-substitution to deca-substitution showed a dramatic reduction in the anti-tumor activity, the activity decreasing as the substitution increases.[15,23] The main influence of alkylation of the Cp rings on the anti-tumor activity is believed to be mainly due to electronic effects rather than steric effects. Alkyl substitution, in addition, reduces the aqueous solubility and therefore may influence transport processes.

Again, the nature of the ion X does not appear to affect significantly the anti-tumor activity of the parent compounds. Reduced activity of complexes with X, for example *p*-nitrophenoxy or 2,4,6-trichlorophenoxy, was attributed to the lack of lability of the Ti—X bond, which is unable to dissociate in solution and generate the active species which was assumed to coordinate to DNA.[22] However, a recent study involving mice inoculated with Ehrlich ascites tumor has shown that the compound $Cp_2Ti^{IV}(NCS)_2$ also exhibits a reduction in anti-tumor efficacy when compared to previous results with the original titanocene dichloride.[45]

In contrast to budotitane, variation of the central metal M is tolerated in the series of Cp_2MX_2 complexes. Optimum cure rates against Ehrlich ascites tumor in mice were obtained for Ti(IV), V(IV), Nb(IV) and Mo(IV). Sporadic cure rates were obtained for Ta(IV) and W(IV), whereas Zr(IV) and Hf(IV) analogues showed no activity.[22]

7.5.2 Reaction with biomolecules

By inductively coupled plasma (ICP) spectroscopy, it was shown that $Cp_2Ti^{IV}Cl_2$ forms adducts of the type $CpTi^{IV}$–DNA at pH = 5.3 and $CpTi^{IV}$–DNA at pH = 7. The titanium–DNA adducts once formed are stable for up to 2 days.[46]

While the exact mechanism of the anti-cancer action of titanocene dichloride seems still unclear, there is some evidence that a Ti(IV) species is formed and stabilized *in vivo* that is transported into cells via blood transport peptides as, for example, glutathione. NMR studies of the metallocenes $Cp_2M^{IV}Cl_2$ (M = titanium, niobium or molybdenum), however, show that with the niobium or molybdenum complex adducts of glutathione are formed, whereas no binding of the titanium complex to glutathione was detected at pH = 2.4 or pH = 6.0.[47] Examples of further studies of the interaction of $Cp_2Ti^{IV}Cl_2$ and derivatives with biomolecules are summarized in Table 7.1.

Table 7.1 Selected examples of the interaction of titanocene dichloride and derivatives with biomolecules

Compound	Biomolecule(s)	Interaction	Ref.
$Cp_2Ti^{IV}Cl_2$	amino acids	carboxyl binding	48, 49
$Cp_2Ti^{IV}Cl_2$	human topoisomerase II	inhibition	50
$Cp_2Ti^{IV}Cl_2$	β-cyclodextrin	inclusion complex	51
$Cp_2Ti^{IV}Cl_2$	glutathione	no interaction	47
$Cp_2Ti^{IV}Cl_2$	nucleotides	phosphate binding	52
$Cp_2Ti^{IV}Cl_2$	6-thiopurines	S(6)/N(7) chelation	53
$Cp_2Ti^{IV}Cl_2$	2-thiocytosine	NH_2/N(3) chelation	53
Ti^{IV}citrate	human apo-transferrin	binding at Fe^{3+} site	29, 54
$Cp_2Ti^{IV}Cl_2$	ATP	phosphate/N(7) binding	27
$Cp_2Ti^{IV}Cl_2$	5′-dAMP	phosphate/N(7) binding	55
$Cp_2Ti^{IV}Cl_2$	DNA	DNA adducts	46

7.5.3 Animal studies

The tumor inhibition influence upon human lung carcinomas xenografted into athymic mice by titanocene dichloride has been studied by Köpf-Maier.[56] In the case of lung adenocarcinoma, titanocene dichloride inhibited tumor growth by more than 50% resulting in treated/control values of 20–50%.

It has been reported that titanocene dichloride suppresses angiogenesis, the formation of blood vessels from preexisting ones, and inhibits biosynthesis of collagenous proteins in the *in vivo* system of the chorioallantoic membrane of the chick embryo.[57] At non-toxic dose regimens, titanocene dichloride retards the growth of Walker 256 carcinosarcoma transplants in rats and reduces the number of seeded implants. These data suggest that the anti-tumor activity of titanocene dichloride may be attributed, at least in part, to its ability to suppress angiogenesis.

In xenografted human renal cell carcinoma in athymic mice, titanocene dichloride showed better activity than the established drugs cyclophosphamide and vinblastine.[58]

Recently, the effect of titanocene dichloride on the activity of natural killer cells in Ehrlich ascites tumor-bearing immunodeficient mice was studied.[59] After inoculation of the tumor, the natural killer cell function declined to subnormal levels. The treatment consisted of intraperitoneal administration of 15 mg/kg/day of titanocene dichloride for 2 days. The treatment significantly enhanced natural killer cell function, and this function was restored to normal values by titanocene dichloride in all stages studied, but not by the established anti-tumor drug cyclophosphamide.

7.5.4 Clinical investigations

As a consequence of the hydrolytic instability of titanocene dichloride at pH >5, a formulation preventing hydrolysis and precipitation reactions had to be developed.[60] Formulations used for clinical studies include titanocene dichloride in a lyophilized powder dissolved in malate buffer at pH = 3.2 and titanocene dichloride in a lyophilized powder dissolved in malic acid at pH = 3.5.[22]

A first clinical phase I protocol, including 40 patients with refractory solid malignancies, evaluated a single-dose administration of titanocene dichloride, repeated every three weeks.[61] The dose-limiting toxicity was renal toxicity, and the recommended dose for phase II studies was 240 mg/m^2. Two minor responses (bladder carcinoma and non-small-cell lung cancer) were observed.

A second phase I study evaluated a weekly administration schedule involving 20 patients.[62] The dose-limiting toxicity again was renal toxicity. The maximum tolerated dose was 140 mg/m^2/week. The dose recommended for phase II studies was also 140 mg/m^2. One patient with adenocarcinoma of unknown primary had a minor response.

In a further phase I clinical trial, a total of 10 patients with progressive advanced cancer were treated with 80 mg/m^2 titanocene dichloride, administered as lyophilized powder, at days 1, 3 and 5 (repeated at day 22).[63] No objective tumor remission was observed. Drug-related side effects observed were renal and hepatic, emesis and metallic taste. The split dose regimen had no advantage with respect to the toxicity profile.

In a phase II trial, 14 patients with metastatic renal-cell carcinoma received 270 mg/m^2 titanocene dichloride every three weeks for six weeks. Although the toxicities and side effects encountered were mild to moderate, no partial or complete response was detectable.[64]

A multicenter phase II clinical trial of titanocene dichloride in 12 patients with metastatic breast cancer has recently been reported by Kröger *et al.*[65] All 12 patients had prior surgery and metastatic disease at study entry. Titanocene dichloride was intravenously administered at a dose of 270 mg/m^2 every three weeks. The conclusion is that titanocene dichloride was not effective in patients with metastatic breast cancer.

7.5.5 Perspectives of titanocene dichloride

One possibility to overcome the limited solubility and low stability in water of neutral titanocene dihalides could be the application of ionic titanocene complexes. Therefore, complexes of the type $[Cp_2Ti^{IV}XL]^+Y^-$ or $[Cp_2Ti^{IV}L_2]^{2+}(Y^-)_2$, where X^- and Y^- are anionic ligands, for example Cl^-, $CF_3SO_3^-$, I^-, $[FeCl_4]^-$ or $[AsF_6]^-$, and L is a neutral ligand, for example phenantroline, CH_3CN, 6-thioguanine, glycine or L-alanine, have been tested. All complexes showed good to moderate anti-tumor activity against fluid Ehrlich ascite tumor but not as active as the parent neutral compounds.[23] Similar results recently have been observed with tin compounds.[66] In order to overcome insolubility and hydrolysis problems, Mokdsi and Harding have prepared the titanocene derivatives $(MeCp)_2TiCl_2$, bis(methylcyclopentadienyl)titanocene dichloride, and the bis-glycine analogue $(MeCp)_2Ti(O_2CCH_2NH_3Cl)_2$ which both exhibit reasonable solubility in water and are relatively stable at pH = 7.0.[55]

Complex formation between β-cyclodextrin and $Cp_2Ti^{IV}Cl_2$ has been proven in solution.[67] The encapsulation of $Cp_2Ti^{IV}Cl_2$ in β-cyclodextrin yields inclusion complexes interesting for pharmaceutical use, owing to the increased aqueous solubility of the drug, better oral absorption, and enhanced chemical and physical stability with respect to oxidation by air, sensitivity to light, rate of disproportionation or polymerization and acidic conditions.[51]

An interesting new approach involving heterobimetallic complexes containing platinum linked to titanium by an organic ligand with hard and soft donors has recently been published.[68] For example, a relatively high *in vitro* cytotoxicity has been found for $Cp_2Ti[\mu\text{-}OSO_2(CH_2)_2Ph_2P]_2PtCl_2{\cdot}4H_2O$, suggesting that treatment of cancer cells with a drug containing two disparate metal centers could be more effective than one containing either metal singly.

At present, transferrin is believed to be a potential carrier of Ti(IV) into the target place. The use of these species as chemotherapeutic agents remains relatively unexplored and awaits future investigation.[23]

Another new development is the combination of two pharmacologically active components in one molecular compound. For example, titanocene complexes with thionucleobases of the type $[Cp_2Ti^{IV}(L)]Cl_2$ (for example L = 6-thioguanine, an established anti-cancer drug) have been synthesized, which could render enhanced activity, since they contain two active agents within the same compound.[53] In fact, synergism between titanocene dichloride and the anti-cancer drug 5-fluorouracil has already been reported.[69] A series of platinum complexes of the anti-cancer drug 6-mercaptopurine, reflecting an analogous strategy to improve platinum drugs, has recently been characterized.[70] This strategy is well known in gold drug development.[71]

7.6 Conclusions

Non-platinum metal-based drugs may be considered as an independent group of anti-tumor agents, and there is hope that some of them may improve the prognosis of patients suffering from certain cancer diseases.[16] Budotitane shows good anti-tumor activity against colon tumor models,[18] and titanocene dichloride shows activity against human adenocarcinomas of the stomach and colon, which are insensitive to common cytostatic agents.[23]

Although budotitane and titanocene dichloride have not shown convincing results in human clinical trials I or II up to now, there is still hope that, by modification of either their chemical configuration or the galenic formulation, there could be a new management of tumors, which so far have not been treatable with chemotherapy.

Acknowledgements

The author thanks the Swiss National Science Foundation and the Alfred Werner Foundation of the University of Zürich for financial support.

References

1. F.A. Cotton, G. Wilkinson, C.A. Murillo and M. Bochmann, *Advanced Inorganic Chemistry*, John Wiley, New York, 1999.
2. E.I. Ochiai, in *General Principles of Biochemistry of the Elements*, E. Frieden (Ed.), Biochemistry of the Elements, Vol. 7, Plenum, New York, 1987.
3. A. Bodner, P. Jeske, T. Weyhermüller *et al.*, *Inorg. Chem.*, **31**, 3737–3748 (1992).
4. F. Caruso, M. Rossi and C. Pettinari, *Expert Opin. Ther. Patents*, **11**, 969–979 (2001).
5. E. Frieden, in *Biochemistry of the Essential Ultratrace Elements*, E. Frieden (Ed.), Biochemistry of the Elements, Vol. 3, Plenum, New York, 1984.
6. H.H. Binder, *Lexikon der chemischen Elemente*, S. Hirzel Verlag, Stuttgart, 1999.
7. G. Manivasagam, U.K. Mudali, R. Asokamani and B. Raj, *Corros. Rev.*, **21**, 125–159 (2003).
8. J. Breme, V. Biehl and A. Hoffmann, *Adv. Eng. Mat.*, **2**, 270–275 (2000).
9. J.P. Fryzek, B. Chadda, D. Marano *et al.*, *J. Occupat. Environ. Med.*, **45**, 400–409 (2003).
10. A. Katzer, S. Hockertz, G.H. Buchhorn and J.F. Loehr, *Toxicology*, **190**, 145–154 (2003).
11. C.X. Cui, Y.T. Shen, Y.J. Xu *et al.*, *Rare Met. Mat. Engi.*, **32**, 627–630 (2003).
12. B. Rosenberg, L. VanCamp, J.E. Trosko and V.H. Mansour, *Nature*, **222**, 385–386 (1969).
13. B. Lippert (Ed.), *Cisplatin: Chemistry and Biochemistry of a Leading Anticancer Drug*, Wiley-VCH, Weinheim, Germany, 1999.
14. P. Köpf-Maier and H. Köpf, *Chem. Rev.*, **87**, 1137–1152 (1987).
15. P. Köpf-Maier and H. Köpf, *Struct. Bonding*, **70**, 103–185 (1988).
16. P. Köpf-Maier, in *Metal Complexes in Cancer Chemotherapy*, B.K. Keppler (Ed.), VCH, Weinheim, 259–296 (1993).
17. P. Köpf-Maier, *Eur. J. Clin. Pharmacol.*, **47**, 1–16 (1994).
18. B.K. Keppler, C. Friesen, H.G. Moritz *et al.*, *Struct. Bonding*, **78**, 97–127 (1991).
19. B.K. Keppler, C. Friesen, H. Vongerichten and E. Vogel, in *Metal Complexes in Cancer Chemotherapy*, B.K. Keppler (Ed.), VCH, Weinheim, 297–323 (1993).

20. M. Hartmann and B.K. Keppler, *Comments Inorg. Chem.*, **16**, 339–372 (1995).
21. B.K. Keppler and E.A. Vogel, in *Platinum and Other Metal Coordination Compounds in Cancer Chemotherapy*, H.M. Pinedo and J.H. Schornagel (Eds), Plenum, New York, 1996.
22. M.M. Harding and G. Mokdsi, *Curr. Med. Chem.*, **7**, 1289–1303 (2000).
23. E. Meléndez, *Crit. Rev. Oncol. Hematol.*, **42**, 309–315 (2002).
24. J. Reedijk, *Chem. Rev.*, **99**, 2499–2510 (1999).
25. P. Köpf-Maier, *J. Struct. Biol.*, **105**, 35–45 (1990).
26. M.J. Clarke and M. Stubbs, in *Metals Ions in Biological Systems*, A. Sigel and H. Sigel (Eds), Vol. 32, 727–780, M. Dekker, New York, 1996.
27. M. Guo and P.J. Sadler, *J. Chem. Soc., Dalton Trans.*, 7–9 (2000).
28. M.L. Guo, H.Z. Sun, H.J. McArdle *et al.*, *Biochemistry*, **39**, 10023–10033 (2000).
29. H. Sun, H. Li and P.J. Sadler, *Chem. Rev.*, **99**, 2817–2842 (1999).
30. P. Yang and M.L. Guo, *Coord. Chem. Rev.*, **186**, 189–211 (1999).
31. A. Fujishima, R.X. Cai, J. Otsuki *et al.*, *Electrochim. Acta*, **38**, 153–157 (1993).
32. N.P. Huang, M.H. Xu, C.W. Yuan and R.R. Yu, *J. Photochem. Photobiol.*, **108**, 229–233 (1997).
33. P. Comba, H. Jakob, B. Nuber and B.K. Keppler, *Inorg. Chem.*, **33**, 3396–3400 (1994).
34. N. Serpone and R.C. Fay, *Inorg. Chem.*, **6**, 1835–1843 (1967).
35. E. Dubler, R. Buschmann and H.W. Schmalle, *J. Inorg. Biochem.*, **95**, 97–104 (2003).
36. J. Mattern, B.K. Keppler and M. Volm, *Arzneim.-Forsch.*, **34**, 1289–1290 (1984).
37. M.R. Berger, M.H. Seelig and A. Galeano, in *Metal Complexes in Cancer Chemotherapy*, B.K. Keppler (Ed.), VCH, Weinheim, 325–349, 1993.
38. M.J. Clarke, F. Zhu and D.R. Frasca, *Chem. Rev.*, **99**, 2511–2533 (1999).
39. B.K. Keppler and D. Schmähl, *Arzneim.-Forsch./Drug Res.*, **36**, 1822–1828 (1986).
40. T. Schilling, B.K. Keppler, M.E. Heim *et al.*, *Invest. New Drugs*, **13**, 327–332 (1996).
41. T. Pieper, K. Borsky and B.K. Keppler, in *Top. Biol. Inorg. Chem.*, M.J. Clarke and P.J. Sadler (Eds), Springer, Berlin, 172–199, 1999.
42. H. Köpf and P. Köpf-Maier, *Angew. Chem., Int. Ed. Engl.*, **18**, 477–478 (1979).
43. G. Wilkinson and J.M. Birmingham, *J. Am. Chem. Soc.*, **76**, 4281–4284 (1954).
44. A. Clearfield, D.K. Warner, C.H. Saldarriaga-Molina *et al.*, *Can. J. Chem.*, **53**, 1622–1629 (1975).
45. M.C. Valadares and M.L.S. Queiroz, *Eur. J. Pharmacol.*, **439**, 35–42 (2002).
46. M.L. McLaughlin, J.M. Cronan, T.R. Schaller and R.D.J. Snelling, *J. Am. Chem. Soc.*, **112**, 8949–8952 (1990).
47. G. Mokdsi and M.M. Harding, *J. Inorg. Biochem.*, **86**, 611–616 (2001).
48. P. Köpf-Maier and I.C. Tornieporth-Oetting, *Biometals*, **9**, 267–271 (1996).
49. T.M. Klapötke, H. Köpf and I.C Tornieporth-Oetting, *Organometallics*, **13**, 3628–3633 (1994).
50. G. Mokdsi and M.M. Harding, *J. Inorg. Biochem.*, **83**, 205–209 (2001).
51. S.S. Braga, I.S. Goncalves, M. Pillinger *et al.*, *J. Organometallics Chem.*, **632**, 11–16 (2001).
52. M. Guo, Z. Guo and P.J. Sadler, *J. Biol. Inorg. Chem.*, **6**, 698–707 (2001).
53. E. Meléndez, M. Marrero, C. Rivera *et al.*, *Inorg. Chim. Acta*, **298**, 178–186 (2000).
54. H. Sun, H. Li, R.A. Weir and P.J. Sadler, *Angew. Chem., Int. Ed.*, **37**, 1577–1579 (1998).

55. G. Mokdsi and M.M. Harding, *J. Organomet. Chem.*, **565**, 29–35 (1998).
56. P. Köpf-Maier, *J. Cancer Res. Clin. Oncol.*, **113**, 342–348 (1987).
57. M. Bastaki, E. Missirlis, N. Klouras *et al.*, *Eur. J. Pharmacol.*, **251**, 263–269 (1994).
58. P. Köpf-Maier, *Anticancer Res.*, **19**, 493–504 (1999).
59. M.C. Valadares, S.I. Klein, A.M.A. Guaraldo and M.L.S. Queiroz, *Eur. J. Pharmacol.*, **473**, 191–196 (2003).
60. B. Mueller, S. Lucks, R. Mueller and W. Mohr, Water-Soluble Pharmaceutical Composition Containing Metallocene Complex. EP Patent 407804 (1991).
61. A. Korfel, M.E. Scheulen, H.J. Schmoll *et al.*, *Clin. Cancer Res.*, **4**, 2701–2708 (1998).
62. C.V. Christodoulou, D.R. Ferry, D.W. Fyfe *et al.*, *J. Clin. Oncol.*, **16**, 2761–2769 (1998).
63. K. Mross, P. Robben-Bathe, L. Edler *et al.*, *Onkologie*, **23**, 576–579 (2000).
64. G. Lummen, H. Sperling, H. Luboldt *et al.*, *Cancer Chemother. Pharmacol.*, **42**, 415–417 (1998).
65. N. Kröger, U.R. Kleeberg, K. Mross *et al.*, *Onkologie*, **23**, 60–62 (2000).
66. M. Gielen, *Appl. Organomet. Chem.*, **16**, 481–494 (2002).
67. I. Turel, A. Demsar and J. Kosmrlj, *J. Inclusion Phenom. Macrocycl. Chem.*, **35**, 595–604 (1999).
68. J.L. Wedgwood, R.A. Kresinski, S. Merry and A.W.G. Platt, *J. Inorg. Biochem.*, **95**, 149–156 (2003).
69. C. Christodoulou, A.G. Eliopoulos, L.S. Young *et al.*, *Br. J. Cancer*, **77**, 2088–2097 (1998).
70. F. Arod, H.W. Schmalle and E. Dubler, *Inorg. Chem.*, to be published (2004).
71. E.R.T. Tiekink, *Crit. Rev. Oncol. Hematol.*, **42**, 225–248 (2002).

8 $_{23}$V Insulin-Mimetic Vanadium-Containing Compounds

Tamás Kiss and Tamás Jakusch

8.1 Chemistry of Vanadium

Vanadium is normally present at very low concentrations ($\sim 10^{-8}$ M) in virtually all cells in plants and animals. *In vivo*, given the constraints of standard physiological conditions (pH 3–7, aerobic aqueous solution and ambient temperature), oxidation states IV and V prevail, with thermodynamically plausible species including the anionic vanadate, a mixture of $[HV^{V}O_4]^{2-}$ and $[H_2V^{V}O_4]^{-}$ ($pK_a \sim 8$[1]) and the cationic oxovanadium $V^{IV}O^{2+}$.[2] Vanadium in these higher oxidation states, as a fairly hard metal ion, readily forms complexes with O-donor ligands and conformably binds N and S donors as well, especially if an O donor also coordinates. Interestingly, the typically soft coordination mode $[V^{IV}(S_4)]$ also occurs in minerals (e.g. patronite)[3] and in thiovanadates.[4]

Vanadium easily switches between the oxidation states V and IV. The redox potential at pH 7 for the couple $H_2VO_4^- + 4H^+ + e^- \rightleftharpoons VO^{2+} + 3H_2O$ is -0.34 V and thus lies in the range where oxovanadium or vanadyl (VO^{2+}) are oxidized to vanadate(V) under aerobic conditions, and vanadate is reduced to oxovanadium(IV) by cellular components such as glutathione and other

Metallotherapeutic Drugs and Metal-Based Diagnostic Agents: The Use of Metals in Medicine
Edited by Gielen and Tiekink

Cys-containing oligopeptides/proteins, ascorbate, NADH and phenolic compounds. The physiological action of vanadate is not restricted to the monovanadate anion, but also extends to condensed species present in the physiological pH range at higher vanadate (V) concentrations: (e.g. divanadate, $H_2V_2O_7^{2-}/HV_2O_7^{3-}$; tetravanadate, $V_4O_{10}^{4-}$; pentavanadate, $V_5O_{12}^{5-}$; and decavanadate, $V_{10}O_{28}^{6-}/HV_{10}O_{28}^{5-}$).[5]

An additional point of interest with respect to the possible general significance of vanadium in living systems is the relation between vanadate, $H_2VO_4^-$, and phosphate, HPO_4^{2-}, at physiological pH and concentrations (i.e. 0.2 μM for vanadate and 30 mM for phosphate in human blood plasma). Vanadate is a well-known inhibitor for phosphate-metabolizing enzymes, e.g. phosphatases and kinases, but can also stimulate specific enzymes such as phosphomutases and phosphoisomerases.[6] A general role for vanadate as a regulator of phosphate-metabolizing enzymes may be anticipated on this basis. Inhibition of a protein-tyrosine-phosphatase has been noted as a possible key step for the insulin-mimetic behaviour of vanadate.

8.2 Biological and Medicinal Aspects of Vanadium

Besides the importance of vanadium in life as a co-factor in vanadate-dependent haloperoxidases[5,7] and vanadium nitrogenases,[8,9] vanadium compounds receive attention because of their potential applications in medicine. For example, various groups of vanadium compounds exhibit an anti-carcinogenic potency towards experimentally induced tumours.[10] These agents include among others organovanadium compounds such as vanadocene dichloride, $[Cp_2V^{IV}Cl_2]$, V^{III} complexes of thiolate-containing ligands, peroxovanadates(V), etc.[11–13] Polyoxovanadates, as generally polyoxometallates, exert anti-viral and anti-tumour activity, mostly on the basis of their interactions with phosphate-metabolizing enzymes.[6,14] However, the most widely studied potential medicinal application of vanadium compounds lies in the treatment of the insulin-dependent or insulin-deficient (IDDM, type 1) and the non-insulin-dependent or insulin-tolerant (NIDDM, type 2) diabetes.[15,16]

8.3 The Role of Insulin in Glucose Metabolism

Diabetes mellitus (DM) is defined as a disease that results in chronic hyperglycaemia due to an absolute or relative lack of insulin and/or insulin resistance, which in turn impairs glucose, protein and lipid metabolism, and finally entrains the characteristic secondary complications.[17] According to the definition of WHO, DM is generally classified as either insulin-dependent (IDDM, type 1) or non-insulin-dependent (NIDDM, type 2).[18] To treat NIDDM several synthetic organic therapeutics have already been clinically used involving sulphonylureas, sulphonamides, biguanides and triglydazone, which has recently been developed. However, IDDM can be controlled only by daily injections of insulin.

Insulin is a signalling hormone, which is essential for the metabolism of both carbohydrate and fat. It is secreted by the pancreas in response to elevated levels of glucose in the bloodstream. The increased insulin level then normally promotes glucose uptake by the liver and gut as well as by peripheral tissues (adipose and muscle), which results in energy production and storage as needed by the organism.[19]

The insulin receptor (IR) is an insulin-activated trans-membrane protein-tyrosine-kinase. Following insulin binding, the receptor undergoes activation by autophosphorylation and subsequently phosphorylates several endogenous proteins on tyrosine moieties. Tyrosyl phosphorylation is linked to a serine/threonine phosphorylation state of key enzymatic systems controlling the glucose and fat metabolism. When insulin is removed, termination occurs at several levels, one of which is dephosphorylation of tyrosyl residues by endogenous protein phosphotyrosine phosphatases (PTPase).[20]

Since vanadate is an inhibitor of phosphatases,[21] it was initially believed that vanadate acts intracellularly by blocking the protein phosphotyrosine phosphatases, acting at the IR and therefore activating it in an insulin-independent manner. However, this turned out not to be the case (see p. 153).

The promotion of glucose uptake in a tissue-specific manner can be enhanced by vanadium. Vanadium compounds, however, can never completely substitute for insulin. (The complete lack of insulin precludes the effectiveness of any vanadium compound.) Nevertheless, many of the observed *in vitro* and *in vivo* effects of vanadium are insulin-like.[21]

8.4 Vanadium Complexes with Biological Activity

After the discovery of the insulin-like effects of simple inorganic vanadium(IV/V) salts, due to the low intestinal absorption of vanadyl(IV) and the toxicity of vanadate(V), intensive research and synthetic work were directed to generate alternative vanadium complexes. Some of them are shown in Figure 8.1.[22,23]

In accordance with the inorganic chemistry of vanadium (see p. 143), many of the biologically tested vanadium complexes involve O coordination. Pyrone and pyridinone derivatives are the best, most-studied compounds. The *trans*-$VO(IV)(Cys\text{-}Me)_2$, Figure 8.1, was the first orally active vanadium(IV) complex to treat DM in rats.[24] One complex from this group, bismaltolatooxovanadium(IV),

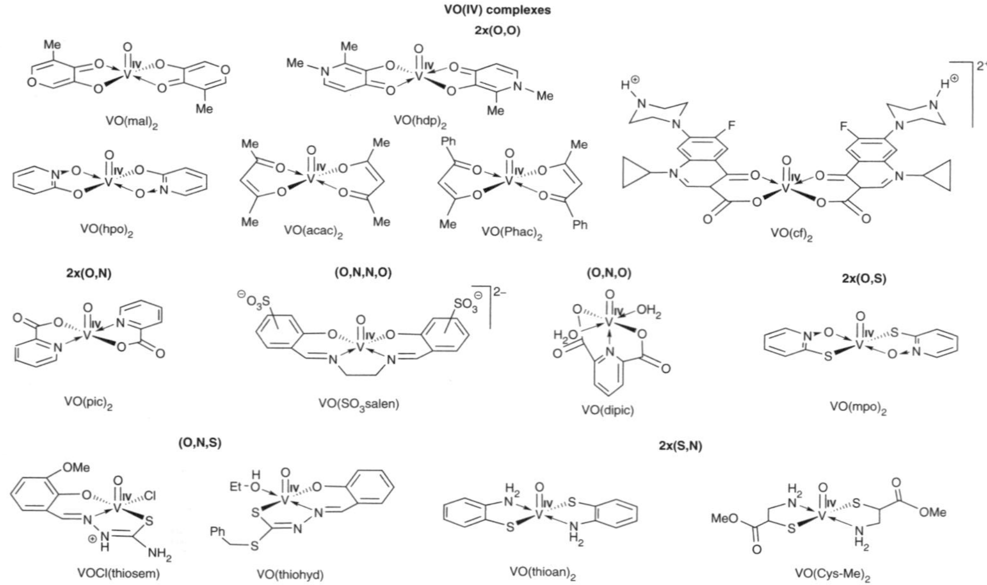

Figure 8.1 Structural formulae of several vanadium complexes prepared and tested for their toxicity and biological activity (based on Ref. 20) $VO(mal)_2$: $V^{IV}O$[bis-maltolate], $VO(hdp)_2$: $V^{IV}O$[bis(1,2-dimethyl-3-hydroxy-4-pyridinone)], $VO(hpo)_2$: $V^{IV}O$[bis(2-hydroxy-pyridine-*N*-oxide)], $VO(acac)_2$: $V^{IV}O$[bis(acetylacetonate)], $VO(Phac)_2$: $V^{IV}O$[bis(benzoylacetonate)], $VO(cf)_2$: $V^{IV}O${bis[1-cyclopropyl-6-fluoro-1,4-dihydroxo-4-oxo-7-(1-piperazinyl)-3-qinolinecarboxylate]}, $VO(pic)_2$: $V^{IV}O$[bis(picolinate)], $VO(SO_3salen)$: $V^{IV}O$[*NN′*-ethylenebis (salicylideneimine-5-sulphonate)], VO(dipic): VIVO-dipicolinate, $VO(mpo)_2$: $V^{IV}O$[bis(2-mercapto-pyridine-*N*-oxide)], VOCl(thiosem): $V^{IV}OCl$(3-methoxysalicylidene-thiosemicarbazonate), VO(thiohyd): $V^{IV}O(OEt)$[salicylidene-(benzylmercaptothiocarbonyl)hydrazonate], $VO(thioan)_2$: $V^{IV}O$[bis(2-aminophenolate)], $VO(Cys\text{-}Me)_2$: $V^{IV}O$[bis(cysteinemethyl ester)], $V(hydraz)_2$: {V^{IV}[bis(phenylacetyla-cetonato-benzoylhydrazone)]}, $(VO)_2(cit)_2$: {$[V^{IV}{}_2O_2(\mu\text{-citrate})_2]$}$^{4-}$, $(VO)_2(OH)_2(UDPglu)_2$: $V^{IV}O\text{-}(\mu\text{-}OH)_2$-(uridylglucose pyrophosphate)$_2$, VO(thiohyd): $V^VO(OEt)$[salicylidene(benzylmercapto-thiocarbonyl)hydrazonate]$^-$, Vanadate: $[H_2V^VO_2]^-$, VO(van-His-Ser): {V^VO (H_2O)[*N*-2-oxido-3-methoxysalicylidene]-histidylserine$^{(2-)}$}, VO_2 (dipic): $[V^VO_2(\text{dipicolinate})]^-$, $(VO)_2(nap\text{-}tea)_2$: $[V^V{}_2O_2${naphthalylidene[hydroxymethyl-bis-(oxymethyl)]aminomethane}$_2]$, VO(py-tris): [V^VO{pyridylidene-tris(methoxy)methylamine}], $VO(O_2)$cmaa: $V^VO(O_2)$[carboxymethyl-aspartate]$^-$, $VO(O_2)_2$Ala-His: $[V^VO$(bis-peroxo)-alanylhistidine$]^-$

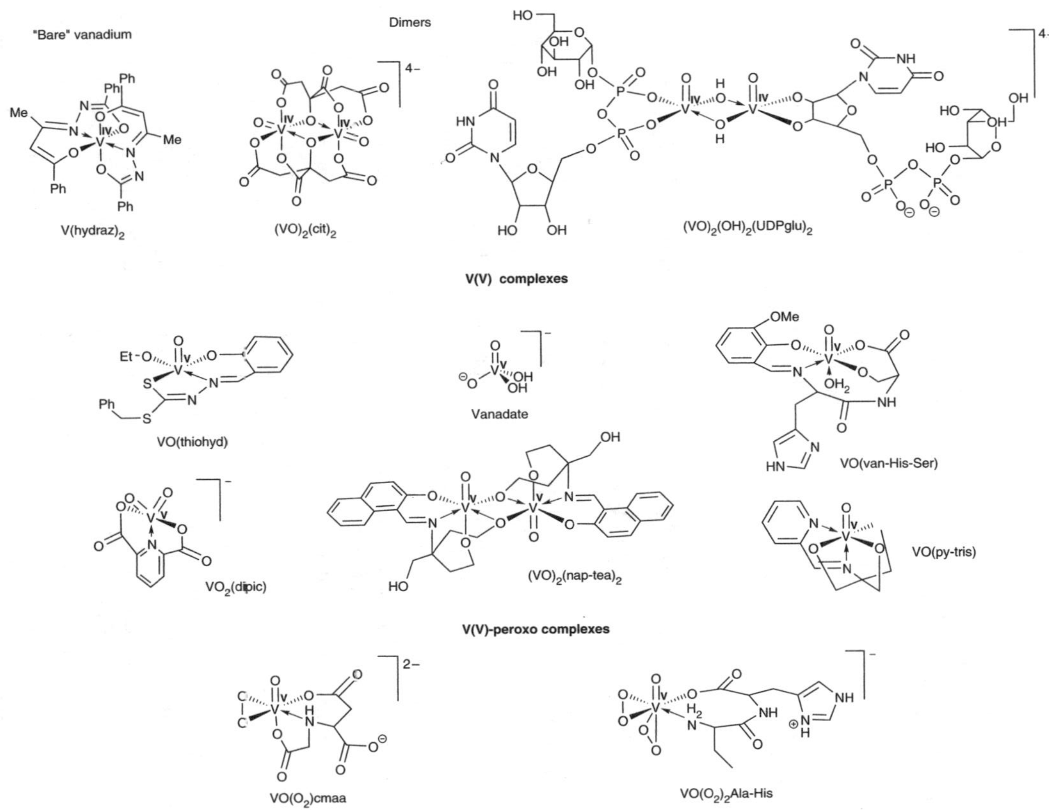

Figure 8.1 (Continued)

Figure 8.1, successfully passed the phase I clinical test.[19] Picolinic acid and its derivatives with N,O coordination,[25] and dipicolinate with O,N,O coordination are also promising ligands, the latter having been successfully applied orally to diabetic cats.[26] Surprisingly, N,S-coordinated complexes have also been found to be efficient in *in vivo* tests,[22] but some of them could not be detected in solution when the components were mixed in a test tube. This may mean that if these complexes are dissolved in water, they rapidly decompose and cannot keep their integrity in the stomach or intestine, where they are expected to be absorbed. The V(V) hydroxylamido complex, an excellent inhibitor of protein tyrosine phosphatase,[27] is an efficient insulin mimetic. In order to be effective,[19] a vanadium compound should fulfil a number of preconditions. Among others, hydrophilicity and lipophilicity should be balanced by an appropriate design of the ligand in order to allow absorption and transport in the bloodstream. It should have a low molecular mass (LMM) in order to cross the cell membrane easily. Of course, an additional demand is low toxicity.

8.5 Biological Activity and Toxicity of Various Vanadium(IV/V) Compounds

Different *in vitro* methods on cell cultures of various types and tissue homogenates have been applied in order to assess the effects of vanadium on glucose[22] and/or fatty acid uptake[28] in comparison with insulin.[29] The results of one of the most comprehensive tests are depicted in Figure 8.2, relating to the findings on mice fibroblast cells.[22]

It can be seen in Figure 8.2 that all the tested complexes (and not only those shown in Figure 8.1) increase glucose uptake with a comparable or higher efficiency than that of insulin in the physiologically or pharmacologically relevant concentration level (0.1–1.0 μM). In this concentration range, none of the tested compounds showed toxicity. VO(IV) complexes (which are generally more toxic than V(V)) tend to be more effective insulin mimetics than V(V) complexes. These observations, together with the very similar biological efficacy of all the tested complexes, corroborate the assumption that the complexes undergo transformation during transport and/or within the cell, thus giving rise to active species different from those originally employed (see p. 151). These *in vitro* studies may be useful for initial screening, but they do not take into account oral bioavailability, which can be tested only *in vivo*. For *in vivo* testing, a number of experimental models of diabetes in rodents have been used. Most widely accepted is the streptozotocin (STZ)-induced diabetic rat.[30] Moderate to good results have been obtained *in vivo* too, with many of the V(IV/V) complexes at oral doses of between 0.1 and 0.7 $mmol\,kg^{-1}\,day^{-1}$.[17,28] As an illustration,[31] the *in vivo* biological effects of several VO(IV) and V(V) compounds on STZ-diabetic rats are given in Figure 8.3.

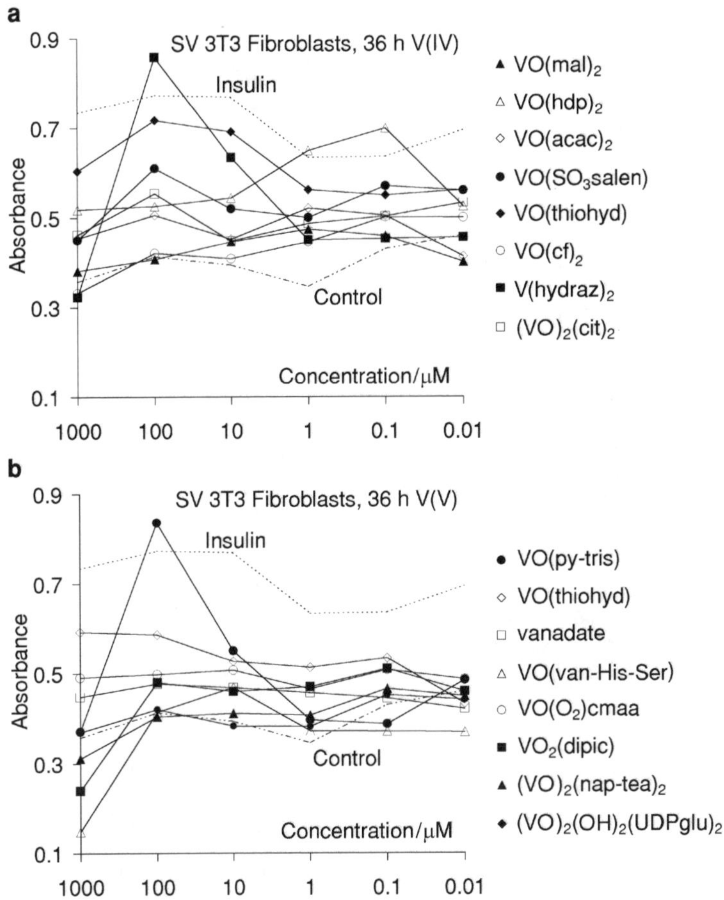

Figure 8.2 Glucose intake by siman virus transformed 3T3 mice fibroblasts in the presence of vanadium compounds. The ordinate is a measure for the reduction equivalents originating from glucose. Data are presented for V(IV) (a) and V(V) complexes (b) separately for cells kept without insulin for 72 h (based on Ref. 22). For abbreviations of the complexes see Figure 8.1

8.6 Speciation of VI(IV) Complexes in Biological Fluids

Due to parallel protonation processes of the metal-binding sites of the coordinating ligands, these complexes certainly decompose in the acidic pH range, e.g. at the pH (~2) of the gastric juice. This is demonstrated in Figure 8.4, in

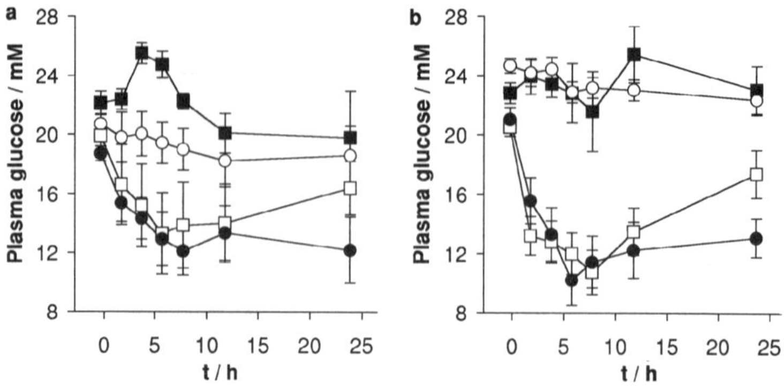

Figure 8.3 Plasma glucose levels for $V^{IV}O^{2+}$ (○), $[V^{V}O_2(mal)_2]^-$ (■), $V^{IV}O(mal)_2$ (●) and $V^{IV}O$(kojic acid)$_2$ (□) following acute (a) oral gavage administration to STZ-diabetic rats at a dose of 0.55 mmol kg^{-1} and (b) intraperitoneal injection in STZ-diabetic rats at a dose of 0.06 mmol kg^{-1} (Reproduced from Ref. 31 with permission from Elsevier)

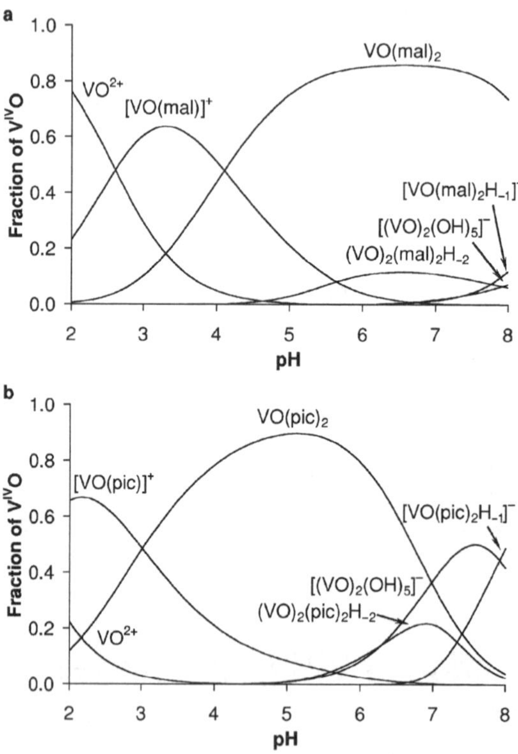

Figure 8.4 Species distribution curves for (a) $V^{IV}O$-maltolato and (b) $V^{IV}O$-picolinato systems as a function of pH, at 1:2 metal ion to ligand ratio, c($V^{IV}O$) = 1 mM (based on data reported in Refs 33–37)

which the species distribution in two well-studied systems, the VO(IV)-maltolato and the VO(IV)-picolinato systems, is depicted as a function of pH.

Accordingly, all other exogenous and endogenous biomolecules present in the stomach or intestine, where the complexes are absorbed, may play a role in VO(IV) binding. This certainly has to be taken into account in the formulation of the drug (e.g. by encapsulation techniques, these problems may well be overcome). The recent results of Sakurai *et al.*[32] support this prediction. After absorption, during their transport in the bloodstream, complex formation with the serum components as active VO(IV) binders also has to be considered. The interactions of several potential insulin-mimetic drugs with the high molecular mass (HMM) protein constituents, e.g. albumin and transferrin (Tf), and some of the LMM constituents (the most potent binders), e.g. lactate, phosphate, oxalate and citrate, have been studied in detail.[33–41] The results of model calculations for serum conditions[41] are depicted in Figure 8.5.

It can be seen in Figure 8.5 that (i) only the pyridinone derivative is a strong enough carrier to preserve a significant proportion of the VO(IV) in the original complex;[37,39] in the other cases, the carrier ligands are displaced by serum components. Accordingly, the most important role of the carrier ligand seems to be to facilitate the absorption of VO(IV). Second, of the two important HMM binders, Tf is much more efficient than albumin and will displace 30–70% of the original carrier from the complex. (At the biologically more

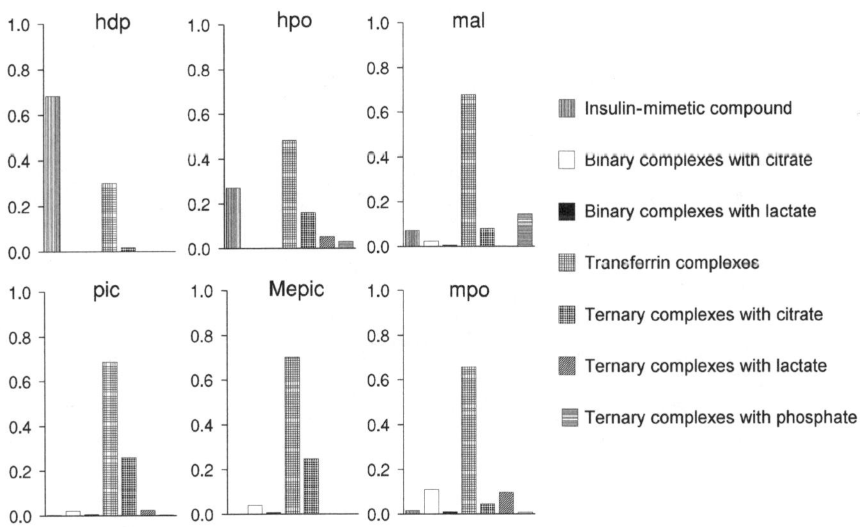

Figure 8.5 Speciation of various insulin-mimetic VO(IV) compounds (100 μM) in serum at pH 7.4 (mal: maltolate, pic: picolinate, Mepic: 6-Me-picolinate, hpo: 2-HO-pyridine-*N*-oxide, mpo: 2-HS-pyridine-*N*-oxide, hdp: 1,2-dimethyl-3-hydroxy-4-pyridinone) (based on data reported in Refs 33–37)

relevant VO(IV) concentrations (<5 mM), practically all the VO(IV) is bound to Tf.) Third, among the LMM binders, citrate is the only 'active' component able to influence the solution state of these insulin mimetics. At physiological pH, VO(IV) exists mostly as the VO(IV)–citrate binary complex and as the VO(IV)–ligand A–citrate mixed ligand complex, but in different proportions.

The decisive role of Tf in VO(IV) binding can be demonstrated by EPR. It has been found[40–42] that in the presence of apoTf, all carrier complexes showed anisotropic EPR spectra even at room temperature (because of the slow tumbling motion of the VO(IV)–protein complex, the parallel and the perpendicular components could not be averaged), unambiguously indicating interactions between the carrier complex of VO(IV) and the protein. As seen in Figure 8.6, in the case of $VO(pic)_2$, the RT spectrum is practically the same as that of VO(IV)–apoTf, suggesting a complete displacement of the carrier ligand by Tf. At the same time, in the case of the VO(IV)–$(hdp)_2$ complex, signals of the anisotropic carrier complex can also be observed, suggesting only partial displacement of the carrier ligand. These results are in complete agreement with the species distributions depicted in Figure 8.5.

These results strongly indicate that vanadium must be released from the carrier compound in order for it to be pharmacologically active. Perhaps the most convincing point is the different pharmacokinetics of disappearance of vanadium and the carrier ligand.[43] Although, in principle, the ligand should be present at roughly twice the concentration of vanadium in the blood, it is found to be present at a lower concentration compared with vanadium from the

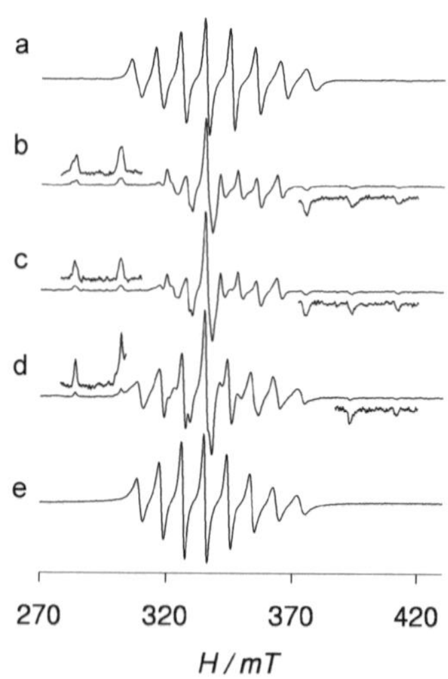

Figure 8.6 Room temperature EPR spectra of (a) VO(IV)$(pic)_2$, (b) VO(IV)-apoTf, (c) VO(IV)$(pic)_2$-apoTf, (d) VO(IV)$(hdp)_2$-apoTf and (e) VO(IV)$(hdp)_2$ at pH 7.4, hdp: 1,2-dimethyl-3-hydroxy-4-pyridinone

1-hour time point onwards. Thus, a number of recent studies have demonstrated fairly conclusively that insulin-mimetic vanadium complexes dissociate rapidly once ingested or injected. In other words, the eventual *in vivo* metabolic fate for chelated vanadium complexes differs little, if at all, from that of non-complexed vanadium compounds, when they are administered orally.[41]

8.7 Possible Mechanism for *in vivo* Vanadium Action

Numerous studies have demonstrated that vanadium mimics nearly all the documented actions of insulin in all insulin-responsive cells and tissues (Table 8.1).

The main questions concerning how vanadate may influence the cellular metabolism are as follows: (i) What are the active forms and effective concentrations of intracellular vanadium? (ii) What is the evidence that vanadium acts as an insulin mimetic through its effects on protein tyrosine phosphorylation?

NMR studies by Sakurai *et al.*[44] show that almost independently of the oxidation state of the vanadium compound, the metal is transported in the blood serum in the IV oxidation state. The binding of VO(IV) to ligands (mostly Tf[37]) prevents its oxidation to vanadate(V), which would otherwise occur rapidly at intracellular pH. Nevertheless, oxidation may occur resulting in the formation of a limited amount of free vanadate. Accordingly, vanadium may be assumed to enter the cell, either in oxidation state IV, through the Tf receptor following the iron pathway, or in oxidation state V, through the phosphate or sulphate pathway. In the intracellular medium, reducing agents can redox-interact with vanadate(V). A frequently discussed candidate for the reduction is glutathione (GSH),[45] although it is a rather ineffectual reducing agent.[46,47] To what extent such redox interactions take place largely depends on

Table 8.1 Some insulin-like actions mediated by vanadium in various insulin-responsive tissues (based on Ref. 8)

Activity	Direction of effect	Target tissue
Hexose transport	Stimulation	skeletal muscle rat adipocytes
Lipogenesis	Stimulation	rat adipocytes
Glucose oxidation	Stimulation	rat adipocytes
Lipolysis	Inhibition	rat adipocytes
Glycogen synthase	Stimulation	skeletal muscle rat adipocytes rat hepatocytes

the stabilization of vanadium in the oxidation state V or IV through the complexation of cell constituents such as GSH, GSSG, ATP, etc. A high intracellular excess of GSH increases the formation possibility of VO(IV) and its complexation with either GSH or GSSG. Both have been shown to be reasonably potent binders for VO(IV).[45,48,49] Other effective reducing agents, such as NADH or ascorbate, may cause the formation of V(III) species.[50,51] Further hydrolytic degradation of VO(IV) may be responsible for the re-oxidation to vanadate(V). Via the redox and complexation reactions, the finely tuned speciation of vanadium may lead to the efficiency of the metal and thus to mimic the effects of insulin. The question of whether the original complex may remain partially intact, i.e. whether it keeps the original carrier ligand at least partially bound to vanadium(IV,V), or whether the endogenous binders of the cells and biological fluids completely displace them, still remains unanswered, although the speciation results discussed in the previous section strongly suggest the former possibility. The possible speciation pathways of vanadium compounds are illustrated in Figure 8.7.

It is commonly accepted that the effects of vanadium (mostly vanadate) on the tyrosine phosphorylation of cellular proteins may account for its biological actions.[50] Because of the importance of protein tyrosine phosphorylation in the actions of insulin (initiated by the receptor tyrosine kinase), it is reasonable to believe the PTPase activity plays a crucial role. Indeed, numerous studies on intact cells and in *in vivo* investigations have shown that the administration of vanadium greatly enhances cellular protein tyrosine phosphorylation.[52–55] However, how PTPase inhibition may lead specifically to some (but not all)

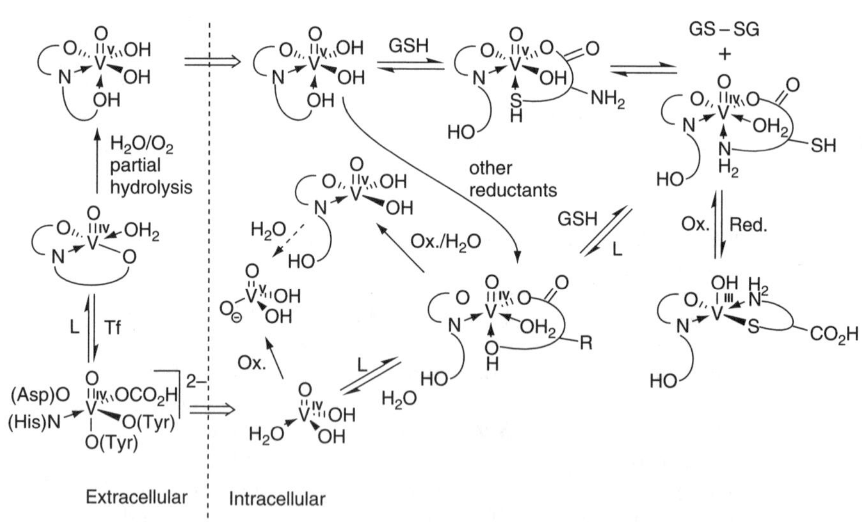

Figure 8.7 The possible speciation pathways of vanadium compounds (based on Ref. 22)

insulin-like metabolic effects *in vivo* without affecting other cell functions that are also controlled by protein tyrosine phosphorylation mechanisms[20] remains unclear. First, it was hypothesized that vanadium exerts its effects via the IR.[56] However, later, it was found that the stimulatory effects of vanadium on glucose uptake and lipogenesis in rat adipocytes were unaffected by IR tyrosine kinase inhibition, which fully blocked effects mediated by insulin.[57] IR-independent effects of vanadium may be explained by the general phosphatase inhibition, but other possibilities have also been suggested. Nonetheless, there remains a need to establish which, if any, of the key intracellular protein tyrosine phopshorylation events are reproduced by vanadium *in vivo* and whether a novel and selective vanadium-sensitive protein tyrosine kinase or PTPase plays important roles.[20] Recently, the possible contribution of nitric oxide (NO) to the action of vanadium has been proposed, in an effort not only to improve the condition of diabetics, but to prevent the onset of DM.[58] This observation might open a new insight into understanding the mechanism of vanadium *in vivo*.

8.8 Conclusion

Both inorganic and chelated vanadium compounds have been demonstrated by a great many *in vitro* and *in vivo* observations to have insulin-mimetic activity. Vanadium compounds have been used to improve STZ-induced IDDM in experimental animals, and vanadyl sulphate has been shown to improve human NIDDM.[59,60] This encourages researchers in the field of vanadium chemistry, biochemistry and physiology to investigate this important issue. Despite the potential of vanadium compounds in the oral treatment of both types of diabetes, only one compound, the VO(IV)–maltolato complex, has been introduced in clinical tests on humans.[7] There are several reasons why pharmaceutical companies have so far shown only marginal interest in vanadium compounds for the treatment of diabetes. Among the obvious reasons is the toxicity of vanadium compounds. There is a general feeling that drugs containing heavy metals will not be accepted by the market (cf. gold and platinum complexes in practical use in medicine). It is also objected that vanadium acts in a non-specific way (as a phosphatase inhibitor), and thus cannot be accepted as a drug. However, there are many drugs in daily use that act in similar non-specific ways (e.g. the already mentioned gold or platinum complexes). However, the main obstacle for a successful clinical introduction appears to be a possible interference and/or competition with other, though more or less ineffectual developments in the area of anti-diabetic treatments.[61]

The most pressing research needs for an improvement of the pharmacological utility of vanadium-containing compounds lie in the areas of comparative tissue uptake, distribution and excretion for a variety of vanadium complexes, and a better knowledge of their metabolic fate. This information will lead to more focused efforts in the developments of new vanadium-containing insulin-mimetic compounds. These results may also change the attitude of clinicians and pharmaceutical experts towards these heavy-metal-containing drugs. As written in one of Orvig's review papers:[19]

> In conclusion, the desirable features of vanadium containing insulin-mimetic agents, including oral activity and potential targeted tissue uptake, tend to outweigh the potential drawbacks, e.g. tissue accumulation and uncertainty of initial dose determination, such that, for the foreseeable near future (10–20 years), these compounds present an opportunity for substantial advance over available diabetic therapies. Whether alone or in combination with other metals, vanadium containing anti-diabetic agents are set to revolutionize the way we look at diabetes.

Acknowledgements

The authors are grateful to Professors H. Sakurai and A. Odani for reading and commenting on this chapter. The work was supported by Hungarian Science Research Fund (OTKA T037385/2002) and the Research Grant of Nagoya University.

References

1. K. Elvingson, A. Gonzales Baró, L. Pettersson, *Inorg. Chem.*, **35**, 3388 (1996).
2. D.W. Boyd, K. Kustin, *Inorg. Biochem.*, **6**, 311 (1984).
3. N.N. Greenwood, A. Earnshow, *Chemistry of the Elements*, 2nd edn, Butterworth-Heinemann, Oxford, 1997, p. 977.
4. N.N. Greenwood, A. Earnshow, *Chemistry of the Elements*, 2nd edn, Butterworth-Heinemann, Oxford, 1997, p. 988.
5. D. Rehder, *Met. Ions Biol. Syst.*, **31**, 1–44 (1995).
6. P.J. Stankiewicz, A.S. Tracey, *Met. Ions Biol. Syst.*, **31**, 249–286 (1995); P.J. Stankiewicz, A.S. Tracey, D.C. Crans, *Met. Ions Biol. Syst.*, **31**, 287–324 (1995).
7. A. Butler, A.H. Baldwin, *Struct. Bond.*, **89**, 109–195 (1997).
8. R. Renirie, W. Hemrika, R. Wever, *J. Biol. Chem.*, **275**, 11650–11654 (2000).
9. R.R. Eady, *Chem. Rev.*, **96**, 3013–3076 (1996).
10. M.J. Dilworth, A.R. Glenn (Eds), *Biology and Biochemistry of Nitrogen Fixation*, Elsevier, Amsterdam, 1991.
11. A.M. Evangelou, *Crit. Rev. Oncol. Hematol.*, **42**, 249–265 (2002).
12. P.Köpf-Maier, W. Wagner, B. Hesse, H. Köpf, *Eur. J. Cancer*, **17**, 665–671 (1981).
13. C. Djordjevic, G.L. Wampler, *J. Inorg. Biochem.*, **25**, 51–58 (1985).
14. M.T. Pope, A. Müller (Eds), *Polyoxometalates: From Platonic Solids to Anti-Retroviral Activity*, Kluwer Academic Publ, Dordrecht, 1994.

15. E.L. Tolman, E. Barris, M. Burns *et al.*, *Life Sci.*, **25**, 1159–1162 (1979).
16. Y. Sechter, S.J.D. Karlish, *Nature*, **284**, 556 (1980).
17. M.A. Atkinson, N.K. Maclaren, *Sci. Am.*, **260**, 42–49 (1990).
18. Diabetes mellitus: Reports of a WHO study group, *WHO Technical Report Series*, **727**, 1985, p. 10.
19. K.H. Thompson, J.H. McNeill, C. Orvig, *Chem. Rev.*, **99**, 2885–2891 (1999).
20. I. Goldwaser, D. Gefel, E. Gershonov *et al.*, *J. Inorg. Biochem.*, **80**, 21–25 (2000).
21. M.C. Cam, R.W. Brownsey, J.H. McNeill, *Can. J. Physiol. Pharmacol.*, **78**, 829–847 (2000).
22. D. Rehder, J. Costa Pessoa, C.F.G.C. Geraldes, M.M.C.A. Castro, T. Kabanos *et al.*, *J. Biol. Inorg. Chem.*, **7**, 384–396 (2002).
23. H. Sakurai, H. Yasui, Y. Adachi, *Expert Opin. Invest. Drug*, **12**, 1189–1203 (2003).
24. H. Sakurai, T. Tsuchiya, M. Nukatsuka, *J. Clin. Biochem. Nutr.*, **8**, 193–200 (1990).
25. H. Sakurai, H. Yasui, *J. Trace Elem. Exp. Med.*, **16**, 269–280 (2003).
26. D.C. Crans, *J. Inorg. Biochem.*, **80**, 123–131 (2000).
27. C. Cuncic, S. Desmarais, N. Detich *et al.*, *Biochem. Pharmacol.*, **58**, 1859–1867 (1999).
28. H. Sakurai, *Chem. Rec.*, **2**, 237–248 (2002).
29. I.G. Fantus, F. Ahmad, G. Deragon, *Diabetes*, **43**, 35–42 (1994).
30. R.H.J. Bell, R.J. Hye, *J. Surg. Res.*, **35**, 433–441 (1983).
31. V.G. Yuen, P. Caravan, N. Glover *et al.*, *J. Inorg. Biochem.*, **68**, 109–116 (1997).
32. H. Sakurai, J. Fugono, H. Yasui, *Mini-Rev. Med. Chem.*, **4**, 41–48 (2004).
33. T. Kiss, E. Kiss, G. Micera, D. Sanna, *Inorg. Chim. Acta*, **283**, 202–210 (1998).
34. P. Buglyo, E. Kiss, I. Fabian *et al.*, *Inorg. Chim. Acta*, **306**, 174–183 (2000).
35. E. Kiss, K. Petrohan, D. Sanna *et al.*, *J. Inorg. Biochem.*, **78**, 97–108 (2000).
36. T. Kiss, E. Kiss, E. Garribba, H. Sakurai, *J. Inorg. Biochem.*, **80**, 65–73 (2000).
37. P. Buglyo, T. Kiss, D. Sanna, G. Micera, *J. Chem. Soc., Dalton Trans.*, 2275–2282 (2002).
38. W.R. Harris, *Clin. Chem.*, **38**, 1809–1818 (1992).
39. H. Sakurai, A. Tamura, J. Fugano *et al.*, *Coord. Chem. Rev.*, **245**, 31–37 (2003).
40. H. Yasui, Y. Kunori, H. Sakurai, *Chem. Lett.*, **32**, 1032–1033 (2003).
41. T. Kiss, T. Jakusch, S. Bouhsina (in preparation).
42. K.H. Thompson, Y. Tsukuda, Z. Xu *et al.*, *Biol. Trace Elem. Res.*, **86**, 31–44 (2002).
43. K.H. Thompson, B.D. Liboiron, G.R. Hanson, C. Orvig, *ACS Monograph* (in press).
44. H. Sakurai, S. Shimomura, K. Fukazawa, K. Ishizu, *Biochem. Biophys. Res. Commun.*, **96**, 293–298 (1980).
45. T. Goda, H. Sakurai, T. Yashimura, *Nippon Kagaku Kaishi*, 654–661 (1998).
46. J. Li, G. Elberg, D.C. Crans, Y. Sechter, *Biochemistry*, **35**, 8314–8318 (1996).
47. I.G. Macara, K. Kustin, L.C. Cantley, *Biochim. Biophys. Acta*, **629**, 95–106 (1980).
48. J. Costa Pessoa, I. Tomaz, T. Kiss, P. Buglyo, *J. Inorg. Biochem.*, **84**, 259–270 (2001).
49. J. Costa Pessoa, I. Tomaz, T. Kiss *et al.*, *J. Biol. Inorg. Chem.*, **7**, 225–240 (2002).
50. A. Stern, A.J. Davison, Q. Wu, J. Moon, *Arch. Biochem. Biophys.*, **299**, 125–128 (1992).
51. K. Kanamori, Y. Kinebuchi, H. Michibata, *Chem. Lett.*, **26**, 423–424 (1997).
52. I.G. Fantus, E. Tsaini, *Mol. Cell. Biochem.*, **182**, 109–119 (1998).
53. Y.R. Hadari, E. Tzahar, O. Nadiv *et al.*, *J. Biol. Chem.*, **267**, 17483–17486 (1992).

54. P. Defilippi, S.F. Retta, C. Olivo *et al.*, *Exp. Cell Res.*, **221**, 141–152 (1995).
55. U.O. Wenzel, B. Fuoqueray, P. Biswas *et al.*, *J. Clin. Invest.*, **95**, 1244–1252 (1995).
56. I.G. Fantus, F. Ahmad, G. Deragon, *Endocrynology*, **127**, 2715–2725 (1990).
57. A. Shiseva, Y. Shechter, *Biochemistry*, **31**, 8059–8063 (1992).
58. A. Tsuji, H. Sakurai, *Biochem. Biophys. Res. Commun.*, **226**, 506–511 (1996).
59. N. Cohen, N. Halberstam, P. Shlimovich *et al.*, *J. Clin. Invest.*, **95**, 2501–2509 (1995).
60. M. Halberstam, N. Cohen, P. Shlimovich *et al.*, *Diabetes*, **45**, 659–666 (1996).
61. D. Rehder, *Inorg. Chem. Commun.*, **6**, 604–617 (2003).

9 $_{25}$Mn Manganese Metallotherapeutics

Jeanne H. Freeland-Graves, Tanushree Bose and Abbass Karbassian

Manganese (Mn) is a transition metal whose name is derived from the Greek word 'magic.' This element is a conundrum of nature in that it is both essential and toxic.[1] The dichotomy is illustrated by its critical role in growth and metabolism versus the debilitating consequences of high exposures. Concern over toxicity and the difficulty in measuring its miniscule concentrations in tissues has limited therapeutic uses until recently.

9.1 Prevalence in the Environment

Manganese is present in the environment as the 12th most prevalent element on earth. The most common of the 11 oxidation states is +2, although it can range from −3 to +7. Natural forms are in sedimentary deposits, iron ores, crude oil and volcanic dust. It is not found as a free metal, but in salts such as oxides, carbonates and silicates. An oxide, pyrolusite (MnO_2), is the most widespread form.[2]

Commercially, manganese is a key component in the metallurgical manufacturing of steel, iron and aluminum and the production of glass, ceramics, dyes, leather and textiles, dry-cell batteries and fertilizer. Newer uses include an anti-fungus compound for foliage (manganese ethylene-bis-dithiocarbamate), an enhancing agent for magnetic resonance imaging, and an anti-knock additive

Metallotherapeutic Drugs and Metal-Based Diagnostic Agents: The Use of Metals in Medicine
Edited by Gielen and Tiekink

[methylcyclopentadienyl manganese tricarbonyl (MMT)] in gasoline.[3] The United States (US) consumes over 500 000 tons of manganese annually, greater than any country in the world.[4]

9.2 Diet and Water

Typical diet intakes of manganese in the US range from 0.7 to 10.9 mg/day, with vegetarians consuming even higher levels.[5,6] Current US recommendations are an adequate intake of 2.3 mg for men and 1.8 mg for women, and a tolerable upper limit of 11 mg.[5]

Foods high in manganese that provide ≥1 mg per serving include nuts, seeds, whole grains, pineapple and tea.[7] Environmental Protection Agency (EPA) levels for manganese in drinking water are set at 0.05 mg/l for aesthetics of taste and stain, rather than health concerns.[8]

9.3 Functions

In contrast to the environment, quantities of manganese in mammalian tissue are scant. Yet this mineral is critical for bone mineralization, synthesis of polysaccharides, amelioration of oxidative stress, and protein and energy metabolism.[9,10] It is essential as a co-factor for manganese superoxide dismutase (SOD), arginase, glutamine synthetase and phosphoenolpyruvate decarboxylase.[5] In addition, manganese is an activator or regulator for a multiplicity of enzymes such as transferases, hydrolases, kinases and decarboxylases,[9,10] but other metals such as magnesium may substitute. This chapter will focus primarily on the MnSOD enzyme, as it is the manganese-containing enzyme with the greatest clinical significance.

9.4 MnSOD

Manganese superoxide dismutase (SOD2) scavenges toxic superoxide radicals that are formed in normal cellular respiration, as well as those generated by polymorphonuclear leukocytes, endothelial cells and mitochondrial electron flux.[11] The protein is located in the mitochondrial matrix, where it catalyzes the dismutation of superoxide O_2^- in the following reactions:

$$O_2^- + Mn^{n+1} \rightarrow O_2 + Mn^{n+}$$

$$H^+ + HO_2 + Mn^{n+} \rightarrow H_2O_2 + Mn^{n+1}$$

Mn^{n+1} is the oxidized metalloenzyme; Mn^{n} is the reduced form. Oxidized MnSOD reacts with superoxide to convert it to hydrogen peroxide (H_2O_2) and molecular oxygen. The resultant H_2O_2 is converted to water and molecular oxygen by glutathione peroxidase or catalase.[12] Other isoforms of SOD also exist: copper zinc (CuZn) SOD (SOD1) is localized in the cytoplasm and nucleus, and CuZn SOD (SOD3) in extracellular spaces. However, it appears that it is the Mn form of SOD that is critical for mammals. Experiments with transgenic mice provide evidence that only the MnSOD knockout is fatal. Reasons for mortality were cardiomyopathy, liver failure, mitochondrial abnormalities, metabolic acidosis and oxidative DNA damage.[13]

The production of superoxide can be beneficial as it promotes survival by protecting from invading organisms, but an overabundance can lead to a variety of pathological consequences. In healthy individuals concentrations of superoxide are kept in balance by spontaneous and SOD-catalyzed dismutations. Stimuli that regulate SOD2 (cytokines, protein kinase stimulating agents, platelet-derived growth factors and manganese ions) in breast cancer are reviewed by Zelko *et al.*[14] In disease states, the SODs are overwhelmed or inactivated, and fail to sustain a normal cell burden of radical oxygen species (ROS). Oxidative stress results when the balance between the generation and removal of the ROS is altered.

In metabolism SOD2 is inactivated by the action of beneficial nitric oxide (NO). This reaction causes the production of cytotoxic peroxynitrite (ONOO—) which, in turn, nitrates SOD2 and renders it inactive. Peroxynitrites also deactivate steroid receptors by nitration and induce lipid peroxidation, DNA damage and cytotoxicity.[11] Pathophysiologies occur when the production of superoxide exceeds the capacity of SOD to remove it, and oxidative damage occurs from the elevated ROS.[14] SOD2 is the initial defense against the detrimental effects of superoxide radicals and modulates the consequences of nitric oxide.[15] Its role in pathologies appears to be related primarily to the oxidation of lipids, DNA and proteins.[13]

9.5 Deficiency

9.5.1 Animals

Deficiencies of manganese in animals are diverse, involving impaired reproduction, testicular degeneration and growth; skeletal abnormalities of the bone and cartilage; altered lipids with reductions in high density lipoprotein (HDL) cholesterol; irregularities of carbohydrate metabolism, particularly related to insulin;[16] defects in ocular tissue;[17] and altered insulin-like growth factor and protein metabolism. Additional information is found in excellent reviews by Keen *et al.*[10,18]

9.5.2 Humans

Manganese deficiency in humans is uncommon but it has been reported in several laboratory-induced experiments.[16] The first was by Doisey who inadvertently fed a low-manganese diet (0.34 mg/day) for 3.5 months to a subject on a semi-purified vitamin K-deficient diet. Symptoms that were not ameliorated by restoration of vitamin K were reddening of a black beard and hair, diminished hair and nail growth, hypocholesterolemia and a scaly, transitory dermatitis.

The last two symptoms were observed also in a metabolic study by the author.[19] Seven college men were fed a semi-purified diet containing 135 mg cholesterol and 2.59 mg Mn/day for 3 weeks, which then decreased to 0.1 mg Mn. After 35 days, 5 of the 7 subjects developed *Miliaria crystallina*, a scaling dermatitis covering the upper torso, groin and lower legs.[20] This dermatitis vanished with manganese repletion. It was theorized that manganese may have been insufficient for the activation of glycosyltransferases, enzymes that add sugars to protein in the formation of collagen. Total cholesterol also declined, presumably related to the requirement of manganese at sites in the biosynthesis of cholesterol.[21] Finally, serum calcium (Ca), phosphorus and alkaline phosphatase increased, suggesting that manganese was being mobilized from stores in bone. The loss of manganese from bone is not surprising as skeletal abnormalities have been observed in manganese deficiencies in chickens, swine and ruminants,[18] and in one child on total parenteral nutrition (TPN) without manganese for 4 years.[22]

Poor manganese status has been implicated in a variety of disease states and metabolic disorders, including osteoporosis, congenital malformations, TPN, Perthes' disease (necrosis of femur head), senile cataracts, epilepsy, amyotrophic lateral sclerosis (degenerative disease of motor neurons), acromegaly (pituitary disorder of excess growth of extremities and face), hip abnormalities, multiple sclerosis,[23] phenylketonuria (congenital defect causing mental retardation) and maple syrup urine disease (neurodegenerative disease of amino acids in urine).[8] One exacerbating factor is the iron intake and status of an individual, as both iron and manganese share a common transport protein in the blood (transferrin) and cell (divalent metal transporter: DMT1). In young women, a supplement of 60 mg Fe/day for 4 months decreased activity of leukocyte manganese.[24] Also, manganese absorption was lowered in women with high ferritin, the iron storage protein.[25]

9.6 Toxicity

Manganism, or manganese neurotoxicity from inhalation of Mn-contaminated dust, has been observed in miners, welders and workers in alkaline battery and ferro-alloy production plants. Initially this Parkinson-like syndrome exhibits fatigue, poor memory, hallucinations, anorexia, impotence and muscular pain in the lower extremities. This is followed by speech disorders, clumsiness in movements and loss of facial expressions. Finally, muscular rigidity, staggering gait and tremor of the upper limbs occur.[26] The affinity of manganese to the basal ganglia in the brain suggests that this mineral might have a role in the pathogenesis of hepatic encephalopathy.[27] Another effect of prolonged exposure might be impotence, as sperm motility and sperm counts were reduced in mice given high manganese doses.[28]

Individuals who may be sensitive to toxicity include those with liver dysfunction and iron or calcium deficiencies or on TPN, as well as pregnant women, the aged, neonates and children with immature organs.[9,25] No evidence exists that diets high in manganese-containing foods induce toxicity or that it is carcinogenic.[2] A review by Finley concluded that the risk from low environmental exposure is weak and does not justify the concern, although others may not agree.[2,29]

9.7 Therapeutic Manganese-Related Agents

The first known therapeutic use of manganese was the treatment of psoriasis, in combination with vitamin B_1 in 1966.[30] Since that time, manganese salts, enzymes, related genes and synthetic antioxidants have been used to treat disorders involving arthritis, cancer, dermatitis, osteoporosis and ischemia/reperfusion injury (lack/return of blood) as reviewed below. Other possible therapeutic targets associated with manganese drugs in the future may involve treatment of cardiovascular diseases, diabetes, epilepsy, acquired immuno deficiency syndrome (AIDS), peritoneal adhesions (scars of membrane surrounding the abdominal cavity) and inflammatory pain.

Initial research focused on the use of manganese-rich herbs and manganese salts in disorders associated with lower manganese status. But the purity, stability and/or poor reaction rates of simple complexes limit their usefulness. Investigations then focused on the role of SOD2 in pathologies as it became clear that oxidative stress is a common contributor to numerous disease states. However, the native form of the enzyme had major pitfalls for clinical applications, including insolubility, proteolytic digestion, defective transport into cells or the brain and immune reactivity.[31] For example, a CuZn isoform of SOD (SOD1), Orgotein®, was tested in clinical trials. Early results showed positive effects in rheumatoid- and osteoarthritis, and reduced side effects of chemotherapy and radiation therapy.[11] However, adverse immunological events occurred, presumably due to the bovine source of the enzyme. Thus, Orgotein® was withdrawn from clinical use except in Spain. Since then lower molecular weight, synthetic forms called SOD mimetics (SODm) have been developed and used for therapeutic purposes.

The three general types of SODm-containing manganese are the Mn(III) metalloporphyrins, Mn(III) salens and Mn(II) (pentaazamacrocyclic ligands)-based complexes. A fourth type, nitroxide, is outside the scope of this chapter as it does not contain manganese. Metalloporphyrin complexes contain manganese instead of iron at the core. These compounds are very stable and are potent inhibitors of oxidative stress that results from lipid peroxidation.[15] Mn(III) salens are free radical scavengers that remove superoxide, but also exhibit catalase activity. Salens investigated in pathologies related to oxidative stress include manganese *N*,*N*-bis(salicylidene)ethylenediamine chloride (EUK-8) and manganese 3-methoxy-*N*,*N*-bis(salicylidene)ethylenediamine chloride (EUK-134) (Figure 9.1). These dual-functioning SODm reduce oxidative injury from superoxide and H_2O_2 effectively.[32] Protective effects of Mn salens have been reported in models of respiratory distress syndrome, neurodegenerative diseases, amyotrophic lateral sclerosis, cardiomyopathies (diseases of heart muscle) and ischemia/reperfusion.[11] Yet a limitation of salen compounds is their non-selectivity for ROS, such that the relative influence of the specific types of ROS cannot be discerned.

A more selective type of SODm is the low molecular weight Mn(II) (pentaazamacrocyclic ligands)-based complex. An example of the non-peptide SOD mimetic is M40403, a bis (cyclo-hexylpyridine-substituted) macrocyclic ligand as shown in Figure 9.2.[33] This artificial antioxidant is more potent than natural forms of other antioxidants, with enhanced pharmacokinetics and stability. Its advantage is its selectivity, in that it scavenges superoxide anions without removing other compounds of interest such as NO, H_2O_2 or hypochlorite.[34]

The disadvantage of native SOD2 is that the protein cannot bind to the surface of the endothelial cell. This adherence to the cell surface is important for the protective effect of SOD on the membrane-associated proteins (NOX). The NOX proteins produce superoxide and H_2O_2 as a defense mechanism, and for cell signaling in smooth muscle and endothelial cells.[35] Thus, both inflammatory cells and vascular cells create superoxide ions that lead to oxidative damage. Recently a chimeric recombination superoxide dismutase was constructed as a fusion gene expressed in *Escherichia coli* that combines the best features of SOD2 and SOD3. The chimera SOD2/3 consists of the coding

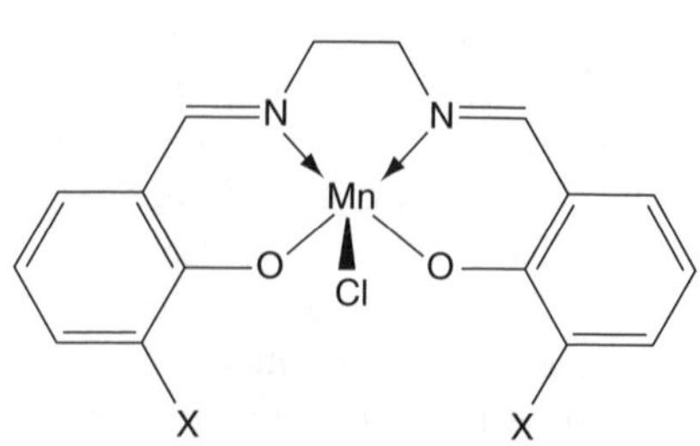

Figure 9.1 SOD mimetic Mn(III) salens. In EUK 8, X is H; in EUK-134, X is OCH_3

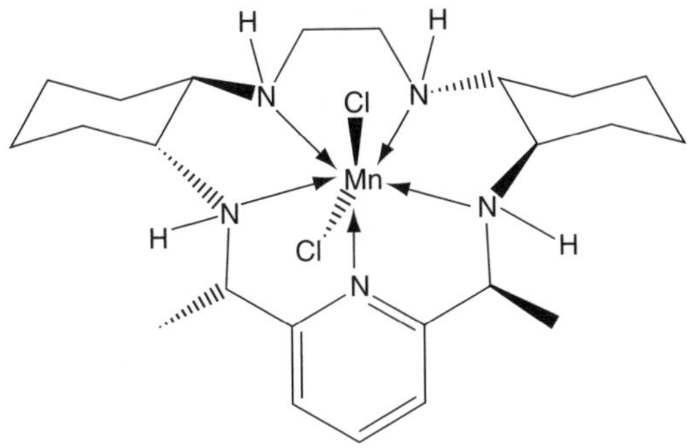

Figure 9.2 SOD mimetic Mn(II) pentaazamacrocylic ligand M40403

sequence from SOD2 (with mitochondrial targeting signal deleted) coupled with the COOH-terminal tail from SOD3. The tail of SOD3 has the important attribute of binding to the endothelial cell surface.[36]

9.8 Therapeutic Uses

A disease approach of the outcomes of selected studies in which various manganese-related therapeutic agents were tested is provided below.

9.8.1 Arthritis

Osteoarthritis is a slow degeneration of the articular cartilage and bone proliferation in joint margins that progressively worsens over time. Rheumatoid arthritis is a chronic inflammatory disease in which the synovial membranes of joints and other tissues are degraded[37] via an autoimmune reaction. In both diseases pro-inflammatory cytokines (tumor necrosis factor alpha, interleukin-1) are up-regulated and induce focal loss of cartilage. In osteoarthritis a major source of oxidative stress is related to nitric oxide; in rheumatoid arthritis the radical oxygen intermediates are more prominent.[38]

In the arthritis diseases, the collagen matrix of the joints is formed from chondroitin sulfate and its precursor, glucosamines, which aggregate to form proteoglycans. Supplements of chondroitin sulfate enhance proteoglycan production as this compound is anti-inflammatory[39]and diminishes the destruction of cultured articular cartilage cells.[40] Glucosamine (GCM) also is effective as it stimulates the activity of human osteoarthritic chondrocytes.[41] The combination of these compounds is synergistic and improves efficacy as compared to either alone.[42]

Manganese is used as a therapeutic agent for both conditions because it activates glycosyltransferases.[16] The chondroitin sulfates that are formed become side chains of the proteoglycans.[43] Also, manganese is necessary as a transcription and posttranslational activator for SOD.[44,45] Infusion of manganese into plasma of conscious horses increased the superoxide scavenging ability in a dose-related manner.[46] More recently, a molecular profiling at the proteome level of synovial tissue via Western blots found that SOD2 was reproducibly overexpressed with rheumatoid arthritis, as compared to those with osteoarthritis.[47]

The combination of manganese ascorbate with chondroitin sulfate and GCM is efficacious as it attenuates symptoms of knee osteoarthritis. An ascorbate form was given in order to strengthen the cross-linking of the collagen fibers. In 34 Navy men, a 4-month study found relief of symptoms according to disease summary and physical examination scores.[48] Das Jr and Hammad also found reduction of severity of symptoms in 72 patients who were classified as being radiographically mild or moderate, as compared to placebo.[49]

Although rheumatoid arthritis is vastly improved by treatment with Enbrel™ (etanercept, a genetically engineered protein), there is a risk of serious infections, including sepsis, which might preclude its use in those predisposed to infections, such as diabetics.[50] Other drugs that target inflammation, such as Kineret™ (anakinra, a blocker of IL-1), may be cost-prohibitive (>$1000 US per month). In a model of collagen-induced autoimmune arthritis with rats treated with GCM, Beren *et al.* found that two evaluations of arthritis, the arthritis index and histologic index, decreased 54% ($p = 0.001$) as compared to controls.[37] However, there was no change in the antibody response to bovine collagen or proliferation of T-cells. Thus, GCM appeared to increase the ability of the joint to withstand an inflammatory attack and might be considered as adjunct therapy.

9.8.2 Cancer

Cancer is a malignant tumor that occurs as a result of abnormal and uncontrolled cellular divisions. Manganese may be critical in the development of some cancers as tumor cells usually have low concentrations of SOD2 and ROS play a role in tumor formation. Also, transfection with SOD2 enhances cellular differentiation and inhibits malignant transformation.[12,51]

A Mn salen mimetic of superoxide dismutase and catalase (EUK-135) has been found to exert pharmacologic efficacy in cell survival following UVB irradiation.[52] UVB irradiation produces both oxidative stress and DNA damage, and induces N-terminal phosphorylation of p53 protein, an important oncogene for human cancers. Pretreatment with EUK-135 before exposure to UVB reduced the concentration of p53, in a dose-dependent manner, and inhibited the mitogen-activated protein kinase (MAPK) pathway response to oxidative stress.

The loss of the malignant phenotype in melanoma from transfection of plasmid cDNA SOD2 was initially reported by Church *et al.*[53] Since then this technique has produced similar positive findings in mouse fibrosarcoma and human cancer cells of the breast, lung, central nervous system, prostate and oral cavity (squamous).[12] In human prostate cells, transfection of the cDNA increases the expression of SOD2 approximately sixfold. Although this increment is less than the normal tenfold, the elevation is sufficient to decrease tumors. Transfection of SOD2 induced apoptosis, a G1 delay in the cell cycle, and diminished tumor volume.[54] The suppression of tumor growth in prostate and other cancers suggests that SOD2 is a tumor-suppressor gene.[12]

Recently, the transfection efficiency of SOD2 has been improved with an adenoviral vector (adenovirusMnSOD), rather than a plasmid. This complex was administered to hamster and human oral cancer cells in combination with 1,3-bis-(2-chloroethyl)-1-nitrosourea (BCNU), a compound that inhibits glutathione peroxidase (GPX). GPX is the enzyme that catalyzes the breakdown of H_2O_2 that is formed by the dismutation of SOD2 on superoxide. An inhibitor of GPX was added since it is believed that it would inhibit the removal of H_2O_2 and the higher levels of this ROS would potentiate the killing of cancer cells.[12] The combination therapy caused 80 and 50% decreases in cell viability for human and hamster oral cancer cells, respectively. The increase in cures indicates that this treatment may be effective for solid oral cavity tumors that are accessible for injection.

Evidence suggests that another manganese complex, manganese *N*-nicotinoyl-*N'*-*o*-hydroxythiobenzhydrazide [Mn(Notbh)(H_2O)], may also be a therapeutic agent for lymphoma and other cancers. This manganese complex (as well as a copper form) prolonged survival in mice exhibiting tumors. The mechanism was believed to be the dual action of a direct annihilation of the tumor cell and a reversal of the immunosuppression observed with tumor growth.[55]

9.8.3 Cardiovascular diseases

An excessive generation of superoxide may be a central component of a number of cardiovascular diseases including hypertension, atherosclerosis and hyperlipidemia.[56] One of the mechanisms by which this free radical contributes to hypertension is the inactivation of NO generated from endothelial cells. Endogenous endothelium-derived NO regulates blood pressure by vasodilation. The coupling of superoxide ions with NO decreases its availability; thus, overproduction of superoxide diminishes blood vessel tone and promotes hypertension.[34] In spontaneously hypertensive rats, treatment with the SOD mimetic M40403 removed superoxide anions and prompted the return to near-normal blood pressure. It was unclear whether this restoration of function was due to prolonged half-life of NO or inhibited formation of peroxynitrite.

Superoxide and its product, peroxynitrite, also participate in the development of atherosclerotic lesions (hardening of the arteries) by stimulating lipid peroxidation. In addition, a lack of nitric oxide due to overproduction of superoxide may promote proliferation of smooth muscle cells and release of pro-inflammatory molecules in the vascular wall. All of these events advance the progression of atherosclerosis and restenosis (a recurrent narrowing of heart valve or artery after corrective surgery).[56] A major generator of superoxide ions in vascular tissues is β-nicotinamide adenine dinucleotide phosphate [NAD(P)H] oxidase; an up-regulation of this enzyme has been implicated in the etiology of cardiovascular diseases. Jiang *et al.*[56] tested the impact of the SODm M40403 on superoxide generated from this oxidase in aorta from apolipoprotein (E)-deficient ($apoE^0$) mice, animals that develop atherosclerosis, hypertension and hyperlipidemia. The removal of superoxide generated from NAD(P)H oxidase reversed the endothelial dysfunction in these $apoE^0$ mice.

These results suggest a clinical application for this mimetic in diseases related to superoxide and endothelial dysfunction, such as hypertension and diabetes, inflammation and ischemia/reperfusion. One caution, however, may be that the mimetic may increase the level of H_2O_2 which also potentiates vascular pathology.[31]

9.8.4 Dermatitis

One unusual therapeutic suggestion utilizing manganese as a remedy is for acute atopic dermatitis.[57] Historically, balneotherapy or bathing in hot springs was used in the treatment for leprosy and syphilis in Japan. This bathing has been shown to be effective for patients with acute atopic dermatitis, who exhibit exacerbations associated with increased density of *Staphylococcus aureus*. The bactericidal activity of the water, as determined by growth of bacteria on soybean casein digest agar plates, was due to the presence of manganese and iodide ions at the pH and concentration found in the hot baths.

9.8.5 Diabetes

Diabetes was first related to manganese by a report that a diabetic man had reductions in blood glucose in response to doses of 5–10 mg manganese and elevated urinary excretion as determined by doses of ^{54}Mn.[58] Also, protracted hypoglycemia following a glucose load was observed in workers with manganese toxicity.[59] Yet a clinical trial in which subjects were given varying manganese doses (to 10 mg) showed no influence on blood glucose.[10] Nonetheless, several studies have found that insulin mRNA is decreased in manganese-deficient rats, suggesting a role for manganese in its transcription or turnover.[10] Another possible association could be the protective role that SOD2 plays in minimizing the oxidative damage from overproduction of ROS in diabetic hyperglycemia. In the three main pathways of pathogenesis of hyperglycemic damage, normalizing levels of SOD2 in bovine aortic endothelial cells prevented the pathogenesis.[60] These results suggest that therapy with SOD mimetics warrant further investigation.

9.8.6 Epilepsy

Lupine is an herb rich in manganese that was prescribed for devil sickness or seizures in Anglo-Saxon England.[61] Since then manganese has continued to be associated with idiopathic epilepsy. Carl *et al.* found low blood manganese in patients with epilepsy due to unknown reasons, as compared to those patients with a history of trauma (that may have triggered the epilepsy).[62] Rats that were genetically prone to epilepsy also exhibited low blood levels of manganese, but seizure onset and severity, brain monoamines and activity of glutamine synthetase were all not responsive to manganese supplements.[63] Glutamine synthetase was of interest as this enzyme has two manganese ions as co-factors and high levels of extracellular glutamate may induce seizures. Eid *et al.* documented that activity of glutamine synthetase was 38% lower in samples from patients with mesial temporal lobe epilepsy (MTLE) than those with other forms, and glutamine synthetase was notably diminished in astrocytes of the MTLE hippocampus.[64] A deficiency of this manganese-containing enzyme was suggested to be a molecular basis for seizures in this type of epilepsy. Whether SOD2 mimetics or glutamine synthetase therapies would be effective in the treatment of this form of epilepsy is unknown.

9.8.7 Human immunodeficiency virus

Although mammalian studies have not been conducted, there is potential for a manganese transporter gene to be a therapeutic strategy for human immunodeficiency virus (HIV) in the prevention of AIDS. Bolton *et al.* reported that increasing the cellular content of manganese in yeast blocks the activity of reverse transcriptase (RT).[65] This key enzyme is the basis by which HIV and other retroviruses transcribe DNA from RNA when they infect normal cells. In HIV, the transcription process is the reverse of the usual copying of RNA from DNA; hence, the enzyme is named 'reverse transcriptase'. The RTs that were investigated each had docking places for two metals – magnesium and manganese. Failure of these metals to dock or attach at these specific sites disrupted the functional ability of the RT. Ordinarily the cellular concentrations of magnesium are several thousand times that of manganese, but raising the manganese content only three times to that of normal reduced the activity of the RT by 50%. The elevated manganese levels were due to a defect in the gene

PMR1, a P-type ATPase that codes for the protein that transports manganese (and calcium) out of cells. Adding trace amounts of Mn^{2+} also inhibited the HIV-1 RT *in vitro* in the presence of magnesium cations. Thus, targeting this gene or its transporter protein to increase intracellular manganese could be a novel and safe approach for anti-retroviral therapy for diseases such as AIDS.

9.8.8 Inflammatory pain and response

A newly discovered role for the use of the SOD mimetic M40403 is as a reliever from inflammatory pain. Wang *et al.* have proposed that superoxide anions generated in response to inflammatory stimuli are mediators of pain.[66] This manganese-containing agent modulates hyperalgesia or excessive sensitiveness to pain via a nociceptive (nerve) signaling cascade (Figure 9.3). Rats were injected with carrageenan (a colloid derived from moss and algae) to produce peripheral inflammation in the paws and other markers of oxidative stress. When a SOD mimetic was administered via intravenous or intrathecal injection, all measures of inflammation and hyperalgesia were inhibited and the

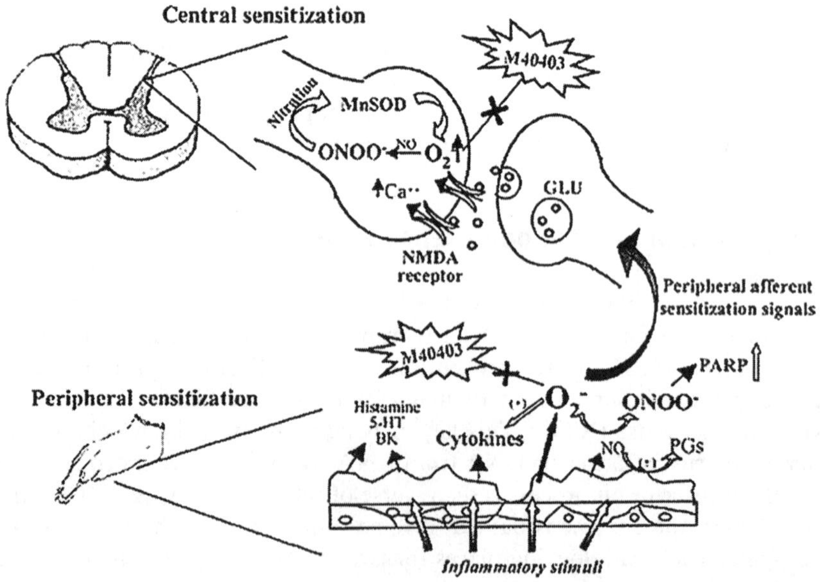

Figure 9.3 Inflammatory stimuli generate superoxide that plays a major role in pain. The MnSOD mimetic, M40403, inhibits the effects of superoxide at both peripheral and central sensitization (based on Ref. 66). Reproduced by permission of the American Society for Pharmacology and Experimental Therapeutics©

endogenous SOD2 was deactivated via nitration. Subsequently, the dysfunctional SOD2 was unable to eradicate the superoxide ions, suggesting their involvement in the nociceptive (nervous) pathway. Clearly this new therapeutic approach could lead to a new array of treatments for hyperalgesic response.

Another way to treat inflammation from paw edema resulting from carrageenan injection is the administration of the chimeric recombinant SOD2/3.[36] Whereas an injection of the native SOD2 produced a modest, non-significant response, SOD2/3 decreased edema by 62%. Also, SOD2/3 was compared to the native SOD in a rat model of acute lung injury. When interleukin (IL) 1α was administered by intratracheal instillation, SOD2/3 totally blocked the accumulation of neutrophils and lung leak. Thus this recombinant of SOD appears to be a powerful anti-inflammatory agent for future clinical applications.

9.8.9 Ischemia and reperfusion injury

Ischemia is a lack of blood flow to a tissue or organ, usually caused by a thrombus which occludes an artery. Initiation of reperfusion (return of the blood supply) within a short period of time is important for reinstating normal function. However, reperfusion also leads to a series of biochemical and cellular responses that produce neurological impairment in the brain and necrotic cardiac tissue.[32,67] The etiology of the destruction is multi-factorial, but it is related to an inflammatory response that includes release of ROS, leukocyte adhesions and infiltration and disintegration of neuronal membranes (also creating additional ROS).[68] One pharmacological strategy to minimize tissue damage is to inhibit the generation or enhance the degradation of ROS. In transgenic mice that overexpressed SOD2, SOD1 or GPX, only the overexpression of SOD2 diminished the myocardial injury due to ischemia and reperfusion.[67] In a rat stroke model, the effectiveness of two manganese salen complexes (EUK-8 and EUK-134) (Figure 9.1) was tested for their efficacy in reducing brain injury when administered as a single intravenous injection 3 h after a stroke.[32] The size of the brain infarct was decreased, much more by use of EUK-134 than EUK-8, and further growth was halted. It is believed that the more potent influence of EUK-134 was due to its greater catalase activity. Catalase degrades H_2O_2 to innocuous water; a lack of catalase in the presence of ferrous iron may shift the H_2O_2 into a highly reactive and damaging hydroxyl radical. The ability of EUK-134 to be efficacious so long after the infarct increases its clinical application for treatment of stroke.

9.8.10 Osteoporosis

Osteoporosis is a debilitating disease that has a multi-factorial etiology associated with genetics, dietary, hormonal and physical activity inadequacies. In a variety of species manganese deficiency results in skeletal defects of decreased length and thickened long bones and swollen, enlarged joints.[16] The primary etiological factor is a defect in the synthesis of proteoglycans as a result of diminished glycosyltransferase activities. In manganese-depleted rats given subcutaneous implants of bone particles, both osteogenesis and osteoclastic activities were impaired. In humans, women with osteoporosis have been reported to exhibit low concentrations of the mineral in plasma[16] and an enhanced response to an oral load. Serum levels of manganese also have been correlated with bone density. When supplements containing manganese, calcium, zinc and copper were given for 2 years to postmenopausal women, the combination was more effective in improving spinal bone mineral density than either calcium or trace minerals alone.[69] Also, a case study reported that a young girl on a manganese-free TPN since day 9 after birth exhibited short stature, brittle bones, and low serum manganese; growth and bone density improved with manganese.[16]

9.8.11 Peritoneal adhesions

Peritoneal adhesions that occur following surgery are often associated with an inflammatory response and ischemia of the tissues. Toxic oxygen radicals promote much of the damage associated with inflammation, particularly lipid peroxidation. Of these, hydroxyl anions are the most toxic and these are postulated to be inhibited by binding iron to prevent the Fenton and Haber-Weiss chemical reactions.[70] However, metal chelates of desferrioxamine and other compounds, such as lactate, citrate and succinate, that bind iron are unstable in the presence of albumin and have low rates of reaction. One agent that sequesters iron, but is also stable, is a desferrioxamine chelate containing manganese(III) (DF-Mn). A study of 40 post-operative rats observed that the manganese-Dfx chelate significantly improved the density of adhesions.[70] An advantage of this SOD mimetic is that it is an alternative approach to preventing oxidative damage that is independent of removing superoxide.[71]

9.8.12 Premenstrual syndrome

Premenstrual syndrome (PMS) is a collection of symptoms in women prior to the onset of menstruation such as moodiness/depression, tiredness, irritability and bloating. Manganese has been suggested for treatment of this condition based on a 169-day metabolic study in which women had increased mood and pain symptoms with a lower (1.0 mg) versus higher (5.6 mg) Mn intake.[72] A possible link with PMS was indicated since the change in mood was found only in their premenstrual phase. However, a subsequent study 10 years later by the same researchers found that dietary manganese only minimally influenced psychological measures in women on an 8-week crossover design study of 0.8 or 20 mg Mn/day.[73] Intake of manganese was not related to measures of hostility, anger or aggression, but low manganese was associated with higher self-confidence. Thus, the efficacy of manganese for symptoms of PMS is unclear.

9.8.13 Shortened life span (premature aging)

One of the most provocative uses of a manganese-containing compound is extending the lifespan. A SOD mimetic, manganese 5,10,15,20-tetrakis-(4-benzoic acid) porphyrin (MnTBAP) (Figure 9.4), was administered by intraperitoneal injection to mice with a SOD2 knockout. Use of this synthetic antioxidant in these transgenic mice increased the survival time by 100%, from 8 to 16 days of age.[13] Cardiac pathology and lipid accrual in the liver were reduced but the mice developed neurological disorders at day 12, due to a spongiform encephalopathy. Since the MnTBAP was unable to cross the

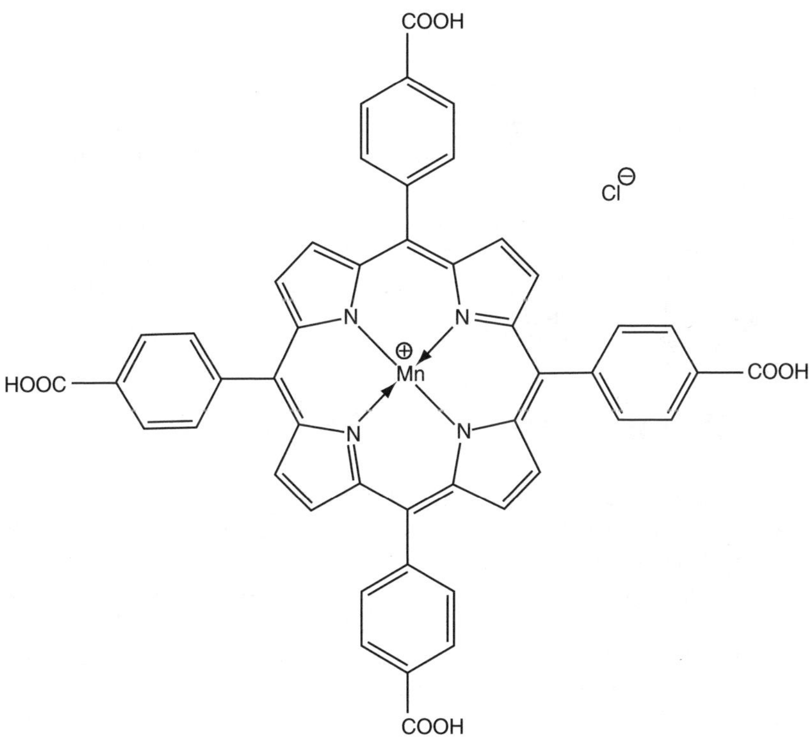

Figure 9.4 SOD mimetic Mn(III) 5,10,15,20-tetrakis-(4-benzoic acid) porphyrin (MnTBAP)

blood-brain barrier, the ultrastructural changes in the brain were attributed to age-related oxidative damage that was allowed to proceed with the alleviation of the cardiomyopathy.

9.8.14 Other disorders

Other potential uses of SOD mimetics may include those conditions for which polymorphisms associated with alterations of the human SOD2 gene have been identified. These include diabetic nephropathy, progeria (premature aging), sporadic motor neuron disease, non-familial idiopathic cardiomyopathy, breast cancer and a diminished tumor-suppressive effect.[14,74] SOD mimetics could also be effective in protecting the lung from cytotoxicity due to oxygen administered in the treatment of acute and chronic lung diseases.[11] For example, Wispe *et al.* found that pulmonary epithelial cells of transgenic mice that overexpressed SOD2 were protected from oxygen injury.[75] Additional therapeutic targets for SOD mimetics are pathophysiological conditions seen in animal models such as gastrointestinal dysfunction associated with alcohol, nonsteroidal anti-inflammatory drugs and *Helicobacter pylori.*[11]

The manganism that occurs from manganese toxicity produces neurological effects that resemble Parkinson's disease as reviewed earlier. One murine study used a pretreatment with manganese chloride ($MnCl_2$) to attenuate the neurotoxicity of 1-methyl-4-phenyl-12,3,6-tetrahydropyridine (MPTP). This neurotoxin induces a Parkinson-like syndrome in humans and primates.[76] In mice given drinking water containing 0.5 or 1.0 mg/ml $MnCl_2$ for 7 days there was a slight prevention the dopamine lowering of MPTP and enhancement of dopamine in lesioned corpus striatum, suggesting that pretreatment with manganese can minimize oxidative stress.

9.9 Magnetic Resonance Imaging (MRI)

Manganese in the brain can be measured without invasive procedures via MRI due to its ferromagnetic properties. In studies of patients with chronic liver disease, MR T-1weighted signal hypersensitivity in the basal ganglia in the brain was related to levels of erythrocyte Mn.[77] Thus, MRI can be used as a diagnostic tool to assess exposure to manganese. The ability of Mn to enhance T-1 weighted images has led to development of Teslascan™ (mangafodipir), an MRI contrast agent containing manganese as a complex with dipyridoxyl (derived from vitamin B_6) (Figure 9.5). Teslascan™ is used in MRI to enhance the detection, localization and characterization of the liver, pancreas and heart. The addition of Teslascan™ improved the correct characterization of lesions with histopathic confirmation by 75%.[78]

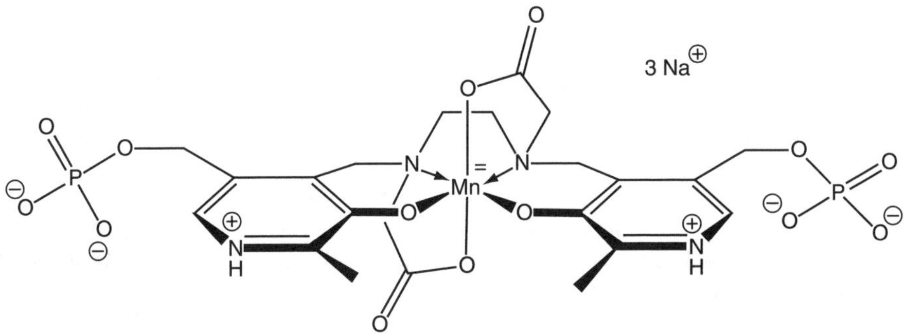

Figure 9.5 Teslascan™ (mangafodipir) is an MRI contrast agent given to diagnose liver problems

9.10 Future Implications

The numerous roles that manganese may play in human biology are just beginning to be appreciated. Low blood values of manganese in several populations suggest that the status of this mineral may be compromised in certain disease states. It is becoming clear that SOD2 is critical for mammalian organisms due to its protective role in preventing cellular and tissue damage by ROS. Imbalance of the SOD enzymes with reactants and its products may lead to a number of pathologies, such as stroke, neurodegeneration and heart failure.[11] Diseases in which genetic variations of the SOD gene have been reported, including shortened life span, sporadic motor neuron disease, non-familial idiopathic cardiomyopathy and breast cancer, may be a focus for new investigations.[13,14] Some success has been shown in treating pathologies such as arthritis, dermatitis, osteoporosis and peritoneal adhesions with manganese salts or simple compounds. However, the use of SOD mimetics and a chimeric recombinant SOD2/3 seems to provide the most opportunities for effective therapies related to diseases associated with manganese and oxidative stress. Whether an up-regulation of this enzyme would benefit those conditions associated with ROS deserves further study. The emergence of new technologies in molecular biology, immunology and chemistry will facilitate the future exploration of potential manganese-based therapies.

Acknowledgements

The authors are indebted to Michael S. Graves for editorial assistance and the Bess Heflin Centennial Professorship for partial support.

References

1. J. Freeland-Graves, in *New Aspects of Trace Element Research*, M. Abdulla, M. Bost, S. Gamon, P. Arnaud and G. Chazot (Eds), Smith-Gordo, London, 1999.
2. World Health Organization (WHO), *Air Quality Guidelines for Europe*, 91, 2nd edn, WHO, Copenhagen, 2000.
3. G. Gerber, A. Leonard and P. Hantson, *Crit. Rev. Oncol. Hematol.*, **42**, 25–34 (2002).
4. The orator. www.theorator.com/bills108/hr2812.html. Accessed June 13, 2004.
5. Institute of Medicine, *Dietary Reference Intakes: Vitamin A, Vitamin K, Arsenic, Boron, Chromium, Copper, Iodine, Iron, Manganese, Molybdenum, Nickel, Silicon, Vanadium, and Zinc*, National Academy Press, Washington, DC, 2002.
6. J. Freeland-Graves, C. Bales and F. Behmardi, in *Nutritional Bioavailability of Manganese*, C. Kies (Ed.), American Chemical Society, Washington, DC, 1987.
7. J. Freeland-Graves, in *Risk Assessment of Essential Elements*, W. Mertz, C. Abernathy and S.S. Olin (Eds) International Life Sciences Institute Press, Washington, DC, 1994.
8. Environmental Protection Agency (EPA), *Health Effects Support Document for Manganese*, US EPA, Washington, DC, 2003.
9. Environmental Protection Agency (EPA), *Drinking Water Health Advisory for Manganese*, US EPA, Washington, DC, 2004.
10. C. Keen, J. Ensunsa and M. Clegg, *Metal Ions Bio. Syst.*, **37**, 89–121 (2000).
11. C. Muscoli, S. Cuzzocrea, D. Riley *et al.*, *Brit. J. Pharmacol.*, **140**, 445–460 (2003).
12. C. Weydert, B. Smith, L. Xu *et al.*, *Free Rad. Bio. Med.*, **34**, 316–329 (2003).
13. S. Melov, *Ann. N.Y. Acad. Sci.*, **959**, 330–340 (2002).
14. I. Zelko, T. Mariani and R. Folz, *Free Rad. Bio. Med.*, **33**, 337–349 (2002).
15. P. Gauuan, M. Trova, L. Gregor-Boros *et al.*, *Bioorg. Med. Chem.*, **10**, 3013–3021 (2002).
16. J. Freeland-Graves and C. Llanes, in *Manganese in Health and Disease*, D. Klimis-Tavantzis (Ed.), CRC Press, Boca Raton, Florida, 1993.
17. H. Gong and T. Amemiya, *Invest. Ophthalmol. Vis. Sci.*, **37**, 1967–1974 (1996).
18. C. Keen, J. Ensunsa, M. Watson *et al.*, *Neurotoxicology*, **20**, 213–224 (1999).
19. B. Friedman, J. Freeland-Graves, C. Bales *et al.*, *J. Nutr.*, **117**, 133–143 (1987).
20. J. Freeland-Graves, *Nutr. Today*, **23**, 13–19 (1988).
21. G. Krishna, H. Whitlock Jr, D. Feldbruegge and J. Proter, *Arch. Biochem. Biophys.*, **114**, 200–215 (1966).
22. N. Norose, M. Terai and K. Norose, *J. Trace Elem. Exp. Med.*, **5**, 100–101 (1992).
23. J. Freeland-Graves and J. Turnlund, *J. Nutr.*, **126**, S2435–S2440 (1996).
24. C. Davis and J. Greger, *Am. J. Clin. Nutr.*, **55**, 747–752 (1992).
25. J. Finley, *Am. J. Clin. Nutr.*, **70**, 37–43 (1999).
26. L. Normandin and A. Hazell, *Metab. Brain Dis.*, **17**, 375–387 (2002).
27. R. Butterworth, *Liver Int.*, **23**, 5–9 (2003).
28. T. Ponnapakkam, K. Bailey, K. Graves and M. Iszard, *Reprod. Toxicol.*, **17**, 547–551 (2003).
29. J. Finley, *Nutr. Rev.*, **62**, 148–153 (2004).
30. Z. Gude and T. Glukhen'kii, *Vestnik Dermatolog. Venero.*, **40**(3), 23–26 (1966).
31. D. Weber and K. Griendling, *Brit. J. Pharmacol.*, **139**, 1059–1060 (2003).

32. K. Baker, C. Marcus, K. Huffman *et al.*, *J. Pharm. Exp. Ther.*, **284**, 215–221 (1998).
33. K. Aston, N. Rath, A. Naik *et al.*, *Inorg. Chem.*, **40**, 1779–1789 (2001).
34. S. Cuzzocrea, E. Mazzon, L. Dugo *et al.*, *FASEB J.*, **18**, 94–101 (2004).
35. A. Gow and H. Ischiropoulos, *Am. J. Physiol. Lung Cell. Mol. Physiol.*, **284**, L915–L916 (2003).
36. B. Gao, S. Flores, J. Leff *et al.*, *Am. J. Physiol. Lung Cell. Mol. Physiol.*, **284**, L917–L925 (2003).
37. J. Beren, S. Hill, M. Diener-West and N. Rose, *Exp. Biol. Med.*, **226**, 144–151 (2001).
38. I. Mazzetti, B. Grigolo, L. Pulsatelli *et al.*, *Clin. Sci.*, **101**, 593–599 (2001).
39. F. Ronca, L. Palmieri, P. Panicucci and G. Ronca, *Osteoarthritis Cartilage*, **6**, 14–21 (1998).
40. D. Uebelhart, E. Thornar, J. Zhang and J. Williams, *Osteoarthritis Cartilage*, **6**, 6–13 (1998).
41. M. Piperno, P. Reboul, M. Hellio le Graverand *et al.*, *Osteoarthritis Cartilage*, **8**, 207–212 (2000).
42. L. Lippiello, J. Woodward, R. Karpman and T. Hammad, *Clin. Orthop. Relat. Res.*, **381**, 229–240 (2000).
43. A. Liu, B. Heinrichs and R. Leach Jr, *Poult. Sci.*, **73**, 663–669 (1994).
44. L. Schrum and H. Hassan, *Biochim. Biophys. Acta*, **1216**, 186–190 (1993).
45. C. Privalle and I. Fridovich, *J. Biol. Chem.*, **267**, 9140–9145 (1992).
46. R. Singh, K. Kooreman, C. Babbs *et al.*, *Am. J. Vet. Res.*, **53**, 1822–1829 (1992).
47. P. Lorenz, P. Ruschpler, D. Koczan *et al.*, *Proteomics*, **3**, 991–1002 (2003).
48. C. Leffler, A. Philippi, S. Leffler *et al.*, *Mil. Med.*, **164**(2), 5–91 (1999).
49. A. Das Jr and T. Hammad, *Osteoarthritis Cartilage*, **8**, 343–350 (2000).
50. Food and Drug Administration, New Warning for Arthritis Drug, Enbrel, www.fda/gov/bbs/topics/ANSWERS/ANS00954.html, Assessed on June 4, 2004.
51. D. St Clair, T. Oberley, K. Muse and W. St Clair, *Free Rad. Bio. Med.*, **16**, 275–282 (1994).
52. D. Decraene, K. Smaers, D. Gan *et al.*, *J. Invest. Dermatol.*, **122**, 484–491 (2004).
53. S. Church, J. Grant, L. Ridnour *et al.*, *Pro. Nat. Acad. Sci. USA*, **90**, 3113–3117 (1993).
54. S. Plymate, K. Haugk, C. Sprenger *et al.*, *Oncogene*, **22**, 1024–1034 (2003).
55. A. Shrivastav, N. Singh and S. Singh, *BioMetab.*, **16**, 311–320 (2003).
56. F. Jiang, Y. Guo, D. Salvemini and G. Dusting, *Brit. J. Pharmacol.*, **139**, 1127–1234 (2003).
57. T. Inoue, S. Inoue and K. Kubota, *Acta Derm. Venereo.*, **79**, 360–362 (1999).
58. A. Rubenstein, N. Levin and G. Elliott, *Lancet*, **2**, 1348–1351 (1962).
59. M. Hassanein, H. Ghaleb, E. Haroun *et al.*, *Brit. J. Indust. Med.*, **23**, 67–70 (1966).
60. T. Nishikawa, D. Edelstein, X. Du *et al.*, *Nature*, **404**, 787–790 (2000).
61. P. Dendle, *Bull. Hist. Med.*, **75**, 91–101 (2001).
62. G. Carl, C. Keen, B. Gallagher *et al.*, *Neurology*, **36**, 1584–1587 (1986).
63. J. Critchfield, G. Carl and C. Keen, *Epilepsy Res.*, **14**, 3–10 (1993).
64. T. Eid, M. Thomas, D. Spencer *et al.*, *Lancet*, **363**, 28–37 (2004).
65. E. Bolton, A. Mildvan and J. Boeke, *Mol. Cell*, **9**, 879–889 (2002).
66. Z. Wang, F. Porreca, S. Cuzzocrea, K. Galen, R. Lightfoot, E. Masini, C. Muscoli, B. Mollace, M. Ndengele, H. Ischiropoulos and D. Salvemini, *J. Pharm. Exp. Ther.*, **309**, 869–878 (2004).

67. S. Jones, M. Hoffmeyer, B. Sharp *et al.*, *Am. J. Physiol. Heart Circ. Physiol.*, **284**, H277–H282 (2003).
68. W. Clarke, Reperfusion Injury in Stroke, www.emedicine.com/neuro/topic602.htm, Accessed June 8, 2004.
69. L. Strause, P. Saltman, K. Smith *et al.*, *J. Nutr.*, **124**, 1060–1064 (1994).
70. G. Soybir, F. Koksoy, F. Ekiz *et al.*, *J. R. Coll. Surg. Edinb.*, **43**, 26–28 (1998).
71. S. Hahn, C. Krishna, A. Samuni *et al.*, *Arch. Biochem. Biophys.*, **288**, 215–219 (1991).
72. J. Penland and P. Johnson, *Am. J. Obstet. Gynecol.*, **168**, 1417–1423 (1993).
73. J. Finley, J. Penland, R. Pettit and C. Davis, *J. Nutr.*, **133**, 2849–2856 (2003).
74. T. Nomiyama, Y. Tanaka, L. Piao *et al.*, *J. Hum. Gen.*, **48**, 138–141 (2003).
75. J. Wispe, B. Warner, J. Clark *et al.*, *J. Bio. Chem.*, **267**, 23937–23941 (1992).
76. P. Rojas and C. Rios, *Neurochem. Res.*, **20**, 1217–1223 (1995).
77. E. Malecki, A. Devenyi, T. Barron *et al.*, *Neurotoxicology*, **20**, 647–652 (1999).
78. Amersham Health. Teslascan™. www.amershamhealth-us.com/teslascan, Accessed on June 10, 2004.

10 $_{26}$Fe The Use of Iron-Based Drugs in Medicine

Xiang Wu and Mei Lin Go

10.1 Introduction

More so than other metals, metallic iron has long been associated with health. Iron-containing chalybeate waters have been vaunted for their healing properties for centuries. Even today, many towns in Europe – Bath in England, Llandrindod Wells in Wales, Spa in Belgium – are renowned for their chalybeate springs. 'Chalybeate', a syrupy brew of iron filings in Rhenish wine was widely prescribed for chlorosis, an anaemia prevalent among adolescent girls in the seventeenth to nineteenth centuries, even before iron was identified as the cause of anaemia.[1] The 'veritable pills of Blaud', a formulation of ferrous sulphate and potassium carbonate, had an unassailable reputation as a cure for a myriad of afflictions in nineteenth-century Europe.[2] In later years, arsenic was added to the pills, prompted by the belief that iron was more effective when combined with arsenic.[3]

It was not until the 1930s that the relationship between iron-deficiency anaemia and the lack of dietary iron was established.[4,5] Today, iron deficiency remains widespread and is considered as 'probably the most frequent nutritional deficiency in the world'.[6] The medical use of iron supplements in the form of oral (ferrous sulphate) or parenteral formulations (iron-dextran, a sodium ferric gluconate complex) for this purpose has been extensively reviewed,[7] and will not be addressed in this chapter.

Metallotherapeutic Drugs and Metal-Based Diagnostic Agents: The Use of Metals in Medicine
Edited by Gielen and Tiekink

10.2 Ferrocene

The scope of the present review will focus primarily on the use of organoiron compounds in therapeutics. Fortunately or otherwise, organoiron chemistry is dominated by a single compound – ferrocene (**1**) or di(η^5-cyclopentadienyl)-iron(II)– whose accidental discovery in 1951 was instrumental in opening up a new area of research in organometallic chemistry. Ferrocene comprises a ferrous ion (Fe^{2+}) coordinated to two cyclopentadienyl rings. The exceptional stability of ferrocene is attributed to the distribution of 18π electrons in the e_{2g} and a_{1g} non-bonding molecular orbitals.

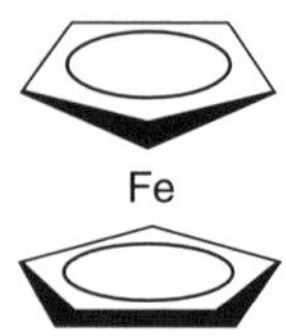

1

Compared to other metallocenes, ferrocene (**1**) has been widely used in drug design, usually as a substitute for the phenyl or alkyl group. This is prompted by several factors. First, ferrocene is more lipophilic than benzene ($\pi_{ferrocene}$ 2.46 compared to $\pi_{benzene}$ 1.96)[8] and its inclusion into a molecule is expected to enhance the membrane permeability and possibly bioavailability of the final compound. Since inappropriate pharmacokinetic properties are the main reasons why compounds fail to transform into effective drugs,[9] the potential utility of the ferrocene moiety in this regard is readily embraced by medicinal chemists. Secondly, novel compounds are obtained when the ferrocene ring is introduced into lead compounds. Ferrocene is an isostere of benzene. It is aromatic but is more electron donating than benzene (σ_{para} of benzene: −0.01; σ_{para} of ferrocene: −0.18).[8] In terms of shape, ferrocene is probably as 'wide' as benzene but it is considerably 'thicker', being as 'thick' as it is 'wide'.

Moreover, ferrocene is widely regarded as a 'safe' metallocene. Acute and chronic toxicity tests of ferrocene and some derivatives attest to the relative safety of ferrocene in many mammalian species (dogs, rats, mice, monkeys).[10–12] However, the issue of whether iron is released from ferrocene-containing compounds during metabolism is not completely resolved and is a matter of concern. Free iron can potentially catalyze free-radical reactions, leading to lipid peroxidation of cellular membranes and the formation of reactive oxygen species via the Fenton reaction.[13] Hanzlik and Soine noted that ferrocene is metabolized by various phase 1 and 2 reactions (hydroxylation, glucuronidation, sulphation) without loss of the metal, implying that the ferrous ion in ferrocene is resistant to removal.[14] Another study showed that more than 60% of ^{59}Fe activity was detected in urine (as ferrocene metabolites or intact ferrocene) after oral administration to rats.[15] In contrast, the ferrous ion appears to be readily released from substituted ferrocenes during metabolism. This is well illustrated in (3,5,5-trimethylhexanoyl) ferrocene (**2**, TMH ferrocene), which is used to deliver iron to biological systems.[16]

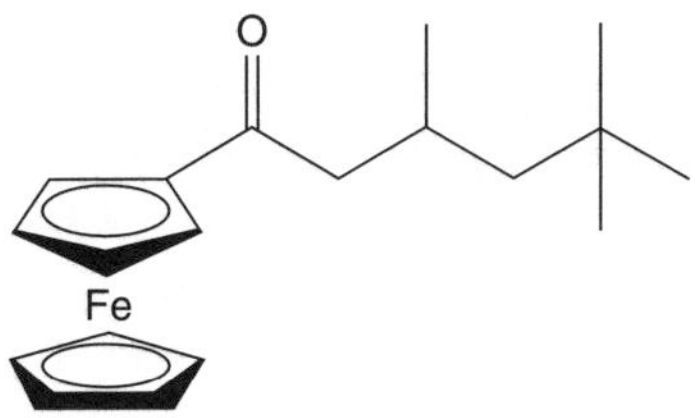

2

10.3 TMH Ferrocene

Special mention must be made of **2** because of its role in the delivery of iron to biological systems. When fed to whole animals or incubated with cultured hepatocytes, TMH ferrocene produces an experimental iron overload that closely mimics hepatocellular iron overload in humans.[15,17] Thus, it serves as a useful experimental model to investigate iron overload in humans. The iron donor properties of TMH ferrocene are unusual in many ways. Unlike dietary iron, the intestinal absorption of TMH ferrocene is not regulated by the amount of iron present in the body. The same amount of TMH ferrocene is absorbed in iron-deficient and iron-loaded rats.[15] Neither is there evidence of dose dependency in its effects. The bioavailability of TMH ferrocene is about six times higher than that of ferrous sulphate. In plasma, TMH ferrocene is not transported by transferrin, the β1 glycoprotein that transfers dietary ferric iron to intracellular sites. Once in the liver, TMH ferrocene is metabolized by a phenobarbital-inducible cytochrome P450 ('TMH ferrocenase') and the iron is released from the hydrocarbon core.[18] The metal is then incorporated into the iron–protein storage complex ferritin and subsequently processed by the body in the same way as dietary iron.[15,17] As the iron in TMH ferrocene is essentially redox-inactive until its release in the liver, the authors anticipated iron-catalyzedoxidative damage to the biological system to be limited.[18]

10.4 Ferrocene in Drug Design

There are many examples in the literature where a phenyl or alkyl group has been replaced by ferrocene as a drug design strategy. However, the anticipated outcome (increase in activity) has not always been encountered. In some instances, the introduction of ferrocene abolished or markedly reduced activity in the resulting compound. In other cases, activity was more or less retained. There are also instances where an increase in activity or an unexpected change in activity profile was detected. Some of these examples are discussed in the following paragraphs.

10.4.1 Examples where introduction of ferrocene has resulted in a loss or no change in activity

The replacement of the aromatic ring in some non-steroidal anti-inflammatory agents (fenclofenac, flurbiprofen, fenbufen, tolmetin) by ferrocene did not improve anti-arthritic or platelet aggregatory activities in the resulting compounds.[19] It was concluded that ferrocene is not an appropriate bioisostere for the phenyl or pyrrole ring. Ferrocene-containing penicillins and cephalosporins[20] and ferrocene-containing rifamycins[21] have been evaluated and found to be no better than the original antibiotic (amoxicillin, carbenicillin, cephalothin, rifamycin) in terms of potency and range of organisms affected. However, it should be noted that in many instances, the synthetic route may not permit a direct substitution of a phenyl or alkyl group by ferrocene. For example, the resemblance between amoxicillin and the ferrocene–amoxicillin derivative **3** is not particularly striking. Thus, a fair comparison of biological activities may not be feasible.

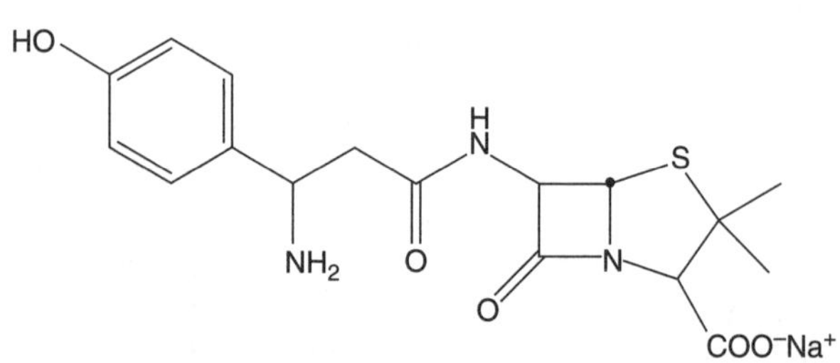

Amoxicillin

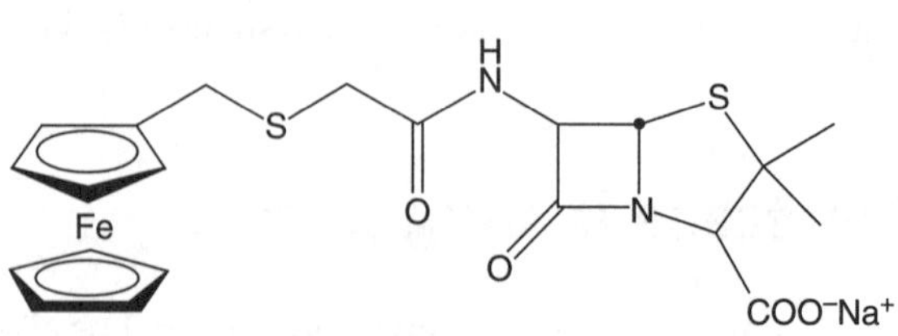

3

The 2,4-diflurophenyl ring of fluconazole is replaced by a ferrocene ring in the ferrocene–fluconazole analogue **4**.[22] The replacement was expected to enhance the permeability of **4** into *Candida* (yeast) cells. In this case, a role was hypothesized for the metal ion. Reports have shown that the growth of *Candida* is dependent on iron: iron deprivation suppressed growth while iron overload had the opposite effect. However, fluconazole showed an increase in its candidacidal activity in the presence of an iron overload.[23] This led to the hypothesis that the inclusion of iron (in the form of ferrocene) in the fluconazole molecule might be an effective way of enhancing its activity against clinically important resistant *Candida* species. This was not found to be the case. When evaluated *in vitro* against several species, the ferrocene analogue actually promoted the growth of *Candida*. When given in a 1:1 combination with fluconazole, the anti-candidacidal activity of the latter was suppressed. The authors concluded that **4** gained ready access into the yeast cells (because of its increased lipophilicity), possibly interacted with the same molecular target as fluconazole (which would explain the inhibition) and may have released the metal ion (thus promoting growth of *Candida*).

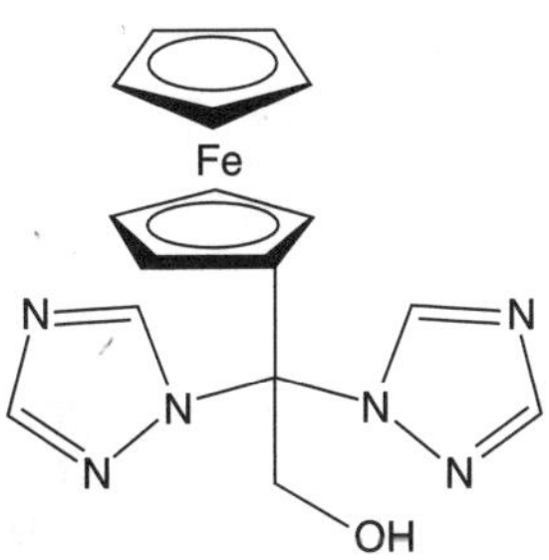

4

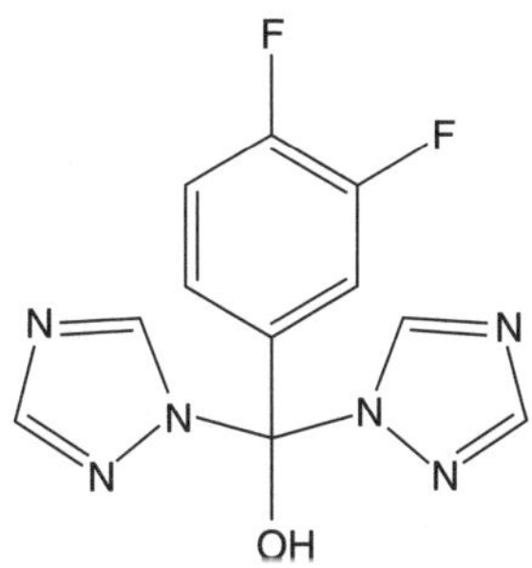

Fluconazole

10.4.2 Examples where introduction of ferrocene has resulted in enhanced activity or a change in activity profile

The replacement of one of the phenyl rings in tamoxifen by ferrocene has given rise to a group of compounds called ferrocifens.[24,25,26] Tamoxifen is a selective estrogen receptor modulator that is used against breast cancer. Its activity is limited to 'estrogen receptor positive' tumours that constitute about 60% of breast cancer cases. The ferrocifens were synthesized in an attempt to overcome the shortcomings of tamoxifen as an anti-tumour agent. The triphenylethylene-like framework of tamoxifen is retained in the ferrocifens in order to ensure recognition for the estradiol receptor. The ferrocene ring was included in order to take advantage of the eventual formation of the ferrocenium ion (in which the iron is in its Fe^{3+} state) on metabolism.[24] The ferrocenium ion is known to be cytotoxic[27] and it was postulated that combining anti-estrogenic and cytotoxic activities in the ferrocifens would be an advantageous strategy.

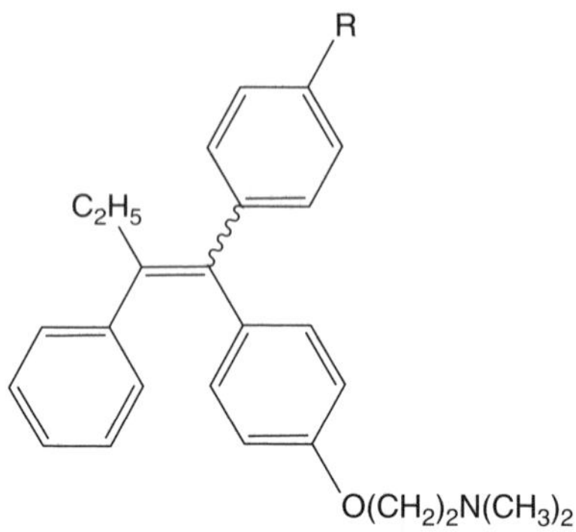

R = H: Tamoxifen (Z + E)

R = OH: Hydroxytamoxifen (Z + E)

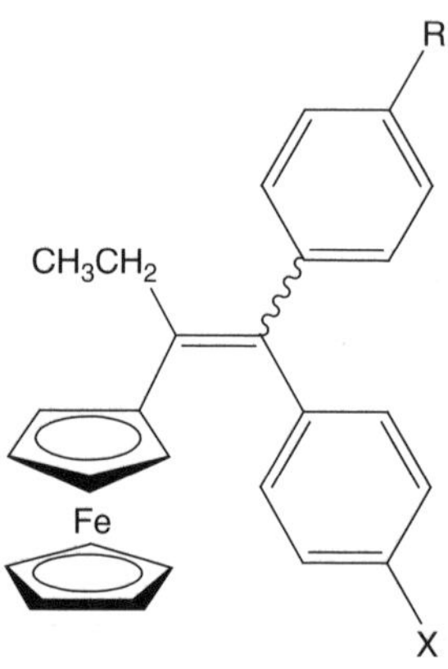

5: R = H; X = $O(CH_2)_2N(CH_3)_2$
6: R = OH; X = $O(CH_2)_2N(CH_3)_2$
7: R = OH; X = $O(CH_2)_3N(CH_3)_2$
8: R = OH; X = $O(CH_2)_4N(CH_3)_2$
9: R = OH; X = H

Compounds **5**–**9** are some of the ferrocifens that have been reported in the literature. The ferrocifen **5** bears the closest resemblance to tamoxifen. The hydroxy ferrocifen **6** is meant to mimic hydroxytamoxifen, the active metabolite of tamoxifen. Compounds **7** and **8** are homologues of **6**, in which the basic dimethylaminoalkoxy side chain is extended from ethyl (**6**, $n = 2$) to propyl (**7**) and butyl (**8**). In ferrocifen **9**, the basic side chain is omitted altogether. Some interesting structure–activity relationships are revealed. First, the presence of the bulky ferrocene ring diminished the affinity of the ferrocifens for the estrogen receptor.[24,25,26] Radioligand displacement assays using tritiated estradiol and sheep uterine cytosol as a source of estrogen receptors showed that binding affinity of hydroxytamoxifen (Z-isomer) decreased from 107% (estradiol = 100%) to 40% (in **6**, Z-isomer) upon introduction of ferrocene.[26] Increasing the chain length of the basic side chain also adversely affected binding affinity: 11.5% for **7** (E + Z) compared to 38.5% for hydroxytamoxifen (E + Z). Second, despite their diminished affinities for the estrogen receptor, the ferrocifens (**6**, **7**, **8**) were able to inhibit the growth of a human breast cancer cell line (MCF-7, estrogen receptor positive) to almost the same extent as tamoxifen.[24,25,26] The observation that estradiol abolished the anti-proliferative effect of **8** suggested that the ferrocifens were interacting at the same site as estrogen in these cells.[25] Moreover, ferrocifen **9** (no basic side chain) and unsubstituted ferrocene had greatly diminished activity, an indication that the tamoxifen template is largely responsible for interaction with the estrogen receptor, as anticipated.[25] Interestingly, the length of the basic side chain determined if the ferrocifens interacted with the receptor as an agonist or antagonist. Thus, ferrocifen **6** caused estrogen-dependent tumour T47-D xenografts in nude mice to increase in size but lengthening the side chain to give higher homologues **7** and **8** restored anti-estrogenic activity.[25] The observation that ferrocifen **6** had estrogenic activity but could still inhibit the growth of breast cancer cells appears to be contradictory. The authors proposed that the addition of the ferrocene ring introduces a genotoxic component that is absent in tamoxifen. Ferrocifens **5** and **6** do indeed induce lesions in the DNA of MCF-7 cells.[26] It is possible that the anti-proliferative effect of **6** is due to its greater genotoxic effect compared to its estrogenic activity. These results support the original hypothesis that the tamoxifen-like framework in ferrocifens confers recognition for the estrogen receptor while the presence of ferrocene induces damage to DNA, possibly via reactive oxygen species generated in a Fenton-like reaction involving the Fe^{2+}/Fe^{3+} couple.[27,28]

Possibly the best known example of a successful ferrocene–phenyl/alkyl substitution is that of the ferrocene analogue of the anti-malarial agent chloroquine–ferrochloroquine, also known as ferroquine. There is a strong likelihood that ferrochloroquine will eventually be used for the treatment of human malaria. Thus the development of ferrochloroquine will be discussed in detail in the following section.

10.5 Ferrochloroquine

In 1997, Biot and coworkers reported the synthesis of four ferrocenyl analogues of chloroquine.[29] In each case, the ferrocene ring is inserted in place of two methylene groups in the butyl side chain of chloroquine. The terminal amino group was varied to give a dimethylamino (**10**), diethylamino (**11**), piperidinyl (**12**) or *N*-methylpiperazinyl (**13**) analogue. The dimethylamino analogue was found to be the most active in mice infected with chloroquine-sensitive and -resistant strains of *Plasmodium*.[29,30] When evaluated against a drug resistant strain of the human *P. falciparum*, its activity against the life stages of the plasmodium in the red blood cell was found to be about 22 times that of chloroquine.[29] Compound **10** was subsequently called ferrochloroquine. Further tests showed that ferrochloroquine had a curative effect on murine *P. vinckei* and protected mice from lethal infection at a daily dose of 8.4 mg/kg when given for 4 days subcutaneously or orally.[30] Furthermore, ferrochloroquine was approximately 300–800-fold more selective against plasmodia than mammalian L5178Y lymphoma cells. Acute toxicity tests in mice showed that ferrochloroquine was more toxic in starved animals.[31] *In vitro* susceptibility tests against *P. falciparum* isolates from Libreville and the Haut-Ogooue region in Gabon confirmed that ferrochloroquine has significant activity against chloroquine-resistant strains.[32,33]

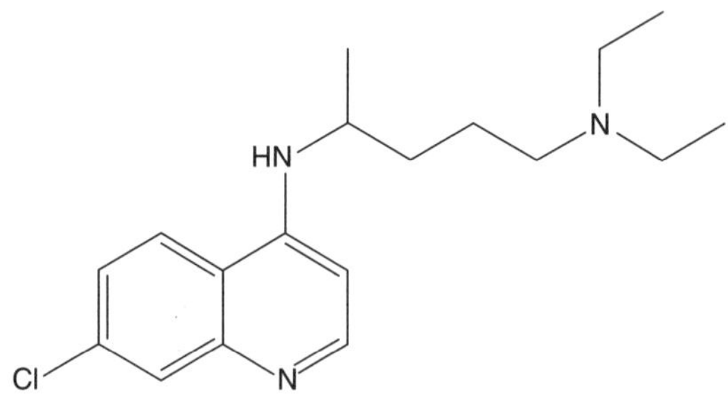

Chloroquine

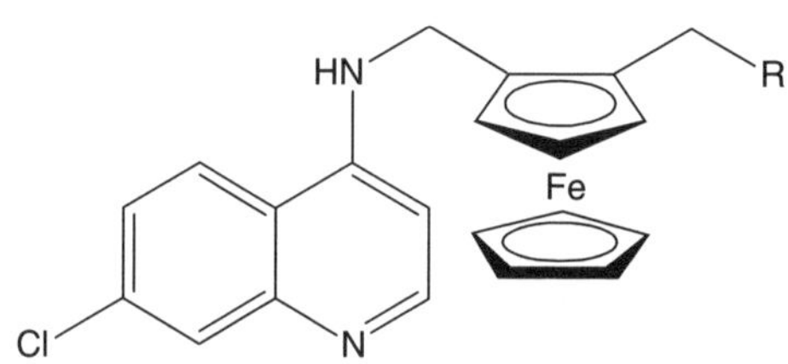

10: R = $N(CH_3)_2$
11: R = $N(C_2H_5)_2$

12: R = 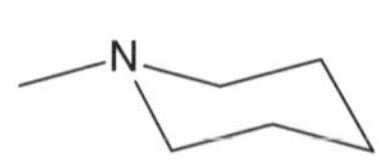

13: R = 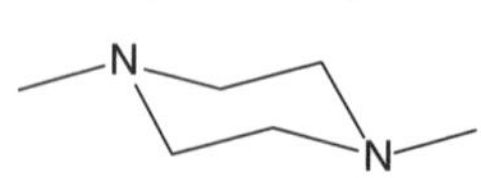

Various structure–activity studies have been reported in the literature. Delhaes and co-workers investigated the importance of chirality to activity.[34] The ferrocene ring in ferrochloroquine is 1,2 substituted. Thus ferrochloroquine has planar chirality and exists in two enantiomeric forms. Both enantiomers (+) (1′R) ferrochloroquine (**14**) and (−) (1′S) ferrochloroquine (**15**) were synthesized and found to be equivalent in anti-plasmodial activity as well as in cytotoxicity against mouse lymphoma cells. When the enantiomers were compared to the racemate, equivalent activity was detected *in vitro* but the racemate had a higher curative effect in infected mice. In view of this observation, racemate ferrochloroquine is the formulation of choice.

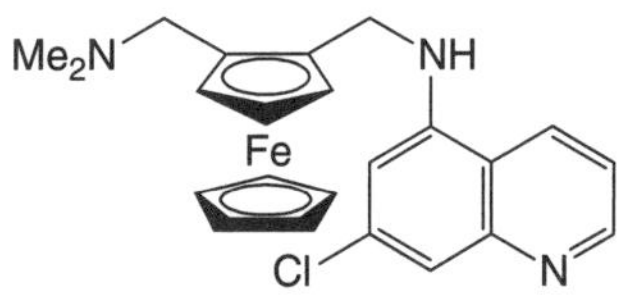

14

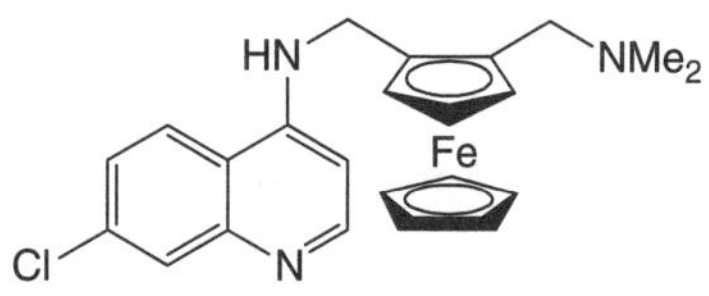

15

Biot and coworkers investigated the attachment of the ferrocene ring to different positions of chloroquine.[35] The attachment of the ferrocene ring to the quinoline ring (**16**) or to the end of the terminal amino function (**17**) caused a decrease in activity. Compound **16** in which ferrocene is linked to the quinolinyl nitrogen gave rise to a salt that is completely inactive, but compound **17** retained activity against chloroquine-sensitive strains. It would seem that the position of ferrocene in ferrochloroquine is optimal.

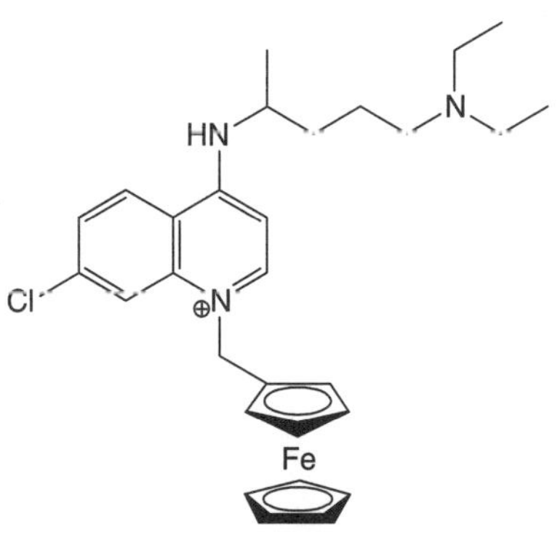

16

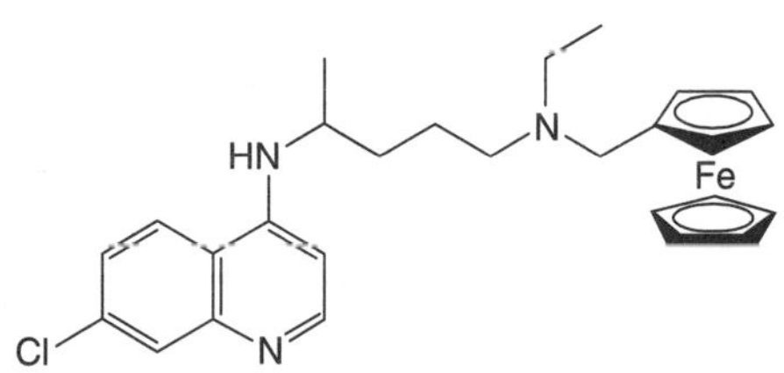

17

Phase I metabolism of chloroquine proceeds by dealkylation of the terminal amino function to give less active mono-desethyl and di-desethyl metabolites. Thus, it was of interest to investigate the activity of equivalent dealkylated derivatives of ferrochloroquine.[36] The desmethyl (**18**) and di-desmethyl (**19**) analogues of ferrochloroquine were synthesized and found to be comparable to ferrochloroquine.

They were more active than chloroquine against sensitive and resistant plasmodia, with **19** being less active than **18** against resistant plasmodia. Thus, if dealkylated compounds like **18** and **19** are formed as metabolites of ferrochloroquine, anti-plasmodial activity is likely to be retained even on metabolism.

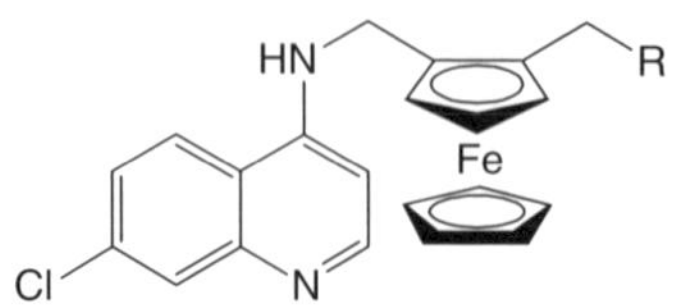

18: R = $NHCH_3$
19: R = NH_2

In ferrochloroquine, the quinoline ring and the terminal basic side chain are attached to the same cyclopentadienyl ring of ferrocene, i.e. the substitution pattern is 1,2-. Beagley and co-workers investigated the effect of attaching the substituents to different cyclopentadienyl rings (i.e. 1,1′-substitution as shown in **20**).[37] Compound **20** was as active as ferrochloroquine against a chloroquine-sensitive strain (D10) of *P. falciparum* but was slightly less active against a resistant plasmodial strain (K1). Another pair of 1,2- and 1,1′-substituted ferrochloroquine analogues (**21, 22**) showed the reverse trend (1,2-substituted **21** > 1,1′-substituted **22**) but both compounds were still more active than chloroquine against resistant plasmodia. Unlike ferrochloroquine, there is an additional basic nitrogen atom in the side chain of **21** and **22**. The presence of this additional basic centre may have influenced anti-plasmodial activity so that a comparison between ferrochloroquine and its 1,2-isomer **20**, and that of **21** and **22** need not result in a similar trend. Thus, the effects of 1,1′- versus 1,2-substitution are probably dependent on the nature of the side chain.

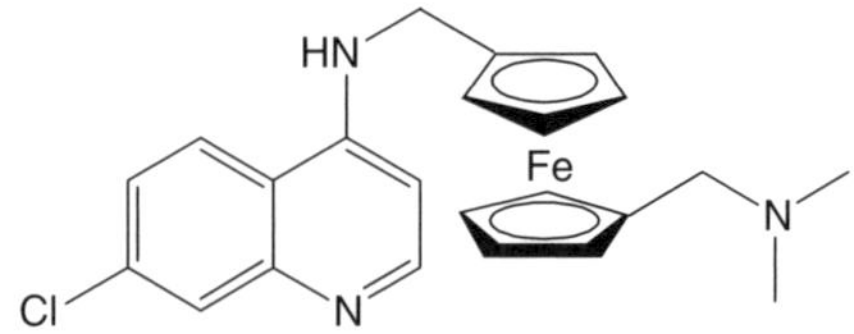

20

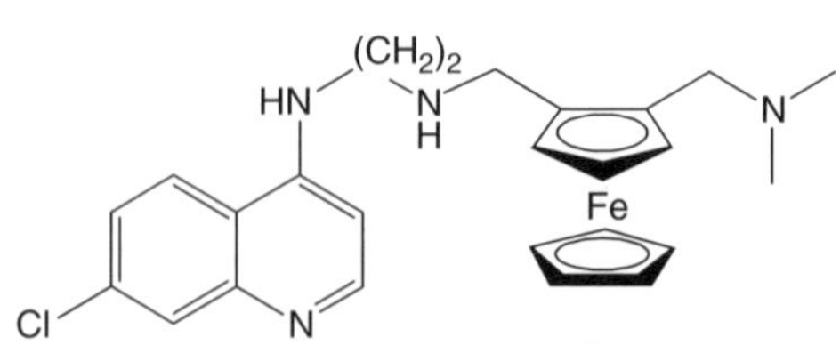

21

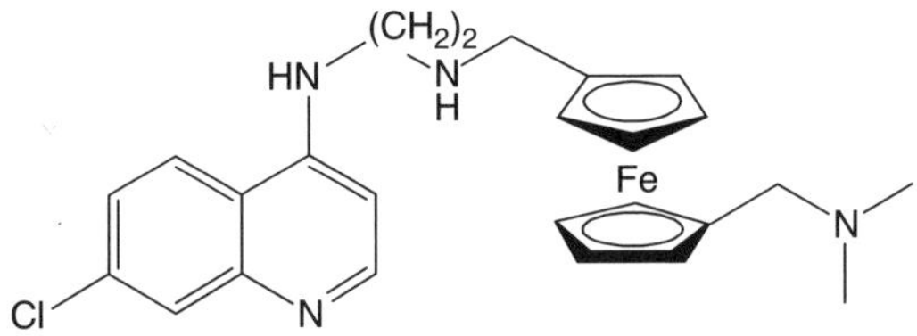

22

Extending the methylene chain in **21** ($n = 2$) to give homologues **23–25** ($n = 3,4,6$) caused a gradual decrease in activity against the chloroquine-sensitive D10 strain, with only **21** retaining the same level of activity as chloroquine.[38] When an amide group is inserted into the secondary amino group of the side chain of **21** and **23–25** to give ureas **26–29** respectively, a rebound in activity was observed (Table 10.1). The increase in activity was consistently observed for all the ureas **26–29**.

However, a different trend is observed when the activity was evaluated against the chloroquine-resistant K1 strain. Here, extending the methylene chain did not have a clear-cut effect on the activities of **21** and **23–25**. The analogue **23** ($n = 3$) is about three times more active than chloroquine while **24** ($n = 4$) is comparable with chloroquine. Among the ureas **26–29**, activity decreased as the methylene chain was extended (**26** > **27** > **28** > **29**), a trend not observed for the amines.

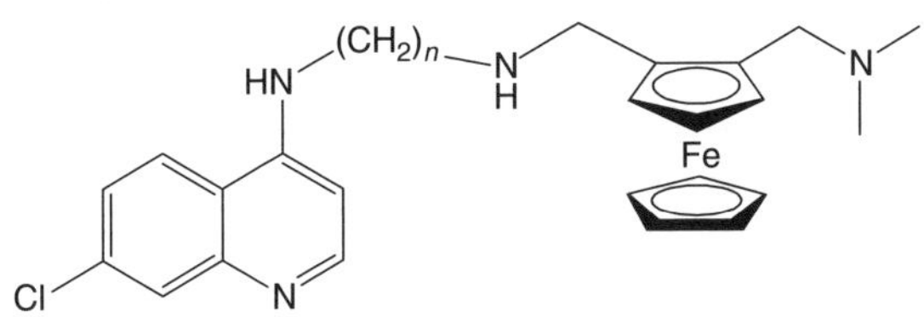

23: $n = 3$
24: $n = 4$
25: $n = 6$

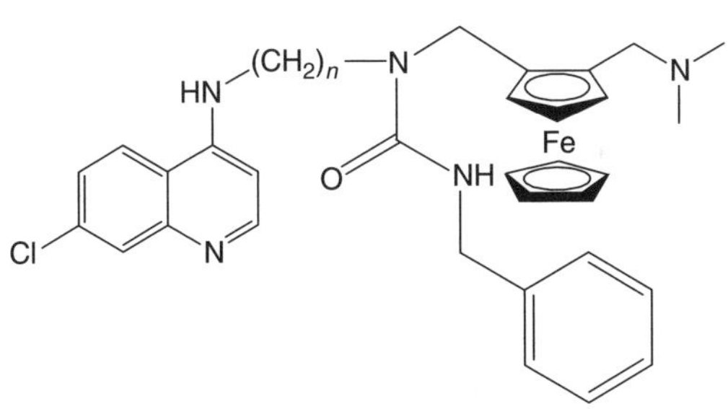

26: $n = 2$
27: $n = 3$
28: $n = 4$
29: $n = 6$

Table 10.1 Anti-plasmodial activity and half-wave potential ($E\frac{1}{2}$) of compounds **21**, amines **23–25** and ureas **26–29**[38]

Compound	n*	IC_{50} D10 (nM)	IC_{50} K1 (nM)	$E^1/_2$ (mV)
Chloroquine	–	41.86	125.38	–
21	2	41.70	73.46	90
23	3	51.37	36.93	66.5
24	4	61.16	111.5	60
25	6	86.92	81.39	–
26	2	21.35	37.50	156.5
27	3	16.20	47.41	131
28	4	16.74	75.23	115.5
29	6	19.01	110.20	122.5

*As shown in structures **21**, **23–25** and **26–29**.

When an amide moiety is attached to the secondary amino function of **21** and **23–25**, a urea is obtained. This has the effect of reducing the basicity of the amino function, as well as diminishing its hydrogen bond donor and acceptor abilities. Bulk is also increased in the urea. These changes in physicochemical properties appear to enhance activity against chloroquine-sensitive plasmodia (**26–29** > **21, 23–25**) but not against resistant plasmodia.

The results in Table 10.1 also emphasize the importance of the methylene side chain for activity. The length of the methylene spacer in 7-substituted 4-aminoquinolines is a major determinant of activity against chloroquine-resistant *P. falciparum*.[39] The same appears to be true for the ferrocene ureas **26–29**. A longer methylene side chain resulted in lower activity against the chloroquine-resistant strain K1 (**26** > **27** > **28** > **29**).

The authors also noted a progressive decrease in the half-wave potential of the ferrocene analogues as the methylene chain is lengthened from $n=2$ to $n=4$.[38] The half-wave potential measures the ease with which Fe^{2+} in ferrocene undergoes oxidation. Comparing the ureas **26** and **28**, the more active **26**, which has a shorter methylene chain ($n=2$), is associated with a larger half-wave potential, indicating that Fe^{2+} is more resistant to oxidation. The significance of this observation was not discussed further by the authors but it may signal a role for Fe^{2+} in anti-plasmodial activity. Not withstanding these observations, Beagley *et al.*, in another study, found equivalent anti-plasmodial activities for ferrochloroquine and its phenyl analogue **30** and concluded that the primary role of ferrocene in ferrochloroquine is that of a hydrophobic spacer.[37]

The mechanism of action of ferrochloroquine in malaria is still under investigation. Ferrochloroquine is a weaker base than chloroquine with pK_a values of 6.99 and 8.19 for the quinoline nitrogen and the tertiary amine, respectively.[31] The mechanism of action of chloroquine has been widely reviewed.[40]

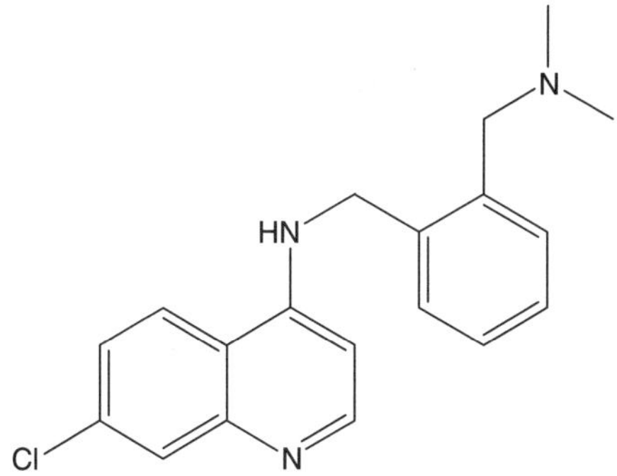

30

The basic properties of chloroquine are believed to contribute in a significant way to its anti-plasmodial activity. It is hypothesized that the diprotonated form of chloroquine is trapped in the acidic food vacuole of the plasmodia (pK_a of chloroquine = 7.94 and 10.03[31]). Because of the poor permeability of the protonated species, exceptionally high concentrations of chloroquine build up in the food vacuole where it is able to disrupt critical processes like the disposal of heme. This mechanism may not apply to ferrochloroquine as it is a weaker base than chloroquine and is also more lipophilic. Ferrochloroquine was originally designed with the intention that the presence of Fe^{2+} would reduce chloroquine resistance.[30] This is indeed found to be true for ferrochloroquine but the question of how, or even if, the ferrous ion contributes to the activity of ferrochloroquine remains unanswered.

10.6 Other Ferrocenyl Anti-plasmodial Agents

The success of ferrochloroquine has encouraged the incorporation of the metallocene into many other anti-malarial agents. Unfortunately, the results have not always been favourable. The ferrocenic mefloquine and quinine analogues (for example, **31**) were less active than the parent compounds,[41] but this may be due to the instability of the ferrocenyl analogues in solution. Various ferrocenic analogues of artemisinin have been synthesized.[42] It was proposed that Fe^{2+} in ferrocene may hasten the cleavage of the peroxide bond which is widely held to contribute to anti-plasmodial activity. Only one analogue (**32**) was found to be comparable to artemisinin. Coincidentally, **32** is also the only analogue with a basic amino function. The amino function enhances solubility through salt formation and this factor may be the reason why anti-plasmodial activity is retained.

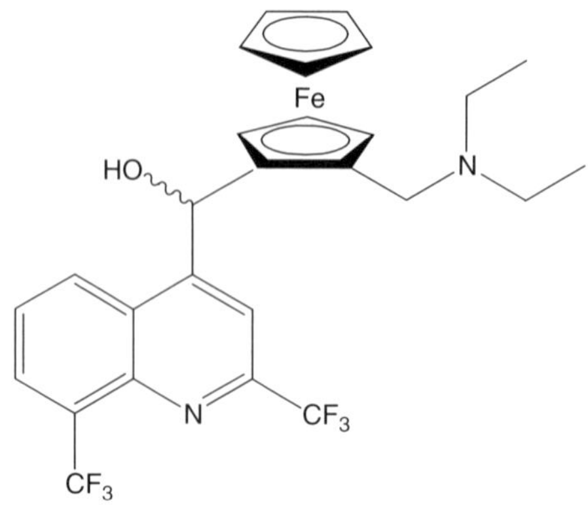

31

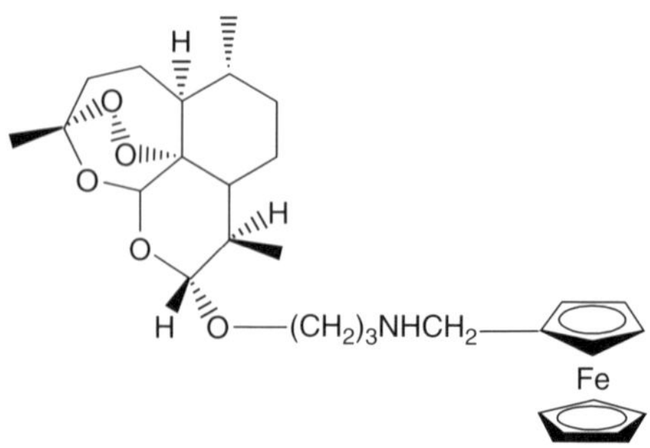

32

In another investigation, ferrocenyl analogues of chalcones were found to exhibit micromolar inhibitory activity against chloroquine-resistant *P. falciparum* K1.[43] The ferrocene ring was inserted into either ring A or ring B of the chalcone template. In general, better activity was observed when ring B was ferrocene. Compared to the chalcones where both rings A and B are phenyl, the ferrocenyl chalcones are generally less active. The most active members, **33** and **34**, inhibited hypoxanthine uptake into *P. falciparum* K1 schizonticides with IC_{50} values of 4.5 μM and 5.1 μM, respectively. Differences in activity are not readily explained by the size or lipophilicity of the ferrocenyl chalcones.

Itoh and coworkers reported the anti-plasmodial activity of ferrocene ellagitannins which contain biphenic acid and ferrocene rings on the glucose core.[44] Omission of ferrocene resulted in less active compounds.

Haworth and Hanlon designed a series of benzimidazolium salts with *N*-ferrocenylmethyl substituents.[45] The authors noted that malaria increases the oxidative stress in red blood cells. In response to this situation, various

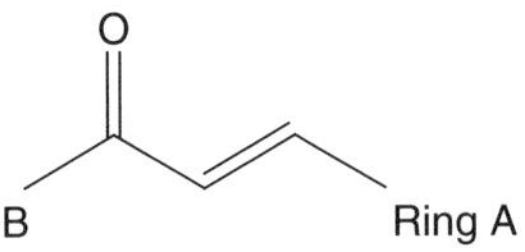

	Ring B	Ring A
33	3-Pyridyl	Ferrocenyl
34	Ferrocenyl	4-Nitrophenyl

physiological processes in the infected host are stepped up, in particular those that supply substrates for reduction which would react with the reactive oxygen species generated by the parasite's presence in the body. They reasoned that exogenous compounds that are reductive targets would also be taken up by infected cells and manifest anti-plasmodial activity. In this context, methylene blue, a cationic redox dye, may exert its anti-plasmodial activity in this way. Since benzimidazolium salts like **35** have a similar reversible redox property, activity might be anticipated in these compounds. The inclusion of the ferrocene ring may be prompted by the fact that the metal ion also participates in a redox reaction (Fe^{2+}/Fe^{3+} equilibrium) and may complement the redox property of the benzimidazolium moiety.

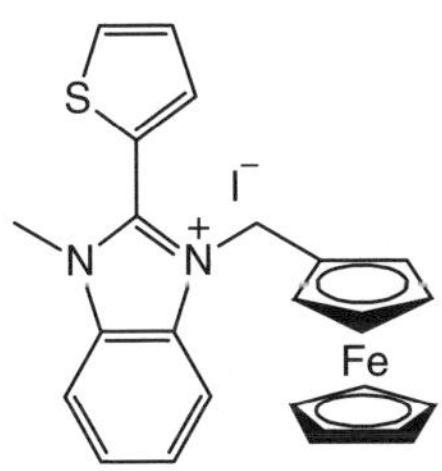

35

10.7 Organoiron as Anti-cancer Agents

The outstanding anti-tumour activity of cisplatin has stimulated a search for other inorganic cytostatics.[46] In the case of iron, attention has focused on ferrocene and ferricenium salts. The current status of these two classes of compounds as cytotoxic agents is reviewed in the following paragraphs. Derivatives in which an alkylating residue is attached to ferrocene (as in the case of ferrocenealdehyde *N*-methyl-*N*-β-chloroethyl hydrazone[47]) are excluded from this discussion.

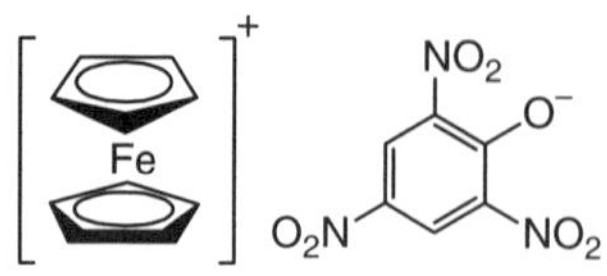

36

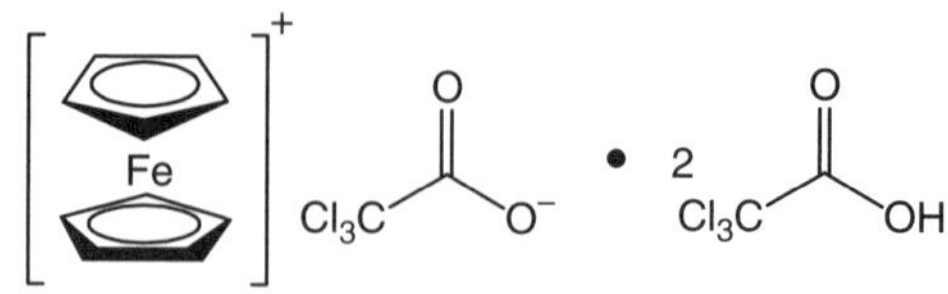

37

Anti-tumour activity in organoiron compounds was first reported for several ferricenium salts (for example, the picrate **36** and trichloroacetate **37**).[28] The inactivity of poorly soluble ferrocene compared to the water-soluble ionic ferricenium salts led to the conclusion that the difference in anti-tumour activities can be traced to solubility. The ferricenium salts (but not the ferrocenes) presumably have the right balance of lipophilicity and hydrophilicity that facilitate permeability. If this is so, then poorly soluble ferrocenium salts should lack anti-tumour activity, and this was corroborated by the absence of activity for the poorly soluble ferricenium heptamolybdate.[28] The corollary would be that appropriate substitution of ferrocene with hydrophilic substituents should result in active compounds. This approach was explored in great detail in a series of polyaspartamide ferrocene conjugates with successful outcomes.[27]

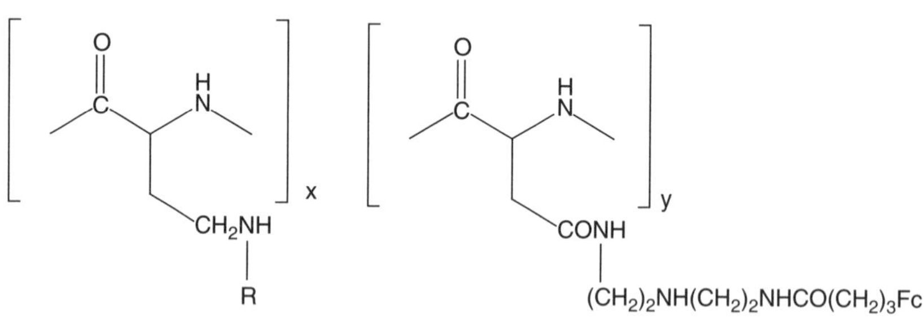

38: R = $-(CH_2)_3N(CH_3)_2$

39: R = $-(CH_2)_2OH$

Compound **38** is an example of a ferrocene (Fc)-containing polyaspartamide conjugate with anti-proliferative activity.[48] The backbone of **38** is formed from a mix of α- and β-/D- and L-aspartamide units linked by amide bonds. The presence of β-peptide and D- configured units prevent premature enzymatic degradation of the polypeptide. The terminal amide group in aspartamide is modified by alkylation. In **38**, a pendant *N*-dimethyaminolpropyl is attached to some amides while 4-ferrocenylbutanoic acid is linked to other amide moieties in the polyaspartamide chain. Like other polyaspartamide conjugates, **38** functions as a water-soluble prodrug, with water solubility conferred by the tertiary amino function which is protonated at physiological pH. The presence of this group may also facilitate the uptake of the conjugate into cells (by adsorptive pinocytosis). The importance of the positively charged amino group is underscored when **38** is compared with its less active analogue **39** which lacks this feature. Once taken up in the cell, the low pH and presence of lysosomal enzymes promote the cleavage of the conjugate to release the active moiety, 4-ferrocenylbutanoic acid.

There has been considerable debate over the identity of the active moiety which exerts the anti-tumour effect. One view is that once internalized in cells, Fe^{2+} in ferrocene exists in equilibrium with Fe^{3+}. Electron transfer between Fe^{2+} and Fe^{3+} occurs rapidly and reversibly.[49] The electrochemical potential for this reaction falls within the range of other biological redox couples.[50] Thus, it is highly conceivable that electron transfer occurs between Fe^{2+}/Fe^{3+} and other biological electron donors/acceptors. Free radicals may be formed in the course of these reactions, and these contribute to anti-proliferative activity.

If this postulate holds true, then the oxidation state of iron (Fe^{2+} or Fe^{3+}) in the metallocene would not be the most important factor determining activity.[49] The more important consideration would be the accessibility of the agent in the cells and the ease with which the Fe^{2+}/Fe^{3+} equilibrium is set up. The latter is determined by the electrochemical potential of the Fe^{2+}/Fe^{3+} couple, which can be influenced by substituents on the metallocene,[51] as well as the conditions in the body compartment.[27] The choice of 4-ferrocenylbutanoic acid as the 'active' ferrocene in the polyaspartamide conjugates is traced to its low formal reduction potential (reflecting ease of oxidation).[52] Among the various ferrocene polyaspartamide conjugates evaluated, anti-proliferative activity was enhanced if structural modifications resulted in a lower reduction potential for the Fe^{2+}/Fe^{3+} couple.[49]

Another view holds that only ferricenium cations, not ferrocene or its derivatives, have anti-proliferative activity. Fe^{3+} in ferrocenium is a stable free radical cation. It can interact with DNA in many ways: covalent bond formation with nucleophilic groups in nucleotides or proteins, electrostatic attraction to negatively charged phosphate residues, charge transfer complex formation and intercalation with nucleotide bases.[28,53] In addition, Osella and co-workers showed that the ferricenium cation can generate the highly reactive hydroxyl radical $OH^{\bullet}$ in aqueous solutions.[53] Using electron spin resonance (ESR), they demonstrated the formation of a radical adduct when ferrocenium carboxylic acid hexafluorophosphate was incubated with a spin trap 5-(diethoxyphosphoryl)-5-methyl-1-pyrroline *N*-oxide (DEPMPO) in 4-(2-hydroxyethyl)-1-piperazineethane sulphonic acid (HEPES) buffer (pH 7.4). Addition of Ehrlich ascites tumour (EAT) cells attenuated the ESR signals indicating that the cells have reacted with the free radicals generated from the redox activity of the ferrocenium salt. They postulated that these free radicals cause the fragmentation of DNA in EAT cells. In a subsequent investigation, they demonstrated the generation of hydroxyl radicals from another ferricenium salt, decamethylferrocenium tetrafluorborate (DEMFc+), which is cytotoxic towards the human breast adenocarcinoma cells (MCF-7).[54] Unlike other ferricenium salts investigated so far, DEMFc+ demonstrates unusual stability in aqueous solutions.

Several investigations have shown that the unsubstituted ferrocene ring has no anti-proliferative activity.[28,53,55] However, this does not hold true for substituted ferrocenes. For example, simple ferrocene derivatives like ferrocenylacetic acid and 3-ferrocenylbutanoic acid have shown modest antiproliferative activity against murine EMT-6 cancer cells.[49] Ong and co-workers noted that the ferrocenyl acridine **40** inhibited the growth of various cancer cell lines (KB, Hela, Colo-205, Hep) with IC_{50} values of 1–2 μg/ml.[56] The mode of action of **40** is proposed to be mediated by the ferrocenium ion which is formed when Fe^{2+} in ferrocene is oxidized in the biological milieu (for example, by peroxidases in the presence of hydrogen peroxide). The ferrocenium cation would then react with the nucleophilic groups in DNA. The acridine ring which is known to be an effective DNA intercalator helps to position the ferrocene moiety in the vicinity of DNA, and facilitates the DNA–ferrocenium reaction. Unfortunately, no experimental evidence was given to support this hypothesis.

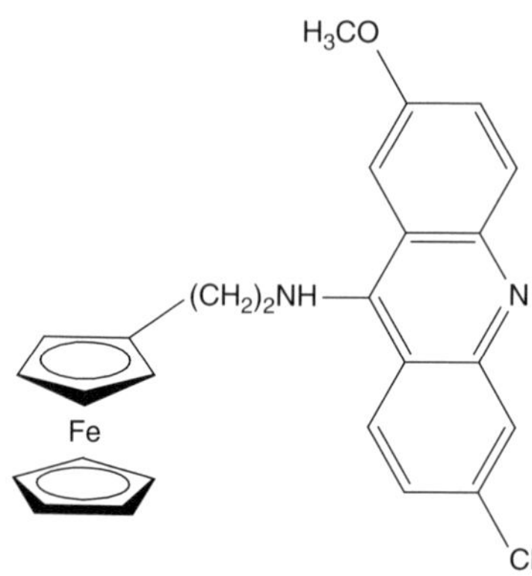

40

The proposal of a ferrocene–DNA interaction has been mentioned earlier in relation to the ferrocifens **5** and **6**. Whereas Ong and coworkers[56] suggested an interaction mediated by the ferrocenium cation, Top *et al.*[26] propose that a Fenton-like reaction involving Fe^{2+} (Equations 10.1–10.3) would explain the genotoxic effects of these ferrocifens.

$$Fe^{2+} + O_2 \rightarrow Fe^{3+} + O_2^{\bullet -} \tag{10.1}$$

$$Fe^{2+} + O_2^{\bullet -} \rightarrow Fe^{3+} + H_2O_2 \tag{10.2}$$

$$Fe^{2+} + H_2O_2 \rightarrow Fe^{3+} + OH^- + OH^{\bullet} \tag{10.3}$$

While these findings may indicate that ferrocenes owe their anti-proliferative activity to 'proxy' species like the ferrocenium radical ion, and reactive oxygen species like the superoxide and hydroxy free radicals, there is evidence to support a more direct role for ferrocene. Mono- and di-substituted acetyl and carboxaldoxime ferrocenes have been reported to have *in vitro* activity against five cancer cell lines.[55] The acetyl derivatives (**41**, **42**) were less active than the carboxaldoxime derivatives (**43**, **44**). Disubstitution of the ferrocene ring further promotes activity (**42** > **41**; **44** > **43**). These compounds were found to be inhibitors of topoisomerase II (Topo II), an enzyme which plays a critical role in the recombination, segregation and DNA repair processes encountered during cell division and replication. Inhibition was not due to interference with the binding or hydrolysis of ATP which is required for the catalytic activity of Topo II. Instead, the ferrocenes interfered with the formation of the DNA–Topo II complex. It was proposed that each of the diacetyl **42** and dicarboxaldoxime **44** derivatives initially interacts with Topo II in the absence of DNA. DNA combines with the drug–Topo II complex and forms a stable ternary complex which promotes the cleavage of DNA and prevents its relegation.

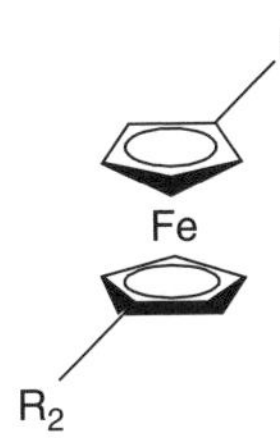

41: R_1 = –$COCH_3$, R_2 = H
42: R_1 = R_2 = –$COCH_3$
43: R_1 = –CH=NOH, R_2 = H
44: R_1 = R_2 = –CH=NOH

Kovjazin and coworkers showed that ferrocene demonstrated an anti-tumour effect when administered intraperitoneally or orally to mice with B-16 melanoma.[57] The anti-tumour effect was observed at very low doses (0.05–0.2 mg/kg) which was about 2000 times less than the LD_{50} of ferrocene. They proposed that the anti-tumour effect of ferrocene was mediated by immune stimulation, possibly against a target like $p21^{ras}$ which is activated by redox signalling. In addition, they noted that immune cells from ferrocene-treated tumour-bearing mice were capable of eliciting an anti-tumour reaction in mice not treated with ferrocene.

10.8 Conclusions

There are several factors that favour the development of organometallics as drug candidates, viz., the fact that the coordination and redox properties of metal ions have been investigated extensively and are well defined, the positively charged metal ions will facilitate binding to negatively charged biomolecules and the novelty introduced by the inclusion of metal ions into a promising molecule.

On the other hand, there is concern over the pharmacokinetic properties of these compounds, in particular the disposition of the metal ion in the body, since accumulation will lead to toxic side effects. The presence of the metal ion also adds a new dimension to the pharmacodynamic activity of the organometallic drug. This is well illustrated with ferrocene, an organoiron moiety that is widely used in drug design. As discussed in the preceding paragraphs, the contribution of the ferrocene moiety remains disputed. The ferrocene moiety could function as an inert hydrophobic spacer, that is, as a convenient isostere of phenyl or alkyl groups whose presence in a molecule would enhance lipophilicity and thus permeability into cells. Alternatively, the chemically stable ferrocene moiety may become redox active in the biological milieu and generate reactive entities like ferrocenium and free radicals that contribute to pharmacodynamic effect. Recent evidence suggests that ferrocenes may act directly, as stimulators of the immune system or as topoisomerase II inhibitors. Whether ferrocene is active *per se* or acts via intermediaries (free radicals, ferricenium cation), the anti-proliferative effect that is sought after in these compounds is not compromised and may even be enhanced by the variety of options. But this may not necessarily be true for other kinds of pharmacological activity. For instance, would anti-inflammatory or anti-malarial activity benefit from the operation of these various pathways? Is it possible to 'switch on' or 'switch off' selected pathways by modifications of the ferrocene ring? These are interesting questions that should be explored in future studies.

References

1. E.M.E. Poskitt, *Br. J. Haemotol.*, **122**, 554–562 (2003).
2. P. Blaud, *Revue Medicale Francaise Etrangere.*, **45**, 341–367 (1832).
3. J. Thomson and L. Findlay, *The Chemical Study of Treatment of Sick Children*, Oliver & Boyd, Edinburgh, 1933.
4. H.M.M. Makay, *Arch. Dis. Childhood,* **3**, 116–144 (1928).
5. H.M.M. Makay, MRC Special Report No. 157, HMSO, London, 1931.
6. L. Hallberg, Iron, in *Human Nutrition and Dietetics*, J.S. Garrow, W.P.I. James and A. Ralph (Eds), Churchill-Livingston, Edinburgh (2000).
7. R.S. Hillman, Hematopoietic agents, in Goodman and Gilman's The *Pharmacological Basis of Therapeutics*, J.G. Hardman, L.E. Limbird and A. Goodman Gilman (Eds), McGraw-Hill, New York (2001).
8. C. Hansch and A. Leo, *Substituent Constants for Correlation Analysis in Chemistry and Biology*, John Wiley & Sons, New York (1979).
9. S. Venkatesh and R.A. Lipper, *J. Pharm. Sci.*, **89**, 145–154 (2000).
10. R.A. Yeary, *Toxicol. Appl. Pharmacol.*, **15**, 666–676 (1969).
11. H.W. Leung, D.W. Hallesy, L.D. Shott *et al.*, *Toxicol. Lett.*, **38**, 103–108 (1987).
12. K.J. Nikula, J.D. Sun, E.B. Barr *et al.*, *Fund. Appl. Toxicol.*, **21**, 127–139 (1993).
13. B. Halliwell and J.M.C. Gutteridge, *Free Radicals in Biology and Medicine*, Oxford University Press, New York (1999).

14. R.P. Hanzlik and W.H. Soine, *J. Am. Chem. Soc.*, **100**, 1290–1291 (1978).
15. P. Nielsen and H.C. Heinrich, *Biochem. Pharmacol.*, **45**, 385–391 (1993).
16. J.W. Halliday and J. Searle, *Biometals*, **9**, 205–209 (1996).
17. E.E. Cable, J.R. Connor and H.C. Isom, *Am. J. Pathol.*, **152**, 781–792 (1998).
18. E.E. Cable and H.C. Isom, *Drug Metab. Dispos.*, **27**, 255–260 (1999).
19. B.E. Maryanoff, S.L. Keeley and F.J. Persico, *J. Med. Chem.*, **26**, 226–229 (1983).
20. D. Scutaru, L. Tataru, I. Mazilu *et al.*, *J. Organomet. Chem.*, **401**, 81–85 (1991).
21. D. Scutaru, I. Mazilu, L. Tataru *et al.*, *J. Organomet. Chem.*, **406**, 183–187 (1991).
22. C. Biot, N. Francois, L. Maciejewski *et al.*, *Bioorg. Med. Chem. Lett.*, **10**, 839–841 (2000).
23. M. Yuriko, E. Brummer and D.A. Stevens. *Myopathology*, **138**, 29–35 (1997).
24. S. Top, J. Tang, A. Vessieres *et al.*, *Chem. Commun.*, 955–956 (1996).
25. G. Jaouen, S. Top, A. Vessieres *et al.*, *Sci. Paris, Ser. IIc, Chim.*, **3**, 89–93 (2000).
26. S. Top, A. Vessieres, C. Cabestaing *et al.*, *J. Organomet. Chem.*, **637–639**, 500–506 (2001).
27. E.W. Neuse, *Macromol. Symposia*, **172**, 127–138 (2001).
28. P. Köpf-Maier, H. Kopf and E.W. Neuse, *J. Cancer Res. Clin. Oncol.*, **108**, 336–340 (1984).
29. C. Biot, G. Glorian, L.A. Maciejewski and J.S. Brocard, *J. Med. Chem.*, **40**, 3715–3718 (1997).
30. O. Domarle, G. Blampain, H. Aganiet *et al.*, *Antimicrob. Agents Chemother.*, **42**, 540–544 (1998).
31. L. Delhaes, H. Abessolo, C. Biot *et al.*, *Parasitol. Res.*, **87**, 239–244 (2001).
32. B. Pradines, T. Fusai, W. Daries *et al.*, *J. Antimicrob. Chemother.*, **48**, 179–184 (2001).
33. C. Atteke, J.M.M. Ndong, A. Aubouy *et al.*, *J. Antimicrob. Chemother.*, **51**, 1021–1024 (2003).
34. L. Delhaes, C. Biot, L. Berry *et al.*, *Chembiochem.*, **3**, 418–423 (2002).
35. C. Biot, L. Delhaes, H. Abessolo *et al.*, *J. Organomet. Chem.*, **589**, 59–65 (1999).
36. C. Biot, L. Delhaes, C.M. N'Diaye *et al.*, *Bioorg. Med. Chem.*, **7**, 2834–2847 (1999).
37. P. Beagley, M.A.L. Blackie, K. Chibale *et al.*, *Dalton Trans.*, 3046–3051 (2003).
38. K. Chibale, J.R. Moss, M. Blackie *et al.*, *Tetrahedron Lett.*, **41**, 6231–6235 (2000).
39. D. De, F.M. Krogstad, L.D. Byers and F.M. Krogstad, *J. Med. Chem.*, **41**, 4918–4926 (1998).
40. L. Tilley, P. Loria and M. Foley, Chloroquine and other quinoline antimalarials, in *Antimalarial Chemotherapy: Mechanisms of Action, Resistance and New Directions in Drug Discovery*, P.J. Rosenthal (Ed.), Humana Press, NJ (2001).
41. C. Biot, L. Delhaes, L.A. Maciejewski *et al.*, *Eur. J. Med. Chem.*, **35**, 707–714 (2000).
42. L. Delhaes, C. Biot, L. Berry *et al.*, *Bioorg. Med. Chem.*, **8**, 2739–2745 (2000).
43. X. Wu, P. Wilairat and M.L. Go, *Bioorg. Med. Chem. Lett.*, **12**, 2299–2302 (2002).
44. T. Itoh, S. Shirakami, N. Ishida *et al.*, *Bioorg. Med. Chem. Lett.*, **10**, 1657–1659 (2000).
45. J. Howarth and K. Hanlon, *Tetrahedron Lett.*, **42**, 751–754 (2001).
46. C.X. Zhang and S.J. Lippard, *Curr. Opin. Chem. Biol.*, **7**, 481–489 (2003).
47. M. Wenzel, M. Schneider and E. Liss, *Z. Naturforsch C.*, **34**, 670–676 (1979).
48. M.T. Johnson, E. Kreft, D.D. N'Da *et al.*, *J. Inorg. Organo. Polym.*, **13**, 255–267 (2003).
49. J.C. Swarts, D.M. Swarts, D.M. Maree *et al.*, *Anticancer Res.*, **21**, 2033–2038 (2001).

50. N.G. Connely and W.E. Geiger, *Chem. Rev.*, **96**, 877–910 (1996).
51. G. Caldwell, M.G. Meirim, E.W. Neuse and C.E.J. van Rensburg, *Appl. Organomet. Chem.*, **12**, 793–799 (1998).
52. J.C. Swarts, E.W. Nuese and G.J. Lambrecht, *J. Inorg. Organo. Polym.*, **4**, 143–153 (1994).
53. D. Osella, M. Ferrali, P. Zanello *et al.*, *Inorg. Chim. Acta*, **306**, 42–48 (2000).
54. G. Tabbi, C. Cassino, G. Cavigiolio *et al.*, *J. Med. Chem.*, **45**, 5786–5796 (2002).
55. Y.N. Vashisht, G.D. Jayaraju and A.K. Kondapi, *Arch. Biochem. Biophys.*, **376**, 229–235 (2000).
56. C.W. Ong, J.Y. Jeng, S.S. Juang and C.F. Chen, *Bioorg. Med. Chem. Lett.*, **2**, 929–932 (1992).
57. R. Kovjazin, T. Eldar, M. Patya *et al.*, *FASEB J.*, **17**, 467–469 (2003).

11 $_{27}$Co Cobalt Complexes as Potential Pharmaceutical Agents

Hui Chao and Liang-Nian Ji

11.1 Introduction

Cobalt is a relatively rare metal that forms approximately 0.001% of the earth's crust. Its vital status was not proved until after World War II, in 1948, when the 'anti-pernicious anemia factor', which came to be called 'Vitamin B_{12}', was finally purified and isolated as crystals.[1] Since then, cobalt complexes have received much attention. The diamagnetic and stable nature of cobalt(III) complexes make them excellent models for metalloenzymes.[2–4] As potential pharmaceutical agents, cobalt complexes have been used for the development of enzyme inhibitors,[5–7] nucleic acid structure probes,[8] DNA-cleaving agents[8] and anti-tumor drugs.[9] In addition, cobalt complexes are also being studied for their utility as drug delivery devices,[10] hypoxic selective agents[11] and positron emission tomography (PET) imaging agents.[12] In this chapter, the use of cobalt in medicine will be outlined.

Metallotherapeutic Drugs and Metal-Based Diagnostic Agents: The Use of Metals in Medicine
Edited by Gielen and Tiekink

11.2 Enzyme Inhibition/Induction

11.2.1 Serine protease inhibitors

The serine proteases comprise a major family of proteolytic enzymes with a diverse range of functions. Serine proteases play key roles in the blood coagulation cascade, in complement activation, bacterial pathogenesis and fibrinolysis.[13] Serine proteases have come into prominence as targets for drug design in anti-viral and anti-bacterial research.[14] Therefore, there is a great deal of interest in the development of inhibitors for serine proteases.

Cobalt(III)-ligated amino acids and dipeptides were reported to inhibit chymotrypsin and trypsin.[15,16] These studies were undertaken to examine the effects of pentaminecobalt(III), which is used as a hydrophilic carboxyl protective group in enzymatic peptide synthesis. It had been previously observed that the presence of this moiety on the amino acid adjacent to the enzyme-susceptible bond blocked catalytic conversion. Recently, cobalt(III) Schiff-base complexes and peptide conjugates of these complexes were shown to effectively inhibit thrombin and thermolysin (Figure 11.1).[5–7] The peptide conjugate selectively inhibited thrombin in a mixture with two other serine proteases. Selectivity is conferred by the peptide component, while inhibition is believed to occur by cobalt(III) coordination to histidines in the active site. Alternatively, the cobalt may be coordinating to histidines near, but not in, the active site to lock the peptide into the substrate-binding pocket. These complexes are also found to bind to histidines in a model system using a zinc finger peptide from the HIV nucleocapsid protein.[17]

11.2.2 Topoisomerase II inhibitors

Topoisomerase II (Topo II) is a nuclear protein known to play crucial roles in DNA replication,[18] condensation and segregation of chromosomes,[19,20] and transcription.[21] Topo II catalyzes a number of ATP-dependent reactions, such

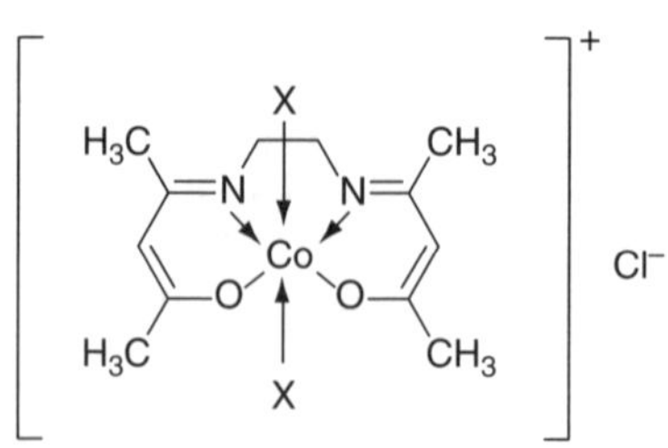

Figure 11.1 R = CH_3 for Co(acacen)X_2; R = COOH for Co(aciden)X_2, peptide attached through COOH group

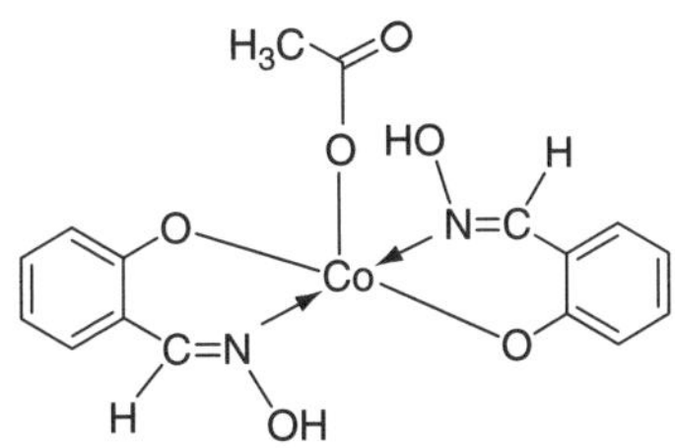

Figure 11.2 Chemical structure of cobalt(III) salicylaldoxime (CoSAL)

as relaxation of negative supercoils, knotting and unknotting, and catenation and decatenation of DNA *in vitro*.[22–24]

Cobalt(III) salicylaldoxime complex (CoSAL) was found to possess anti-proliferative properties (Figure 11.2).[25] Recently, through DNA-binding studies and studies of the action of this complex on Topo II catalytic activities, it was found that Topo II is one of the cellular targets of CoSAL and thereby accounts for the anti-proliferative activity. The inhibition of topo II activity by CoSAL is due to the bidirectional mode of interaction with enzyme and DNA which could aid in forming a drug-mediated cleavage complex.[26] The 2-furfural semicarbazone and thiosemicarbazone cobalt complexes were also found to suppress human DNA topoisomerase II activity with DNA nicks and DNA fragmentation,[27] but they did not alkylate the bases of DNA, cause intercalation between base pairs or cause cross-linking of DNA strands. Cobalt complexes of 2-furaldehyde oximes were also synthesized and examined.[28] These cobalt complexes were inhibitors of DNA polymerase α-activity and *de novo* purine synthesis with marginal inhibition of ribonucleoside reductase and dihydrofolate reductase activities with DNA fragmentation, but they did not inhibit L1210 DNA Topo II activity, except reducing thymidylate synthetase activity.

11.2.3 Heme oxygenase-1 inducers

Heme oxygenase-1 (HO-1) or HSP 32 is the inducible isoform of heme oxygenase (HO),[29] which catalyzes the breakdown of heme into carbon monoxide, iron and biliverdin that is immediately reduced to bilirubin.[30] These products of the HO reaction have important physiological effects: carbon monoxide functions as a signaling molecule, biliverdin and its reduced product bilirubin are potent anti-oxidants, and 'free' iron increases oxidative stress and regulates the expression of many mRNAs. Many studies have suggested that HO-1 acts as an inducible defense against oxidative stress, in models of inflammation, ischemia-reperfusion, hypoxia and hyperoxia-mediated injury.[31–35]

Cobalt protoporphyrin has been well characterized as inducer of HO-1 and increased the expression of HO-1 at the level of transcription.[36,37] This was extended using primary cultures of chick embryo liver cells and the leghorn male hepatoma (LMH) line of avian hepatoma cells.[38] Cobalt protoporphyrin strongly induced luciferase expression in both types of transfected cells by a mechanism fundamentally different from that of sodium arsenite and other stress inducers. Deletional analysis showed that cobalt porphyrin-dependent up-regulation of gene expression was found to require the first 5.6 kb of the HO-1 upstream promoter, but not the consensus AP-1 binding elements found in this region. Further investigations show that an undefined regulatory region in the 5′-UTR of the chick HO-1 gene confers up-regulation of reporter gene (luciferase) expression in the presence of heme and other selected metalloporphyrins.[39]

11.3 Nucleic Acid Binding and Cleavage

11.3.1 Cobalt(III) polypyridyl complexes

Over the past decades, the potential application of polypyridyl transition metal complexes as non-radioactive nucleic acid probes, which can recognize and cleave DNA, has stimulated intensive research.[8,40–43] In comparison to other complexes, the octahedral polypyridyl complexes are particularly suitable for these applications because they are coordinatively saturated, inert to substitution, rigid and structurally well defined. In addition, these complexes can interact non-covalently with nucleic acids by intercalation, groove binding or electrostatic binding. When the size, shape and chirality of the complexes are suited to the DNA structure, the complexes may approach closely to and intercalate the base pairs of DNA; and the spectroscopic properties, such as electronic absorption and steady-state emission, and the thermodynamic properties, such as viscosity, will display change. On the other hand, upon irradiation, these complexes can induce single- or double-stranded cleavage of plasmid DNA. Although they have not received as much attention as the ruthenium system, cobalt complexes have attracted some interest due to their unique electronic structure and redox properties.

Maiya *et al.* reported the interaction between DNA and cobalt(III) complexes-containingdipyrido[3,2-*a*:2′,3′-*c*]phenazine (dppz) and its derivatives (Figure 11.3).[44,45] This study demonstrated that these cobalt(III) complexes bind to DNA in an intercalative mode and affect the photocleavage of DNA (supercoiled pBR 322 DNA). Ji and co-workers discovered that cobalt(III) complexes-containing imidazo[4,5-*f*][1,10]-phenanthroline derivatives are avid binders of calf-thymus DNA (Figure 11.4).[46–49] Cobalt(III) complexes have been found to promote the single-stranded cleavage of plasmid pBR 322 DNA upon irradiation. Mechanisms for photocleavage were also investigated. Recently,

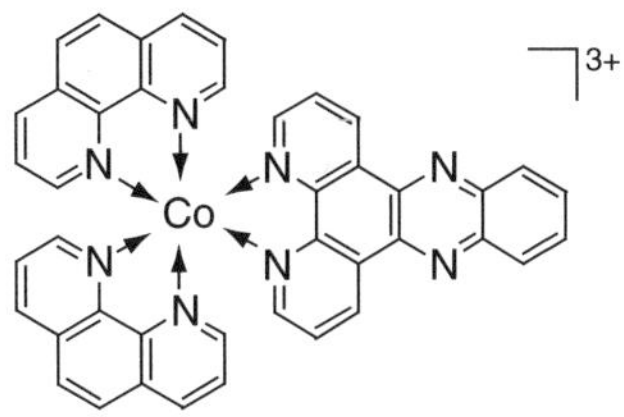

$[Co(phen)_2(dppz)]^{3+}$

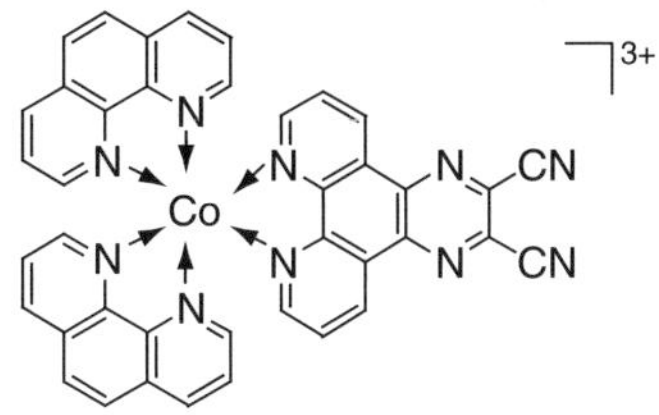

$[Co(phen)_2(dicnq)]^{3+}$

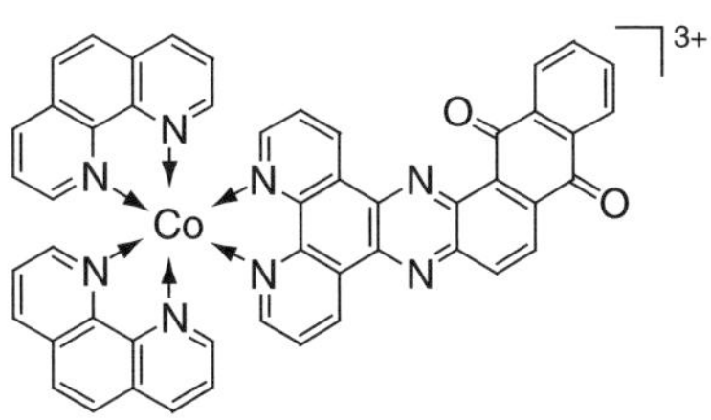

$[Co(phen)_2(qdppz)]^{3+}$

Figure 11.3 Cobalt(III) complexes-containing dppz and its derivatives

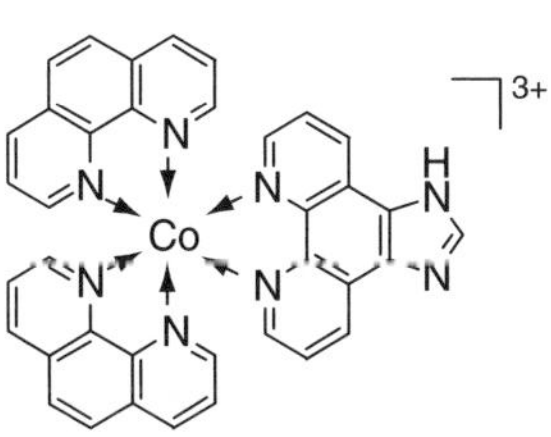

$[Co(phen)_2(IP)]^{3+}$

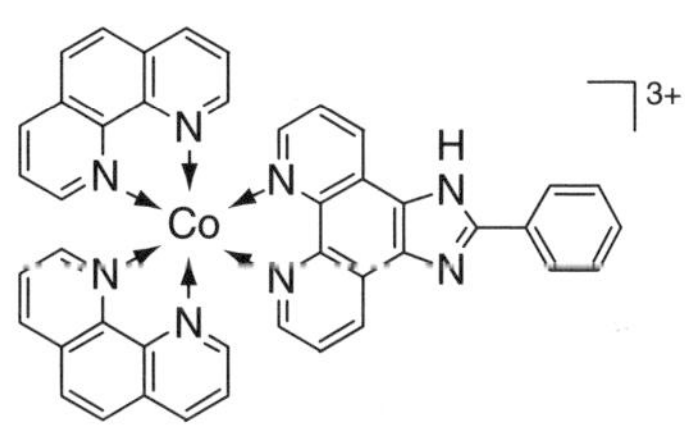

$[Co(phen)_2(PIP)]^{3+}$

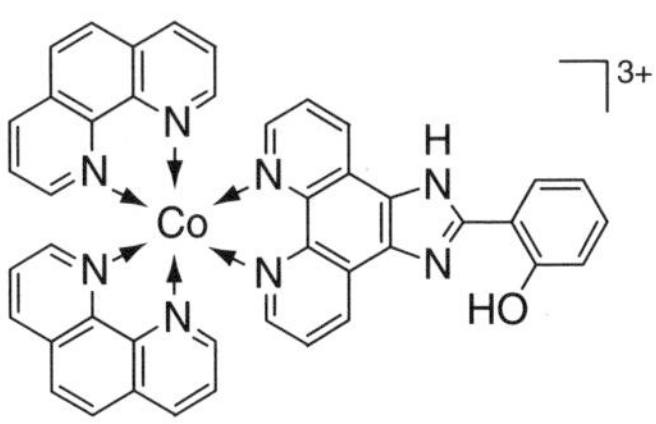

$[Co(phen)_2(HPIP)]^{3+}$

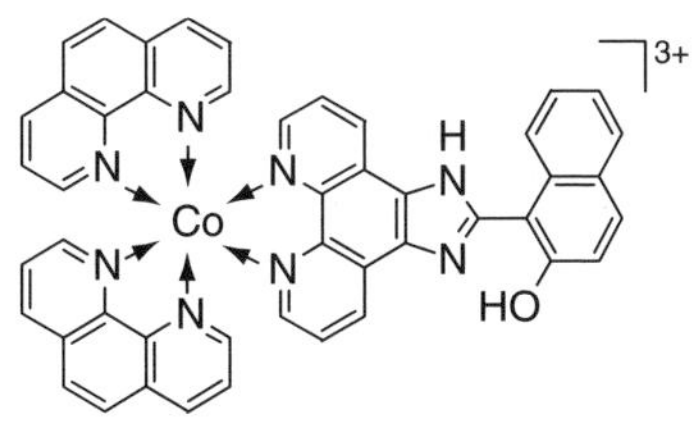

$[Co(phen)_2(HNAIP)]^{3+}$

Figure 11.4 Cobalt(III) complexes with imidazo[4,5-*f*][1,10]phenanthroline derivatives

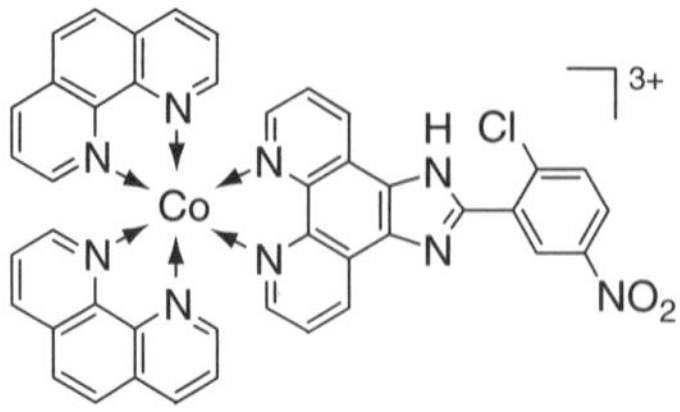

$[Co(phen)_2(CNOIP)]^{3+}$

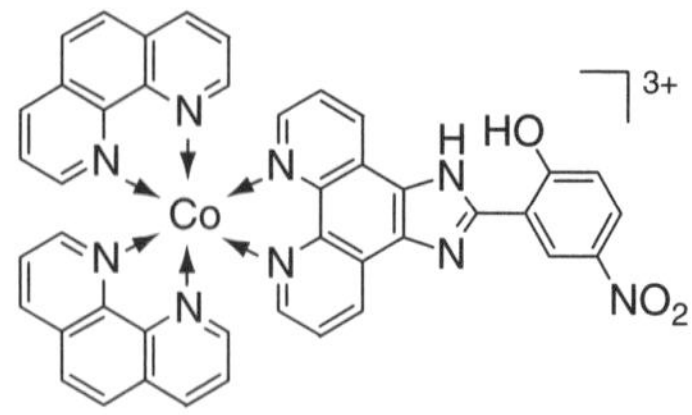

$[Co(phen)_2(HNOIP)]^{3+}$

Figure 11.4 (Continued)

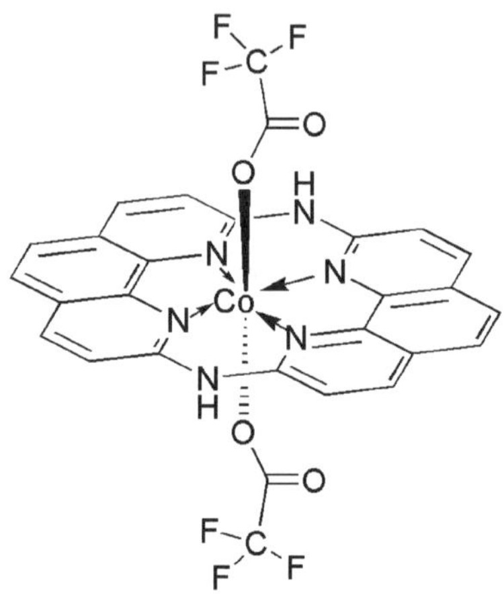

Figure 11.5 Chemical structure of $[Co(tfa)_2(happ)]$

Cheng and co-workers have synthesized a novel octahedral cobalt(II) complex $[Co(tfa)_2(happ)]$ (tfa = trifluoroacetate; happ = hexaazaphenanthroline-cyclophane) (Figure 11.5). This complex exhibits low affinity toward double-stranded DNA and low cleavage reactivity toward single-stranded DNA. However, this complex was found to act as a specific probe for the DNA bulge with single-stranded DNA cleavage activity in the presence of H_2O_2.[50,51]

In the last few years, some other ligands have been complexed to cobalt(III) complexes.[52,53] Nonetheless, no significant improvements have been obtained with respect to the above-mentioned examples.

11.3.2 Cobalt(III) bleomycin complexes

The bleomycins (Blms) are a group of glycopeptide antibiotics isolated from *Streptomyces verticillus*, which are used clinically for treating cancers of the head, neck and testes, as well as certain lymphomas.[54] These antibiotics require a metal ion like iron(II) and dioxygen as co-factors to produce their drug action.[9,55–58] When iron(II) reacts with bleomycin in the presence of O_2, activated bleomycin HOO—Fe(III)Blm is formed which is proposed to have a square bipyramidal metal ion coordination.[59,60] This species is particularly difficult to study because it is unstable and reacts with itself in the absence of DNA.[60–62] At the same time, HOO—Fe(III)Blm, Fe(III)- and Fe(II)Blm are also paramagnetic, making nuclear magnetic resonance (NMR) structural investigations difficult.

The Co(II)Blm complex has been used as a model for FeBlm as it can be aerobically oxygenated in a fashion reminiscent of Fe(II)Blm to create HOO—Co(III)Blm (green) and H_2O—Co(III)Blm (brown), respectively.[63,64] The diamagnetic and stable nature of Co(III)Blm have made them excellent subjects for NMR structural investigations.

Recently, the structure of Co(III)Blm bound to oligonucleotide substrates has been studied in detail by NMR spectroscopy.[2,65–70] Examination of HOO—Co(III)Blm associated with DNA oligomers containing a specific binding site of 5′-guanine-pyrimidine-3′ reveals that the metal domain and linker are folded into a unit that interacts with the minor groove edge of the guanine. The drug's bithiazole moiety partially intercalates DNA and hydrogen bonding occurs between the pyrimidine moiety and minor groove of the G residue proximate to the intercalation site. In the process of association, the hydroperoxide ligand becomes oriented toward the C4′-hydrogen of the deoxyribose. These NMR studies also provided a number of observations to support the hypothesis that HOO—Co(III)Blm is an excellent structural model of activated FeBlm.[2,65,69]

Although Co(III)Blms strongly bind to DNA, previous studies indicate that they do not cause damage to DNA under aerobic conditions. However, the situation was changed in 1982 when Chang and Meares first observed a significant DNA photocleavage by a cobalt(III)–bleomycin complex.[71] This renewed interest in the photochemistry of cobalt(III) complexes with bleomycin. Subsequently, Saito and co-workers studied the photocleavage characteristics of HOO—Co(III)Blm (green).[72] When the self-complementary dodecamer 5′-CGCTTTAAAGCG-3′ was used as the target, selective cleavage was observed after alkali treatment at positions C-3 and C-11. Free cytosine and alkali-labile lesions were detected as the products of the reaction by high performance liquid chromatography (HPLC) and attributed to cleavage at C-3 initiated by hydrogen atom abstraction from C-4′ of deoxyribose. These results clearly demonstrate that the site of attack by cobalt(III)Blm is deoxyribose, but they do not offer insight into the oxidizing species or for its mechanism of formation.

To elucidate the mechanism of the light-induced DNA strand cleavage reaction, Mascharak and co-workers[73,74] have synthesized a series of Co(III)-PMAH complexes as simplified Co(III)Blm analogues that lack the DNA-binding bithiazole moiety of the natural ligand. DNA cleavage and spin-trapping experiments demonstrate that the UV light-induced DNA cleavage reaction by

Co(III)-PMAHs is neither a consequence of photoreduction of the complexes to cobalt(II) nor are they initiated by singlet oxygen. They should be attributed to the formation of the C/N-based radical on the ligand framework, which rapidly reacts with water to produce hydroxyl radicals near to the DNA helix with consequent cleavage. Recently, Nishida *et al.* have suggested a different mechanism for visible light-induced DNA cleavage reaction by HOO—Co(III)Blm.[75] The cleavage process is related to the peroxide adduct. When the system is irradiated by visible light, it seems quite likely that a direct hydroxylation reaction may occur at the C4′ position associated with a concerted heterolytic O—O cleavage reaction without formation of a C4′ radical.

11.3.3 Cobalt(III) polyamine complexes

Hydrolysis of the phosphodiester bond is of critical importance at several stages in a cell cycle (including DNA repair and excision, integration and signal transduction). Metal-containing enzyme-mediated nucleic acid hydrolysis is therefore a common catalytic pathway in nucleic acid biochemistry.[76] Cobalt(III) polyamine complexes have long been known to be among the most effective catalysts for the hydrolysis of phosphate esters.[77–79] Chin *et al.*[80] have demonstrated the particular capabilities of 1,4,7,10-tetraazacyclododecane (cyclen) or tris(3-aminopropyl)amine (trpn) complexes (Figure 11.6), in which cobalt(III) complexes with an enforced *cis*-configuration at the labile coordination sites are proved as one of the most key elements for artificial nucleases. Recently Chin *et al.* have prepared a dimeric cobalt(III) complex and have proposed a mechanism in which the phosphate is doubly Lewis acid activated and subsequently cleaved by attack of a cobalt(III)-bound hydroxide ion.[81]

Vance and Czarnik have reported that the dinuclear cobalt(III) cyclen complexes based on both a very rigid anthracene spacer and a very flexible alkyl spacer promoted the hydrolysis of the phosphate diester BNPP (bis(nitrophenyl)phosphate).[82] The rigid complexes are more reactive than the flexible complexes, whereas no cooperativity between the metal ions is observed for the

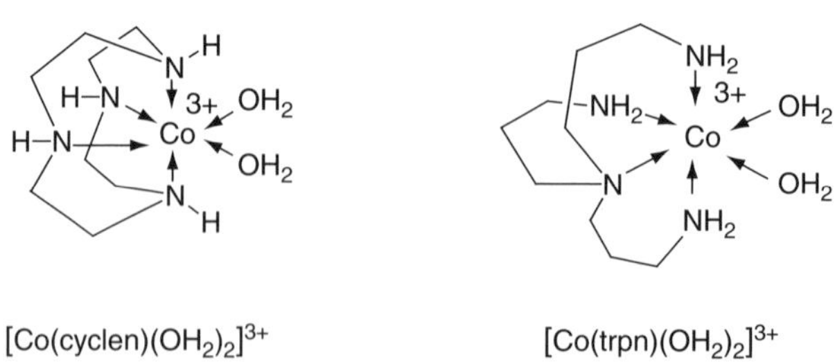

Figure 11.6 Cobalt(III) complexes of cyclen or trpn

BNPP hydrolysis. Hettich and Schneider *et al.* have increased the DNA affinity of dinuclear cyclen-Co(III) by modification with positively charged peralkylammonium ions that can interact with the anionic phosphate ester linkages in DNA.[83] The presence of additional positive charges has small effects on the hydrolysis rates of BNPP ester but leads to substantially enhanced cleavage of plasmid DNA. The highest activity was observed for a complex in which two cobalt(III) centers can act simultaneously on the DNA grooves.

11.4 Miscellaneous

11.4.1 Cobalamin conjugates as drug delivery devices

Most of the front-line chemotherapeutic drugs such as doxorubicin, taxol, etoposide and the alkylating agents have a dose-limiting toxicity that arises from interaction of the drug with an organ system that is remote from the site of cancer. If these cytotoxic drugs could be delivered selectively to cancer cells, their therapeutic index would be greatly increased and the serious side-effects such as bone marrow suppression, cardiotoxicity, nephrotoxicity, hepatotoxicity, neurotoxicity, hair loss and nausea might be eliminated, or at least reduced.

Cobalamin (Figure 11.7) is an essential micronutrient that is required for human health and, more importantly, is required in large quantities by cells that are replicating DNA prior to cell division. Cancer cells have an increased ability to transport and to sequester cobalamin in large excess over the amount required for normal cellular metabolism and cell replication.[84,85] In some patients with tumors, up to a 50-fold increase in the major cobalamin transport proteins has been observed. Collins and Hogenkamp also showed that rapidly dividing cells up-regulate cobalamin-binding receptors during DNA replication.[10,86,87] By taking advantage of the body's own machinery to concentrate and to maintain cobalamin levels *in vivo*, combined with the up-regulation of cobalamin-binding receptors in neoplastic cells, cobalamin has the potential to be a site-selective drug delivery vehicle for the treatment, imaging, and diagnosis of cancer.

Numerous bioconjugate derivatives of cobalamin with chemotherapeutic agents and radionuclides have been synthesized and some of them have shown promise in initial biological tests. For instance, Hogenkamp *et al.* reported tumor imaging via indium 111-labeled diethylenetriaminepentaacetate adenosylcobalamin.[10,87] The same group also used vitamin B_{12} (cyanocobalamin) to deliver a boron-rich nido-carborane to tumors.[88] In their study, the biological activity of cyanocobalamin-nido-carborane derivatives is relatively high – 92.93%, when compared with ^{57}Co cyanocobalamin. These ^{11}B cobalamin conjugates might be useful agents in treating malignant tumors via neutron capture therapy. Recently, Grissom *et al.* have prepared some fluorescent derivatives of cobalamin by linking

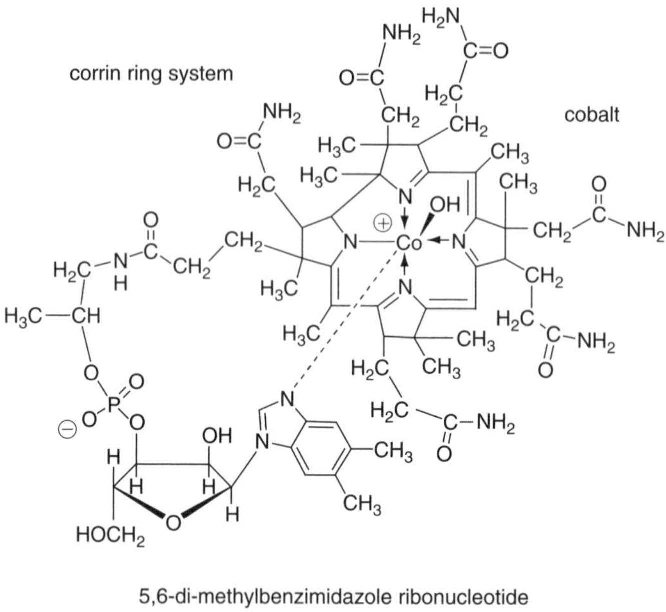

Figure 11.7 Chemical structure of cobalamin

fluorophores to cobalamin through a propylamide spacer (Figure 11.8).[89] Unlike the cryptofluorescent and profluorescent cobalamin conjugates synthesized previously, the compounds reported herein exhibit less quenching of the fluorophore excited state by cobalamin. The CobalaFluors are overtly fluorescent molecules

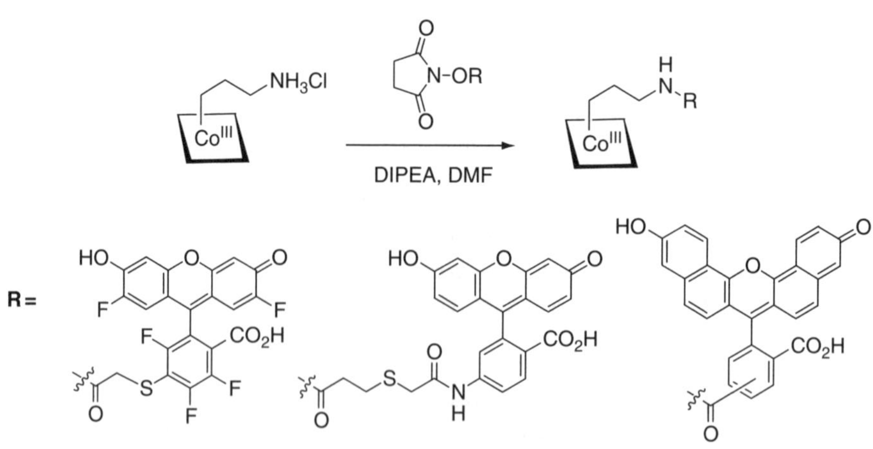

Figure 11.8 Fluorescent derivatives of cobalamin

that are potentially suitable for the *in vitro* and *in vivo* imaging of transcobalamin receptors on cancer cells.

Due to the light sensitivity of the carbon—cobalt bond,[90] the inert cobalamin–drug bioconjugate can be activated to cleave the Co—C bond by visible or near-UV light, thereby releasing the active drug in the irradiated tissue. Cobalamin bioconjugates can also be activated by ultrasound to cleave the Co—C bond, allowing release of the drug deep within a tissue that is not as easily accessible with red light.[91] Sonolysis of aqueous solutions produces a high concentration of hydroxyl radicals and hydrogen atoms. These, respectively, reactive oxidizing and reducing species are responsible for initiating most reactions in aqueous solvents, including cleavage of the Co—C bond and activation of the prodrug to release the active chemotherapeutic agent.

11.4.2 Hypoxic selective agents

It is well established that hypoxic cells exist in solid tumors and that these cells are resistant to the cytotoxic effects of ionizing radiation.[92,93] The non-cycling status of many hypoxic cells[94] and the difficulty of achieving adequate drug concentrations in regions distant from blood vessels[95] indicate that hypoxic cells may also be resistant to conventional chemotherapeutic agents. Therefore, there is a great interest in the design of drugs that are selectively cytotoxic under hypoxic conditions.[96,97] The basic approach is to utilize net one-electron bioreduction at a transition intermediate as the activation step. The properties of the Co(II)/Co(III) redox couple offer distinct advantages in this quest.

Ware *et al.* reported the interesting hypoxia-selective cytotoxic behaviors of cobalt(III) complexes containing coordinated nitrogen mustards.[11,98] Two series of cobalt(III) complexes of the bidentate mustard ligands *N*,*N'*-bis(2-chloroethyl)ethylenediamine (BCE) and *N*,*N*-bis(2-chloroethyl)ethylenediamine (DCE) have been synthesized (Figure 11.9).[99] The complexes exhibited hypoxia-selective cytotoxicity against Chinese hamster ovary (CHO) fibroblasts and UV4 cells *in vitro*. In both of the BCE and DCE series, the pattern of cytotoxicity of the cobalt complexes paralleled that of the free ligands, suggesting that the cytotoxicity of the complexes is due to the release of the free ligands. The unsymmetrical DCE–cobalt complexes were an order of magnitude more cytotoxic than the corresponding free DCE compounds. The most effective complex was $[Co(Meacac)_2(DCE)]^+$, which was approximately 30-fold selective toward hypoxic EMT6 cell in culture, and which showed enhanced activity against EMT6 cells in multicellular spheroids.[100] This work showed that metal complexes of nitrogen mustards have significant selectivity toward hypoxic mammalian cells and that they represent a promising new class of hypoxia-selective cytotoxins.

The kinetic basis for the *in vitro* sensitizing activity of $[Co(Meacac)_2(DCE)]^+$ has also been investigated by pulse radiolysis.[100] The rate constants for the

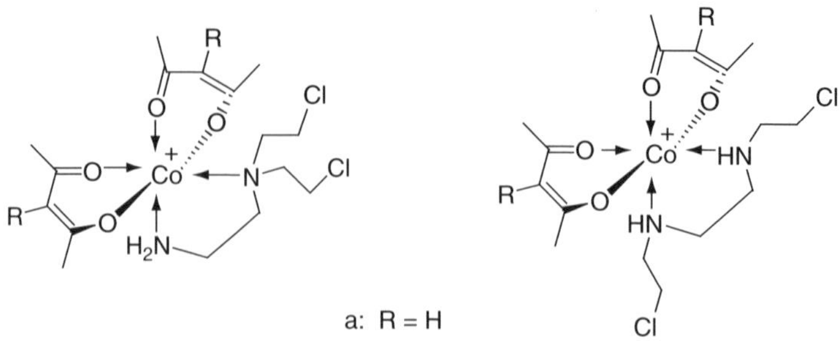

a: R = H
b: R = CH_3
c: R = C_2H_6
d: R = C_3H_7

Figure 11.9 $[Co(Racac)_2(DCE)]^+$ (left) and $[Co(Reacac)_2(BCE)]^+$ (right)

one-electron reduction of $[Co(Meacac)_2(DCE)]^+$ by model reductants exhibited a marked dependency on the reduction potential of the reductant, with values up to several orders of magnitude lower than those reported for misonidazole. To clarify the mechanism of hypoxic selectivity, other series of potentially radiosensitizing cobalt(III) complexes, derived from the lead complex $[Co(Meacac)_2(DCE)]^+$ by replacing the $Racac^-$ ligand with monoanionic, bidentate dithiocarbamato or tropolonato (2-hydroxy-2,4,6-cycloheptatrien-1-one) ligands, were prepared (Figure 11.10).[101] These results imply that the dithiocarbamato and tropolonato complexes are not suitable as bioreductive anti-cancer drugs. Ware *et al.* recently synthesized cobalt(III) complexes containing the tridentate nitrogen mustard ligand *N*,*N*-bis(2-chloroethyl)diethylenetriamine (DCD)

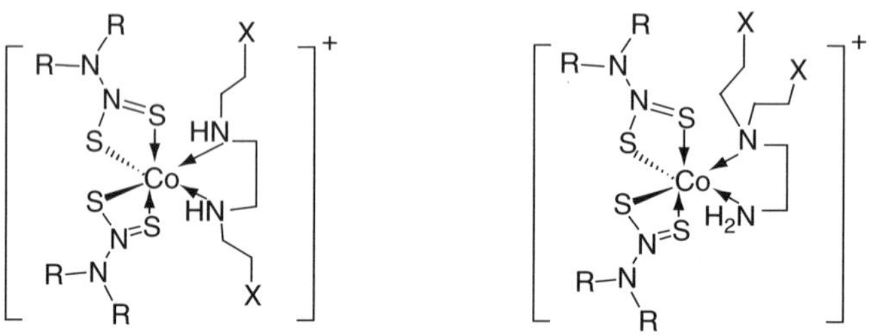

a: X = H, R_2 = Me_2
b: X = Cl, R_2 = Me_2
c: X = Cl, R_2 = $(CH_2)_4$

a: X = H, R_2 = Me_2
b: X = H, R_2 = Et_2
c: X = H, R_2 = $(CH_2)_4$
d: X = Cl, R_2 = Me_2
e: X = Cl, R_2 = Et_2

Figure 11.10 $[Co(R_2dtc)_2(BCE)]^+$ (left) and $[Co(R_2dtc)_2(DCE)]^+$ (right)

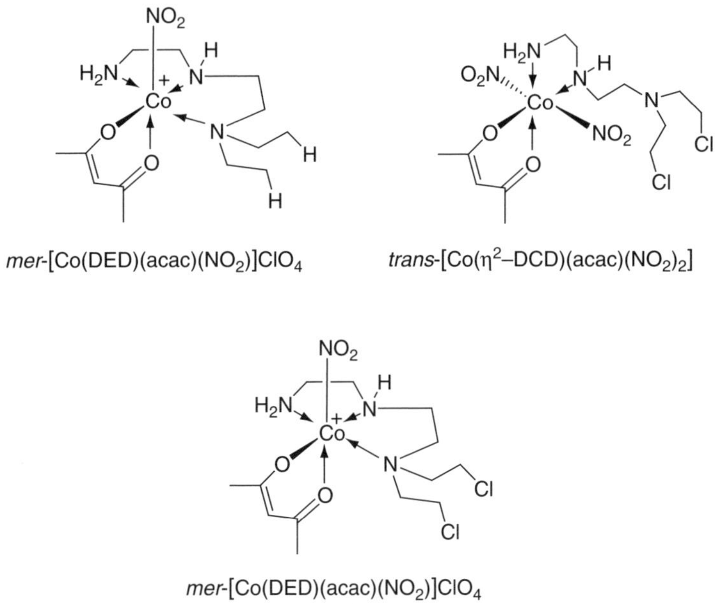

Figure 11.11 Cobalt(III) complexes containing tridentate ligands DCD and DED

(Figure 11.11).[102] Two DCD complexes were characterized by X-ray crystal structure determinations, which confirm the tridentate coordination with a long cobalt–tertiary nitrogen bond distance in the former, and bidentate coordination with the mustard nitrogen as a pendant arm in the latter. The bidentate complex, which contains a free nitrogen mustard, had cytotoxicity similar to that of the free ligand, but cytotoxicity was successfully masked in the tridentate complex, which showed modest hypoxic selectivity (fivefold) in a clonogenic assay.

11.4.3 PET imaging agents

Positron emission tomography is a technology that relies upon the injection of short-lived radionuclide-labeled tracers to examine metabolic processes in the body.[103] Once injected, these harmless radioactive tracers (which have a short useful lifetime) allow the functioning of the metabolic processes under investigation to be examined with a PET camera. During the last decades, platelets radiolabeled with various radionuclides were successfully used in nuclear medicine for imaging of thrombosis and/or atherosclerosis.[104] However, short-lived PET radionuclides such as ^{68}Ga (68 min) have limited their clinical use in cell labeling. The ^{55}Co complexes are medium half-life PET radionuclides with a relative long half-life of 18.2 h. This may enable quantitative uptake and cell kinetic studies with radiolabeled blood cellular elements.

Ellis and Sharma prepared cobalt(II) oxine (8-hydroxyquinoline) and cobalt(III) tropolonate with ^{57}Co as models for the ^{55}Co nucleus.[12] Results show that the uptake of ^{57}Co(II) oxine into erythrocytes was only 37%, and 65% of the activity eluted from the cells in cell-free plasma within 30 min. By contrast, high leukocyte and erythrocyte labeling efficiencies (>90%) were obtained with a ^{57}Co(III) tropolonate complex and the elution in cell-free plasma over 4 h was <8%. This suggests that cobalt(III) tropolonate may be useful for radiolabeling leukocytes for PET investigations. Schmaljohann and co-workers investigated the platelet uptake of ^{57}Co(II) oxine, ^{57}Co(III) tropolonate and ^{57}Co MPO (MPO = mercapopyridine-*N*-oxide) *in vitro*.[105] Their experimental results show that these ^{57}Co complexes are not ideal for platelet labeling and the eventually following PET application. The reason could be a higher stability of the ^{57}Co complexes, similar as postulated for ^{68}Ga-MPO[106] and ^{111}In complexes.[107,108] Thus, after diffusion into the platelets the strongly bound cobalt complexes can more easily diffuse back across the cell membrane prior to dissociation and trapping of the ^{57}Co within the platelet.

11.5 Conclusions

Because of their unique electronic structures and redox properties, cobalt complexes have the potential to be used for the development of enzyme inhibition or induction, DNA binding and cleavage, drug delivery devices, hypoxic selective agents and PET imaging agents. Some cobalt complexes have already shown great promise. The use of cobalt-based drugs in medicine assures continued research and development. One can confidently predict increasing research in this area because of the enormous potential cobalt presents in this context.

Acknowledgements

We are grateful to the National Natural Science Foundation of China, the National Science Foundation of Guangdong Province and Research Fund of Royal Society of Chemistry, UK, for financial support.

References

1. G. Dodson, J.P. Glusker and D. Sayre (Eds), *Structural Studies on Molecules of Biological Interest – A Volume in Honour of Dorothy Hodgkin*, Clarendon, Oxford, 1981.
2. R.X. Xu, D. Nettesheim, J.D. Otvos and D.H. Petering, *Biochemistry* **33**, 907–916 (1994).

3. J.C. Noveron, M.M. Olmstead and P.K. Mascharak, *J. Am. Chem. Soc.* **121**, 3553–3554 (1999).
4. M.A. Pennella, J.E. Shokes, N.J. Cosper *et al.*, *Proc. Natl. Acad. Sci. USA* **100**, 3713–3718 (2003).
5. A. Böttcher, T. Takeuchi, K.I. Hardcastle *et al.*, *Inorg. Chem.* **36**, 2498–2504 (1997).
6. T. Takeuchi, A. Böttcher, C.M. Quezada *et al.*, *J. Am. Chem. Soc.* **120**, 8555–8556 (1998).
7. T. Takeuchi, A. Böttcher, C.M. Quezada *et al.*, *Bioorg. Med. Chem.* **6**, 1–5 (1999).
8. L.-N. Ji, X.-H. Zou and J.-G. Liu, *Coord. Chem. Rev.* **216–217**, 513–536 (2001).
9. R.M. Burger, *Chem. Rev.* **98**, 1153–1169 (1998).
10. D.A. Collins, H.P.C. Hogenkamp and M.W. Gebhard, *Mayo Clin. Proc.* **74**, 687–691 (1999).
11. D.C. Ware, W.R. Wilson, W.A. Denny and C.E.F. Rickard, *J. Chem. Soc., Chem. Commun.* 1171–1173 (1991).
12. B.L. Ellis and H.L. Sharma, *Nucl. Med. Commun.* **20**, 1017–1021 (1999).
13. E. Tsilikounas, C. Kettner and W. Bachovchin, *Biochemistry* **31**, 12839–12846 (1992).
14. B. Duffy, C. Schwietert, A. France *et al.*, *Biol. Trace Elem. Res.* **64**, 197–213 (1998).
15. S. Bagger, *J. Inorg. Biochem.* **52**, 165–171 (1993).
16. S. Bagger and K. Wagner, *J. Pept. Res.* **52**, 273–282 (1998).
17. A.Y. Louie and T.J. Meade, *Proc. Natl. Acad. Soc. USA* **95**, 6663–6668 (1998).
18. S.J. Brill, S. Dinardo, K. Voelkel-Meimanam and R. Sternglanz, *Nature* **326**, 414–416 (1987).
19. G.C. Uemura and M. Yanagida, *EMBO J.* **5**, 1003–1010 (1986).
20. J. Newport and T. Spann, *Cell* **48**, 219–230 (1987).
21. J.C. Wang and A.S. Lynch, *Curr. Opin. Genet. Dev.* **3**, 764–768 (1993).
22. G. Capranico and F. Zunino, *Eur. J. Cancer* **28A**, 2055–2066 (1992).
23. J. Roca and J.C. Wang, *Cell* **71**, 833–840 (1992).
24. M.P. Watt and D.I. Hickson, *Biochem. J.* **303**, 681–695 (1994).
25. P. Lumme and H. Elo, *Inorg. Chim. Acta* **92**, 241–251 (1984).
26. D. Jayaraju, Y.N. Vashisht Gopal and A.K. Kondapi, *Arch. Biochem. Biophys.* **369**, 68–77 (1999).
27. I.H. Hall, C.B. Lackey, T.D. Kistler *et al.*, *Pharmazie* **55**, 937–941 (2000).
28. I.H. Hall, K.F. Bastow, A.E. Warren *et al.*, *Appl. Organomet. Chem.* **13**, 819–828 (1999).
29. K.K. Elbirt, H.L. Bonkovsky, *Proc. Assoc. Am. Phys.* **111**, 438–447 (1999).
30. M.D. Maines, *Annu. Rev. Pharmacol. Toxicol.* **37**, 517–554 (1997).
31. S. Ryter, L.E. Otterbein, D. Morse and A.M. Choi, *Mol. Cell. Biochem.* **234–235**, 249–263 (2002).
32. G. Sass, M.C.P. Soares, K. Yamashita *et al.*, *Hepatology* **38**, 909–918 (2003).
33. D. Willis, A.R. Moore, R. Frederick and D.A. Willoughby, *Nat. Med.* **2**, 87–90 (1996).
34. L.E. Otterbein, J.K. Kolls, L.L. Mantell *et al.*, *J. Clin. Invest.* **103**, 1047–1054 (1999).
35. M.D. Maines, V.S. Raju and N. Panahian, *J. Pharmacol. Exp. Ther.* **291**, 911–919 (1999).
36. M.L. Tomaro, J. Frydman and R.B. Frydman, *Arch. Biochem. Biophys.* **286**, 610–617 (1991).
37. K. Mitani, H. Fujita, Y. Fukuda *et al.*, *Biochem. J.* **290**, 819–825 (1993).

38. Y. Shan, J. Pepe, T.H. Lu *et al.*, *Arch. Biochem. Biophys.* **380**, 219–227 (2000).
39. Y. Shan, J. Pepe, R.W. Lambrecht and H.L. Bonkovsky, *Arch. Biochem. Biophys.* **399**, 159–166 (2002).
40. A. Sigel and H. Sigel (Eds), *Metal Ions in Biological Systems* Vol. 33, Marcel Dekker, New York, 1996.
41. K.E. Erkkila, D.T. Odom and J.K. Barton, *Chem. Rev.* **99**, 2777–2796 (1999).
42. I. Ortmans, C. Moucheron and A. Kirsch-De Mesmaeker, *Coord. Chem. Rev.* **168**, 233–271 (1998).
43. L.-N. Ji, Q.L. Zhang and H. Chao, *Chinese Sci. Bull.* **46**, 1332–1337 (2001).
44. S. Arounaguiri and B.G. Maiya, *Inorg. Chem.* **35**, 4267–4270 (1996).
45. C.V. Sastri, D. Eswaramoorthy, L. Giribabu and B.G. Maiya, *J. Inorg. Biochem.* **94**, 138–145 (2003).
46. Q.L. Zhang, J.G. Liu, H. Chao *et al.*, *J. Inorg. Biochem.* **83**, 49–55 (2001).
47. Q.L. Zhang, J.G. Liu, J. Liu *et al.*, *J. Inorg. Biochem.* **85**, 291–296 (2001).
48. Q.L. Zhang, J.G. Liu, H. Xu *et al.*, *Polyhedron* **20**, 3049–3055 (2001).
49. Q.L. Zhang, J.G. Liu, J.Z. Liu *et al.*, *Inorg. Chim. Acta* **339**, 34–40 (2002).
50. C.C. Cheng, Y.N. Kuo, K.S. Chuang *et al.*, *Angew. Chem. Int. Ed.* **38**, 1255–1257 (1999).
51. C.C. Cheng, W.C. Huang-Fu, K.C. Hung *et al.*, *Nucleic Acids Res.* **31**, 2227–2233 (2003).
52. Y.F. Song and P. Yang, *Polyhedron* **20**, 501–506 (2001).
53. V.G. Vaidyanathan and B.U. Nair, *J. Inorg. Biochem.* **94**, 121–126 (2003).
54. R.H. Blum, S.K. Carter and K.A. Agre, *Cancer* **31**, 903–914 (1973).
55. J. Stubbe and J.W. Kozarich, *Chem. Rev.* **87**, 1107–1136 (1987).
56. J.C. Dabrowiak, *Adv. Inorg. Biochem.* **4**, 69–113 (1983).
57. S.M. Hecht, *Acc. Chem. Rev.* **19**, 383–391 (1986).
58. J. Stubbe, J.W. Kozarich, W. Wu and D.E. Vanderwall, *Acc. Chem. Rev.* **29**, 322–330 (1996).
59. R.M. Burger, J. Peisach and S.B. Horwitz, *J. Biol. Chem.* **256**, 11636–11644 (1981).
60. A. Kuramuchi, K. Takahashi, T. Takita and U. Umezawa, *J. Antibiot.* **34**, 576–582 (1981).
61. M. Nakamura and J. Peisach, *J. Antibiot.* **41**, 638–647 (1988).
62. J. Templin, L. Berry, S. Lyman *et al.*, *Biochem. Pharmacol.* **43**, 615–623 (1992).
63. R.X. Xu, W.E. Antholine and D.H. Petering, *J. Biol. Chem.* **267**, 944–949 (1992).
64. R.X. Xu, W.E. Antholine and D.H. Petering, *J. Biol. Chem.* **267**, 950–955 (1992).
65. W. Wu, D.E. Vanderwall, S.M. Liu *et al.*, *J. Am. Chem. Soc.* **118**, 1268–1280 (1996).
66. W. Wu, D.E. Vanderwall, C.J. Turner *et al.*, *J. Am. Chem. Soc.* **118**, 1281–1294 (1996).
67. P. Fulmer, C. Zhao, W. Li *et al.*, *Biochemistry* **36**, 4367–4374 (1997).
68. S.T. Hoehn, H.D. Junker, R.C. Bunt *et al.*, *Biochemistry* **40**, 5894–5905 (2001).
69. C. Zhao, C. Xia, C. Mao *et al.*, *J. Inorg. Biochem.* **91**, 259–268 (2002).
70. C. Xia, F.H. Försterling and D.H. Petering, *Biochemistry* **42**, 6559–6564 (2003).
71. C.H. Chang and C.F. Meares, *Biochemistry* **21**, 6332–6334 (1982).
72. I. Saito, T. Morii, H. Sugiyama *et al.*, *J. Am. Chem. Soc.* **111**, 2307–2308 (1989).
73. J.T. Tan, S.E. Hudson, S.J. Brown *et al.*, *J. Am. Chem. Soc.* **114**, 3841–3853 (1992).

74. J.T. Tan, E.T. Farinas, S.S. David and P.K. Mascharak, *Inorg. Chem.* **33**, 4295–4308 (1994).
75. Y. Nishida, M. Kunita and S. Nishino, *Inorg. Chem. Commun.* **2**, 156–157 (1999).
76. J.A. Cowan, *Chem. Rev.* **98**, 1067–1088 (1998).
77. J. Chin, *Acc. Chem. Res.* **24**, 145–152 (1991).
78. D.S. Sigman and A. Mazumder, D.M. Perrin, *Chem. Rev.* **93**, 2295–2316 (1993).
79. A. Sreedhara and J.A. Cowan, *J. Biol. Inorg. Chem.* **6**, 337–347 (2001).
80. J. Chin, M. Banaszczyk, V. Jubian and X. Zou, *J. Am. Chem. Soc.* **111**, 186–190 (1989).
81. N.H. Williams, B. Takasaki, M. Wall and J. Chin. *Acc. Chem. Res.* **32**, 485–493 (1999).
82. D.H. Vance and A.W. Czarnik. *J. Am. Chem. Soc.* **115**, 12165–12166 (1993).
83. R. Hettich and H. Schneider, *J. Am. Chem. Soc.* **119**, 5638–5647 (1997).
84. H. Flodh, *Acta Radiol.* Suppl. **284**, 55–60 (1968).
85. H.P.C. Hogenkamp, D.A. Collins, C.B. Grissom and F.G. West, in *Chemistry and Biochemistry of* B_{12}, R. Banerjee (Ed.), John Wiley & Sons: New York, 385–410 (1999).
86. D.A. Collins and H.P.C. Hogenkamp, *J. Nucl. Med.* **38**, 717–723 (1997).
87. D.A. Collins, H.P.C. Hogenkamp, M.K. O'Connor *et al.*, *Mayo Clin. Proc.* **75**, 568–580 (2000).
88. H.P.C. Hogenkamp, D.A. Collins, D. Live *et al.*, *Nucl. Med. Biol.* **27**, 89–92 (2000).
89. C.C. Smeltzer, M.J. Cannon, P.R. Pinson *et al.*, *Org. Lett.* **3**, 799–801 (2001).
90. D. Dolphin (Ed.), *Vitamin* B_{12}, John Wiley & Sons: New York, 1982.
91. W.A. Howard, A. Bayomi, E. Natarajan *et al.*, *Bioconjug. Chem.* **8**, 498–502 (1997).
92. R.A. Gatenby, H.B. Kessler, J.S. Rosenblum *et al.*, *Int. J. Radiat. Oncol. Biol. Phys.* **14**, 831–838 (1988).
93. J. Overgaard, H.S. Hansen, A.P. Andersen *et al.*, *Int. J. Radiat. Oncol. Biol. Phys.* **16**, 1065–1068 (1989).
94. P.L. Olive and R.E. Durand, *Int. J. Radiat. Oncol. Biol. Phys.* **16**, 755–761 (1989).
95. D.J. Kerr and S.B. Kaye, *Cancer Chemother. Pharmacol.* **19**, 1–5 (1987).
96. W.A. Denny, W.R. Wilson and M.P. Hay, *Br. J. Cancer* **74** (Suppl. 27), S32 (1996).
97. A. Hasrat and J.E. van Lier, *Chem. Rev.* **99**, 2379–2450 (1999).
98. D.C. Ware, B.D. Palmer, W.R. Wilson and W.A. Denny, *J. Med. Chem.* **36**, 1839–1846 (1993).
99. R.F. Anderson, W.A. Denny, D.C. Ware and W.R. Wilson, *Br. J. Cancer* 74 (Suppl. 27), S48–S51 (1996).
100. W.R. Wilson, J.W. Moselen, S. Cliffe *et al.*, *Int. J. Radiat. Oncol. Biol. Phys.* **29**, 323–327 (1994).
101. D.C. Ware, H.R. Palmer, F.B. Pruijn *et al.*, *Anticancer Drug Des.* **13**, 81–103 (1998).
102. D.C. Ware, P.J. Brothers, G.R. Clark *et al.*, *J. Chem. Soc., Dalton Trans.* 925–932 (2000).
103. A. Alavi and M. Reivich, *Positron Emission Tomography*, New York: Alan R. Liss, Inc., 1985.
104. M. Rodrigues and H. Sinzinger, *Thromb. Res.* **76**, 399–432 (1994).

105. J. Schmaljohann, G. Karanikas and H. Sinzinger, *J. Labelled Cpd. Radiopharm.* **44**, 395–403 (2001).
106. C.J. Mathias, W.A. Heaton, M.J. Welch *et al.*, *Int. J. Appl. Radiat. Isot.* **32**, 651–656 (1981).
107. R.D. Abeysinghe, B.L. Ellis and J.B. Porter, *Eur. J. Nucl. Med.* **21**, 1141–1147 (1994).
108. C. Mathias and M. Welch, *J. Nucl. Med.* **20**, 659 (1979).

12 $_{29}$Cu Chemotherapeutic Copper Compounds

Francisco González-Vílchez and Rosario Vilaplana

12.1 Introduction

For years, medicinal chemistry was almost exclusively based on organic compounds and natural products. Although the fundamental importance of life dependence on metal ions has been recognized from oldest times,[1,2] scarce attention has been given to inorganic substances. However, during the past three decades, transition metal complexes have been isolated and studied as potential cancer chemotherapeutic agents.[3–6] Concretely, the success of *cis*-dichlorodiammine platinum complex (cisplatin) in humans and the remarkable anti-tumour activity of cisplatin-like compounds have clearly provided great impetus for search towards finding more active and selectively cytotoxic inorganic complexes.[7–11] Over the years a large number of complexes containing platinum or non-platinum metals have been found to be effective against a variety of experimental tumours.[12–18]

Much research on anti-tumour compounds has been carried out synthesizing and characterizing new non-platinum metal complexes, where the nature of both the metal and the ligand were varied. It is remarkable that investigations in this area are focused on the use of biologically active complexes formed by essential ions, such as copper.[19] Since any essential metal which escapes its normal metabolic pathways can be very toxic to the organism (for example,

Metallotherapeutic Drugs and Metal-Based Diagnostic Agents: The Use of Metals in Medicine
Edited by Gielen and Tiekink

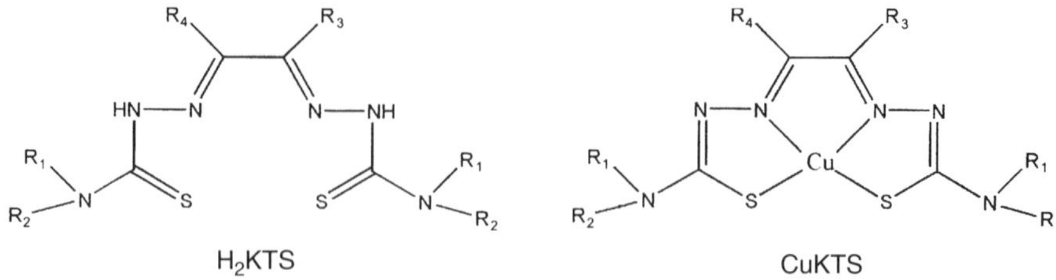

(H_2KTS: 3-ethoxy-2-oxobutyraldehyde, bTSC; R_4 = ethoxyethyl, R_1–R_3 = H)

Figure 12.1 General structures of bis(thiosemicarbazones) (bTSC) and their copper complexes

copper in Wilson disease), complexes of such metals may serve as effective cytotoxic agents.[2,20] Indeed, it is well known for many years that a large number of bis(thiosemicarbazones) (bTSC) H_2KTS and several groups of their copper complexes CuKTS, whose generalized structure is shown in Figure 12.1, show promising anti-tumour activities.[21] The anti-cancer action of CuKTS has been clearly demonstrated in *in vitro* and *in vivo* experiments.[19]

In this chapter, the biological properties of recently obtained selected compounds of the most important groups of chemotherapeutic copper complexes will be discussed. Special emphasis would be focused on bTSC-type copper complexes, the most investigated group of biologically active substances.

12.2 Copper-Purine Derivatives Complexes

The synthesis, characterization and biological (anti-tumour and anti-proliferative) activity of new mononuclear or binuclear copper(II) complexes with 6-(2-chlorobenzyl-amino)purine (HL_1) (**A**) and 6-(3-chlorobenzylamino)purine (HL_2) (**B**) (Scheme 12.1) have been reported by Travnýcĕk *et al.*,[22] the composition of the complexes depending on the molar ratio of the reactants and the pH of the medium. The mononuclear complex $[Cu(H^+L_2)_2Cl_3]Cl \cdot H_2O$ (**1**) was isolated and green-blue crystals suitable for single-crystal X-ray analysis were obtained. On the other hand, binuclear complexes of composition $[Cu_2(\mu\text{-}Cl)_2(\mu\text{-}L_1)_2(H_2O)_2]$ (**2**) and $[Cu_2(\mu\text{-}HL_1)_2(\mu\text{-}Cl)_2(HL_1)_2Cl_2]$ (**3**) were characterized and studied. From all the obtained data and comparison with crystallographic parameters observed for similar complexes,[23] their structures were proposed.

Biological activity testing against mouse melanoma cell line *B*16-*F*0, human malignant melanoma cell line *G*361, human osteogenic sarcoma cell line *HOS* and human breast adenocarcinoma cell line *MCF*7 were performed for the three complexes, using a calcein AM assay, in which the incubation of cells with

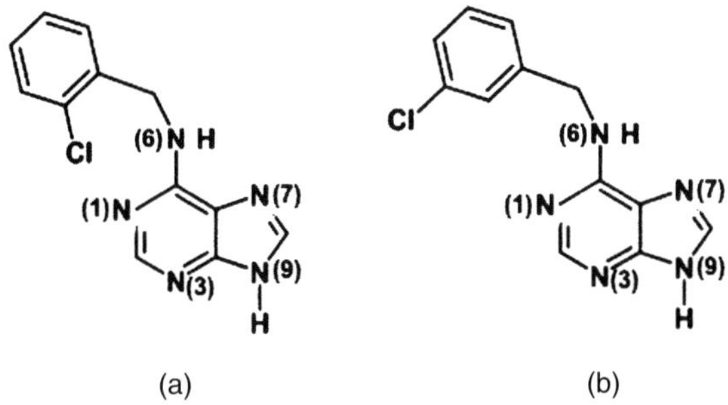

Scheme 12.1

the tested complexes lasted for 72 h at 37 °C in 5% CO_2 atmosphere. At the end of this period, the cells were incubated for 1 h with calcein AM and the fluorescence of the live cells was measured. Inhibition of $p34^{cdc2}$ kinase by the complexes was assayed as described previously.[24] The stability of the tested complexes in the media used was monitored by measurement of UV–V spectra. No significant changes in the maxima of the absorption peaks were observed after 24 h.

Both the anti-tumour and $p34^{cdc2}$ inhibitory activities of the complexes and the starting compounds, i.e. $CuCl_2 \cdot 2H_2O$, HL_1 and HL_2, were assayed in *in vitro* kinase and cytotoxicity assays, respectively. The data obtained from a cytotoxicity assay are presented in Table 12.1. All the tested complexes are shown to be potent growth inhibitors of the different cancer cell lines used. The IC_{50}

Table 12.1 IC_{50} values (μM) for selected complexes assessed by a calcein AM assay of surviving tumour cells compared with the values found for $CuCl_2 \cdot 2H_2O$, HL_1, HL_2 and Bohemin, respectively

	Cell line			
Compound	*B16-F0*	*G361*	*HOS*	*MCF7*
1[a]	17	59	69	54
2[a]	8.5	31	29	30
3[a]	8.2	20	23	24
$CuCl_2 \cdot 2H_2O$	100	>100	>100	>100
HL_1	85	95	>100	>100
HL_2	>100	>100	>100	>100
Bohemin[b]	c	c	46	53

[a] The studied complexes were sufficiently soluble in the solvent mixture (DMSO/water) needed for testing.
[b] 6-(Benzylamino)-2-[(3-hydroxypropyl)amino]-9-isopropylpurine.
[c] Not tested.

values for $CuCl_2 \cdot 2H_2O$, HL_1 and HL_2 obtained on all used cell lines were found to be remarkably higher than those for the appropriate Cu(II) complexes. This indicates that the cytotoxic activity of the ligands strongly increases after the formation of the Cu(II) complexes. The most sensitive to the assayed compounds were *B16-F0* mouse melanoma cells, displaying cytotoxicities with IC_{50} in the range from 8.2 to 19 μM. *G361*, *HOS* and *MCF7* cell lines were usually less sensitive, with IC_{50} values ranging from 20 to 88 μM. Surprisingly, the values obtained for *HOS* and *MCF7* cell lines were in many cases lower than those found for the best-known cytokine-derived cdk inhibitor, Bohemin (Table 12.1). Moreover, for the *G361* and *HOS* cell lines, the mononuclear complex **1** was found to be several times less potent than tested binuclear Cu(II) species **2** and **3**.

The inhibition of crude $p34^{cdc2}$ kinase by complexes **2** and **3** was also estimated. The data obtained are shown in Table 12.2. The IC_{50} values are significantly lower than those previously reported for the two structurally similar Ni(II) complexes, for which larger inhibitory values (80 and 110 μM, respectively) were found.[24] In conclusion, the prepared complexes exhibit considerable anti-cancer activity, and studies are continuing to explore their possible clinical use.[22]

Travnýcěk *et al.*[25] prepared also a new dimeric Cu(II) complex with the ligand HL_2 (**B**) (see Scheme 12.1), of composition $[Cu_2(\mu\text{-}Cl)_2(\mu\text{-}HL_2)_2Cl_2]$ (**4**) and with a structure similar to that of complex **2**; two chloride ions replacing two water molecules, and two ligands HL_2 connecting the central atoms through N(3) and N(9) atoms of HL_2.

Anti-tumour activity of complex **4** was determined using the cytotoxicity assay mentioned above. Obtained data (Table 12.3) show that this complex is a potent growth inhibitor of the same cancer cell lines tested for previous copper complexes and the human chronic myelogenous leukaemia *K562*. Surprisingly, formation of Cu(II) complexes had even stronger impact on anti-tumour activity than the non-toxic compound HL_2: the IC_{50} values found for complex **4** were in the range from 32 to 39 μM, whereas the lowest IC_{50} value determined for the starting compounds HL_2 and $CuCl_2 \cdot 2H_2O$ was 100 μM.

Table 12.2 IC_{50} values for tested compounds added to crude $p34^{cdc2}$ kinase[a]

Compound	IC_{50} (μM)
Bohemin[b]	1.2
2	15
3	10
BAP[c]	200

[a] IC_{50} values were subtracted from the dose–response curves.
[b] 6-(Benzylamino)-2-[(3-hydroxypropyl)amino]-9-isopropylpurine.
[c] 6-Benzylaminopurine.

Table 12.3 IC_{50} values (μM) assessed by a calcein AM assay of surviving tumour cells (complex **4**) compared with the values found for $CuCl_2 \cdot 2H_2O$ and HL_2

Compound	Cell line			
	*K*562	*G*361	*HOS*	*MCF*7
4	32	36	38	39
$CuCl_2 \cdot 2H_2O$	100	>100	>100	>100
HL_2	>100	>100	>100	>100

12.3 Copper-Thiosemicarbazone Complexes

The anti-tumour potential of quinonoidal compounds is known since 1974 and several 1,2-napthaquinone derivatives behave as anti-cancer compounds.[26–28] Recently, the synthesis and structural characterization of metal conjugates of 1,2-naphthaquinone (NQTS, Scheme 12.2) along with their anti-tumour activities against the *MCF*7 breast cancer cell line have been studied.[29] Thiosemicarbazonates have been intensively studied due to their inhibitory action on the DNA enzyme ribonucleotide diphosphate reductase and their selectivity towards hormone-responsive cancers.[30] All synthesized compounds were evaluated for their anti-proliferative activities against the human breast cancer cell line *MCF*7 and their IC_{50} values (μM) determined. From the results obtained it is clear that the addition of the TSC-pharmacophore to the parent quinonoidal moiety considerably enhances the anti-proliferative activity of the quinone, and the Cu(II) compound has a low IC_{50} value (4.02 mM). This cytotoxic action could be attributed to its topoisomerase II inhibitory activity.

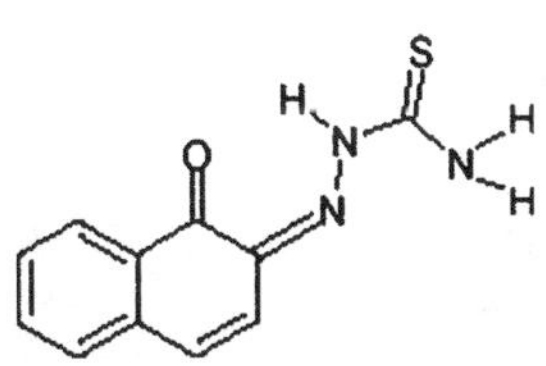

NQTS

Scheme 12.2

A new salicylaldehyde thiosemicarbazone (H_2-STSC, Scheme 12.3) has been synthesized as the main hydrolysis product of the parent compound helicin-thiosemicarbazone (H-TSC).[31] The new H_2STSC ligand, which bears a chain with the nucleophilic S, N centres and has anti-tubercolostatic activity, was characterized by IR and NMR spectroscopies and X-ray diffractometry, and its biological activity tested, in comparison with H-TSC. Although complexation assays of H_2-STSC with several metals were performed, only the reaction of this ligand with $CuSO_4$ was successful, leading to the solid complex [Cu(H-STSC)(OH_2)]$_2SO_4 \cdot 2DMSO \cdot 6H_2O$ (**I**).

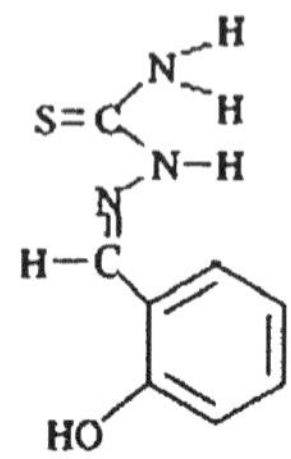

H_2-STSC = salicyladehide thiosemicarbazone

Scheme 12.3

The complex **I** has been tested *in vitro* on human leukaemic U937 cell lines, focusing the experiments on the activity with respect to cell proliferation inhibition and apoptosis induction. The results indicate that the complex inhibited about 40% of the cell proliferation at 0.3 and 0.5 μg/ml, but DNA fragmentation and apoptosis were not observed at these concentrations (Figure 12.2a).

A possible explanation of the unusual biological behaviour of the complex **I** can be found in the copper coordination geometry: the metal centre is substantially square planar, with the Cu atom displaced only 0.07 Å from the coordination plane and an angle between the chelation planes of 7.6°. Instead, the

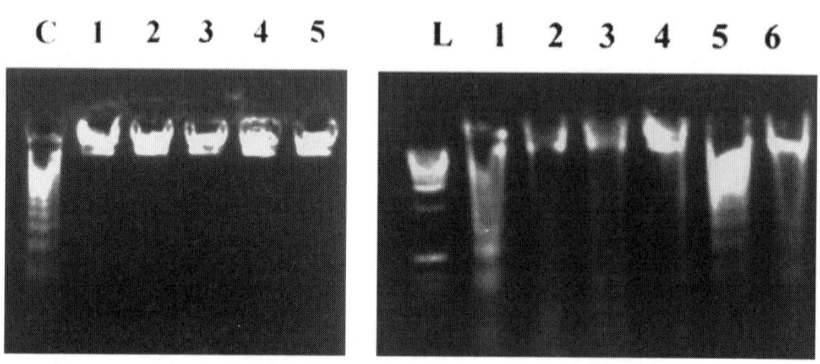

Figure 12.2 (a) Apoptosis assays on complex **I**. Lane **C**: positive control; lane **1**: untreated control cells; lanes **2** and **3**: H-TSC (20 and 30 μg/ml, respectively); lanes **4** and **5**: complex **I** (0.3 and 0.5 μg/ml, respectively). (b) Lane **L** ladder; lane **1**: cycloeximide; lane **2**: untreated control cells; lane **3**: DMSO; lane **4**: [Cu(H_3ut)Cl_2]$\cdot 2H_2O$; lane **5**: [Cu(H_2ut)(OH_2)Cl]$\cdot 2H_2O$; lane **6**: [Cu(H_2L)(OH_2)Cl]Cl

monomeric copper complexes [Cu(H_2L)(OH_2)Cl]Cl (**II**; H_2L = pyridoxal-thiosemicarbazone) and [Cu(H_2ut)(OH_2)Cl]·2H_2O (**III**; H_3ut = 5-formyluracil-thiosemicarbazone), previously synthesized and able to induce apoptosis (Figure 12.2b), presented a different geometry.[32] These and other compounds characterized in the same research inhibited cell proliferation but only compounds **II** and **III** caused DNA fragmentation when used at 40 μg/ml, inducing apoptosis.

There is a common fact that the two complexes **II** and **III**, active in inducing apoptosis, share: the copper coordination geometry. While in the complex **I**, as explained above, the metal centre is substantially square planar with the copper atom displaced 0.07 Å from the coordination plane, complexes **II** and **III** showed the Cu atom displaced by 0.24 and 0.25 Å, respectively, from the coordination plane, with interplanar angles of 10.2° and 11.1°, respectively. These important structural features are postulated as responsible of the apoptotic properties of the last two mentioned complexes.

Carboxamidrazones (**a**) (Scheme 12.4) bear structural similarities to heterocyclic thiosemicarbazones (**b**), and their copper conjugates show remarkable anti-tumour activities towards the hormone-responsive cancers. Thus, copper complexes of *N*′-(2-benzyloxy-benzylidiene) pyridine-2-carboxamidrazone (**pcm**) have shown to be highly effective against the human breast cancer cell line *MCF*7.[33] The corresponding Cu(II)-**pcm** complex (**1**) was prepared as a green solid and suitable X-ray quality crystals were obtained.

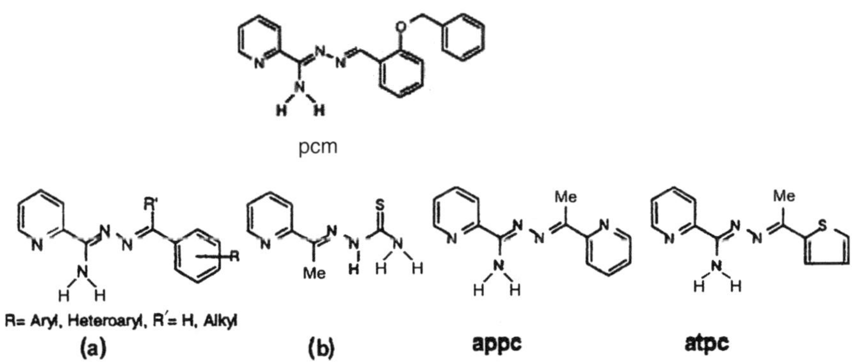

Scheme 12.4

The ligand and the copper complex were tested for their anti-tumour activity against the human breast cancer cell line *MCF*7. While the cells treated with **pcm** were viable, those treated with complex (**1**) showed cell death at comparatively low concentrations inducing a direct cytotoxic action. The IC_{50} value for the copper complex is 3 μM, which is four times less than that of the ligand. Compared with earlier data and with those of analogous thiosemicarbazones,[30] the present copper compound is found to exhibit higher anti-tumour activity which may be due to the generation of cytotoxic Cu(I) species through intracellular enzymatic reduction.[34,35] Alternatively, the copper complexes have been shown to undergo affinity binding with estrogen receptors in the case of breast cancer cells producing anti-proliferative effects observed in the present case.[36]

Following research on the anti-cancer properties of carboxamidrazone derivatives, the biological action of two copper conjugates of the new ligands **appc** and **atpc** (Scheme 12.4) [Cu(appc)Cl_2] (**c**) and [Cu(atpc)Cl_2] (**d**), against the hormone-responsive cancer cell line, mouse melanoma B16F10, was studied.[37] The complex is monomeric with **appc** molecule acting as a planar tridentate ligand. During metal complexation, the ligand utilizes two of its pyridyl nitrogens which are at the apex of the bipyramid and one diazine nitrogen which is part of the trigonal plane. Two unidentate chloride ions occupy the remaining positions of the trigonal plane in *cis* configuration.

Figure 12.3 shows the bar diagram of the anti-proliferative activities of copper complexes against the B16F10 cell line. The incubation of the cells with the parent ligands for 48 h is found to result in a decrease in the metabolic activity of these cells but not in a complete stop of the cell proliferation even at the highest dose employed (10 μM). The copper complexes on the other hand show a dose-dependent inhibition of the cell proliferation in the same concentration range indicating higher anti-proliferative activities. Complex **c** was found to be more potent ($IC_{50} = 6.8$ μM) than complex **d** ($IC_{50} = 10$ μM), suggesting that substitution on the N1 nitrogen of pyridyl-2-carbox-midrazone is only effective when the heteroatom is a nitrogen in a six-member aryl ring. For thiosemicarbazone ligands and their metal complexes, the presence of a thiocarbonyl sulphur in the side chain is known to increase the cytotoxicity.[38]

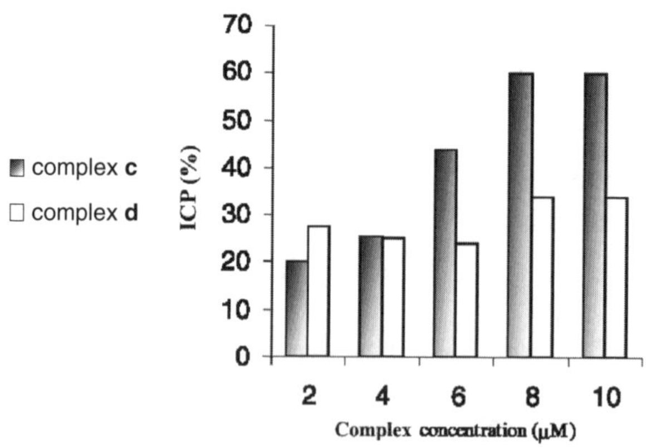

Figure 12.3 Inhibitory effect of complexes **c** and **d** on cell line B16F10 (*y*-axis: inhibition of cell proliferation, ICP)

The results, however, clearly indicate that the heterocyclic sulphur in the carboxamidrazone side chain offers no advantage in terms of the anti-proliferative activity against B16F10 mouse melanomal cells. Thus, the five-coordinate copper complexes are consistently found to exhibit enhanced anti-proliferative activities against B16F10 mouse melanoma cells. On the other hand, copper conjugation to the carboxamidrazone ligands may facilitate their intracellular transportation resulting in their estrogen receptor blocking action as previously suggested.[36]

Other interesting thiosemicarbazone derivatives have been studied in a search for new anti-tumour active substances. Thus, copper complexes of glucosone-bTSC (Scheme 12.5) were isolated and their anti-tumour activity studied.[39] The complexes appeared as crystalline substances of red colour, soluble in DMSO, DMF and some other organic solvents.

It was established that the synthesized copper complexes **Ib–Id**, possess low acute toxicities on white mongrel mice ($LD_{100} \geq 2000$ mg/kg), while complexes **If** and **Ig** are characterized by pronounced toxicity ($LD_{100} = 80$ and 100 mg/kg, respectively).

According to the chemotherapy tests developed on mice and rats, bTSC do not offer any protection against Walker's carcinosarcoma (WCS) and sarcoma 45. However, complexes **Ib** and **Ic**, in contrast to their ligands, exhibited somewhat more pronounced anti-tumour properties under the same conditions of testing with both sarcoma 45 {TGI (Tumour Growth Inhibition) = 54 and 53%, respectively, $p < 0.05$} and WCS (TGI = 47 and 50%, respectively, $p < 0.05$). Compound **Ib** retained the effect when tested on sarcoma 180 (TGI = 55%, $p < 0.05$). Complex **Ig** inhibited WCS growth (TGI = 34%, $p = 0.05$).

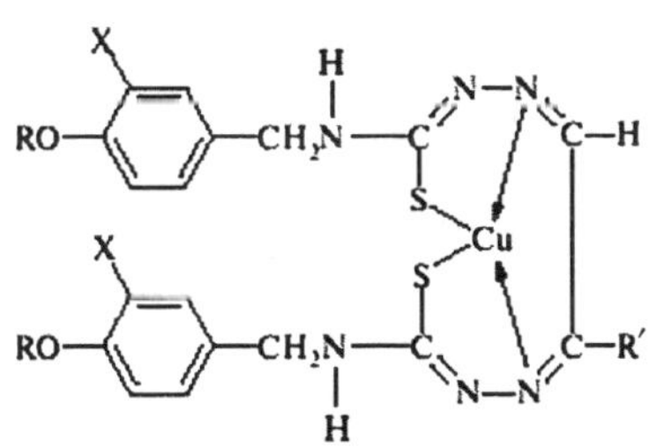

Ia–Ig

I: R = C_3H_7 (a), *i*-C_3H_7 (b), C_4H_9 (c), CH_3 (d), C_2H_5 (e, g), C_4H_9 (f);
R′ = CH_3 (a–e), $(CHOH)_3CH_2OH$ (f, g); X = H (a–c, f), NO_2 (d), Cl (e, g)

Scheme 12.5

A comparison of the results of these experiments to those for a bromobenzyl analogue tested on the same strains shows that the substitution of bromine for chlorine does not change the therapeutic effect of the ligand, but significantly decreases the anti-tumour effect of the complex, in particular for sarcoma 45.

Thiosemicarbazones of heterocyclic aldehydes possess anti-tumour properties and can inhibit DNA synthesis.[40] In the search for new anti-tumour drugs of this type, thiosemicarbazones of cyclodimers of 2-alkoxyacroleins, such as 2-formyl-2,5-dimethoxy-2,3-dihydro-4H-pyran thiosemicarbazones (Ligand 12.1), and its copper complex (**Cu–I**) were synthesized. The complex formation reaction was accompanied by the reduction of Cu(II) to Cu(I) under the action of the initial thiosemicarbazone. The acute toxicity (LD_{50} = 62.5 mg/kg, ip) of **Cu–I** was determined on intact SHK mice. The DNA-inhibiting activity of the copper complex was studied *in vivo* on C_3Ha mice with rapidly growing hepatoma 22I. In addition, its anti-tumour activity was studied *in vitro* on a CV-1 grivet kidney cell culture, using 6-mercaptopurine as the reference compound. In the *in vivo* tests, **Cu–I** (40 mg/kg) was introduced into mice and allowed to act for 3, 6, 9 or 24 h before analysis. For evaluating the selectivity of drug action, the DNA synthesis inhibition was determined both in tumour cells and in organ tissues (spleen and large intestine). The copper complex inhibited the inclusion of ^{14}C-thymidine into tumour cells by 90 and 74% after 9 and 24 h, respectively; at the same time, DNA synthesis in the splenic and intestinal cells did not exceed 47% for a 24 h exposure. Therefore, the presence of copper accounts for an increase in both efficacy and selectivity of the anti-tumour action, through an increase of the number of single-strand scissions of DNA chains and decreased DNA–protein interactions in cell nuclei.

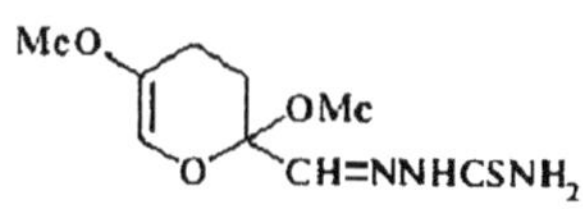

Ligand 12.1

The results of the anti-tumour tests with compound **Cu–I** in a dose of 20–25 mg/kg indicate suppression of the growth of both RShM-5 and LLC tumours by 99% without any loss of the experimental animals. The pronounced anti-tumour activity of **Cu–I** was observed not only for intraperitoneal injection. Moreover, a significant effect was retained upon peroral administration of the drug: treatment with a 35 mg/kg dose of the complex inhibited the growth of RShM-5 by 81%.

Initial interest in 2-formylpyridine thiosemicarbazone (**HL**) and its substituted derivatives arose from their marked anti-tumour properties.[41] More recently it has been shown that a dimer Cu(II) complex of HL is also active.[42] In view of the interest in these substances, three modified ligands based on the HL type, 6-methyl-2-formylpyridine thiosemicarbazone (**6HL**), 2-formylpyridine-2′-methyl thiosemicarbazone (**2′L**) and (2-formylpyridine-4′-methyl thiosemicarbazone) (**4′HL**) have been obtained (Scheme 12.6) and a selection of their Cu(II) complexes have been synthesized with N-donor ligands [2,2′-bipyridyl (bipy) and 4-dimethylaminopyridine (dmap)], as well as an S-donor ligand [pentafluorothiophenolate (pftp)] and O-donor ligands [paratrityl-phenolate (ptp) and the monohydrogen phosphate ion (HPO_4^{2-})]. The complexes have been characterized by physical and spectroscopic techniques.[43]

Cytotoxicity testing (IC_{50}) for selected compounds was carried out on the cell line HCT-8, a human colon adenocarcinoma as previously described, cells of which are generally resistant to intercalating drugs (relative to other cell lines).[44] Results evidence that CuL^+ is less active ($IC_{50} = 14\,\mu M$) than the free ligand **HL** ($IC_{50} = 4.7\,\mu M$). As ethylenediaminetetraacetic acid (EDTA) can quantitatively remove the Cu(II) from CuL^+, the cytotoxicities of both **HL** and CuL^+ were re-evaluated with EDTA present to ensure the activity of **HL** was not due also to the formation of cytotoxic metal chelates in the culture medium. The IC_{50} value of $3.8\,\mu M$ for **HL** + EDTA is not significantly changed from the average value of $4.7\,\mu M$ for **HL**; thus, the activity is due to the uncomplexed **HL** and not to metal complexes formed in the assay medium by accessible metal ions and **HL** or EDTA. The addition of EDTA to CuL^+ has, however, increased the cytotoxicity when compared with CuL^+ alone (IC_{50} values 7.6 and $14\,\mu M$, respectively), again showing the uncomplexed ligand to be more active than its Cu(II) complex. In contrast, but using WCS, the cytotoxicity of H_2KTS (Figure 12.1) has been shown to be due to the *in vivo* formation of the complex CuKTS, as H_2KTS has only no activity in animals maintained on a copper-deficient diet.[45]

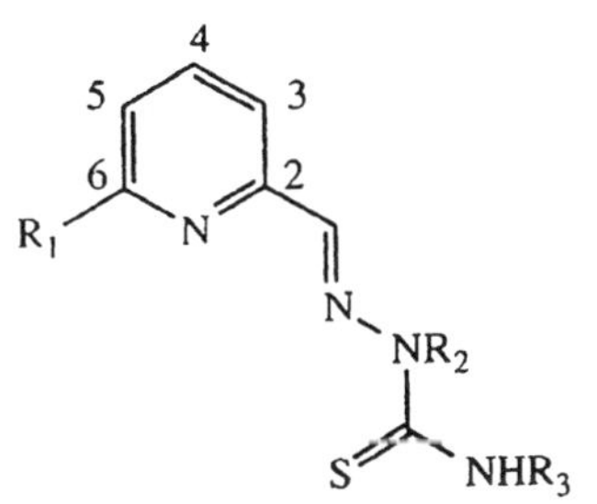

HL	$R_1 = R_2 = R_3 = H$
6HL	$R_1 = CH_3$, $R_2 = R_3 = H$
2′L	$R_2 = CH_3$, $R_1 = R_3 = H$
4′HL	$R_3 = CH_3$, $R_1 = R_2 = H$

Scheme 12.6

The ligand **2′L** and its complex Cu $(2'L)^{2+}$ are also both inactive. As published, *in vitro* and *in vivo* activity for **2′L** was lost compared to **HL**.[46] The addition of the 2′-methyl group effectively traps the ligand in the thione form, rendering impossible the overall conjugation. The loss of activity for **2′L** may therefore be due to the ligand being unable to have this delocalized form and/or its inability to coordinate as an anionic moiety.

The ligand 6-methyl-2-formylpyridine thiosemicarbazone, **6HL**, and its complex, Cu $(6L)^+$, are both inactive ($IC_{50} \geq 20\,\mu M$). The general decrease in activities for the 6-methylated ligands was thought to be a steric effect due to the interference with the ring N chelation and partly electronic due to changes in the conjugated ligand system.

12.4 Copper-Benzohydroxamic Acid Complexes

Several hydroxamic compounds, i.e. benzohydroxamic acid (BHA), have been synthesized and tested for their anti-tumour properties, most of them showing significant activity. Moreover, substituted BHA has also been prepared to enhance the anti-tumour effect of BHA.[47,48] With the same objective in mind, new Cu(II)-BHA complexes were isolated and evaluated not only for their efficacy in tumour cell growth inhibition and enhancement of longevity in tumour-bearing mice, but also as correctors of the perturbed biological parameters produced by the action of oxygen free radicals.[49] It is well known that a treatment with free radical scavengers eliminates such effects and causes the killing of tumour cells using the cell-damaging action of the named scavengers.[50]

The complex $[Cu\{C_6H_5(O)CNHO\}_2]$ (**1**) was tested against Ehrlich ascitic carcinoma bearing mice. LD_{50} value of complex **1** was 125 mg/kg. Treatment with complex **1** (25 mg/kg, ip dose/day, 4 days) resulted in a significant cancer growth inhibition (65%) and increased life span of tumour-bearing mice (150%). Additionally, the perturbed haematological parameters were restored. The effect of this treatment on the transplantability of the neoplastic cells was noted by the reduction of intraperitoneal tumour burden in mice inoculated with cells treated with complex **1**, with respect to control. A 56% reduction of tumour cell burden was observed. Free radicals generated by administration of complex **1** were estimated by the presence of the 'TBA reactive substance' in the serum of control mice treated with the complex and are responsible for killing tumour cells and inhibiting ribonucleotide reductase.[47] All the above observations show the efficacy of complex **1** at the mentioned dose, having no adverse side effect on the host. The complex could be utilized in combination with other suitable drugs to protect the host haematological parameters. However, in such use, systemic toxicity of **1**, if any, should be properly evaluated.

12.5 Copper-Imidazole Derivatives Complexes

The strong anti-tumour activity of the *trans*-bis (acetato) bis (imidazole) copper(II) complex $[Cu(O_2CMe)_2(HIm)_2]$, against the mouse cancer cell line B16 melanoma is well known.[51] A similar complex $[Cu(O_2CMe)_2L_2]$ (**1**), where L = 1-methyl-4,5-diphenylimidazole, has been synthesized and its effects on the induction of cell division delays, on Sister Chromatid Exchanges (SCEs) and on the suppression of Mitotic Indices (MIs) were examined as well.[52] SCEs have been proposed as a very sensitive method for detecting mutagens and/or carcinogens and furthermore as a method of evaluating chemotherapy *in vitro* and *in vivo*.[53,54]

The crystal structure of complex **1** indicates that two L molecules (via the pyridine-type N atoms) and two acetate ligands (via the carboxylate O atoms) are coordinated to the metal ion, forming a nearly square-planar arrangement. The Cu atom is six-coordinate with the ligand atoms forming a distorted octahedron. As expected, the structure of **1** shows a remarkable similarity to that of the anti-tumour complex $[Cu(O_2CMe)_2(HIm)_2]$.

The effect of complex **1** on the plasmid pKS DNA was examined, looking for the change of the conformational forms of the plasmid from the closed circular supercoiled (form I) to the open circular relaxed (form II). The unwinding effect of complex **1** is clearly observed when the plasmid is incubated with low concentrations (0.1 or 0.2 mM) of the compound (Figure 12.4), as seen from the progressive increase of the amount of the relaxed DNA (form II). Concentrations of the complex above 0.5 mM cause scissions on both forms of DNA, as shown by the appearance of a reduced amount of DNA in the gel.

Complex **1** induced small but non-significant changes in SCEs. By contrast, the complex induced a reduction of the proliferation rate index (PRI), an indicator of the cytostatic effect, and a reduction of MIs, an indicator of the cytotoxic effect, on human lymphocytes in cultures.

The obtained results show that concentrations of complex **1** between 0.77×10^{-7} and 1.54×10^{-6} M produced cytogenetic damage in culture human

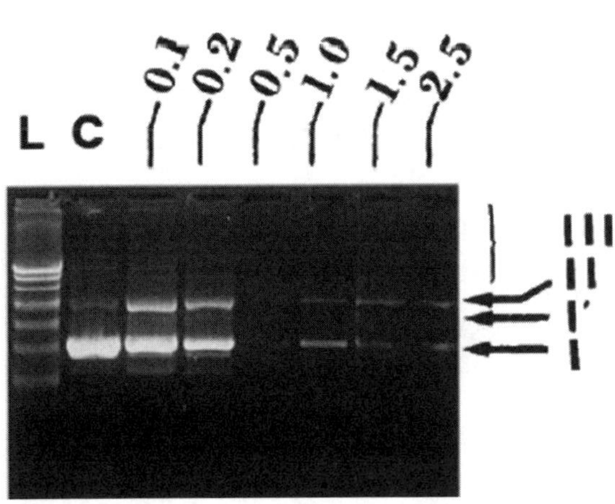

Figure 12.4 Agarose gel electrophoretic patterns of plasmid bluescript KS incubated with different concentrations of **1** (in mM). Lane **L**: kDNA/Hind III markers; lane **C**: intact pKS. Form III: catenanes or knott forms of DNA

lymphocytes. This DNA damage was observed by the induction of SCEs and the reduction of PRIs and MIs, indicating high cytostatic and cytotoxic action of **1**. Thus, the vulnerability of cells to **1** may be interpreted as having some relevance to the treatment of human cancer.

Chelating donor ligands as *N,N'*-1,10-phenanthroline (phen) are potent *in vitro* inhibitors of the growth of *Candida albicans*, a commensal of the human body considered to be an important fungal pathogen which is often fatal in immuno-compromised patients.[55,56] Phen itself and its metal complexes represent a novel set of highly active anti-fungal agents whose mode of action is significantly different from that of the state-of-the-art polyene and azole prescription. Additionally, benzimidazole and many of its derivatives exhibit a variety of biological actions, including anti-bacterial, -viral, -cancer and -fungal activity.[57]

The compound 2-(4'-thiazolyl)benzimidazole, {thiabendazole (TBZH) (**1**)}, (Scheme 12.7), with structural features common to phen and benzimidazole is non-toxic to humans and used as a fungicide in agriculture. Reactions of **1** with copper(II) acetate, chloride, nitrate and butanedioate yields $[Cu(TBZH)_2(H_2O)_2]$ (**2**), $[Cu(TBZH)_2Cl]Cl{\cdot}H_2O{\cdot}EtOH$ (**3**), $[Cu(TBZH)_2(NO_3)_2]$ (**4**) and $[Cu(TBZH)(O_2C\text{–}CH_2CH_2\text{–}CO_2)]$ (**5**), respectively. The molecular structure of complex **3** comprises a five-coordinate copper centre with the metal bound to two chelating TBZH ligands and one chloride ion. The geometry is best described as trigonal bipyramidal.[58]

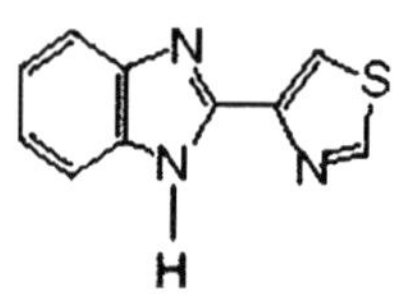

Structure of thiabendazole

Scheme 12.7

The chemotherapeutic potential (IC_{50}) of **1**, **3** and **4** towards the human squamous carcinoma tongue cell line, CAL-27, and the malignant melanoma (melanocyte) skin cell line, SK-MEL-31, was determined. TBZH was capable of killing both cancer lines only at higher concentrations with IC_{50} value of 91.7 and 136.9 μg/ml, for the tongue and skin cell line, respectively. In the case of compounds **3** and **4**, the IC_{50} values were almost identical (33.1 and 31.9 μg/ml, respectively) in the CAL-27 cell line and 39.1 and 30.7 μg/ml, respectively, in the SK-MEL-31 cell line. These doses were found to be significantly lower than that for the metal-free TBZH.

To date TBZH is the first *N,N'*-donor ligand that, *in vitro*, exhibits poor anti-candida activity on its own but when complexed to a copper(II) centre becomes relatively potent.

12.6 Copper-Polycarboxylate Complexes

The sequestering agent 1,2-propylenediamine-*N*,*N*,*N*,*N*′-tetraacetic acid (PDTA) is a good example of a polycarboxylic ligand forming active metal complexes against several human tumour cells.[17,59–61] It is well known that Cu(II) complexes with chelating ligands play an important role as models of the associative complexes involved in the substitution reactions at the copper site of SOD enzymes; indeed strongly donor ligands as PDTA should favour the formation of six-coordinate copper complexes, the coordination number of the associative complex.[62] Based on these arguments, a recent communication deals with the synthesis of the solid compound $[Cu(PDTA\text{-}H_2)(H_2O)_2]\cdot H_2O$ (**1**), anti-cancer properties of which have been evaluated against the established ovarian tumour cell line TG (*in vitro*) and the solid tumour Sarcoma 180 (*in vivo*).[63] Animals were BDF_1 female mice with weights within a 3 g value range and a minimum weight of 17 g. The number of animals was six per test group. The therapeutic activity of the complex was obtained from the T/C percentage which is described as T/C (%) = (100) × mean lifespan of treated mice/mean lifespan of untreated mice. The inhibiting effects of complex **1** are shown in Figure 12.5 which presents the evolution with time of the total number of culture cells in the absence (control) or presence of the complex.

In this way, the growth kinetics of culture treated with doses of 1 μg/ml and 10 μg/ml of complex is significantly lower ($p < 0.001$) than that shown in

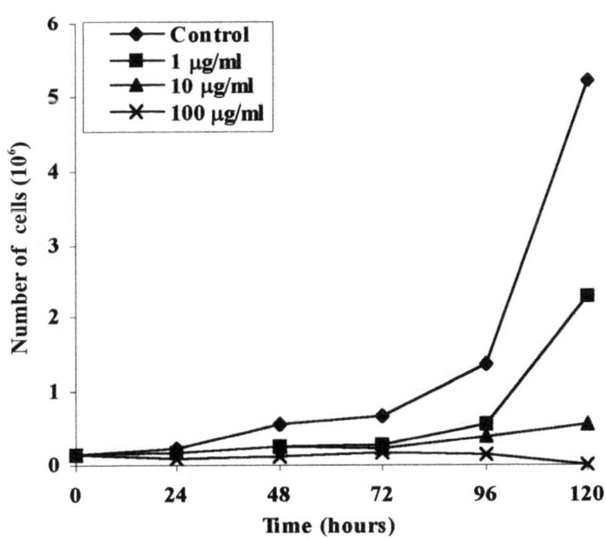

Figure 12.5 Determination of the effects exerted by complex **1** on the growth kinetics of the TG human cancer cell line

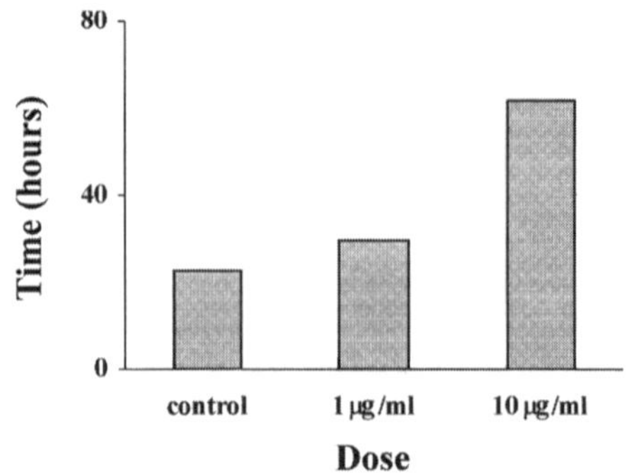

Figure 12.6 Variation of the duration of the cellular cycle of the TG ovarian cancer cell line with a dose of complex **1** (µg/ml, *x*-axis)

control cultures after 48 h of the treatment. Similar results are obtained 24 h after administration of the higher dose of the compound (100 µg/ml). As seen, the growth kinetics of the tumour cells decreases with increasing doses of compound **1**; thus, the proliferation process is almost completely cancelled by this complex after 96 h of treatment even at the lower doses handled.

The effect produced by the copper complex on the vital cellular cycle was also studied. Figure 12.6 shows the duration of the cellular cycle corresponding to treated and control TG cell lines. At doses of 1 µg/mg, an important induced increasing of the TG cellular cycle is observed (30 h) if compared to that of the control cells (23 h). The length of the cycle of treated cells at a dose of 10 µg/mg was also significantly enhanced (62 h). However, a dose of 100 µg/ml was strongly cytotoxic and the death of the treated cells was observed. Thus, obtained results demonstrate that significant enhancement of the vital cellular cycle is reached at all doses lower than 100 µg/ml.

In conclusion, this chapter included representative complexes of copper formed with different groups of compounds that have been recently tested for their anti-cancer activity. Thus, it has been established that the copper ion chelation plays a definite role in the biological activity of most of the selected organic ligands, enhancing the cytotoxicity and the anti-tumour activity of the metal-free substances on several tumour cell lines, *in vivo* and *in vitro*. The tested complexes are strong inhibitors of the enzyme ribonucleotide reductase, an obligatory enzyme in the pathway of synthesis of precursors of DNA.

Acknowledgements

Authors greatly appreciate financial support from MCYT, Spain (Grants PPQ2000-0035-P4 and BQU2001-2455) and EC (COST Chemistry Projects D20/005 and D20/009).

References

1. A.V. Xavier (Ed.), *Frontiers in Bioinorganic Chemistry*, VCH, Weinheim, 1986.
2. H. Sigel and A. Sigel, *J. Indian Chem. Soc.*, **77**, 501–509 (2000).
3. H. Sigel (Ed.), *Metal Ions in Biological Systems*, M. Dekker, Inc., New York, Vol. 11, 1980.
4. B.K. Kepler (Ed.), *Metal Complexes in Cancer Chemotherapy*, VCH, Weinheim, 1993.
5. A. Sigel and H. Sigel (Eds), *Metal Ions in Biological Systems*, M. Dekker, Inc., New York, Vol. 32, 1996.
6. Z. Guo and P.J. Sadler, *Angew. Chem. Int. Ed.*, **38**, 1512–1531 (1999).
7. M.P. Hacker, E.B. Douple and I.H. Krakoff (Eds), *Platinum Coordination Complexes in Cancer Chemotherapy*, Nijhoff, Boston, 1984.
8. S.L. Bruhn, J.H. Toney and S.J. Lippard, *Prog. Inorg. Chem.*, **38**, 477–532 (1990).
9. M. Van Bensichem and N. Farrell, *Inorg. Chem.*, **31**, 634–639 (1992).
10. J.A.R. Navarro, J.M. Salas, M.A. Romero *et al.*, *J. Med. Chem.*, **41**, 332–338 (1998).
11. B. Lippert (Ed.), *Cisplatin: Chemistry and Biochemistry of a Leading Anticancer Drug*, VHCA, Zurich/VCH, Weinheim, 1999.
12. M. Gielen, *Metal-Based Drugs*, **1**, 213–219 (1994); **2**, 99–103 (1995); *Coord. Chem. Rev.*, **151**, 41–51 (1996).
13. M. Gielen, R. Wilem, H. Dalil *et al.*, *Metal-Based Drugs*, **5**, 83–90 (1998).
14. E. Wong and C.M. Giandomenico, *Chem. Rev.*, **99**, 2451–2466 (1999).
15. E. Gallori, C. Vettori, E. Alessio *et al.*, *Arch. Biochem. Biophys.*, **376**, 156–162 (2000).
16. J. Reedijk, *Proc. Natl. Acad. Sci. USA*, **100**, 3611–3616 (2003).
17. M.J. Clarke, *Coord. Chem. Rev.*, **236**, 209–233 (2003).
18. R.A. Vilaplana, A. Castiñeiras and F. González-Vílchez, *Bioinorg. Chem. Appl.*, in press.
19. D.H. Petering, in *Metal Ions Biol. Syst.*, **11**, 197–229 (1980).
20. J.J.R. Fráusto da Silva and R.J.P. Williams, *The Biological Chemistry of the Elements*, Clarendon, Oxford, 1991.
21. F.A. French and E.J. Blanz Jr, *Cancer Res.*, **25**, 1454–1458 (1965); **26**, 1638–1640 (1966).
22. Z. Travnýcěk, M. Maloň, Z. Šindelář *et al.*, *J. Inorg. Biochem.*, **84**, 23–32 (2001).
23. A. Marzotto, A. Ciccarese, D.A. Clemente and G. Valle, *J. Chem. Soc., Dalton Trans.*, 1461–1468 (1995).
24. Z. Travnýcěk, M. Maloň, M. Biler *et al.*, *Transition Met. Chem.*, **25**, 265–269 (2000).
25. M. Maloň, Z. Travnýcěk, M. Maryško *et al.*, *Inorg. Chim. Acta*, **323**, 119–129 (2001).
26. C.J. Li, L.J. Zhang, B.J. Dezube *et al.*, *Proc. Natl. Acad. Sci. USA*, **90**, 1839–1842 (1993).
27. S.M. Planchon, S. Wuerzberger, B. Frydman *et al.*, *Cancer Res.*, **55**, 3706–3711 (1995).
28. B. Frydman, L.J. Marton, J.S. Sun *et al.*, *Cancer Res.*, **57**, 620–627 (1997).
29. Z. Afrasiabi, E. Sinn, J. Chen *et al.*, *J. Inorg. Biochem.*, **357**, 271–278 (2004).
30. A. Murugkar, B. Unnikrishnan, S. Padhye *et al.*, *Metal-Based Drugs*, **6**, 177–182 (1999).

31. M. Belicchi Ferrari, S. Capacchi, G. Pelosi *et al.*, *Inorg. Chim. Acta*, **286**, 134–141 (1999).
32. M. Belicchi Ferrari, G. Gasparri Fava, E. Leporati *et al.*, *J. Inorg. Biochem.*, **70**, 145–154 (1998).
33. N. Gokhale, S. Padhye, D. Rathbone *et al.*, *Inorg. Chem. Commun.*, **4**, 26–29 (2001).
34. Y. Li and M.A. Trush, *Carcinogenesis*, **14**, 1303–1311 (1993).
35. C.C. Winterbourn and R.W. Carrell, *Biochem. J.*, **165**, 141–148 (1977).
36. P.F. Predki and B. Sarkar, *J. Biol. Chem.*, **267**, 5842–5846 (1992).
37. N.H. Gokhale, S.S. Padhye, S.B. Padhye *et al.*, *Inorg. Chim. Acta*, **319**, 90–94 (2001).
38. D. West, S. Padhye and P. Sonawane, *Struct. Bonding*, **76**, 1 (1991).
39. E.R. Dilanyan, T.R. Ovsepyan and G.M. Stepanyan, *Pharm. Chem. J.*, **34**, 408–410 (2000).
40. N.A. Keiko, L.G. Stepanova, M.G. Voronkov *et al.*, *Pharm. Chem. J.*, **36**, 407–409 (2002).
41. B.A. Booth, K.C. Agrawal, E.C. Moore and A.C. Sartorelli, *Cancer Res.*, **34**, 1308–1314 (1974).
42. W.E. Antholine, B. Kalyanaraman and D.H. Petering, *Environ. Health Perspect.*, **64**, 19–35 (1985).
43. E.W. Ainscough, A.M. Brodie, W.A. Denny *et al.*, *J. Inorg. Biochem.*, **70**, 175–185 (1998).
44. G.J. Finlay, B.C. Baguley and W.R. Wilson, *Anal. Biochem.*, **139**, 272–277 (1984).
45. H.G. Petering, H.H. Buskirk and J.A. Crim, *Cancer Res.*, **27**, 1115–1121 (1967).
46. F.A. French, E.J. Blanz Jr, S.C. Shaddix and R.W. Brockman, *J. Med. Chem.*, **17**, 172–181 (1974).
47. S. Natelson, P. Pantazis and E.A. Natelson, *Clin. Chim. Acta*, **229**, 133–145 (1994).
48. I.H. Hall, R.Izydore, E.S. Hall, M.C. Miller *et al.*, *Anticancer Drugs*, **3**, 273–280 (1992).
49. J.A. Khanam, S.P. Bag, B. Sur and P. Sur, *Indian J. Pharmacol.*, **29**, 157–161 (1997).
50. T. Yoshikawa, S. Kokura, K. Tainaka *et al.*, *Cancer Res.*, **55**, 1617–1620 (1995).
51. H. Tamura, H. Imai, J. Kuwahara and Y. Sugiura, *J. Am. Chem. Soc.*, **109**, 6870–6871 (1987).
52. C.P. Raptopoulou, S. Paschalidou, A.A. Pantazaki *et al.*, *J. Inorg. Biochem.*, **71**, 15–27 (1998).
53. D.F. Deen, L.E. Kendall, L.J. Marton and P.J. Tofilon, *Cancer Res.*, **46**, 1599–1602 (1986).
54. D. Mourelatos, J. Dozi-Vassiliades, A. Kotsis and C. Gourtsas, *Cancer Res.*, **48**, 1129–1131 (1988).
55. M. Geraghty, M. McCann, M. Devereux *et al.*, *Metal-Based Drugs*, **6**, 41–48 (1999).
56. M. Devereux, M. McCann, V. Leon *et al.*, *Polyhedron*, **22**, 3187–3194 (2003).
57. N.S. Habib, S.M. Rida, E.A.M. Badawey *et al.*, *Pharmazie*, **52**, 346–349 (1997).
58. M. Devereux, M. McCann, D.O. Shea *et al.*, *J. Inorg. Biochem.*, in press.
59. R. Vilaplana, M.A. Romero, M. Quirós *et al.*, *Metal-Based Drugs*, **2**, 211–219 (1995).
60. M. Carballo, R. Vilaplana, G. Márquez *et al.*, *Biochem. J.*, **328**, 559–564 (1997).
61. F. González-Vílchez, R. Vilaplana, G. Blasco and L. Messori, *J. Inorg. Biochem.*, **71**, 45–51 (1998).
62. Z. Lu, C. Duan, Y. Tian *et al.*, *Inorg. Chem.*, **35**, 2253–2258 (1996).
63. S. Kamah, R. Vilaplana, J. Moreno *et al.*, *Metal-Based Drugs*, **7**, 219–224 (2000).

13 $_{30}$Zn The Role of Zinc as a Metallotherapeutic Agent

Jane V. Higdon and Emily Ho

13.1 Introduction

The role of zinc in a wide range of cellular processes, including cell proliferation, reproduction, immune function, and defense against free radicals, has been well established.[1,2] Zinc is considered to be the most abundant trace intracellular element, and there exists increasing evidence that zinc plays an important role in both genetic stability and function.[3] About 25% of the total zinc present in rat liver is found within the nucleus.[4] Zinc is a component of over 300 zinc metalloenzymes and over 2000 zinc-associated transcription factors including DNA-binding proteins with zinc fingers.[5–7] Thus, zinc plays an important role in several aspects of health. This chapter will focus on the function of zinc, recommended intakes, deficiency signs, potential therapeutic uses of zinc in health, and the influence of zinc status on DNA integrity and cancer.

13.2 Functions

Zinc is the most abundant intracellular trace element, and plays an integral role in numerous diverse biochemical processes.[8] The biochemical functions of zinc have been broadly classified as catalytic, structural and regulatory.[9]

Metallotherapeutic Drugs and Metal-Based Diagnostic Agents: The Use of Metals in Medicine
Edited by Gielen and Tiekink

13.2.1 Catalytic functions

Zinc is required for the catalytic activity of more than 300 specific enzymes.[10] Unlike iron and copper, zinc does not participate in redox reactions, but functions as a Lewis acid to accept a pair of electrons.[11] Some well-studied zinc metalloenzymes include alcohol dehydrogenase, carbonic anhydrase, alkaline phosphatase and ribonucleic acid polymerases.

13.2.2 Structural functions

Zinc also plays a structural role by stabilizing the tertiary structure of zinc metalloenzymes and other critical proteins.[9] For example, copper is required for the catalytic activity of copper–zinc superoxide dismutase (CuZnSOD), but zinc is required to stabilize the enzyme.[12] The coordination of zinc by four amino acid side chains to form zinc finger motifs facilitates stable protein folding to form biologically active proteins. The zinc finger motif is the most common recurring structure in transcription factors, including Specificity protein 1 (Sp1), retinoic acid receptors and vitamin D receptors.[10] Zinc finger motifs are also found in signal transduction factors, such as protein kinases.[9]

13.2.3 Regulatory functions

Although less studied than its catalytic and structural functions, zinc is also known to play a direct role in the regulation of gene expression. Transcription of the metallothionein gene is regulated by metal response elements (MRE) in its promoter.[13] The MRE-binding transcription factor-1 (MTF-1) is a multiple zinc finger-containing protein that binds to MRE sequences in DNA and activates gene transcription. MTF-1 binding to MRE is activated by zinc but not other transition metals, and may be related to cellular zinc status. Although MTF-1 regulation of other genes has not been established, the fact that a null mutation for MTF-1 is lethal during embryonic development in mice suggests other genes are regulated by MTF-1.[14] Zinc status has also been found to modulate the expression of genes related to zinc transport, growth, energy utilization, stress responses and signal transduction.[15]

13.3 Zinc Deficiency

The study of zinc deficiency in animals and humans points to critical roles for zinc in growth, neurological development, immune function, wound healing and reproduction. Despite the identification of multiple biochemical functions for zinc, the precise biochemical mechanisms for most symptoms of zinc deficiency are not yet clear.[8]

13.3.1 Severe zinc deficiency

Severe zinc deficiency has been observed in experimental human zinc depletion,[16] in adults and children on total parenteral nutrition solutions lacking zinc[17,18] and in patients with acrodermatitis enteropathica – a rare autosomal recessive disorder of zinc uptake associated with mutations in an intestinal zinc transporter gene.[19] The clinical manifestations of severe zinc deficiency in humans include slowing or cessation of growth and development, delayed sexual maturation and impotence, characteristic skin rashes, alopecia, chronic and severe diarrhea, immune system deficiencies, impaired wound healing, diminished appetite, impaired taste sensation, night blindness, corneal changes and behavioral disturbances.[8,20] Although not a useful indicator of mild zinc deficiency, plasma zinc concentrations, which normally range from 75 to 120 μg/dL (11.5–18.5 μmol/L), have been found to fall below 70 μg/dL with experimental zinc depletion.

13.3.2 Mild zinc deficiency

Although the lack of a sensitive indicator of marginal zinc deficiency in individuals hinders its study, controlled trials of moderate zinc supplementation have demonstrated that milder zinc deficiency states contribute to impaired growth and development and increased susceptibility to life-threatening infections in young children.[20]

13.3.3 Growth retardation

Significant delays in linear growth and weight gain are common features of mild zinc deficiency in children. In the 1970s and 1980s, several randomized placebo-controlled trials of zinc supplementation in young children with significant growth delays were conducted in Denver, Colorado. Modest zinc supplementation (5.7 mg/day) resulted in increased growth rates compared to placebo.[21] Since then, numerous studies in developing countries have observed similar results with modest zinc supplementation. A meta-analysis of 33 controlled intervention trials conducted throughout the world found that supplementation of children with zinc (3–20 mg/day) resulted in highly significant increases in linear growth and weight gain.[22] These findings indicate that growth-limiting zinc deficiency in children is widespread, particularly in developing countries.

13.4 Increased Susceptibility to Infectious Disease

13.4.1 Diarrhea

The adverse effects of zinc deficiency on immune system function are likely to increase the susceptibility of children to infectious diarrhea, while persistent diarrhea contributes to zinc deficiency and malnutrition. The results of intervention trials in developing countries indicate that zinc supplementation is effective in preventing and treating infectious diarrhea in children. Six out of nine supplementation trials found that the incidence of diarrhea was significantly (12–55%) lower in children who were supplemented with zinc.[23] Pooled analysis of those trials revealed an overall decrease of 18% in the incidence of diarrhea in zinc-supplemented children.[24] At least seven published trials found that zinc supplementation shortened the duration of an episode of acute diarrhea in children.[23] A meta-analysis of five trials of zinc supplementation in persistent diarrhea (episodes lasting at least 14 days) found that zinc supplementation reduced the rate of treatment failure or death by 42%, leading the World Health Organization (WHO) to recommend the use of zinc in the treatment of persistent diarrhea.[25]

13.4.2 Pneumonia

The results of at least five intervention trials indicate that zinc supplementation can prevent childhood pneumonia in developing countries. A pooled analysis of those trials found that the risk of pneumonia was 41% lower in children who were supplemented with zinc.[24]

13.4.3 Malaria

Although more limited, there is some evidence from supplementation trials that zinc supplementation can reduce the incidence of malaria in children. Two intervention trials found that supplementing children with zinc decreased visits to health facilities for the treatment of malaria by about one-third.[26,27]

13.5 Risk Factors for Zinc Deficiency

Conditions that increase the risk of zinc deficiency include protein-energy malnutrition, severe or persistent diarrhea, malabsorption syndromes (celiac disease and short bowel syndrome), inflammatory bowel disease (Crohn's disease and ulcerative colitis), anorexia nervosa, alcoholic liver disease and sickle cell anemia.[8] Life-stage groups at increased risk of zinc deficiency include infants and children, pregnant and lactating women, and the elderly.

13.5.1 Dietary factors that decrease zinc absorption

High dietary intakes of phytate (myo-inositol hexaphosphate), common in those whose major food staples are grains and legumes, substantially reduce the absorption of zinc.[10] Clinical zinc deficiency in humans was first described in 1961, when the consumption of diets with low zinc bioavailability due to high phytate content was associated with 'adolescent nutritional dwarfism' in the Middle East.[28] Zinc absorption from a meal with a phytate: zinc molar ratio that exceeds 15 may be 50% lower than from a meal where the phytate: zinc molar ratio is less than 5.[29] Calcium in combination with phytate reduces zinc absorption still further. This effect is particularly relevant to populations consuming a diet that is highly dependent on tortillas made with lime (calcium oxide).[8] The bioavailability of zinc from foods containing animal protein is generally high due to a relative absence of inhibitors of zinc absorption and the presence of sulfur-containing amino acids that enhance zinc absorption.

13.5.2 Intake recommendations

The latest recommended dietary allowances (RDA) for zinc (Table 13.1) were established by the Food and Nutrition Board of the Institute of Medicine in 2001 using a factorial approach to estimate the amount of absorbed zinc required to replace endogenous losses and maintain normal growth.[10] The RDA of 11 mg/day for adult men and 8 mg/day for adult women is based on data indicating that the bioavailability of zinc from typical US diets is 41% for men and 48% for women. The WHO Expert Committee established recommended nutrient intakes (RNI) for zinc that differ depending on whether a low-, moderate-, or high-bioavailability diet is consumed (Table 13.2). In a high-bioavailability diet with a phytate:zinc molar ratio of less than 5, the bioavailability of zinc is assumed to be about 50%, while in a low-bioavailability diet with a phytate:zinc molar ratio greater than 15, the bioavailability of zinc is assumed to be about 15%.[30]

Table 13.1 Recommended dietary allowances (RDA) and tolerable upper levels of intake (UL) for zinc[10]

		RDA		UL
Life stage	Age	Males (mg/day)	Females (mg/day)	Males and females (mg/day)
Infants	0–6 months	2*	2*	4
Infants	7–12 months	3	3	5
Children	1–3 years	3	3	7
Children	4–8 years	5	5	12
Children	9–13 years	8	8	23
Adolescents	14–18 years	11	9	34
Adults	≥19 years	11	8	40
Pregnant	≤18 years	–	12	34
Pregnant	≥19 years	–	11	40
Lactating	≤18 years	–	13	34
Lactating	≥19 years	–	12	40

* Adequate intake (AI).

Table 13.2 WHO/FAO recommended nutrient intakes (RNIs) for dietary zinc from diets ranging from high zinc bioavailability (assumed 50%) to low zinc bioavailability (assumed 15%)[30]

Life stage	Age	Males (mg/day)	Females (mg/day)
Infants	0–6 months	1.1–6.6*	1.1–6.6
Infants	7–12 months	2.5–8.4	2.5–8.4
Children	1–3 years	2.4–8.3	2.4–8.3
Children	4–6 years	2.9–9.6	2.9–9.6
Children	7–9 years	3.3–11.2	3.3–11.2
Adolescents	10–18 years	5.1–17.1	4.3–14.4
Adults	≥19 years	4.2–14.0	3.0–9.8
Pregnant, 1st trimester	All ages	–	3.4–11.0
Pregnant, 2nd trimester	All ages	–	4.2–14.0
Pregnant, 3rd trimester	All ages	–	6.0–20.0
Lactating, 0–3 months	All ages	–	5.8–19.0
Lactating, 4–6 months	All ages	–	5.3–17.5
Lactating, 6–12 months	All ages	–	4.3–14.4

* The lower number represents recommended zinc intake from high-bioavailability diets (phytate:zinc molar ratio <5). The higher number represents recommended zinc intake from low-bioavailability diets (phytate:zinc molar ratio >15).

13.6 Therapeutic Uses of Zinc

13.6.1 Wilson disease

Wilson disease is a rare genetic copper overload disorder characterized by progressive liver disease and/or neurological deterioration.[31] Oral zinc supplements are used in the treatment of Wilson disease to decrease copper absorption. Pharmacologic doses of zinc induce metallothionein synthesis by intestinal mucosal cells. Metallothionein has higher affinity for copper than zinc and preferentially binds copper in the intestine. Copper bound to metallothionein is not absorbed but eliminated in the feces during normal turnover of intestinal mucosal cells.[32]

13.6.2 Age-related macular degeneration

Age-related macular degeneration (AMD) is the most common cause of blindness and vision impairment in Americans over the age of 60 years.[33] Zinc is present in relatively high concentrations in the retina, and lower retinal and serum zinc levels have been associated with macular degeneration in primates.[34] However, scientific evidence that zinc intake is associated with the development or progression of AMD is limited. Observational studies have not demonstrated clear associations between dietary zinc intake and the incidence of AMD.[35–37] A randomized controlled trial found that supplementation with 200 mg/day of zinc sulfate (81 mg/day of elemental zinc) for 2 years reduced the loss of vision in patients with AMD.[38] However, a later trial using the same dose and duration found no beneficial effect in patients with a more advanced form of AMD in one eye.[39] A large randomized controlled trial of daily anti-oxidant (500 mg of vitamin C, 400 IU of vitamin E and 15 mg of β-carotene) and high-dose zinc (80 mg of zinc and 2 mg of copper) supplementation found that the anti-oxidant combination plus high-dose zinc and high-dose zinc alone significantly reduced the risk of advanced macular degeneration compared to placebo in individuals with moderate to severe signs of the disease in at least one eye.[40] At present, there is little evidence that zinc supplementation is beneficial for people with early signs of macular degeneration, but additional randomized controlled trials are warranted.[41]

13.6.3 Human immunodeficiency virus (HIV) infection

The essential role of zinc in immune system function has generated considerable scientific interest in the potential for adequate zinc status to slow the progression of HIV disease and decrease the incidence of opportunistic infections.

Decreased serum zinc levels have been associated with more advanced disease and increased mortality in HIV patients.[42,43] However, it is unclear whether low plasma zinc levels in HIV patients are related to zinc deficiency or disease progression.[44] In the largest intervention trial conducted in AIDS patients, supplementation with 45 mg/day of zinc for one month reduced the incidence of opportunistic infections compared to placebo.[45] However, the HIV virus also requires zinc, and excessive zinc intake may stimulate the progression of HIV infection. In an observational study of HIV-infected men, increased zinc intake was associated with more rapid disease progression, and any intake of zinc supplements was associated with poorer survival.[46,47] These results indicate that further research is necessary to determine optimal zinc intakes for HIV-infected individuals.[48]

13.6.4 Type 1 diabetes

Type 1 (insulin-dependent) diabetes results from the autoimmune destruction of the insulin-producing pancreatic β-cells. Supplementation studies in animals indicate that increasing pancreatic zinc levels reduce the severity of chemically induced type 1 diabetes.[49] Evidence that higher zinc intakes are protective against type 1 diabetes in humans is very limited. A case-control study in Sweden[50] and an ecological study in the UK[51] found that higher concentrations of zinc in the drinking water were associated with a significantly lower incidence of type 1 diabetes, but an ecological study in Finland found no such relationship.[52] Increased urinary zinc excretion (hyperzincuria) and decreased plasma zinc concentrations have been observed in patients with type 1 diabetes in some[53,54] but not all studies.[55,56] However, there is no evidence in humans that higher zinc intakes are beneficial in the management of type 1 diabetes. In fact, one study found that supplementation with 50 mg/day of zinc for one month increased hemoglobin A1c levels in type 1 diabetics.[57]

13.6.5 Wound healing

Impaired wound healing is a long-recognized consequence of zinc deficiency.[58,59] However, the results of numerous clinical trials of zinc supplementation on healing of pressure ulcers, vascular ulcers, surgical incisions and other skin lesions are mixed.[60–62] In general, zinc supplementation has been found to improve wound healing in zinc-deficient populations and individuals with low plasma zinc concentrations, but evidence of benefit in zinc-sufficient populations is weak.[63] Moreover, limited data in humans and animals suggest that very high doses of supplemental zinc may have adverse effects on immune responses essential for wound healing.[64,65]

13.7 Common Cold

13.7.1 Oral zinc

The use of zinc lozenges within 24 h of the onset of cold symptoms and continued every 2–3 h while awake until symptoms resolve has been advocated for reducing the duration of the common cold. At least ten controlled trials of zinc gluconate lozenges for the treatment of common cold in adults have been published. Five trials found that zinc lozenges reduced the duration of cold symptoms,[66–70] while five trials found no difference between zinc lozenges and placebo lozenges with respect to the duration or severity of cold symptoms.[71–75] Of the two placebo-controlled trials that examined the effect of zinc acetate lozenges, one found that their use significantly shortened the duration of cold symptoms,[76] while the other found no difference between zinc acetate and placebo lozenges.[71] Despite numerous well-controlled clinical trials, evidence that zinc lozenges are effective in reducing the duration or severity of common cold symptoms is still lacking.[77]

13.7.2 Intranasal zinc

Intranasal zinc preparations designed to be applied directly to the nasal epithelium are also marketed as over-the-counter cold remedies. While two placebo-controlled trials found that intranasal zinc gluconate modestly shortened the duration of cold symptoms when started 24–48 h after the onset of symptoms,[78,79] two other placebo-controlled studies of intranasal zinc found it to be of no benefit.[80,81] Of concern are several case reports of persistent loss of the sense of smell (anosmia) in individuals who used intranasal zinc as a cold remedy.[82,83] Since zinc-associated anosmia may be irreversible, zinc nasal gels and nasal sprays should be avoided.

13.8 Zinc Status and Its Relevance to Cancer

Zinc deficiency affects many cellular systems in the host because of zinc's essential role in many aspects of cellular metabolism. Zinc deficiency can occur in populations with low dietary zinc intake and high concentration of phytate, a powerful chelator of divalent metals. Although severe zinc deficiency is rare, mild deficiency is highly prevalent even in developed countries. Populations that are at high risk include individuals at early stages of the life cycle (i.e. infancy and childhood) when requirements for zinc are high. In addition, the elderly have an increased risk of zinc depletion because zinc absorption may be impaired and they tend to consume low zinc diets.[84] Foods rich in zinc include red meat and seafood. Several plant sources such as whole grains and legumes also contain zinc, but the zinc in these sources is much less bioavailable. Thus, vegetarians may also be at risk of zinc deficiency.

The role of zinc in cancer has received increasing attention. The link between zinc deficiency and cancer has been established in human, animal and cell culture studies. Zinc status is compromised in cancer patients compared to healthy controls.[85,86] Zinc deficiency causes oxidative DNA damage[87] and chromosome breaks have been reported in animals fed a zinc-deficient diet.[88] In rats, dietary zinc deficiency causes an increased susceptibility to tumor development when exposed to carcinogenic compounds.[89–93] Zinc deficiency has also been suggested to be a contributor to the development of esophageal tumors in rats.[91,92,94] *In vitro*, cell culture studies have also shown that zinc deficiency can lead to increased oxidative damage to testicular cell DNA.[95] Together, these data strongly suggest that zinc deficiency itself compromises the integrity of DNA and may increase one's risk of developing cancer. However, the exact molecular mechanism by which zinc deficiency impacts DNA integrity and cancer risk remains unclear and is likely a complex multi-layered process. Understanding the basic molecular mechanisms behind human nutrient deficiencies will help us better understand the role of proper nutrition in the prevention of many chronic disease states such as cancer.

13.8.1 Zinc deficiency and oxidative stress

A significant portion of the North American population does not consume adequate zinc. Approximately 10% of the US population consumes less than half the recommended level for zinc.[96] Importantly, zinc deficiency results in an increased sensitivity to oxidative stress[97] and may, in part, account for the mechanism by which zinc deficiency increases the risk for cancer development. There is now increasing evidence that oxidative stress plays an important contributing factor in several chronic degenerative diseases such as cancer.[98] Zinc's anti-oxidant function may be related to several factors. First, zinc is an essential component of CuZnSOD, one of the cell's first lines of defense against reactive oxygen species (ROS), which functions to remove the superoxide anion. Zinc is also an important factor in regulating metallothionein metabolism. Metallothionein, a small molecular-weight protein, is important in zinc homeostasis and has potent anti-oxidant activity. Zinc can induce the expression of the cysteine-rich metallothionein protein that acts as a potent scavenger of hydroxyl radicals. Loss of this protein can impair cellular anti-oxidant defenses and contribute to the cell's sensitivity to oxidative stress. Other potential mechanisms for zinc's anti-oxidant effects are the antagonism of redox-active transition metals, such as iron or copper, and the prevention of oxidation of sulfhydryl groups within proteins. Zinc can limit hydroxyl radical formation from H_2O_2 by competing with both iron and copper in Fenton reactions and appears to stabilize sulfhydryl groups in several proteins.[99–104]

Compromised zinc status clearly has a significant impact on anti-oxidant capacity of the cell. Several laboratories, including our own, have shown increased oxidative stress with zinc deficiency. Increased oxidant production has been shown in zinc-deficient cell cultures[105] and increases in oxidative protein and DNA damage have been shown in zinc-deficient rats.[87,95,97,106,107] Zinc-supplementation strategies have also been shown to be beneficial against oxidant damage and the progression of ROS-induced diseases such as diabetes.[49,108] Addition of zinc also protects cells from UV-induced DNA damage and apoptosis.[109,110] In humans, zinc supplementation appears to help patients with slowing the progression of AMD.[111,112] Thus, the anti-oxidant role of zinc could be an important mechanism in maintaining DNA integrity in the cell by preventing oxidative DNA damage in the cell.

Reactive oxygen species are commonly produced during normal cellular metabolism. However, under certain conditions of stress, such as poor nutrition, an increase in the production of ROS may overwhelm host anti-oxidant defenses, resulting in oxidative damage. There is increasing evidence that the pathology and disease development associated with oxidative stress may not be due simply to increases in oxidative damage.[113] Rather, ROS may also act as signaling molecules that trigger distinct pathways to induce pathology. Thus, the role of zinc deficiency in the development of chronic diseases such as diabetes and cancer may be far more complex than simply causing oxidative damage. Instead, zinc status may also affect redox-sensitive signals and ultimately alter signal pathways involved in stress response and DNA repair.

13.8.2 Zinc and DNA repair

The effects of zinc on DNA integrity may not be limited to zinc's anti-oxidant properties. Zinc plays a critical role in the regulation of transcription and replication of DNA through zinc finger proteins. Additionally, many DNA repair mechanisms involve zinc. Many proteins involved in both base-excision repair and nucleotide-excision repair are zinc finger or zinc-associated proteins. For example, the tumor suppressor protein p53 plays an important role in coordinating events leading to appropriate DNA repair. The p53 protein plays a role in modulating cell cycle progression, apoptosis, DNA repair and cell proliferation/differentiation.[114] Also, p53 is most well known for its ability to induce G_1 arrest in the cell cycle, allowing the cell to induce adequate repair of DNA before cellular division.[115] Notably, the prevalence of p53 mutations in tumors is very high. In fact, over 50% of human malignancies contain a mutation in p53.[116] The majority of these mutations are found in the region of the gene that encodes for the DNA-binding region of p53.[117,118] This binding region also contains the zinc-binding domain. Several lines of evidence suggest that p53 acts, in part, as a DNA-binding protein.[119] Thus, to coordinate the events related to DNA repair, p53 must be able to bind to specific DNA-binding domains to transcriptionally activate downstream targets involved in DNA repair. An increase in p53 protein expression with zinc deficiency was found in several different cell lines.[120,121] An upregulation in p53 expression is seen both in chemically induced zinc deficiency (using an intracellular zinc chelator, *N*,*N*,*N*′,*N*′-tetrakis(2-pyridylmethyl)ethylenediamine TPEN) and by feeding cells zinc-deficient media. This upregulation of p53 expression is most likely in response to DNA damage induced by zinc deficiency. Although there is an increase in p53 expression with zinc deficiency, there is evidence that p53 may be dysfunctional and, hence, DNA repair would be severely compromised. A marked decrease in the ability of p53 to bind to downstream DNA targets has been found with zinc deficiency.[105] Because the p53 protein is a transcription factor, the ability of p53 to bind to DNA promoter regions is critical for coordinating the events that control DNA repair and apoptosis. Thus, despite increases in p53 protein level in zinc-deficient cells, this p53 is likely dysfunctional, and activation of gene transcription of specific genes needed for DNA repair could be compromised.

The DNA-binding activity of p53 is largely mediated by a conformation-sensitive structure in the central portion of the protein (residues 102–292).[118] Mutations in this region cause an 'unfolding' of this structure and a loss of binding activity. Other researchers have also found that the removal of zinc, by either chemical chelation or adding zinc-deficient media, alters the expression of p53.[122,123] Direct chemical chelation also appears to reversibly alter p53 conformation, with the loss of DNA-binding activity.[124] Thus, zinc deficiency may render the p53 protein to adopt a 'mutant-like' conformation that will alter the cell's ability to appropriately respond to DNA damage.

The effect of zinc deficiency was also investigated on the expression of apyrimidic endonuclease (APE), an important endonuclease in base-excision repair.[125] DNA base-excision repair is a major pathway responsible for the repair of both cellular alkylation and oxidative DNA damage. A critical step in this pathway involves the cleavage of damaged sites in DNA by APE. APE (which is also known as Ref. 1) is a multifunctional protein that not only repairs AP sites, but also controls DNA-binding activity, via redox mechanisms, of numerous transcription factors that are involved in cancer promotion and progression (such as AP-1, NFκB, p53).[126] In addition, APE levels appear to be elevated in a number of cancers.[127–129] Zinc deficiency increases the expression of APE, most likely in response to DNA damage induced by zinc deficiency.[105] These transcription factors play important roles in controlling oxidative stress responses and cell proliferation.[130,131] Perturbations in their binding and signal pathways could significantly impair the cell's ability to handle oxidative stress and ultimately be very detrimental to the cell. We have also found a marked decrease in binding of AP1 and NFκB with zinc deficiency.[105] Other investigators have also detected alterations in NFκB signaling with zinc deficiency.[132] Therefore, the impact of zinc status on DNA integrity is most likely a multi-layered process involving both increases in oxidative stress and combined perturbations in DNA repair and other oxidant-sensitive signal pathways.

13.8.3 Zinc and prostate cancer

Etiologic factors that initiate and enhance the progression of prostate malignancy are beginning to emerge. One major research focus is the role of diet and nutrition. Dietary factors possibly linked to prostate cancer are numerous; however, the most provocative data focus attention upon those nutrients and phytochemicals involved in oxidative defense. A clear understanding of how anti-oxidants may protect the prostate from the genetic damage that is associated with tumor development remains unclear. It is well established that the consumption of fruit and vegetables is associated with reduced risk of many cancers.[133,134] These strong and reproducible epidemiological associations have led to an interest in determining the specific components within whole foods responsible for this observed reduced risk. However, supplementation with certain anti-oxidant compounds has not been successful. The anti-oxidant β-carotene which when provided as a dietary supplement to populations at high risk of lung cancer, in two independent trials, actually resulted in the early termination of these studies due to significantly increased incidence in the supplemented populations.[135,136] As a result of the strikingly negative outcome of the β-carotene trials, our own approach has shifted away from evaluating the health benefits of single dietary supplements to an approach with whole foods that contain an array of nutrients that are targeted to specific biological functions. Small clusters of cells representing the earliest stages of prostate cancer, in an apparently latent form, reside in millions of men (estimates up to 50%) of every race and culture of the world.[137] Therefore, it can be deduced that initiating factors leading to changes in the local prostate environment can shift the balance in individuals whether to develop 'active' prostate cancer. Because prostate cancer is a disease of aging, hypotheses have been formulated that oxidative stress plays a potentially important role in the etiology of this disease, and thus most of the dietary intervention studies to date have focused on enhancing anti-oxidant status. This has been done by evaluating supplements of vitamin E,[138] selenium[139] or β-carotene[138,140] on risk of developing prostate cancer. Currently, a large chemoprevention trial, the selenium and Vitamin E Cancer Prevention Trial (SELECT), is underway to evaluate the effects of selenium and vitamin E either alone or in combination on prevention of prostate cancer.[141] This study is based entirely on anti-oxidant function of these two nutrients using dietary supplementation. The outcome of this study will not be known for several years.

Zinc is a nutrient of particular interest in prostate cancer. Zinc plays an important role in protecting DNA from damage as an anti-oxidant and a component of many DNA repair proteins. Zinc is also unique in that it bears anti-oxidant,[1] anti-inflammatory[142] and pro-apoptotic activity.[143,144] Thus, zinc supplementation has the potential to target multiple points of the carcinogenesis cascade (Figure 13.1). Normal human prostate accumulates the highest levels of zinc of any soft tissue in the body.[145,146] However, a marked decrease in zinc content is associated with prostate cancer.[147–149] Zinc deficiency also compromises hormone status and reproductive function in men.[150–153] Several studies have implicated changes in zinc accumulation in the development and progression of prostate malignancy.[154,155] There also exists some evidence that

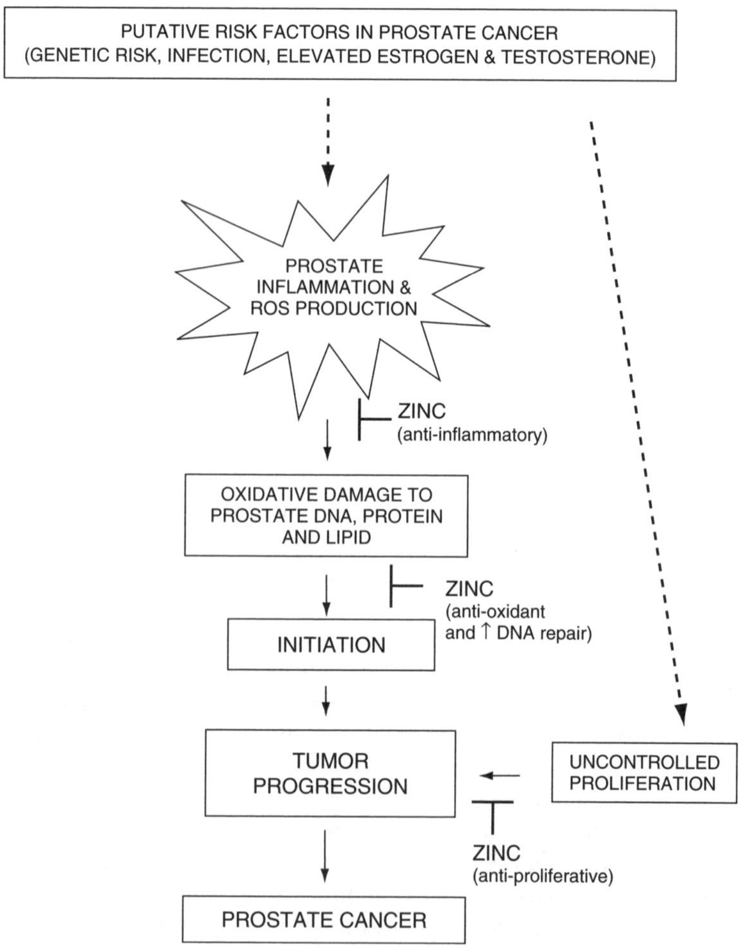

Figure 13.1 Potential inhibitory actions of zinc in prostate carcinogenesis

increased dietary zinc is associated with a decrease in the incidence of prostate cancer.[156] Currently it is unknown why the prostate accumulates high zinc concentrations. However, this phenomenon may render the prostate sensitive to changes in zinc intake. We have shown in various other cell types that changes in intracellular zinc have a dramatic effect on DNA damage and repair, and hence risk of cancer.[105] It is possible that dietary zinc deficiency will increase an individual's risk of oxidative DNA damage in the prostate and prostate cancer. In addition, there appears to be a loss of zinc during prostate cancer. Thus, zinc need may be enhanced in prostate cancer patients. Zinc supplementation strategies may not only aid in the prevention of cancer, but could also play an important role in limiting its malignancy.

There is some controversy regarding the efficacy of zinc supplements for the prevention of prostate cancer. Several studies have shown that high cellular zinc levels inhibit prostate cancer cell growth.[154,157,158] However, a recent epidemiological study has shown an increase in risk for prostate cancer with high zinc supplement use.[159] Increased risk was seen in subjects with very high dose zinc supplement use (more than 100 mg/day) or long-term zinc supplement use (10 or more years). The current upper level of intake for zinc is 40 mg/day. Thus it is possible that these subjects could have been in the 'toxic' range for zinc intake. As with most therapeutics, higher doses do not always equate with an increase in efficacy.

13.9 Safety

13.9.1 Toxicity

Isolated outbreaks of acute zinc toxicity have occurred as a result of the consumption of food or beverages contaminated with zinc released from galvanized containers. Signs of acute zinc toxicity are abdominal pain, diarrhea, nausea and vomiting. Single doses of 225–450 mg of zinc usually induce vomiting.[10] Milder gastrointestinal distress has been reported at doses of 50–150 mg/day of elemental zinc. Metal fume fever has been reported after the inhalation of zinc oxide fumes. Profuse sweating, weakness, and rapid breathing may develop within 8 h of zinc oxide inhalation and persist 12–24 h after exposure is terminated.[8] Total zinc intakes of 60 mg/day (50 mg/day supplemental and 10 mg/day dietary zinc) have been found to result in signs of copper deficiency.[160] In 2001, the Food and Nutrition Board of the Institute of Medicine established tolerable upper intake levels (UL) for zinc based on the prevention of copper deficiency (Table 13.1). The UL for zinc in adults is 40 mg/day.[10] Intranasal zinc is known to cause a loss of the sense of smell (anosmia) in laboratory animals.[161] Zinc sulfate nasal spray was reported to cause persistent anosmia in humans in the 1930s,[162] and there have been several recent case reports of individuals who developed anosmia after using intranasal zinc gluconate.[82,83]

13.9.2 Drug interactions

Concomitant administration of zinc supplements with tetracycline and quinolone classes of antibiotics may decrease antibiotic absorption with potential reductions in efficacy.[163] The therapeutic use of metal chelating agents, such as penicillamine for Wilson disease and diethylenetriamine pentaacetate (DTPA) for iron overload, has resulted in severe zinc deficiency. Anticonvulsant drugs, especially sodium valproate, may also precipitate zinc deficiency. Prolonged use of diuretics may increase urinary zinc excretion, resulting in increased loss of zinc. The tuberculosis medication ethambutol has metal chelating properties and has been shown to increase zinc loss in rats.[8]

13.10 Conclusions

Zinc clearly plays an important role in maintaining human health. A significant portion of the population could be at risk for zinc deficiency and may benefit from zinc-supplementation strategies. Zinc may also be beneficial in the treatment of several disorders, such as several pro-inflammatory conditions and cancer. However, issues considering dose and toxicity also need to be addressed.

References

1. T.M. Bray and W.J. Bettger, *Free Rad. Bio. Med.*, **8**, 281–291 (1990).
2. S.R. Powell, *J. Nutr.*, **130**, 1447S–1454S (2000).
3. I.E. Dreosti, *Mutat. Res.*, **475**, 161–167 (2001).
4. R.J. Cousins, *Proc. Nutr. Soc.*, **57**, 307–311 (1998).
5. A.S. Prasad, *J. Am. Coll. Nutr.*, **17**, 542–543 (1998).
6. A.S. Prasad and O. Kucuk, *Cancer Metastasis Rev.*, **21**, 291–295 (2002).
7. A.S. Prasad, *BMJ*, **326**, 409–410 (2003).
8. J.C. King and C.L. Keen, *Nutrition in Health and Disease*, M. Shils, J.A. Olson, M. Shike and A.C. Ross (Eds), pp. 223–239, Williams & Wilkins, Baltimore, 1999.
9. M.J. Dibley, *Present Knowledge in Nutrition*, B.A. Bowman and R.M. Russell (Eds), pp. 329–351, ILSI Press, Washington, DC, 2001.
10. Food and Nutrition Board of the Institute of Medicine, *Dietary Reference Intakes for Vitamin A, Vitamin K, Boron, Chromium, Copper, Iodine, Iron, Manganese, Molybdenum, Nickel, Silicon, Vanadium, and Zinc*, pp. 442–501, National Academy Press, Washington, DC, 2001.

11. K.A. McCall, C. Huang and C.A. Fierke, *J. Nutr.*, **130**, 1437S–1446S (2000).
12. B. Halliwell and J.M.C. Gutteridge, *Free Radicals in Biology and Medicine*, pp. 107–112, Oxford University Press, New York, 1999.
13. S.R. Davis and R.J. Cousins, *J. Nutr.*, **130**, 1085–1088 (2000).
14. C. Gunes, R. Heuchel, O. Georgiev *et al.*, *EMBO J.*, **17**, 2846–2854 (1998).
15. R.J. Cousins, R.K. Blanchard, J.B. Moore *et al.*, *J. Nutr.*, **133**, 1521S–1526S (2003).
16. M.T. Baer and J.C. King, *Am. J. Clin. Nutr.*, **39**, 556–570 (1984).
17. T. Arakawa, T. Tamura, Y. Igarashi *et al.*, *Am. J. Clin. Nutr.*, **29**, 197–204 (1976).
18. R.G. Kay, C. Tasman-Jones, J. Pybus *et al.*, *Ann. Surg.*, **183**, 331–340 (1976).
19. S. Kury, M. Kharfi, R. Kamoun *et al.*, *Hum. Mutat.*, **22**, 337–338 (2003).
20. M. Hambidge, *J. Nutr.*, **130**, 1344S–1349S (2000).
21. P.A. Walravens, K.M. Hambidge and D.M. Koepfer, *Pediatrics*, **83**, 532–538 (1989).
22. K.H. Brown, J.M. Peerson, J. Rivera and L.H. Allen, *Am. J. Clin. Nutr.*, **75**, 1062–1071 (2002).
23. R.E. Black, *J. Nutr.*, **133**, 1485S–1489S (2003).
24. Z.A. Bhutta, R.E. Black, K.H. Brown *et al.*, *J. Pediatr.*, **135**, 689–697 (1999).
25. Z.A. Bhutta, S.M. Bird, R.E. Black *et al.*, *Am. J. Clin. Nutr.*, **72**, 1516–1522 (2000).
26. C.J. Bates, P.H. Evans, M. Dardenne *et al.*, *Br. J. Nutr.*, **69**, 243–255 (1993).
27. A.H. Shankar, B. Genton, M. Baisor *et al.*, *Am. J. Trop. Med. Hyg.*, **62**, 663–669 (2000).
28. A.S. Prasad, J.A. Halsted and M. Nadimi, *Am. J. Med.*, **31**, 532–546 (1961).
29. J.R. Turnlund, J.C. King, W.R. Keyes *et al.*, *Am. J. Clin. Nutr.*, **40**, 1071–1077 (1984).
30. Food and Agriculture Organization, Human Vitamin and Mineral Requirements: Report of a Joint FAO/WHO Expert Consultation, http://www.fao.org/DOCREP/004/Y2809E/y2809e0m.htm#bm22, 2002.
31. W.A. Gahl, *Cecil Textbook of Medicine*, L. Goldman and J.C. Bennett (Eds), pp. 1130–1132, W.B. Saunders Co., Philadelphia, 2000.
32. D.W. Cox and E.A. Roberts, *Sleisenger & Fordtran's Gastrointestinal and Liver Disease*, M. Feldman, L.S. Friedman and M.H. Sleisenger (Eds), pp. 1269–1277, W.B. Saunders, St Louis, 2002.
33. National Eye Institute, Statistics/Data, http://www.nei.nih.gov/eyedata/, 2004.
34. B.H. Grahn, P.G. Paterson, K.T. Gottschall-Pass and Z. Zhang, *J. Am. Coll. Nutr.*, **20**, 106–118 (2001).
35. G.M. VandenLangenberg, J.A. Mares-Perlman, R. Klein *et al.*, *Am. J. Epidemiol.*, **148**, 204–214 (1998).
36. W. Smith, P. Mitchell, K. Webb and S.R. Leeder, *Ophthalmology*, **106**, 761–767 (1999).
37. E. Cho, M.J. Stampfer, J.M. Seddon *et al.*, *Ann. Epidemiol.*, **11**, 328–336 (2001).
38. D.A. Newsome, M. Swartz, N.C. Leone *et al.*, *Arch. Ophthalmol.*, **106**, 192–198 (1988).
39. M. Stur, M. Tittl, A. Reitner and V. Meisinger, *Invest. Ophthalmol. Vis. Sci.*, **37**, 1225–1235 (1996).
40. Age-Related Eye Disease Study Research Group, *Arch. Ophthalmol.*, **119**, 1417–1436 (2001).
41. J.R. Evans, *Cochrane Database Syst. Rev.*, CD000254 (2002).
42. H. Lai, S. Lai, G. Shor-Posner, F. Ma *et al.*, *J. Acquir. Immune. Defic. Syndr.*, **27**, 56–62 (2001).

43. N. Wellinghausen, W.V. Kern, W. Jochle and P. Kern, *Biol. Trace Elem. Res.*, **73**, 139–149 (2000).
44. R. Kupka and W. Fawzi, *Nutr. Rev.*, **60**, 69–79 (2002).
45. E. Mocchegiani, S. Veccia, F. Ancarani *et al.*, *Int. J. Immunopharmacol.*, **17**, 719–727 (1995).
46. A.M. Tang, N.M. Graham, A.J. Kirby *et al.*, *Am. J. Epidemiol.*, **138**, 937–951 (1993).
47. A.M. Tang, N.M. Graham and A.J. Saah, *Am. J. Epidemiol.*, **143**, 1244–1256 (1996).
48. M.K. Baum, G. Shor-Posner and A. Campa, *J. Nutr.*, **130**, 1421S–1423S (2000).
49. E. Ho, N. Quan, Y.H. Tsai *et al.*, *Exp. Biol. Med. (Maywood)*, **226**, 103–111 (2001).
50. B. Haglund, K. Ryckenberg, O. Selinus and G. Dahlquist, *Diabetes Care*, **19**, 873–875 (1996).
51. H.X. Zhao, M.D. Mold, E.A. Stenhouse *et al.*, *Diabet. Med.*, **18**, 709–717 (2001).
52. E. Moltchanova, M. Rytkonen, A. Kousa *et al.*, *Diabet. Med.*, **21**, 256–261 (2004).
53. J. Honnorat, M. Accominotti, C. Broussolle *et al.*, Orgiazzi, *Biol. Trace Elem. Res.*, **32**, 311–316 (1992).
54. T. Isbir, L. Tamer, A. Taylor and M. Isbir, *Diabetes Res.*, **26**, 41–45 (1994).
55. C. Ruiz, A. Alegria, R. Barbera *et al.*, *J. Trace Elem. Med. Biol.*, **12**, 91–95 (1998).
56. A.H. Zargar, M.I. Bashir, S.R. Masoodi *et al.*, *Saudi Med. J.*, **23**, 539–542 (2002).
57. J.J. Cunningham, A. Fu, P.L. Mearkle and R.G. Brown, *Metabolism*, **43**, 1558–1562 (1994).
58. A. Rahmat, J.N. Norman and G. Smith, *Br. J. Surg.*, **61**, 271–273 (1974).
59. H.H. Sandstead, V.C. Lanier Jr, G.H. Shephard and D.D. Gillespie, *Am. J. Clin. Nutr.*, **23**, 514–519 (1970).
60. A.B. Lansdown, *Lancet*, **347**, 706–707 (1996).
61. M. Gray, *J. Wound Ostomy Continence Nurs.*, **30**, 295–299 (2003).
62. E.A. Wilkinson and C.I. Hawke, *Arch. Dermatol.*, **134**, 1556–1560 (1998).
63. E.A. Wilkinson and C.I. Hawke, *Cochrane Database Syst. Rev.*, CD001273 (2000).
64. Y. Lim, M. Levy and T.M. Bray, *J. Nutr.*, **134**, 811–816 (2004).
65. S. Houston, J. Haggard, J. Williford Jr *et al.*, *J. Am. Geriatr. Soc.*, **49**, 1130–1132 (2001).
66. B.H. McElroy and S.P. Miller, *Am. J. Ther.*, **10**, 324–329 (2003).
67. J.C. Godfrey, B. Conant Sloane, D.S. Smith *et al.*, *J. Int. Med. Res.*, **20**, 234–246 (1992).
68. S.B. Mossad, M.L. Macknin, S.V. Medendorp and P. Mason, *Ann. Intern. Med.*, **125**, 81–88 (1996).
69. G.A. Eby, D.R. Davis and W.W. Halcomb, *Antimicrob. Agents Chemother.*, **25**, 20–24 (1984).
70. W. Al-Nakib, P.G. Higgins, I. Barrow *et al.*, *J. Antimicrob. Chemother.*, **20**, 893–901 (1987).
71. R.B. Turner and W.E. Cetnarowski, *Clin. Infect. Dis.*, **31**, 1202–1208 (2000).
72. D.S. Smith, E.C. Helzner, C.E. Nuttall Jr *et al.*, *Antimicrob. Agents Chemother.*, **33**, 646–648 (1989).
73. K. Weismann, J.P. Jakobsen, J.E. Weismann *et al.*, *Dan. Med. Bull.*, **37**, 279–281 (1990).
74. M.L. Macknin, M. Piedmonte, C. Calendine *et al.*, *JAMA*, **279**, 1962–1967 (1998).
75. B.M. Farr, E.M. Conner, R.F. Betts *et al.*, *Antimicrob. Agents Chemother.*, **31**, 1183–1187 (1987).

76. A.S. Prasad, J.T. Fitzgerald, B. Bao *et al.*, *Ann. Intern. Med.*, **133**, 245–252 (2000).
77. J.L. Jackson, E. Lesho and C. Peterson, *J. Nutr.*, **130**, 1512S–1515S (2000).
78. S.B. Mossad, *QJM*, **96**, 35–43 (2003).
79. M. Hirt, S. Nobel and E. Barron, *Ear Nose Throat J.*, **79**, 778–780, 782 (2000).
80. R.B. Turner, *Clin. Infect. Dis.*, **33**, 1865–1870 (2001).
81. E.A. Belongia, R. Berg and K. Liu, *Am. J. Med.*, **111**, 103–108 (2001).
82. B.W. Jafek, M. Linschoten and B.W. Murrow, Zicam Induced Anosmia, American Rhinologic Society 49th Annual Fall Scientific Meeting, pp. 48–49, 2003.
83. C.A. DeCook and A.R. Hirsch, *Chem. Senses*, **25**, 659 (2000).
84. K.M. Hambidge, N.F. Krebs and L. Miller, *Am. J. Clin. Nutr.*, **68**, 410S–413S (1998).
85. A. Federico, P. Iodice, P. Federico *et al.*, *Eur. J. Clin. Nutr.*, **55**, 293–297 (2001).
86. A.S. Prasad, F.W. Beck, T.D. Doerr *et al.*, *J. Am. Coll. Nutr.*, **17**, 409–418 (1998).
87. P.I. Oteiza, M.S. Clegg, M.P. Zago and C.L. Keen, *Free Rad. Bio. Med.*, **28**, 1091–1099 (2000).
88. M.S. Golub, M.E. Gershwin, L.S. Hurley *et al.*, *Am. J. Clin. Nutr.*, **42**, 1229–1239 (1985).
89. L.Y. Fong, A. Sivak and P.M. Newberne, *J. Natl. Cancer Inst.*, **61**, 145–150 (1978).
90. L.Y. Fong, J.X. Li, J.L. Farber and P.N. Magee, *Carcinogenesis*, **17**, 1841–1848 (1996).
91. L.Y. Fong, K.M. Lau, K. Huebner and P.N. Magee, *Carcinogenesis*, **18**, 1477–1484 (1997).
92. L.Y. Fong and P.N. Magee, *Cancer Lett.*, **143**, 63–69 (1999).
93. L.Y. Fong, V.T. Nguyen, A.E. Pegg and P.N. Magee, *Cancer Epidemiol. Biomarkers Prev.*, **10**, 191–199 (2001).
94. P.M. Newberne, T.F. Schrager and S. Broitman, *Pathobiology*, **65**, 39–45 (1997).
95. P.I. Oteiza, K.L. Olin, C.G. Fraga and C.L. Keen, *J. Nutr.*, **125**, 823–829 (1995).
96. P. Wakimoto and G. Block, *J. Gerontol. A Biol. Sci. Med. Sci.*, **56**, 65–80 (2001).
97. C.G. Taylor, W.J. Bettger and T.M. Bray, *J. Nutr.*, **118**, 613–621 (1988).
98. L. Castro and B.A. Freeman, *Nutrition*, **17**, 161, 163–165 (2001).
99. R.E. Kelly, M.I. Mally and D.R. Evans, *J. Biol. Chem.*, **261**, 6073–6083 (1986).
100. M.W. Washabaugh and K.D. Collins, *J. Biol. Chem.*, **261**, 5920–5929 (1986).
101. D. Conte, S. Narindrasorasak and B. Sarkar, *J. Biol. Chem.*, **271**, 5125–5130 (1996).
102. J.E. Hesketh, *Int. J. Biochem.*, **15**, 743–746 (1983).
103. Y.C. Tse-Dinh and R.K. Beran-Steed, *J. Biol. Chem.*, **263**, 15857–15859 (1988).
104. H.W. Fu, J.F. Moomaw, C.R. Moomaw and P.J. Casey, *J. Biol. Chem.*, **271**, 28541–28548 (1996).
105. E. Ho and B.N. Ames, *Proc. Natl. Acad. Sci. USA*, **99**, 16770–16775 (2002).
106. K.L. Olin, M.K. Shigenaga, B.N. Ames *et al.*, *Proc. Soc. Exp. Biol. Med.*, **203**, 461–466 (1993).
107. P.L. Oteiza, K.L. Olin, C.G. Fraga and C.L. Keen, *Proc. Soc. Exp. Biol. Med.*, **213**, 85–91 (1996).
108. E. Mocchegiani, M. Muzzioli and R. Giacconi, *Biogerontology*, **1**, 133–143 (2000).
109. M.T. Leccia, M.J. Richard, A. Favier and J.C. Beani, *Biol. Trace Elem. Res.*, **69**, 177–190 (1999).
110. I.R. Record, M. Jannes and I.E. Dreosti, *Biol. Trace Elem. Res.*, **53**, 19–25 (1996).
111. R.A. Mittra, *Minn. Med.*, **86**, 40–46 (2003).

112. Age-Related Eye Disease Study Research Group, *Arch. Ophthalmol.*, **119**, 1417–1436 (2001).
113. K. Datta, S. Sinha and P. Chattopadhyay, *Natl. Med. J. India*, **13**, 304–310 (2000).
114. A.J. Levine, *Cell*, **88**, 323–331 (1997).
115. D.P. Lane, *Nature*, **358**, 15–16 (1992).
116. M. Hollstein, D. Sidransky, B. Vogelstein and C.C. Harris, *Science*, **253**, 49–53 (1991).
117. R.S. Bedwal and A. Bahuguna, *Experientia*, **50**, 626–640 (1994).
118. N.P. Pavletich, K.A. Chambers and C.O. Pabo, *Genes Dev.*, **7**, 2556–2564 (1993).
119. W.D. Funk, D.T. Pak, R.H. Karas *et al.*, *Mol. Cell. Biol.*, **12**, 2866–2871 (1992).
120. B.N. Ames, *Mutat. Res.*, **475**, 7–20 (2001).
121. E. Ho, C. Courtemanche and B.N. Ames, *J. Nutr.*, **133**, 2543–2548 (2003).
122. J.C. Fanzo, S.K. Reaves, L. Cui *et al.*, *Am. J. Physiol. Cell Physiol.*, **281**, C751–757 (2001).
123. C. Meplan, M.J. Richard and P. Hainaut, *Oncogene*, **19**, 5227–5236 (2000).
124. P. Hainaut and M. Hollstein, *Adv. Cancer Res.*, **77**, 81–137 (2000).
125. G. Fritz, *Int. J. Biochem. Cell Biol.*, **32**, 925–929 (2000).
126. A.R. Evans, M. Limp-Foster and M.R. Kelley, *Mutat. Res.*, **461**, 83–108 (2000).
127. Y. Xu, D.H. Moore, J. Broshears *et al.*, *Anticancer Res.*, **17**, 3713–3719 (1997).
128. B.G. Thomson, R. Tritt, M. Davis *et al.*, *Anticancer Res.*, **20**, 4153–4157 (2000).
129. F. Puglisi, F. Barbone, G. Tell *et al.*, *Oncol. Rep.*, **9**, 11–17 (2002).
130. U. Siebenlist, G. Franzoso and K. Brown, *Annu. Rev. Cell Biol.*, **10**, 405–455 (1994).
131. E. Shaulian and M. Karin, *Oncogene*, **20**, 2390–2400 (2001).
132. P.I. Oteiza, M.S. Clegg and C.L. Keen, *J. Nutr.*, **131**, 21–26 (2001).
133. G. Block, B. Patterson and A. Subar, *Nutr. Cancer*, **18**, 1–29 (1992).
134. K.A. Steinmetz and J.D. Potter, *J. Am. Diet. Assoc.*, **96**, 1027–1039 (1996).
135. The Alpha-Tocopherol Beta Carotene Cancer Prevention Study Group, *N. Engl. J. Med.*, **330**, 1029–1035 (1994).
136. G.S. Omenn, G.E. Goodman, M.D. Thornquist *et al.*, *N. Engl. J. Med.*, **334**, 1150–1155 (1996).
137. W.G. Nelson, T.L. DeWeese and A.M. DeMarzo, *Cancer Metastasis Rev.*, **21**, 3–16 (2002).
138. O.P. Heinonen, D. Albanes, J. Virtamo *et al.*, *J. Natl. Cancer Inst.*, **90**, 440–446 (1998).
139. L.C. Clark, G.F. Combs and B.W. Turnbull, *JAMA*, **276**, 1957–1963 (1997).
140. N.R. Cook, M.J. Stampfer, J. Ma *et al.*, *Cancer*, **86**, 1783–1792 (1999).
141. E.A. Klein, I.M. Thompson, S.M. Lippman *et al.*, *J. Urol.*, **166**, 1311–1315 (2001).
142. P. Connell, V.M. Young, M. Toborek *et al.*, *J. Am. Coll. Nutr.*, **16**, 411–417 (1997).
143. M. Provinciali, A. Donnini, K. Argentati *et al.*, *Free Rad. Bio. Med.*, **32**, 431–445 (2002).
144. A.Q. Truong-Tran, J. Carter, R.E. Ruffin and P.D. Zalewski, *Biometals*, **14**, 315–330 (2001).
145. R. Lahtonen, *Prostate*, **6**, 177–183 (1985).
146. A.B. Kar and A.R. Chowdhury, *J. Urol.*, **96**, 370–371 (1966).
147. N.K. Dhar, T.C. Goel, P.C. Dube *et al.*, *Exp. Mol. Pathol.*, **19**, 139–142 (1973).
148. E.A. Platz and K.J. Helzlsouer, *Epidemiol. Rev.*, **23**, 93–101 (2001).

149. Z.M. Bataineh, I.H. Bani Hani and J.R. Al-Alami, *Saudi Med. J.*, **23**, 218–220 (2002).
150. G.B. Martin, C.L. White, C.M. Markey and M.A. Blackberry, *J. Reprod. Fertil.*, **101**, 87–96 (1994).
151. A.S. Prasad, C.S. Mantzoros, F.W. Beck *et al.*, *Nutrition*, **12**, 344–348 (1996).
152. S.A. Hamdi, O.I. Nassif and M.S. Ardawi, *Arch. Androl.*, **38**, 243–253 (1997).
153. C.D. Hunt, P.E. Johnson, J. Herbel and L.K. Mullen, *Am. J. Clin. Nutr.*, **56**, 148–157 (1992).
154. L.C. Costello and R.B. Franklin, *Prostate*, **35**, 285–296 (1998).
155. L.C. Costello and R.B. Franklin, *Oncology*, **59**, 269–282 (2000).
156. A.R. Kristal, J.L. Stanford, J.H. Cohen *et al.*, *Cancer Epidemiol. Biomarkers Prev.*, **8**, 887–892 (1999).
157. P. Feng, J.Y. Liang, T.L. Li *et al.*, *Mol. Urol.*, **4**, 31–36 (2000).
158. R.G. Uzzo, P. Leavis, W. Hatch *et al.*, *Clin. Cancer Res.*, **8**, 3579–3583 (2002).
159. M.F. Leitzmann, M.J. Stampfer, K. Wu *et al.*, *J. Natl. Cancer Inst.*, **95**, 1004–1007 (2003).
160. M.K. Yadrick, M.A. Kenney and E.A. Winterfeldt, *Am. J. Clin. Nutr.*, **49**, 145–150 (1989).
161. K. McBride, B. Slotnick and F.L. Margolis, *Chem. Senses*, **28**, 659–670 (2003).
162. F.F. Tisdall, A. Brown and R.D. Defries, *J. Pediatr.*, **18**, 60–62 (1938).
163. S.S. Hendler, and D.R. Rorvik (Eds), *PDR for Nutritional Supplements*, pp. 539–542, Medical Economics Company, Inc., Montvale, 2001.

14 $_{31}$Ga Therapeutic Gallium Compounds

Lawrence R. Bernstein

14.1 Introduction

Gallium was described before it was ever observed. This bold feat was accomplished by Dmitri Mendeleev,[1] who in 1871 postulated that 'eka-aluminum' would fill the gap below aluminum in his newly created periodic table of the elements, and would chemically resemble its neighbor. The element was actually discovered in 1875 by Paul-Émile Lecoq de Boisbaudran, not in association with aluminum, but rather as a trace element in sphalerite (zinc sulfide) from southern France. Lecoq de Boisbaudran named the element gallium 'in honor of France (Gallia)',[2] though there have been lingering suspicions that the name also may be derived from 'gallus,' Latin for 'le coq' (rooster).

The first published investigation of gallium's therapeutic potential was that by Levaditi *et al.* in 1931.[3] Gallium tartrate eradicated experimental syphilis in rabbits with a single dose of 30–45 mg Ga/kg intramuscularly or 15 mg Ga/kg intravenously. At 225 mg Ga/kg it was found to eliminate infection by *Trypanosoma evansi* in mice. Despite the reported efficacy and lack of toxicity (750 mg Ga/kg was tolerated in mice), there was no published follow-up to this research over the next several decades.

A period of intensive research on the toxicity and tissue distribution of gallium occurred from about 1949 to 1952. The earliest of these studies showed that

Metallotherapeutic Drugs and Metal-Based Diagnostic Agents: The Use of Metals in Medicine
Edited by Gielen and Tiekink

gallium was poorly absorbed from oral ingestion of common gallium salts or chelates,[4] and that injected gallium (generally as the citrate) tended to concentrate in bone, liver and kidney tissue.[5] Attention quickly turned to gallium radioisotopes, particularly ^{72}Ga, as these could be easily used to trace the distribution of gallium in experimental animals and humans.[6] Radiogallium was soon found to concentrate in regions of high bone turnover, particularly in some bone tumors.[7]

Based on these findings, a preclinical and clinical study (34 patients) was undertaken to explore the use of radioactive ^{72}Ga to treat primary and metastatic bone cancers.[8] The guiding hypothesis was that ^{72}Ga would become highly concentrated in tumors, which would then be destroyed by the localized radiation. Results indicated that due to its short half-life (14.1 h), by the time the ^{72}Ga became concentrated in the target tumors, the radiation level had already decreased so much as to be ineffective, while the patient had been subjected to undesirable levels of whole body radiation.[8] In addition, since ^{72}Ga was produced by neutron irradiation of stable ^{71}Ga, from which it could not be chemically separated, ^{72}Ga was by necessity administered with large amounts of non-radioactive 'carrier' gallium, which was felt to pose a toxicity risk. The use of non-radioactive gallium as a therapeutic agent was apparently not contemplated at this time, and no further work on the potential therapeutic use of gallium appeared until the 1970s. The studies of the early 1950s did, nevertheless, produce data on gallium tissue distribution in experimental animals and humans, and ultimately led to the now widespread use of radioactive gallium (mainly ^{67}Ga, with a half-life of about 78 h) as a diagnostic agent for some cancers, infections and inflammatory diseases.

Research conducted from the early 1970s to the present has demonstrated that non-radioactive gallium has several therapeutic activities, including (a) decreasing accelerated bone mineral resorption and lowering associated elevated plasma calcium levels; (b) inhibiting neoplastic proliferation; and (c) treating certain infectious diseases, particularly those caused by intracellular organisms. In animal models, gallium also has shown selective immunomodulating activity, particularly relating to allograft rejection and autoimmune diseases. Recent reviews[9–11] have covered much of this research and have discussed in detail what is known of gallium's mechanisms of action. Although we are a bit ahead of Mendeleev when he described gallium before it had been observed, we still have only taken a glimpse into gallium's therapeutic potential.

14.2 Chemistry and Mechanisms of Action

14.2.1 Aqueous biochemistry

The chemistry of gallium as it pertains to physiologic solutions has been reviewed,[9,12–14] and will be only briefly summarized here.

Under physiological conditions, gallium is trivalent in aqueous solution (Ga^{3+}).[15] The small and highly charged Ga^{3+} ion (ionic radius 0.62 Å (octahedral) and 0.47 Å (tetrahedral)[16]) is hydrolyzed nearly completely over a wide pH range, forming various hydroxide species, particularly $Ga(OH)_4^-$ (gallate). Concomitant with hydroxide formation in aqueous solutions is the formation of H_3O^+, with resulting lowered pH. If the pH is raised, highly insoluble amorphous $Ga(OH)_3$ precipitates. This $Ga(OH)_3$ tends to convert on aging to the apparently stable crystalline phase GaO(OH).[15] Thus, 'free' Ga^{3+}, which is actually tightly coordinated to six water molecules, has low solubility in most aqueous solutions. At pH 7.4 and 25 °C, overall gallium solubility in equilibrium with crystalline GaO(OH) is approximately 1 μM; there is essentially no unbound Ga^{3+}, and 98.4% of the dissolved gallium is present as $Ga(OH)_4^-$ and 1.6% as $Ga(OH)_3$.[15,17] Both $Ga(OH)_3$ and GaO(OH) display the amphoteric properties first predicted by Mendeleev,[1] becoming increasingly soluble with increasing acidity or basicity, though even at pH 2 their solubility is only approximately 10^{-2} M, and at pH 10 it is only approximately $10^{-3.3}$ M.[15]

These factors of gallium solution chemistry have major implications regarding gallium's therapeutic use. When gallium salts, such as the chloride or nitrate, are dissolved in water (or in other aqueous solutions such as normal saline), most of the gallium ions are hydrolyzed as described above, leaving the resulting solution highly acidic. Over time, such solutions, unless they are extremely dilute or are further acidified, are not stable, and some precipitation of gallium hydroxides will occur. Being acidic, the solutions are not appropriate for parenteral administration by injection. To overcome these problems, gallium solutions for injection are usually prepared with citrate, which chelates the gallium, preventing hydrolysis and improving stability.

Further in this regard, it has been repeatedly observed that low gallium absorption occurs when gallium salts are administered orally.[4,18,19] This low absorption is likely due in large part to the formation of poorly soluble gallium hydroxides in the gastrointestinal tract.

14.2.2 Gallium and iron

The medicinal chemistry of gallium appears to be dominated by the striking similarity in chemical behavior between Ga^{3+} and ferric iron (Fe^{3+}). This similarity, as has been noted,[9] is due to the close correspondence between the two ions in a number of parameters, including ionic radius and factors relating to bond formation (such as electronegativity, ionization potential and electron affinity). Gallium is thus expected to follow many of the same chemical pathways as ferric iron in the body, and to be able to occupy the ferric iron site in some proteins and chelates.

It is the differences between the two ions, however, that allow for gallium's therapeutic potential and that minimize its toxicity. The first major difference is that Ga^{3+} is essentially irreducible under physiological conditions, whereas Fe^{3+} is readily reducible to Fe^{2+} (a much more soluble and significantly larger ion). This difference means that, *in vivo*, Ga^{3+} does not enter Fe^{2+}-bearing molecules such as heme (a fundamental component of hemoglobin and myoglobin, as well as cytochromes and numerous other enzymes) and thus does not interfere with oxygen transport and other vital functions. It also means that when Ga^{3+} substitutes for Fe^{3+} in a redox-active enzyme, it cannot participate in redox reactions, making the enzyme non-functional. Furthermore, Ga^{3+} will not participate in Fenton-type redox reactions, in which hydroxyl and other highly reactive oxygen-bearing free radicals are produced; such reactions make unbound iron ions highly toxic when present in blood plasma. The other major difference is that Fe^{3+} is even less soluble in neutral aqueous solutions than is Ga^{3+}: at pH 7.4 and 25 °C, the solubility of Fe^{3+} (in equilibrium with FeO(OH)) is only approximately 10^{-18} M, compared to the analogous figure of approximately 10^{-6} M for Ga^{3+}.[14] This difference means that whereas essentially all Fe^{3+} is protein-bound or chelated in blood plasma, small but potentially significant amounts of gallate ($Ga(OH)_4^-$) can exist at equilibrium, which can participate in activities not available to the protein-bound metal.

Because unbound iron ions are highly toxic, iron is maintained bound to a series of proteins and small molecules when it is absorbed and transported throughout the body.[20] In blood plasma, iron exists predominately as Fe^{3+} bound to transferrin (TF), which is the major iron transport protein. Similarly, at equilibrium, nearly all Ga^{3+} in blood plasma is bound to TF.[9,13] Metal-bearing TF, particularly when saturated with two metal ions, can bind to transferrin receptor (TFR) on cell membranes; this complex is taken up by endocytosis, the metal is released from TF as the endosome is acidified to pH 5.5, and the TF and TFR are recycled.[20] TFR is expressed in all nucleated cells, but at the highest levels in many neoplastic cells, as well as in normal hepatocytes; Kupffer cells; erythroid precursors; and cells of the basal epidermis, endocrine pancreas, seminiferous tubules and mucosal epithelium.[21–23] These cells all have a high need for iron: rapidly proliferating cells, particularly neoplastic cells, must manufacture the Fe^{3+}-bearing enzyme ribonucleotide reductase, which is essential for DNA synthesis; erythroid precursors (mainly in the marrow) reduce Fe^{3+} to Fe^{2+} to produce hemoglobin and related molecules; and liver cells as well as macrophages absorb and store iron to regulate overall iron levels.

The iron-binding capacity of TF in blood plasma is normally approximately 3.3 μg of Fe^{3+}/ml (referred to as the total iron-binding capacity), though normally only about 33% of the available metal-binding sites (two per TF molecule) are occupied by Fe^{3+}.[24] Thus, at normal iron saturation levels, plasma TF has the capacity to bind as much as approximately 2.7 μg/ml (40 μM) of Ga^{3+}; if Ga^{3+} concentrations exceed this level, then significant amounts of gallate will form, together with traces of $Ga(OH)_3$ and gallium citrate.[9]

Gallium also binds to the Fe^{3+} sites of lactoferrin, a protein closely related structurally to transferrin. Lactoferrin binds Fe^{3+} and Ga^{3+} more avidly than does TF, and can remove Ga^{3+} from TF.[25] Apolactoferrin exerts anti-microbial activity by locally sequestering iron, an essential nutrient, and occurs in amounts of approximately 0.5–1 mg/ml in epithelial secretions such as milk, tears, seminal fluid and nasal discharge.[26,27] It is also secreted at sites of infection and inflammation.[28–30]

Ferritin, a very large (440 kDa) protein used for iron storage, will also bind gallium. Gallium can be transferred to ferritin from transferrin or lactoferrin, with ATP and other phosphate-bearing compounds acting as mediators.[31,32] Ferritin is present in most cell types, but is concentrated in Kupffer cells of the liver and other tissue macrophages, and in duodenal mucosal cells, particularly when there is abundant iron in the diet.[20,24]

The binding of gallium to transferrin, lactoferrin and ferritin accounts for much of the tissue distribution observed when gallium radioisotopes are administered. ^{67}Ga is found to concentrate most highly in many tumors and at sites of inflammation and infection.[33–35] Many tumors overexpress TFR, and the avidity of tumors for ^{67}Ga has been correlated to transferrin receptor 1 (TFR1; CD71) expression.[14,36–39] Exceptions to this generalization exist however, and gallium appears to enter some ^{67}Ga-avid tumors by TFR1-independent mechanisms.[40–42] A second membrane-bound transferrin receptor (TFR2α) has been described,[43] which is most highly expressed in hepatocytes, some erythroid cells and the crypt cells of the duodenum (which sense plasma iron levels and then, when they mature into villi enterocytes, regulate dietary iron absorption).[43–45] Some of the reported non-transferrin-receptor-mediated gallium uptake is likely through TFR2α. Little information is available on the direct uptake by cells of gallium bound to low molecular weight (LMW) molecules; it is noted, however, that Ga^{3+}, as well as Fe^{3+} and other trivalent or tetravalent metal ions, greatly upregulates the uptake of LWM-bound Fe by monocytes, macrophages, neutrophils and myeloid cells, as well as the binding of transferrin and lactoferrin to cell membranes.[46] The mechanism of LMW-bound ferric iron absorption may apply to LMW-bound gallium.

The concentration of ^{67}Ga at sites of inflammation and infection likely stems from its binding to lactoferrin, as well as from uptake by some leukocytes[35,47,48] and, when present, bacteria.[49,50]

Although gallium is avidly taken up by a wide range of proliferating cancer cells, it is not concentrated by rapidly proliferating normal cells, such as gastrointestinal mucosal cells, hematopoietic cells of the bone marrow and elsewhere, and the transient cells of hair follicles; significant ^{67}Ga accumulations do not occur at these sites. The reasons for the lack of accumulation in normal proliferating tissue have not been explored, but are likely due in part to local recycling of iron, so that little new uptake of iron (or gallium) from plasma occurs.

14.2.3 Mechanisms of action

The mechanisms for the observed therapeutic activities of gallium have been reviewed,[9] so will be only briefly summarized here.

Much of gallium's therapeutic activity derives from its ability to mimic Fe^{3+} and yet not to participate in the redox reactions available for Fe^{3+}. This mimicry leads to the concentration of gallium at sites in the body where Fe^{3+} is taken up from plasma, including proliferating cancer cells; infected cells, particularly macrophages; and proliferating bacteria and parasites. When gallium reaches these sites it will compete with Fe^{3+} and will interfere with its absorption, metabolism and activity.

Iron is essential to cell division, largely because it is present in the active site of ribonucleotide reductase, an enzyme that catalyzes the production of the deoxyribonucleotides required for DNA. By competing with iron, gallium can interfere with ribonucleotide reductase activity and inhibit DNA synthesis.[51] Furthermore, Ga^{3+} can substitute for Fe^{3+} in the M_2 subunit of ribonucleotide reductase, deactivating the enzyme through a conformational change.[52,53] If DNA synthesis is inhibited in proliferating cells, apoptosis may result: in human leukemic CCRF-CEM cells deprived of iron, by exposure to either deferoxamine or 12.5–100 μM gallium nitrate, apoptosis is induced.[54] Similarly, exposure of human peripheral blood mononuclear cells to 50–100 μM gallium nitrate induces apoptosis.[55]

Cancer cells that take up gallium become iron-deprived, which causes upregulation of TFR1.[56] Increased TFR1 promotes increased Ga-TF uptake, leading to increased iron deprivation, and this cycle continues until apoptosis of the cell results. Ga-TF may cause additional iron deprivation by preventing sufficient acidification of Fe-TF-containing endosomes to allow for the intracellular release of Fe.[56]

Several *in vitro* studies[57,58] have found that Ga^{3+} can directly bind to DNA, and may compete with magnesium in this regard. As these studies were done using isolated DNA at pH values of 4–5, their relevance to *in vivo* gallium activity is not clear.

Gallium nitrate and Ga-TF were found to potently inhibit protein tyrosine phosphatase (PTPase) from Jurkat human T-cell leukemia cells and HT-29 human colon cancer cells (IC_{50} value of 2–6 μM).[59] This activity did not, however, correlate with growth inhibition in these cells, and a relationship between the observed PTPase inhibition and gallium's anti-tumor activity has not been established.

As mentioned, gallium inhibits the proliferation of some pathogenic microorganisms. It may be particularly effective in treating some intracellular pathogens, such as species of *Mycobacterium*.[60] Infected cells (particularly macrophages) take up Ga-TF; the infecting organisms then take up gallium instead of iron and, as with cancer cells, cannot synthesize sufficient DNA for replication, and ultimately die.

In addition to these anti-proliferative activities, gallium is potently anti-resorptive to bone mineral, appears to have anabolic (formation-stimulating) effects on bone, inhibits some T-cell and macrophage activation and has other selective immunomodulatory activities.[9]

14.3 Therapeutic Gallium Compounds

At present, only citrated gallium nitrate (Ganite™) is approved for therapeutic use (in the United States, for cancer-related hypercalcemia). Based on publicly available information, the only other potentially therapeutic gallium compounds that have been in clinical trials are gallium chloride, gallium maltolate and gallium 8-quinolinolate. Gallium-transferrin and a gallium–transferrin–doxorubicin conjugate have been administered to a small number of cancer patients, and several other gallium compounds have undergone preclinical testing.

14.3.1 Gallium nitrate and citrated gallium nitrate

Most data on the biological and therapeutic activities of gallium derive from investigations using gallium nitrate in aqueous solution. The term 'gallium nitrate,' however, has not been used consistently: it has been used to describe (a) chelator-free gallium nitrate solutions, employed for most of the *in vitro* and some of the animal studies and (b) gallium nitrate solutions containing citrate as a chelator, employed for all of the clinical studies and some of the animal studies. Chelator-free gallium nitrate solutions contain ionic gallium (mostly as the pH-dependent hydroxide species discussed previously), whereas citrate-containing solutions at neutral pH contain gallium citrate as a coordination complex. The commercially available product for injection (Ganite™) contains 97.8 mM of both gallium and citrate at neutral pH,[61] which allows essentially all the gallium to bind to citrate, with very little gallium remaining in other forms. This solution is here referred to as 'citrated gallium nitrate' or 'CGN,' whereas gallium nitrate without citrate or other chelators is abbreviated 'GN.' Dose levels for both GN and CGN are given in terms of anhydrous gallium nitrate ($Ga(NO_3)_3$).

In vitro *and non-human* in vivo *studies*

Anti-tumor activity of GN was first reported in 1971:[62] efficacy was demonstrated against Walker 256 ascites carcinosarcoma in rats ($GaCl_3$ and $Ga(SO_4)_3 \cdot 18H_2O$ had very similar effects), and >90% growth inhibition was observed for six of eight solid tumors implanted subcutaneously (sc) in rodents. Complete tumor regression occurred in several animals. Unchelated, acidic GN solutions were administered intraperitoneally (ip) daily for 10 days; LD_{10} (at 30 days) was 63 mg/kg (mouse), 50 mg/kg (rat); LD_{50} was 80 mg/kg (mouse), 67.5 mg/kg (rat); and the effective dose varied from 30 to 60 mg/kg. *In vitro*, GN gave an ID_{50} of 355 μM (anhydrous basis) against Walker 256 carcinosarcoma cells.[63] *In vivo* efficacy against implanted human medulloblastoma (Daoy cells) was observed in nude mice that received ip GN (50 mg/kg/day for 10 or 15 days): the growth rate of macroscopic tumors was reduced or reversed[64] and the progression of microscopic disease delayed.[65]

The finding of an anti-hypercalcemic effect in humans, which results from inhibited bone resorption rather than increased urinary calcium excretion,[66,67] stimulated considerable research on gallium's bone-related activity (reviews: [9,68]). Gallium concentrates in skeletal tissue, particularly at sites of rapid bone mineral deposition such as active metaphyseal growth plates and healing fractures.[5,6,69,70] Concentration on the endosteal and periosteal surfaces of diaphyseal bone also occurs, but to a lesser extent.[70] Several studies (using bone fragments *in vitro* or implanted in rodents) found that gallium adsorbs to bone surfaces and then inhibits osteoclastic bone resorption;[71–73] gallium-treated bone is also less soluble in acetate buffer and less readily resorbed by monocytes.[74] Gallium does not enter the crystal lattice of hydroxylapatite, but rather deposits in part on bone mineral surfaces, possibly as gallium phosphate.[75] At anti-resorptive concentrations (up to at least 15 μM, and possibly >100 μM), GN does not affect osteoclast morphology or viability,[67,71,73] or act as an osteoclastic metabolic inhibitor, as do bisphosphonates.[76] Gallium is found to directly inhibit the osteoclast vacuolar-class ATPase proton pump.[76] In young female rats, administration of ip GN resulted in bone containing more calcium, having a larger average hydroxylapatite crystallite size, and having a higher density relative to that in untreated individuals.[69,74]

Some *in vitro* data indicate that GN may stimulate bone formation. Experiments with rat osteogenic sarcoma (ROS) and normal rat osteoblast cells found that GN can decrease osteocalcin (OC) and OC mRNA levels[77,78] and increase c-fos mRNA levels;[79] both activities are associated with bone formation. *In vivo* and clinical studies have not shown clear, consistent anabolic activity, though elevated plasma alkaline phosphatase (a marker of bone formation) was observed in post-menopausal women treated with CGN.[80]

Gallium nitrate and other gallium salts have shown selective *in vitro* and *in vivo* immunomodulating activity. This activity appears to stem primarily from inhibition of some T-cell activation and proliferation[48,80–82] and inhibition of inflammatory cytokine secretion by activated macrophages,[80,83,84] without generalized cytotoxicity to lymphocytes or macrophages. Some of the selective anti-inflammatory activity may be due to pro-inflammatory T-helper type 1 (Th-1) cells being much more sensitive to inactivation by iron deprivation than anti-inflammatory, pro-antibody Th-2 cells.[85] Synoviocyte matrix metalloproteinase activity (elevated in inflammatory joint diseases) was also shown to be dose-dependently inhibited by GN at concentrations of 10–100 μM.[86] In animal models, GN has shown efficacy in suppressing adjuvant-induced arthritis,[80] experimental encephalomyelitis,[81] experimental autoimmune uveitis,[87] type 1 diabetes,[88] endotoxic shock,[89,90] and allograft rejection.[91]

Clinical experience

Parenterally administered CGN has demonstrated single-agent clinical efficacy in cancer-related hypercalcemia (for which it is approved in the United States), Paget's disease of bone, and several types of cancer, including lymphoma (43% response in relapsed malignant non-Hodgkin's lymphoma),[66,92] advanced refractory urothelial carcinoma (17% response),[93] advanced bladder carcinoma (40% average response in two studies),[94,95] advanced or recurrent epithelial ovarian carcinoma resistant to cisplatin (12% response),[96] metastatic or advanced non-squamous cell cervical carcinoma (12% response),[97] advanced or recurrent squamous cell carcinoma (8% response)[98] and metastatic prostate cancer (2 of 23 patients had partial response, and 7 had reduction in bone pain after treatment for only 7 days).[99]

Early anti-cancer clinical trials employed short (generally $\leq$30 min) iv bolus infusions at doses up to 1350 mg/m^2;[100] subsequently, only prolonged (generally $\geq$5 days) iv infusions or sc injections have been used, due to dose-limiting renal toxicity from bolus iv dosing.

For cancer-related hypercalcemia, CGN is administered as a continuous iv infusion for 5 days at 200 mg/m^2/day ($\sim$5 mg/kg/day).[101,102] Studies comparing CGN to other anti-hypercalcemic drugs found that the proportion of patients achieving normocalcemia was higher for those receiving CGN than for those receiving pamidronate,[103] calcitonin[101] or etidronate.[104]

A small study[105,106] found that CGN may be particularly efficacious against multiple myeloma. Thirteen relapsed multiple myeloma patients (11 at stage III) were treated for 6 or 12 months with sc CGN at 30 mg/m^2/day, administered for two weeks followed by two weeks without therapy, together with the M-2 chemotherapy protocol. A 5-day iv infusion of 100 mg/m^2/day CGN was administered every other month. These patients were matched with 167 similar patients who received only M-2 therapy. Patients receiving CGN showed an increase in total body calcium (decreased bone resorption), stabilized measures of bone density, reduction in vertebral fractures and decreases in measures of pain versus those not receiving CGN. Significantly, mean survival in the CGN-treated group was 87+ months, with several long-term survivors (including stage III patients (at presentation) alive at 137+ and 144+ months, and a stage IIIB patient alive at 96+ months in complete remission), compared to mean survival of 48 months in the 167 patients who received M-2 chemotherapy alone, with no long-term survivors.

Markers of bone turnover were significantly reduced in drug-resistant patients with Paget's disease of bone (a disease characterized by localized, greatly accelerated bone remodeling) who received relatively low doses of CGN (iv at 100 mg/m^2/day for 5 days[107] or sc at 0.25 or 0.5 mg/kg/day in two non-consecutive 14-day cycles).[108] No drug-related adverse events were reported other than moderate, transient reductions in hemoglobin and serum iron-binding capacity.

Pharmacokinetics

Data on the pharmacokinetics of CGN are sparse and show significant individual variability;[109–112] much of the variability may be due to differences in renal function or the extent of metastatic disease (because neoplastic tissue and associated areas of inflammation may take up significant amounts of gallium). Steady-state plasma gallium levels of 1.2–1.5 μg/ml were achieved within 48 h in eight hypercalcemic patients who received 200 mg/m^2 CGN as a continuous infusion for 5 days, and levels of 1.0–1.3 μg/ml were achieved in six patients who received 100 mg/m^2 on the same schedule (plus one patient whose plasma gallium level never exceeded 0.45 μg/ml).[112] A gallium plasma level of 1 μg/ml was considered therapeutic for cancer-related hypercalcemia in this study. The non-linear dose/plasma concentration relationship was not discussed. Elimination of gallium following administration of iv CGN is considered biphasic, with an initial elimination half-life of 0.15 to 1.5 h and a terminal half-life of 6 to 196 h; more than half of the administered dose is generally excreted in the urine within 24 h.[109–111] In a single patient who received CGN at 0.5 mg/kg (20 mg/m^2) sc daily for 14 days, maximal gallium plasma concentrations of 0.95 to 1.85 μg/ml occurred 1–2 h after injection; 19% of the administered gallium was renally excreted on day 1, and 28% by day 7.[113]

Toxicity

The major toxicity associated with iv CGN is renal, as indicated by proteinuria, increases in blood urea nitrogen (BUN) and serum creatinine, and a decrease in creatinine clearance. Rats given ip CGN at 100 mg/kg develop renal tubule damage, which is caused at least in part by precipitation of gallium and calcium phosphates within the tubule lumina.[114] In humans, renal toxicity is consistently observed following a 30-min infusion of $\geq$750 mg/m^2, though it is ameliorated by concomitant mannitol diuresis.[110] At lower doses or with longer infusion times, renal toxicity is reduced: in well-hydrated patients receiving 300–400 mg/m^2/day for 7 days, markers of renal tubule damage are not significantly elevated,[115] and well-hydrated hypercalcemia patients receiving 200 mg/m^2/day for 5 days rarely show significant impairment of creatinine clearance.[101,104,112]

Another dose-related toxicity observed is anemia, which is generally mild (<3 g/dl decrease in hemoglobin) and has only been clinically significant at high bolus iv doses.[100,110] When CGN is administered as prolonged infusions at 200–300 mg/m^2, decreases in hemoglobin are usually no more than 1–2 g/dl.[66,105,108] The anemia, when observed, is similar to that arising from iron deficiency, with hypochromia and microcytosis; iron-binding capacity is slightly decreased, while serum iron, TF and ferritin levels are normal.[116]

Other adverse events occasionally observed following slow iv infusion of CGN that may be drug-related include nausea and vomiting,[96] mild respiratory alkalosis,[116] hypophosphatemia,[116] hypocalcemia [116] and optic neuritis.[117]

14.3.2 Gallium chloride

Gallium chloride ($GaCl_3$; GC) has shown anti-cancer activity in animal studies; activity was greatest during the exponential growth phase of tumors.[118] Animal studies found strong preferential uptake by tumor cells when GC was administered orally,[119] without the high renal uptake and toxicity observed following ip administration.[120]

Oral GC was administered to 18 lung cancer patients for 15 days at doses ranging from 100 to 1400 mg/day.[18] Bioavailability was low and serum gallium levels did not increase at doses over 400 mg/day. After 15 days, a steady-state gallium serum level of 371 ± 142 ng/ml was reached; the level was significantly lower in patients with metastases than in those without. Mild reductions in hemoglobin and magnesium were observed.[121] The maximum accessible serum concentrations were deemed too low for GC to be used as a single agent in lung cancer.[18] Possible potentiation of cisplatin and etoposide therapy by GC in lung cancer patients was reported.[122]

14.3.3 Gallium 8-quinolinolate

Gallium 8-quinolinolate (tris(8-quinolinolato)gallium(III); GQ; Figure 14.1) was developed to provide high oral gallium bioavailability for use in cancer treatment.[123] GQ inhibited the *in vitro* proliferation of A549 human lung adenocarcinoma cells: the IC_{50} was 2.5 μM, similar to that for GN and about a tenth the value for GC.[124] In rats transplanted with Walker carcinosarcoma 256, tumor reduction of >50% was observed following oral gavage at 24 mg/kg/day on days 3–9 after transplantation; an equimolar oral dose of GN had no anti-tumor effect.[125] An anti-hypercalcemic effect was also observed. A dose of 62.5 mg/kg/day for two weeks was well tolerated in healthy Swiss mice, whereas doses of 125 mg/kg/day proved toxic (including leucopenia, but not anemia) and sometimes fatal.[123] In the animals that received 62.5 mg/kg/day, gallium was concentrated in the bone (7(3) μg/g), liver (4(2) μg/g), spleen (2(1) μg/g) and

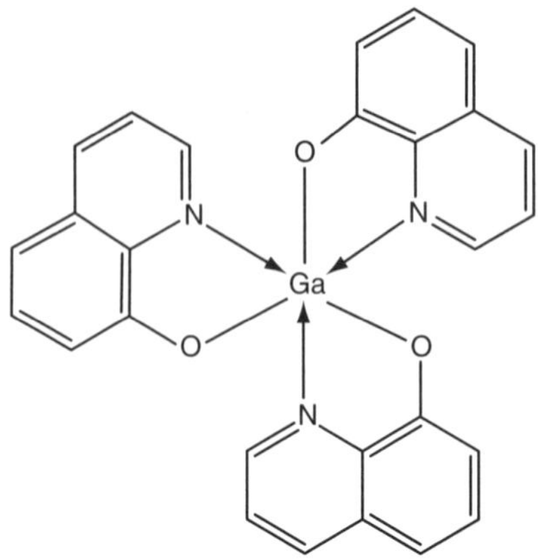

Figure 14.1 Structure of tris(8-quinolinolato)gallium(III) (gallium 8-quinolinolate; GQ)

kidneys (1.8(2) μg/g), with <0.03 ng/g gallium found in the brain, lungs, testes, and ovaries (serum concentration not given).[123] This compound entered human clinical trials in 2004 (B. Keppler, personal communication, 2004).

14.3.4 Gallium maltolate

Gallium maltolate (tris(3-hydroxy-2-methyl-4*H*-pyran-4-onato)gallium(III); GM; Figure 14.2) is an orally active complex that is moderately soluble in both aqueous and lipidic solutions (octanol/water partition coefficient of 0.41(8)) and is stable over a pH range of about 5.5–8.0.[126] The compound has shown oral gallium bioavailability of at least 25–57% in healthy human subjects, with roughly linear absorption and elimination kinetics following single doses of 100–500 mg and an average maximum serum gallium level (C_{max}) of 569(17) ng/ml at the 500-mg dose.[126] Similar results were obtained in elderly subjects with Paget's disease of bone; in these patients it was found that C_{max} was 1138(532) ng/kg after 3 days of dosing at 600 mg/day.[127] Only about 2% of administered gallium was excreted in the urine 72 h after dosing, and no renal toxicity was observed.[126] These results strongly suggest that plasma gallium following oral GM administration is nearly all bound to transferrin; in contrast, much of the plasma gallium following iv CGN administration is in the form of gallate, which is rapidly excreted in the urine and may cause renal toxicity.[126] The drug appears well tolerated at doses of up to 600 mg/day for 3 days.[126,127]

In animal studies, oral GM has shown efficacy in some rat models of rheumatoid arthritis, in which inflammation and bone degradation were

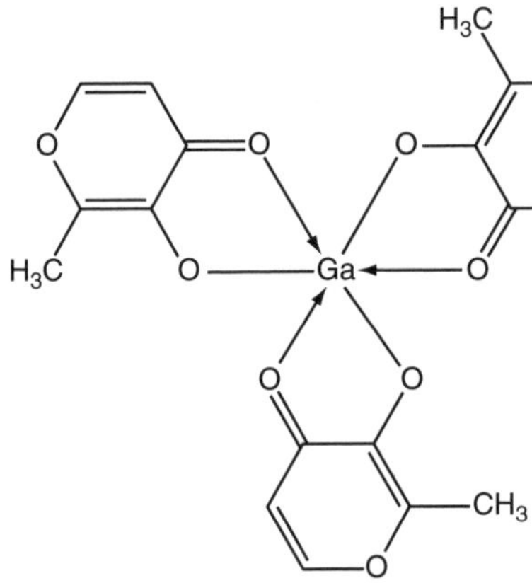

Figure 14.2 Structure of tris(3-hydroxy-2-methyl-4*H*-pyran-4-onato)gallium(III) (gallium maltolate; GM)

suppressed (A. Bendele, personal communication, 2004); tuberculosis in guinea pigs;[128] and *Rhodococcus equi* infection in mice.[129]

14.3.5 Other gallium compounds

A gallium–doxorubicin–transferrin conjugate was 100 times more inhibitory to a doxorubicin-resistant MCF-7 human breast cancer cell line than doxorubicin alone, and 10 times more inhibitory than an iron–doxorubicin–transferrin conjugate.[130] Ga-TF in conjunction with doxorubicin and other chemotherapeutic agents, sometimes with a preliminary dose of deferoxamine, has shown anti-tumor activity and low toxicity following iv administration to breast cancer patients (J. Head, personal communication, 2004).

Gallium chelated to pyridoxal isonicotinoyl hydrazone (PIH; an iron chelator that can enter cells through a non-TFR-mediated mechanism) (Figure 14.3) was found to be growth inhibitory in a human T-lymphoblastic leukemic CCRF-CEM cells resistant to gallium alone, and was more inhibitory than PIH alone.[131]

A complex of gallium with 2-acetylpyridine ^{4}N-dimethylthiosemicarbazone and chloride (Figure 14.4) showed IC_{50} values in the low nanomolar range against ovarian (41M), mammary (SK-BR-3) and colon (SW480) carcinoma cells, though the values were similar to those of the ligand alone.[132] Further testing is needed to determine whether the gallium helps target the compound to gallium-avid cancer cells.

Gallium protoporphyrin IX (GaPPIX; Figure 14.5) at concentrations of about 1 µg/ml showed *in vitro* growth-inhibitory activity against many pathogenic

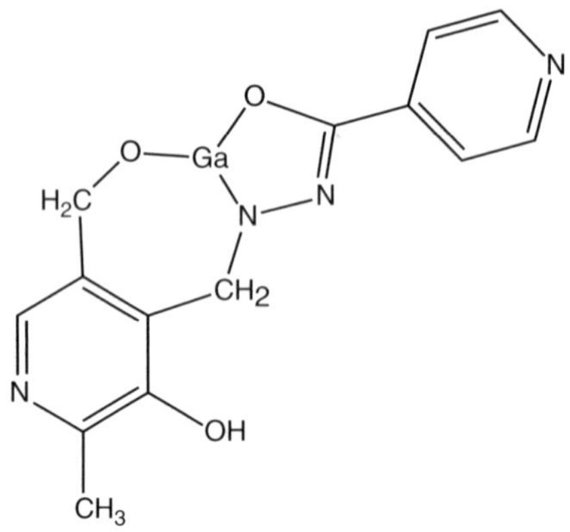

Figure 14.3 Structure of pyridoxal isonicotinoyl hydrazone gallium(III)

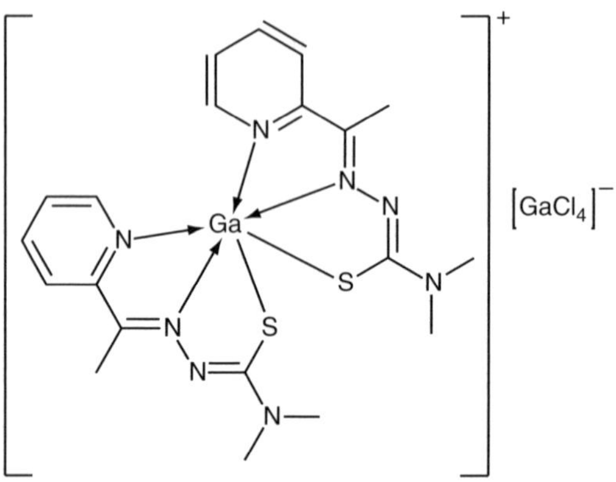

Figure 14.4 Structure of bis(2-acetylpyridine ^4N-dimethylthiosemicarbazone)gallium(III), gallium(III) tetrachloride

bacteria, including Gram-positive and -negative species and mycobacteria,[133] and against the malarial parasite *Plasmodium falciparum*.[134] Protoporphyrin IX or gallium alone had approximately one-hundredth the anti-bacterial activity of intact GaPPIX against the most sensitive organisms.[133] GaPPIX appears to enter bacteria primarily through heme uptake systems; toxicity may occur by incorporation of the molecule into cytochrome/quinol oxidases, which leads to the generation of reactive oxidative species in locally cytotoxic concentrations.[133] Human primary fibroblasts exposed to 100 μg/ml GaPPIX for 48 h displayed no decrease in viability, and mice given 25–30 mg/kg ip, followed by four daily ip doses of 10–12 mg/kg, showed no acute toxic effects.[133]

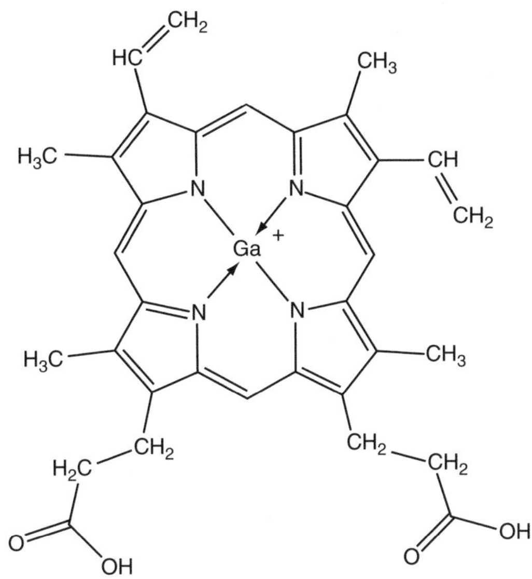

Figure 14.5 Structure of gallium(III) protoporphyrin IX (GaPPIX)

Abbreviations Used

ATP, adenosine triphosphate; CGN, citrated gallium(III) nitrate aqueous solution; GaPPIX, gallium(III) protoporphyrin IX; GC, gallium(III) chloride; GM, gallium(III) maltolate; GN, gallium(III) nitrate; GQ, gallium(III) 8-quinolinolate; ip, intraperitoneal; iv, intravenous; PIH, pyridoxal isonicotinoyl hydrazone; PTPase, protein tyrosine phosphatase; sc, subcutaneous; TF, transferrin; TFR, transferrin receptor.

References

1. D. Mendeleev, *Zhurnal Russkoe Fiziko-Khimicheskoe Obshchestvo*, **3**, 25 (1871).
2. P.-É. Lecoq de Boisbaudran, *Annales de Chimie*, Series 5, **10**, 100–141 (1877).
3. C. Levaditi, J. Bardet, A. Tchakirian and A. Vaisman, *Comptes Rendus de l'Académie des Sciences*, Series D, **192**, 1142–1143 (1931).
4. H.C. Dudley and M.D. Levine, *J. Pharmacol. Exp. Ther.*, **95**, 487–493 (1949).
5. H.C. Dudley, J.I. Munn and K.E. Henry, *J. Pharmacol. Exp. Ther.*, **98**, 105–110 (1950).
6. H.C. Dudley and G.E. Maddox, *J. Pharmacol. Exp. Ther.*, **96**, 224–227 (1949).
7. H.C. Dudley, G.W. Imirie Jr and J.T. Istock, *Radiology*, **55**, 571–578 (1950).
8. M. Brucer, G.A. Andrews and H.D. Bruner, *Radiology*, **61**, 534–536 (1953). [Also subsequent papers in this volume, pp. 537–613.]

9. L.R. Bernstein, *Pharmacol. Rev.*, **50**, 665–682 (1998).
10. G. Apseloff, *Am. J. Ther.*, **6**, 327–339 (1999).
11. P. Collery, B. Keppler, C. Madoulet and B. Desoize, *Crit. Rev. Oncol. Hematol.*, **42**, 283–296 (2002).
12. M.A. Green and M.J. Welch, *Nucl. Med. Biol.*, **16**, 435–448 (1989).
13. G.E. Jackson and M.J. Byrne, *J. Nucl. Med.*, **37**, 379–386 (1996).
14. R.E. Weiner, *Nucl. Med. Biol.*, **23**, 745–751 (1996).
15. C.F. Baes Jr and R.E. Mesmer, *The Hydrolysis of Cations*, Wiley, New York, 1976.
16. R.D. Shannon, *Acta Cryst.*, **A32**, 751–767 (1976).
17. W.R. Harris and V.L. Pecoraro, *Biochemistry*, **22**, 292–299 (1983).
18. P. Collery, H. Millart, O. Ferrand *et al.*, *Anticancer Res.*, **9**, 353–356 (1989).
19. D.H. Ho, J.R. Lin, N.S. Brown and R.A. Newman, *Eur. J. Pharmacol.*, **183**, 1200 (1990).
20. R.R. Crichton, S. Wilmet, R. Legssyer and R.J. Ward, *J. Inorg. Biochem.*, **91**, 9–18 (2002).
21. K.C. Gatter, G. Brown, I.S. Trowbridge *et al.*, *J. Clin. Pathol.*, **36**, 539–545 (1983).
22. T.F. Byrd and M.A. Horwitz, *J. Clin. Med.*, **91**, 969–976 (1993).
23. H.A. Huebers and C.A. Finch, *Physiol. Rev.*, **67**, 520–582 (1987).
24. G.M. Brittenham, *Hematology, Basic Principles and Practice*, R. Hoffman, E.J. Benz Jr, S.J. Shattil, B. Furie, and H.J. Cohen (Eds), pp. 327–349, Churchill Livingstone, New York, 1991.
25. W.R. Harris, *Biochemistry*, **25**, 803–808 (1986).
26. P.L. Masson, J.F. Heremans and C. Dive, *Clin. Chim. Acta*, **14**, 735–739 (1966).
27. S.M. Larson and G.L. Schall, *JAMA*, **218**, 257 (1971).
28. P.L. Masson, J.F. Heremans and E. Schonne, *J. Exp. Med.*, **130**, 643–658 (1969).
29. R. Bennett and T. Kokocinski, *Br. J. Haematol.*, **39**, 509–521 (1978).
30. H.S. Birgens, *Dan. Med. Bull.*, **38**, 244–252 (1991).
31. R.E. Weiner, G.J. Schreiber, P.B. Hoffer and J.T. Bushberg, *J. Nucl. Med.*, **26**, 908–916 (1985).
32. R.E. Weiner, *J. Nucl. Med.*, **30**, 70–79 (1989).
33. B. Nelson, R.L. Hayes, C.L. Edwards *et al.*, *J. Nucl. Med.*, **13**, 92–100 (1972).
34. P. Hoffer, *J. Nucl. Med.*, **21**, 484–488 (1980).
35. M.F. Tsan, *J. Nucl. Med.*, **26**, 88–92 (1985).
36. S.M. Larson, J.S. Rasey, D.R. Allen *et al.*, *J. Natl. Cancer Inst.*, **64**, 41–53 (1980).
37. C.R. Chitambar and Z. Zivkovic, *Cancer Res.*, **4**, 3929–3934 (1987).
38. W. Feremans, W. Bujan, P. Neve *et al.*, *Am. J. Hematol.*, **36**, 215–216 (1991).
39. Y. Tsuchiya, A. Nakao, T. Komatsu *et al.*, *Chest*, **102**, 530–534 (1992).
40. M.-H. Sohn, B.J. Jones, J.H. Whiting Jr *et al.*, *J. Nucl. Med.*, **34**, 2135–2143 (1993).
41. C.R. Chitambar and D. Sax, *Blood*, **80**, 505–511 (1992).
42. R.E. Weiner, I. Avis, R.D. Neumann and J.L. Mulshine, *J. Cell. Biochem.*, **24** (Suppl.), 276–287 (1996).
43. H. Kawabata, R. Yang, T. Hirama *et al.*, *J. Biol. Chem.*, **274**, 20826–20832 (1999).
44. H. Kawabata, T. Nakamaki, P. Ikonomi *et al.*, *Blood*, **98**, 2714–2719 (2001).
45. W.J. Griffiths and T.M. Cox, *J. Histochem. Cytochem.*, **51**, 613–624 (2003).
46. O. Olakanmi, G.T. Rasmussen, T.S. Lewis *et al.*, *J. Immunol.*, **169**, 2076–2084 (2002).
47. E.E. Camargo, H.N. Wagner Jr and M.F. Tsan, *Nucl. Med.*, **19**, 147–150 (1979).

48. W.R. Drobyski, R. Ul-Haq, D. Majewski and C.R. Chitambar, *Blood*, **88**, 3056–3064 (1996).
49. S. Menon, H.N. Wagner Jr and M.F. Tsan, *J. Nucl. Med.*, **19**, 44–47 (1978).
50. T. Emery and P. Hoffer, *J. Nucl. Med.*, **21**, 935–939 (1980).
51. D.W. Hedley, E.H. Tripp, P. Slowiaczek and G.J. Mann, *Cancer Res.*, **48**, 3014–3018 (1988).
52. J. Narasimhan, W.E. Antholine and C.R. Chitambar, *Biochem. Pharmacol.*, **44**, 2403–2408 (1992).
53. C.R. Chitambar, J. Narasimhan, J. Guy *et al.*, *Cancer Res.*, **51**, 6199–6201 (1991).
54. R.U. Haq, J.P. Wereley and C.R. Chitambar, *Exp. Hematol.*, **23**, 428–432 (1995).
55. K.L. Chang, W.T. Liao, C.L. Yu *et al.*, *Toxicol. Appl. Pharmacol.*, **193**, 209–217 (2003).
56. C.R. Chitambar and P.A. Seligman, *J. Clin. Invest.*, **78**, 1538–1546 (1986).
57. H.A. Tajmir-Riahi, M. Naoui and R. Ahmad, *Metal Ions in Biology and Medicine*, **2**, 98–101, J. Anastassopoulou, P. Collery, J.C. Etienne, T. Theophanides *et al.* (Eds), John Libbey Eurotext, Paris, 1992.
58. M. Manfait and P. Collery, *Magnesium Bull.*, **4**, 153–155 (1984).
59. M.M. Berggren, L.A. Burns, R.T. Abraham and G. Powis, *Cancer Res.*, **53**, 1862–1866 (1993).
60. O. Olakanmi, B.E. Britigan and L.S. Schlesinger, *Infect. Immun.*, **68**, 5619–5627 (2000).
61. Genta Inc., Ganite^{TM} package insert (1993).
62. M.M. Hart, C.F. Smith, S.T. Yancey and R.H. Adamson, *J. Natl. Cancer Inst.*, **47**, 1121–1127 (1971).
63. R.H. Adamson, G.P. Canellos and S.M. Sieber, *Cancer Chemother. Rep. Pt. 1*, **59**, 599–610 (1975).
64. H.T. Whelan, M.H. Schmidt, G.S. Anderson *et al.*, *Pediatr. Neurol.*, **8**, 323–327 (1992).
65. H.T. Whelan, M.B. Williams, D.M. Bajic *et al.*, *Pediatr. Neurol.*, **11**, 44–46 (1994).
66. R.P. Warrell Jr, C.J. Coonley, D.J. Straus and C.W. Young, *Cancer*, **51**, 1982–1987 (1983).
67. R.P. Warrell Jr, R.S. Bockman, C.J. Coonley *et al.*, *J. Clin. Invest.*, **73**, 1487–1490 (1984).
68. R. Bockman, *Semin. Oncol.*, **30**(2), Suppl. 5, 5–12 (2003).
69. R.S. Bockman, A.L. Boskey, N.C. Blumenthal *et al.*, *Calcif. Tissue Int.*, **39**, 376–381 (1986).
70. R.S. Bockman, M.A. Repo, R.P. Warrell Jr *et al.*, *Proc. Nat. Acad. Sci. USA*, **87**, 4149–4153 (1990).
71. T.J. Hall and T.J. Chambers, *Bone Miner.*, **8**, 211–216 (1990).
72. R. Donnelly, R.S. Bockman, S.B. Doty and A.L. Boskey, *Bone Miner.*, **12**, 167–179 (1991).
73. H.C. Blair, S.L. Teitelbaum, H.-L. Tan and P.H. Schlesinger, *J. Cell. Biochem.*, **48**, 401–410 (1992).
74. M.A. Repo, R.S. Bockman, F. Betts *et al.*, *Calcif. Tissue Int.*, **43**, 300–306 (1988).
75. L.R. Bernstein and R.S. Bockman, *Materials Res. Soc. Fall Meeting 1988, Program and Abstracts*, p. 320 (1988).
76. P.H. Schlesinger, S.L. Teitelbaum and H.C. Blair, *J. Bone Miner. Res.*, **6**, Suppl. 1, S127 (1991).

77. P.T. Guidon Jr, R. Salvatori and R.S. Bockman, *J. Bone Miner. Res.*, **8**, 103–112 (1993).
78. L.G. Jenis, C.E. Waud, G.S. Stein *et al.*, *J. Cell. Biochem.*, **52**, 330–336 (1993).
79. P.T. Guidon Jr and R.S. Bockman, *Clin. Res.*, **38**, 328A (1990).
80. V. Matkovic, G. Apseloff, D.R. Shepard *et al.*, *Curr. Ther. Res.*, **50**, 247–254 (1991).
81. C. Whitacre, G. Apseloff, K. Cox *et al.*, *J. Neuroimmunol.*, **39**, 175–182 (1992).
82. E.H. Huang, D.M. Gabler, M.E. Krecic *et al.*, *Transplantation*, **58**, 1216–1222 (1994).
83. N. Makkonen M.-R. Hirvonen, K. Savolainen *et al.*, *Inflamm. Res.*, **44**, 523–528 (1995).
84. D. Mullet, X. Bian, G.W. Cox *et al.*, *FASEB J.*, **9**, A944 (1995).
85. J.A. Thorson, K.M. Smith, F. Gomez *et al.*, *Cell. Immunol.*, **134**, 126–137 (1991).
86. F.S. Panagakos, E. Kumar, C. Venescar and P. Guidon, *Biochimie*, **82**, 147–151 (2000).
87. M.C. Lobanoff, A.T. Kozhich, D.I. Mullet *et al.*, *Exp. Eye Res.*, **65**, 797–801 (1997).
88. J.O. Flynn, D.V. Serreze, N. Gerber and E.H. Leiter, *Diabetes*, **41**, 38A (1992).
89. M.E. Krecic, D.R. Shepard, G. Apseloff *et al.*, *FASEB J.*, **9**, A944 (1995).
90. M.E. Krecic-Shepard, D.R. Shepard, D. Mullet *et al.*, *Life Sci.*, **65**, 1359–1371 (1999).
91. C.G. Orosz, E. Wakely, S.D. Bergese *et al.*, *Transplantation*, **61**, 783–791 (1996).
92. D.J. Straus, *Semin. Oncol.*, **30**(2), Suppl. 5, 25–33 (2003).
93. A.D. Seidman, H.I. Scher, M.H. Heinemann *et al.*, *Cancer*, **68**, 2561–2565 (1991).
94. E.D. Crawford, J.H. Saiers, L.H. Baker *et al.*, *Urology*, **38**, 355–357 (1991).
95. P.A. Seligman, P.L. Moran, R.B. Schleicher and E.D. Crawford, *Am. J. Hematol.*, **41**, 232–240 (1992).
96. J.H. Malfetano, J.A. Blessing and M.D. Adelson, *Am. J. Clin. Oncol.*, **14**, 349–351 (1991).
97. J.H. Malfetano, J.A. Blessing and H.D. Homesley, *Am. J. Clin. Oncol.*, **18**, 495–497 (1995).
98. J.H. Malfetano, J.A. Blessing, H.D. Homesley and P. Hanjani, *Invest. New Drugs*, **9**, 109–111 (1991a).
99. H.I. Scher, T. Curley, N. Geller *et al.*, *Cancer Treat. Rep.*, **71**, 887–893 (1987).
100. A.Y. Bedikian, M. Valdivieso, G.P. Bodey *et al.*, *Cancer Treat. Rep.*, **62**, 1449–1453 (1978).
101. R.P. Warrell Jr, R. Israel, M. Frisone *et al.*, *Ann. Intern. Med.*, **108**, 669–674 (1988).
102. B. Leyland-Jones, *Semin. Oncol.*, **30**(2), Suppl. 5, 13–19 (2003).
103. F. Bertheault-Cvitkovic, J.-P. Armand, M. Tubiana-Hulin *et al.*, *Proc. 9th EORTC/National Cancer Inst. Symposium on New Drugs in Cancer Chemotherapy*, p. 140 (1996).
104. R.P. Warrell Jr, W.K. Murphy, P. Schulman *et al.*, *J. Clin. Oncol.*, **9**, 1467–1475 (1991).
105. R.P. Warrell Jr, D. Lovett, F.A. Dilmanian *et al.*, *J. Clin. Oncol.*, **11**, 2443–2450 (1993).
106. R. Niesvizky, *Semin. Oncol.*, **30**(2), Suppl. 5, 20–24 (2003).
107. V. Matkovic, G. Apseloff, D.R. Shepard and N. Gerber, *Lancet*, **335**, 72–75 (1990).

108. R.S. Bockman, F. Wilhelm, E. Siris *et al.*, *J. Clin. Endocrinol. Metab.*, **80**, 595–602 (1995).
109. S.W. Hall, K. Yeung, R.S. Benjamin *et al.*, *Clin. Pharmacol. Ther.*, **25**, 82–87 (1979).
110. I.H. Krakoff, R.A. Newman and R.S. Goldberg, *Cancer*, **44**, 1722–1727 (1979).
111. D.P. Kelsen, N. Alcock, S. Yeh *et al.*, *Cancer*, **46**, 2009–2013 (1980).
112. R.P. Warrell Jr, A. Skelos, N.W. Alcock and R.S. Bockman, *Cancer Res.*, **46**, 4208–4212 (1986).
113. R.S. Bockman, R.P. Warrell Jr, B. Bosco *et al.*, *J. Bone Miner. Res.*, **4**, S167 (1989).
114. R.A. Newman, A.R. Brody and I.H. Krakoff, *Cancer*, **44**, 1728–1740 (1979).
115. B. Leyland-Jones, R.B. Bhalla, F. Farag *et al.* Jr, *Cancer Treat. Rep.*, **67**, 941–942 (1983).
116. R.P. Warrell Jr and R.S. Bockman, *Important Advances in Oncology 1989*, V.T. DeVita, S. Hellman and S.A. Rosenberg (Eds), pp. 205–220, J.B. Lippincott, Philadelphia, 1989.
117. K.G. Csaky and R.C. Caruso, *Am. J. Ophthalmol.*, **124**, 567–568 (1997).
118. Y. Carpentier, F. Liautaud-Roger, F. Labbe *et al.*, *Anticancer Res.*, **7**, 745–748 (1987).
119. P. Collery, H. Millart, J.P. Simoneau *et al.*, *Trace Elem. Med.*, **1**, 159–161 (1984).
120. P. Chappuis, P. Collery, F. Labbe *et al.*, *Acta Pharm. Biol. Clin.*, **3**, 148–151 (1984).
121. P. Collery, H. Millart, D. Lamiable *et al.*, *Magnesium*, **8**, 56–64 (1989).
122. P. Collery, M. Morel, H. Millart *et al.*, *Metal Ions in Biology and Medicine*, **1**, 437–442, P. Collery, L.A Poirier, M. Manfait and J.C. Etienne (Eds), John Libbey Eurotext, Paris, 1990.
123. P. Collery, J.L. Domingo and B.K. Keppler, *Anticancer Res.*, **16**, 687–692 (1996).
124. P. Collery, F. Lechenault, A. Cazabat *et al.*, *Anticancer Res.*, **20**, 955–958 (2000).
125. M. Thiel, T. Schilling, D.C. Gey *et al.*, *Contrib. Oncol.*, **54**, 439–443 (1999).
126. L.R. Bernstein, T. Tanner, C. Godfrey and B. Noll, *Metal-Based Drugs*, **7**, 33–48 (2000).
127. B.L. Lum, A. Gottlieb, R. Altman *et al.*, *J. Bone Miner. Res.*, **18**, Suppl. 2, S391 (2003).
128. L. Schlesinger, unpublished data (2000).
129. J.R. Harrington, R.J. Martens, N.D. Cohen *et al.*, Gallium Therapy: A Novel Metal-Based Antimicrobial Strategy for Control of *Rhodococcus equi* Foal Pneumonia, *Am. Assoc. Equine Pract. Proc.* (2004).
130. F. Wang, X. Jiang, D.C. Yang *et al.*, *Anticancer Res.*, **20**, 799–808 (2000).
131. C.R. Chitambar, P. Boon and J.P. Wereley, *Clin. Cancer Res.*, **2**, 1009–1015 (1996).
132. V.B. Arion, M.A. Jakupec, M. Galanski *et al.*, *J. Inorg. Biochem.*, **91**, 298–305 (2002).
133. I. Stojiljkovic, V. Kumar and N. Srinivasan, *Mol. Microbiol.*, **31**, 429–442 (1999).
134. K. Begum, H.S. Kim, V. Kumar *et al.*, *Parasitol. Res.*, **90**, 221–224 (2003).

15 $_{32}$Ge Biological Activity of Organogermanium Compounds

Edmunds Lukevics and Luba Ignatovich

15.1 Introduction

The German chemist Winkler,[1,2] working at the Freiberg Academy of Mines, was able to isolate during the investigation of the then-discovered mineral argyrodite (Ag_6GeS_5, containing 5 to 7% germanium) a new element in its free state. He called the new element *germanium* in honour of his motherland.

Germanium (atomic number 32, atomic weight 72.59, melting point 937.4, boiling point 2830 °C), a greyish-white lustrous crystalline metal, belongs to family 14 of periodic table (formerly called group IVA), so that its physical and chemical properties resemble those of the non-metal silicon and, to a lesser extent, tin. Germanium is usually classified as semi-metal and has semiconductor properties. In nature, germanium is widely, albeit sparsely, distributed and constitutes approximately 7 ppm of the earth's crust. It is associated with sulfide ores of other elements, particularly with those of copper, zinc, lead, tin and antimony.[3] Germanium is highly concentrated in some coals at about 500 ppm. The highest reservoirs, worldwide, of germanium are found in Zaire with concentrations reaching 1000 ppm. The concentration of germanium in

Metallotherapeutic Drugs and Metal-Based Diagnostic Agents: The Use of Metals in Medicine
Edited by Gielen and Tiekink

sea water was estimated at 0.05 μg l^{-1}. Germanium is generally found in the +4 oxidation state as the oxide or in solution as germanic acid, $Ge(OH)_4$.

Up to the middle of the twentieth century, organogermanium derivatives were the least understood among the analogous compounds of the silicon subgroup elements. The first organogermanium compound, i.e. tetraethylgermane, was synthesized for the first time by Winkler in 1887 by the reaction of tetrachlorogermane and diethylzinc.[4,5] Its properties were consistent with those predicted by D. Mendeleev for 'ekasilicon'.

The major use of germanium is as an optical material (60% infrared systems, 8% fibre optics, 9% γ-ray, X-ray and infrared detectors), 10% semiconductors, transistors, diodes and rectifiers, 13% other applications (catalysts, metallurgy and chemotherapy).[6]

Edible plant foods usually contain less than 1 μg g^{-1} although some plants used for Chinese herbs, medicines and vegetables such as ginseng, garlic, oats, soya beans, Shiitake mushroom, aloe, litchi contain appreciably higher amounts of germanium.[7,8] Germanium is very well absorbed after ingestion (>90%). Germanium in plants is present in organic forms with Ge—O bonds, but the actual chemical structure of these germanium compounds is not known.[9] It has been hypothesized that germanium plays an important role in the photo-electrochemical process of photosynthesis, the metabolism and self-defence (protection from invading viruses) processes of these germanium-bearing plants.[7,8] These questions regarding the role of germanium in plant metabolism and protection are undoubtedly an important research topic requiring investigation.

The mean concentrations of germanium in normal human tissues are: lymph node: 0.9 μg kg^{-1}; skeletal muscle: 3.0 μg kg^{-1}; liver: 0.04 mg kg^{-1}; lung: 0.09 mg kg^{-1}; brain: 0.1 mg kg^{-1}; blood: 0.2 mg kg^{-1}; testes: 0.5 mg kg^{-1}; kidney: 9.0 mg kg^{-1} wet weight.[9]

In the 1970s, dietary germanium supplements became very popular due to the alleged therapeutic value of germanium (stimulation of iron consumption and haemoglobin production). In 1994, the first organogermanium pharmaceutical propagermanium was launched in Japan under the trade name Serocion® (Sanwa Kagaku Kenkyusho Co. Ltd). Its biological spectrum of activity includes protection against viruses, immunostimulation and hepatoprotection. Propagermanium, belonging to the class of germsesquioxanes, has low toxicity.

This achievement has stimulated further investigations of the biological activity not only of germsesquioxanes but also of other classes of low-toxic organogermanium compounds.

Today, numerous organogermanium compounds possessing anti-tumour, immunomodulating, interferon-inducing, radioprotective, hypotensive and neurotropic properties have been synthesized. The most intensively investigated organogermanium compounds are germsesquioxanes of general formula $[(GeCHR^1CHR^2COOH)_2O_3]_n$ (R^1, R^2 = H, alkyl, aryl, hetaryl); **1**, $R^1 = R^2 = H$ 2-carboxyethylgermanium sesquioxide, repargermanium, rexagermanium, proxigermanium, propagermanium, Serocion®; **2** 2-(3-dimethylaminopropyl)-

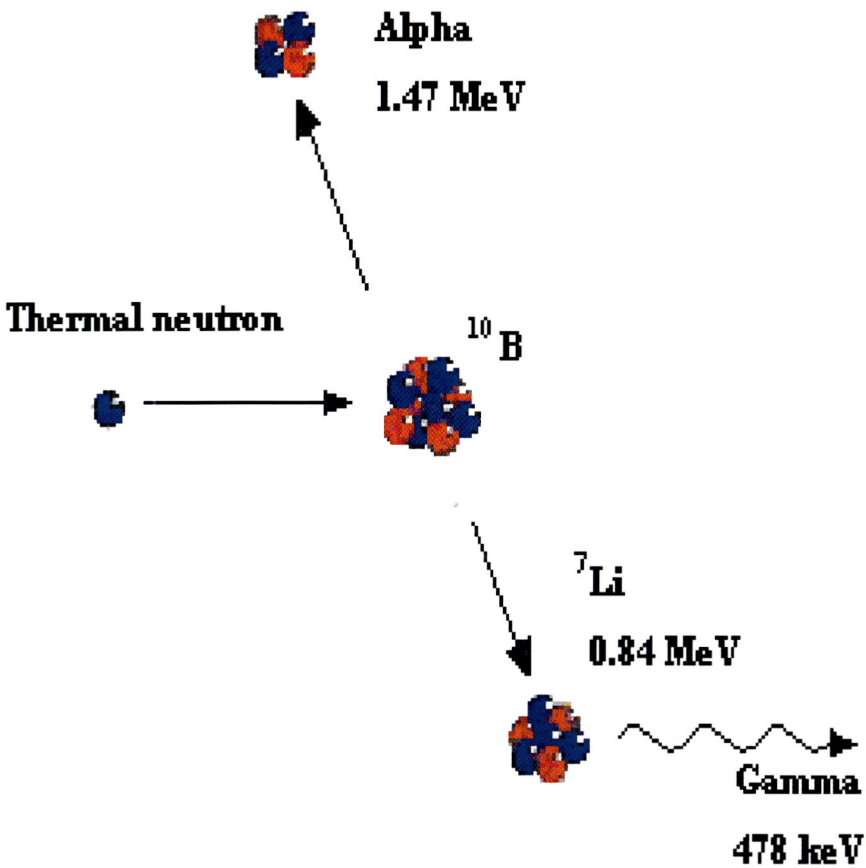

Plate 1 (Figure 2.1) The mechanism of boron neutron capture therapy (BNCT)

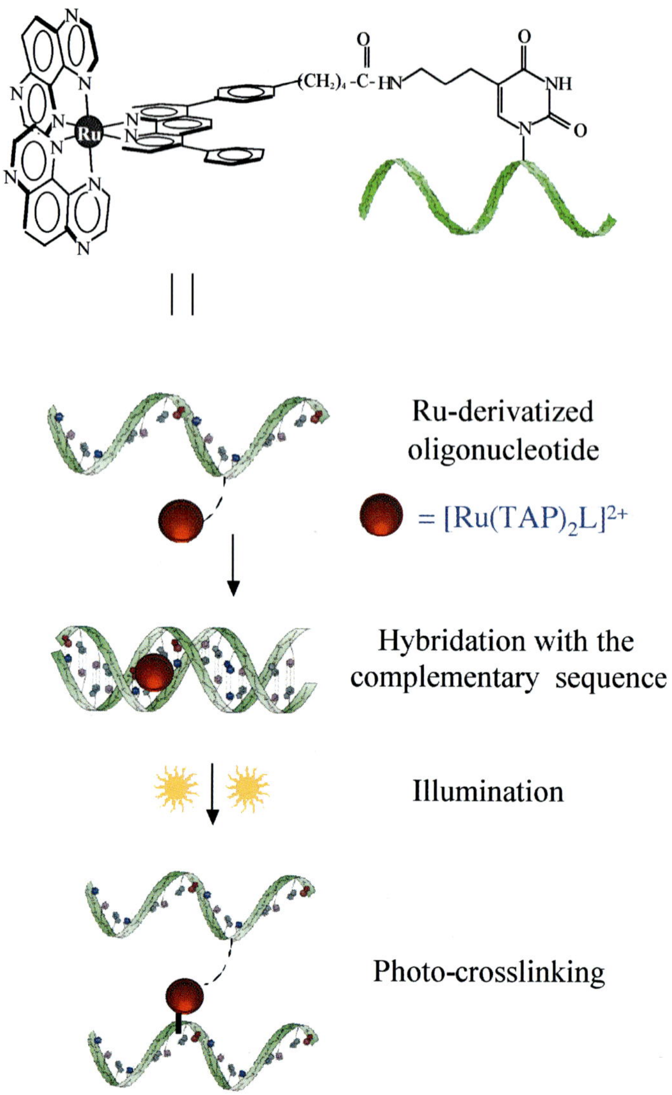

Plate 2 (Figure 19.12) The $[Ru^{II}(TAP)_2\ dip']^{2+}$ complex anchored onto a synthetic oligonucleotide at the level of the 5 position of a thymine

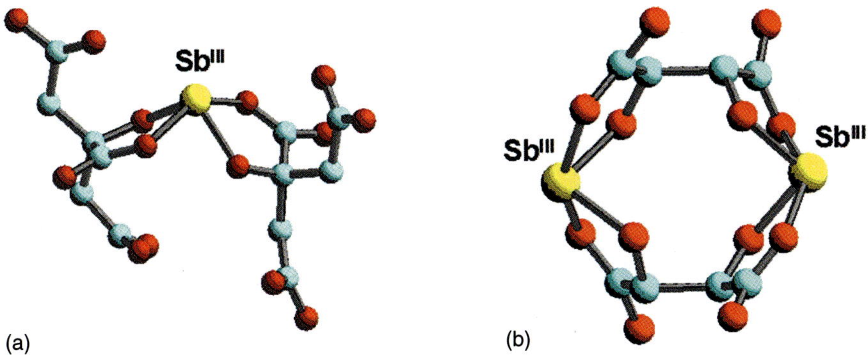

Plate 3 (Figure 23.1) X-ray crystal structures of antimony(III) citrate (a) and potassium antimony(III) tartrate (PAT, $K_2Sb_2(C_4H_2O_6)_2$) (b)

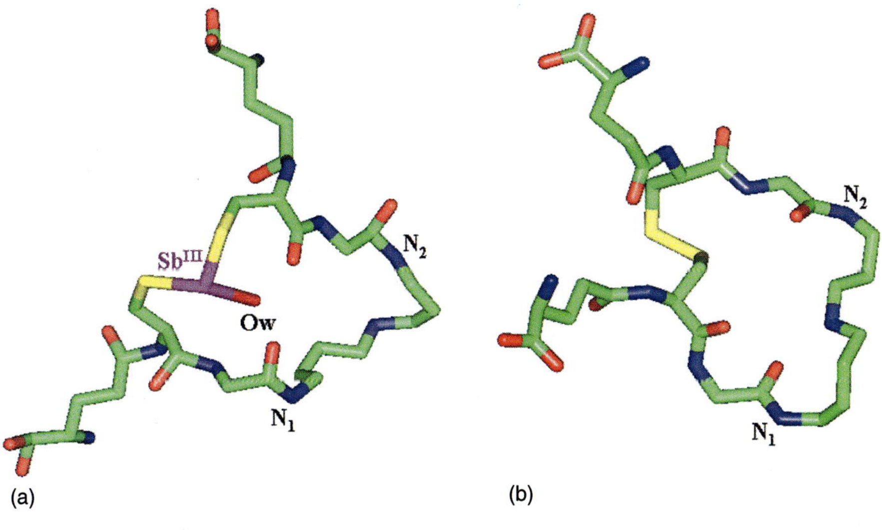

Plate 4 (Figure 23.4) Structures of $Sb(T(S)_2)$ complex together with the oxidized form of trypanothione (adapted with permission from Ref. 90)

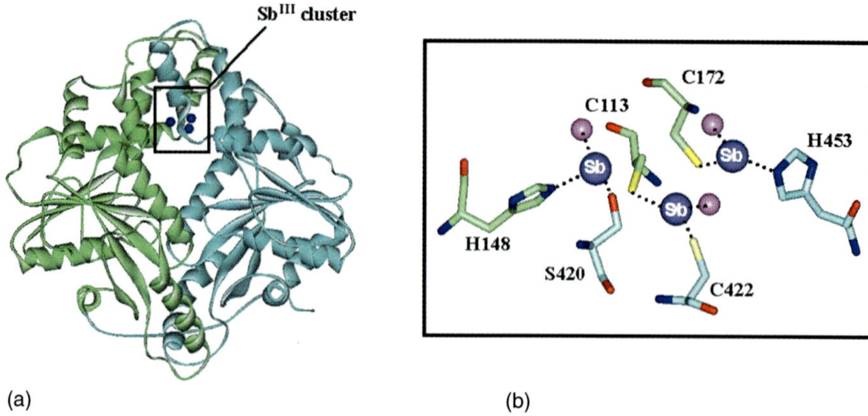

Plate 5 (Figure 23.5) Overall structure of the ArsA ATPase (a) and coordination of Sb^{III} at the metal binding site (b) (coordinates from PDBID: 1F48). Residues His^{148}, Cys^{113} and Cys^{172} are from domain A1 while Ser^{420}, Cys^{422} and His^{453} are from domain A2. The purple balls represent the bound chlorides (Cl^-)

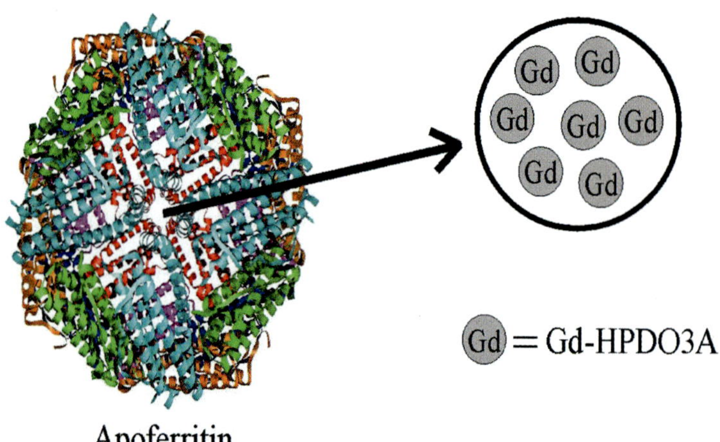

Plate 6 (Figure 28.13) Schematic representation of Gd-HPDO3A entrapped into the Apoferritin cavity

8,8-diethyl-2-aza-8-germaspiro[4,5]decane, spirogermanium; **3**, germatranes; **4**, germylporphyrines; **5**, monomeric and air-stable decaphenylgermanocene; **6**, germanium-modified triazoles and many of germyl-substituted heterocycles (mainly derivatives of furan, thiophene, isoxazoline and uracil).

$(GeCH_2CH_2COOH)_2O_3$

1

2-Carboxyethylgermanium sesquioxide (*anti-viral, immunomodulation, anti-tumour, anti-arthritic*)

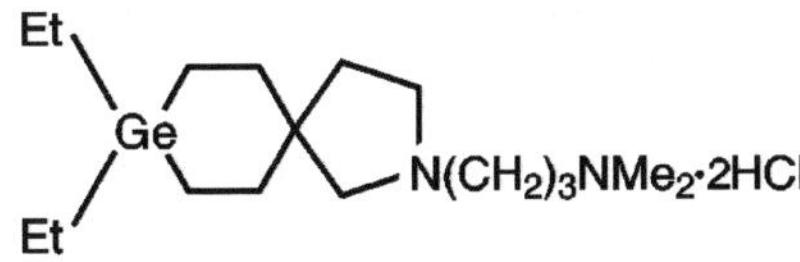

2

Spirogermanium (*anti-tumour, anti-malarial, hepatoprotection*)

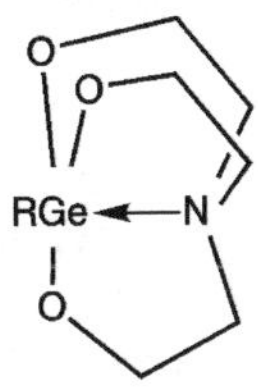

3

Germatranes (*anti-tumour, neurotropic*)

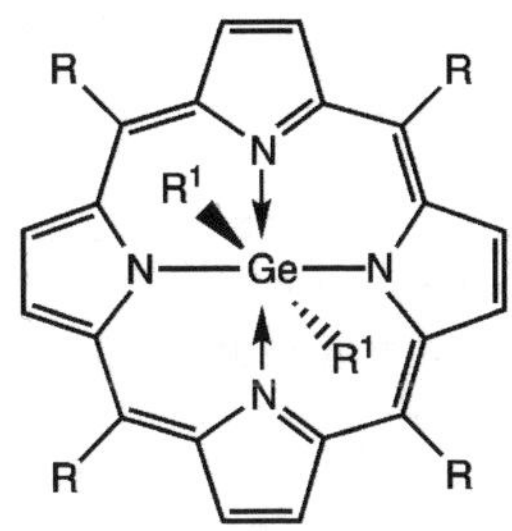

4

Germanium(IV) porphyrines (*anti-tumour*)

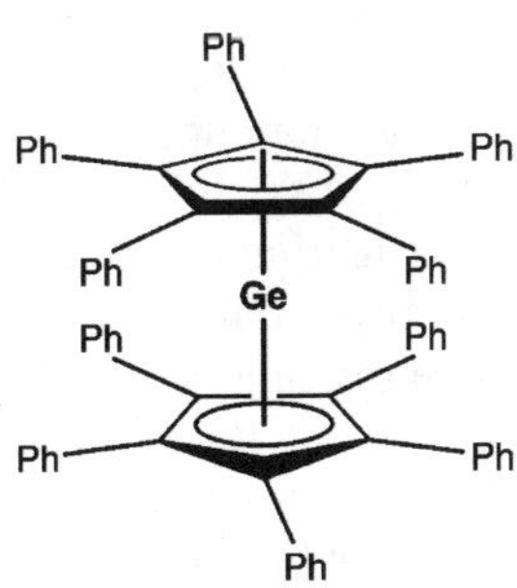

5

Decaphenylgermanocene (*anti-tumour*)

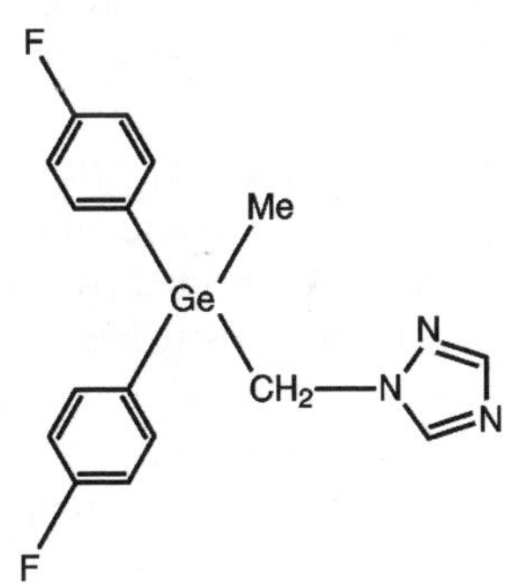

6

Germa-flusilazole (*fungicidal activity*)

Several reviews covering the biological properties of organogermanium compounds have appeared owing to the numerous biological investigations on these compounds.[7,8,10–22]

15.2 Biological Activity

15.2.1 2-Carboxyethylgermanium sesquioxide

2-Carboxyethylgermanium sesquioxide (**1**) received much attention in the nineties. The biochemical significance of this crystalline structure consisting of Ge_6O_6 rings[23] will become apparent in discussing organic germanium compounds' ability to scavenge free radicals, protect against radiation, enrich the body's oxygen supply and rid the body of heavy metals. The synthesis is carried out by hydrolysis of an organogermanium trichloride[10]

$$CH_2{=}CHR + HGeCl_3 \longrightarrow Cl_3GeCH_2CH_2R \longrightarrow (O_{1.5}GeCH_2CH_2COOH)_n$$

$R = COOH,\ COCl,\ CN$ ***2-Carboxyethylgermanium sesquioxide***

A large number of other organogermanium sesquioxides have been synthesized for biological testing.[10]

2-Carboxyethylgermsesquioxane exhibits acute toxicity for mice and rats with a mean value $LD_{50} = 6000–10\,000\ mg\,kg^{-1}$ (p.o., i.p.) and 4500–5700 $mg\,kg^{-1}$ (i.v.).[24] Subchronic oral toxicity of $[(HOOCCH_2CH_2Ge)_2O_3]_n$ (in rats $1\ g\,kg^{-1}\ day^{-1}$ during 6 months) has been reported.[25,26] Daily oral administration of $120\ mg\,kg^{-1}day^{-1}$ of 2-carboxyethylgermsesquioxane, hereafter Ge-132, for 24 weeks showed no toxic effects in mice.[27]

Oral administration of proxigermanium for a year at $750\ mg\,kg^{-1}\ day^{-1}$ induced diarrhoea in rats. However, it has been found that a dose of $83\ mg\,kg^{-1}\ day^{-1}$ is not toxic. Proxigermanium administered in a dose of 15–240 mg $person^{-1}$ has not affected physiological functions in healthy volunteers.[28]

The compound is not embriotoxic, teratogenic,[29–31] mutagenic, carcinogenic or antigenic.[20,21,32] The oral administration of proxigermanium does not affect fertility at doses 350, 700 and $1400\ mg\,kg^{-1}\ day^{-1}$ 60 days before and at mating in male rats and 14 days before, at and 7 days after mating in female rats.[28]

The histological investigation revealed no significant renal toxic effects of propagermanium when administered at a high dose for eight weeks. Propagermanium administration was not associated with any alteration in the changes induced by adriamycin or mercuric chloride. This confirms that propagermanium may be a safe compound for use by individuals with damaged kidneys.[33]

On the other hand, case reports on the nephrotoxicity of organogermanium preparations (high doses, long-term administration) have been published.[9,18,20,34–41] One-month oral subacute toxicity of propagermanium has been studied.[42]

The anti-tumour activity of Ge-132 and its derivatives has been studied. 2-Carboxyethylgermsesquioxane has been revealed to possess anti-tumour,[7,8,10,17,43] interferon-inducing[10,17,44–46] immunomodulating (normalizing the ratio of T-lymphocyte subpopulations),[10,47–53] and anti-viral[10,54] properties. The anti-tumour activities of $(O_{1.5}GeCH_2CH_2COOH)_n$ and $(O_{1.5}GeCH_2CH_2CONH_2)_n$[7,8] were first reported by Asai.[7,8] Mice with the Ehrlich ascites tumour received 20 μg ml^{-1} $(O_{1.5}GeCH_2CH_2COOH)_n$ and 20 and 40 μg ml^{-1} of $(O_{1.5}GeCH_2CH_2CONH_2)_n$ daily for 7 days. The inhibition of the tumour cells' growth was higher than in control by 26, 47 and 31%, respectively. 2-Carboxyethylgermsesquioxane exhibited some anti-tumour activity in adenocarcinoma LA-795, in the Lewis lung carcinoma 3LL,[55–58] melanoma B16[59] and leukaemia L-1210[60] and had a limited anti-metastatic effect in mice.[56] Complete remission of pulmonary spindle cell carcinoma has been observed after oral administration of germanium sesquioxide.[61] In some cases 2-carboxyethylgermsesquioxane reinforces the effect of bleomycine and 5-fluorouracil.[62] It has been shown that $(O_{1.5}GeCH_2CH_2COOH)_n$,[63,64] $(O_{1.5}GeCH_2CH_2COONa)_n$[55] and $(O_{1.5}GeCH_2CH_2CONH_2)_n$[65] prolong life of mice with implanted ascitic hepatomas AH-44 and AH-66 and also of rats with syngenic bladder carcinoma BC47.[66]

Germanium-132 was not cytotoxic to carcinoma cells growing *in vitro* and had its anti-tumour effect via the stimulation of host-mediated, immunopotentiating mechanisms,[14,56,67] leading to the augmentation of natural killer cells' activity and activation of macrophages in mice.[56] Carboxyethylgermanium sesquioxide has some preventive effects on the precancerous lesions in rat glandular stomach induced by *N*-methyl-*N′*-nitrosoguanidine.[68]

The Ge-132 administered p.o. activated mouse murine macrophage-mediated tumour cytotoxicity. When Ge-132-activated macrophages were treated with ganglioside *in vitro*, synergistic activity appeared.[69] 2-Carboxyethylgermsesquioxane, 2-carbamoylethylgermsesquioxane and α,β-dicarboxyethylgermsesquioxane significantly stimulated mouse peritoneal macrophages and human monocytoid cells by oral administration of a single dose of 100 mg kg^{-1}.[70]

The anti-tumour activity of various organogermanium sesquioxanes [$(O_{1.5}GeCH_2CHRCOX)_n$, R = H, Me, X = OH, NH_2][59] and sesquisulphides[71] has been studied. The most active was the methacrylic acid derivative (80% life prolongation in melanoma B16 and 72% life prolongation in mice with Lewis lung carcinoma).

The sesquioxanes $(O_{1.5}GeCH_2CHRCOX)_n$[72,73] and $(4\text{-}RC_6H_4GeO_{1.5})_n$ (R = Hal, CN)[74–76] exhibited anti-tumour activity and prolonged the lives of animals with Ehrlich ascites tumour.

Organogermanium sesquioxanes containing uracil or 5-fluorouracil (5-FU) moieties possess an anti-tumour activity against IMC carcinoma in mice.[77,78] 1-[*p*-[Bis(β-chloroethylamino)phenyl]-2-amino-2-carboxyethylgermanium sesquioxane (LD_{50} 1765 mg kg^{-1}) inhibited the growth of sarcoma S-180 by 78% (in mice, i.p.), whereas a 65% inhibition was achieved with 5-FU under the same experimental conditions.[79] Some germanyl heterocyclic (indolyl or furyl) amino acid derivatives, such as 1-(3′-indolyl)-2-amino-2-carboxyethylgermsesquioxane, and corresponding sesquisulfide have anti-tumour activity in sarcoma S-180 comparable to that of 5-FU.[80]

The results of bioassay showed that the organogermanium sesquioxanes containing the α-aminophosphonate group $O_{1.5}GeCHR'CHR''C(O)NHCH(R)P(O)(OPh)_2$ exhibit cytotoxicity *in vitro*.[81,82]

Preliminary pharmacological investigations of 2-(2-germaoxa-3H-benzofuran-3-yl)-2-acetylaminoglycine and 1-(2-hydroxyphenyl)-2-amino-2-carboxyethyl germanium sesquioxane showed that the sesquioxane derivative had a low toxicity (LD_{50} >10 g kg^{-1} for mice) and inhibited the growth of S-180 in mice by 51.6% (orally). The activity of this compound was slightly higher than that of 5-FU under the same experimental conditions.[83]

Glucopyranosyl derivatives of carboxyethylgermanium sesquioxane such as 2,4-di(O-acetyl)-1,3,6-tri-O-(carboxyethyl)germanyl-β-D-glucopyranose showed EAC(Erlich Ascites Carcinoma) growth inhibition in mice (i.p.) by 62%.[84]

Germsesquioxanes $(GeCR^1R^2CR^3HCOX)_2O_3$ (R^1, R^2, R^3 = H or lower alkyl, Ph; X = OH, O-lower alkyl group, amino group or a negative charge) have been proposed as therapeutic agents for brain aging.[85] Ge-132 can antagonize the damage of neural behaviour caused by the heavy element lead.[86]

Germanium-132 can inhibit bone resorption by osteoclasts in a concentration-dependant manner.[87] The therapeutic effect of Ge-132 for experimental osteoporosis has been studied using ovariectomized rats maintained on a low calcium-containing diet. Ge-132 prevented decreased bone strength, and affected the femur cortical bone index and bone mineral mass caused by osteoporosis.[88] The peripheral quantitative computed tomography measurements of the femur demonstrated that a high correlation existed between the transverse strength and total bone mineral density or bone mineral content.[89]

Recently, the effects of Ge-132 on the low density lipoprotein oxidation and atherosclerosis in Kurosawa and Kusanagi hypercholesterolemic (KHC) rabbit have been studied.[90] Treatment with Ge-132 resulted in decreases in the oxidation rate and in the formation rate of thiobarbituric acid-reactive substances following copper-induced oxidation of low-density lipoprotein. Ge-132 is suggested to possess anti-oxidative properties, but this did not lead to any attenuation of atherosclerotic progression in the KHC rabbits.

The non-enzymic glycosylation (Maillard reaction) of amino acids with glucose has been markedly suppressed in the presence of Ge-132 in the range of 1–10 mmol l^{-1}. These results demonstrate the anti-diabetic properties of Ge-132.[91] Some derivatives of Ge-132 also inhibit the Maillard reaction.[92,93] As protein modification, such as glycation, may play a role in initiating changes that lead to cataract development, the effect of Ge-132 on galactose-induced cataractagenesis has been studied. It has been found that Ge-132 acted as an antiglycation agent and delayed cataract formation. This compound was effective in maintaining Na(+)-K(+)-ATPase.[94,95]

15.2.2 Spirogermanium

The organogermanium compound attracting greatest attention is 2-(3-dimethylaminopropyl)-8,8-diethyl-2-aza-8-germaspiro[4,5]decane (**2**), hereafter spirogermanium. The synthesis of this compound is complicated and involved nine stages:[96]

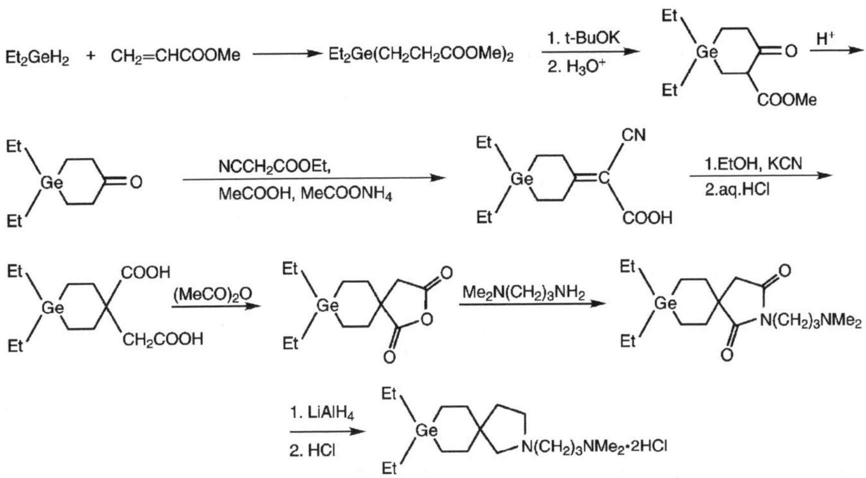

spirogermanium

Preclinical toxicological evaluation of spirogermanium in white mice and beagle dogs confirmed the lack of bone-marrow toxicity and pointed to a dose-limiting toxic effect of spirogermanium on the central nervous system (CNS), manifesting in dogs by pilo-erection (muscle abridgement with hair movement), nystagmus (involuntary oscillatory movement of eyeballs) and tremor, and lethal doses generalized seizures.[17,97,98] The neurotoxicity of spirogermanium at doses 32–60 mg m^{-2} has been observed in clinical trials as well.[10] The dose-limiting toxicity has been observed in patients.[99] Spirogermanium is rapidly removed from the plasma and appears not to be accumulated in tissues.[20]

Spirogermanium (**2**) was the first organogermanium compound tested clinically as an anti-tumour agent.[10,11,17,98,100–110] There were many reports on its anti-tumour effectiveness on a large number of human tumour types, e.g. ovarian carcinoma, cervix carcinoma, breast carcinoma, renal cell carcinoma, metastatic prostate cancer, melanoma, non-Hodgkin lymphoma, non-small-cell lung cancer and colorectal carcinoma.

The preliminary results obtained from *in vitro*[111–113] and *in vivo*[100,101] screenings have suggested that spirogermanium might be useful in the treatment of human cancers. Passing through toxicological studies, spirogermanium has entered clinical trials on humans bearing cancerous tumours.[100,101] The mode of action is thought to involve the inhibition of protein synthesis with a secondary suppression of RNA and DNA synthesis. It has been reported that spirogermanium also has anti-arthritic and suppressor cell-inducing activity.[114] A Phase I clinical study[99,115,116] showed good drug tolerance. Further evaluation in Phase II revealed consistent neurotoxicity, ranging from dizziness to disorientation, as well as mild haematological and hepatic toxicity. Modest activity was demonstrated in ovarian cancer. No responses were seen in patients with refractory colorectal carcinoma or non-small-cell lung cancer.[109,117–119] Low response and relatively high toxicity have also been found in the Phase II treatment of gastric carcinoma.[120]

Spirogermanium has some anti-malarial activity,[15] but no effect on the *Babesia microti* infection in the hamster model.[121]

15.2.3 Germatranes

The next group of biologically active organogermanium compounds are *germatranes* (**3**) – tricyclic organogermanium derivatives of triethanolamine (1-germa-2,8,9-trioxa-5-azatricyclo[3.3.3.0^{1,5}]undecane) containing a hypervalent germanium atom with a transannular bond to nitrogen (bond length 2.01–2.29 Å) which is responsible for their chemical stability.[122] Several methods for the preparation of germatranes have been elaborated. The reaction of tetra-alkoxy- or trialkoxy-germatranes with triethanolamine occurring under mild conditions in the presence of a catalyst or without catalyst giving 70–90% yield is considered the simplest and most available:[123]

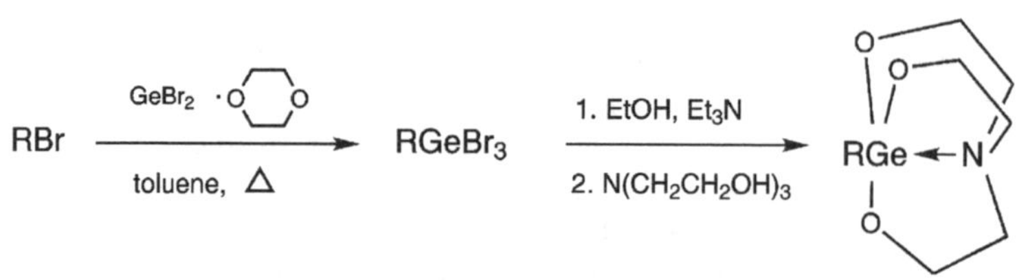

R = alkyl, aryl, hetaryl ***Germatranes***

The acute toxicity of germatranes depends on the substituent at the germanium atom.[22] Their mean LD_{50} at intraperitoneal (i.p.) administration varies from 16.5 to 10 000 mg kg^{-1}.[16,124]

All furylgermatranes are less toxic than the corresponding thienylderivatives.[16] The position of the bromine atom and germatranyl group in the thiophene ring dramatically influences the toxicity of the compounds, varying from the very toxic 5-bromo-2-thienylgermatrane to the non-toxic 3-bromo-4-thienylgermatrane. 2-Isomers belonging to the thiophene series appear to be the most toxic while, in the furan series, the 2-derivatives are less toxic than 3-isomers. Phenylgermatrane[125] is two times less toxic than 2-thienylgermatrane ($16.5\,mg\,kg^{-1}$) but still exhibits high toxicity ($35.5\,mg\,kg^{-1}$). *p*-Fluorophenylgermatrane exhibits a comparable toxicity (LD_{50} $42.2\,mg\,kg^{-1}$). The introduction of a bromine atom or a methyl group in the *para*-position of the phenyl ring decreases the toxicity by two times (LD_{50} 65.5 and $70\,mg\,kg^{-1}$, respectively). *p*-(2-Thienyl)phenylgermatrane (LD_{50} $324\,mg\,kg^{-1}$) is less toxic (~9 times) in comparison with the unsubstituted phenylgermatrane. Introduction of methylene-, siloxy- or germoxy-group between the aromatic ring and the germatranyl cage substantially lowers the toxicity of germatranes.[22]

The high toxicity of phenyl- and thienyl-germatranes can be explained neither by the presence of the tricyclic germatrane ring with a penta-coordinated germanium atom nor by the presence of a π-electron system in the substituent. Alkylgermatranes (LD_{50} ranging from 355 to $3000\,mg\,kg^{-1}$),[22] *p*-biphenyl-, *p*-(2-thienyl)phenyl- or 2,2′-bithienyl-germatranes (LD_{50} 282, 324 and $447\,mg\,kg^{-1}$, respectively) containing the same germatrane system and vinylgermatrane (π-bond) (LD_{50} $5600\,mg\,kg^{-1}$)[22] have low toxicity or are non-toxic. Hydrolysis of the Ge—C bond decreases the toxicity by more than 500 times. Both the products of the Ge—O bond hydrolysis – germanic acid (LD_{50} ~$2000\,mg\,kg^{-1}$) and triethanolamine (LD_{50} $1450\,mg\,kg^{-1}$) – are also less toxic than the starting 2-thienylgermatrane. As the mechanism of arylgermatrane's biological activity is not known yet, one can speculate that both parts of the molecule, the germatrane system (for binding to the receptor?) and aryl group bound directly to the germanium atom (for toxic bioarylation; 2-thienylmethyl- and benzylgermatranes are less toxic – 325 and >$1000\,mg\,kg^{-1}$), are important for the expression of high toxicity.

The neurotropic activity of germatranes **3** has been studied.[10,11,16,22,126–132] In most cases the heteroarylgermatranes $R(CH_2)_nGe(OCH_2CH_2)_3N$ with the germatranyl group directly connected to the carbon atom ($n = 0$) possess a higher toxicity and a higher CNS activity.[16] Hydrogermatrane $HGe(OCH_2CH_2)_3N$ exhibits the highest depressant activity (ED_{50} $0.0015\,mg\,kg^{-1}$) on the CNS.[16] Thus, the therapeutic index for this compound is rather high (>200 000). The transition from hydrogermatrane to hydroxygermatrane $HOGe(OCH_2CH_2)_3N$ demonstrated that the CNS-depressant activity is noticeably decreased.[16,22] The substitution of the hydroxyl group in the germatranol molecule by a trimethylsiloxy, triphenylsiloxy or triphenylgermoxy group leads to the complete loss of the CNS-depressant activity. Arylgermatranes were more active than hetaryl derivatives in memory improvement tests. Phenylgermatrane, *p*-tolylgermatrane, *p*-fluorophenylethynylgermatrane, benzylgermatrane, and *o*- and *p*-bromobenzylgermatranes completely prevented animals from retrogradal amnesia caused by an electric shock. 4-(Dimethylamino)phenylgermatrane has been studied more thoroughly by its administration into the stomach.[16] The neurotropic action of this germatrane is characterized by a serotonin-blocking as well as by M-cholinemimetic and GABA-ergic mechanisms.

The anti-tumour activity of two types of germatranes $RGe(OCH_2CH_2)_3N$ and $R_3MOGe(OCH_2CH_2)_3N$ (R = alkyl, aryl, furyl, thienyl, phenyl; M = Si, Ge) has been studied.[22] The germatranes $R_3MOGe(OCH_2CH_2)_3N$ possess higher cytotoxicity on HT-1080, MG-22A, B16 and Neuro 2A tumour cells (IC_{50} 0.9–6.1 $\mu g\,ml^{-1}$) than $RGe(OCH_2CH_2)_3N$ (IC_{50} ~100 $\mu g\,ml^{-1}$). Hydroxygermatrane had the highest cytotoxic activity on HT-1080 tumour cells. 1-Triphenylgermoxygermatrane (IC_{50} 3.2 $\mu g\,ml^{-1}$) suppressed the growth of melanoma B16 cells three times more effectively than 1-triphenylsiloxygermatrane (IC_{50} 10 $\mu g\,ml^{-1}$). 1-Trimethylsiloxy- and 1-trimethylgermoxygermatranes inhibited the growth of Sarcoma 37 in mice more effectively than the parent 1-hydroxygermatrane (by 75, 60 and 40% correspondingly).[11] Administration of 3-(1-germatranyl)propionic acid to mice afflicted with Ehrlich ascites tumour caused a 78% increase in survival time.[59] Germatranes $N(CH_2CH_2O)_3Ge(CHR)_2CONH_2$ (R = H, Ph) showed some activity against IMC carcinoma in mice.[133]

15.2.4 Germanium modified organic compounds

Besides spirogermanium and germanium sesquioxanes, numerous other organogermanium compounds have shown anti-tumour activity against experimental tumours. They include the octahedral complex dialkyl-5,10,15,20-tetra-[3′,5′-bis(*t*-butyl)phenyl]-porphynato-germanium(IV) (**4**)[134] and some deca-phenylgermanocene (**5**) derivatives.[135] The dimethyl derivative **4** (R^1 = Me) was cytotoxic *in vitro* towards HeLa cells, inhibited the solid form of B16 melanoma and IMC carcinoma (33 and 39% inhibition, respectively) and produced a complete remission of tumours in 50% of rats bearing a Walker 256 carcinosarcoma.[134] The monomeric air-stable decaphenylgermanocene **5** has been synthesized. This compound produced 40–80% cure rates in mice bearing the fluid Ehrlich ascites tumour.[135,136]

1,1-Bis(1,3-dithian-2-yl)-1-germa-3-cyclopentene has been reported to be more potent than 2-carboxyethylgermanium sesquioxane against IMC carcinoma in mice.[137]

The cytotoxicity of germyl derivatives of hetaryldiketones have been tested on HT-1080 and MG-22A tumour cell cultures.[138] 2,2-Dimethyl-5-(5′-trimethylgermyl-2′-furfurylidene)-1,3-dioxane-4,6-dione and 2,2-dimethyl-5-(5′-trimethylgermyl-2′-thienylidene)-1,3-dioxane-4,6-dione (IC_{50} ~100 $\mu g\,ml^{-1}$) have a limited influence on the HT-1080 cell line but their cytotoxicity considerably increases in MG-22A cells (IC_{50} 1–10 $\mu g\,ml^{-1}$). The germanium derivative of Meldrum's acid exhibits a low toxicity and the highest anti-Corazol activity (262%) and depresses the pharmacological effects of phenamine by 55.5%.[138]

Germyl- and silyl-trifluoroacetylderivatives of furan possess a wide range of toxicity (LD_{50} 11.2–2240 $mg\,kg^{-1}$).[139] In this type of compound, germanium derivatives are more toxic than the corresponding silicon derivatives. Triethylgermylderivative is a highly toxic compound: LD_{50} 11.2 mg kg^{-1}. It is interesting to note that the trimethylderivatives of germanium and silicon analogues have comparable toxicities, but the substitution of the methyl groups to ethyl groups dramatically changes the toxicity: the germanium derivative becomes 200 times more toxic than the silicon analogue. The germyl-substituted compounds possessed the highest neurotropic activity.[139]

1-Triphenylgermyl-4-propionyl semicarbazides, thiosemicarbazides $Ph_3GeCHR'CH_2CONHNHC(X)NHR''$ ($R' = H$, Ph; $R'' = Ph$, *p*-tolyl; $X = O$, S) and their heterocyclic derivatives have been found to possess inhibitory effects on gastric carcinoma MGC-803 *in vitro*.[140]

Germanium and silicon derivatives of furfural semicarbazone and thiosemicarbazone exhibited similar anti-tumour activities against the melanoma B16 in mice (40–48% inhibition of the growth).[141]

2-(5′-Triethylgermyl-3′-isoxazolinyl)pyridine hydrochloride, 2-(5′-triethylgermylmethyl-3′-isoxazolinyl)pyridine hydrochloride, (3-methyl-4-triphenylgermylisoxazolinyl-2)-5-carboxylic acid ethyl ester and (3-ethyl-4-triphenylgermyl-isoxazolinyl-2)-5-carboxylic acid ethyl ester had a borderline activity on HT-1080 and MG-22A tumour cell cultures, and a low activity on B16. The most active cytotoxic substance was (3-ethyl-4-triphenylgermylisoxazolinyl-2)-5-carboxylic acid ethyl ester. Germylisoxazolin-2-yl derivatives are stronger tumour-growth inhibitors and NO-inducers than their silyl analogues.[142] The cytotoxic activity of germyl-substituted 4,4-dioxo-3a, 6a-dihydrothieno[2,3-d]isoxazolines-2 has been studied.[143] Germanium-containing compounds had a cytotoxic effect on Neuro 2A and B16 cell lines. 4,4-Dioxo-3-phenyl-5-trimethylgermyl-3a,6a-dihydrothieno[2,3-d]isoxazoline-2 showed the highest tumour-cell growth inhibition.

Water soluble derivatives of $n\text{-}Bu_2Sn(O_2CCHR^1CHR^2GeR^3)_2 \cdot H_2O$ (type I: $R^3 = (OCH_2CH_2)_3N$, $R^1 - H$, and R^2 is H, CH_3, C_6H_5, 4 ClC_6H_4, 2 ClC_6H_4, 4-$CH_3C_6H_4$, 2-$CH_3C_6H_4$, 2,4-ClC_6H_3, 4-$CH_3OC_6H_4$, 4-$O_2NC_6H_4$, type II: $R^3 = Ph_3$, $R^1 = H$ and R^2 is CH_3, C_6H_5, 4-ClC_6H_4, 4-$CH_3C_6H_4$, 4-$CH_3OC_6H_4$) were screened *in vitro* for their cytotoxicity activity against KB cells, HCT-8 cells and Bel7402.[144] All compounds showed some activity. The germatrane-substituted derivatives were more active than the Ph_3Ge-substituted derivatives. Methyl-substituted analogues $Me_2Sn(O_2CCHR^1CHR^2GeR^3)_2$ where $R^1 = Me$ and H, $R^2 = H$, CH_3, C_6H_5, 4-$CH_3OC_6H_4$, 4-ClC_6H_4, 4-$CH_3C_6H_4$ were found to have a potential activity against bacteria.[145] The mechanism involved in the bactericidal effect of these compounds has not been discussed.

5-Trimethylgermyluracil, its silicon counterpart and their 1-(2-tetrahydrofuryl) derivatives display similar cytotoxicities to melanoma B16 cells.[11] 5-Trimethylgermyl and 5-trimethylsilyl derivatives of 2′-deoxyuridine are similar in their anti-metabolic properties: β-anomers possess a weak biological action, and α-anomers inhibit the replication of the herpes simplex virus HSV-1 and reveal cytotoxic properties in *in vitro* experiments on cell cultures of the human ovary carcinoma CaOv.[146,147]

The germanium analogue of *N*-neopentyl-*N*-nitrosourea exhibited a considerably higher cytotoxicity to leukaemia L1210 cells than the carbon analogue.[148] Some trimethylgermyl derivatives of retinobenzoic acid show a high retinoidal activity in human promyelotic leukaemia cells HL-60.[149]

β-(Trimethylgermyl)alanine and β-(trimethylsilyl)alanine have been synthesized and used to prepare germa- and sila-containing decapeptides. Their binding affinities and antagonistic properties for the human gonadotropin-releasing hormone receptor were similar to those of the carbon analogue but the testosterone and luteinizing hormone suppression lasted significantly longer.[150,151]

Germanium analogues of anti-muscarinic drugs have been prepared and it has been shown that there exists a Ge/Si/C bioisosterism.[152]

Several classes of organogermanium derivatives (germathiazolidines, germadithioacetals, germatranes and germylated sulfides) possess a remarkable radioprotective activity.[153–163]

The germanium analogue **6** of the leading agricultural fungicide flusilazole has been synthesized.[164] The fungicidal properties of flusilazole and of bis(4-fluorophenyl)methyl(1H-1,2,4-triazol-1-yl-methyl)germane have been compared and found to be similar. A dose-dependent fungicidal activity has been demonstrated for the germanium derivative of *o*-chlorobenzamide.[165]

References

1. C.A. Winkler, *Ber. Dtsch. Chem. Ges.*, **19**, 210–211 (1886).
2. C.A. Winkler, *J. Prakt. Chem.*, **34**, 177–229 (1886).
3. P.S. Gleim, in *McGraw-Hill Encyclopedia of Science and Technology*, vol. 8, 6th edn, McGraw-Hill, New York, 1987, pp. 76–83.
4. C.A. Winkler, *J. Prakt. Chem.*, **36**, 177–209 (1887).
5. C.A. Winkler, *Ber. Dtsch. Chem. Ges.*, **20**, 677–680 (1887).
6. T.O. Liewellyn, in *Minerals Yearbook* – 1990, vol. 1 *(Metals and Minerals), US Department of Interrior Bureau of Mines, US Government Printing Office*, Washington, 1993, pp. 491–494.
7. K. Asai, *Organic Germanium: A Medical Godsend*, Ed. L. Kagakusha, Tokyo, 1977.
8. K. Asai, *Miracle Cure: Organic Germanium*, Japan Publ. Inc., Tokyo, 1980.
9. G.A. Schauss, *Biol. Trace Elem. Res.*, **29**, 267–280 (1991).
10. E. Lukevics, T.K. Gar, L.M. Ignatovich and V.F. Mironov, *Biological Activity of Germanium Compounds*, Zinatne, Riga, 1990 (in Russian).
11. E. Lukevics and L.M. Ignatovich, *Appl. Organomet. Chem.*, **6**, 113–126 (1992).
12. G. Atassi, *Rev. Silicon, Germanium, Tin and Lead*, **8**, 219–235 (1985).
13. J.S. Thayer, *Appl. Organomet. Chem.*, **1**, 227–234 (1987).
14. R.R. Brutkiewicz and F. Suzuki, *In Vivo*, **1**, 189–204 (1987).
15. S. Goodman, *Med. Hypotheses*, **26**, 207–215 (1988).
16. E. Lukevics, S. Germane and L. Ignatovich, *Appl. Organomet. Chem.*, **6**, 543–564 (1992).
17. P. Köpf-Maier, *Eur. J. Clin. Pharmacol.*, **47**, 1–16 (1994).

18. E. Lukevics and L.M. Ignatovich, in *The Chemistry of Organic Germanium, Tin and Lead Compounds*, Ed. S. Patai, John Wiley & Sons Ltd, 1995, pp. 857–864.
19. S. Maeda, in *The Chemistry of Organic Germanium, Tin and Lead Compounds*, Ed. S. Patai, John Wiley & Sons Ltd, 1995, pp. 871–910.
20. G.B. Gerber and A. Leonard, *Mutat. Res.*, **387**, 141–146 (1997).
21. S.-H. Tao and P.M. Bolder, *Regul. Toxicol. Pharmacol.*, **25**, 211–219 (1997).
22. E. Lukevics and L. Ignatovich, in *The Chemistry of Organic Germanium, Tin and Lead Compounds*, Ed. Z. Rappoport, John Wiley & Sons Ltd, 2002, pp. 1653–1683.
23. M. Tsutsui, N. Kakimoto, D.D. Axtell *et al.*, *J. Am. Chem. Soc.*, **98**, 8287–8289 (1976).
24. S. Nakayama, T. Tsuji and K. Usami, *Showa Ikagakkai Zhasshi*, **46**, 227–235 (1986), *Chem. Abstr.*, **106**, 334 (1987).
25. F. Anger, J.P. Anger, L. Guillou *et al.*, *J. Toxicol. Clin. Exp.*, **11**, 421–436 (1991).
26. F. Anger, J.P. Anger, L. Guillou and A. Papillon, *Appl. Organomet. Chem.*, **6**, 267–272 (1992).
27. T. Sanai, S. Okuda, K. Onoyama *et al.*, *Kidney International*, **40**, 882–890 (1991).
28. A. Hoshi, *Drugs of the Future*, **18**, 905–908 (1993).
29. Y. Sugiya, S. Sakamaki and H. Satoh, *Oyo Yakuri*, **31**, 1191–1200 (1986), *Chem. Abstr.*, **105**, 164632 (1986).
30. Y. Sugiya, K. Yoshida, K. Eda *et al.*, *Oyo Yakuri*, **32**, 139–152 (1986), *Chem. Abstr.*, **105**, 164634 (1986).
31. Y. Sugiya, K. Yoshida, K. Eda *et al.*, *Oyo Yakuri*, **32**, 123–138 (1986), *Chem. Abstr.*, **105**, 164633 (1986).
32. M. Kagoshima and M. Suzuki, *J. Med. Pharm. Sci.*, **15**, 1497–1502 (1986).
33. K. Asano, M. Yamano, K. Haruyama *et al.*, *J. Toxicol. Sci.*, **19**(II), 131–143 (1994).
34. M. Nagata, T. Yoneyama, K. Yanagida *et al.*, *J. Toxicol. Sci.*, **10**, 333–341 (1985).
35. S. Okada, S. Kijama, Y. Oh *et al.*, *Curr. Ther. Res.*, **141**, 265–273 (1987).
36. A.G. Schauss, *Ren. Fail.*, **13**, 1–4 (1991).
37. J.I. Van der Spoel, B.H.C. Stricker, M.R. Esseveld and M.E.I. Schipper, *Lancet*, **336**, 117 (1990).
38. A. Takeuchi, N. Yoshizawa, S. Oshima *et al.*, *Nephron*, **60**, 436–442 (1992).
39. B. Hess, J. Raisin, A. Zimmermann *et al.*, *Am. J. Kidney Dis.*, **21**, 548–552 (1993).
40. A. Shinohara, M. Chiba and Y. Inaba, *J. Anal. Toxicol.*, **23**, 625–631 (1999).
41. H. Brandenbergen and M. Roth, in *Analytical Toxicology for Clinical, Forensic and Pharmaceutical Chemists*, Eds. H. Brandenbergen, R.A.A. Maes and W. de Gruyter, 1997, pp. 285–344.
42. S. Nakai, S. Chittrakarn and T. Togashi, *Oyo Yakuri*, **56**, 1–16 (1998), *Chem. Abstr.*, **130**, 20310 (1999).
43. H. Fukazawa, Y. Ohashi, S. Sekiyama *et al.*, *Head Neck Surg.*, **16**, 30–38 (1994).
44. H. Aso, F. Suzuki, T. Yamaguchi *et al.*, *Microbiol. Immunol.*, **29**, 65–74 (1985).
45. S. Arai, Y. Tomita, T. Munakata *et al.*, *Int. J. Immunother.*, **3**, 97–103 (1987).
46. T. Munakata, S. Arai, K. Kuwano *et al.*, *J. Interferon Res.*, **7**, 69–76 (1987).
47. F. Suzuki, R.B. Brutkiewicz and R.B. Pollard, *Int. J. Immunother.*, **2**, 239–245 (1986).
48. K. Miyao and N. Tanaka, *Drugs of the Future*, **13**, 441–453 (1988).
49. L. Pronai and S. Arimori, *Biotherapy*, **4**, 1–8 (1992).
50. M. Kuwabara, *J. Vet. Sci.*, **55**, 471–473 (1993).

51. K. Ikemoto, M. Kobayashi, T. Fukumoto *et al.*, *Experientia*, **52**, 159–166 (1996).
52. S. Yokochi, Y. Ishiwata, H. Hashimoto *et al.*, *Scand. J. Immunol.*, **48**, 183–191 (1998).
53. H. Kobayashi, H. Aso, N. Ishida *et al.*, *Immunopharm. Immunotoxicol.*, **14**, 841–864 (1992).
54. H. Aso, F. Suzuki, T. Ebina and N. Ishida, *J. Biol. Response Mod.*, **8**, 180 (1989).
55. T. Sato, A. Ishikawa and Y. Ishida, Pat. 834794 Belg. (1976), *Chem. Abstr.*, **86**, 37752y (1977).
56. N. Kumano, T. Ishikawa, S. Koinumaru *et al.*, *Tohoku J. Exp. Med.*, **146**, 97–104 (1985).
57. N. Kumano, Y. Nakai, T. Ishikawa *et al.*, *Am. Soc. Microbiol.*, 1380–1382 (1981).
58. N. Kumano, Y. Nakai, T. Ishikawa *et al.*, *Am. Soc. Microbiol.*, 1525–1527 (1980).
59. E. Lukevics, S.K. Germane, A.A. Zidermane *et al.*, *Khim-Farm Zh.*, **18**, 154–159 (1984), *Chem. Abstr.*, **101**, 130791y (1984) .
60. S. Arimori, K. Watanabe, M. Yoshida and T. Nagao, in *Immunomodulation by microbial product and related synthetic compounds, Proc. Intern. symp., Amsterdam*, 536–539 (1981).
61. M.G. Mainwaring, C. Poor, D.S. Zander and E. Harman, *Chest*, **117**, 591–593 (2000).
62. H. Kobayashi, T. Komuro and H. Furue, *Jap. J. Cancer Chemother.*, **13**, 2588–2593 (1986).
63. A. Ishikawa, Y. Ishida, S. Ikegami *et al.*, Pat. 54-21992 Jpn (1979), *Chem. Abstr.*, **92**, 644n (1980).
64. N. Kakimoto, Pat. 54-147932 Jpn (1979), *Chem. Abstr.*, **92**,185931v (1980).
65. N. Kakimoto and K. Miyao, Pat. 54-160742 Jpn (1979), *Chem. Abstr.*, **93**, 80060c (1980).
66. H. Sato and A. Iwaguchi, *Jap. J. Cancer Chemother.*, **6**, 79–83 (1979).
67. F. Suzuki and R.B. Pollard, *J. Interferon Res.*, **14**, 223–233 (1994).
68. X. Ming, H. Yin and Z. Zhu, *Chung Hua Wai Ko Tsa Chih*, **34**, 221–223 (1996), www.ncbi.nlm.nih.gov:80; PMID: 9387686.
69. Y. Liu, K. Ma, Y. Wang and Z. Chui, *Zhongguo Mianyixue Zazhi*, **11**, 207–210 (1995), *Chem. Abstr.*, **124**, 164456 (1996).
70. Y. Wang, Y. Liu and Z. Chui, *Shengwu Huaxue Yu Shengwu Wuli Jinzhan*, **24**, 39–44 (1997), *Chem. Abstr.*, **128**, 201 (1998).
71. N. Kakimoto, M. Matsui, T. Takada and M. Akiba, *Heterocycles*, **23**, 2681–2684 (1985).
72. H. Aso, E. Shibuya, F. Suzuki *et al.*, *Jap. J. Cancer Chemother.*, **12**, 2345–2351 (1985).
73. T. Sato, Pat. 55-1677222 Jpn (1980), *Chem. Abstr.*, **94**, 185729b (1981).
74. S. Kato and M. Okamoto, Pat. 59-130814 Jpn (1984), *Chem. Abstr.*, **101**, 216415r (1984).
75. S. Kato and M. Okamoto, Pat. 60-01192 Jpn (1985), *Chem. Abstr.*, **102**, 204108a (1985).
76. Tokyo Soda Co., Ltd., Pat. 60-1191 Jpn (1985), *Chem. Abstr.*, **102**, 204107z (1985).
77. K. Takakusaki and H. Kakimoto, Pat. 59-31785 Jpn (1984), *Chem. Abstr.*, **101**, 91237a (1984).

78. K. Takakusaki and H. Kakimoto, Pat. 59-95283 Jpn (1984), *Chem. Abstr.*, **101**, 152092y (1984).
79. F. Jiang and C. Chen, *Zhongguo Yaowu Huaxue Zazhi*, **4**, 1–3 (1994), *Chem. Abstr.*, **123**, 9879j (1995).
80. F. Jiang, M. Liu and Z. Zlao, *Zhongguo Yaowu Huaxue Zazhi*, **5**, 202–204 (1995), *Chem. Abstr.*, **124**, 232624p (1996).
81. Q. Wang, Z. Chen, Q. Zeng and R. Huang, *Heteroatom Chem.*, **10**, 209–212 (1999).
82. Q. Wang, Q. Zeng and Z. Chen, *Heteroatom Chem.*, **10**, 5–8 (1999).
83. M. Jiang, M. Fengchao, K. Lixia and M. Zeng, *Zhongguo Yaowu Huaxue Zazhi*, **7**, 280–282 (1997), *Chem. Abstr.*, **130**, 81583v (1999).
84. F. Jiang and M. Chen, *Guangdong Weiliang Yuansu Kexue*, **5**, 41–44 (1998), *Chem. Abstr.*, **130**, 47172 (1999).
85. K. Kamura, N. Kakimoto and M. Akiba, Pat. 98-0003287 Jpn (1998), *Chem. Abstr.*, **130**, 163195 (1999).
86. C. Han, L. Li, S. Cui *et al.*, *Weisheng Dulixue Zazhi*, **12**, 159–160 (1998), *Chem. Abstr.*, **130**, 263260 (1999).
87. M. Kumegawa and N. Kakimoto, Pat. 94-258811 Jpn (1994), *Chem. Abstr.*, **125**, 49328 (1996).
88. A. Fujii, N. Koboyama, J. Yamane *et al.*, *Gen. Pharmac.*, **24**, 1527–1532 (1993).
89. H. Matsumoto, G.-Z. Jiang, T. Hashimoto *et al.*, *Int. J. Oral-Med. Sci.*, **1**, 10–16 (2002).
90. Y. Wakabayashi, *Biosci. Biotechnol. Biochem.*, **65**, 1893–1896 (2001).
91. J. Chen, G. Zvang and S. Li, *Yingyong Huaxue*, **15**, 55–58 (1998), *Chem. Abstr.*, **129**, 225551 (1998).
92. K. Nakamura, K. Nomoto, K. Kariya *et al.*, *Amino Acids*, **1**, 263–278 (1991).
93. N. Kakimoto, T. Yoshihara, H. Akao and M. Akiba, Pat. 94-657078 Jpn (1994), *Chem. Abstr.*, **124**, 176539 (1996).
94. N.J. Unakar, M. Johnson, J. Tsui *et al.*, *Exp. Eye Res.*, **61**, 155–164 (1995).
95. N.J. Unakar, J. Tsui and M. Johnson, *Curr. Eye Res.*, **16**, 832–837 (1997).
96. L.M. Rice, J.W. Wheeler and C.F. Geschickter, *J. Heterocycl. Chem.*, **11**, 1041–1047 (1974).
97. M.C. Henry, E. Rosen, C.D. Port and B.S. Levine, *Cancer Treat. Rep.*, **64**, 1207–1210 (1980).
98. J.H. Saiers, B. Blumenstein, M. Slavik *et al.*, *Cancer Treat. Rep.*, **71**, 1305–1306 (1987).
99. J. Harvey, M. McFadden, F.P. Smith *et al.*, *Invest. New Drugs*, **8**, 53–56 (1990).
100. M. Slavik, O. Blanc and J. Davis, *Invest. New Drugs*, **1**, 255–234 (1983).
101. M. Slavik and J.H. Saiers, *Cancer*, **11**, 25–30 (1981).
102. M. Slavik, L. Elias, J. Mrema and J.H. Saiers, *Drugs Exp. Clin. Res.*, **8**, 379–385 (1982).
103. J.J. Kavanagh, P.B. Saul, L.J. Copeland *et al.*, *Cancer Treat. Rep.*, **69**, 139–140 (1985).
104. N. Vogelzang, D. Gesme and B. Kennedy, *Am. J. Clin. Oncol.*, **8**, 341–344 (1985).
105. F.H. Dexeus, C. Logothetis, M.L. Samuels and B. Hassan, *Cancer Treat. Rep.*, **70**, 1129–1130 (1986).
106. K.J. Padya, A. Kramar, R.F. Asbury and D.G. Haller, *Am. J. Clin. Oncol.*, **11**, 496–498 (1988).

107. C.K. Mirabelli, A.M. Badger, C.P. Sung *et al.*, *Anticancer Drug Des.*, **3**, 231–242 (1989).
108. E.A. Monnot, C.G. Kindberg, T.S. Johnson *et al.*, *Int. J. Pharm.*, **60**, 41–52 (1990).
109. B.F. Kimler, *J. Neurooncol.*, **20**, 103–109 (1994).
110. R.F. Asbry, A. Cnaan, L. Johnson *et al.*, *Am. J. Clin. Oncol.*, **17**, 166–169 (1994).
111. B.T. Hill, S.A. Whatley, A.S. Bellamy *et al.*, *Cancer Res.*, **42**, 2852–2856 (1982).
112. S.J. Yang and S. Rafla, *Am. J. Clin. Oncol.*, **5**, 331–337 (1983).
113. B.T. Hill and R.D. Whelan, *Invest. New Drugs*, **4**, 359–365 (1986).
114. A.M. Badger, D.A. Schwartz, D.H. Picker *et al.*, *J. Med. Chem.*, **33**, 2963–2970 (1990).
115. P.J. Byrne, P.S. Schein, P. Maguire *et al.*, *Proc. Am. Assoc. Cancer Res., Am. Soc. Clin. Oncol.*, **21**, 351 (1980).
116. P.S. Schein, M. Slavik, T. Smyth *et al.*, *Cancer Treat. Rep.*, **64**, 1051–1056 (1980).
117. T.E. Lad, R.R. Blought, M. Evrard *et al.*, *Invest. New Drugs*, **7**, 223–224 (1989).
118. D.S. Ettinnger, D.M. Finkelstein, M.D. Abeloff *et al.*, *Invest. New Drugs*, **8**, 183–185 (1990).
119. D.S. Ettinnger, D.M. Finkelstein, R.C. Donehower *et al.*, *Med. Pediatr. Oncol.*, **17**, 197–201 (1989).
120. Y. Novik, L.M. Ryan, D.G. Haller *et al.*, *Med. Oncol.*, **16**, 261–266 (1999).
121. L.M. Weiss, M. Wittner, S. Wasserman *et al.*, *J. Infect. Dis.*, **168**, 1289–1292 (1993).
122. E. Lukevics, L. Ignatovich and S. Belyakov, *J. Organomet. Chem.*, **588**, 222–230 (1999).
123. S.S. Karlov and G.S. Zaitseva, *Chem. Heterocycl. Compd.*, **37**, 1325–1357 (2001).
124. E. Lukevics, L.M. Ignatovich and S.K. Germane, *Chem. Heterocycl. Compd.*, **31**, 1241–1248 (1995).
125. E. Lukevics, S.K. Germane, M.A. Trushule *et al.*, *Khim.-Farm. Zh.*, **21**, 1070–1074 (1987), *Chem. Abstr.*, **108**, 68300b (1988).
126. E. Lukevics, L. Ignatovich, N. Porsiurova and S. Germane, *Appl. Organomet. Chem.*, **2**, 115–120 (1988).
127. E. Lukevics, S. Germane, M. Trushule *et al.*, *Latv. PSR Zinat. Akad. Vestis*, **N5**, 79–82 (1988), *Chem. Abstr.*, **109**, 104165t (1988).
128. E. Lukevics, L. Ignatovich, N. Shilina and S. Germane, *Appl. Organomet. Chem.*, **6**, 261–266 (1992).
129. E. Lukevics and L. Ignatovich, *Main Group Met. Chem.*, **17**, 133–144 (1994).
130. L. Ignatovich, S. Belyakov, J. Popelis and E. Lukevics, *Chem. Heterocycl. Compd.*, **36**, 603–606 (2000).
131. E. Lukevics, L. Ignatovich, T. Shul'ga and S. Belyakov, *Main Group Met. Chem.*, **25**, 463–464 (2002).
132. E. Lukevics, L. Ignatovich, T. Shul'ga *et al.*, *J. Organomet. Chem.*, **659**, 165–171 (2002).
133. K. Sato and H. Kakimoto, Pat. 59-31784 Jpn (1984), *Chem. Abstr.*, **101**, 91238b (1984).
134. T.K. Miyamoto, N. Sugita, Y. Matsumoto *et al.*, *Chem. Lett.* 1695–1698 (1983).
135. P. Köpf-Maier, *Cancer Chemother. Pharmacol.*, **23**, 225–230 (1989).
136. P. Köpf-Maier, C. Janik and H. Schumann, *Inorg. Chim. Acta*, **152**, 75–76 (1988).

137. K. Sato and H. Kakimoto, Pat. 59–193897 Jpn (1984), *Chem. Abstr.*, **103**, 22788k (1985).
138. E. Lukevics, L. Ignatovich and I. Shestakova, *Appl. Organomet. Chem.*, **17**, 898–905 (2003).
139. L. Ignatovich, Dz. Zarina, I. Shestakova *et al.*, *Metal-Based Drugs*, **8**, 211–214 (2002).
140. F. Li, Z. Zhang and H. Gao, *Metal-Based Drugs*, **3**, 241–242 (1996).
141. E.Ya. Lukevics, L.M. Ignatovich, A.A. Zidermane and A.Zh. Dauvarte, *Latv.PSR. Zinat. Akad. Vestis, ser. chim.*, 483–486 (1984), *Chem. Abstr.*, **101**, 211331a (1984).
142. E. Lukevics, P. Arsenyan, S. Germane and I. Shestakova, *Appl. Organomet. Chem.*, **13**, 795–798 (1999).
143. E. Lukevics, P. Arsenyan, I. Shestakova *et al.*, *Metal-Based Drugs*, **7**, 63–66 (2000).
144. S. Xuequing, Y. Zhiqiang, X. Qinglan and L. Jinshan, *J. Organomet. Chem.*, **566**, 103–110 (1998).
145. M.A. Choudhary, M. Mazhar, S. Ali *et al.*, *Metal-Based Drugs*, **8**, 275–281 (2002).
146. S.Ya. Melnic, A.A. Bakhmedova, T.P. Nedorezova *et al.*, *Bioorg. Khim.*, **11**, 1248–1252 (1985).
147. S.Ya. Melnic, A.A. Bakhmedova, T.D. Miniker *et al.*, *Bioorg. Khim.*, **10**, 1645–1654 (1984).
148. S. Ninomiya, F. Liu, H. Nakagawa *et al.*, *Chem. Pharm. Bull.*, **34**, 3273–3278 (1986).
149. T. Yamakawa, H. Kagechika, E. Kawachi *et al.*, *J. Med. Chem.*, **33**, 1430–1437 (1990).
150. M. Merget, K. Gunter, M. Bernd *et al.*, *J. Organomet. Chem.*, **628**, 183–194 (2001).
151. R. Tacke, M. Merget, R. Bertermann *et al.*, *Organometallics*, **19**, 3486–3497 (2000).
152. R. Tacke, T. Heinrich, T. Korner *et al.*, *Phosphorus Sulfur Silicon*, **150/151**, 69–87 (1999).
153. J. Satgé, A. Gazes, M. Bouchaut *et al.*, *Eur. J. Med. Chem.-Chem. Ther.*, **17**, 433–436 (1982).
154. M. Fatome, H. Sentenac-Roumanou, C. Lion *et al.*, *Eur. J. Med. Chem.*, **19**, 119–122 (1984).
155. M. Fatome, H. Sentenac-Roumanou, C. Lion *et al.*, *Eur. J. Med. Chem.-Chem. Ther.*, **23**, 257–266 (1988).
156. J. Satgé, G. Rima, M. Fatome *et al.*, *Eur. J. Med. Chem.*, **24**, 48–54 (1989).
157. G. Rima, J. Satgé, M. Fatome *et al.*, *Eur. J. Med. Chem.*, **26**, 291–297 (1991).
158. G. Rima, J. Satgé, H. Sentenac-Roumanou *et al.*, *Appl. Organomet. Chem.*, **10**, 113–123 (1996).
159. G. Rima, J. Satgé, H. Sentenac-Roumanou *et al.*, *Main Group Met. Chem.*, **20**, 255–267 (1997).
160. G. Rima, J. Satgé, R. Dagiral *et al.*, *Metal-Based Drugs*, **5**, 139–146 (1998).
161. G. Rima, J. Satgé, R. Dagiral *et al.*, *Metal-Based Drugs*, **6**, 49–60 (1999).
162. B. Célaries, G. Rima, L. Court *et al.*, *Metal-Based Drugs*, **8**, 199–210 (2001).
163. B. Célaries, C. Amourette, C. Lion, G. Rima, *Appl. Organomet. Chem.*, **17**, 135–138 (2003).
164. R. Tacke, B. Becker, D. Berg *et al.*, *J. Organomet. Chem.*, **438**, 45–55 (1992).
165. Monsanto Co., Pat. 538231 EP (1993), *Chem. Abstr.*, **119**, 160256S (1993).

16 $_{33}$As Metallotherapeutic Arsenic Compounds

Paul C. Ho

16.1 Introduction

Arsenic ($_{33}$As) is one of the most toxic heavy elements derived from the natural environment. The name 'arsenic' is derived from the Greek word 'arsenikon' meaning 'potent'.

Arsenicals (compounds that contain arsenic) have long been known as poisons and paradoxically used as therapeutic agents in the history of medicine.[1] As early as 2000 BC, arsenic trioxide (As_2O_3), obtained from smelting copper, was used as a drug. Hippocrates (460 to 377 BC) used orpiment (As_2S_3) and realgar (As_2S_2) as escharotics (corrosive agents) to tumors or cancerous ulcers. Aristotle (384 to 322 BC) and Pliny the Elder (AD 23 to 79), documented the medicinal properties of arsenicals. Galen (AD 130 to 200) recommended a paste of arsenic sulfide for the treatment of ulcers. Paracelsus (1493 to 1541) used elemental arsenic extensively. He broke the rigid tradition of Galenic medicine and administered realgar internally to treat cancer-like tumors. Paracelsus gave cogency to the belief that there was a therapeutic window for toxic compounds in which clinical benefits can be achieved without extreme toxicity. In the eighteenth century, Fowler's solution (1% potassium arsenite) was used for the relief of various ailments. Arsenicals remained as the mainstay of medicine in the *Materia Medica* in the nineteenth century[2] and were used as

Metallotherapeutic Drugs and Metal-Based Diagnostic Agents: The Use of Metals in Medicine
Edited by Gielen and Tiekink

the remedies for such diverse diseases or complaints as anorexia and other nutritional disturbances, neuralgia, rheumatism, asthma, tuberculosis, diabetes, intermittent fever, skin disorders, hematologic abnormalities, malaria and syphilis.[2,3] At that time, Fowler's solution was so popular that it was mocked in cartoons. Fowler's solution was the primary therapy for chronic myelogenous leukemia (CML) until it was replaced by radiation and cytotoxic chemotherapy in the twentieth century.

In the early twentieth century, physicians still commonly used arsenicals to treat pellagra and malaria.[1] In 1905, Fritz Swchaudinn proved the cause of syphilis to be *Treponema pallidum*. Soon later in 1910, Paul Ehrlich synthesized the arsenical compound Salvarsan or 606 (as the 606th compound tested in his laboratory) and its derivative that became the standard effective therapeutic agents for the treatment of syphilis for nearly 40 years until the introduction of penicillin.[4]

Arsenicals were also administered as vaginal inserts or powder for insufflation for treating vaginal discharge of any origin.[1] They were apparently effective against *Trichomonas*, and were only subsequently replaced by metronidazole in the mid-twentieth century.

Although safer and more effective therapeutic agents have, over the years, replaced most medicinal uses of arsenicals, arsenicals remain important in the chemotherapy of trypanosomal infections.[5] Melarsoprol is the arsenic drug currently approved for the treatment of late-stage African trypanosomiasis.

Arsenicals have been used in Chinese traditional medicine for more than 1000 years. Arsenic trioxide (As_2O_3), also called *Pi shuang*, was recorded in the *Compendium of Materia Medica* (edited by Li Shizhen, 1518–1593). *Pi shuang* and *Xiong Huang* (realgar, As_2S_2) have been used for the treatment of cancer and other conditions.[6] Chinese physicians from Harbin led the revival in the medicinal use of arsenicals as anti-cancer agents in the early 1970s, when they used arsenic trioxide for treating cancers including esophageal carcinoma, malignant lymphoma and leukemia.[7] In particular, notable success was observed in the treatment of acute promyelocytic leukemia (APL) with the characteristic chromosome translocation t(15;17).[8,9] Long-term clinical trials showed that low plasma concentrations of As_2O_3 (1–2 μM) are effective in the treatment of APL,[10] even in the patients resistant to all *trans*-retinoic acid (ATRA) or conventional chemotherapy.[11] Complete remission rates (CRR) in the newly diagnosed and relapsed patients with APL were between 85 and 93%.[9,11–13] More importantly, the side effects associated with arsenic were mild and responded to symptomatic treatment or resolved with dose reduction, while other severe toxic effects, such as myelosuppression commonly associated with conventional chemotherapy, were absent.[9,11–13] After the successful clinical trials, the Food and Drug Administration (FDA) of USA approved arsenic trioxide injections (Trisenox®) in the year 2000 for the treatment of APL in adult patients who fail to respond to other chemotherapy or have relapsed disease.[14] The success of using arsenic in the treatment of APL has prompted researchers to evaluate the potential of arsenic in treating other malignant disorders, including chronic lymphocytic leukemia (CLL),[15] multiple myeloma (MM),[15,16] and other solid tumors such as neuroblastoma,[17] gastric[18] and cervical tumors.[19] This chapter focuses on the uses of arsenic for the treatment of malignancies.

16.2 Chemistry of Arsenic

To comprehend the activity, toxicology and metabolism of arsenic, it is essential to understand its chemistry. Arsenic (atomic number 33, atomic weight 75) is a metalloid of the nitrogen group of the periodic table. It can exist in three different valency states, including elemental (zero oxidation), trivalent (As^{III}) and pentavalent (As^{V}). Arsenic forms alloys with metals and also readily reacts with carbon, oxygen and hydrogen, forming covalent bonds.[1] Trivalent arsenicals are more toxic than the pentavalent ones. The toxicity of an arsenical also varies with the physical state of the compound, and the rate of absorption and elimination. The pentavalent arsenic compounds include derivatives of the arsenic acid, H_3AsO_4 (the arsenates). The trivalent inorganic arsenic compounds are derivatives of the arsenious acid, $H_2As_2O_3$ (the arsenites) and arsenic trioxide (As_2O_3). The organic arsenicals that contain (a) covalent bond(s) between arsenic and carbon are often formed from inorganic arsenic compounds through metabolism. They are generally excreted more rapidly from the body than the inorganic compounds.[1] The chemical structures of the trivalent and pentavalent inorganic arsenicals and the methylated metabolites are shown in Table 16.1.

Table 16.1 Inorganic As^{III} and As^{V} compounds and the methylated metabolites

Chemicals with symbolic name	Structure	pK_a
iAs^{III} (arsenite)	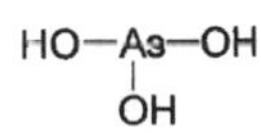	9.2 13.5
iAs^{V} (arsenate)	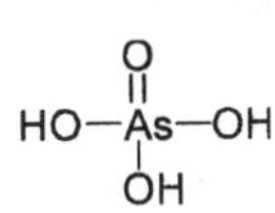	2.3 6.9 11.4
MMA (monomethylarsonic acid)	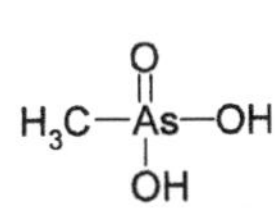	3.6 8.2
DMA (dimethylarsinic acid)	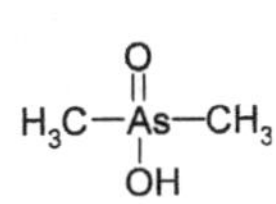	9.3

16.3 Mechanisms of Action

Arsenic causes cellular alterations via numerous pathways, including induction of apoptosis, inhibition of proliferation, stimulation of differentiation, and inhibition of angiogenesis[20] (Figure 16.1). The fundamental biological effects of arsenic (mainly the trivalent forms, arsenite and As_2O_3) may be mediated by reactions with closely spaced cysteine residues on critical cell proteins.[21] The subsequent anti-carcinogenic or carcinogenic actions of arsenic could be due to combined molecular mechanisms that are related to length of arsenic exposure (e.g. acute vs chronic), level of exposure (low vs high doses) and/or species of arsenic exposure (e.g. arsenite, arsenate, MMA, DMA, etc.). Mechanisms of the anti-cancer activities could also depend on the cell types.[22] In recent years, when arsenic was reintroduced to the modern medicine for the treatment of APL, much attention has been focused on the mechanisms of action for this leukemia. *In vitro* and *in vivo* experiments suggested that induction of apoptosis and partial differentiation of APL cells are likely to constitute the cellular effects of As_2O_3. The induction of apoptosis and partial differentiation by arsenic are mediated via multiple and complex pathways.

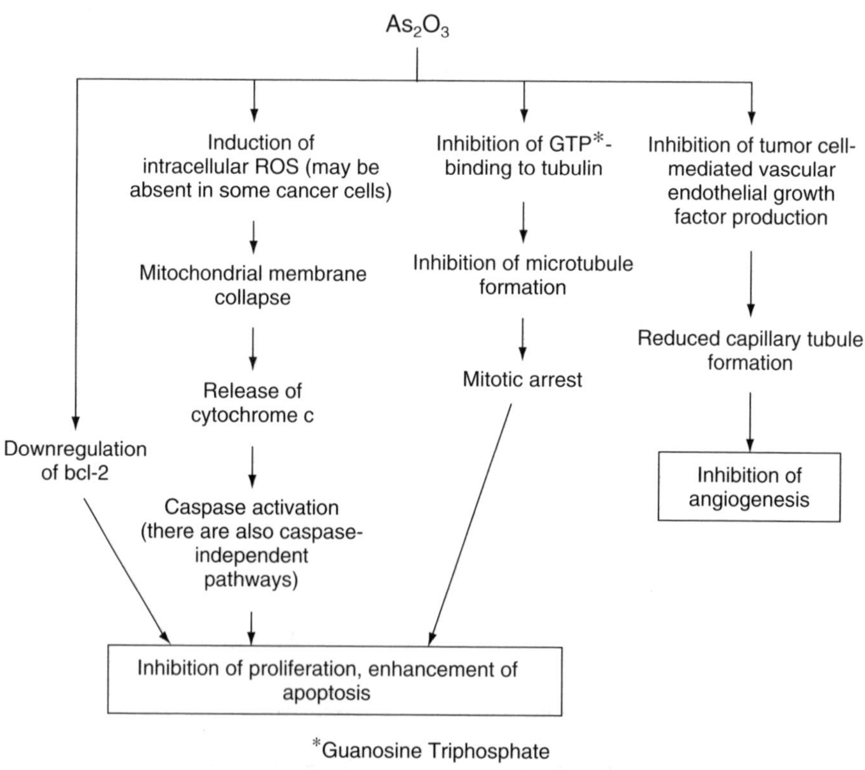

Figure 16.1 Mechanisms of action of As_2O_3 in the treatment of malignancies

16.3.1 Modulation of PML and PML–RARα genes

The vast majority of cases of APL are characterized by the t(15; 17) translocation. This translocation generates a fusion between the *PML* (promyelocytic leukemic) gene and the *RARα* (retinoic acid receptor alpha) gene, which encodes a transcription factor.[23] PML is a tumor suppressor involved in complex functions including growth arrest and apoptosis induction. The PML–RARα fusion protein inhibits PML-dependent apoptotic pathways and blocks myeloid differentiation by direct transcriptional inhibition of retinoic acid (RA) target genes.[24] The block in myeloid differentiation by PML–RARα can be released by treatment with pharmacological levels of RA, providing the basis for ATRA therapy for APL. ATRA is used in treatment of APL to induce differentiation of leukemic promyelocytes.[25] As_2O_3 has been proposed as an alternative to treatment with ATRA, because it can induce complete remission in both ATRA-sensitive and ATRA-resistant APL patients.[26] The anti-leukemic effects of As_2O_3 were mediated by its ability to induce the relocation and degradation of PML, and the degradation of PML–RARα in APL cells.[27–29] Sequence analysis of the *PML* gene has shown the presence of a cysteine-rich region that may be a potential site for interaction with trivalent arsenic. Evidence suggested that in APL patients, RA induces differentiation, whereas arsenic induces both a partial differentiation and apoptosis. Their mechanisms of action are believed to be distinct, but both appear to induce degradation of the PML–RARα fusion protein and their effects may be synergistic.[30]

Although both RA and arsenic induce degradation of PML–RARα, several differences in their action on APL cells have been reported. *In vitro*, ATRA induces differentiation of APL along a granulocytic pathway, whereas, as identified from both *in vitro* and *in vivo* studies, arsenic induces dual differentiation and apoptosis of APL at low concentrations (0.0–0.5 μM), and only apoptosis at high clinically achievable concentrations (1–2 μM).[28,31–33] In APL cells sensitive to ATRA, *in vitro* treatment with arsenic antagonizes ATRA-induced differentiation, whereas treatment with ATRA decreases arsenic-induced apoptosis.[28,34] However, in several models of cells resistant to differentiation by ATRA alone, sub-apoptotic concentrations of arsenic can synergize with ATRA to induce differentiation.[11,34] Gianni *et al.*[35] isolated an As_2O_3-resistant NB4 subline (NB4-AsR). These cells failed to undergo apoptosis, but maintained the partial differentiation response to arsenic, whereas ATRA-triggered differentiation and apoptosis were accelerated in the As_2O_3-treated NB4-AsR cells. In a clinical study, As_2O_3 was found able to induce differentiation in patients with APL who have relapsed after ATRA treatment.[36] This is consistent with a previous *in vitro* study suggesting that APL cell growth arrest and differentiation could occur without PML–RARα degradation,[37] and that preservation of the malignant phenotype may not require a detectable expression of the fusion protein in some APL clones.[38]

One potential important difference in the mechanism by which As_2O_3 and ATRA induce degradation of the PML–RARα fusion protein is that As_2O_3 targets the PML portion of the protein, whereas ATRA targets the RARα.[27] In non-APL cells, arsenic accelerates the targeting of PML onto nuclear bodies (NB) and induces degradation of the PML protein.[27] PML has been shown to suppress growth[39] and induce apoptosis.[40] As_2O_3 alters the traffic of PML proteins, enhancing their association with NB as well as their apoptotic properties.[27,41] NB/PML localization appears to be required for the arsenic-induced anti-proliferative and apoptotic effect. Furthermore, As_2O_3 increases covalent modifications of PML by SUMO-1, a small ubiquitin-like protein. The SUMO-1 polymodified forms of PML are compartmentalized entirely in the PML-NBs in non-APL cells, whereas the unmodified form of PML is found in the soluble nucleoplasmic fraction.[41,42]

16.3.2 Induction of intracellular reactive oxygen species (ROS)

Induction of apoptosis by As_2O_3 could be caused by the production of intracellular ROS.[43] The generation of ROS appears to occur consistently after arsenic exposure,[10,44–46] and the generated ROS subsequently initiates the programmed cell death.[45] The generation of ROS from the As_2O_3-induced apoptosis in U937 myeloid cells could be prevented by the anti-oxidant *N*-acetyl-L-cysteine (NAC).[44] The ROS generated in response to arsenic exposure leads to the accumulation of intracellular hydrogen peroxide (H_2O_2) through activation of flavoprotein-dependent superoxide-producing enzymes such as NADPH oxidase.[45] H_2O_2 acts as a mediator to induce apoptosis through release of cytochrome c to cytosol, activation of CPP32 protease (caspase 3), and PARP (poly (ADP-ribose) polymerase) degradation.[45] H_2O_2 has been suggested to mediate As_2O_3-induced apoptosis in APL-derived NB4 cells. In these cells, apoptosis was preceded by mitochondrial membrane potential disruption, increased H_2O_2 content, and followed by cytochrome c release and caspase-3 activation.[46] Elevating the intracellular ROS levels would sensitize the cells to As_2O_3-induced apoptosis, and the inherent intracellular ROS levels would also determine the tumor cells' susceptibility to As_2O_3.[47,48] Nevertheless, there are controversies in the literature regarding the role of ROS in the arsenic-induced apoptosis. It has been reported that the increased intracellular peroxide levels accompanied with As_2O_3-induced apoptosis could be significantly inhibited by NAC (a thiol-containing anti-oxidant), diphenylene iodonium (an inhibitor of NADPH oxidase), 4,5-dihydro-1,3-benzene disulfonic acid (a selective scavenger of O_2^-) and catalase.[45] However, another study showed that dithiothreitol (DTT) a reducing agent blocked the As_2O_3-induced apoptosis, but did not antagonize the As_2O_3-induced ROS production.[49] Butylated hydroxyanisole, a potent anti-oxidant that can inhibit the production of free radicals, has been shown to exert no effects on the As_2O_3-mediated cytotoxicity in ovarian and prostate cancer cells. It was then suggested that As_2O_3-mediated cytotoxicity in these cancer cells is not related to As_2O_3-mediated superoxide radical generation.[50]

16.3.3 Collapse of the mitochondrial membrane potential

Similar to other chemotherapeutic drugs, As_2O_3 can induce an early mitochondrial membrane potential ($\Delta\varphi m$) collapse in intact APL cells,[51] malignant lymphocytes[49] and carcinoma cell lines.[52,53] Evidence suggests that arsenic might act on mitochondria directly to induce apoptosis.[54,55] As_2O_3 induces a decrease in the $\Delta\varphi m$ via its action on mitochondrial permeability transition (MPT) pores.[56] The permeability transition pore complex (PTPC) is a multi-protein complex that interacts with the apoptosis-regulating protein bcl-2/bax family.[54,57] During apoptosis, the PTPC of the mitochondrial membrane interacts with the bcl-2/bax complex, leading to a progressive permeabilization, with dissipation of the $\Delta\varphi m$.[54] Unlike conventional chemotherapeutic agents, As_2O_3 acts directly on mitochondria to destroy the mitochondrial inner trans-membrane potential to promote apoptosis. DTT can effectively block the As_2O_3-induced $\Delta\varphi m$ collapse and apoptosis in APL cells.[10,49,51] It is known that arsenite at low concentrations selectively binds to the closely spaced (vicinal) thiol (SH) groups in proteins. The function of the adenine nucleotide translator, a fundamental component of the MPT multi-protein complex, is regulated by the redox state of the vicinal SH groups. When these SH groups form a disulfide bond, the MPT pores are open and the $\Delta\varphi m$ is disrupted, whereas reduction of the disulfide bond returns the MPT pores to a closed conformation. Therefore, it is presumed that As_2O_3-induced $\Delta\varphi m$ collapse and apoptosis are associated with dithiol oxidation or cross-linking.

Arsenic trioxide could also disrupt mitochondria through other indirect mechanisms. As_2O_3 treatment of malignant cells results in increased intracellular levels of H_2O_2, which lowers the mitochondrial membrane potential and leads to cytochrome c release and subsequent activation of the caspase pathway.[45]

16.3.4 Roles of glutathione

A decrease in the glutathione (GSH) level that increases the intracellular level of ROS predisposes cells to DNA damage. It has been reported that arsenic-induced apoptosis appeared to be associated with changes in GSH levels.[49,51,55] GSH appears to be a key player in determining the consequences of arsenic exposure including whether apoptosis will take place. The sensitivity of various tumor cell lines to As_2O_3-induced growth inhibition and apoptosis has been shown to be inversely related to the intracellular GSH concentration.[10] As arsenic binds to sulfhydryl group–containing GSH, in cells with a low GSH content, the capacity to eliminate ROS will be decreased. This results in the increase in ROS level, which causes the activation of caspase-3.[45,58] The antioxidant NAC, which increased the synthesis of GSH, completely suppressed As_2O_3-induced apoptosis.[17] In contrast, buthionine sulfoximine (BSO), a depleter of cellular GSH, and ascorbic acid, which decreased the intracellular GSH content, substantially enhanced the effect of As_2O_3.[10,46,49,51] In addition, the increased resistance to arsenic displayed by NB4/As cells may be related to a high intracellular GSH level because reduction of the GSH level by BSO completely restored the sensitivity of these cells to arsenic.[55] Therefore, the intracellular GSH level could be a good indicator for the sensitivity of various cancer cells to arsenic.

16.3.5 Down-regulation of bcl-2

It has been proposed that As_2O_3 promotes apoptosis through the modulation of bcl-2. In the NB4 cell line, As_2O_3 downregulated bcl-2 expression at both the mRNA and the protein levels, but did not influence *bax*, *bcl-x*, *c-myc* and *p53* gene expressions.[31] Exposure to As_2O_3 also downregulated bcl-2 RNA and protein expression in other cell lines.[31,59–61] In HL60 and some breast cancer cell lines, in which As_2O_3 failed to promote apoptosis, no modification of bcl-2 protein levels has been detected.[62] Bcl-2 overexpression attenuated As_2O_3-induced apoptosis effect in U937 cells[44] and HeLa cells,[19] suggesting the involvement of bcl-2 in the regulation of arsenic-mediated apoptosis. However, the overexpression of bcl-2 in the bcl-2-transfected ARP-1 and 8226 cell lines and the high bcl-2 expressing myeloma cells (U266 and ARH-77) did not protect these cells against As_2O_3-induced apoptosis, indicating that As_2O_3 is a potent inducer of apoptosis in myeloma cells and working in a bcl-2 independent fashion.[63]

16.3.6 Involvement of p53

Although As_2O_3 does not increase p53 or activity in most cells, a marked increase in p53 protein level was seen during As_2O_3-induced apoptosis in a human gastric cancer cell line, while co-incubation with the p53 anti-sense oligonucleotide suppressed both p53 overexpression and apoptosis induced by As_2O_3.[64] In MBC-1, a B-cell lymphoma line, As_2O_3 exposure caused upregulation of p53 expression, resulting in caspase activation and, ultimately, apoptosis.[65] p53 accumulation was also implicated in the mechanism by which As_2O_3 treatment induced apoptosis and G_1-phase arrest in human T-cell lymphotropic virus type 1-infected cells.[66]

However, the role of p53 in arsenic-related cellular effects is still obscure. No difference in sensitivity to arsenic was found between fibroblast bearing p53 and fibroblast deficient in this gene, strongly suggesting that p53 is not involved.[67] Hence, the relationship of p53 and arsenic in carcinogenesis or in cancer therapy could be cell- and tissue-type-specific and depends on dose and cellular environment.[22]

16.3.7 Activation of caspases

Apoptosis is mediated by caspases, a class of cysteine proteases that cleave key cellular proteins to induce morphological features of nuclear condensation and fragmentation. Arsenic exposure causes the activation of caspases both *in vitro* and *in vivo*. Clinical trials with relapsed APL patients showed that responses to As_2O_3 were accompanied by enhanced expression of proenzymes for caspases 2 and 3, and activation of caspases 1 and 2.[36] The activation of caspase 3 and cleavage of its specific substrates such as poly (ADP-ribose) polymerase (PARP) could be seen in most cells treated with As_2O_3.[46,51,68,69] In addition to caspase 3, As_2O_3 also activated caspase 8 and Bid in a GSH concentration-dependent manner in NB4 cells.[55] Inhibition of caspase 8 blocked not only the activation of caspase 3 but also the loss of $\Delta\varphi m$. However, caspase 3 was not activated in 1 μM As_2O_3-treated BJAB (B-cell lymphoma), Molt-4 (acute lymphocytic leukemia) and SKW-3 (chronic lymphocytic leukemia) cells, despite substantial apoptosis.[49] This observation implied that different downstream cell-death effector molecules of the $\Delta\varphi m$ collapse may contribute to As_2O_3-induced apoptosis, and caspase-3 is not always necessarily involved.

16.3.8 Other mechanisms of arsenic-induced apoptosis

As_2O_3 may additionally affect tumor-cell growth by inhibiting angiogenesis, the formation of blood vessels that support the tumor growth.[70] In mice with methylcholanthrene-induced fibrosarcomas, a single dose of As_2O_3 was sufficient to induce preferential vascular shutdown in tumor tissue and massive necrosis in the central portion of the tumor, whereas skin, muscle and kidney vasculature remained relatively unaffected.[71]

The cytoskeleton has also been suggested as a potential cellular target for arsenic because its major constituent, tubulin, has a relatively high sulfhydryl content. Arsenic-induced growth inhibition appeared to be related to changes in the cytoskeleton, leading to a decreased production or secretion of interleukin-2 (IL-2).[72] Disruption of microtubule assembly and spindle formation during mitosis can promote apoptosis. In myeloid leukemic cells, As_2O_3 markedly inhibited GTP-induced polymerization of monomeric tubulin for microtubule formation.[73] Although As_2O_3 effectively induced apoptosis in ovarian and prostate cancer cells, addition of the strong anti-oxidant butylated hydroxyanisole did not rescue the cells from As_2O_3-induced apoptosis.[50] This showed that the cytotoxic effects of As_2O_3 were not mediated by superoxide generation in these cells. It was proposed that apoptosis is induced in these cells via an alternate mechanism, perhaps involving interactions with tubulin or other cytoskeletal elements.[50]

16.4 Pharmacokinetic Profiles

16.4.1 Absorption, distribution and excretion

Arsenicals are readily absorbed orally. The rate of absorption depends on the solubility and the chemical species of the arsenic compound. In rat, the oral absorption of As_2O_3 is mediated through an electrogenic process in the small intestine, involving a proton gradient.[74] The best pH for arsenic absorption is 5.0.[75] In patients with relapsed acute myeloid leukemia (AML) or APL, the systemic availability of oral As_2O_3 was found comparable to that after iv dosing.[76] In APL patients treated with intravenous infusion of As_2O_3 (10 mg/day) over 2–3 h, the maximal plasma concentration (C_{max}) was between 5.54 and 7.30 μmol/l and the time to maximal plasma concentration (t_{max}) was around 4 h.[11,77] After the maximal concentration, plasma arsenic was rapidly eliminated. The plasma distribution half-life ($t_{1/2\alpha}$) and the elimination half-life were 0.89 ± 0.29 h and 12.13 ± 3.31 h, respectively.[77] The apparent distribution volume (V_c) was 3.83 ± 0.45 l. The systemic clearance (CL_s) was 1.43 ± 0.17 l/h. Continuous administration of As_2O_3 did not alter its pharmacokinetic behavior. The 24-h arsenic content in urine accounted to 1–8% of the daily dose (10 mg). During As_2O_3 treatment, arsenic accumulation in hair and nail increased continuously; the peak concentration could be fivefold to sevenfold higher than that of pre-treatment, reaching 2.5–2.7 μg/g tissue at CR.[11,77] After withdrawal of the drug, arsenic contents in both urine and hair or nail declined gradually. No bone marrow suppression or severe organ impairment was observed.

16.4.2 Metabolism

The absorbed arsenic undergoes hepatic biomethylation to form monomethylarsonic acid (MMA) and dimethylarsinic acid (DMA) that are less toxic but not totally innocuous.[78,79] Methylation of arsenic involves a stepwise process of two-electron reduction of the pentavalent arsenic species (e.g. As^V, MMA^V and DMA^V) to the trivalent arsenic species (e.g. As^{III}, MMA^{III} and DMA^{III}), followed by the oxidative addition of a methyl group to the trivalent arsenic species.[80] The postulated pathway for biomethylation of arsenic is shown in Figure 16.2. GSH acts as the reducing agent, and S-adenosyl methionine (SAM) as the methyl donor. DMA^V is the dominant urinary metabolite (55–80%) compared with MMA^V (10–20%) and inorganic arsenic species (10–30%).[81–84]

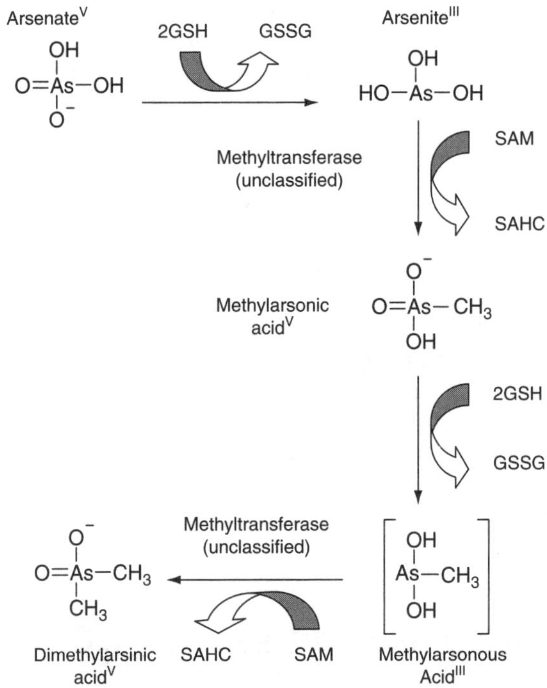

Figure 16.2 Possible metabolic pathways of inorganic arsenic. GSH – Glutathione; GSSG – Oxidized glutathione; SAM – S-adenosyl methionine; SAHC – S-adenosyl homocysteine

After acute poisoning, the highest concentration of arsenic is in the kidney and liver;[85] whereas after chronic arsenic ingestion, arsenic accumulates in the liver, kidneys, heart and lungs, and smaller amounts in muscles, nervous system, gastrointestinal tract and spleen.[85] Though most arsenic is cleared from these sites after withdrawal, residual amounts remain in the keratin-rich tissues such as nails, hair and skin.

Trivalent arsenicals, including sodium arsenite and the more soluble As_2O_3, inhibit many enzymes by reacting with biological ligands that possess available sulfur groups. In all organs, As^{III} is the main form of arsenic bound to tissues, compared with As^{V} and the methylated metabolites, which show very weak affinity for tissues and therefore are rapidly eliminated in bile, blood and urine.[85,86]

16.5 Toxic Side Effects of As_2O_3

In the treatment of APL patients, the side effects of As_2O_3 appear to be mild. In contrast to many other chemotherapeutic agents, it causes no myelosuppression.

Instead, As_2O_3 induces leukocytosis in about 50% of patients.[11,13,36] The leukocytosis can resolve in all cases without chemotherapy.[87] The APL patients on As_2O_3 can also develop retinoic acid syndrome (RAS)-like symptoms such as fever, skin rash and edema, which can be readily relieved by steroid administration.[88] Other mild effects were reported in about 40–50% of relapsed patients, including fatigue, fever, edema, nausea, anorexia, diarrhea, emesis, headache, insomnia, cough, dyspnea, dermatitis, tachycardia, pain, hypokalemia, hypomagnesemia and hyperglycemia. The most common (>10%) Grade 3 and 4 adverse events were abdominal pain, epistaxis, dyspnea, hypoxia, bone pain, thrombocytopenia, neutropenia, hypokalemia and hyperglycemia.[11,36,89] In a clinical trial, prolonged QT intervals (the time intervals for the contraction of the ventricle from the beginning of the Q wave to the end of T wave; a prolonged QT interval indicates cardiac toxicity) were observed in all patients during induction therapy with As_2O_3 and ventricular premature contractions were noticed during eight of the twelve courses of therapy.[90] QT prolongation was also observed in 38% of patients in another clinical study.[89] The prolonged QT intervals could return to baseline following cessation of As_2O_3.

16.6 Indications and Uses of As_2O_3

The FDA has approved As_2O_3 (Trisenox®, 1 mg/ml in 10 ml) for the treatment of patients with APL who have not responded to, or have relapsed following the use of ATRA and anthracycline-based chemotherapy, which is considered as the first-line therapy. For induction treatment, Trisenox® should be administered intravenously at a dose of 0.15 mg/kg daily until bone marrow remission. Total induction dose should not exceed 60 doses. Consolidation treatment should begin three to six weeks after completion of induction therapy. Trisenox® should be administered intravenously at a dose of 0.15 mg/kg daily for 25 doses over a period up to 5 weeks.[91]

16.7 Summary and Conclusions

In the last decade, the efficacy of As_2O_3 has been established in both newly diagnosed and relapsed APL patients. Its common adverse effects are mild and myelosuppression is minimal. The main mechanisms of action of As_2O_3 in APL include induction of apoptosis, promotion of differentiation and inhibition of angiogenesis. Currently, As_2O_3 is the recommended treatment for APL patients who have relapses after or are refractory to ATRA therapy. Recent studies indicate that As_2O_3 could also be potentially efficacious in other hematologic neoplasms, including chronic myelogenous leukemia (CML),[92] multiple myeloma

(MM)[16] and myelodysplastic syndrome (MDS).[93] Further clinical trials are required to establish the efficacy of As_2O_3 in these malignancies.

Acknowledgements

The research work on arsenicals was supported by a grant (RP148-000-032-112) from the National University of Singapore.

References

1. D.M. Jolliffe, *J. Royal Soc. Med.*, **86**, 287–289 (1993).
2. S. Waxman, K.C. Anderson, *Oncologist*, **6**(S2), 3–10 (2001).
3. M.S. Tallman, *Blood Rev.*, **15**, 133–142 (2001).
4. A.S. Lyon, R.J. Petrucelli, *Medicine, An Illustrated History*, Macmillan Co. Aust., Pty. Ltd, Australia, 1979.
5. B. Bouteille, O. Oukem, S. Bisser, M. Dumas, *Fundam. Clin. Pharmacol.*, **17**, 171–181 (2003).
6. Jiang Su New Medical College, *Encyclopedia of Chinese Medicine*, pp. 1620–1622. Shanghai, China, Shanghai Scientific Publishing House, 1986.
7. Z. Wang, *Cancer Chemother. Pharmacol.*, **48**(S1), S72–S76 (2001).
8. R.A. Larson, K. Kondo, J.W. Vardiman *et al.*, *Am. J. Med.*, **76**, 827 (1984).
9. P. Zhang, S.Y. Wang, L.H. Lu *et al.*, *Chin. J. Hematol.*, **17**, 58–62 (1996).
10. J. Dai, R.S. Weinberg, S. Waxman, Y. Jing, *Blood*, **93**, 268–277 (1999).
11. Z.X. Shen, G.Q. Chen, J.H. Ni *et al.*, *Blood*, **89**, 3354–3360 (1997).
12. H.D. Sun, L. Ma, X.C. Hu, T.D. Zhang, *Clin. J. Comb. Trad. Med. West Med.*, **12**, 170–171 (1992).
13. C. Niu, H. Yan, T. Yu *et al.*, *Blood*, **94**, 3315–3524 (1999).
14. D. Bradley, *PSTT*, **3**, 401 (2000).
15. A.J. Murgo, *Oncologist*, **6**(S2), 22–28 (2001).
16. N.J. Bahlis, J. McCafferty-Grad, I. Jordon-McMurry *et al.*, *Clin. Cancer Res.*, **8**, 3658–3668 (2002).
17. Y. Akao, H. Yamada, Y. Nakagawa, *Leuk. Lymphoma*, **37**, 53–63 (2000).
18. T.C. Zhang, E.H. Cao, J.F. Li *et al.*, *Eur. J. Cancer*, **35**, 1258–1263 (1999).
19. S.H. Woo, I.C. Park, M.J. Park *et al.*, *Int. J. Oncol.*, **21**, 57–63 (2002).
20. W.H. Miller Jr, H.M. Schipper, J.S. Lee *et al.*, *Cancer Res.*, **62**, 3893–3903 (2002).
21. E.T. Snow, *Pharmacol Ther.*, **53**, 31–65 (1992).
22. M.J. McCabe, K.P. Singh, S.A. Reddy *et al.*, *J. Pharmcol. Exp. Ther.*, **295**, 724–733 (2000).
23. J. Borrow, A.D. Goddard, D. Sheer, E. Solomon, *Science*, **249**, 1577–1580 (1990).
24. J.L. Slack, M.E. Rusiniak, *Ann. Hematol.*, **79**, 227–238 (2000).
25. M.B. Hansen, C.C. Sigman, F. Andreola *et al.*, *Carcinogenesis*, **21**, 1271–1279 (2000).
26. E.M. Rego, L.Z. He, R.P. Warrell Jr *et al.*, *Proc. Natl. Acad. Sci. USA*, **97**, 10173–10178 (2000).

27. J. Zhu, M.H. Koken, F. Quignon *et al.*, *Proc. Natl. Acad. Sci. USA*, **94**, 3979–3983 (1997).
28. W. Shao, M. Fanelli, F.F. Ferrara *et al.*, *J. Natl. Cancer Inst.*, **90**, 124–133 (1998).
29. A.T. Look, *J. Natl. Cancer Inst.*, **90**, 86–88 (1998).
30. V. Lallemand-Breitenbach, M.C. Guillemin, A. Janin *et al.*, *J. Exp. Med.*, **189**, 1043–1052 (1999).
31. G.Q. Chen, J. Zhu, X.G. Shi *et al.*, *Blood*, **88**, 1052–1061 (1996).
32. G.Q. Chen, X.G. Shi, W. Tang *et al.*, *Blood*, **89**, 3345–3353 (1997).
33. K. Kinjo, M. Kizaki, A. Muto *et al.*, *Leukemia*, **14**, 431–438 (2000).
34. Y. Jing, L. Wang, L. Xia *et al.*, *Blood*, **97**, 264–269 (2001).
35. M. Gianni, M.H. Koken, M.K. Chelbi-Alix *et al.*, *Blood*, **91**, 4300–4310 (1998).
36. S.L. Soignet, P. Maslak, Z.G. Wang *et al.*, *N. Engl. J. Med.*, **339**, 1341–1348 (1998).
37. C. Nervi, F.F. Ferrara, M. Fanelli *et al.*, *Blood*, **92**, 2244–2251 (1998).
38. M. Fanelli, S. Minucci, V. Gelmetti *et al.*, *Blood*, **93**, 1477–1481 (1999).
39. Z.M. Mu, K.V. Chin, J.H. Liu *et al.*, *Mol. Cell. Biol.*, **14**, 6858–6867 (1994).
40. F. Quignon, F. De Bels, M. Koken *et al.*, *Nat. Genet.*, **20**, 259–265 (1998).
41. S. Muller, M.J. Matunis, A. Dejean, *EMBO J.*, **17**, 61–70 (1998).
42. T. Sternsdorf, E. Puccetti, K. Jensen *et al.*, *Mol. Cell. Biol.*, **19**, 5170–5178 (1999).
43. K. Iwama, S. Nakajo, T. Aiuchi, K. Nakaya, *Int. J. Cancer*, **92**, 518–526 (2001).
44. Y.J. Choi, J.W. Park, S.I. Suh *et al.*, *Int. J. Oncol.*, **21**, 603–610 (2002).
45. Y.C. Chen, S.Y. Lin-Shiau, J.K. Lin, *J. Cell. Physiol.*, **177**, 324–333 (1998).
46. Y. Jing, J. Dai, R.M. Chalmers-Redman *et al.*, *Blood*, **94**, 2102–2111 (1999).
47. J. Yi, F. Gao, G. Shi *et al.*, *Apoptosis*, **7**, 209–215 (2002).
48. J. Yi, F. Gao, G. Shi *et al.*, *Chin. Med. J.*, **115**, 603–606 (2002).
49. X.H. Zhu, Y.L. Shen, Y.K. Jing *et al.*, *J. Natl. Cancer Inst.*, **91**, 772–778 (1999).
50. R. Uslu, U.A. Sanli, C. Sezgin *et al.*, *Clin. Cancer Res.*, **6**, 4957–4964 (2000).
51. X. Cai, Y.L. Shen, Q. Zhu *et al.*, *Leukemia*, **14**, 262–270 (2000).
52. Z.Y. Shen, L.J. Tan, W.J. Cai *et al.*, *Int. J. Mol. Med.*, **4**, 33–37 (1999).
53. J.G. Seol, W.H. Park, E.S. Kim *et al.*, *Int. J. Oncol.*, **18**, 249–255 (2001).
54. G. Kroemer, H. de Thé, *J. Natl. Cancer Inst.*, **91**, 743–745 (1999).
55. K. Kitamura, Y. Minami, K. Yamamoto *et al.*, *Leukemia*, **14**, 1743–1750 (2000).
56. A. Roulston, C. Reinhard, P. Amiri, L.T. Williams, *J. Biol. Chem.*, **273**, 10232–10239 (1998).
57. P. Costantini, E. Jacoto, D. Decaudin, G. Kroemer, *J. Natl. Cancer J.*, **92**, 1042–1053 (2000).
58. A.M. DiPietrantonio, T. Hsieh, J.M. Wu, *Biochem. Biophys. Res. Commun.*, **255**, 477–482 (1999).
59. Y. Akao, H. Mizoguchi, S. Kojima *et al.*, *Br. J. Haematol.*, **102**, 1055–1060 (1998).
60. W.H. Park, J.G. Seol, E.S. Kim *et al.*, *Cancer Res.*, **60**, 3065–3071 (2000).
61. A. Konig, L. Wrazel, R.P. Warrell Jr *et al.*, *Blood*, **90**, 562–570 (1997).
62. M. Lu, J. Levin, E. Sulpice *et al.*, *Exp. Hematol.*, **27**, 845–852 (1999).
63. Y. Gazitt, D. Douer, Q. Liu, *Blood*, **98**, 159a–160a (2001).
64. X.H. Jiang, B.C. Wong, S.T. Yuen *et al.*, *Int. J. Cancer*, **91**, 173–179 (2001).
65. Z.Y. Shen, J. Shen, W.J. Cai *et al.*, *Int. J. Mol. Med.*, **5**, 155–158 (2000).
66. K. Ishitsuka, S. Hanada, K. Uozumi *et al.*, *Leuk. Lymphoma*, **37**, 649–655 (2000).
67. C. Huang, W.Y. Ma, J. Li, Z. Dong, *Cancer Res.*, **59**, 3053–3058 (1999).

68. Y. Akao, Y. Nakagawa, K. Akiyama, *FEBS Lett.*, **455**, 59–62 (1999).
69. S. Huang, C.F. Huang, T. Lee, *Biochem. Pharmacol.*, **60**, 771–780 (2000).
70. G.J. Roboz, S. Dias, G. Lam *et al.*, *Blood*, **96**, 1525–1530 (2000).
71. Y.S. Lew, S.L. Brown, R.J. Griffin *et al.*, *Cancer Res.*, **59**, 6033–6037 (1999).
72. L. Vega, P. Ostrosky-Wegman, T.I. Fortoul *et al.*, *Immunopharmacol. Immunotoxicol.*, **21**, 203–220 (1999).
73. Y.M. Li, J.D. Broome, *Cancer Res.*, **59**, 776–780 (1999).
74. M.J. Gonzalez, M.V. Aguilar, M.C. Martinez, *J. Trace Elem. Med. Biol.*, **11**, 239–247 (1997).
75. R.N. Ratnaike, *Postgrad. Med. J.*, **79**, 391–396 (2003).
76. C.R. Kumana, W.Y. Au, N.S.L. Lee *et al.*, *Eur. J. Clin. Pharmcol.*, **58**, 521–526 (2002).
77. J.H. Ni, G.Q. Chen, Z.X. Shen *et al.*, *Chin. Med. J.*, **111**, 1107–1110 (1998).
78. D.J. Thompson, *Chem. Biol. Interact.*, **88**, 89–114 (1993).
79. H.V. Aposhian, *Ann. Rev. Pharmacol. Toxicol.*, **37**, 397–419 (1997).
80. K.T. Suzuki, T. Tomita, Y. Ogra, M. Ohmichi, *Chem. Res. Toxicol.*, **14**, 1604–1611 (2001).
81. J.P. Buchet, R. Lauwerys, in *Arsenic Exposure and Health*, W.R. Chappell, C.O. Abernathy, C.R. Cothern (Eds), Science and Technology Letters, Norwood, 1994.
82. J.P. Buchet, R. Lauwerys, H. Roels, *Int. Arch. Occup. Environ. Health*, **48**, 71–79 (1981).
83. M. Vahter, in *Arsenic Exposure and Health Effects*, W.R. Chappell, C.O. Abernathy, R.L. Calderon (Eds), Elsevier, Amsterdam, 1999.
84. C. Hopenhayn-Rich, A.H. Smith, H.M. Goeden, *Environ. Res.*, **60**, 161–177 (1993).
85. L. Benramdane, M. Accominotti, L. Fanton *et al.*, *Clin. Chem.*, **45**, 301–306 (1999).
86. G.M. Bogdan, A. Sampayo-Reyes, H.V. Aposhian, *Toxicol.*, **93**, 175–193 (1994).
87. L.H. Camacho, S.L. Soignet, S. Chanel *et al.*, *J. Clin. Oncol.*, **18**, 2620–2625 (2000).
88. C.P. Lin, M.J. Huang, I.Y. Chang, W.Y. Lin, *Leuk. Lymphoma*, **38**, 191–194 (2000).
89. M.H. Cohen, S. Hirschfeld, F.H. Susan *et al.*, *Oncologist*, **6**, 4–11 (2001).
90. K. Ohnishi, H. Yoshida, K. Shigeno *et al.*, *Ann. Intern. Med.*, **133**, 881–885 (2000).
91. C.K. McEvoy, *AHFS Drug Information*, American Society of Health-System Pharmacy, Bethesda, 2003.
92. C. Perkins, C.N. Kim, G. Fang, K.N. Bhalla, *Blood*, **95**, 1014–1022 (2000).
93. A. List, M. Beran, J. DiPersio *et al.*, *Leukemia*, **17**, 1499–1507 (2003).

17 $_{34}$Se The Use of Selenium-Based Drugs in Medicine

Michael Carland and Tahli Fenner

17.1 Introduction to Selenium and its Biochemistry

First isolated in 1817 by J.J. Berzelius,[1] selenium (Se) is a group 16 metalloid, lying directly below sulfur in the periodic table. Like sulfur, selenium may be found in the oxidation states of 0, −2 (selenide), +4 (selenite) and +6 (selenate).[2–4] It is the 73rd most common element in the earth's crust,[5] making it the least abundant element that has been shown to possess a defined role in human biology.[4]

The toxic effects of selenium had been reported as early as 1295. In China, Marco Polo described the symptoms of acute selenosis (blind staggers)[4] that was suffered by livestock grazing on plants now known to be selenium accumulators.[6] Despite this, selenium was shown to be an essential trace element in mammals in 1957[7] although it was 16 years before its roles in mammalian biology were defined.

It is now understood that a healthy human diet requires around 55–70 µg of selenium per day[8] although as much as 10 times this amount may be ingested

Metallotherapeutic Drugs and Metal-Based Diagnostic Agents: The Use of Metals in Medicine
Edited by Gielen and Tiekink

without ill-effect.[9] A daily intake in excess of 900 μg per day has been shown to trigger the onset of chronic selenosis, the symptoms of which may include gastrointestinal upsets, hair loss, malformed nails and mild nerve damage.[10,11]

As its position on the periodic table suggests, selenium chemistry has a great deal in common with sulfur chemistry. This fact is exploited by living systems, which often use the same enzymes as those used for inorganic sulfur metabolism to metabolise inorganic selenium.[6] Mammals reduce selenite to the key selenium metabolite selenide (H_2Se) with the enzymes glutathione reductase or thioredoxin reductase (TrxR). Excess selenide may then be successively methylated to yield the volatile (and malodorous) methylselenol and dimethylselenide (Figure 17.1).[12] These compounds are excreted through the lungs, leading to pungent breath in people with excessive selenium in their diets.[4] Further methylation produces the trimethylselenonium ion, which is soluble in water and may be excreted in urine.[12,13] These metabolic pathways enable the removal of excess selenium from the body.

The amino acid selenocysteine (SeCys) is the principal form by which mammals exploit selenium chemistry for biological activity,[6,14] and its widespread presence across most living kingdoms has seen it dubbed the 21st amino acid.[15] Birringer *et al.*[6] provide an excellent discussion of selenium-containing metabolites and pathways in a recent review.

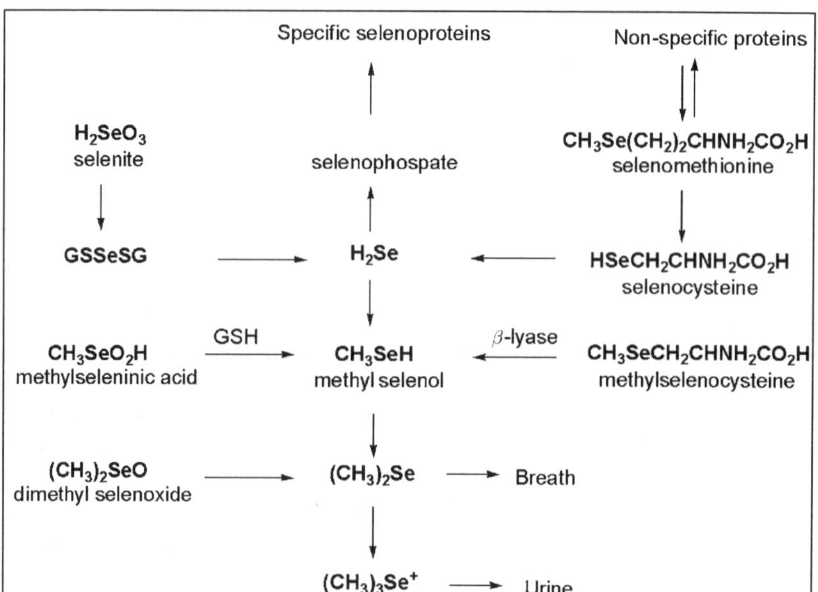

Figure 17.1 Pathways of selenium metabolism

The first defined role for selenium in mammals was in the enzyme glutathione peroxidase (GPx), which was described in 1973.[16] The chief role of GPx is to catalyse the reduction of potentially harmful lipid peroxides to alcohols where glutathione is the reducing agent.[17] Since this discovery, some 18 selenium-containing proteins have been detected in humans and the roles of many of them are the subject of much debate.[6] Of those whose roles have been elucidated, GPx, TrxR[18] and iodothyronine deiodoninases[19] are among the best understood in terms of biological function. Mammalian TrxR is associated with a number of important reactions including the reduction of potentially harmful peroxides, disulfide bonds in peptides, regeneration of ascorbic acid (vitamin C) and the reduction of selenite to selenide.[20] In all cases selenium is present in the form of SeCys.

Less well-understood selenoproteins include those with uninspiring names such as selenoprotein P and selenoprotein W. While most do not have universally accepted roles in biology, many are implicated in various processes. Selenoprotein P, for example, is an extraordinary water-soluble glycoprotein containing up to 12 SeCys residues.[6] Several theories regarding its biological function have been postulated, including as a heavy-metal binding agent,[21] blood-borne anti-oxidant,[22] cell death inhibitory factor[23] and an unusual form of selenium transport.[24]

This chapter aims to provide an insight into the therapeutic potential of natural and synthetic selenium-containing compounds that exhibit interesting and beneficial biological activity. It is beyond the scope of this chapter to cover all applications or therapeutics; however, we have selected some interesting examples to focus on. In addition, a number of excellent review papers are available (and cited where appropriate) for more in-depth critiques.

17.2 Diseases Associated with Selenium Deficiency

Dietary deficiency in selenium has been associated with a wide range of conditions, from the development of cardiomyopathy involved in Keshan disease to mood disorders and Parkinson's disease.[25–28] Considering the role of selenium in anti-oxidant enzymes such as GPx, it is not surprising that selenium deficiency is implicated in conditions associated with oxidative stress, such as atherosclerosis.[23,26,27,29] Selenium dietary supplements, sodium selenite and selenium-enriched yeast are available to counter potential deficiency and a number of excellent reviews address the consequences of inadequate selenium intake.[26,27,30] Whilst both sodium selenite and selenomethionine (SeMet) (present in Se-rich yeast) are available for biosynthesis of selenoproteins, SeMet is advocated by some as a preferred source of selenium as it is the form commonly found in food and is less toxic than sodium selenite.[26,31] Excess SeMet can be stored in non-specific body proteins in place of methionine; thus SeMet appears to increase selenium levels more efficiently than sodium selenite (Figure 17.1). However, selenium from SeMet is less bioavailable and SeMet stored in proteins must undergo a longer metabolic pathway to generate H_2Se for use in selenoprotein synthesis than sodium selenite.[26,31]

17.2.1 Asthma

A symptom of asthma is lung inflammation, which has been associated with the generation of hydrogen peroxide as a by-product of increased metabolic activity.[32] Hydrogen peroxide activates a nuclear transcription factor for inflammatory cytokines, NF-κB.[32,33] GPx appears to have a role in the regulation of NF-κB activity. Comparison of selenium-deficient and selenium-supplemented cells showed upregulation and inhibition of NF-κB, respectively. Whether these results are purely due to removal of hydrogen peroxide by GPx remains unclear.[17] Studies have found evidence of a relationship between dietary selenium intake and asthma, and subjects supplemented with dietary sodium selenite showed decrease in asthma symptoms.[26] Recent research by Kim and co-workers established that GPx levels increased and the activity of NF-κB was reduced in allergen-sensitized mice treated with sodium selenite. Additionally, selenium-pretreated mice were less affected by exposure to the allergen.[32]

17.2.2 Keshan disease

Keshan disease is a cardiomyopathy that manifests in Chinese populations having low selenium intake. After the late 1970s, when the relationship between Keshan disease and selenium was established, selenium supplementation of susceptible individuals substantially reduced the incidence of Keshan disease. However, selenium deficiency did not fully explain the occurrence of the disease, and it was later found that a viral co-factor, coxsackievirus, was necessary for development of the disease.[28] Beck and co-workers demonstrated that selenium-deficient mice infected with a harmless strain of coxsackievirus developed inflammation of the heart muscle, whereas selenium-replete mice did not. This was suspected to result from decreased GPx activity, as selenium-deficient mice had a fivefold reduction in GPx activity.[34] Confirmation of the role of GPx came through the use of GPx-1 knockout mice, displaying optimal activity of all other selenoenzymes. Infection of the knockout mice with the same strain of coxsackievirus also induced myocarditis, as was found in selenium-deficient mice.[35] Interestingly, when selenium-replete mice were infected with the virus isolated from selenium-deficient mice, they too developed myocarditis. The virus was shown to mutate to a more virulent strain in selenium-deficient mice. In contrast, no nucleotide changes were found in virus isolated from selenium-replete mice.[28]

The possibility that oxidative stress may induce changes in viral RNA has implications for many other diseases. Beck and colleagues also examined the effect of selenium on influenza virulence, and found that genes associated with viral replication had at least 29 nucleotide changes in selenium-deficient mice, compared with no changes in selenium-replete mice.[36]

17.2.3 Human immunodeficiency virus (HIV)

Declining selenium levels have been shown to accompany the progression of HIV and loss of $CD4^+$ T-cells. Studies have also shown that selenium-deficient individuals have a substantial increase in mortality from HIV-related illnesses compared with selenium-replete individuals.[26,27] Evidence implicating chronic oxidative stress and the progression of HIV has been found.[37] HIV patients show membrane damage indicative of oxidative stress and higher amounts of lipid peroxidation by-products.[37] If oxidative stress facilitates the progression of HIV, then it is not surprising that anti-oxidant selenoenzymes may have a role in slowing the decline of disease sufferers. Many studies also show that adequate selenium is necessary for healthy immune function.[27] Kiremidjian-Schumacher found that supranutritional doses of selenium enhanced immune function, even in individuals with adequate selenium.[38,39]

17.3 Therapeutic Uses of Supranutritional Doses of Selenium

Some debate exists regarding the recommended daily allowance of selenium, in particular if it should be increased. Proponents of supranutritional doses of selenium cite numerous studies, primarily in the area of cancer prevention, expounding the benefits of taking doses of selenium exceeding that required for maximal selenoenzyme function.[27,30] Whilst the doses advocated are not in the toxic range for humans, caution should be taken when making such recommendations as there is substantial evidence that at high concentrations selenium acts as a pro-oxidant.

17.3.1 Arsenism

Arsenism is not uncommon in undeveloped areas of Asia having contaminated water sources.[40,41] Diabetes, heart disease, many types of cancer and skin lesions are just some of the effects of chronic arsenic exposure. At present there is no treatment available to assist recovery after the source of arsenic has been removed.[41] The interaction between dietary selenium and arsenic has been documented for some time.[42–44] Work by Gailer has shown that selenium metabolites such as H_2Se and glutathione react with arsenic *in vivo* to generate the seleno-bis(S-glutathionyl) arsinium ion $[(GS)_2AsSe]^-$ which can be excreted in bile.[21] Recently, a study of farmers suffering from arsenism in Mongolia showed that selenium supplementation significantly improved skin lesions and liver function.[40,41] In addition, the arsenic concentrations in hair and blood of individuals reduced twofold more than in control groups.[41] Although this trial was small, it certainly offers a potentially simple method to assist sufferers of arsenism.

17.3.2 Cancer prevention

Studies as early as 1969 suggested that there was an inverse correlation between dietary selenium levels and cancer mortality rates.[45] Accompanying the discovery of the enzyme GPx, the chemopreventative action of selenium was initially attributed to its presence in anti-oxidant enzymes such as GPx and other selenoproteins. In 1996 Clark found that selenium supplementation, 200 μg/day in the form of selenium-enriched yeast, reduced the risk of prostate, colon and lung cancer by up to 48% in subjects with a history of skin cancer. The dose of selenium used was higher than the current American 'recommended dietary allowance', and higher than the 90 μg/day required for maintenance of maximum-plasma and whole-blood GPx activities, as established by Duffield.[27,46–48] The anti-oxidant explanation for the chemopreventative action of selenium lost popularity as it appeared that supranutritional doses of selenium were necessary for anti-cancer activity.[47] Additional evidence that the mechanism of selenium anti-cancer activity was discrete from its nutritional role was provided by studies that showed selenium compounds that were only poorly able to be used for the biosynthesis of selenoproteins also had comparable anti-cancer activities.[49–51]

Much work has been directed at elucidating the mechanism of selenium anti-cancer activity. Studies have used different forms of selenium, commonly sodium selenite and SeMet, which are used in dietary supplements. Observed differences in the type of activity for each form and studies showing contradictory results have hindered finding a simple answer. Generally it is considered that, at nutritional levels, selenium in selenoproteins acts as an anti-oxidant, enhancing cell growth and immune activity. At supranutritional levels, selenium metabolites behave as pro-oxidants and inhibit cell growth, and at even higher concentrations they become toxic.[47,48] Numerous mechanisms have been proposed to account for the chemopreventative properties of selenium. These include induction of apoptosis and cell-cycle arrest by low-weight selenium metabolites, such as methylselenol;[47,52–55] activation of Ref. 1 by selenomethionine – which leads to increased DNA repair;[56] inhibition of enzymes such as

TrxR;[48,57,58] and newly discovered selenoproteins with unknown functions, all of which may be shown to play a part in cancer prevention.[48]

Selenium metabolites

A large number of studies have shown that the generation of low molecular weight selenium metabolites from selenium-containing compounds is critical for their anti-cancer activity.[48–51,59,60] Excess natural inorganic and organic selenium is metabolized for excretion in animals by methylation of H_2Se; Figure 17.1.[12,49] Methylselenol is considered to be the species responsible for the chemopreventative activity of supranutritional doses of dietary supplements.[48–52,59] Dimethylselenide was excluded as a candidate for anti-cancer activity as dimethylselenoxide, a dimethylselenide precursor, had negligible anti-cancer activity. The fully methylated metabolite, trimethylselenonium ion, was found to have no anti-cancer activity, possibly as it is too rapidly excreted.[47,49,59] Methylselenol is too reactive to be administered directly. However, alternative precursors that generate methylselenol endogenously have provided evidence of the importance of methylselenol to anti-cancer activity.

Methylselenocysteine (SeMC) is converted to methylselenol by β-lyase and is a more direct source of methylselenol than sodium selenite or SeMet (Figure 17.1).[52,59] SeMC is a more potent chemopreventative agent than SeMet or sodium selenite, presumably due the more efficient production of methylselenol.[50,51,59] Additionally, SeMC is more available than SeMet for conversion to methylselenol as it cannot be stored in general proteins (replacing methionine).[59] L-Se-methylselenocysteine (L-SeMC) has shown much promise as a chemopreventative agent, with lower toxicity than sodium selenite or SeMet, and has been patented for use as a nutriceutical by Pharmase Inc., USA.[61]

Another source of methylselenol is methylseleninic acid (MSeA, Figure 17.1). This has the advantage of metabolism to methylselenol through reaction with reduced glutathione (GSH); thus it is an effective methylselenol source in cell types that have low concentrations of β-lyase.[52,60] A recent study found that methylselenol generated from both MSeA and SeMet (after exposure to methioninase, METase) displayed almost identical actions, that being induction of cell-cycle arrest at G_1 and apoptosis (programmed cell death), in Du145 human prostate cancer cells.[54] Apoptosis is considered to be the primary cause of cell death by anti-cancer drugs.[62] Lu found that SeMet in the absence of METase had no effect on human prostate cancer cells at concentrations as high as 100 μM. In contrast, methylselenol generated from METase-treated SeMet significantly increased apoptosis at concentrations as low as 1 μM SeMet.[54]

Ip found that MSeA treatment of human premalignant breast cells caused cell-cycle arrest at the G_o/G_1 stage and increased the rate of apoptosis by five times. Ip suggests that methylselenol may induce apoptosis by upregulating the expression of a number of proteins known to mediate programmed cell death. Selenols exist *in vivo* as the selenoate anion; these have the potential to form selenylsulfide linkages (—S—$SeCH_3$) that could affect the activity of cysteine-rich proteins.[52,63]

It has also been proposed that the generation of superoxide from selenoate anions may cause oxidative stress and apoptosis.[55,64,65] Spallholz used chemiluminescence to detect superoxide formation from various selenium-containing compounds.[55] They found that putative selenol precursors, such as MSeA and diselenides, generated superoxide whereas monoselenides did not. Interestingly, SeMet alone did not generate superoxide; however, superoxide was detected when SeMet was tested in the presence of METase.[55] This gives further credence to the importance of methylselenol in cancer prevention.

DNA repair

A new hypothesis regarding the chemopreventative action of dietary selenium that has stimulated much interest has emerged from the work of Smith and co-workers.[56,66] Smith has found that treatment of human lung cells with SeMet appeared to increase repair of DNA damaged by UV radiation and chemical carcinogens such as carboplatin.[56] The results suggest that SeMet activates a protein, redox factor Ref. 1, which in turn activates the tumor-suppressor protein p53. In cells lacking the *p53* gene there was no observed induction of DNA repair by SeMet. Interestingly, in contrast with many studies, no apoptosis or cell-cycle arrest was observed with SeMet treatment. The tumor-suppressor protein p53 has roles in both induction of apoptosis and enhancement of DNA repair depending on cell type, and often functions poorly in human cancer cells.[62,67] Further research to elucidate the activity of selenium compounds in improving DNA repair will certainly be of interest, particularly with the aim of treating people for whom inadequate DNA repair places them at risks for certain cancers.

Inhibition of thioredoxin reductase (TrxR)

The activity of selenoprotein TrxR impacts upon a large range of biological processes.[20] Relevant to cancer prevention is the role of TrxR in apoptosis and protection against oxidative stress. Increased activity of TrxR leads to increased levels of reduced thioredoxin (Trx). Reduced Trx is able to bind to apoptosis signalling kinase (ASK-1), resulting in inhibition of apoptosis. On the other hand, inhibition of TrxR can lead to increased ASK-1 activity and subsequent apoptosis.[20,68] There are a number of clinically used chemotherapeutic compounds that inhibit TrxR, and evidence that selenium was also an inhibitor of TrxR led to speculation that this may hold the key to its anti-cancer activity.[20,48] The findings of a number of studies[48,57,68,69] have presented conflicting evidence regarding this hypothesis and it is clear that more research is required to ascertain exactly how selenium acts as a cancer preventative.

17.4 Synthetic Selenium-Containing Therapeutics

17.4.1 Anti-oxidants and anti-inflammatory agents

Ebselen

Reactive oxygen species (ROS) are involved in many biological processes including the inflammatory response.[33,70,71] Ebselen (2-phenyl-1,2-benzisoselenazol-3(2*H*)-one) (**1**) is an anti-oxidant and anti-inflammatory that has undergone Phase III clinical trials as a neuroprotective agent and is soon to become the first synthetic organoselenium therapeutic released on the market. Selenium is not liberated from Ebselen **1**, and as such it is relatively non-toxic. The anti-inflammatory action of Ebselen **1** is attributed to its ability to interfere with a number of pathways to inflammation, either through its anti-oxidant capacity or through direct inhibition of enzymes.[1,72,73]

Ebselen **1** catalytically reduces peroxynitrite, a strong oxidant that is involved in inflammation.[74–78] Peroxynitrite is formed from the reaction of nitric oxide with superoxide. Under physiological conditions peroxynitrite can lead to the formation of highly reactive $^{\bullet}NO_2$ and $^{\bullet}OH$ radical species. A recent study found that Ebselen **1** was a potent inhibitor of nitric oxide-induced apoptosis, possibly through generation of the Ebselen oxide after reaction with peroxynitrite.[79] This study suggested that the Ebselen oxide can inactivate a key regulator of apoptosis, the protein kinase JNK1, by oxidizing the sulfhydryl moiety.[79]

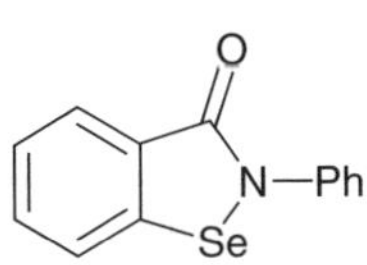

1

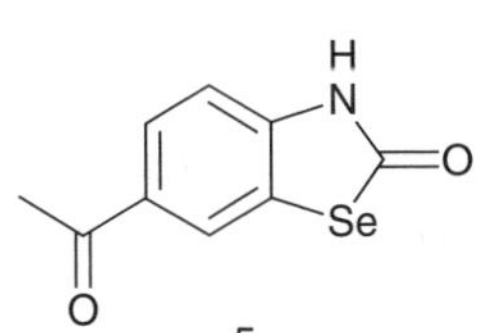

5

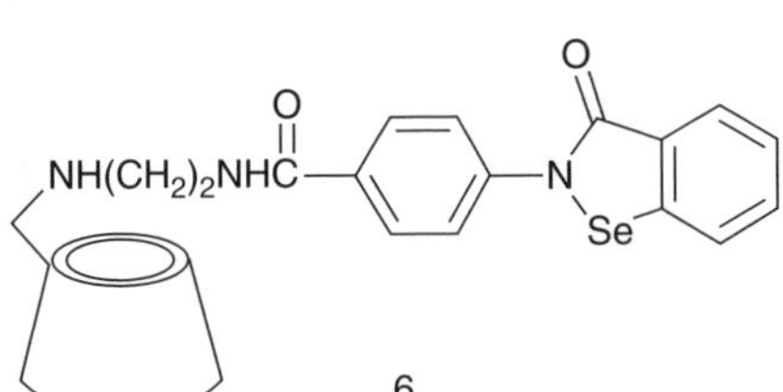

6

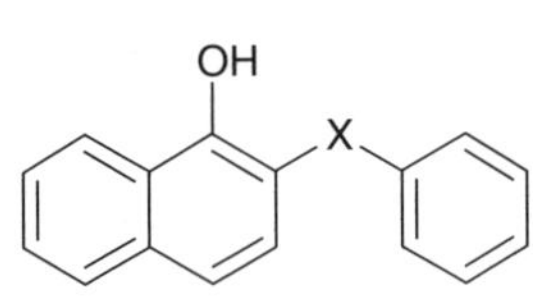

7a X = Se
7b X = CH_2

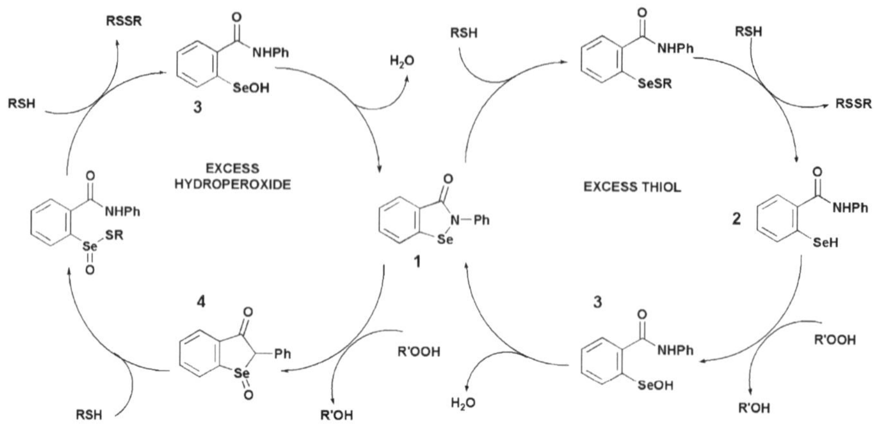

Figure 17.2 Ebselen – mechanism of GPx-like activity

Ebselen **1** acts as a GPx mimic by reducing hydroperoxides to water or the corresponding alcohol (Figure 17.2).[72] There are a number of mechanisms proposed for this process. When the concentration of the thiol exceeds that of the hydroperoxide, Ebselen **1** first reacts with thiol to generate a selenol **2**. This can then reduce hydroperoxides to yield seleninic acid **3**, which, after loss of water, regenerates Ebselen **1**.[72,80] Evidence of the formation of selenol **2** was found by trapping with 1-chloro-2,4-dinitrobenzene and inhibition of activity by iodoacetate.[81–83] An alternate mechanism appears to operate when hydroperoxide concentrations are higher, whereby Ebselen **1** is first oxidized by hydroperoxide to the oxide **4** before regeneration of Ebselen **1** by two equivalents of thiol.[72,80] It is noteworthy that the sulfur analogue of Ebselen **1** is inactive and compounds such as **5**, lacking direct Se—N bonds, are also inactive.[80,84,85] Numerous compounds have been prepared as potential GPx mimics that contain Se—N bonds, and are discussed in an excellent review.[1]

Recent work by Zhao and Holmgren demonstrates that Ebselen **1** is also able to catalytically reduce hydroperoxides through reaction with the Trx system[86] (Figure 17.3). Ebselen **1** is a substrate for TrxR and a rapid oxidant of reduced thioredoxin. Selenol **2** can be formed directly by TrxR/NADPH or from reaction with reduced Trx. Selenol **2** rapidly reduces hydroperoxides to form seleninic acid **3**. Ebselen **1** is regenerated after spontaneous loss of water.[86] Comparison of the interaction of Ebselen **1** with the Trx system and GSH showed that the rate of reduction of hydrogen peroxide was eight times higher with TrxR (50 nM) than GSH (1 mM). When the complete TrxR/Trx system was tested the hydrogen peroxide reductase activity of Ebselen **1** increased 13-fold.[86] Given the ubiquitous nature of the Trx system this work suggests that much of the hydroperoxide-reducing ability of Ebselen **1** could result from the interaction between the two.

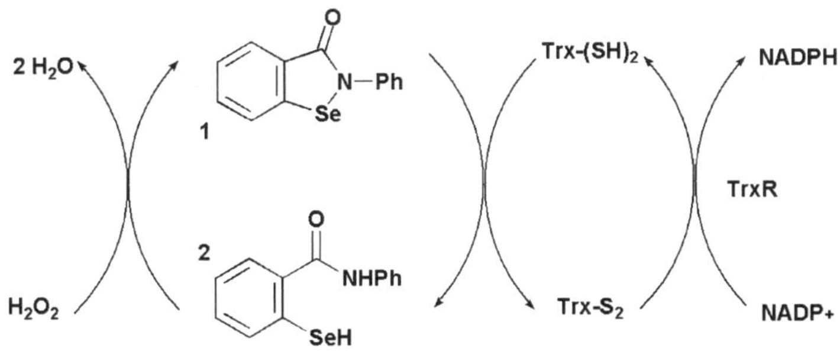

Figure 17.3 Ebselen – interaction with TrxR/Trx system

The role of hydroperoxides in inflammation is well known. Hydroperoxides are essential for the activation of inflammatory enzymes known as the lipoxygenases (LOX) and cyclooxygenases (COX).[17,87] Of the lipoxygenases, 5-LOX is considered to be critical to inflammation as it catalyses the biosynthesis of pro-inflammatory leukotrienes from arachidonic acid.[88] LOX and COX are attractive targets for inhibition in the pursuit of novel anti-inflammatories.[1] The capacity for Ebselen **1** to act as a LOX inhibitor is critical to its anti-inflammatory activity. The inhibition of LOX by Ebselen **1** may be as a result of its anti-oxidant activity or through the direct interaction of Ebselen **1** and LOX.

Other anti-oxidants and anti-inflammatories

The efficacy of Ebselen **1** is somewhat limited by its poor solubility in water. In the pursuit of preparing GPx mimics with increased water solubility, Liu and co-workers prepared β-cyclodextrins with an Ebselen **1** moiety tethered to the primary ring such as **6**.[89] These compounds had excellent aqueous solubility and cyclodextrin **6** displayed GPx activity on par with that of Ebselen **1**.[89]

Engman and colleagues prepared the selenium analogue **7a** of a known 5-LOX inhibitor, 2-benzyl-1-naphthol **7b**.[90] Phenylselenenyl-1-naphthol **7a** showed no anti-oxidant activity in the presence of hydrogen peroxide and a thiol. In a lipid peroxidation model, anti-oxidant activity was poorer than the parent compound.

However, the selenium analogue **7a** was a fivefold more potent inhibitor of leukotriene B4 synthesis than the parent and consequently a potentially efficient anti-inflammatory.[90]

Several research groups including those of Engman and Schiesser have prepared selenium-containing analogues of a number of biologically relevant anti-oxidants including tocopherols[91,92] and carbohydrates[93] in a bid to increase the solubility (and thus, efficacy). Recently Back and Moussa have prepared some interesting cyclic seleninate esters **8** that have GPx-like activity.[94,95] The

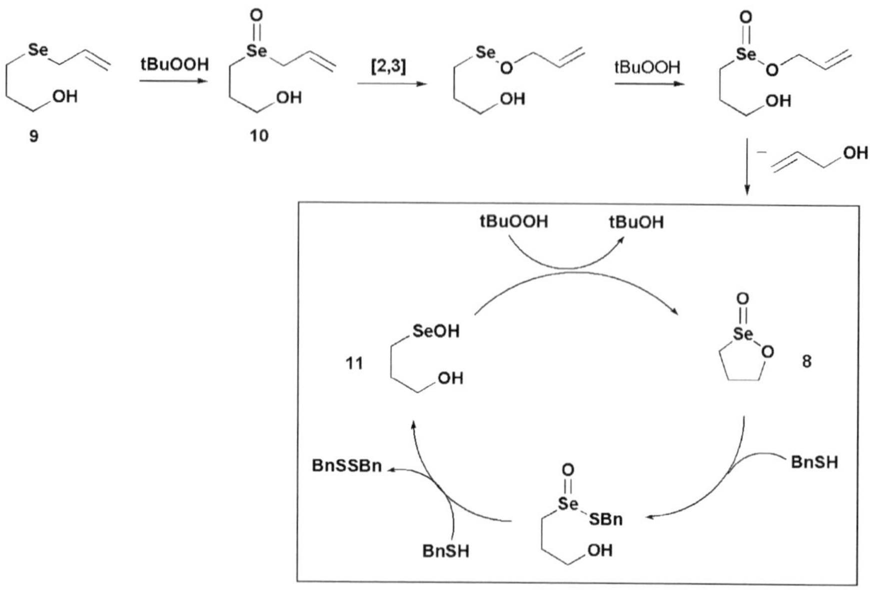

Figure 17.4 Cyclic seleninate ester **8** – generation and mechanism of GPx-like activity

cyclic seleninate **8** was generated *in situ* from allyl-3-hydroxypropyl selenide **9** or (Figure 17.4). After oxidation of allyl selenide **9** with *t*-butylhydroperoxide the generated oxide **10** underwent a [2,3] sigmatropic rearrangement. Further oxidation and cyclization gave the cyclic seleninate ester **8**.[94,95] Seleninate ester **8** is believed to be the species acting as the GPx mimic. Reaction with two equivalents of thiol gives the seleninic acid **11** which is able to reduce the hydroperoxide to regenerate the cyclic seleninate ester.[94,95] Using the *t*-butylhydroperoxide and benzyl thiol model, to test for GPx-like activity, cyclic seleninate ester **8** had substantially higher catalytic activity than Ebselen **1**.[94] Perhaps compounds containing a Se—O heterocycle merit the level of attention given to Se—N-containing compounds.

17.4.2 Anti-cancer

Despite very promising research, to date, no synthetic organoselenium compounds are in clinical use as anti-cancer agents. Early research generally involved the substitution of sulfur for selenium in sulfur-containing compounds of known activity, often resulting in little improvement in activity or toxicity.[1] However, many organoselenium compounds showing potent anti-cancer activity have been developed, such as triphenylselenonium ion **12**,[49] selenocyanates **13**,[1,49] selenazofurin **14a**,[1,15,96] 1,3-selenazine derivative **15**[1,97] and selenocysteine conjugates **16** and **17**.[1,98–101]

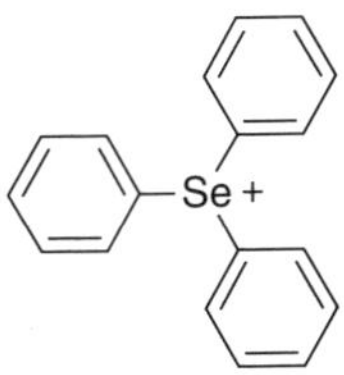

12

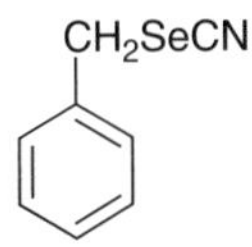

13

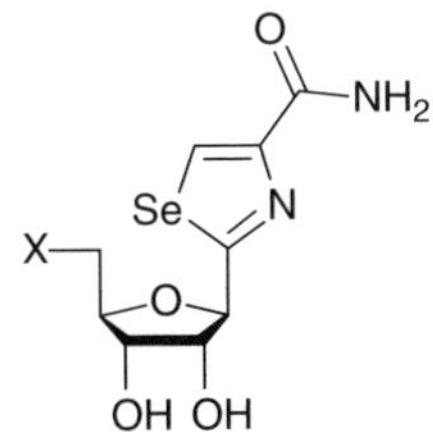

14a X = OH
14b X = PO_3H_2

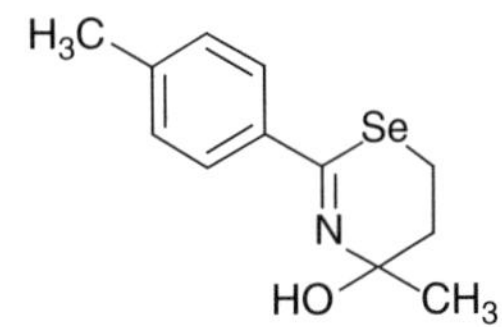

15

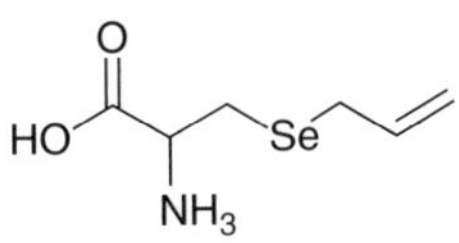

16

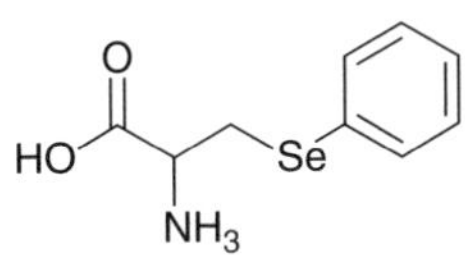

17

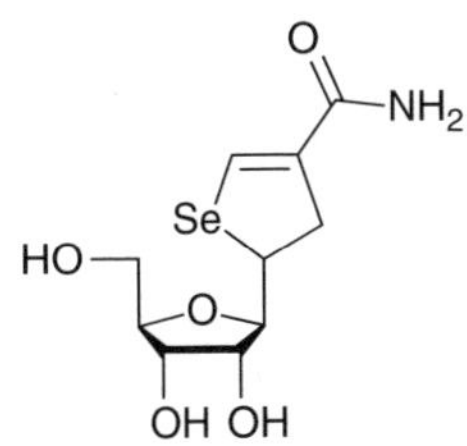

18

Selenazofurin

Selenazofurin **14a** is a *c*-nucleoside that is an inosine monophosphate dehydrogenase (IMPDH) inhibitor.[96] IMPDH catalyzes the oxidation of inosine monophosphate (IMP) to xanthosine monophosphate (XMP), which is a key step in the synthesis of purines. In rapidly multiplying cells IMPDH activity is increased; thus IMPDH inhibition is the target of some potential anti-cancer agents. Due to increased IMPDH inhibition, selenazofurin **14a** has 5–10-fold higher activity than the sulfur analogue.[1,96] Analogues of selenazofurin have also been prepared including the monophosphate derivative **14b** and selenophenfurin **18** which also shows IMPDH inhibition.[102]

Selenocysteine conjugates

As discussed earlier (section 'selenium metabolites') methylselenocysteine (MSeC) undergoes β-elimination by β-lyases to generate methylselenol, a critical metabolite in the anti-cancer activity of dietary selenium. Unnatural selenocysteine conjugates have been prepared, which are also metabolized by β-lyase to selenols, ammonia and pyruvate[99,100] (Figure 17.5).

16
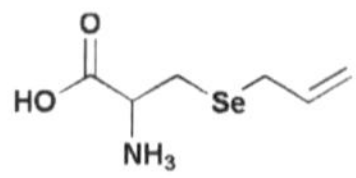

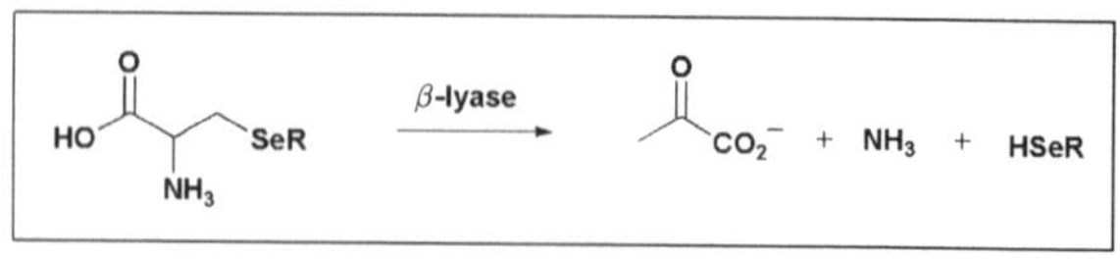

17
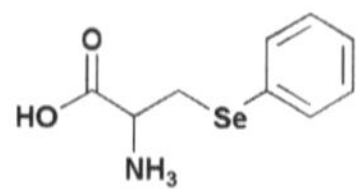

Figure 17.5 Selenocysteine conjugates – prodrugs with anti-cancer activity

Ip found that Se-allyl-DL-selenocysteine **16** displayed excellent anti-tumor activity against mammary tumors in rats.[101] Research has shown that the selenocysteine conjugates are much better substrates for β-lyase than the corresponding sulfur analogue, showing rates of β-elimination up to two orders of magnitude higher.[100,101] Selenocysteine (Se-Cys) conjugates, particularly those containing phenyl substituents such as **17**, show promise as prodrugs able to deliver selenols selectively to kidneys, as β-lyase is present in high concentrations in the kidneys and liver.[99] An additional use for Se-Cys conjugates is to ameliorate the toxic effects of cisplatin.[98] Cisplatin is severely nephrotoxic and there are currently no compounds used clinically to protect against its toxicity. Rooseboom and co-workers found that Se-Cys conjugates improved viability of cisplatin-treated cells having normal β-lyase activity whereas the corresponding cells having very low β-lyase activity showed no improvement in cell viability.[98] Owing to the selective delivery of selenol by Se-Cys conjugates it is thought that these compounds may provide protection against the toxic effects of cisplatin without adversely affecting its efficacy against cancer. Excessive production of ROS has been implicated in the toxicity of cisplatin; thus one possible mechanism for the protective effect of Se-Cys conjugates involves the induction of anti-oxidant enzymes.[98] An alternative hypothesis is that selenol binds to the platinum contained in cisplatin, deactivating it. There is evidence that methylselenol is capable of binding to platinum, giving credence to this hypothesis.[103]

Selenocyanates

Selenocyanates such as **13**, particularly bis-selenocyanates, exhibit anti-cancer activity in a number of different models.[1] Rao found evidence that bis-selenocyanate inhibits protein kinase C (PKC) activity, which is intimately involved with tumor development.[104] It is believed that selenocyanates are metabolized to the corresponding selenol (in the same manner as are thiocyanates), by glutathione and glutathione transferases.[49] Indeed, selenols appear to be a frequently seen metabolite of organoselenium compounds with anti-cancer activity.

17.4.3 Anti-hypertensives

May and co-workers have developed a series of phenylaminoethylselenides (PAESe) **19** that have anti-hypertensive properties.[105–107] PAESe's anti-hypertensive activities appear to result from inhibition of dopamine β-monooxygenase (DBM).[105] DBM catalyzes the conversion of dopamine to noradrenalin and is dependent on ascorbate for catalytic activity (Figure 17.6). PAESes inhibit DBM by depleting ascorbate in the vesicles of chromaffin cells. DBM oxidizes PAESe **19** to the selenium oxide **20** using reduced ascorbate as a co-factor. The reduced PAESe **19** is regenerated by reduced ascorbate.

A cycle is established which leads to the exhaustion of local supplies of reduced ascorbate, the essential co-factor for DBM activity. Of the PAESes, 4-hydroxy-α-methyl-phenyl-2-aminoethyl selenide **21** was found to be an orally active anti-hypertensive, with little central nervous system permeability.[106] PAESes also have anti-oxidant activity. Specifically, they are able to reduce peroxynitrite at a rate more than four orders of magnitude higher than the rate of reduction of hydrogen peroxide.[108]

17.4.4 Anti-virals and anti-bacterials

Anti-virals

Selenium-containing anti-virals, usually in the form of nucleoside synthetase inhibitors where a lower heteroatom such as oxygen or sulfur has been substituted,[1] include selenazofurin **14a**. While bearing a broad spectrum of anti-viral activity, unfortunately selenazofurin is highly toxic at therapeutic concentrations making it an unsuitable therapeutic.[109] Compounds such as **22** show *in vitro* activity against HIV at nM concentrations, below the level at which toxicity is observed.[110]

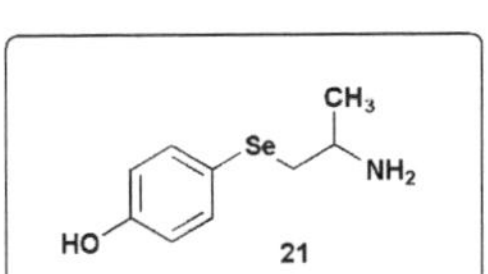

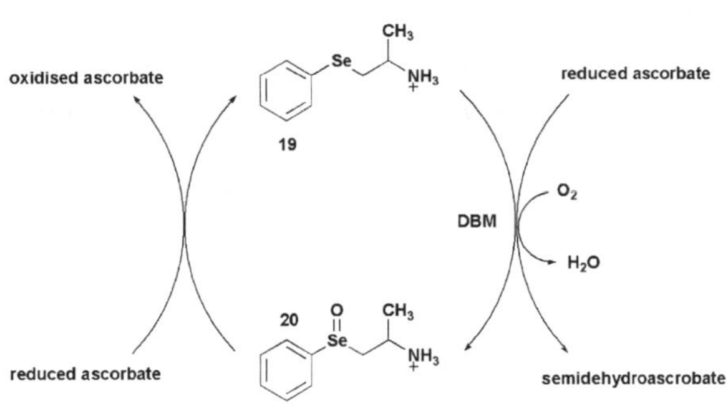

Figure 17.6 Mechanism of anti-hypertensive activity of PAESe

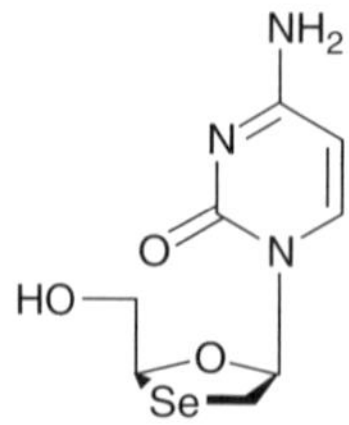

22

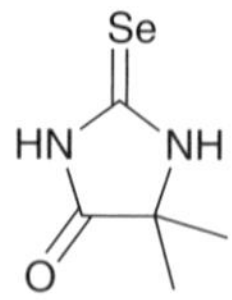

23

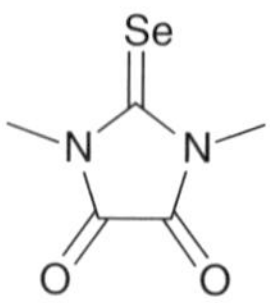

24

Anti-bacterials

While the substitution of sulfur for selenium in traditional β-lactam antibiotics has been met with limited success,[111–113] several selenium-containing heterocyclic compounds, for example 1,3 selenazine **15**, have shown strong inhibitory activity against several pathogenic strains including *Staphylococcus aureus*.[1,114] It is interesting to note that the sulfur homologue in this case displayed little analogous activity, indicating the selenium atom was a key factor in their anti-bacterial properties. A similar result was observed in the anti-mycobacterial activity of seleno-azole derivatives **23** and **24** and the thio-azole analogues.[115] Ebselen **1** has also been shown to possess anti-bacterial activity *in vitro* against *S. aureus* and it is thought that this activity comes about through the selenium atom interfering with the action of an essential thiol group.[116]

17.5 Conclusion

While there can be no question that selenium is a key player in the body's ability to maintain itself, the precise nature of many of its roles remains an area of much speculation. No doubt much will be clarified over the coming years as researchers unravel the mysteries of its functions. Selenium-containing therapeutics have been slow to find a place in medicine, presumably because of the historical association of selenium with toxicity. The beneficial roles of selenium having been established, it now seems only a matter of time before selenium-containing compounds join the pool of therapeutics used to treat disease.

References

1. G. Mugesh, W.-W. du Mont and H. Sies, *Chem. Rev.*, **101**, 2125–2179 (2001).
2. C.G. Jacob, G.I. Giles, N.M. Giles and H. Sies, *Angew. Chem. Int. Ed.*, **42**, 4742–4758 (2003).
3. A. Krief and L. Heven, *Organoselenium Chemistry I: Functional Group Transformations*, Springer-Verlag, Berlin, 1988.

4. P.A. Cox, *The Elements on Earth*, Oxford University Press, Oxford, 1995.
5. N.N. Greenwood and A. Earnshaw, *Chemistry of the Elements*, Pergamon Press, Oxford, 1984.
6. M. Birringer, S. Pilawa and L. Flohe, *Nat. Prod. Rep.*, **19**, 693–718 (2002).
7. K.F. Schwarz, C.M. Foltz, *J. Am. Chem. Soc.*, **79**, 3292–3293 (1957).
8. S.W. May, *Exp. Opin. Invest. Drugs*, **11**, 1261–1269 (2002).
9. O.A. Levander, *Biomed. Environ. Sci.*, **10**, 214–219 (1997).
10. J. Alexander, J. Hogberg, Y. Thomassen and J. Aaseth, Selenium, *Handbook of Toxicity of Inorganic Compounds*, H.G. Seiler, H. Sigel and A. Sigel (eds), Marcel Dekker, New York, 1988.
11. D.G. Barceloux, *J. Toxicol. Clin. Toxicol.*, **37**, 145–172 (1999).
12. H.E. Ganther, D.G. Hafeman, N.G. Hoekstra *et al.*, Selenium metabolism and function in man and animals, *Trace Element Analytical Chemistry in Medicine and Biology*, P. Bratter and P. Schramel (eds), Walter de Gruyter, New York, 1984.
13. I.S. Palmer, R.P. Gunsalus, A.W. Halverson and O.E. Olson, *Biochim. Biophys. Acta*, **208**, 260–266 (1970).
14. R.S. Glass, W.P. Singh, W. Jung *et al.*, *Biochemistry*, **32**, 12555–12559 (1993).
15. S.W. May, *Exp. Opin. Invest. Drugs*, **8**, 1017–1030 (1999).
16. H.E. Ganther, D.G. Hafeman, N.G. Hoekstra *et al.*, *Science*, **179**, 588–590 (1973).
17. R. Brigelius-Flohe, *Free Rad. Biol. Med.*, **27**, 951 (1999).
18. A. Lescure, D. Gautheret, P. Carbon and A. Krol, *J. Biol. Chem.*, **274**, 38147–38154 (1999).
19. L.B. Banu, M.J. Berry and P.R. Larsen, *Nature*, **349**, 438–440 (1991).
20. J. Nordberg and E.S.J. Arner, *Free Rad. Biol. Med.*, **31**, 1287–1312 (2001).
21. J. Gailer, *Appl. Organometal. Chem.*, **16**, 701–707 (2002).
22. R.F. Burk, *Curr. Opin. Gastroen.*, **17**, 162–166 (2001).
23. M. Hirashima, T. Naruse, H. Maeda *et al.*, *Biol. Pharm. Bull.*, **26**, 794–798 (2003).
24. K.E. Hill, J. Zhou, W.J. McMahan *et al.*, *J. Biol. Chem.*, **278**, 13640–13646 (2003).
25. K.S. Zafar, A. Siddiqui, I. Sayeed *et al.*, *J. Neurochem.*, **84**, 438–446 (2003).
26. M.P. Rayman, *The Lancet*, **356**, 233–241 (2000).
27. M.P. Rayman, *Proc. Nutr. Soc.*, **61**, 203–215 (2002).
28. J.R. Arthur, R.C. McKenzie and G.J. Beckett, *J. Nutr.*, **133**, 1457S–1459S (2003).
29. R.W. Alexander, *Hypertension*, **25**, 155–161 (1995).
30. J. Neve, *Curr. Opin. Clin. Nutr.*, **5**, 659–663 (2002).
31. G.N. Schrauzer, *J. Nutr.*, **130**, 1653–1656 (2000).
32. D.-W. Jeong, M.-H. Yoo, T.S. Kim *et al.*, *J. Biol. Chem.*, **277**, 17871–17876 (2002).
33. W. Droge, *Physiol. Rev.*, **82**, 47–95 (2001).
34. M.A. Beck, P.C. Kolbeck, F.H. Rohr *et al.*, *J. Med. Virol.*, **43**, 166–170 (1994).
35. M.A. Beck, R.S. Eswothy, Y.S. Ho and F.F. Chu, *FASEB J.*, **12**, 1143–1149 (1998).
36. M.A. Beck, H.K. Nelson, Q. Shi *et al.*, *FASEB J.*, **15**, 1481–1483 (2001).
37. G.W. Pace and C.D. Leaf, *Free Rad. Biol. Med.*, **19**, 523–528 (1995).
38. L. Kiremidjian-Schumacher, R. Martin, H.I. Wishe *et al.*, *Biol. Trace Elem. Res.*, **41**, 115–127 (1994).
39. L. Kiremidjian-Schumacher and M. Roy, *BioFactors*, **14**, 161–168 (2001).
40. W. Wang, L. Yang, S. Hou *et al.*, *Curr. Sci.*, **81**, 1215–1218 (2001).
41. L. Yang, W. Wang, S. Hou *et al.*, *Environ. Geochem. Hlth.*, **24**, 359–374 (2002).
42. O.A. Levander and C.A. Baumann, *Toxicol. Appl. Pharmacol.*, **9**, 106–115 (1966).
43. O.A. Levander and C.A. Baumann, *Toxicol. Appl. Pharmacol.*, **9**, 98–105 (1966).

44. O.A. Levander and L.C Argrett, *Toxicol. Appl. Pharmacol.*, **14**, 308–314 (1969).
45. R.J. Shamberger and D.V. Frost, *Can. Med. Assoc. J.*, **104**, 82 (1969).
46. A.J. Duffield, C.D. Thomson, K.E. Hill and W. Sheila, *Am. J. Clin. Nutr.*, **70**, 896–903 (1999).
47. G.F. Combs and W.P. Gray, *Pharmacol. Ther.*, **79**, 179–192 (1998).
48. H.E. Ganther, *Carcinogenesis*, **20**, 1657–1666 (1999).
49. H.E. Ganther and J.R. Lawrence, *Tetrahedron*, **53**, 12299–12310 (1997).
50. C. Ip, R.M. Budnick and H.E. Ganther, *Cancer Res.*, **51**, 595–600 (1991).
51. C. Ip and H.E. Ganther, *Cancer Res.*, **50**, 1206–1211 (1990).
52. C. Ip, Y. Dong and H.E. Ganther, *Cancer Met. Rev.*, **21**, 281–289 (2002).
53. H.-M. Shen, W.-Z. Ding and C.-N. Ong, *Free Rad. Biol. Med.*, **33**, 552–561 (2002).
54. Z. Wang, C. Jiang and J. Lu, *Mol. Carcinogenesis*, **34**, 113–120 (2002).
55. J.E. Spallholz, V.P. Palace and T.W. Reid, *Biochem. Pharmacol.*, **67**, 547–554 (2004).
56. Y.R. Seo, M.R. Kelley and M.L. Smith, *Proc. Natl. Acad. Sci. USA*, **99**, 14548 (2002).
57. H.E. Ganther and C. Ip, *J. Nutr.*, **131**, 301–304 (2001).
58. S. Gromer and J.H. Gross, *J. Biol. Chem.*, **277**, 9701–9706 (2002).
59. C. Ip, *J. Nutr.*, **128**, 1845–1854 (1998).
60. C. Ip, H.J. Thompson, Z. Zhu and H.E. Ganther, *Cancer Res.*, **60**, 2882–2886 (2000).
61. J.E. Spallholz, T.W. Reid and R.D. Walkup, *A Method of Using Synthetic L-Se-Methylselenocysteine as a Neutriceutical*, EP 1205471, Eur. Pat. Appl., 2002.
62. J.M. Brown and B.G. Wouters, *Cancer Res.*, **59**, 1391–1399 (1999).
63. P.R. Harrison, J. Lanfear, L. Wu *et al.*, *Biomed. Environ. Sci.*, **10**, 235–245 (1997).
64. J.E. Spallholz, *Biomed. Environ. Sci.*, **10**, 260–270 (1997).
65. M.S. Stewart, J.E. Spallholz, K.H. Neldner and B.C. Pence, *Free Rad. Biol. Med.*, **26**, 42–48 (1999).
66. R. Longtin, *J. Natl. Cancer Inst.*, **95**, 98–100 (2003).
67. M.L. Smith, J.M. Ford, M.C. Hollander *et al.*, *Mol. Cell. Biol.*, **20**, 3705–3714 (2000).
68. K. Jonsson-Videsater, L. Bjorkhem-Bergman, A. Hossain *et al.*, *Biochem. Pharmacol.*, **67**, 513–522 (2004).
69. M.M. Berggren, J.F. Mangin, J.R. Gasdaska and G. Powis, *Biochem. Pharmacol.*, **57**, 187–193 (1999).
70. E.M. Conner and M.B. Grisham, *Nutrition*, **12**, 274–277 (1996).
71. F.A. Fitzpatrick, *Int. Immunopharmacol.*, **1**, 1651–1667 (2001).
72. T. Schewe, *Gen. Pharmac.*, **26**, 1153–1169 (1995).
73. Y. Nakamura, Q. Feng, T. Kumagai *et al.*, *J. Biol. Chem.*, **277**, 2687–2694 (2002).
74. H. Sies and G.E. Arteel, *Free Rad. Biol. Med.*, **28**, 1451–1455 (2000).
75. H. Sies and H. Masumoto, *Adv. Pharmacol.*, **38**, 229–246 (1997).
76. G.E. Arteel, K. Briviba and H. Sies, *Chem. Res. Toxicol.*, **12**, 264–269 (1999).
77. D.G. Musaev, Y.V. Gelemtii, C.L. Hill and K. Hirao, *J. Am. Chem. Soc.*, **125**, 3877–3888 (2003).
78. D.G. Musaev and K. Hirao, *J. Phys. Chem.*, **107**, 1563–1573 (2003).
79. K.P. Sarker, K.K. Bismas, J.L. Rosales *et al.*, *J. Neurochem.*, **87**, 1345–1353 (2003).
80. G. Mugesh and H.B. Singh, *Chem. Soc. Rev.*, **29**, 347–357 (2000).

81. I.A. Cotgreave, R. Morgenstern, L. Engman and J. Ahokas, *Chem. Biol. Interact.*, **84**, 69–76 (1992).
82. M. Maiorino, A. Roveri, M. Coassin and F. Ursini, *Biochem. Pharmacol.*, **37**, 2267–2271 (1988).
83. R. Morgenstern, I.A. Cotgreave and L. Engman, *Chem. Biol. Interact.*, **84**, 77–84 (1992).
84. I.A. Cotgreave, U. Johansson, G. Westergren *et al.*, *Agent. Action.*, **24**, 313–319 (1988).
85. V. Galet, J.L. Bernier, J.P. Henichart *et al.*, *J. Med. Chem.*, **37**, 2903–2911 (1994).
86. R. Zhao and A. Holmgren, *J. Biol. Chem.*, **277**, 39456–39462 (2002).
87. H. Kuhn and A. Borchert, *Free Rad. Biol. Med.*, **33**, 154–172 (2002).
88. B. Haribabu, M.W. Verghese, D.A. Steeber *et al.*, *J. Exp. Med.*, **192**, 433–438 (2000).
89. Y. Liu, B. Li, L. Li and H.-Y. Zhang, *Helv. Chim. Acta*, **85**, 9–18 (2002).
90. L. Engman, D. Stern, H. Frisell *et al.*, *Bioorg. Med. Chem.*, **3**, 1255–1262 (1995).
91. J. Malmstrom, M. Jonsson, I.A. Cotgreave *et al.*, *J. Am. Chem. Soc.*, **123**, 3434–3440 (2001).
92. T. Fenner and C.H. Schiesser, *Molecules*, **9**, 472–479 (2004).
93. M.A. Lucas, O.T.K. Nguyen, C.H. Schiesser and S. Zheng, *Tetrahedron*, **56**, 3995 (2000).
94. T.G. Back and Z. Moussa, *J. Am. Chem. Soc.*, **125**, 13455–13460 (2003).
95. T.G. Back and Z. Moussa, *J. Am. Chem. Soc.*, **124**, 12104–12105 (2002).
96. G. Gebeyehu, V.E. Marquez, A. Van Cott *et al.*, *J. Med. Chem.*, **28**, 99–105 (1985).
97. M. Koketsu, H. Ishihara, W. Wu *et al.*, *Eur. J. Pharmac. Sci.*, **9**, 157–161 (1999).
98. M. Rooseboom, G. Schaaf, J.N.M. Commandeur *et al.*, *J. Pharmacol. Exp. Ther.*, **301**, 884–892 (2002).
99. M. Rooseboom, N.P.E. Vermeulen, I. Andreadou and J.N.M. Commandeur, *J. Pharmacol. Exp. Ther.*, **294**, 762 (2000).
100. J. Andreadou, W.M.P.B. Mange, J.N.M. Commandeur *et al.*, *J. Med. Chem.*, **39**, 2040–2046 (1996).
101. C. Ip, Z. Zhu, H.J. Thompson *et al.*, *Anticancer Res.*, **19**, 2875–2880 (1999).
102. P. Franchetti, L. Cappallecci, G.A. Sheikha *et al.*, *J. Med. Chem.*, **40**, 1731–1737 (1997).
103. G.S. Baldew, J.G.J. Mol, F.J.J. de Kanter *et al.*, *Biochem. Pharmacol.*, **41**, 1429–1437 (1991).
104. C.V. Rao, B. Simi, Y. Hirose *et al.*, *Int. J. Oncol.*, **16**, 519–527 (2000).
105. S.W. May, K. Wimalasena, H.H. Herman *et al.*, *J. Med. Chem.*, **31**, 1066–1068 (1988).
106. S.W. May, L. Wang, M.M. Gill-Woznichak *et al.*, *J. Pharmacol. Exp. Ther.*, **283**, 470–477 (1997).
107. J.D. Overcast, A.E. Ensley, C.J. Buccafusco *et al.*, *J. Cardiovas. Pharmacol.*, **38**, 337–346 (2001).
108. M.M. Woznichak, J.D. Overcast, K. Robertson *et al.*, *Arch. Biochem. Biophys.*, **379**, 314–320 (2000).
109. M.J. Parhnam and E. Graf, *Biochem. Pharmacol.*, **36**, 3095–3102 (1987).
110. C.K. Chu, L. Ma, S. Olgen *et al.*, *J. Med. Chem.*, **43**, 3906–3912 (2000).
111. M. Alpegiani, A. Bedeschi, E. Perrone and G. Franceschi, *Tetrahedron Lett.*, **27**, 3041–3044 (1986).

112. M.W. Carland, R.L. Martin and C.H. Schiesser, *Tetrahedron Lett.*, **42**, 4737–4739 (2001).
113. G.A. Brown, K.M. Anderson, M. Murray *et al.*, *Tetrahedron*, **56**, 5579–5587 (2000).
114. M. Koketsu, H. Ishihara and M. Hatsu, *Res. Commun. Mol. Pathol. Pharmacol.*, **101**, 179–186 (1998).
115. D. Dieidda, G. Lampis, C. Maullu *et al.*, *Pharmacol. Res.*, **36**, 193–197 (1997).
116. R. Nozawa, T. Yokota and R. Fujimoto, *Antimicrob. Agents Chemother.*, **33**, 1388–1390 (1989).

18 $_{43}$Tc Technetium in Medicine

Oyebola O. Sogbein and John F. Valliant

18.1 Overview

Technetium plays an important role in modern healthcare because one of its isotopes, technetium-99m (^{99m}Tc), is the most widely used radionuclide in diagnostic medicine. Its prevalent use is a result of the isotope's nearly ideal nuclear properties, widespread availability and relatively low cost. Tc-labeled compounds are currently used to image different organs and a wide range of different disease states. These include cardiac and kidney functions and the spread of cancer to bone. Hospitals that have a nuclear medicine department are routinely performing Tc-imaging procedures on a daily basis.

This chapter contains an overview of the chemistry of technetium with particular emphasis on areas that relate to its use in nuclear medicine. The document was written principally for newcomers to the field. For this reason, a historical look at the discovery and uses of Tc and a brief description of some of the most widely used agents in clinical nuclear medicine are included. An introduction to some of the research that is currently underway in technetium radiopharmaceutical chemistry is also highlighted.

Metallotherapeutic Drugs and Metal-Based Diagnostic Agents: The Use of Metals in Medicine
Edited by Gielen and Tiekink

18.2 Technetium: Historical Accounts and Properties Relevant to Nuclear Medicine

The element technetium ($Z=43$), which is derived from the Greek word '*technikos*' meaning artificial, is located in a central position in the periodic table and possesses 25 known isotopes ranging from 86 to 118 atomic mass units. All of the isotopes of technetium are radioactive having a wide range of different half-lives. These include short-lived isotopes such as ^{86m}Tc ($t_{1/2}=1.11\,\mu s$), ^{93m}Tc ($t_{1/2}=43.5$ min) and ^{94}Tc ($t_{1/2}=53$ min) and those with half-lives surpassing thousands of years like ^{97}Tc ($t_{1/2}=2.6\times10^6$ y), ^{98}Tc ($t_{1/2}=4.2\times10^6$ y) and ^{99}Tc ($t_{1/2}=2.1\times10^5$ y).

The existence of technetium was predicted by Mendeleev in 1869 on the basis of trends in the periodic table. Mendeleev reported that the unknown element should be similar to manganese and gave it the name '*ekamanganese*'. In 1925, an erroneous claim of the discovery of element 43, which was named masurium, was published.[1] The actual discovery was published in 1937, when Perrier and Segrè reported the isolation of element 43 from a sample of molybdenum that was bombarded with deuterons from the Berkeley cyclotron.[2] The official naming of element 43 as technetium occurred just after the end of World War II.[3]

The discovery of this 'artificial' element prompted a search for natural sources of technetium. In 1956, Merrill showed that ^{99}Tc existed in stellar matter prompting new theories into the production of heavy elements in stars.[4] Later, in 1961, a terrestrial source of ^{99}Tc was isolated in minute quantities from African pitchblende (a uranium-rich ore) by Kenna and Kuroda.[5] The low natural occurrence of technetium impeded investigation of its chemistry because long-lived isotopes were only available at an exorbitant cost. Currently ^{99}Tc, which is isolated in kilogram quantities as a by-product of ^{235}U fission, is available at a cost of less than \$100 (US) per gram.

The first report of nuclear isomerism by element-43 was by Segrè and Seaborg in 1938.[6] That isotope, ^{99m}Tc ($t_{1/2}=6.02$ h, $E_\gamma=141$ keV), possesses ideal properties for medical radioimaging because the γ emission is sufficiently energetic to allow for visualization of sites deep within the human body by scintillation cameras without exposing patients to high levels of ionizing radiation. The daughter of ^{99m}Tc, the long-lived β^- emitter ^{99}Tc ($E_{max}=293.6$ keV), also does not contribute significantly to the overall radiation dose. A further attraction to ^{99m}Tc is the half-life of 6 h, which provides sufficient time for labeling, administration and biodistribution without having to use excessive amounts of radioactivity to compensate for losses due to radioactive decay. The long half-life also provides the opportunity to perform protracted imaging studies, which can take up to several hours, without having to significantly increase the injected dose.

An important invention that made ^{99m}Tc widely available to all hospitals is the ^{99}Mo/^{99m}Tc generator, which was developed at the Brookhaven National

Laboratory in the late 1950s.[7] ^{99}Mo/^{99m}Tc generators are based on the decay of the parent ^{99}Mo to ^{99m}Tc which occurs with 87.5% efficiency,[8] the other principal pathway being direct decay to the isomer, ^{99}Tc. ^{99}Mo/^{99m}Tc generators contain high specific activity ^{99}Mo ($E_{\beta} - \max = 1.36$ MeV, $t_{1/2} = 66.0$ h), which is adsorbed onto an alumina (Al_2O_3) column as molybdate (MoO_4^{2-}). Flushing the column with a 0.9% saline (0.15 M NaCl) solution results in the selective elution of ^{99m}Tc as $^{99m}TcO_4^-$. The parent $^{99}MoO_4^{2-}$ remains bound to the column. The attractiveness of the generator lies in the fact that it provides a closed, self-shielded system for the production of sterile, non-pyrogenic and isotonic solutions of $^{99m}TcO_4^-$. Generators are routinely shipped, typically on a weekly basis, to nuclear medicine departments around the world.

Over the useful lifetime of a generator (typically 1–2 weeks), the amount of ^{99m}Tc that can be eluted is affected by several factors including column elution efficiency, the initial activity of the ^{99}Mo, and the elapsed time between elutions. After one half-life of ^{99m}Tc, approximately 47% of the maximum amount of technetium that is available can be eluted from a generator while after four half-lives (approximately 24 h), 88% of the available activity can be obtained. It is therefore convenient to 'milk' generators on a daily basis.

The first clinical use of ^{99m}Tc was reported by Harper in 1964 at the University of Chicago.[9] $^{99m}TcO_4^-$ eluted from a generator was used to probe the anatomical definition of various organs including the brain, the liver and the thyroid gland. These seminal studies led to the premise that ^{99m}Tc complexes of alternative chemical forms could enable imaging of different sites and/or biological processes because different ^{99m}Tc complexes would be expected to exhibit unique distributions *in vivo*.

In order to prepare complexes of ^{99m}Tc, TcO_4^- is reacted with a ligand in the presence of a reducing agent. A wide range of different reducing agents have been investigated with $SnCl_2$ being the most widely employed for preparing complexes of Tc(V) and Tc(I) while boron hydrides are used to prepare organometallic Tc(I) complexes (discussed later). The reduction potential of TcO_4^- to TcO_2 in acidic aqueous solutions is +0.738 V, which is comparable to the standard reduction potential for Fe^{3+}/Fe^{2+} (+0.771 V).[10]

The development of technetium 'instant kits' in 1968 was another important discovery that further propelled the clinical use of ^{99m}Tc.[11] A kit consists of a pre-mixed formulation of reducing agent and metal-complexing agent in a sterile vial to which generator eluent containing $^{99m}TcO_4^-$ is added. The desired product is then formed, usually after the application of heat, in high radiochemical yield and purity without having to perform complex synthetic or purification procedures.

An interesting aspect of the early development of ^{99m}Tc radiopharmaceuticals is that the structures of compounds produced at the tracer level were often not well understood. Determining the structures of ^{99m}Tc complexes unambiguously is complicated by the miniscule amounts of the isotope that is present at the tracer level (10^{-7}–10^{-10} M), which is below the detection limits of standard structural characterization techniques. This problem is overcome by using the long-lived isotope ^{99}Tc, which can be handled safely in large quantities (typically 10–250 mg per reaction) with only nominal shielding. Reactions with ^{99}Tc provide sufficient material to characterize new Tc complexes, which in turn serve as reference standards for those compounds produced at the tracer level. In the absence of a license to handle the long-lived ^{99}Tc isotope, which is considered a disposal problem, researchers can use rhenium (Re) to prepare well-characterized reference standards. This latter approach must be done cautiously because there are numerous examples of where the chemistry of the two congeners differs significantly.

18.3 Technetium Radiopharmaceuticals

$^{99m}TcO_4^-$ can be used to image the thyroid, salivary glands and ectopic gastric mucosa. To target Tc to different sites it is necessary to prepare complexes of the isotope to affect its physicochemical properties. The first generation of Tc radiopharmaceuticals were based on relatively simple coordination complexes of Tc. The distribution of these compounds, which are referred to as Tc-essential compounds, depends upon the charge, size, stability and lipophilicity of the metal complexes. With the increasing demand for compounds that target specific receptor systems, there has been a shift toward tagging biomolecules with Tc as a more rational means of achieving site selectivity. This latter class of compounds are referred to as Tc-tagged radiopharmaceuticals.

18.3.1 Tc-essential compounds

There are a significant number of different Tc complexes that have been employed as radiotracers.[12,13] Compounds of Tc(V) are particularly prominent as they are readily prepared in water from TcO_4^- via $SnCl_2$ reduction. Ceretec™, for example, is a Tc(V) complex of the hexamethylpropyleneamineoxime (HMPAO) ligand, which is used to image cerebral blood flow (Figure 18.1a).[14] The HMPAO ligand forms a neutral square pyramidal complex with the $\{TcO\}^{3+}$ core, which is a prominent feature in Tc(V) chemistry.[15] Neurolite™ (Figure 18.1b) is similarly used to image cerebral blood flow and consists of a chelate, 1-ethylcysteinate dimer (ECD), that is made up of two cysteine ethyl ester units which form a neutral complex with the $\{TcO\}^{3+}$ core.[16] The square pyramidal complex freely crosses the blood–brain barrier whereupon an ester group of one of the isomers of Tc-ECD is hydrolyzed by an esterase, which in turn prevents premature efflux of the imaging agent from the brain.

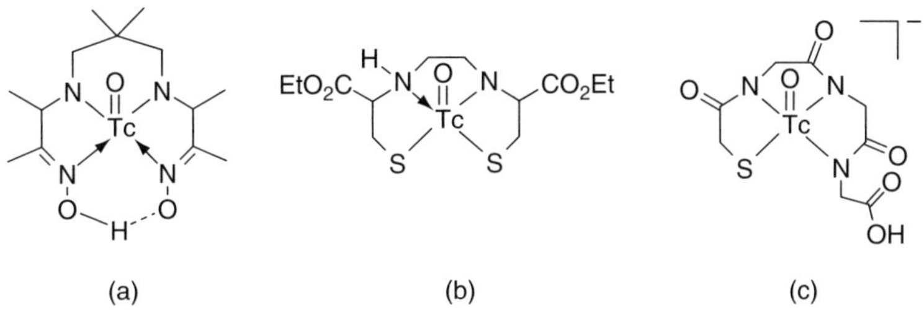

Figure 18.1 Tc(V) Chelate complexes

The ^{99m}Tc complex of mercaptoacetyltriglycine (MAG_3) (Figure 18.1c) is used as an agent to assess kidney function.[17,18] Like Ceretec™ and Neurolite™, $TcMAG_3$ forms a square pyramidal complex with Tc(V) with the basal plane consisting of the three nitrogen and one sulfur donor atoms. The carboxylic acid group, which can be used as a site to link targeting agents, does not coordinate to the metal center and is believed to help facilitate excretion via the kidneys.

One of the most important roles Tc radiopharmaceuticals play in diagnostic medicine is in helping to assess cardiac function. The structures of approved cardiac imaging agents vary significantly and involve coordination to different Tc cores. These include the technetium dioxo complex $[TcO_2\{P(R)_2CH_2CH_2P(R)_2\}_2]^+$ (Myoview™),[19] which has an approximately octahedral geometry about Tc (Figure 18.2a), and the Tc(III) complex, ^{99m}Tc-teboroxime (Cardiotec™), which contains boronic acid adducts of oxime type ligands. Tc-teboroxime is formally a seven-coordinate-complex containing a labile chloride as the seventh ligand.[20]

Technetium isonitrile complexes of the type $[Tc(CNR)_6]^+$ are also widely used to image heart function.[21] The hexakis-((2-methoxy-2-methyl-1-propyl) isonitrile) complex of Tc(I) (Figure 18.2b, Tc-Sestamibi, Tc-MIBI or Cardiolite™)

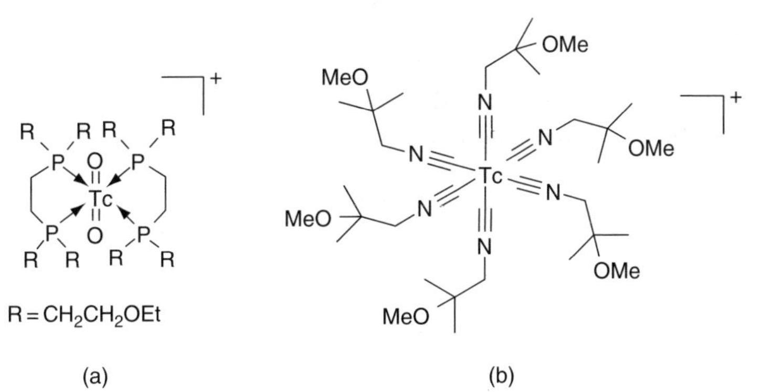

Figure 18.2 (a) ^{99m}Tc-teboroxime (Cardiotec™) (b) Tc-Sestamibi (Cardiolite™)

is used for myocardial perfusion imaging studies and breast cancer imaging (under the name Miraluma™).[22] ^{99m}Tc-Sestamibi is indicated for detecting coronary artery disease by localizing myocardial ischemia and infarction and for evaluating myocardial function. Miraluma™ is used as a second-line screening method after mammography for patients who have had an abnormal mammogram.

The preparation of Tc-isonitrile complexes in water[23] deserves special mention as it is recognized as a milestone in Tc chemistry. $[Tc(CNR)_6]^+$ complexes, unlike most organometallic compounds, can be prepared in high yield in aqueous solutions where the concentration of the metal is extremely low. Tc-Sestamibi is currently prepared using a kit which consists of, amongst other components, a tetrakis(isonitrile) Cu(I) complex as the source of the isonitrile ligands and stannous chloride as the reducing agent. The product can be prepared in high yield by heating the kit with $^{99m}TcO_4^-$ in boiling water for 10 min. Tc-Sestamibi was the first organometallic compound to be approved for routine clinical use as a diagnostic agent.

Another prevalent use of Tc is for the detection of cancer metastases to bone. Tc complexes of phosphonate ligands are routinely used to image increased calcium ion concentrations associated with stress to bone.[24] The mechanism of localization is believed to be associated with the interaction of the ligand hydroxyl groups with free calcium ions. One of the most widely employed agents is based on the methylenediphosphonate (MDP) ligand. The exact structure of the product(s) produced when MDP is labeled with ^{99m}Tc is not known conclusively.[25,26]

18.3.2 Tc-tagged compounds

Because it is difficult to rationally design Tc-essential radiopharmaceuticals that can target specific receptors, there has been a shift toward using biomolecules as a means of directing the biodistribution of Tc. This approach is non-trivial because Tc is a transition metal which complicates the labeling chemistry. In order to tag a biomolecule with Tc, the parent agent must have the appropriate arrangement of donor atoms which can form a stable metal complex or the targeting agent must be derivatized with a ligand that is capable of coordinating the radiometal.

Preparing Tc-tagged compounds requires that labeling not have a detrimental effect on the affinity of the parent targeting agent for the target receptor and that it not significantly alter the pharmacokinetics of the parent in an unfavorable manner. Furthermore, it is essential that the Tc complex remain intact *in vivo* so that the resulting images represent the distribution of the Tc bioconjugate and not a fragment of the parent radiopharmaceutical.

Unlike ^{11}C and ^{18}F, which are commonly used positron-emitting radionuclides, or ^{123}I which (like ^{99m}Tc) is used for single photon emission computed tomography (SPECT), Tc cannot be directly bound to the carbon backbone of a targeting agent by replacing a hydrogen atom. For compounds possessing the appropriate arrangement of good donor atoms such as thiol, amino or amido groups, it is possible to perform a direct labeling with technetium. This approach has been used to tag proteins and monoclonal antibodies; however, the regioselectivity of labeling is typically low, resulting in the formation of multiple species. Direct labeling can also disrupt the targeting ability of the parent compound and is therefore not widely employed.

A more effective approach to tagging a targeting agent with Tc is to use a bifunctional chelate (BFC). A BFC is a ligand that has the appropriate combination of donor atoms and structure to form a well-defined and stable/inert metal (Tc) complex while also possessing an additional functional group for bioconjugation. Labeling can be performed by first forming the chelate–Tc complex and then undertaking a bioconjugation reaction (known as the pre-labeling approach) or by appending the free ligand to the molecule and then labeling with Tc (post-labeling). Post-labeling is the more attractive of the two methods for clinical applications because it reduces the number of steps associated with handling radioactive materials and it is more amenable to the development of instant kits.

There is a virtual cornucopia of BFCs for Tc that have been developed.[27–30] Tc has eight accessible oxidation states which affords the opportunity to design structurally diverse ligand systems. The two main oxidation states that are most widely utilized to produce BFC complexes for nuclear medicine applications are Tc(V) and more recently Tc(I).

Bifunctional ligands of Tc(V)[31] involve, predominately, coordination complexes of the $\{TcO\}^{3+}$ core. This particular core is stabilized by a wide range of donors but has a preference for thiolate, amido and π-donating alkoxide ligands, which help satisfy the high formal charge at the metal center. Chelates are typically tetradentate and form complexes having square pyramidal geometries. These include ligands that contain all-nitrogen donors, like the N_4 propylene diamine dioxime (PnAO) type ligands and the oft-employed mixed nitrogen–sulfur ligands. Examples of the latter are triamidomonothiol (N_3S), monoamine monoamide (N_2S_2) (Figure 18.3a),[32,33] diamidodithiols (Figure 18.3b)[34] and diaminodithiol ligands (Figure 18.3c).[35] Amino acid analogues

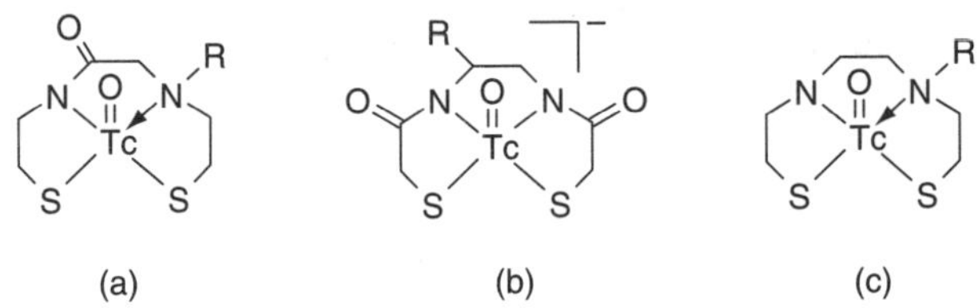

Figure 18.3 Tc-N_2S_2 Chelate complexes

of these compounds have also been prepared; however, once a stereogenic center is introduced into the backbone of the ligand, coordination to Tc results in the formation of a mixture of isomers.[36] These diastereomers, in which substituents about the stereogenic center are located syn or anti to the Tc-oxo bond, are often difficult to separate even at the macroscopic scale. For substituted PnAO[37] and certain amino acid-based monoamide monamino Tc(V) complexes,[38] even when the isomers are separated they have been shown to interconvert in solution.

A large number of different small-molecule–Tc(V)-chelate complexes have been prepared as novel radiotracers.[39] These include derivatives of steroids, which are designed to target estrogen receptors expressed on breast cancers, and biological dyes, which are designed to target amyloid plaques associated with Alzheimer's disease. ^{99m}Tc TRODAT-1 (Figure 18.4),[40] which is a tropane analogue containing a diaminodithiol ligand, deserves special notation as it is the first Tc complex to target specific receptors in the human brain. ^{99m}Tc TRODAT-1 has been used for imaging the central nervous system's (CNS) dopamine transporters (DAT) which is particularly useful for studying patients with Parkinson's disease.[41]

A limitation of tetradentate Tc(V) ligands arises when trying to label proteins and peptides, which are attractive targeting agents for delivering Tc to specific receptor systems (discussed later). Peptides can contain sequences of amino acids which are ideally set up to bind Tc(V) in a manner analogous to the ligands described above. This creates the same set of problems encountered with direct labeling in that the regioselectivity of labeling can be low thereby resulting in the formation of complex mixtures of products.

For tagging proteins and peptides with Tc, substituted hydrazine ligands, specifically hydrazinonicotinamides, were developed.[42] These ligands form stable organohydrazino complexes with Tc (Figure 18.5a)[43] even in the presence of competing thiolate ligands from cysteine-containing amino acids. The remaining coordination sites on Tc are occupied by a co-ligand (L). There are a wide variety of co-ligands that can be employed,[44,45] including tricine, mannitol and trisodium triphenylphosphine-3′,3″,3‴-trisulfonate (TPPTS), which is a convenient means of modifying the hydrophilicity/hydrophobicity of the Tc conjugate.[46]

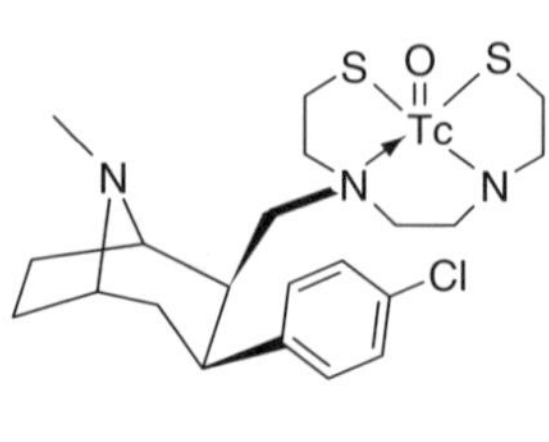

Figure 18.4 ^{99m}Tc TRODAT-1

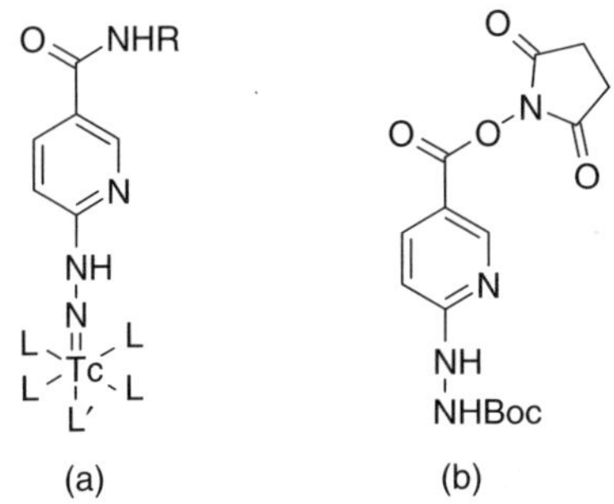

Figure 18.5 (a) Hydrazinonicotinamide complex of Tc-containing co-ligands (L, L′) and (b) active ester derivative of a protected form of the ligand for tagging amino groups

N-hydroxysuccinimidylhydrazinonicotinamide, which is commonly referred to as HYNIC, is the best-known Tc–hydrazine ligand. The active ester of a Boc-protected form of the ligand (Figure 18.5b) provides a convenient means for tagging amino groups on biomolecules. HYNIC conjugates of a variety of targeting agents including chemotactic peptides,[47,48] somatostatin analogues,[49] leukotriene B_4 (LTB_4)[50] and vitronectin receptor antagonists,[51] and large biomolecules like human IgG[52] have been reported. It is therefore somewhat surprising that there are no approved radiopharmaceuticals derived from this particular ligand. One of the drawbacks of HYNIC is that producing macroscopic quantities of reference standards (^{99}Tc or Re complexes) has proven to be difficult, which is a result of the complex coordination chemistry of the ligand.[53] Notwithstanding, HYNIC continues to be one of the most widely employed ligands in Tc radiopharmaceutical chemistry.

Because of their large size, traditional Tc–chelate/ligand complexes can influence the 'homing' ability of targeting agents particularly for low molecular weight compounds. Furthermore, conventional Tc ligands can undergo redox and/or transchelation reactions *in vivo* resulting in premature loss of the radionuclide. These characteristics along with the increasing demand for radiopharmaceuticals that target specific receptor systems has motivated a search for new Tc synthons, which are smaller, more robust and have better-defined structures.

The Tc(I) isonitrile complexes on which Cardiolite™ is based are inert owing in part to their low-spin d^6 electronic configuration. Tc(I) complexes would therefore appear to be ideal synthons from which to prepare receptor-targeted radiopharmaceuticals. Unfortunately the poly-substituted isonitrile complexes are difficult to monosubstitute and they are relatively non-reactive making them unsuitable starting materials from which to prepare Tc-tagged radiopharmaceuticals.

Alberto and co-workers reported the synthesis of $[Tc(CO)_3(OH_2)_3]^+$ as a practical and versatile Tc(I) synthon.[54] In this complex, the three facially oriented water molecules are sufficiently labile that they can be readily displaced by a variety of mono-, bi- and tridentate ligands (discussed later). The synthesis of the Tc(I) precursor involves reduction of TcO_4^- with borohydride in the presence of CO (Scheme 18.1). The $Tc(CO)_3^+$ core can be prepared at the macroscopic scale with ^{99}Tc and at the tracer level using ^{99m}Tc. A further innovation was the development of a kit for $[^{99m}Tc(CO)_3(OH_2)_3]^+$, which contains potassium boranocarbonate ($K_2H_3BCO_2$); a compound that acts as both a reducing agent and a source of CO.[55]

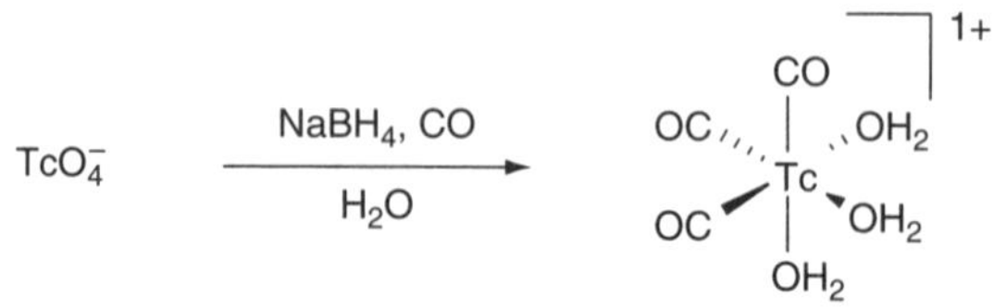

Scheme 18.1

The $[Tc(CO)_3]^+$ core interacts with a broad range of different ligands including thioureas, isonitriles and phosphines. The resulting complexes are for the most part octahedral, d^6-low spin and are therefore typically inert, which is ideal for radiopharmaceutical applications. Aliphatic amines and carboxylates coordinate rapidly with $[Tc(CO)_3]^+$; however, the resulting complexes are more reactive than complexes containing soft donors. Thioethers, for example, form inert complexes but do so at a very slow rate. A compromise between the rapid coordination of hard donors and the inertness of complexes containing soft donors can be found for ligands that contain aromatic amines, including substituted pyridines and imidazoles.[56]

Access to $[Tc(CO)_3(OH_2)_3]^+$ is transforming Tc radiopharmaceutical chemistry because it provides a means to append inert and low molecular weight Tc complexes to targeting vectors.[57,58] There are a number of bifunctional Tc(I) ligands that are now available for preparing bioconjugates (Figure 18.6). These include N_2O, N_3 and NSN type ligands which form neutral or cationic complexes with the $[Tc(CO)_3]^+$ core.[59,60] Schibli and co-workers investigated the reaction of the $[Tc(CO)_3]^+$ core with a variety of bi- and tridentate pyridine- or imidazole-type ligands to gain insight into the optimal chelate type and 'denticity' for radiopharmaceutical development.[61] Their study revealed that

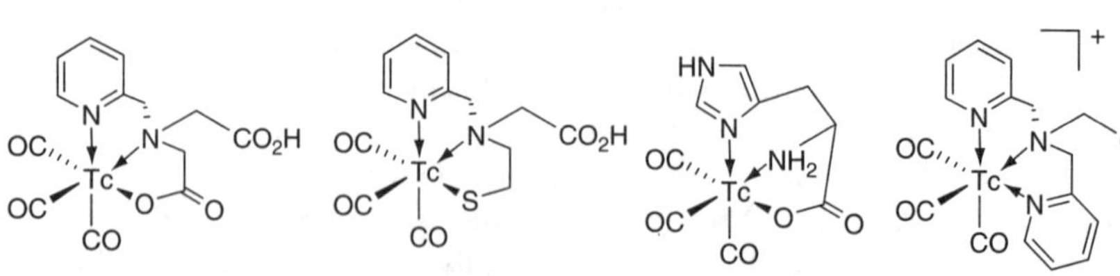

Figure 18.6 Examples of tridentate ligand complexes of Tc(I)

both the bidentate and tridentate chelates bound rapidly with the $[Tc(CO)_3]^+$core on a macroscopic scale and at the tracer level. The tridentate systems were shown to be superior for developing Tc-tagged compounds because they demonstrated enhanced stability in ligand challenge experiments and the complexes remained intact *in vivo*. The bidentate analogues, in contrast, decomposed after being incubated with histidine and cysteine for 24 h and they showed significant binding to plasma after one hour of incubation in human blood.

18.4 Peptide-Targeted Radiopharmaceuticals

Oligopeptides have been actively investigated as vectors for targeting radionuclides to specific receptor systems. For example, ^{111}In-labeled Octreotide, which is a somatostatin analogue, is used clinically to image neuroendocrine tumors.[62] There have been a substantial number of peptides that have been labeled with Tc using both tetradentate $N_xS_{(4-x)}$ and HYNIC-type ligands. These include compounds targeted at the formyl peptide receptor for imaging sites of infection and inflammation and bombesin receptors for tumor imaging. Reviews of peptide–Tc conjugates can be found throughout the literature.[63–66]

Because different amino acid residues can bind technetium, it is essential that the structure of Tc–peptide complexes that are being evaluated as novel radiopharmaceuticals be determined unambiguously. Although this would be best done by X-ray crystallography, the majority of ^{99}Tc-labeled peptides do not form single crystals. As a result, multi-NMR spectroscopy has proven to be a very useful tool for studying the structure of ^{99}Tc–peptide derivatives.

A recent example of the work involved in characterizing a peptide-based metalloradiopharmaceutical was reported by Francesconi and co-workers.[67] [^{99m}TcO] apcitide is a cyclic peptide derivative that is designed to target the GPllb/llla receptor, which is expressed on activated platelets, as a means of imaging deep vein thrombosis in the lower extremities. The peptide as a bis-succinimidomethyl ether-linked dimmer is available as a kit (trade name AcuTect) and is approved for clinical use in the United States.

The peptide is engineered in such a manner that the metal is designed to bind to the three deprotonated amides and thiol of the Gly^{11}-Gly^{12}-Cys^{13}-binding motif (an N_3S ligand). The product produced by labeling the peptide with macroscopic amounts of ^{99}Tc (using $TcOCl_4^-$) showed similar retention time to the product produced at the tracer level. NMR studies of [^{99}TcO] apcitide, following HPLC purification, in H_2O/D_2O mixtures showed the loss of the amide protons associated with coordination of Tc to the deprotonated nitrogen atoms of the Gly-Gly-Cys sequence. This was accompanied by changes in the chemical shift and splitting patterns of the α-methylene protons of the glycine amino acids and the α-methine proton of the cysteine CH group. These shifts, which are similar to those reported for simpler but analogous N_3S chelates, support the belief that the metal binds to the target sequence. It should also be noted that infrared spectroscopy confirmed that the complex contains, as would be expected, the $\{TcO\}^{3+}$ core. The Tc=O stretching vibration in Tc(V) coordination complexes is characteristic and typically appears between 890 and 1020 cm^{-1}.

18.5 Current Areas of Research

18.5.1 Bioorganometallic chemistry

In addition to coordination complexes, there is a significant amount of interest in preparing sandwich complexes of the $Tc(CO)_3^+$ core.[68,69] $CpTc(CO)_3$ (Cp = cyclopentadienide) is an attractive and highly sought-after synthon for developing Tc-tagged radiopharmaceuticals because of the complex's small size and stability. The synthesis of organometallic complexes of ^{99m}Tc is challenging in that they must be prepared in aqueous media (organometallic complexes are traditionally prepared under anhydrous reaction conditions) where the concentration of the metal is very low.[70]

Wenzel was the first to report a method for synthesizing $CpTc(CO)_3$ at the tracer level from $^{99m}TcO_4^-$.[71] The reaction (Scheme 18.2) involved heating $^{99m}TcO_4^-$ in the presence of ferrocene as the source of Cp and $Mn(CO)_5Br$ as the CO reservoir. Spradau and Katzenellenbogen reported an enhanced version of the Wenzel reaction in which $Cr(CO)_6$ was used as the source of CO and $CrCl_3$ added as a reducing agent.[72] The latter reaction conditions have been used as a means of preparing a variety of different sandwich compounds including an organometallic derivative of Octreotide.[73] One of the major limitations of these methods is that they require the use of organic solvents and harsh reaction conditions which precludes routine clinical use and direct labeling of certain classes of biomolecules including peptides.

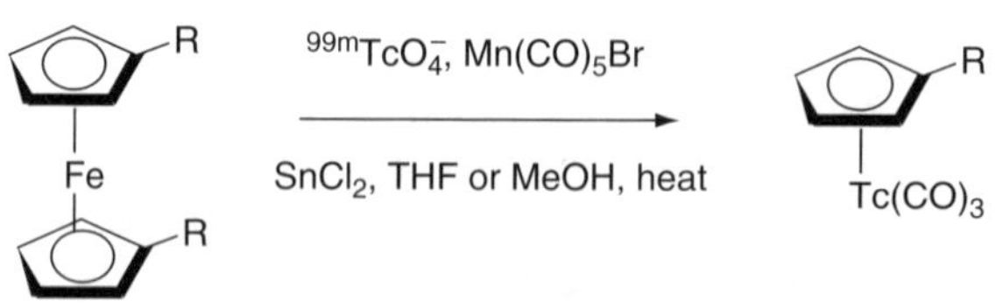

Scheme 18.2

One of the reasons for needing such complex synthetic protocols is that Cp does not react readily with metals in aqueous solution and its conjugate base tends to oligomerize in water particularly when heated. Alberto and co-workers developed an ingenious solution to the problem by derivatizing the Cp ring with an acyl group which increases the stability of Cp thus preventing unwanted side reactions. Reaction of acyl-Cp derivatives with the $Tc(CO)_3^+$ core afforded good to excellent yields of $RC(O)CpTc(CO)_3$ complexes at both the macroscopic and tracer levels (Scheme 18.3).[74] This approach was recently used to prepare a half-sandwich complex of technetium linked to a serotonergic ligand.[75]

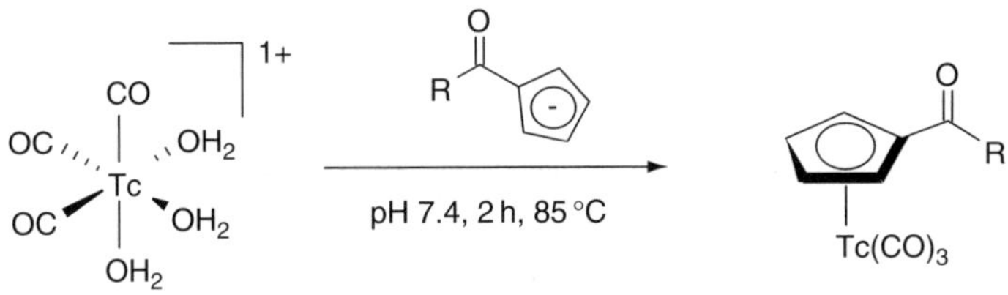

Scheme 18.3

As an alternative to Cp-type ligands, our group developed a means of preparing half-sandwich compounds of Tc using *nido*-carborane ligands of the general type $(R_2C_2B_9H_{10})^-$.[76] When deprotonated, these ligands are isolobal to cyclopentadienide and they form sandwich-type compounds in which the metal is bound to the open face of the cluster. Carboranes are attractive ligands for radiometal chemistry because, unlike traditional organometallic ligands, they are compatible with the aqueous reaction environment and the products are typically more stable than the corresponding Cp complexes. One additional advantage of carboranes over traditional ligands is that they can be readily functionalized with a wide range of different groups at one or more of the carbon and/or boron vertices regioselectively. This affords a tremendous amount of flexibility when designing novel compounds.

Our original method for the synthesis of Tc–carborane complexes in water involved reacting $[Tc(CO)_3(OH_2)_3]^+$ with a *nido*-carborane in the presence of NaOH. The product $[(\eta^5\text{-}C_2B_9H_{11})Tc(CO)_3]^-$, which can also be prepared having a linker arm for bioconjugation, was stable toward histidine and cysteine challenge experiments. The same reaction applied at the tracer level afforded only low yields of the ^{99m}Tc–carborane complexes. More recently, we discovered that fluoride can be used in place of hydroxide to promote the synthesis of Re/Tc-metallocarboranes at both the macroscopic and the tracer levels in water (Scheme 18.4).[77] This is a significant discovery because it creates the opportunity to directly label carborane–biomolecule conjugates under mild reaction conditions. The ^{99m}Tc-labeled carboranes, like the ^{99}Tc analogues, were produced in very high yields and they are stable to ligand challenge experiments.

18.5.2 Tc and solid supports

One of the drawbacks of Tc-tagged compounds is that they are typically designed in such a manner that the bioconjugates are capable of binding to the target

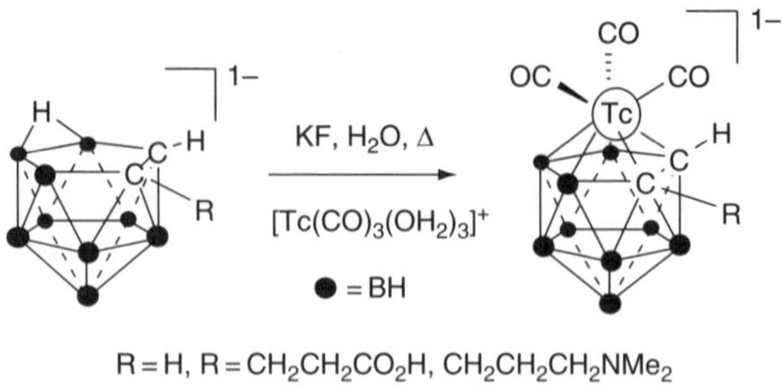

Scheme 18.4

receptor whether or not the radionuclide is present. In view of the fact that during labeling there is large excessive of ligand present compared to Tc (more than 10 000-fold), the unlabeled material must be removed prior to use because it can displace the Tc complex from the target site thereby reducing image quality/ accuracy. This is of course not an issue for Tc-essential compounds where the free ligand and Tc complex have completely different biodistribution profiles.

The simplest solution to this problem is to purify the labeling mixture by HPLC. This approach, although effective, is not suitable for the daily labelings done in hospitals. An alternative approach for the preparation of 'ligand-free' samples is to label compounds that are bound to an insoluble support in such a manner that upon coordination of Tc only the desired complex is released into solution. Unreacted ligand remains bound to the support and is removed by simple filtration.

There have been several approaches for carrying out solid-phase labeling with Tc and/or Re.[78–80] Thornback, Ballinger and co-workers[81] showed that it was possible to label an N_3S chelate that was connected to a series of different solid supports through a succinimidyl linker via the ligand-thiol group. Although these systems afforded high specific activity formulations (specific activity in this instance refers to the amount of activity divided by the amount of free ligand present), radiochemical yields varied significantly (13–80%). Alternatively, the same chelate, derivatized with a peptide, was linked through the thiol to a gold support. Upon labeling with ^{99m}Tc, the desired product was isolated in excellent radiochemical yield (Scheme 18.5).[82]

Alberto and co-workers developed a solid-phase labeling system for the $Tc(CO)_3^+$ core.[83] Tridentate chelates comprised of different combinations of aliphatic and aromatic amines and carboxylates were linked to a TentaGel-type support. This resin, which is a polystyrene/poly(ethylene glycol)-based support, was selected because it swells in water and is stable to elevated reaction temperatures. Cleavage of the tertiary amine linker proceeded in modest to good yields with ^{99}Tc and ^{99m}Tc but interestingly not with Re. For reactions involving $^{99m}Tc(CO)_3^+$, the products produced had very high specific activities (refer to the above definition).

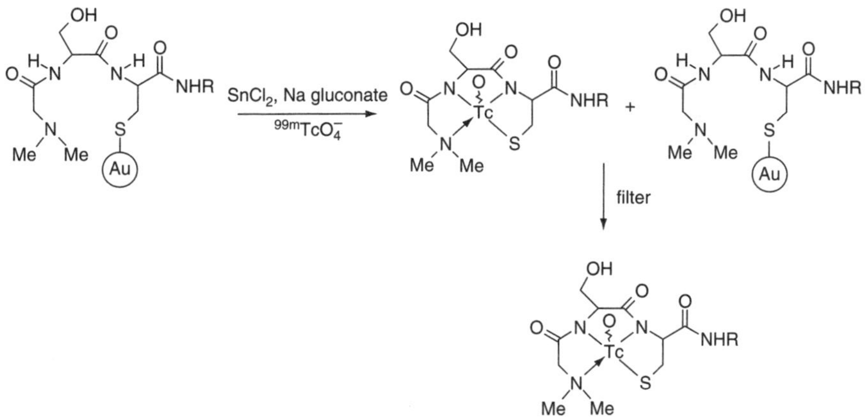

Scheme 18.5

Beyond labeling, solid-phase synthesis (SPS) is also starting to play an increasingly important role in the development of new radiopharmaceuticals. SPS affords the opportunity to prepare libraries of chelate–biomolecule conjugates as a more rapid and efficient means of identifying viable Tc radiopharmaceuticals. The majority of SPS research involving Tc has been focused on the synthesis of peptide conjugates.

The preparation of peptide-targeted Tc radiopharmaceuticals is well suited to SPS because most bifunctional Tc ligands contain amino or carboxylic acid linker groups. Standard automated peptide synthesis protocols can be used to link these bifunctional ligands to resin-bound peptides and to vary the factors which have the greatest impact on receptor-binding affinity and biodistribution of technetium-labeled peptides.[84] These include the site of attachment of the Tc ligand, the presence and type of a spacer group between the ligand and the peptide, and the nature of the Tc–ligand complex (charge, size, lipophilicity, etc.). SPS offers a convenient means of identifying optimal combinations of these factors because libraries of bioconjugates can be prepared in parallel for those instruments that have multiple reaction wells.

Several methods have been developed for linking Tc ligands to peptides on solid supports. Our group showed that N_3S-type chelates and their Re complexes could be conveniently incorporated into peptides when bound to Merrifield-type resins (Scheme 18.6).[85] Hoffman *et al.* used a similar method for linking a chelate and its Re complex, through a variety of different spacer groups, to a peptide which targets bombesin receptors on tumor cells.[86] Gariépy and co-workers described a method for adding an N_2S_2 chelate (a diamidodithiol ligand) to the N-terminus of peptides bound to a support.[87]

Blower and co-workers recently described a novel SPS method in which a HYNIC derivative of 9-fluorenylmethyl carbamate (Fmoc)-protected lysine (Figure 18.7) was used as a metal-binding amino acid analogue.[88] This particular

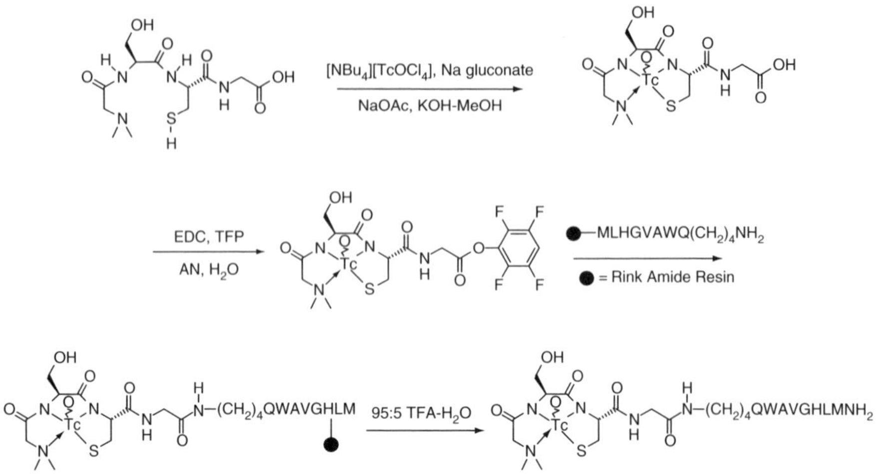

Scheme 18.6

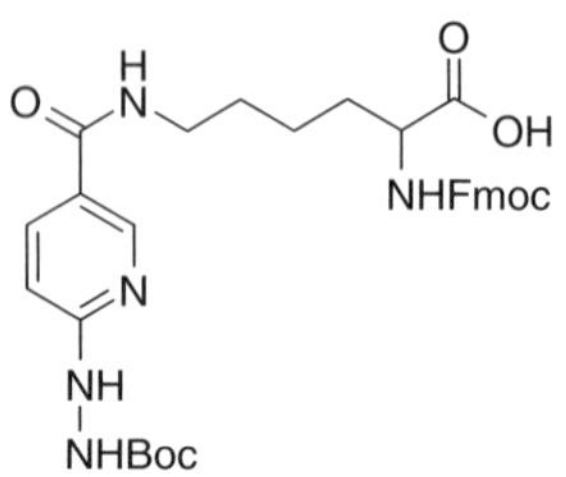

Figure 18.7 HYNIC derivative of 9-fluorenylmethyl carbamate (Fmoc) protected-lysine; a metal-binding amino acid analogue

ligand is attractive because it can be incorporated within the backbone of peptide, as opposed to simple N or C termini conjugation, using standard Fmoc solid-phase peptide chemistry. The lysine–HYNIC ligand was used to prepare a Tc-labeled analogue of salmon calcitonin, a 32-amino acid neuropeptide, as a means of targeting the radionuclide to sites of bone disease and tumors. An analogous system which can bind the $\{TcO\}^{3+}$ core was reported by Hunter and Luyt. They described the synthesis of an Fmoc-protected-lysine-derived bifunctional chelator that can be incorporated into growing peptides.[89]

Our group, as part of a collaborative research effort, recently described the development of a new SPS method for the preparation of Tc(I)- and Re(I)-labeled peptides.[90] We showed that a tridentate chelate derived from lysine (referred to as a single amino acid chelate or SAAC) which binds the $Tc(CO)_3^+$ core can be incorporated into the backbone of peptides using a conventional automated synthesizer (Scheme 18.7). One particular advantage to working

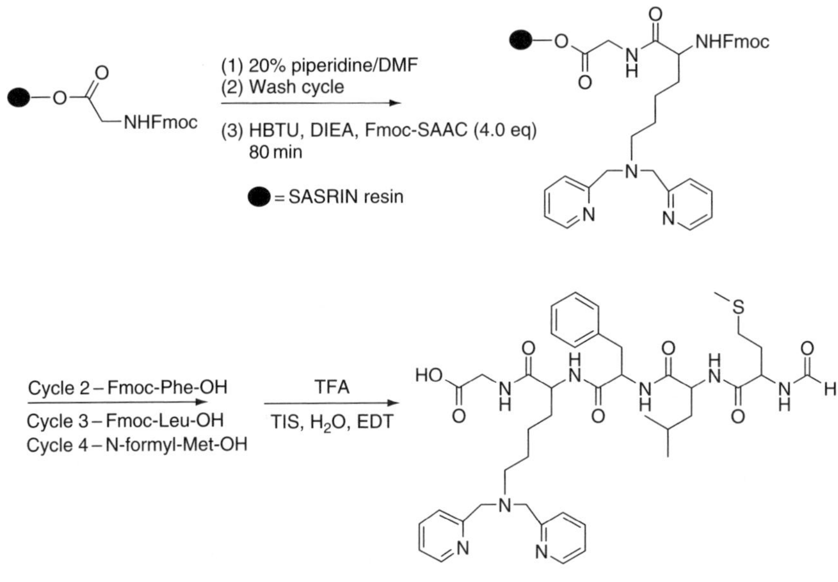

Scheme 18.7

with the $M(CO)_3^+$ core is that the coordination complexes are by and large inert (discussed earlier). In the case of the SAAC ligand, this feature makes it possible to incorporate the Re complex into the peptide using exactly the same procedure that is used to prepare the ligand. This affords a convenient means of preparing well-defined reference standards, for when it is time to label the peptide-ligand conjugates with ^{99m}Tc. As mentioned previously, having a fully characterized product is an essential part of developing and receiving approval for any new radiopharmaceutical.

18.5.3 Technetium-94m

One of the limitations of imaging compounds labeled with ^{99m}Tc by SPECT is that it is difficult to acquire quantitative biodistribution and kinetic data. As a result, there has been a growing interest in producing and using a positron-emitting radionuclide of Tc.

In positron emission tomography (PET), two collinear γ rays which appear as a result of positron annihilation are detected in coincidence by a ring of detectors which surround the subject under study. PET has better resolution and sensitivity than SPECT and because of the coincidence detection, obtaining quantitative data for PET is relatively straightforward, which is not the case for SPECT.

The most promising of the four positron-emitting isotopes of Tc is ^{94m}Tc. ^{94m}Tc has a half-life of 52.5 min and a positron branching ratio of 0.72. ^{94m}Tc can be produced via three main methods which include bombardment of ^{94}Mo with protons (^{94}Mo(p,n)^{94m}Tc), ^{93}Nb with ^{3}He (^{93}Nb(^{3}He,2n)^{94m}Tc), and ^{92}Mo with α particles (^{92}Mo(α,pn)^{94m}Tc).[91] The irradiation of natural or enriched molybdenum targets with protons appears to be the most effective, feasible and widely employed production method. It can be performed on small medical cyclotrons which are becoming increasingly available as the clinical role of PET continues to grow.

The use of ^{94m}Tc for quantitative studies is more complicated than for more traditional positron-emitting isotopes like ^{18}F and ^{11}C.[92] For these isotopes, the major decay pathway is positron emission. The photons produced by annihilation are the only source of γ rays. When ^{94m}Tc decays to ^{94}Mo, positron emission is accompanied by multiple γ ray emissions with the most abundant having an energy of 871 keV (92%). These γ rays, after scattering, can be detected by the PET camera when they are coincident with an annihilation photon which in turn increases the data scatter and complicates the acquisition of quantitative data. Another problem is that the production of ^{94m}Tc is accompanied by the production of ^{94}Tc which also decays via positron and γ ray emissions. This issue is further complicated by the fact that ^{94}Tc and ^{94m}Tc have different half-lives (293 min vs 52.5 min), which means that the ratio of ^{94}Tc to ^{94m}Tc increases with time. The change in ratio must be taken into account during any decay corrections.

Despite these issues, approaches have been developed which enable quantitative data to be obtained from ^{94m}Tc-labeled compounds. Synthesis and biodistribution studies of ^{94m}Tc analogues of established ^{99m}Tc radiopharmaceuticals like Sestamibi[93] and teboroxime[94] have been reported. ^{94m}Tc has also been used to label receptor-targeted radiopharmaceuticals including an estradiol derivative that was designed to target tumors that are ER-positive estrogen receptor.[95] In this case, an iminodiacetic acid-thioether tridentate chelate was linked to the 7α position of the steroid and the bioconjugate labeled with the $^{94m}Tc(CO)_3^+$. The biodistribution of the steroid was subsequently evaluated using a dedicated small animal PET scanner along with the counting of excised tissues.

18.6 Outlook

There are many exciting areas of research associated with Tc that are open to further exploration. These span from basic coordination chemistry, through radiochemistry and radiopharmacy research, to radiopharmaceutical development. Advances in these areas will require contributions from not only inorganic and radiopharmaceutical chemists but also scientists with a broader set of skills including those having expertise in organic chemistry and drug discovery, imaging physics and biology/biochemistry.

Modern drug discovery techniques, which are revolutionizing pharmaceutical research, are only just beginning to have an impact on Tc-radiopharmaceutical chemistry. The number of Tc radiopharmaceuticals approved lately is disappointing when compared to the amount of effort being put into developing new agents.[96] In order to discover new Tc radiopharmaceuticals in a time- and resource-efficient manner it is essential that new methods for library construction, for systems beyond simply peptides, be developed. Furthermore, high throughput screening systems must be developed which allow these libraries to be rapidly evaluated for those factors that are relevant to radiopharmaceutical development.

One of the key steps to achieving the above-stated goal is the continuing exploration of the fundamental coordination and radiochemistry of Tc. The potential impact of new Tc chemistry is illustrated by the effect that the development of the $Tc(CO)_3^+$ synthon is having on modern Tc radiopharmaceutical chemistry. This particular core, amongst other things, enables solid-phase synthetic procedures to be performed, which were not possible with preexisting Tc synthons. New cores will similarly open new avenues for radiopharmaceutical development.

Another area for future research is in combining the diagnostic power of technetium with the therapeutic capability of other radioisotopes. Technetium's congener, Re, has two isotopes (^{186}Re and ^{188}Re) which are available commercially and which can be used to create radiotherapy agents. There is an oft-cited avowal that Tc and Re can be considered a 'matched pair' for imaging and therapy, the idea being that the Tc complex can be used for detecting and staging diseases like cancer while the Re complex, which would in theory posses the same biodistribution profile, would then be used for therapy. Unfortunately the stability of $^{186/188}$Re complexes is often much less than the ^{99m}Tc analogue meaning that the biodistribution profiles are in fact often different. New ligands which are capable of preventing premature decomposition of radio-rhenium complexes is a major obstacle to making the matched-pair dream a reality.

An additional area that deserves special notation is the use of animal imaging.[97] Scanners that are dedicated for animal imaging are now available for both PET[98,99] and SPECT.[100,101] These instruments afford the opportunity to obtain images of animal models at a higher resolution than with the counterparts designed for humans. Furthermore, researchers who have animal scanners are capable of performing multiple imaging studies in the same animal, which increases the accuracy of experiments, without being encumbered by clinical demands on an instrument.

As the pharmaceutical industry continues to recognize the importance of molecular and animal imaging in drug development,[102,103] ^{99m}Tc-labeled compounds will begin to play an important role in drug development and testing. With the ability to incorporate Tc into peptides that target specific receptors, it is conceivable that in the future, lead pharmaceutical candidates identified from large *in vitro* screening studies will subsequently be evaluated for their ability to displace receptor-specific Tc ligands *in vivo*. Screening *in vivo* offers a means of evaluating compounds in a more realistic environment than is presently offered by screens involving isolated cells or receptors.

References

1. (i) W. Noddack, I. Tacke, O. Berg, *Naturwissenschaften* **13**, 567 (1925). (ii) I. Tacke, The discovery of eka-manganeses, *Angew. Chem.* **38**, 1157–1160 (1925).
2. (i) C. Perrier, E. Segrè, Some chemical properties of element 43, *J. Chem. Phys.* **5**, 712–716 (1937). (ii) C. Perrier, E. Segrè, Radioactive isotopes of element 43, *Nature* **140**, 193–194 (1937). (iii) E. Segrè, Radioactive isotopes of element 43, *Nuovo Cemento* **14**, 528 (1937).
3. C. Perrier, E. Segrè, Technetium: The element of atomic number 43, *Nature* **159**, 24 (1947).
4. P.W. Merrill, Technetium in the N-type star, *19 Piscium Publs. Astron. Soc. Pacific* **68**, 70–71 (1956).
5. B.T. Kenna, P.K. Kuroda, Isolation of naturally occurring technetium, *J. Inorg. Nucl. Chem.* **23**, 142–144 (1961).
6. E. Segrè, G.T. Seaborg, Nuclear isomerism in element 43, *Phys. Rev.* **54**, 772 (1938).
7. P. Richards, A survey of the production of Brookhaven National Laboratory of radioisotopes for medical research, *Trans. 5th Nuclear Congress, 7th Int. Electronic Nuclear Symposium* (Rome), 223–244 (1960).
8. J. Varma, C.E. Mandeville, The disintegration of ^{99}Mo, *Phys. Rev.* **94**, 91–94 (1954).
9. P.V. Harper, R. Beck, D. Charleston, K.A. Lathrop, Optimalization of a scanning method using ^{99m}Tc, *Nucleonics* **22**, 50–54 (1964).
10. The CRC Handbook of Chemistry and Physics, 64th Edition, Robert C. Weast (ed.) CRC Press, Cleveland, Ohio, USA, 1984.
11. W.C. Eckelman, P. Richards, Instant ^{99m}Tc-DTPA, *J. Nucl. Med.* **11**, 761–762 (1970).
12. J.R. Dilworth, S.J. Parrott, The biomedical chemistry of technetium and rhenium, *Chem. Soc. Rev.* **27**, 43–55 (1998).
13. K. Schwochau, Technetium radiopharmaceuticals – fundamentals, synthesis, structure and development, *Angew. Chem. Int. Ed.* **33**, 2258–2267 (1994).
14. J.P. Leonard, D.P. Novotnik, R.D. Nierinckx, Technetium-99m-d, 1-HM-PAO: A new radiopharmaceutical for imaging regional brain perfusion using SPECT – a comparison with iodine-123 HIPDM, *J. Nucl. Med.* **27**, 1819–1823 (1986).
15. S. Jurisson, E.O. Schlemper, D.E. Troutner *et al.*, Synthesis, characterization, and X-ray structural determinations of technetium(V)-oxo-tetradentate amine oxime complexes, *Inorg. Chem.* **25**, 543–549 (1986).
16. R.C. Walovitch, T.C. Hill, S.T. Garrity *et al.*, Characterization of technetium-99m-L, L-ECD for brain perfusion imaging, Part 1: Pharmacology of technetium-99m ECD in nonhuman primates, *J. Nucl. Med.* **30**, 1892–1901 (1989).
17. A.R. Fritzberg, S. Kasina, S.D. Eshima, D.L. Johnson, Synthesis and biological evaluation of technetium-99m MAG_3 as a hippuran replacement, *J. Nucl. Med.* **27**, 111–116 (1986).
18. K. Itoh, ^{99m}Tc-MAG_3: Review of pharmacokinetics, clinical application to renal diseases and quantification of renal function, *Ann. Nucl. Med.* **15**, 179–190 (2001).
19. J.D. Kelly, A.M. Forster, B. Higley *et al.*, Technetium-99m-tetrofosmin as a new radiopharmaceutical for myocardial perfusion imaging, *J. Nucl. Med.* **34**, 222–227 (1993).

20. K.E. Linder, M.F. Malley, J.Z. Gougoutas *et al.*, Neutral, seven-coordinate dioxime complexes of technetium(III): Synthesis and characterization, *Inorg. Chem.* **29**, 2428–2434 (1990).
21. B.L. Holman, A.G. Jones, J. Lister-James *et al.*, A new technetium-99m-labeled myocardial imaging agent, hexakis(t-butylisonitrile)-technetium(I) [Tc-99m TBI]: Initial experience in the human, *J. Nucl. Med.* **25**, 1350–1355 (1984).
22. E. Prats, F. Aisa, M.D. Abos *et al.*, Mammography and ^{99m}Tc-MIBI scintimammography in suspected breast cancer, *J. Nucl. Med.* **40**, 296–301 (1999).
23. M.J. Abrams, A. Davison, A.G. Jones *et al.*, Synthesis and characterization of hexakis(alkyl isocyanide) and hexakis(aryl isocyanide) complexes of technetium(I), *Inorg. Chem.* **22**, 2798–2800 (1983).
24. L.M. Lamki, Bone scintigraphy current trends and future prospects, *J. Nucl. Med.* **26**, 312–314 (1985).
25. K. Libson, E. Deutsch, B.L. Barnett, Structural characterization of a technetium-99-diphosphonate complex. Implications for the chemistry of technetium-99m skeletal imaging agents, *J. Am. Chem. Soc.* **102**, 2476–2478 (1980).
26. J.L. Martin Jr, J. Yuan, C.E. Lunte *et al.*, Technetium-diphosphonate skeletal imaging agents: EXAFS structural studies in aqueous solution, *Inorg. Chem.* **28**, 2899–2901 (1989).
27. R. Alberto, Technetium, *Compr. Coord. Chem. II* **5**, 127–270 (2004).
28. P. Thornton, Manganese, technetium and rhenium, *Annual Reports on the Progress of Chemistry, Section A: Inorganic Chemistry* **99**, 191–200 (2003).
29. P.J. Blower, Inorganic pharmaceuticals, *Annual Reports on the Progress of Chemistry, Section A: Inorganic Chemistry* **97**, 587–603 (2001).
30. B.I. Kharisov, M.A. Mendez-Rojas, State-of-the-art coordination chemistry of radioactive elements, *Russ. Chem. Rev.* **70**, 865–884 (2001).
31. B. Johannsen, H. Spies, Technetium(V) chemistry as relevant to nuclear medicine, *Top. Curr. Chem.* **176**, 77–121 (1996).
32. L.M. Gustavson, T.N. Rao, D.S. Jones *et al.*, Synthesis of a new class of technetium chelating agents: N_2S_2 monoaminemonoamide (MAMA) ligands, *Tetrahedron. Lett.* **32**, 5485–5488 (1991).
33. J.P. O'Neill, S.R. Wilson, J.A. Katzenellenbogen, Preparation and structural characterization of monoamine-monoamide bis(thiol) oxo complexes of technetium(V) and rhenium(V), *Inorg. Chem.* **33**, 319–323 (1994).
34. D. Brenner, A. Davison, J. Lister-James, A.G. Jones, Synthesis and characterization of a series of isomeric oxotechnetium(V) diamido dithiolates, *Inorg. Chem.* **23**, 3793–3797 (1984).
35. H.F. Kung, M. Molnar, J. Billings *et al.*, Synthesis and biodistribution of neutral lipid-soluble technetium-99m complexes that cross the blood–brain barrier, *J. Nucl. Med.* **25**, 326–332 (1984).
36. R.A. Bell, B.E. McCarry, J.F. Valliant, The synthesis, NMR spectroscopy, and X-ray structure of a new rhenium N_2S_2 chelate complex, *Inorg. Chem.* **37**, 3517–3520 (1998).
37. J.E. Cyr, D.P. Nowotnik, Y. Pan *et al.*, Technetium(V) oxo complexes of substituted propylene diamine dioxime (PnAO) ligands: Water-dependent interconversion between *syn* and *anti* isomers, *Inorg. Chem.* **40**, 3555–3561 (2001).
38. E. Wong, T. Fauconnier, S. Bennett *et al.*, Rhenium(V) and technetium(V) oxo complexes of an N2N′S peptidic chelator: Evidence of interconversion between the *syn* and *anti* conformations, *Inorg. Chem.* **36**, 5799–5808 (1997).

39. S.S. Jurisson, J.D. Lyden, Potential technetium small molecule radiopharmaceuticals, *Chem. Rev.* **99**, 2205–2218 (1999).
40. H.F. Kung, Development of Tc-99m labeled tropanes: TRODAT-1, as a dopamine transporter imaging agent, *Nucl. Med. Biol.* **28**, 505–508 (2001).
41. P.D. Mozley, J.S. Schneider, P.D. Acton *et al.*, Binding of [99mTc]TRODAT-1 to dopamine transporters in patients with Parkinson's disease and in healthy volunteers, *J. Nucl. Med.* **41**, 584–589 (2000).
42. D.A. Schwarz, M.J. Abrams, M.M. Hansen *et al.*, Preparation of hydrazino-modified proteins and their use of the synthesis of technetium-99m-protein conjugates, *Bioconjug. Chem.* **2**, 333–336 (1991).
43. T. Nicholsen, J. Zubieta, Complexes of rhenium with benzoylazo and related ligands. Crystal and molecular structures of the 'green chelate' benzoylazo complex $[ReCl_2(PPh_3)_2(NNCOC_6H_4\text{-p-Cl})](N_2, O)$, of the analogous 1-azophthalazine chelate complex $[ReCl_2(PPh_3)_2(NNC_8H_5N_2)][N_2N_1]$ and of the *cis*-dichloro organodiazenido complexes of the type $[ReCl_2(PPh_3)_2(NNR)L](L_2NCCH_3, NH_3$, and $C_5H_5N)$. A comparison to the structure of the *trans*-dichloro dimethylformamid derivative $[ReCl_2(PPh_3)_2(NNCO_2CH_3)(Me_2NCHO)]$. The structural characterization of the mixed hydrazido(1-) hydrazideo(2-) complexes $[ReCl_2(PPh_3)_2(NNHR)(NHNHR^1)]$, $(R = R^1 = –COC_6H_5$; $R = –COC_6H_5R$, $R^1 = –CO_2CH_3)$, *Polyhedron* **7**, 171–185 (1988).
44. S. Liu, D.S. Edwards, A.R. Harris, A novel ternary ligand system for ^{99m}Tc-labeling of hydrazino nicotinamide-modified biologically active molecules using imine-N-containing heterocycles as coligands, *Bioconjug. Chem.* **9**, 583–595 (1998).
45. D.S. Edwards, S. Liu, J.A. Barrett *et al.*, New and versatile ternary ligand system for technetium radiopharmaceuticals: Water soluble phosphines and tricine as coligands in labeling a hydrazinonicotinamide-modified cyclic glycoprotein IIb/IIIa receptor antagonist with ^{99m}Tc, *Bioconjug. Chem.* **8**, 146–154 (1997).
46. M. Ono, Y. Arano, T. Mukai *et al.*, Control of radioactivity pharmacokinetics of ^{99m}Tc-HYNIC-labeled polypeptides derivatized with ternary ligand complexes, *Bioconjug. Chem.* **13**, 491–501 (2002).
47. M.J. Abrams, J. Juweid, C.I. ten Kate *et al.*, Technetium-99m-labeled hydrazino nicotinamide derivatized chemotactic peptide analogs for imaging focal sites of bacterial infection, *J. Nucl. Med.* **34**, 1964–1974 (1993).
48. J.W. Babich, W. Graham, S.A. Barrow, A. Fischman, Comparison of the infection imaging properties of a ^{99m}Tc labeled chemotactic peptide with ^{111}In IgG, *Nucl. Med. Biol.* **22**, 643–648 (1995).
49. C. Decristoforo, W. Cholewimski, E. Donnemiller *et al.*, Detection of somatostatin receptor-positive tumours using the new ^{99m}Tc-tricine-HYNIC-*D*-Phe^1-Tyr^3-octreotide: First results in patients and comparison with ^{111}In-DTPA-*D*-Phe^1-octreotide, *Eur. J. Nucl. Med.* **27**, 1580–1580 (2000).
50. S. Liu, A.R. Harris, N.E. Williams, D.S. Edwards, ^{99m}Tc-Labeling of a hydrazinonicotinamide-conjugated LTB4 receptor antagonist useful for imaging infection and inflammation, *Bioconjug. Chem.* **13**, 881–886 (2002).
51. S. Liu, D.S. Edwards, M.C. Ziegler *et al.*, ^{99m}Tc-Labeling of a hydrazinonicotinamide-conjugated vitronectin receptor antagonist useful for imaging tumors, *Bioconjug. Chem.* **12**, 624–629 (2001).

52. E.T. Dams, W.J. Oyen, O.C. Boerman *et al.*, Technetium-99m labeled to human immunoglobulin G through the nicotinyl hydrazine derivative: A clinical study, *J. Nucl. Med.* **39**, 119–124 (1998).
53. J.W. Babich, W. Graham Coco, S. Barrow *et al.*, ^{99m}Tc-labeled chemotactic peptides: Influence of coligand on distribution of molecular species and infection imaging properties. Synthesis and structural characterization of model complexes with the $\{Re(\eta^2\text{-}HNNC_5H_4N)(\eta^2\text{-}NNC_5H_4N)\}$ core, *Inorg. Chim. Acta* **309**, 123–136 (2000).
54. R. Alberto, R. Schlibi, A. Egli, P.A. Schubiger, Novel organometallic aqua complex of technetium for the labeling of biomolecules: synthesis of $[^{99m}Tc(OH_2)_3(CO)_3]^+$ from $[^{99m}TcO_4]^-$ in aqueous solution and its reaction with a bifunctional ligand, *J. Am. Chem. Soc.* **120**, 7987–7988 (1998).
55. R. Alberto, K. Ortner, N. Wheatley *et al.*, Synthesis and properties of boranocarbonate: A convenient in situ CO source for the aqueous preparation of $[Tc(H_2O)_3(CO)_3]^+$, *J. Am. Chem. Soc.* **123**, 3135–3136 (2001).
56. R. Alberto, R. Schibli, R. Waibel *et al.*, Basic aqueous chemistry of $[M(OH_2)_3(CO)_3]^+$ (M = Re, Tc) directed towards radiopharmaceutical application, *Coord. Chem. Rev.* **190–192**, 901–919 (1999).
57. R. Waibei, R. Alberto, J. Willude *et al.*, Stable one-step technetium-99m labeling of His-tagged recombinant proteins with a novel Tc(I)-carbonyl complex, *Nat. Biotechnol.* **17**, 897–901 (1999).
58. A. Egli, R. Alberto, L. Tannahill *et al.*, Organometallic ^{99m}Tc-aquaion labels peptide to an unprecedented high specific activity, *J. Nucl. Med.* **40**, 1913–1917 (1999).
59. S.R. Banerjee, M.K. Levadala, N. Lazarova *et al.*, Bifunctional single amino acid chelates for labeling of biomolecules with the $\{Tc(CO)_3\}^+$ and $\{Re(CO)_3\}^+$ cores. Crystal and molecular structures of $[ReBr(CO)_3(H_2NCH_2C_5H_4N)]$, $[Re(CO)_3\{(C_5H_4NCH_2)_2NH\}]Br$, $[Re(CO)_3\{(C_5H_4NCH_2)_2NCH_2CO_2H\}]Br$, $[Re(CO)_3\{X(Y)NCH_2CO_2CH_2CH_3\}]Br$ (X = Y = 2-pyridylmethyl; X = 2-pyridylmethyl, Y = 2-(1-methylimidazolyl)methyl; X = Y = 2-(1-methylimidazolyl)methyl), $[ReBr(CO)_3\{(C_5H_4NCH_2)NH(CH_2C_4H_3S)\}]$, and $[Re(CO)_3\{(C_5H_4NCH_2)N(CH_2C_4H_3S)(CH_2CO_2)\}]$, *Inorg. Chem.* **41**, 6417–6425 (2002).
60. J.K. Pak, P. Benny, B. Spingler *et al.*, N^{ε} functionalization of metal and organic protected L-histidine for a highly efficient, direct labeling of biomolecules with $[Tc(OH_2)_3(CO)_3]^+$, *Chem. Eur. J.* **9**, 2053–2061 (2003).
61. R. Schibli, R. La Bella, R. Alberto *et al.*, Influence of the denticity of ligand systems on the in vitro and in vivo behavior of ^{99m}Tc(I)-tricarbonyl complexes: A hint for the future functionalization of biomolecules, *Bioconjug. Chem.* **11**, 345–351 (2000).
62. D. Kwekkeboom, E.P. Krenning, M. de Jong, Peptide receptor imaging and therapy, *J. Nucl. Med.* **41**, 1704–1713 (2000).
63. S. Liu, D.S. Edwards, ^{99m}Tc-Labeled small peptides as diagnostic radiopharmaceuticals, *Chem. Rev.* **99**, 2235–2268 (1999).
64. C. Van de Wiele, A. Signore, R.A. Dierckx, Peptide receptor imaging: Advances in the diagnosis of pulmonary diseases, *Am. J. Resp. Med.* **1**, 177–183 (2002).
65. J. Blum, H. Handmaker, N.A. Rinne, Technetium labeled small peptide radiopharmaceuticals in the identification of lung cancer, *Curr. Pharm. Des.* **8**, 1827–1836 (2002).

66. S.M. Okarvi, Recent developments in ^{99m}Tc-labelled peptide-based radiopharmaceuticals: An overview, *Nucl. Med. Commun.* **20**, 1093–1112 (1999).
67. L.C. Francesconi, Y. Zheng, J. Bartis *et al.*, Preparation and characterization of [^{99}TcO] Apcitide: A technetium labeled peptide, *Inorg. Chem.* **43**, 2867–2875 (2004).
68. F. Le Bideau, M. Salmain, S. Top, G. Jaouen, New and efficient routes to biomolecules substituted with cyclopentadienyltricarbonylrhenium and -technetium derivatives, *Chem. Eur. J.* **7**, 2289–2294 (2001).
69. N. Metzler-Nolte, Labeling of biomolecules for medicinal applications-bioorganometallic chemistry at its best, *Angew. Chem. Int. Ed.* **40**, 1040–1043 (2001).
70. G. Jaouen, S. Top, A. Vessières, R. Alberto, New paradigms for synthetic pathways inspired by bioorganometallic chemistry, *J. Organomet. Chem.* **600**, 23–36 (2000).
71. M. Wenzel, Technetium-99m labeling of cymantrene analogs with various substituents. A new preparation of technetium-99m radiodiagnostics, *J. Labelled Compd. Radiopharm.* **31**, 641–650 (1992).
72. T.W. Spradau, J.A. Katzenellenbogen, Preparation of cyclopentadienyltricarbonylrhenium complexes using a double ligand-transfer reaction, *Organometallics* **17**, 2009–2017 (1998).
73. T.W. Spradau, W.B. Edwards, C.J. Anderson *et al.*, Synthesis and biological evaluation of Tc-99m-cyclopentadienyltricarbonyltechnetium-labeled octreotide, *Nucl. Med. Biol.* **26**, 1–7 (1999).
74. J. Wald, R. Alberto, K. Ortner, L. Candreia, Aqueous one-pot synthesis of derivatized cyclopentadienyl-tricarbonyl complexes of ^{99m}Tc with an in situ CO source: Application to a serotonergic receptor ligand, *Angew. Chem. Int. Ed.* **40**, 3062–3066 (2001).
75. J. Bernard, K. Ortner, B. Spingler *et al.*, Aqueous synthesis of derivatized cyclopentadienyl complexes of technetium and rhenium directed toward radiopharmaceutical application, *Inorg. Chem.* **42**, 1014–1022 (2003).
76. J.F. Valliant, P. Morel, P. Schaffer, J.H. Kaldis, Carboranes as ligands for the preparation of organometallic Tc and Re radiopharmaceuticals. Synthesis of $[M(CO)_3(\eta^5\text{-}2,3\text{-}C_2B_9H_{11})]^-$ and rac-$[M(CO)_3(\eta^5\text{-}2\text{-}R\text{-}2,3\text{-}C_2B_9H_{10})]^-$ (M = Re, ^{99}Tc; R = $CH_2CH_2CO_2H$) from $[M(CO)_3Br_3]^{2-}$, *Inorg. Chem.* **41**, 628–630 (2002).
77. O.O. Sogbein, P. Merdy, P. Morel, J.F. Valliant, Preparation of Re(I)- and ^{99m}Tc(I)-metallocarboranes in water under weakly basic reaction conditions, *Inorg. Chem.* **43**, 3032–3034 (2004).
78. J.B. Arterburn, K.V. Rao, M.C. Perry, Solid-supported hydrazine substrate for labeling estradiol ligands with rhenium, *Angew. Chem. Int. Ed.* **39**, 771–772 (2000).
79. R. Dunn-Dufault, A. Pollak, J. Fitzgerald *et al.*, A solid-phase technique for preparation of no-carrier-added technetium-99m radiopharmaceuticals: Application to the streptavidin/biotin system, *Nucl. Med. Biol.* **27**, 803–807 (2000).
80. J.A. Bravo, A. Gibson, K. Loughran, M. Bradley, Solid-phase synthesis of oxo (mercaptoacetylglycylglycylglycine)rhenate(V), *Chem. Commun.* **9**, 837–838 (2001).
81. R. Dunn-Dufault, A. Pollak, J.R. Thornback, J.R. Ballinger, Convenient preparation of no-carrier-added technetium-99m radiopharmaceuticals using solid-phase technology, *Bioconjug. Chem.* **10**, 832–837 (1999).
82. A. Pollak, D.G. Roe, C.M. Pollock *et al.*, A convenient method of preparing high specific activity technetium complexes using thiol-containing chelators adsorbed on gold, *J. Am. Chem. Soc.* **121**, 11593–11594 (1999).

83. S. Mundwiler, L. Candreia, P. Haefliger *et al.*, Preparation of no-carrier-added technetium-99m complexes via metal-assisted cleavage from a solid phase, *Bioconjug. Chem.* **15**, 195–202 (2004).
84. A.J. Fischman, J.W. Babich, H.W. Strauss, A ticket to ride: Peptide radiopharmaceuticals, *J. Nucl. Med.* **34**, 2253–2263 (1993).
85. J.F. Valliant, R.W. Riddoch, D.W. Hughes *et al.*, The solid-phase synthesis and NMR spectroscopy of a ^{99}Tc chelate-bombesin derived peptide conjugate, *Inorg. Chim. Acta* **325**, 155–163 (2001).
86. C.J. Smith, H. Gali, G.L. Sieckman *et al.*, Radiochemical investigations of ^{99m}Tc-N_3S-X-BBN[7-14]NH_2; An in vitro/in vivo structure-activity relationship study where X = 0, 3, 5, 8, and 11-carbon tethering moieties, *Bioconjug. Chem.* **14**, 93–102 (2003).
87. J. Gariépy, S.Rémy, X. Zhang *et al.*, A simple two-step approach for introducing a protected diaminedithiol chelator during solid-phase assembly of peptides, *Bioconjug. Chem.* **13**, 679–684 (2002).
88. W.E.P. Greenland, K. Howland, J. Hardy *et al.*, Solid-phase synthesis of peptide radiopharmaceuticals using Fmoc-N-ε-(HYNIC-Boc)-lysine, a technetium-binding amino acid: Application to Tc-99m-labeled salmon calcitonin, *J. Med. Chem.* **46**, 1751–1757 (2003).
89. D.H. Hunter, L.G. Luyt, Lysine conjugates for the labelling of peptides with technetium-99m and rhenium, *J. Labelled Compd. Radiopharm.* **43**, 403–412 (2000).
90. K.A. Stephenson, J. Zubieta, S.R. Banerjee *et al.*, A new strategy for the preparation of peptide-targeted radiopharmaceuticals based on an fmoc-lysine-derived single amino acid chelate (SAAC). Automated solid-phase synthesis, NMR characterization, and in vitro screening of fMLF(SAAC)G and fMLF[(SAAC-Re$(CO)_3)^+$]G, *Bioconjug. Chem.* **15**, 128–136 (2004).
91. S.M. Qaim, Production of high purity ^{94m}Tc for positron emission tomography studies, *Nucl. Med. Biol.* **27**, 323–328 (2000).
92. M.F. Smith, M.E. Daube-Witherspoon, P.S. Plascjak *et al.*, Device-dependent activity estimation and decay correction of radionuclide mixtures with application to Tc-94m PET studies, *Med. Phys.* **28**, 36–45 (2001).
93. C.K. Stone, B.T. Christian, R.J. Nickles, S.B. Perlman, Technetium 94m-labeled methoxyisobutyl isonitrile: Dosimetry and resting cardiac imaging with positron emission tomography, *J. Nucl. Cardiol.* **1**, 425–433 (1994).
94. J.R. Nickles, A.D. Nunn, C.K. Stone, B.T. Christian, Technetium-94m-teboroxime: synthesis, dosimetry and initial PET imaging studies, *J. Nucl. Med.* **34**, 1058–1066 (1993).
95. L.G. Luyt, H. Bigott, M.J. Welch, J.A. Katzenellenbogen, 7α- and 17α-Substituted estrogens containing tridentate tricarbonyl rhenium/technetium complexes: synthesis of estrogen receptor imaging agents and evaluation using microPET with technetium-94m, *Bioorg. Med. Chem.* **11**, 4977–4989 (2003).
96. M.J. Welch, J.S. Lewis, $[Tc(CO)_3]^+$ Chemistry: A promising new concept for SPECT? *Eur. J. Nucl. Med.* **30**, 1302–1304 (2003).
97. S.R. Cherry, In Vivo molecular and genomics imaging: New challenges for imaging physics, *Phys. Med. Biol.* **49**, R13–R48 (2004).
98. M.E. Phelps, Positron emission tomography provides molecular imaging of biological processes, *Proc. Natl. Acad. Sci. USA* **97**, 9226–9233 (2000).

99. D.J. Rowland, J.S. Lewis, M.J. Welch, Molecular imaging: The application of small animal positron emission tomography, *J. Cell. Biochem.* (Suppl. 39), 110–115 (2002).
100. A. Wirrwar, N. Schramm, H. Vosberg, H.W. Muller-Gartner, High resolution SPECT in small animal research, *Rev. Neurosci.* **12**, 187–193 (2001).
101. F.G. Blankenberg, H.W. Strauss, Nuclear medicine applications in molecular imaging, *J. Magn. Reson. Imaging* **16**, 352–361 (2002).
102. M.S. Berridge, D.L. Heald, Z. Lee, Imaging studies of biodistribution and kinetics in drug development, *Drug Dev. Res.* **59**, 208–226 (2003).
103. S.J. Gatley, N.D. Volkow, J.S. Fowler *et al.*, Positron emission tomography and its use to image the occupancy of drug binding sites, *Drug Dev. Res.* **59**, 194–207 (2003).

19 $_{44}$Ru Perspectives of Ruthenium Complexes in Cancer Therapy

Olivier Lentzen, Cécile Moucheron and Andrée Kirsch-De Mesmaeker

19.1 Introduction

In 1827, the Russian scientist Gottfried Wilhelm Osann (1797–1866) announced the discovery of three new metals among which was ruthenium. However, when the corresponding samples were sent to Berzelius for confirmation, he found nothing new after analysis. Osann retracted his claims in 1829, and the discovery of ruthenium was finally attributed to Karl Karlovitch Klaus (1796–1864). Klaus obtained six grams of pure ruthenium in 1844 while studying insoluble residues of platinum ores. He also sent a sample of ruthenium to Berzelius, who confirmed this time the presence of a new element. Klaus decided to keep the name 'ruthenium' in recognition of Osann's work and in honor of his native Russia (*Ruthenia* is the mediaeval Latin name of Russia).[1–3]

The main earth stocks of ruthenium are localized in the Ural Mountains (Russia) and in both South and North America. Ruthenium is a hard white metal which can be isolated from ores of the platinum group metals by numerous chemical processes.[4] Ruthenium is mainly consumed in metallurgy for alloy production. Indeed, small amounts of this metal are used to increase the hardness of platinum and palladium or to improve the resistance of titanium to corrosion.[1,4]

Metallotherapeutic Drugs and Metal-Based Diagnostic Agents: The Use of Metals in Medicine
Edited by Gielen and Tiekink

Ruthenium exists in all possible oxidation states from 0 to VIII.[1] Compounds with high oxidation states of the metal can be used as oxidants for organic molecules, but the most stable derivatives in aqueous solution are generally in the oxidation state II or III.[1,5] These ions, as well as those of other transition elements, have high affinities for lone pairs of electrons and are thus often complexed by nitrogen atoms of high-field polyazaaromatic ligands. As reported in different recent reviews, the resulting octahedral coordination complexes have found several medical applications in the past 20 years.[5–7] For example, the radiophysical properties of ^{97}Ru and ^{103}Ru complexes can be exploited in radiodiagnostic imaging.[8] Other ruthenium complexes are also employed as immunosuppressants, NO (nitrosyls) delivery/scavenger tools, vasodilator/vasoconstrictor agents or drugs for cancer chemotherapy.[9] These metallic compounds are particularly attractive for these applications, mainly due to (i) their tendency to selectively bind biomolecules; (ii) well-developed synthesis processes; and (iii) the possibility to adjust complex properties (ligand substitution rate, redox properties, etc.) by varying the ligands coordinated to the metal center.

In this contribution, instead of reporting an exhaustive list of ruthenium's medical applications, we have decided to present in succession two different approaches commonly used for the discovery of active anti-cancer compounds and applied also with Ru complexes. The first one consists in testing directly *in vitro* and *in vivo* complexes chosen on the basis of their analogy with other molecules known for their anti-tumor activity. If this approach allowed the discovery of the majority of the active Ru compounds, its drawback, especially for a chemist, is that their chemical mechanisms of action are not necessarily established before their testing and remain unclear even after detection of an activity. The second approach, maybe more attractive for a chemist, is based on thorough studies of the properties of some particular complexes. The chemical, photochemical, physical and photophysical properties led, step by step, to the development of novel derivatives as potential drugs in anti-cancer therapy. This 'chemical' approach is more rigorous but, necessarily, requires much more time to develop an active compound. Finally, in the last part of this chapter, a new therapeutic approach is described and discussed. It is based on oligonucleotides derivatized by active ruthenium complexes with the aim of specifically targeting a sequence of a desired gene. This strategy is promising in medicine for the design of drugs with specific activities.

19.2 Ruthenium Complexes that Mimic Platinum Drugs

The search for metal-based compounds able to bind irreversibly to DNA is generally considered as one of the most attractive directions of research exploited to develop anti-cancer drugs.[7,10–12] Indeed, since its clinical introduction in the

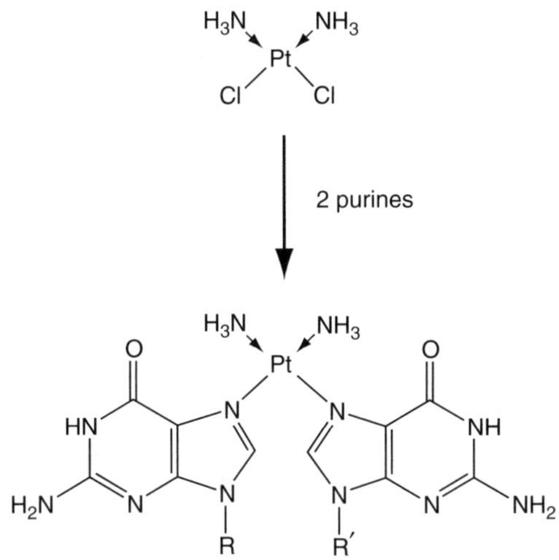

Figure 19.1 Cisplatin and its adduct with two guanine residues

late 1970s, cisplatin [*cis*-diamminedichloro-platinum(II)] remains one of the most efficient anti-cancer drugs. The cytotoxicity of this complex has been studied extensively and was shown to originate mainly from the formation of covalent cross-links that result from the binding of cisplatin to two purines of DNA (Figure 19.1).[13,14] Such a substitution product (usually called adduct) is able to inhibit DNA and RNA synthesis and thereby induce cell death.[15] Despite its important activity against various human cancers, cisplatin exhibits insufficient efficiency against some of the most frequent human cancers such as breast or colon tumors.[16] Moreover, patients treated with cisplatin often undergo acquired resistance and heavy side effects.[17]

Both the advantages and the drawbacks of cisplatin chemotherapy have motivated scientists to search for other metal-based drugs with better therapeutic properties. Besides searching for platinum analogues, a further approach consists in developing anti-tumoral agents containing other metal ions such as ruthenium. In this section, we present different ruthenium complexes designed to mimic the action of platinum complexes and presenting either an *in vitro* or an *in vivo* anti-tumor activity. Thus, they belong to the first approach for the discovery of active compounds as described earlier.

19.2.1 Chloro-ammino derivatives

The potential anti-cancer activity of ruthenium compounds was first examined by Clarke and co-workers who studied chloro-ammino derivatives of the general formula $[Ru(NH_3)_{6\text{-}X}Cl_X]^{Y+}$.

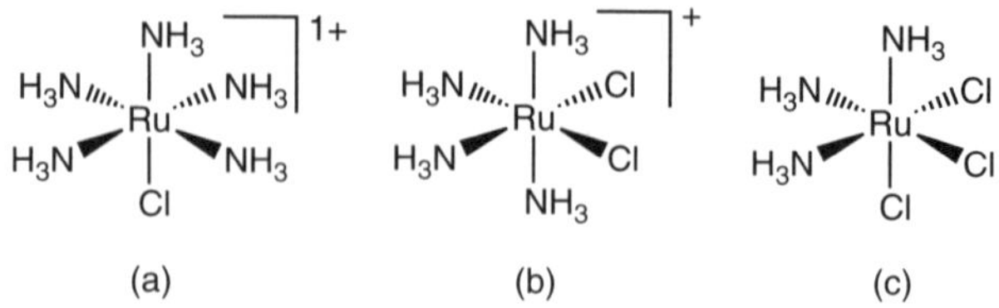

Figure 19.2 Chloro-ammino derivatives. (a) $[Ru^{II}(NH_3)_5Cl]^{1+}$, (b) *cis*-$[Ru^{III}(NH_3)_4Cl_2]^+$, (c) *fac*-$[Ru^{III}(NH_3)_3Cl_3]$

These compounds were chosen due to their similarities with cisplatin. Indeed, they contain chloride ligands that can be substituted in order to allow the binding of the complex with nucleic acids. This binding has been widely studied with the complex $[Ru^{II}(NH_3)_5Cl]^+$ (Figure 19.2a) and its aqua analogue $[Ru^{II}(NH_3)_5(H_2O)]^{2+}$.[18–20] Indeed, these complexes were shown to bind rather rapidly to DNA with an affinity constant of about 5000 M^{-1} (thanks mainly to the ionic attraction between the negatively charged DNA and the positively charged complex). The principal binding product of the complex with DNA involves the loss of the chloride/aqua ligand to the benefit of the coordination with a guanine N^7 nitrogen that is fairly accessible in the DNA major groove. Nevertheless, despite their binding to DNA, penta-ammineruthenium complexes do not behave as efficient anti-cancer agents.

In contrast, *cis*-$[Ru^{III}(NH_3)_4Cl_2]^+$ (Figure 19.2b) and *fac*-$[Ru^{III}(NH_3)_3Cl_3]$ (Figure 19.2c), which were tested against various tumors, have shown *in vivo* activity comparable to that of cisplatin.[11,21–23] Their cytotoxicity has been correlated to DNA binding although it is usually accepted that most Ru^{III} complexes are almost inert to ligand substitution and thus bind more slowly to DNA than their Ru^{II} analogues. In order to rationalize the efficiency of these Ru^{III} compounds, an 'activation by reduction' mechanism has been suggested. This mechanism involves the *in situ* reduction of the inactive Ru^{III} complex into the corresponding Ru^{II} drug, which is then able to lose chloride ligands much more rapidly. The possibility of such an *in vivo* reduction by cellular components has been demonstrated.[22,24] Moreover, it is generally accepted that tumors are reducing environments that could favor reduction of Ru^{III} into Ru^{II} and thereby allow ruthenium binding and, thus, accumulation in tumor tissues. Among these chloro-ammino derivatives, the best anti-tumor activity was found for *fac*-$[Ru^{III}(NH_3)_3Cl_3]$.

19.2.2 Dimethyl-sulfoxide complexes

The solubility of ruthenium complexes can be highly improved by substituting the NH_3 ligands by dimethylsulfoxide (DMSO) molecules. The initial *in vitro* and *in vivo* studies with DMSO complexes were performed by Mestroni, Alessio

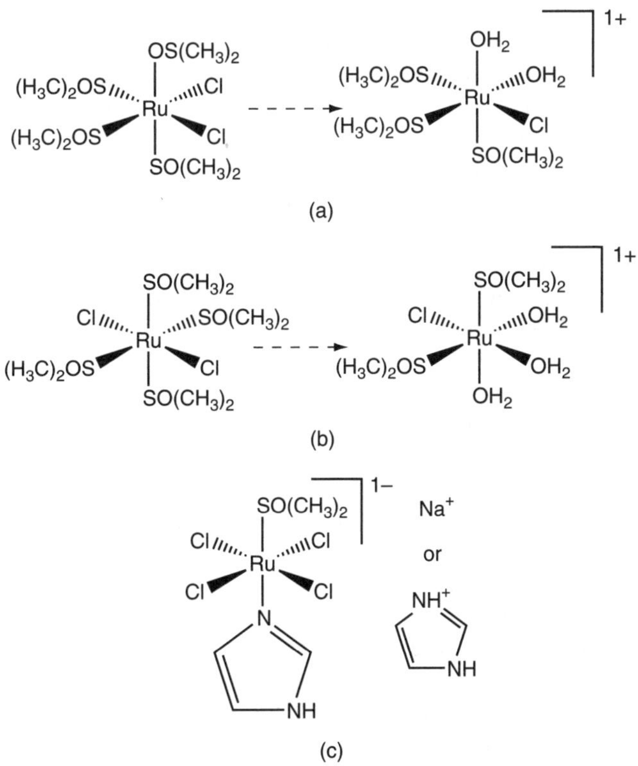

Figure 19.3 Dimethyl-sulfoxide complexes. (a) *cis*-[Ru^{II}(DMSO)$_4$Cl$_2$] and the corresponding aqua species, (b) *trans*-[Ru^{II}(DMSO)$_4$Cl$_2$] and the corresponding aqua species, (c) Na{*trans*-[Ru^{III}(DMSO)(Im)Cl$_4$]$^-$} (NAMI) and its imidazolium salt (NAMI-A)

and co-workers on the *cis*- and *trans*-[Ru^{II}(DMSO)$_4$Cl$_2$] complexes (Figure 19.3a and 3b).[25,26]

Although both isomers were shown to be able to coordinate N^7 positions of guanine residues of DNA, different studies have shown that *trans*-[Ru^{II}(DMSO)$_4$Cl$_2$] binds much more rapidly to DNA than the *cis*-isomer.[25,27–29] Moreover, only monoadducts of the *cis*-isomer on DNA are observed, whereas the *trans*-complex, similarly to cisplatin, is able to form DNA biadducts leading to intra- and inter-strand cross-links.[25,30] The difference of behavior between the two isomers has been explained by differences in the DMSO coordination sites. Indeed, in the *trans*-[Ru^{II}(DMSO)$_4$Cl$_2$], the four DMSO ligands are coordinated via the S atom whereas the *cis*-isomer contains three S-bonded DMSO molecules and the fourth one is coordinated via the oxygen atom. In solution, due to the important *trans* labilizing effect induced by a S-bonded DMSO on another DMSO ligand, the *cis*-[Ru^{II}(DMSO)$_4$Cl$_2$] loses rapidly its O-bonded DMSO in preference to an aqua molecule, whereas the *trans*-isomer loses two DMSO ligands

(Figure 19.3). After this first step, both isomers slowly lose one chloride ligand, resulting in a species containing three reactive groups in the case of the *trans*-isomer but containing only two reactive sites in the case of the *cis*-isomer (Figure 19.3).[29] Moreover, the reaction of the complex on DNA is facilitated thanks to a weaker steric hindrance in the case of the *trans* compound which contains less remaining DMSO ligands than the *cis*-isomer.

cis- And *trans*-[$Ru^{II}(DMSO)_4Cl_2$] have been tested *in vitro* and *in vivo* against different tumor cells, and the *trans*-isomer was shown to be almost 16 times more active than the *cis*-isomer, confirming a mechanism including DNA binding.[28,31,32] On the other hand, if *trans*-[$Ru^{II}(DMSO)_4Cl_2$] is less active than cisplatin against P388 leukemia for example, the reverse tendency is observed when the ruthenium complex is tested against a Pt-resistant P388 leukemia tumor line.[31] Moreover, *trans*-[$Ru^{II}(DMSO)_4Cl_2$] presents a good anti-metastatic activity that is particularly interesting since metastatic cancers are particularly difficult to cure.[32,33]

The encouraging results obtained with *trans*-[$Ru^{II}(DMSO)_4Cl_2$] are at the origin of the search for other *trans*-DMSO ruthenium complexes as anti-cancer agents. The more promising compounds of this kind are the Ru^{III} complexes Na{*trans*-[$Ru^{III}(DMSO)(Im)Cl_4$]} (NAMI) and its imidazolium salt NAMI-A (Figure 19.3c) which is a more stable and reproducible solid than NAMI. (NAMI has two DMSO molecules of crystallization that can be randomly replaced by water or acetone molecules of the crystallization solvent.)[34,35] NAMI-A is currently being tested in clinical phase I studies and presents an interesting anti-metastatic activity on a broad range of tumors among which is lung metastasis.[35–37] Such a complex could be particularly useful in minimizing the growth of undetected micro-metastases following surgery or radiotherapy.

However, the mechanism by which NAMI and NAMI-A exert their anti-metastatic activity is not yet elucidated, but could require *in vivo* reduction. It has been shown that NAMI loses two of its chloride ligands and is transformed into the corresponding, more reactive, aqua species that can bind to various biomolecules among which is DNA.[38] NAMI is also able to inhibit DNA and RNA polymerases, but not to form significant DNA interstrand cross-links.[6,39] Thus, the anti-metastatic activity of NAMI and NAMI-A seems to arise from other mechanisms than in the case of cisplatin, such as protein–DNA cross-links for example.[39–41]

19.2.3 Complexes with mixed chloride and heterocyclic ligands

Several Ru^{II} complexes of the type *cis*-[$Ru^{II}(L)_2Cl_2$] (where L is a bidentate heteroaromatic ligand) have also been prepared and are currently investigated for their anti-tumor properties on the basis of their similar reactivity as compared to cisplatin.

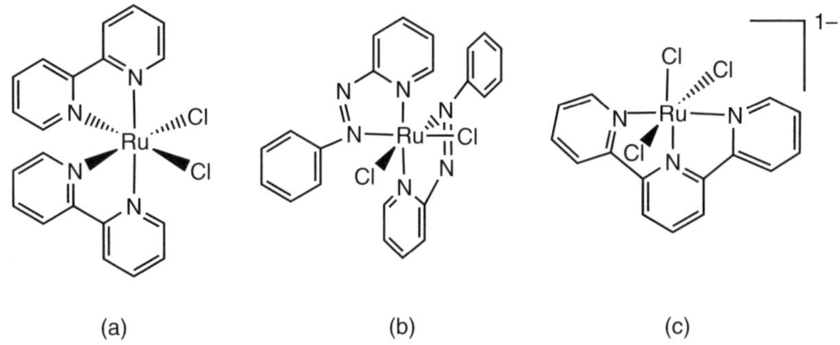

Figure 19.4 Complexes with chlorides and heterocyclic ligands. (a) *cis*-[$Ru^{II}(bpy)_2Cl_2$], (b) *cis*-[$Ru^{II}(azpy)_2Cl_2$], (c) *mer*-[$Ru^{II}(terpy)Cl_3$]$^-$

The complex *cis*-[$Ru^{II}(bpy)_2Cl_2$] (bpy = 2,2′-bipyridine; Figure 19.4a) has been tested on different cell lines but is mostly inactive as anti-tumor agent.[27] In contrast, one of the isomers (the one with C_2 symmetry) of the *cis*-[Ru^{II} $(azpy)_2Cl_2$] (azpy = 2-phenylazopyridine; Figure 19.4b) exhibits a high cytotoxicity that is comparable or even better than that of cisplatin against several fast-growing cell lines.[27,42,43] The higher activity of *cis*-[$Ru^{II}(azpy)_2Cl_2$] as compared to *cis*-[$Ru^{II}(bpy)_2Cl_2$] seems to be due to a higher flexibility of the azpy ligand which allows an easier substitution of the chloride ligands and thus the binding of the complex to two bases.[42]

Another example is the complex *mer*-[$Ru^{II}(terpy)Cl_3$]$^{1-}$, where terpy is the tridentate ligand 2,2′:6′,2″-terpyridine (Figure 19.4c). This complex presents a high cytotoxic activity that seems to be related to the possibility of interstrand DNA cross-links formation.[27,44] Indeed, this complex is able to form a number of interstrand cross-links which is almost identical to that formed under the same conditions by cisplatin. This cross-linking should be favored, as compared to the case of the *cis*-[$Ru^{II}(L)_2Cl_2$] compounds, thanks to the presence of the three chloride ligands that should increase the probability of adduct formation and decrease the hindrance resulting from a biadduct formation.

19.3 Design of New Anti-cancer Compounds based on the Photoreactivity of Polyazaaromatic Ruthenium(II) Complexes

In this section, we present the development of new potential anti-cancer agents, which is based upon some special properties that have been determined for some Ru^{II} complexes. More specifically, the photoreactivity of these ruthenium complexes has guided this investigation. The use of light to cure certain diseases is called phototherapy.[45] Usually, in phototherapy, the primary process is the absorption of light by a photosensitizer which, in a second step, reacts with a targeted endogenous molecule (O_2 or DNA) via energy or electron transfer.

$$\text{Photosensitizer}^* + \text{Target} \longrightarrow \text{photosensitizer} + \text{Target}^* \quad \text{(energy transfer)}$$

or

$$\text{Photosensitizer}^* + \text{Target} \longrightarrow \text{photosensitizer}^{\bullet -} + \text{Target}^{\bullet +} \quad \text{(electron transfer)}$$

Metal compounds such as polyazaaromatic ruthenium(II) complexes are particularly good candidates as photosensitizers.[46] Indeed, their properties can be modulated by varying the ligands chelated to the metal center as shown by their different photophysical and photochemical behaviors in the absence and in the presence of DNA.[47–51] The interaction of ruthenium(II) complexes with DNA enables their reaction with the nucleic acids. Three interaction geometries are possible: the complex can be externally associated in the electrostatic environment of the DNA double helix, adsorbed in the DNA grooves or intercalated between two DNA base pairs via one of its ligands. Finally, their reaction in the electronic excited state is favored (i) by their strong absorption in the visible part of the spectrum, which increases the probability of population of this excited state and (ii) by a rather long lifetime of this excited state (i.e. longer than 100 ns).

The photophysics of most Ru^{II} polypyridyl complexes can be described on the basis of the photophysical scheme established for the reference complex $[Ru^{II}(bpy)_3]^{2+}$ (Figure 19.5).[47,48] The photon absorption populates a singlet metal to ligand charge transfer excited state (^{1}MLCT), which deactivates rapidly by intersystem crossing (ISC) to the triplet MLCT state (^{3}MLCT) with a quantum yield of 1.[52] This triplet can deactivate to the ground state by radiative deactivation, radiationless deactivation or thermal activation to a ^{3}MC (metal centered) state. This latter is a bit higher in energy than the ^{3}MLCT

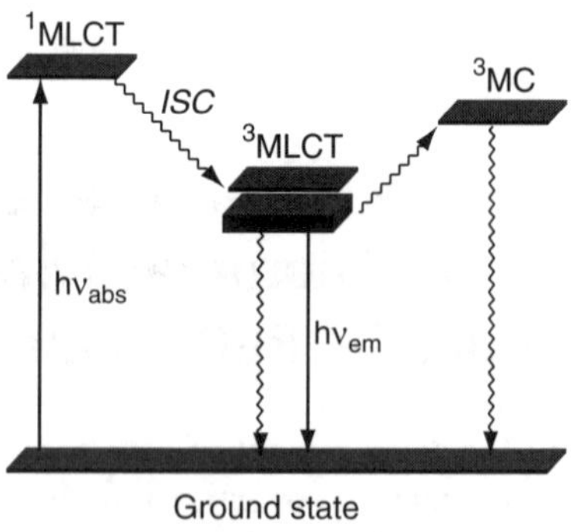

Figure 19.5 Photophysical scheme for $[Ru^{II}(bpy)_3]^{2+}$

state and deactivates without emission to the ground state. In the presence of DNA, the ^{3}MLCT excited state of the complex generally gives rise to electron or energy transfer processes.[48,49]

19.3.1 Energy transfer processes leading to photo-cleavages

This type of DNA damages arises from an energy transfer between a triplet excited photosensitizer (porphyrin, metal complex, etc.) and an oxygen molecule, leading to the production of singlet oxygen ($^1O_2$).[46,53]

$$[Ru^{II}(bpy)_3]^{2+*} + {}^3O_2 \longrightarrow [Ru^{II}(bpy)_3]^{2+} + {}^1O_2{}^*$$

Singlet oxygen is a very reactive species which may induce, among other processes, formation of oxidizing agents such as superoxide radicals ($O_2^{\bullet-}$, $OOH^\bullet$) or hydroxyl radicals ($OH^\bullet$). These species are able to damage DNA, leading mainly to oxidation of the guanine moiety into 8-oxoguanine and to DNA cleavages.[53] These cleavages have been demonstrated by gel electrophoresis experiments with plasmid DNA. Indeed, a single strand break, in one of the two strands of the DNA double helix of a supercoiled closed circular form of plasmid DNA, causes its conversion into the open circular form which exhibits a different migration velocity on an agarose gel.

The area in which the generation of singlet oxygen by photosensitizers is applied in medicine in order to destroy cancer cells is called photodynamic therapy (PDT).[54] Several hundreds of molecules have already been shown to photosensitize singlet oxygen production. In this context, polypyridyl Ru^{II} complexes are interesting due to, as mentioned above, their important absorption in the visible part of the spectrum and their long excited-state lifetimes. For example, the excitation of $[Ru^{II}(bpy)_3]^{2+}$ (Figure 19.6a) or $[Ru^{II}(phen)_3]^{2+}$ (phen = 1,10-phenanthroline) (Figure 19.6b) was shown to induce production of singlet oxygen and DNA single strand breaks.[55–57] Rather high quantum yields of singlet oxygen production (from 0.1 to 1.0 depending on the complex and on the solvent) have been determined for different Ru^{II} complexes.[56,58]

Pauly and co-workers have recently shown that singlet oxygen production by the complex $[Ru^{II}(bpy)_2(phen)]^{2+}$ is able to block partially the activity of a bacteriophage RNA-polymerase.[59] Moreno and co-workers have encapsulated the complex $[Ru^{II}(dip(SO_3Na)_2)_3]^{2+}$ ($dip(SO_3Na)_2$ = sodium salt of disulfonated 4,7-diphenyl-1,10-phenanthroline; Figure 19.6c) into polyacrylamide nanoparticles.[60] They found that the polyacrylamide matrix does not quench the $^1O_2{}^*$ production, allowing it to reach the external solution and to react with a targeted molecule. These particles may be modified with tumor-recognizing groups and be used in photodynamic therapy. The main drawbacks of the *in vivo* treatments in photodynamic therapy are collateral damages to healthy cells, acquired resistance, and limitation of light penetration in tissues.[60]

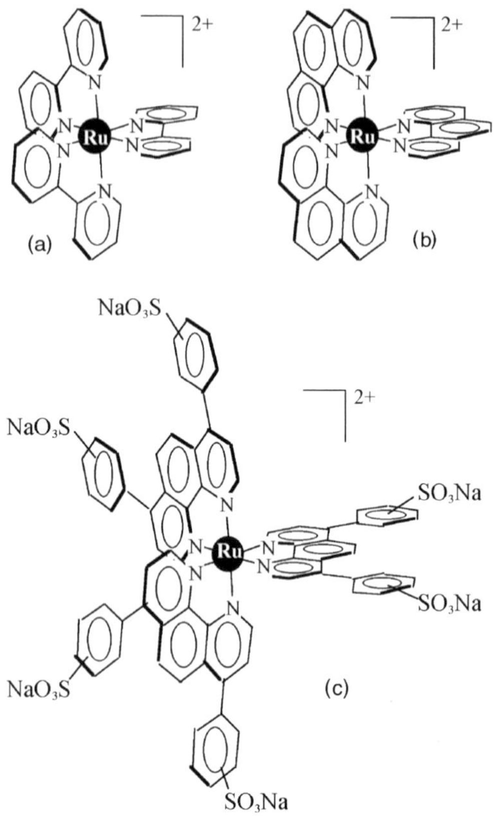

Figure 19.6 Complexes that generate singlet oxygen. (a) $[Ru^{II}(bpy)_3]^{2+}$, (b) $[Ru^{II}(phen)_3]^{2+}$, (c) $[Ru^{II}(dip(SO_3Na)_2)_3]^{2+}$ ($dip(SO_3Na)_2$ = sodium salt of disulfonated 4,7-diphenyl-1, 10-phenanthroline)

19.3.2 Photoelectron transfer processes leading to DNA cleavages

We have limited this section to the discussion of DNA cleavages resulting from direct photoelectron transfer processes from a DNA base to the excited state of a photosensitizer.[46] By modulating the photosensitizer's absorption spectrum, the photo-oxidation can occur with genetic material situated at different depths in tissues. For example, some authors studied derivatives of Ru^{II}-2,3-naphthalocyanine (Figure 19.7) that are characterized by a huge absorption in the 'phototherapeutic window' (650–850 nm) where the depth of light penetration into mammalian tissues is maximum.[45] Ru^{II}-2,3-naphthalocyanine compounds were shown to be active *in vitro* against cancer cells, most probably via an electron transfer process since singlet oxygen was not detected in solution.[61]

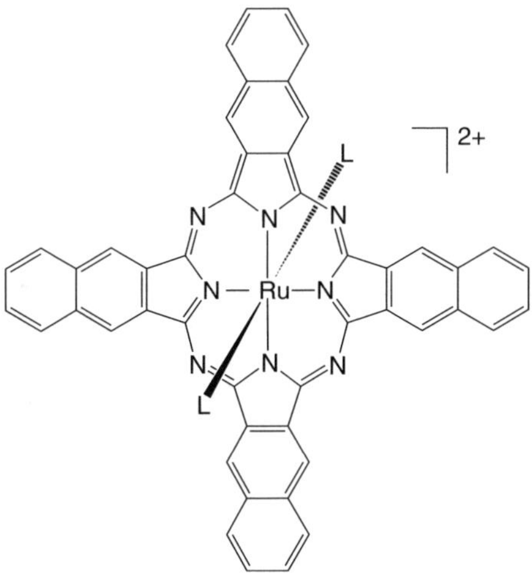

Figure 19.7 Ru^{II}-2,3-naphthalocyanine derivatives. L can be different ligands such as benzonitrile, pyridinesulfonate, aminoethanesulfonate, alanine, etc. The choice of these ligands influences the solubility of the complex as well as its physical and biological properties

Besides, it has been demonstrated that a photo-induced electron transfer takes place from a guanine (electron donor) to the ^{3}MLCT excited state of some Ru^{II} complexes containing π-deficient ligands such as TAP (1,4,5,8-tetraazaphenanthrene) (Figure 19.8a), HAT (1,4,5,8,9,12-hexaazatriphenylene) (Figure 19.8b) or BPZ (2,2′-bipyrazine) (Figure 19.8c) (electron acceptors).[48,49,62–64] Indeed, these ligands are known to stabilize the HOMO (highest occupied molecular orbital)

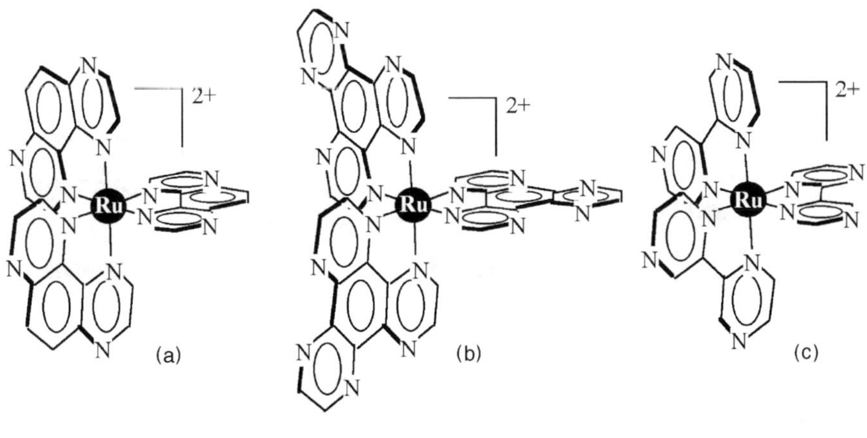

Figure 19.8 Photo-oxidizing Ru^{II} complexes. (a) $[Ru^{II}(TAP)_3]^{2+}$, (b) $[Ru^{II}(HAT)_3]^{2+}$, (c) $[Ru^{II}(BPZ)_3]^{2+}$

and LUMO (lowest unoccupied molecular orbital) orbitals of the complex, which makes the complex much more oxidizing in the excited state.

For example, the reduction potential of $[Ru^{II}(TAP)_3]^{2+}$ (Figure 19.8a) in its ground state (addition of one electron to the LUMO orbital of the complex) is negative versus the saturated calomel electrode (SCE) (rather high in energy) and obviously not positive enough (not low enough in energy) to abstract an electron from a guanine moiety (Figure 19.9).

In contrast, in the excited state, the reduction potential of the complex becomes positive because the electron from the guanine moiety can be added in the semi-occupied HOMO of the complex. In this way, the charge transfer process becomes exergonic and the excited state of the complex is thus able to abstract an electron from the guanine donor (Figure 19.9). In agreement with this, the luminescence of the complex is quenched by the GMP (guanosine-5′-monophosphate).[65] The electron transfer corresponding to this luminescence extinction has been demonstrated by laser flash photolysis experiments. Indeed, the differential absorption spectrum obtained after the pulsed excitation (i.e. after reduction of the excited complex by GMP) is similar to that obtained when the complex is reduced by pulse radiolysis or electrochemically.[65]

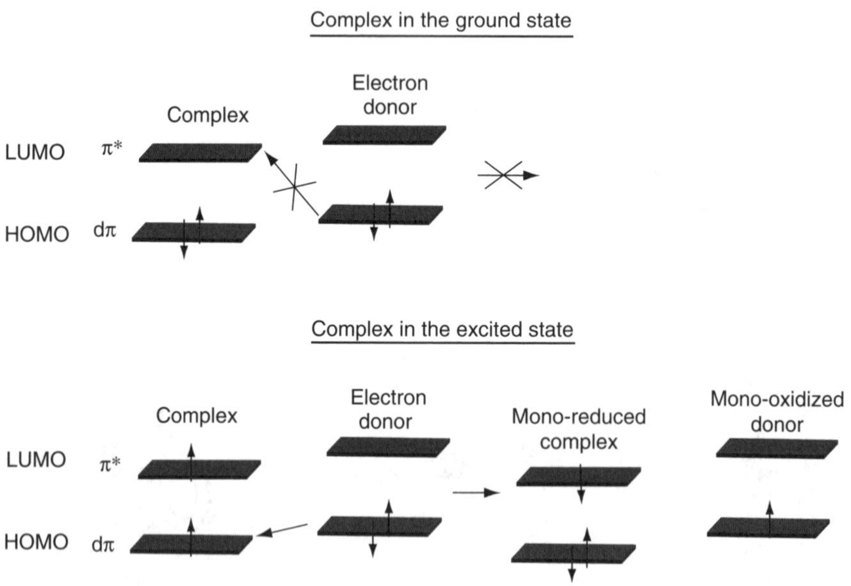

Figure 19.9 Schematic representation of the HOMO/LUMO orbitals of a Ru^{II} complex and an electron donor (GMP for example). When the complex is in the ground state, the electron transfer is endergonic; in the excited state the electron transfer becomes exergonic and leads to the formation of the mono-reduced complex and the mono-oxidized donor ($GMP^{\bullet+}$ for example)

$$[Ru^{II}(TAP)_3]^{2+} + GMP \xrightarrow{h\nu} [Ru^{II}(TAP)_2(TAP^{\bullet-})]^{+} + GMP^{\bullet+}$$

Such an electron transfer has also been observed in the presence of DNA and with other complexes containing at least two π-deficient ligands.[48,49,63] As in the case of singlet oxygen production, gel electrophoresis studies with plasmid DNA after illumination with this type of complex have shown that the formation of the radical cation on the guanine is the reason for efficient DNA cleavages.[46,50,51,66]

19.3.3 Photoelectron transfer processes leading to photoadduct formation

Surprisingly, direct DNA strand break is not the major process observed after the illumination of $[Ru^{II}(TAP)_3]^{2+}$ or complexes containing at least two TAP ligands in the presence of CT-DNA (calf thymus-DNA) or oligonucleotides. As demonstrated by gel electrophoresis in denaturing conditions, absorption spectroscopy, dialysis experiments and ElectroSpray Mass Spectrometry (ESMS), the illumination of photo-oxidizing Ru^{II} complexes in the presence of CT-DNA and oligonucleotides leads to the formation of a covalent adduct between the complex and the nucleic acid.[63,67] The production of this photoadduct is initiated by a photoelectron transfer process from a guanine base to the excited state of the complex as described in Figure 19.9, probably accompanied by a proton transfer.[68] The proposed mechanism implies, after the electron and proton transfer, the recombination of the two radicals (protonated radical anion of one TAP ligand and deprotonated radical cation of the guanine base), which gives rise to a covalent bond between both species. The TAP ligand which behaved as the electron acceptor in the complex is then rearomatized in a last step.

$$[Ru^{II}(TAP)_3]^{2+*} + GMP \longrightarrow [Ru^{II}(TAP)_2(TAP^{\bullet-})]^{1+} + GMP^{\bullet+}$$

$$[Ru^{II}(TAP)_2(TAP^{\bullet-})]^{1+} + GMP^{\bullet+} \longrightarrow [Ru^{II}(TAP)_2(TAPH^{\bullet})]^{2+} + GMP(-H)^{\bullet}$$

$$[Ru^{II}(TAP)_2(TAPH^{\bullet})]^{2+} + GMP(-H)^{\bullet} \longrightarrow \text{photoadduct}$$

The photoadduct formed by visible light absorption of different Ru–TAP complexes in the presence of CT-DNA or GMP has been isolated and characterized by ESMS and NMR spectroscopy.[67,69] These analyses revealed that a covalent bond has been formed between the C^2 position of one TAP ligand and the N^2 position of the guanine (Figure 19.10a).

As in the cases of the other metal complexes forming adduct with DNA (such as the Pt compounds), this new type of photo-adduct could be exploited to design new anti-tumor agents that are able to perturb DNA function or

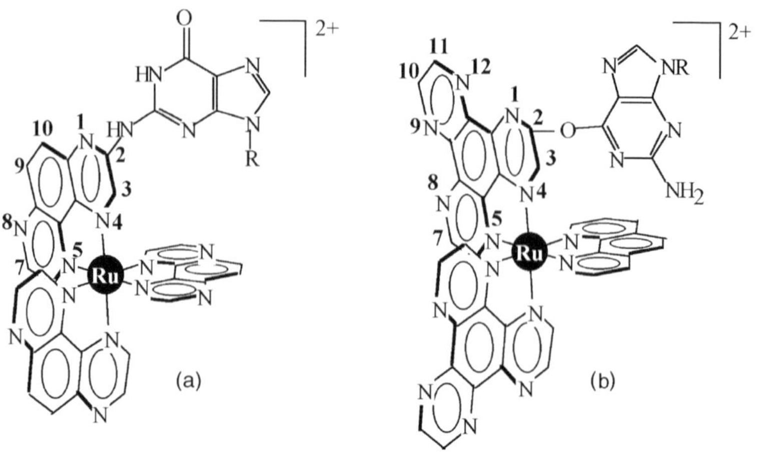

Figure 19.10 Photoadduct obtained after continuous illumination of $[Ru^{II}(TAP)_3]^{2+}$ (a) $[Ru^{II}(HAT)_2phen]^{2+}$ (b) in the presence of GMP or DNA. R = H after acid hydrolysis

enzymes involved in gene transcription and replication. Pauly and coworkers studied the influence of the photoadduct formed with $[Ru^{II}(TAP)_2(phen)]^{2+}$ on the activity of a RNA-polymerase.[59] The results show that the illumination of 10^{-5} M of the complex in the presence of plasmid DNA induces a 50% reduction of the *in vitro* transcription of a bacteriophage RNA-polymerase. The inhibition is almost complete for a complex concentration of 5×10^{-5} M.

The $[Ru^{II}(BPZ)_3]^{2+}$ (Figure 19.8c) and other complexes with at least two BPZ ligands are also able to react via photo-induced electron transfer with GMP and DNA.[64,70] This reaction leads to plasmid DNA cleavages and photo-addition of the complex probably at the level of DNA guanines. However, although the covalent nature of this adduct has been proven by dialysis and electrophoresis experiments, its structure has yet to be reported.

Photo-oxidizing Ru^{II} complexes with HAT ligands are less studied than their TAP analogues. Nevertheless, the occurrence of photo-induced electron transfer has also been proven with $[Ru^{II}(HAT)_3]^{2+}$ (Figure 19.8b), $[Ru^{II}(HAT)_2(phen)]^{2+}$ and $[Ru^{II}(HAT)_2(bpy)]^{2+}$.[63] Very recently, photoadduct formation between $[Ru^{II}(HAT)_2(phen)]^{2+}$ and GMP, CT-DNA or oligonucleotides has been evidenced by absorption spectroscopy, ESMS and NMR.[71] In this case, the covalent bond seems to be formed between the C^2 or C^7 position of a HAT ligand and the exocyclic O^6 position of the guanine (Figure 19.10b). Interestingly, the $[Ru^{II}(HAT)_2(phen)]^{2+}$ is able to form a second adduct, corresponding to the addition of two GMP molecules on the complex.[71] The formation of such a biadduct is very promising for a possible use of this complex as anti-cancer agent since this photoreaction could result in DNA cross-links at the level of guanine units in close proximity, as in the case of cisplatin.

19.4 Targeting Base Residues of Specific Sequences

The reactions or photoreactions of metal complexes with DNA described previously usually occur randomly at guanine sites of nucleic acids. Recently, different authors studied the possibility of directing the photoreaction of Ru^{II} complexes to a specific gene. Oligonucleotides can indeed be used to target specifically a sequence of mRNA (called anti-sense strategy) or double-stranded DNA (called anti-gene strategy) with triple-helix formation.[72–75] The derivatization of an oligonucleotide by a metallic complex should allow the combination of the specific recognition of the sequence by the oligonucleotide with the (photo)-reactivity of a ruthenium complex. Such a system would allow the targeting of a specific gene (the complementary sequence of the derivatized oligonucleotide) and the induction of the (photo)reaction on a specific base unit of this gene.

This approach was used by Ossipov and co-workers who studied oligonucleotides derivatized by the photoreactive and intercalating $[Ru^{II}(terpy)(dppz)(CH_3CN)]^{2+}$ complex (dppz = dipyrido [3,2-a:2′,3′-c] phenazine, Figure 19.11).[76] In this case, the photoreaction is not initiated by a photoelectron transfer. Instead, the visible illumination of this compound induces the loss of the CH_3CN ligand, giving rise to the corresponding aqua complex $[Ru^{II}(terpy)(dppz)(H_2O)]^{2+}$ which forms an adduct on DNA similar to the Pt aqua complexes. The $[Ru^{II}(terpy)(dppz)(CH_3CN)]^{2+}$ complex has thus been anchored to different synthetic oligonucleotides. After hybridization with the complementary sequence, the illumination of the corresponding derivatized duplexes results in the photo-addition of the aqua complex on a guanine residue of the complementary strand. This sequence-specific photoreaction inducing the cross-linking of the duplex strands could be interesting in the development of anti-cancer agents or tools for DNA diagnostic studies.

In order to combine the photo-redox properties of a Ru–TAP complex described previously, with the sequence specificity of an oligonucleotide, Kirsch-De Mesmaeker and co-workers chose a system comprising a

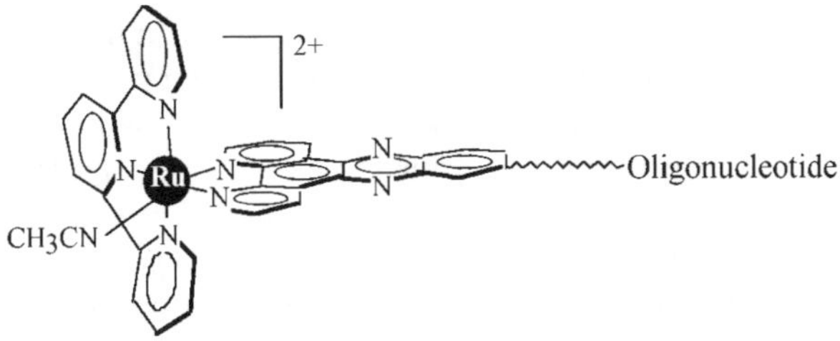

Figure 19.11 The photoreactive $[Ru^{II}(terpy)(dppz)(CH_3CN)]^{2+}$ complex anchored via the dppz ligand onto an oligonucleotide

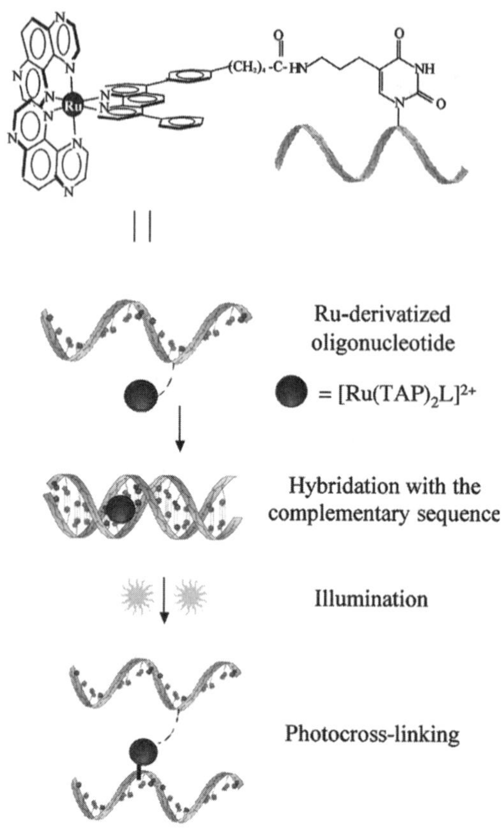

Figure 19.12 The $[Ru^{II}(TAP)_2dip']^{2+}$ complex anchored onto a synthetic oligonucleotide at the level of the 5-position of a thymine (see Plate 2)

$[Ru^{II}(TAP)_2dip']^{2+}$ (dip′ = dip-$(CH_2)_4COOH$) complex anchored to different synthetic 17 mer oligonucleotides (Figure 19.12).[77–79] $[Ru^{II}(TAP)_2dip']^{2+}$ contains two π-deficient TAP ligands in order to make the complex photo-oxidizing and a dip′ ligand for the derivatization. The derivatized probe strands are free of guanine residues (to prevent intramolecular reactions) and were hybridized to their complementary sequences containing various numbers of guanine units located in different positions. It was demonstrated that an electron transfer occurs from the guanine moieties of the complementary strand to the excited complex anchored to the probe strand. The recombination of the radicals generated by this primary photoprocess as described above leads to the formation of a photoadduct of the tethered complex on a guanine of the complementary strand. As evidenced by denaturing gel electrophoresis and by ESMS, this adduct is responsible for the photocross-linking of the two oligonucleotide strands (Figure 19.12).[80,81] Interestingly, this photocross-linking

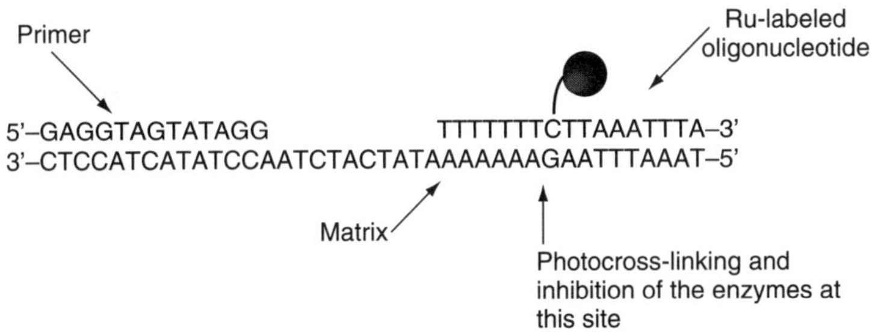

Figure 19.13 Representation of the system studied by Kirsch-De Mesmaeker and co-workers, comprising a 13 mer DNA primer and a 17 mer Ru^{II} labeled oligonucleotide, both hybridized to a 40 mer matrix

is able to block *in vitro* the action of different enzymes such as a digestion enzyme (exonuclease III) and different DNA-polymerases, with 100% efficiency.[81] Thus, a 17 mer Ru^{II}-labeled oligonucleotide was hybridized to its complementary sequence located on the 5′ extremity of a 40 mer matrix. After illumination, the elongation of the 13 mer DNA primer hybridized to the 3′ extremity of the same matrix was completely blocked at the position corresponding to the formation of the photocross-linking (Figure 19.13).

In conclusion, these data obtained with Ru-labeled oligonucleotides indicate quite interestingly that the cross-linking of the two strands induced under illumination could be developed for future applications, either in the frame of the anti-sense or anti-gene strategy, or for DNA diagnostic studies.

19.5 Conclusions

As mentioned in the introduction, two approaches in the search of novel anti-cancer drugs based on ruthenium complexes exist in the literature and each presents different advantages and drawbacks.

In the first approach, rather simple $Ru^{II/III}$ complexes have been prepared and directly evaluated for their potential anti-cancer activity by *in vivo* and *in vitro* tests. Their selection is based on the knowledge acquired from research on the chemistry of Pt complexes and derivatives, used presently as anti-cancer agents and acting mainly at the level of DNA. Thus, Ru^{II} and Ru^{III} complexes 'mimic' in some way the activity of the Pt complexes that have been screened. These compounds react by substitution of a monodentate ligand by a water molecule and the resulting aqua complex further reacts with the DNA bases. Among the most interesting results from these studies is the fact that some of these Ru complexes have revealed important anti-metastatic activity which is not the case for the Pt complexes.

In the second approach, more sophisticated Ru complexes for which the synthesis and characterization are not always obvious have been prepared. The photophysical and photochemical properties of these samples have guided their development into novel derivatives for applications as molecular tools in DNA diagnostics or as potential anti-cancer drugs. These photoactive Ru compounds have revealed two main types of photoreactivity. They can behave as photosensitizers of oxygen with generation of singlet excited oxygen. As such, they are being developed for photodynamic therapy, but in this area the ruthenium complexes seem to be less promising than some organic dyes or porphyrins. More interestingly, some Ru complexes with polyazaaromatic ligands behave as photo-oxidants and give rise to photo-induced electron transfer processes that lead to DNA damage, i.e. cleavage or adduct formation with DNA. The most promising application thus far is probably when these compounds are tethered to oligonucleotides that can be photocrosslinked to their complementary sequence and in this way inhibit the normal function of DNA enzymes.

References

1. E.A. Seddon and K.R. Seddon, The chemistry of ruthenium, in *Topics in Inorganic and General Chemistry, Monograph 19*, Amsterdam Elsevier, 1984.
2. G.W. Osann, *Philosophical Magazine*, **2**, 391–392 (1827).
3. K.K. Klaus, *Philosophical Magazine*, **27**, 230–231 (1845).
4. D.R. Lide, *Handbook of Chemistry and Physics, 77th edition*, London CRC Press, 1996–1997.
5. M.J. Clarke, Ruthenium complexes, in *Metal Complexes in Cancer Chemotherapy*, Weinheim B.K. Keppler (ed.), New York: VCH, 1993.
6. M.J. Clarke, *Coord. Chem. Rev.*, **236**, 209–233 (2003).
7. C.X. Zhang and S.J. Lippard, *Curr. Opin. Chem. Biol.*, **7**, 481–489 (2003).
8. S.C. Srivastava, L.F. Mausner and M.J. Clarke, *Prog. Clin. Biochem. Med.*, **10**, 111–149 (1989).
9. P.C. Ford, J. Bourassa, K. Miranda *et al.*, *Coord. Chem. Rev.*, **171**, 185–202 (1998).
10. B.K. Keppler, *New J. Chem.*, **14**, 389–403 (1990).
11. B.K. Keppler, *Metal Complexes in Cancer Chemotherapy*, New York: VCH Weinheim, 1993.
12. M.J. Clarke and P.J. Sadler, Metallopharmaceuticals: DNA interaction, in *Topics in Biological Inorganic Chemistry*, Volume 1, Berlin Springer, 1999.
13. S.M. Cohen and S.J. Lippard, *Prog. Nucleic Acid Res. Mol. Biol.*, **67**, 92–130 (2001).
14. W.I. Sundquist and S.J. Lippard, *Coord. Chem. Rev.*, **100**, 293–322 (1990).
15. E.R. Jamieson and S.J. Lippard, *Chem. Rev.*, **99**, 2467–2498 (1999).
16. E. Wong and C.M. Giandomenico, *Chem. Rev.*, **99**, 2451–2466 (1999).
17. V. Brabec and J. Kasparkova, *Drug Resist. Updat.*, **5**, 147–161 (2002).
18. M.J. Clarke and M. Buchbinder, *Inorg. Chim. Acta*, **27**, L87–L88 (1978).
19. M.J. Clarke, B. Jansen, K.A. Marx and R. Kruger, *Inorg. Chim. Acta*, **124**, 13–28 (1986).

20. V.M. Rodriguez-Bailey, K.J. LaChance-Galang, P.E. Doan and M.J. Clarke, *Inorg. Chem.*, **36**, 1873–1883 (1997).
21. J.R. Durig, J. Danneman, W.D. Behnke and E.E. Mercer, *Chem. Biol. Interact.*, **13**, 287–294 (1976).
22. D. Frasca, J. Ciampa, J. Emerson *et al.*, *Metal-Based Drugs*, **3**, 197–209 (1996).
23. M.J. Clarke, F. Zhu and D.R. Frasca, *Chem. Rev.*, **99**, 2511–2533 (1999).
24. M.J. Clarke, S. Bilter, D. Rennert *et al.*, *J. Inorg. Biochem.*, **12**, 79–87 (1980).
25. E. Alessio, G. Mestroni, G. Nardin *et al.*, *Inorg. Chem.*, **27**, 4099–4106 (1988).
26. G. Mestroni, E. Alessio, G. Sava *et al.*, The development of tumor-inhibiting ruthenium dimethylsulfoxide complexes, in *Metal Complexes in Cancer Chemotherapy*, Weinheim B.K. Keppler (ed.), New York: VCH, 1993.
27. O. Novakova, J. Kasparkova, O. Vrana *et al.*, *Biochemistry*, **34**, 12369–12378 (1995).
28. G. Sava, S. Pacor, F. Bregant *et al.*, *Pharmacology (Life Sci. Adv.)*, **9**, 79–84 (1990).
29. O. Novakova, C. Hofr and V. Brabec, *Biochemistry Pharmacol.*, **60**, 1761–1771 (2000).
30. G. Esposito, S. Cauci, F. Fogolari *et al.*, *Biochemistry*, **31**, 7094–7103 (1992).
31. M. Coluccia, G. Sava, F. Loseto *et al.*, *Eur. J. Cancer*, **29**, 1873–1879 (1993).
32. G. Mestroni, E. Alessio, M. Calligaris *et al.*, *Prog. Clin. Biochem. Med.*, **10**, 71–87 (1989).
33. G. Sava, S. Pacor, F. Bregant *et al.*, *Anticancer Drugs*, **1**, 99–108 (1990).
34. A. Bergamo, M. Cocchietto, I. Capozzi *et al.*, *Anticancer Drugs*, **7**, 697–702 (1996).
35. G. Sava and A. Bergamo, *Int. J. Oncol.*, **17**, 353–365 (2000).
36. G. Sava, I. Capozzi, K. Clerici *et al.*, *Clin. Exp. Metastasis*, **16**, 371–379 (1998).
37. S. Zorzet, A. Sorc, C. Casarsa *et al.*, *Metal-Based Drugs*, **8**, 1–7 (2001).
38. E. Gallori, C. Vettori, E. Alessio *et al.*, *Arch. Biochem. Biophys.*, **376**, 156–162 (2000).
39. A. Barca, B. Pani, M. Tamaro and E. Russo, *Mut. Res.*, **423**, 171–181 (1999).
40. B. Sanna, M. Debidda, G. Pintus *et al.*, *Arch. Biochem. Biophys.*, **403**, 209–218 (2002).
41. L. Messori, P. Orioli, D. Vullo *et al.*, *Eur. J. Biochem.*, **267**, 1206–1213 (2000).
42. A.H. Velders, H. Kooijman, A.L. Spek *et al.*, *Inorg. Chem.*, **39**, 2966–2967 (2000).
43. A.C.G. Hotze, A.H. Velders, F. Ugozzoli *et al.*, *Inorg. Chem.*, **39**, 3838–3844 (2000).
44. P.M. van Vliet, S.M.S. Toekimin, J.G. Haasnoot *et al.*, *Inorg. Chim. Acta*, **231**, 57–64 (1995).
45. G. Stochel, A. Wanat, E. Kulis and Z. Stasicka, *Coord. Chem. Rev.*, **171**, 203–220 (1998).
46. B. Armitage, *Chem. Rev.*, **98**, 1171–1200 (1998).
47. B. Durham, J.V. Caspar, J.K. Nagle and T.J. Meyer, *J. Am. Chem. Soc.*, **104**, 4803–4810 (1982).
48. C. Moucheron, A. Kirsch-De Mesmaeker and J.M. Kelly, *J. Photochem. Photobiol., B.*, **40**, 91–106 (1997).
49. A. Kirsch-De Mesmaeker, J.-P. Lecomte and J.M. Kelly, *Top. Curr. Chem.*, **177**, 25–76 (1996).
50. I. Ortmans, C. Moucheron and A. Kirsch-De Mesmaeker, *Coord. Chem. Rev.*, **168**, 233–271 (1998).
51. C. Moucheron, A. Kirsch-De Mesmaeker and J.M. Kelly, Photophysics and photochemistry of metal polypyridyl and related complexes with nucleic acids, in *Structure and Bonding, vol 92*, Springer-Verlag, Berlin, Heidelberg, 1998.

52. J.N. Demas and M.F. Taylor, *Inorg. Chem.*, **18**, 3177–3179 (1979).
53. J. Piette, *J. Photochem. Photobiol., B.*, **11**, 241–260 (1991).
54. I.J. MacDonald and T.J. Dougherty, *J. Porphyr. Pthalocya.*, **5**, 105–129 (2001).
55. C.V. Kumar, J.K. Barton and N.J. Turro, *J. Am. Chem. Soc.*, **107**, 5518–5523 (1985).
56. C. Tanielian, C. Wolff and M. Esch, *J. Phys. Chem.*, **100**, 6555–6560 (1996).
57. J.W. Dobrucki, *J. Photochem. Photobiol., B.*, **65**, 136–144 (2001).
58. F. Wilkinson, W.P. Helman and A.B. Ross, *J. Phys. Chem. Ref. Data*, **22**, 113–262 (1993).
59. M. Pauly, I. Kayser, M. Schmitz *et al.*, *J. Chem. Soc. Chem., Commun.*, **N10**, 1086–1087 (2002).
60. M.J. Moreno, E. Monson, R.G. Reddy *et al.*, *Sensors and Actuators*, **90**, 82–89 (2003).
61. J.F. Vollano, G.E. Bossard, S.A. Martellucci *et al.*, *J. Photochem. Photobiol., B.*, **37**, 230–235 (1997).
62. J.M. Kelly, M.M. Feeney, L. Jacquet *et al.*, *Pure Appl. Chem.*, **69**, 767–772 (1997).
63. J.-P. Lecomte, A. Kirsch-De Mesmaeker, M.M. Feeney and J.M. Kelly, *Inorg. Chem.*, **34**, 6481–6491 (1995).
64. P. Vicendo, S. Mouysset and N. Paillous, *Photochem. Photobiol.*, **65**, 647–655 (1997).
65. J.-P. Lecomte, A. Kirsch-De Mesmaeker, J.M. Kelly *et al.*, *Photochem. Photobiol.*, **55**, 681–689 (1992).
66. N. Paillous and P. Vicendo, *J. Photochem. Photobiol., B.*, **20**, 203–209 (1993).
67. M.M. Feeney, J.M. Kelly, A.B. Tossi *et al.*, *J. Photochem. Photobiol., B.*, **23**, 69–78 (1994).
68. I. Ortmans, B. Elias, C. Moucheron *et al.*, *J. Chem. Soc., Dalton Trans.*, **4**, 668–676 (2004).
69. L. Jacquet, D.R. Davies, A. Kirsch-De Mesmaeker and J.M. Kelly, *J. Am. Chem. Soc.*, **119**, 11763–11768 (1997).
70. E. Gicquel, H. Nierengorben, M. Luhmer, J.F. Constant, E. Defrancq, P. Dumy, A. van Dorsselaer, N. Paillous and P. Vicendo, *Photochem. Photobiol.*, **72**, 583–589 (2000).
71. R. Blasius, C. Moucheron and A. Kirsch-De Mesmaeker, *Chem. Eur. J.*, accepted (2004).
72. M.P. Knauert and P.M. Glazer, *Hum. Mol. Genet.*, **10**, 2243–2251 (2001).
73. K.M. Vasquez and P.M. Glazer, *Q. Rev. Biophys.*, **35**, 89–107 (2002).
74. E. Uhlmann and A. Peyman, *Chem. Rev.*, **90**, 543–584 (1990).
75. I. Tamm, B. Dörken and G. Hartmann, *Lancet*, **358**, 489–497 (2001).
76. D. Ossipov, S. Gohil and J. Chattopadhyaya, *J. Am. Chem. Soc.*, **121**, 13416–13433 (2002).
77. I. Ortmans, S. Content, N. Boutonnet *et al.*, *Chem. Eur. J.*, **5**, 2712–2721 (1999).
78. D. Garcia-Fresnadillo, N. Boutonnet, S. Schumm *et al.*, *Biophys. J.*, **82**, 978–987 (2002).
79. S. Schumm, M. Prévost, D. Garcia-Fresnadillo *et al.*, *J. Phys. Chem.*, **106**, 2763–2768 (2002).
80. O. Lentzen, J.F. Constant, E. Defrancq *et al.*, *Chembiochem.*, **4**, 195–202 (2003).
81. O. Lentzen, E. Defrancq, J.F. Constant *et al.*, *J. Biol. Inorg. chem.*, **9**, 100–108 (2004).

20 $_{45}$Rh Rhodium in Medicine

Florian P. Pruchnik

20.1 Introduction

The first reports on the use of inorganic compounds in therapy appeared about 3500 years ago when a suspension of rust was administered as a cure for impotence, and iron compounds have often been used since for the treatment of anemia. Inorganic compounds have been used as anti-cancer drugs for 140 years and systematic studies of their cytotoxicity began about 70 years ago. However, very intense investigations of the anti-tumor activity of inorganic and organometallic compounds started after the recognition in 1969 that *cis*-$[PtCl_2(NH_3)_2]$ is an efficient anti-cancer agent.[1–7]

20.2 Anti-tumor Activity of Rhodium Complexes

Rhodium complexes belong to the very promising class of anti-tumor agents. Cytostatic and anti-cancer activity characterize mainly Rh(I), Rh(II) and Rh(III) coordination and organometallic compounds containing nitrogen, sulfur and oxygen ligands. The anti-tumor activity of rhodium complexes was expected because Rh(I), Rh(II) and Rh(III) complexes have d^8, d^7 and d^6

Metallotherapeutic Drugs and Metal-Based Diagnostic Agents: The Use of Metals in Medicine
Edited by Gielen and Tiekink

electronic configuration and are isoelectronic with Pt(II), Pt(III) and Pt(IV) complexes; thus chemical properties of rhodium and platinum complexes are expected to be, to some extent, similar.

20.2.1 Rhodium(I) complexes

Rhodium(I) square planar complexes [RhCl(COD)(NH_3)], [RhCl(COD)(piperidine)] and [Rh(acac)(COD)] (COD = 1,5-cyclooctadiene) are efficient agents against leukemia L1210, Ehrlich ascites carcinoma, sarcoma 180 and Lewis lung carcinoma, and inhibit the development of metastases. Acetylacetonato complexes are significantly more active than complexes containing piperidine, 2,2′-bipyridine and 1,10-phenanthroline. The anti-tumor activity of these complexes is interesting because it is comparable with that of cisplatin. Furthermore, they are only marginally nephrotoxic.[8–10] The 1,5-cyclooctadiene complexes show better anti-tumor properties than coordination complexes with norbornadiene (NBD), and complexes with 1,5-hexadiene are not active anti-tumor agents. The cationic complexes [Rh(COD)(N—N—R)]Cl, where N—N—R = 2-pyridinalmethylimine, 2-pyridinalethylimine and 2-pyridinalisopropylimine, exhibit activity against Lewis lung carcinoma, P388 lymphocytic leukemia and a mammary carcinoma (MC) and on lung metastatic tumor.[11–13] The anti-neoplastic properties of these complexes do not seem related to a cytotoxic action. It is more probable that modifications occurring at the primary tumor level, most likely different from lethal effects directed to tumor cells, lead to the reduction of spontaneous lung metastasis in the treated animals. Complexes possessing nitrogen leaving-groups show similarly pronounced effects on the primary tumor and on lung metastases, while the dimeric rhodium(I) complexes [$Rh_2Cl_2(COD)_2$], [$Rh_2Cl_2(NBD)_2$] and [$Rh_2Cl_2(CH_2{=}CHCH_2CH_2CH{=}CH_2)_2$] display negligible effects on subcutaneous tumor growth, showing instead a remarkable anti-metastatic effect.[8–14]

The square-planar rhodium(I) carbonyl complexes [$RhX(CO)_2$], where X = sulfonamide derivatives, were prepared, characterized and assayed as cytotoxic and anti-tumor agents *in vitro* against KB oral carcinoma and *in vivo* against P388, Ehrlich ascites and advanced B16 melanoma. Assays against three *Trypanosoma* strains were also performed. The coordination complex [Rh(sulfamethoxydiazine)($CO)_2$] appeared to be active in all mentioned biological systems. This examined complex did not show nephrotoxicity. Relationships between biological activity and π-electronic charge localization on the nitrogen atom of the amidic group were also discussed.[15] Dithiocarbamate and xanthate rhodium(I) carbonyl complexes – [$Rh(S_2CNR_2)(CO)_2$] and [$Rh(S_2COR)(CO)_2$] – were also synthesized and assayed as cytostatic and anti-tumor agents *in vitro* against KB cells and *in vivo* against P388 leukemia, Ehrlich ascites carcinoma, Sarcoma 180 ascites and ADJ/PC6A solid tumor as well as against five *Trypanosoma* strains. The dicarbonyl(diphenyldithiocarbamato)-

rhodium(I) complex $[Rh(S_2CNPh_2)(CO)_2]$ was active in all aforenamed biological systems without showing evident nephrotoxicity.[16] The NBD xanthate complexes $[Rh(S_2COR)(NBD)]$ show activity against Ehrlich and Landschutz ascitic tumors.[17,18]

The cationic complexes $[Rh(NBD)(N{-}N)]ClO_4$, where N—N = 3,3′-dimethylene-2,2′-1,8-naphthyridine and 3,3′-trimethylene-2,2′-1,8-naphthyridine, were assayed against ascitic tumors and primary solid tumors in the form of pure complexes and coordination complexes adsorbed on polymers (polyethylene glycols). These complexes appear to be promising drugs because of their solubility in polymer dissolved in water, which make them easier to use in comparison with neutral complexes. These complexes are similar to cisplatin in reducing tumor growth and in increasing the survival lifespan of mice.[19]

20.2.2 Rhodium(II) complexes

Binuclear rhodium(II) complexes belong to the most promising anti-tumor complexes. Physicochemistry and reactivity of these complexes were very thoroughly investigated.[20] Properties of many complexes containing less than four bridging ligands have also been characterized.[20,21] The chemical structures of some of these complexes are presented in Figure 20.1.

In 1972 Bear and co-workers reported the anti-tumor activity of rhodium(II) carboxylates $[Rh_2(OOCR)_4]$.[22] These complexes were shown to be active against oral carcinoma, Ehrlich ascites, leukemia P388 and L1210 leukemia.[22–30] Anti-tumor activity of $[Rh_2(OOCR)_4]$ depends on the hydrophobic properties of the R group: $CH_3 < CH_3CH_2 < CH_3CH_2CH_2 < CH_3CH_2CH_2CH_2 > CH_3CH_2CH_2CH_2CH_2$. Other rhodium(II) complexes with four bridges: $[Rh_2(OOCR)_4L_2]$ (L – isonicotinic acid, HSA),[31,32] $[Rh_2\{O(HN)CCF_3\}_4]$,[33,34] $[Rh_2(OOCCF_3)_4]$ and $[Rh_2(OOCCF_3)_4L_2]$ (L = sulfadiazine),[35] rhodium(II) citrate,[36,37] rhodium(II) ketogluconate and glucoronate and their adducts with cyclophosphamide[38] have also

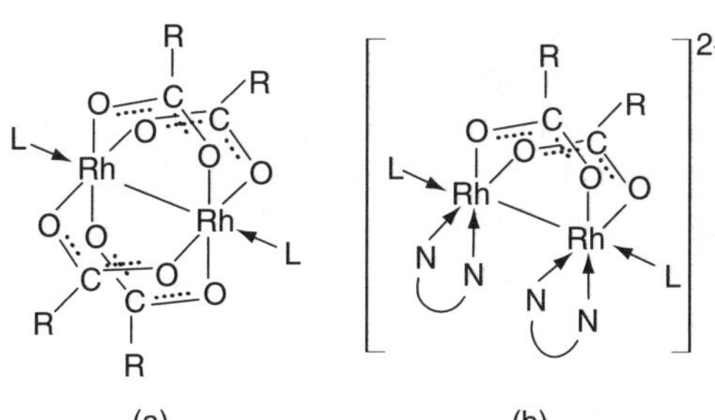

Figure 20.1 Structure of $[Rh_2(OOCR)_4]$ (a) and $[Rh_2(OOCR)_2(N\text{-}N)_2L_2]X_2$ (b) complexes. N-N = bipy, phen and their derivatives, L = Lewis base, X = $RCOO^-$, BF_4^-, etc.

been investigated. The *in vitro* activity of $[Rh_2(OOCR)_4L_2]$ (R = Me, Et, Pr) adducts with isonicotinic acid against K562 human leukemic cell line and investigations of LD_{10} (poisoning with 10% lethality) of these complexes *in vivo* in mice indicate that, in the presence of cellular membrane or blood lipids, isonicotinic acid adducts dissociate giving the parent complexes which therefore enter the cells more easily.[31] Thus, adducts with isonicotinic acid can be used as carriers of rhodium(II) carboxylates to cells. The same was suggested in the case of the reaction products between HSA (human serum albumin) and $[Rh_2(OOCR)_4]$ or $[Rh_2\{O(HN)CCF_3\}_4]$.[32] Promising anti-tumor properties characterizes the amidate complex $[Rh_2\{O(HN)CCF_3\}_4]$ because it is more active than *cis*-$[PtCl_2(NH_3)_2]$, and LD_{50} values for these complexes are comparable.[33,34] The activity of rhodium(II) trifluoroacetate and its adduct with sulfadiazine tested in mice against Ehrlich ascites carcinoma was much higher than that of rhodium(II) acetate.[35] Interesting anti-tumor properties also characterize rhodium(II) citrate, ketogluconate and glucoronate which are soluble in water.[36–38] Their toxicity is low and their anti-tumor activity is comparable with that of cisplatin. They can form inclusion or association compounds with cyclodextrin and hydroxyapatite. The associates allow for the controlled release of pure rhodium complexes into tumor cells.[37] Rhodium(II) complexes containing different bridging ligands are generally more reactive. This is caused by differences in the trans-influence of ligands. The formamidinato carboxylato complex $[Rh_2(form)_2\{O(HN)CCF_3\}_2(H_2O)_2]$ (form = *N,N'*-di-*p*-tolylformamidinate, *p*-$MeC_6H_4NCHNC_6H_4Me$-*p*) shows activity against the Yoshida ascites sarcoma and the T8 sarcoma of Guerin.[39] This complex reacts with 9-ethylguanine (9-EtGh) giving the complex $[Rh_2(form)_2\{9\text{-EtGH}\}_2](CF_3COO)_2$ containing two 9-ethylguanine bridging ligands coordinated via O- and N7-atoms.[40] The complexes $[Rh_2(RCOO)_2(N\text{-}N)_2L_2](RCOO)_2$ and $[Rh_2Cl_2(RCOO)_2(N\text{-}N)_2]$ (R = H, Me, Et, *n*-Pr, *n*-Bu, Ph, PhCHOH, MeCHOH, N-N = 2,2′-bipyridine, 4,4′-dimethyl-2,2′-bipyridine, 1,10-phenanthroline, 2,9-dimethyl-1,10-phenanthroline, 4,7-diphenyl-1,10-phenanthroline) (Figure 20.1b) constitute another group of rhodium(II) complexes showing interesting biological activities.[41–47] They are effective against the human carcinoma KB cell line,[41,42] against bacteria[43–45] and against the alga *Chlorella vulgaris*.[46] Efficient cytostatic agents against the KB carcinoma, Hul703 (bladder cancer), SW707 (colon adenocarcinoma) and T47D (breast cancer) are acetato complexes containing bridging orthometallated methoxyphenylphosphine ligands:[48] $[Rh_2(OAc)_2\{\mu\text{-}P(C_6H_3OMe\text{-}3)(C_6H_4OMe\text{-}3)_2\}_2(HOOCMe)(H_2O)]$,[48] $[Rh_2(OAc)_3\{\mu\text{-}P(C_6H_4O\text{-}2)(C_6H_4OMe\text{-}2)_2\}(HOOCMe)]$ and $[Rh_2(OAc)_3\{\mu\text{-}P(C_6H_3(OMe\text{-}6)O\text{-}2)(C_6H_3(OMe)_2\text{-}2,6)_2\}(HOOCMe)]$ (Figure 20.2).[49] Other acetato mono-orthometallated and di-orthometallated complexes, $[Rh_2(OAc)_3\{\mu\text{-}P(C_6H_3(OMe\text{-}3)(C_6H_4OMe\text{-}3)_2\}(HOOCMe)_2]$, $[Rh_2(OAc)_3\{\mu\text{-}P(C_6H_3OMe\text{-}4)(C_6H_4OMe\text{-}4)_2\}(HOOCMe)_2]$ and $[Rh_2(OAc)_3\{\mu\text{-}P(C_6H_3SO_3Na\text{-}3)(C_6H_4SO_3Na\text{-}3)_2\}(HOOCMe)_2]$, exhibit lower anti-tumor activity or are biologically inactive. Thus cytostatic activity of these complexes strongly depends on structure and properties of the phosphine-bridging ligand.

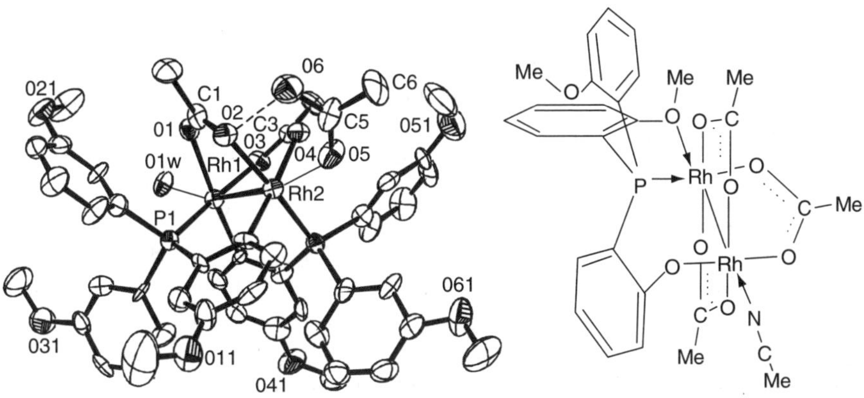

Figure 20.2 Structure of $[Rh_2(OAc)_2\{\mu\text{-}P(C_6H_3OMe\text{-}3)(C_6H_4OMe\text{-}3)_2\}_2(HOOCMe)(H_2O)]$ (Rh–Rh: 2.4913(18) Å)[48] and $[Rh_2(OAc)_3\{\mu\text{-}P(C_6H_4O\text{-}2)(C_6H_4OMe\text{-}2)_2\}(NCMe)]$ (Rh–Rh: 2.421(1) Å)[49]

Rhodium binuclear complexes can very easily coordinate ligands along the Rh–Rh axis. However, the substitution of equatorial carboxylato ligands is much more difficult. Thus, $[Rh_2(OOCR)_4]$ complexes form labile axial adducts with N- and O-donors. Substitution of bridging RCOO ligands by other ligands including amino acids also occurs but this reaction is considerably slower, especially in the presence of strong acids.[50] Thus, anti-tumor activity can follow both from formation of adducts with ligands coordinated in axial positions and coordination of ligands in equatorial coordination sites. The sulfhydryl (SH) compounds such as cysteine and glutathione can react to liberate carboxylate ligands and bind irreversibly to the rhodium.[28] Rhodium(II) carboxylates *in vivo* inhibit synthesis of DNA, but minimal inhibition of RNA synthesis has been observed.[25,27]

Investigations of the influence of rhodium(II) carboxylates on the activity of enzymes revealed that all those having SH groups in or near their active site were irreversibly inhibited, while the enzymes without SH essential groups were not affected. The reaction of rhodium(II) carboxylates with enzymes containing SH groups closely parallels the toxicity and anti-tumor activity of these complexes.[28] The rhodium(II) carboxylates bind to a variety of proteins and irreversibly inhibit those with cysteines in the active site. The mechanism of action of dirhodium(II) complexes can be similar to that of cisplatin. Recently, the ditolylformamidinato complex $[Rh_2\{\mu\text{-}HC(NC_6H_4Me\text{-}4)_2\}_2(9\text{-}EtGH)_2(MeCN)][BF_4]_2$ with bridging 9-ethylguanine ligands coordinated via O- and N7-atoms (head-to-head configuration) and the analogous 9-ethyladenine complex (head-to-tail configuration) have been prepared and structurally characterized[40] (Figure 20.3). In the acetato complex containing 2,2′-bipyridine, 9-ethylguanine is coordinated as a terminal equatorial ligand[51] (Figure 20.3).

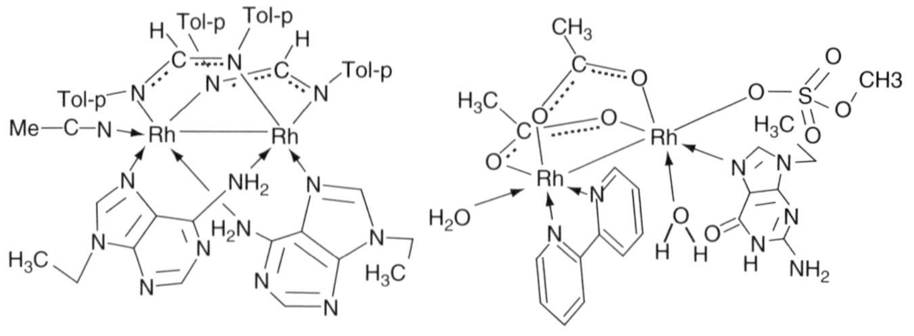

Figure 20.3 Structure of $[Rh_2\{\mu\text{-}HC(NC_6H_4Me\text{-}4)_2\}_2(9\text{-}EtAH)_2(MeCN)][BF_4]_2$ (Rh–Rh: 2.510(3) Å) and $[Rh_2(\mu\text{-}OOCMe)_2(bpy)(9\text{-}EtGH)(H_2O)_2(O_3SOMe)](O_3SOMe)\cdot H_2O$ (Rh–Rh: 2.5112(7) Å)

These complexes were prepared from dirhodium(II) complexes containing two or three bridging ligands. The structure and the Rh–Rh distances in these complexes suggest that compounds containing both bridging and terminal purine nucleotide ligands can be formed *in vivo* from $[Rh_2(OOCR)_4]$ or other tetrabridged complexes. The complexes of this type can be responsible for the anti-tumor activity of the binuclear rhodium(II) complexes. These data indicate that dirhodium complexes can coordinate not only to guanine but also to adenine, while cisplatin forms bonds mainly with guanine. This was confirmed by a combination of matrix-assisted laser desorption ionization mass spectrometry (MALDI MS) and enzymatic digestion experiments of $Rh_2(\mu\text{-}OOCMe)_2$(TCTCTAATCTC) and $Rh_2(\mu\text{-}OOCMe)_2$(CCTCTGGTCTCC). It has been established that the $Rh_2(\mu\text{-}OOCMe)_2$ unit forms an adduct with AA and GG sites.[52] In the presence of oxygen, the dimers $[Rh_2(\mu\text{-}OOCMe)_2(bpy)_2(MeCN)_2][BF_4]_2$ and $[Rh_2(\mu\text{-}OOCMe)_2(phen)_2(MeCN)_2][BF_4]_2$ react with 2-aminothiophenol giving rhodium(III) octahedral complexes: $[Rh\{C_6H_4S(NH_2)\text{-}N,S\}\{C_6H_4S(NH_2)\text{-}S\}_2(phen)]$, $[Rh_2\{\mu\text{-}C_6H_4S(NH_2)\text{-}N,S\}_2\{C_6H_4S(NH_2)\text{-}S\}_2(bpy)_2]^{2+}$ and $[Rh\{C_6H_4S(NH_2)\text{-}N,S\}_2(bpy)]$.[53] In the reaction of $[Rh_2(\mu\text{-}OOCMe)_2(bpy)_2(MeCN)_2][BF_4]_2$ with PhSH in the absence of O_2, a new binuclear rhodium(II) complex with two bridging and two terminal C_6H_5S ligands is formed: $[Rh_2(\mu\text{-}C_6H_5S)_2(C_6H_5S)_2(bpy)_2]$.[54] Thus, sulfido ligands are capable of stabilizing Rh(II) complexes. However, the complexes $[Rh_2(\mu\text{-}OOCMe)_2(N\text{-}N)_2(H_2O)_2](CH_3COO)_2$ in alcohols and water immediately react with SH compounds (glutathione, cysteine, coenzyme A) giving polynuclear complexes with $(Rh_2^{3+})_n$ chains.[55] Formation of complexes with rhodium in non-integral mean oxidation states +1.25, +1.33, +1.5 and +1.75 is also possible because $[Rh_2(\mu\text{-}OOCMe)_2(N\text{-}N)_2(H_2O)_2]^{2+}$ complexes, even in water–alcohol solutions, are readily reduced giving rhodium wires containing infinite chains of Rh–Rh bonds.[56–59] Recently, the structure of the complex $\{[Rh_4(\mu\text{-}OOCH)_4(bpy)_4](BF_4)\}_n\cdot 0.5nC_4H_8O_2$ with a Rh_4^{5+} core has been determined. The mass spectra of $[Rh_2(\mu\text{-}OOCR)_2(N\text{-}N)_2(H_2O)_2](RCOO)_2$ in different alcohol matrices revealed their reduction to the di-, tetra-, hexa- and even octanuclear rhodium complexes in which the mean oxidation states of rhodium change in the range of 1.25–1.75. In solution, rhodium wires dissociate with the formation of mainly tetranuclear (and probably octanuclear) complexes. The latter have 30 and 31 *d* electrons and are isoelectronic with platinum blues[60] which show promising anti-tumor activity.[61] Thus, during the cellular reactions of $[Rh_2(\mu\text{-}OOCR)_2(N\text{-}N)_2(H_2O)_2]^{2+}$, species similar to the platinum blues can be probably formed. The binding of the $[Rh_2(\mu\text{-}OOCMe)_4(H_2O)_2]$ and $[Rh_2(\mu\text{-}OOCMe)_2(phen)_2(H_2O)_2]^{2+}$ complexes to DNA has been investigated. The binding constants of these complexes to calf-thymus DNA were determined from electronic spectra recorded during titrations. The equilibrium constants are equal to $4.6 \times 10^2\ M^{-1}$ and $1.7 \times 10^4\ M^{-1}$ for $[Rh_2(\mu\text{-}OOCMe)_4(H_2O)_2]$ and $[Rh_2(\mu\text{-}OOCMe)_2(phen)_2(H_2O)_2]^{2+}$, respectively.[62] The electronic spectra indicate that DNA is coordinated in the axial coordination site immediately upon mixing. Both Rh(II) complexes inhibit transcription more effectively than cisplatin, plausibly owing to their binding to the enzyme T7-RNA polymerase.

Interaction of rhodium(II) complexes with HSA has also been investigated. The spectrophotometric titration of HSA with $[Rh_2(OOCR)_4]$ has shown that albumin binds eight equivalents of rhodium dimer.[32,47,63–65] The complex is bound to histydyl imidazole groups and is influencing the conformation of HSA.

20.2.3 Rhodium(III) complexes

As rhodium(III) coordination complexes have a d^6 electronic configuration, they are isoelectronic with Ru(II) and Pt(IV) complexes which are interesting anti-tumor complexes, e.g. (OC-6-43)-$[PtCl_2(OOCMe)_2(CyNH_2)(NH_3)]$ (Cy = cyclohexyl) (JM216), an orally active anti-tumor agent. Rhodium(III) complexes generally have an octahedral structure and are inert complexes; nevertheless, many of the Rh(III) coordination complexes tested show considerable anti-tumor and anti-microbial activity. The meridional complex (OC-6-21)-$[RhCl_3(NH_3)_3]$ is remarkably active against sarcoma 180.[66,67] The complexes containing dimethyl sulfoxide (OC-6-21)-$[RhCl_3(Me_2SO\text{-}S)_2(L)]$ (L = NH_3, imidazole) show a cytostatic activity against A2780, A2780/cp8, LoVo and Calu cells similar to that of cisplatin.[68] In contrast, the anionic complexes (OC-6-11)-$Na[RhCl_4(Me_2SO\text{-}S)_2]$, (OC-6-11)-$Na[RhCl_4(L)(Me_2SO\text{-}S)]$ and (OC-6-11)-$LH[RhCl_4(L)_2]$ (L = imidazole) are inactive. An anti-neoplastic activity *in vivo* against rhabdomyosarcoma characterizes a rhodium(III) chloride complex with oxalyl homocysteine thiolactone.[69] Administration of the rhodium(III) complex $[RhCl_2L_4]Cl$ (L = sulfaquinoxaline) to rats treated earlier with thioacetamide (inducing liver tumor) caused a significant restoration of the liver function.[70,71]

The cytostatic activity of polypyridyl and pyrazole rhodium(III) complexes against HCV29T tumor cells increases in the series $[RhCl_2(Hpz)_4][RhCl_4(Hpz)_2]$ **1** < $[RhCl_3(tpta)]{\cdot}H_2O$ **2** < $[RhCl_3(tpy)]$ **3** < *cis*-$[PtCl_2(NH_3)_2]$ < $[Rh(tpy)_2(Him)]Cl_3{\cdot}3H_2O$ **4** (Hpz = pyrazole, Him = imidazole, tpy = 2,2′:6′,2″-terpyridine, tpta = 2,4,6-tris(2-pyridyl)-1,3,5-triazine) (Figure 20.4).[72]

The cytostatic activity of the complexes may be a result of their interaction with DNA. The effectiveness of the restraining of DNA migration in the presence of complexes in an 8% native polyacrylamide increased in the order: **1** (20%) < **4** (40%) < **2** (55%) < **3** (90%). This order is consistent with the cytostatic activity of complexes apart from the most active complex **4**. It is possible that complex **4** relatively effectively interacts with DNA, but the rate of migration of the interaction product is not very strongly lowered because its structure is considerably different than the structures of DNA adducts with the complexes **1**, **2** and **3**. Interaction of complex **2** with DNA has been also investigated using DNase I footprinting. Investigations indicated that **2** is the most effective to change the DNase I footprinting of the ATGCGCT fragments of DNA. The obtained results suggest that the cytostatic activity of polypyridyl complexes can result from their interaction with DNA, most likely via formation

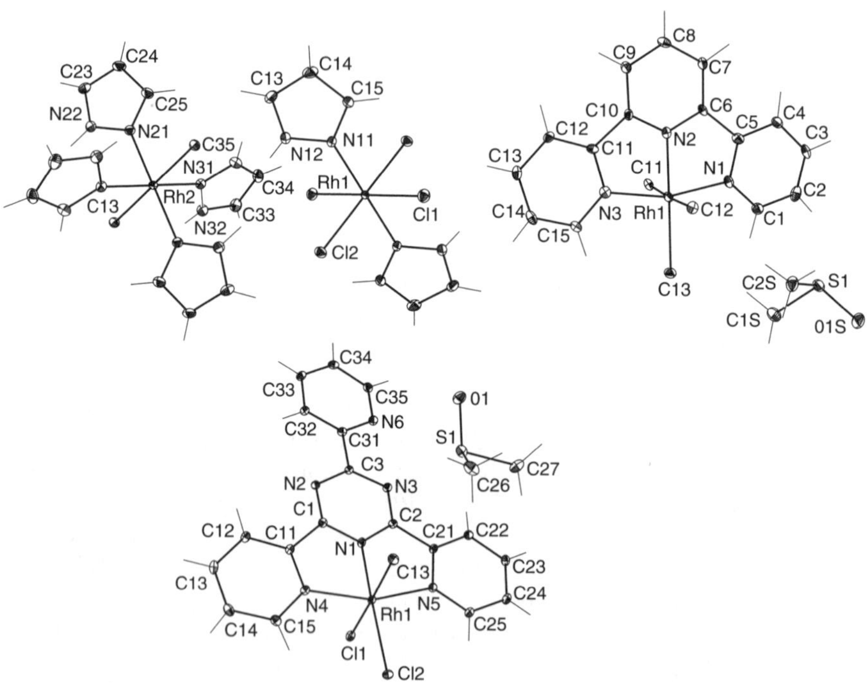

Figure 20.4 Chemical structures of $[RhCl_2(Hpz)_4][RhCl_4(Hpz)_2]$, $[RhCl_3(tpy)]{\cdot}(CH_3)_2SO$, $[RhCl_3(tpta)]{\cdot}(CH_3)_2SO$

of a Rh–N coordination bond with guanine or adenine bases. However, intercalation of tpy or tpta ligands between nucleotide bases cannot be excluded.

Rhodium(III) complexes with phenanthrenequinone diimine (phi) and 1,10-phenanthroline and its derivatives specifically bind DNA (Figure 20.5).[73–77] The phi ligand intercalates between the DNA base pairs in the major groove site of the DNA and orients the complex with respect to the helix. In these intercalations, the long axis of the phi ligand is parallel to the long axis of the base pair. The complex $[Rh(phen)_2(phi)]^{3+}$ binds to DNA with $K \geq 10^7\,M^{-1}$ and on photo-excitation, cleaves duplex DNA. These complexes bind DNA selectively in the more accessible major groove sites owing to steric clashes between pyrimidines and the H2 and H9 hydrogens of phenanthroline ligands.

The $[Rh(phen)_2(phi)]^{3+}$ complex cleaves DNA selectively at 5′-pyr-pyr-pur-3′ sites, e.g. 5′-CCAG-3′, whereas cleavage at 5′-pur-pur-pyr-3′ sites is suppressed. Other rhodium(III) complexes containing the phi ligand, with a less sterically demanding coordination sphere, are not selective, e.g. $[Rh(bpy)_2(phi)_2]^{3+}$. The phenanthroline complex containing the 5,6-chrysenequinone diimine ligand (chrysi) (Figure 20.5) is too large to intercalate into standard base pair steps. Therefore, this complex can recognize and cleave a single mismatch site within a 2725 base pair sequence of plasmid DNA. Interactions of rhodium(II) complexes $[Rh_2(OOCR)_4]$ (R = Me, Et, Pr, CF_3) and $[Rh_2\{O(NH)CR\}_4]$ (R = Me,

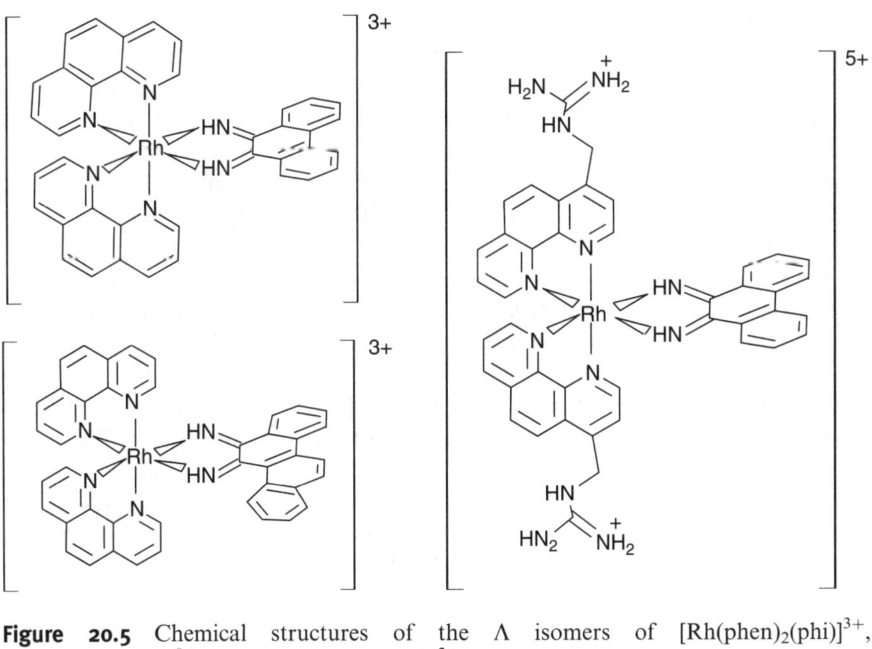

Figure 20.5 Chemical structures of the Λ isomers of $[Rh(phen)_2(phi)]^{3+}$, $[Rh(MGP)_2(phi)]^{5+}$ and $[Rh(phen)_2(chrysi)]^{3+}$

CF_3) with DNA were investigated using electrochemical methods.[78] The analysis of the suppressions of the anodic peak currents for guanine and adenine bases indicate that these complexes interact mainly with adenine.

20.3 Anti-bacterial and Anti-parasitic Activity

Many Rh(I), Rh(II) and Rh(III) complexes are active against malaria, leishmaniasis, trypanosomes, bacteria and algae. The anti-malarial activity of [RhCl(COD)(CQ)] (CQ = chloroquine) is similar to that of chloroquine diphosphate.[79,80] However, chloroquine complexes with other metals (ruthenium and gold) are more active.[79] Numerous rhodium complexes were investigated for the treatment of leishmaniasis.[79,81–83] Rhodium(III) cationic complexes containing 2-hydroxybenzothiazole are very efficient anti-leishmanial agents:[81] they totally inhibit the growth of *Leishmania donovani*. The rhodium(III) complex with mepacrine has an ED_{50} (effective dose generating 50% inhibition) of less than 1 μM. Ultrastructural studies (investigation of structure by means of electron microscope) suggest that the amastigote kinetoplast-mitochondrion complex is the primary site of action of the rhodium complex.[82] The complexes: $[RhBr_2L_4]Br$ (L = 2-amino-6-ethoxybenzothiazole or 2-bromothiazole) and $[RhCl_2L_4]Cl$ (L = mefloquine, 2-mepacrine or oxamniquine) exhibited high cytotoxicity against the macrophage cell line J-774.[83] The complex with 2-amino-6-ethoxybenzothiazole exhibited high activity *in vivo* in parasitized Wistar rats.[83] The $[Rh_2(OOCR)_4L_2]$ adducts with anti-malarial drugs (Plasmaquine, Chloroquine, Primaquine, Mepacrine, Pentaquine, Amodiaquine; Figure 20.6) show both anti-tumor and trypanocidal activities. The activity of rhodium complexes against trypanosomes was thoroughly investigated.[17,18,84–87] Trypanocidal effects (*Trypanosoma evansi*, *T. equiperdum*, *T. congolense*, *T. cruzi*) of [RhX(NBD)] (X = alkyl and aryl xanthates) were investigated.[17]

Of this series, the most active is the complex containing cyclopentyl xanthate. The activities of $[Rh_2L_2(COD)_2]X_2$ (L = anti-parasitic drug 2-hydroxystilbamidine, COD = 1,3-cyclooctadiene, $X^- = Cl^-, NO_3^-, ClO_4^-, BPh_4^-$) and $[Rh(COD)L_2]X$ (L = anti-parasitic drug: Benznidazole = 2-nitro-*N*-(phenylmethyl)-1*H*-imidazole-1-acetamide, Nifurtimox = 3-methyl-*N*-[(5-nitro-2-furanyl)methylene]-4-thiomorpholine-1,1-dioxide, Niridazole = 1-(5-nitrothiazol-2-yl)-2-imidazolidinone) were assayed on rats infected with *T. evansi*, and *T. congolense*.[18] The most active was the complex with benznidazole. Rhodium complexes with diamidine and phenanthridinium trypanocides show higher therapeutic properties than the parent drugs.[85] Another group of complexes showing activity against *T. cruzi* are $[RhX_2L_4]X$ complexes, where L is a derivative of thiazole.[86] It has been found that the rhodium(III) complexes do not interact with native DNA but some of them inhibit the precursor uptake and the

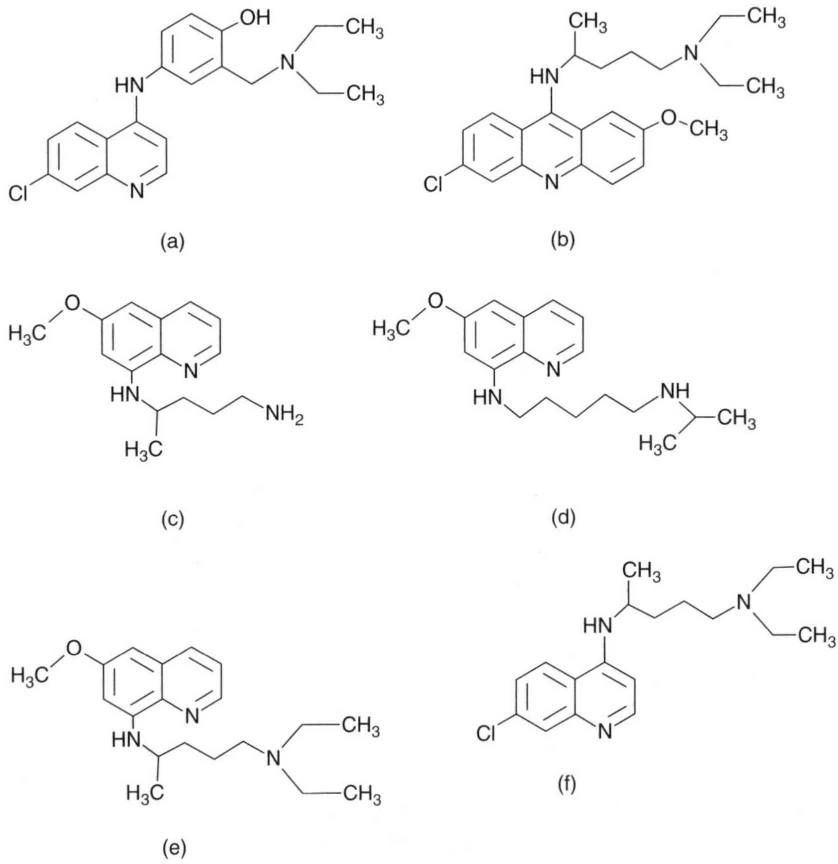

Figure 20.6 Chemical structures of anti-malarial drugs used as ligands in syntheses of rhodium complexes: (a) Amodiaquine, (b) Mepacrine, (c) Primaquine, (d) Pentaquine, (e) Plasmaquine and (f) Chloroquine

viability of metacyclic forms.[86] The rhodium(I) complex with clotrimazole, (1-[(2-chlorophenyl)diphenylmethyl]-1*H*-imidazole), [RhCl(COD)(CTZ)] efficiently inhibits *in vitro* cultures of *T. cruzi*.[87] Complexes of Rh(III) and Rh(II) with 2-amino-6-ethoxybenzothiazole, 2-aminonaphtothiazole, 2-chloropyridine and benzimidazole dithiocarbamate were evaluated for their anti-filarial activity against *Molinema dessetae* and *Brugia pahangi*.[88]

The binuclear rhodium(II) complexes $[Rh_2(RCOO)_2(N\text{-}N)_2(H_2O)_2](RCOO)_2$, $[Rh_2Cl_2(RCOO)_2(N\text{-}N)_2]$ and $[Rh_2Cl_2(MeCOO)(tpy)_2](H_3O)Cl_2{\cdot}9H_2O$ (R = H, Me, Bu, Ph, PhCHOH; N-N = 2,2′-bipyridine (bpy), 1,10-phenanthroline (phen), 2,9-dimethyl-1,10-phenanthroline (dmp) and 6,7-dimethyl-2,3-di(2-pyridyl)quinoxaline (dmpq); tpy = 2,2′:6′,2″-terpyridine) are effective anti-bacterial agents against *Staphylococcus aureus* and *Escherichia coli*. The most active anti-bacterial agents against *S. aureus* were $[Rh_2(PhCOO)_2(phen)_2(H_2O)_2](PhCOO)_2$, $[Rh_2(PhCOO)_2(dmpq)_2(H_2O)_2](PhCOO)_2$, $[Rh_2(BuCOO)_2(phen)_2(H_2O)_2](BuCOO)_2$ and $[Rh_2(BuCOO)_2(bpy)_2(H_2O)_2](BuCOO)_2$. These complexes show rather low activity against *E. coli*. The most active anti-bacterial agents against *E. coli* were $[Rh_2(RCOO)_2(N\text{-}N)_2(H_2O)_2](RCOO)_2$ and $[Rh_2Cl_2(RCOO)_2(N\text{-}N)_2]$ (R = H and Me). These complexes exhibit low activity against *S. aureus*. The activity of the complexes $[Rh_2(RCOO)_2(N\text{-}N)_2(H_2O)_2](RCOO)_2$ against *E. coli* decreases in the series: $R = H \cong CH_3 > C_2H_5 > C_3H_7 \cong C_4H_9$. The reverse order was found in the case of *S. aureus*. The complexes were considerably more active than the uncoordinated nitrogen ligands.[43–45] The $[Rh_2(PhCOO)_2(phen)_2(H_2O)_2](PhCOO)_2$ complex is active against many *Staphylococcus* strains resistant to commonly used antibiotics.[43] The complexes $[Rh_2(RCOO)_2(N\text{-}N)_2(H_2O)_2](RCOO)_2$ are effective cytostatic agents against the synchronously cultivated green alga *C. vulgaris*. For the $[Rh_2(MeCOO)_2(dpp)_2(H_2O)_2](MeCOO)_2$ (dpp = 4,7-diphenyl-1,10-phenanthroline) and $[Rh_2(PhCHOHCOO)_2(phen)_2(H_2O)_2](PhCHOHCOO)_2$ complexes, ED_{50} values equal to 10^{-6} M and 8×10^{-6} M, respectively, were found.[46] The complexes inhibited the synthesis of DNA, protein and pigments (chlorophyll *a*, chlorophyll *b*, carotenes, luteine and violaxantine). However, the synthesis of RNA at low concentration of the complexes increased. A further well-defined series of metal complexes showing anti-bacterial activity is that of rhodium(III) complexes with pyridine: *trans*-$[RhX_2(py)_4]Y$.[89] They are more active against Gram-positive bacteria than against Gram-negative. This perhaps is correlated with penetration because the highest anti-bacterial activity is exhibited by the most lipophilic complexes. The increase in activity corresponded to the increase in the number of C atoms in the alkyl group bound with pyridine ring. The complexes with pyridine derivatives substituted at the 3-, 4- or 5-positions had equal effects. The complex *cis*-$[RhCl_2(bpy)_2]^+$ is inactive. Many rhodium(III) complexes, such as $RhCl_3$, $(NH_4)_3[RhCl_6]$, (OC-6–21)-$[RhCl_3(NH_3)_3]$, were effective in the filamentation of bacteria.

20.4 Radiorhodium Agents

The utility of radiation in the destruction of cancerous tissues was recognized and exploited since the time of M. Sklodowska-Curie. Radiometals possess a wider range of nuclear properties and show diverse chemistries; therefore they are very valuable pharmaceuticals.[90] Desirable radionuclide characteristics for radiotherapy are high specific activity, low γ-emission, a decay half-life similar to the half-life of residence at the target, high linear energy transfer and ranges of particles comparable with the site of deposition. The radionuclide ^{105}Rh [(35.4 h, 0.567 MeV β^- (70%), 0.247 MeV β^- (30%), 0.320 MeV γ)] is an attractive β^--emitter for radiotherapy because of its advantageous properties. Complexes used for radiotherapy should be inert, to minimize loss of the rhodium complex from its carrier molecule. The rhodium(III) complexes of d^6 low-spin electronic configuration belong to the most inert complexes. The complexes with multidentate, chelating ligands are more inert than the complexes with appropriate monodentate ligands. Therefore very frequently, the complexes with multidentate ligands are used as radiotherapeutic agents. In order to bind a metallic complex with a carrier, it is important to prepare a suitable bifunctional chelate that will act both as a ligand to the radionuclide and as a linker to the carrier molecule. Examples of such molecules are thiobis[ethylenenitrilo(tetra-acetic acid)], $S\{CH_2CH_2N(CH_2COOH)_2\}_2$ (L)[91] and 1,5,9,13-tetrathiacyclo-hexadecane-3,11-diol (L′)[92] which form octahedral Rh(III) complexes [RhL] and $[RhCl_2L']^+$, respectively. The ^{105}Rh(III) complexes with $HOOCCH_2S(CH_2CH_2S)_3CH_2COOH$ (2,5,8,11-tetrathiadodecane-1,12-dicarboxylic acid) and analogous tetrathiatridecane- and tetrathiatetradecane-dicarboxylic acids were evaluated for their stability under physiological conditions and *in vivo* in normal mice. The results indicated that the *cis*-$[RhCl_2\{HOOCCH_2(SCH_2CH_2)_3SCH_2COOH\}]^+$ complex showed the most promising *in vivo* characteristics. Thus, this complex is a potential radiopharmaceutical.[93] The stability of the ^{105}Rh-bleomycin complex in plasma is sufficient to allow targeted delivery of the radioisotope.[94] The biodistribution of ^{105}Rh complexes with trithiaazacyclotetradecanes and tetrathiacyclohexadecanes (1,4,7-trithia-11-azacyclotetradecane, 1,5,9-trithia-13-azacyclohexadecane, 1,5,9,13-tetrathia-cyclohexadecane-3-ol, etc.) has been investigated in rats. The results revealed that the *cis*-$[RhCl_2\{14\}ane]^+$ complexes exhibited less kidney retention than the *trans*-$[RhCl_2\{16\}ane]^+$ complexes.[95] The ^{103m}Rh complexes have also been proposed for targeted radiotherapy of small tumors.[96] Rhodium trichloride was used as catalyst for regiospecific deuteriation and tritiation of various drugs.[97] The ^{106}Ru–^{106}Rh radioisotopes are used for the treatment of choroidal melanoma.[7] Brachytherapy with ^{106}Ru–^{106}Rh plaques can be used for small medium-sized choroidal melanomas.[98,99] The treatment using ^{106}Ru–^{106}Rh radionuclides gives frequently satisfactory results.

20.5 Photochemotherapy

Photochemotherapy has been revealed to kill tumor cells most frequently by inactivating nucleic acids. However, the interaction of photosensitizers with serum proteins, annexins, etc., can also be important. The $[Rh(phen)_2(phi)]^{3+}$ complex (Figure 20.5) binds selectively to DNA duplexes by intercalation into the major groove but shows no detectable interaction with A-conformational RNA duplexes. However, the RNA folds of bovine (BIV) and human (HIV) immunodeficiency viruses have remarkable structures in which intramolecular base triples are formed and the major grooves are widened. At these sites, transcription is activated. The $[Rh(phen)_2(phi)]^{3+}$ complex targets sites of triple base interactions which are cleaved upon photoactivation. The primary site cleaved upon photoactivation in the BIV by the rhodium complex is U24, a base participating in the novel base triple with A13 and U10 bases.[100] The Δ-$[Rh(phen)_2(phi)]^{3+}$ complex binds to this site with an affinity of $2 \times 10^6\,M^{-1}$. Thus, the rhodium complex is able to compete for RNA target with the BIV trans-activating (Tat) peptide.[100]

A rhodium complex, *rac*-$[Rh(bpy)_2(phzi)]^{3+}$ (phzi = Benzo[*a*]phenazine-5,6-quinone diimine), intercalates as a sterically demanding intercalator to destabilized mismatched sites in double-helical DNA and, upon photoactivation, promotes direct strand scission at single base mismatch sites.[101] Unlike the $[Rh(bpy)_2(chrysi)]^{3+}$ complex, the phzi complex intercalates and cleaves with a high affinity and efficiency. The specific binding constants for CA, CC and CT mismatches are 0.3, 1 and $6 \times 10^7\,M^{-1}$, respectively. *cis*-$[RhCl_2(phen)_2]^+$ and related complexes, in photochemically induced reactions, can form covalent bonds with calf-thymus DNA, nucleotides and nucleosides.[102] The *cis*-$[RhCl_2(phen)_2]^+$ and *cis*-$[RhCl_2(phen)(phi)]^+$ complexes form, upon photoexcitation (320–400 nm), cross-inks in nucleic acids through the involvement of purines.[103] The former complex, despite the negligible absorption beyond the 320–400 nm region, when irradiated with red light, populates a highly reactive excited state that forms covalent bonds with DNA.[104] A series of $[Rh_2(RCOO)_4]$ complexes (R = H, Me, Et, CF_3) was examined for enhancement of efficiency of X-ray-induced killing of *Salmonella typhimurium*.[105,106] A range of *cis*-Pt(II) and $[Rh_2(RCOO)_4]$ complexes have been examined for their ability to increase the radiation sensitivity of aerobic and hypoxic V-79 cells *in vitro*. The platinum complexes show considerably smaller levels of sensitization than rhodium(II) carboxylates.[107]

20.6 Toxicity of Rhodium Complexes

There are no data on rhodium involvement in either human or animal biochemical systems. Studies of the $^{105}RhCl_3$ metabolism in rats administered via the oral, intravenous and intramuscular routes revealed that rhodium was poorly absorbed. Initial rapid elimination occurred in the urine and later via the gastrointestinal tract. Rhodium was mainly accumulated in kidney and liver, but minor fractions were found in many soft tissues. The biological half-life of ^{105}Rh varied between 4.3 and 16.5 days. The few studies on rhodium toxicity indicate that Rh salts are slightly toxic by oral ingestion and are mild skin irritants.[108–112] The rhodium trichloride fed to mice at 5 ppm in drinking water produced tumors: adenocarcinoma and lymphoma-leukemia. Thus, $RhCl_3$ at these dosage levels is slightly carcinogenic. The oral LD_{50} values for rats are 500–2000 mg/kg for $RhCl_3$ and $Na_3[RhCl_6]$, >5000 mg/kg for RhI_3,

>500 mg/kg for $[Rh(NH_3)_5Cl]Cl_2$ and $[RhCl(PPh_3)_3]$, and >2000 mg/kg for $Rh(acac)_3$. An examination of the frequencies of the positive prick test and patch test reactions to the chlorinated complexes of platinum-group metals revealed that the present concentration of these metals in the environment does not increase the incidence of reactions to these salts in patients with dermatitis and/or urticaria.[112] Rhodium complexes show relatively low mutagenicity. The Ames tests indicate that the $[Rh_2(RCOO)_2(N\text{-}N)_2(H_2O)_2](RCOO)_2$ complexes (R = H, Me, Et, Ph; N-N = bpy, phen, dmpq) at MIC (minimal inhibitory concentration) do not show any mutagenic activity.[45] Rhodium trichloride, at a concentration 300 μM, induced mutations in V-79 cells.[113] Apart from the test of Ames, the bacterial SOS chromotest is used for the assessment of genotoxicity. The latter is a rapid test giving results within hours. In this assay, the *lacZ* gene of the tester strain *E. coli* PQ37 is fused to the bacterial *sfiA* SOS operon and *lacZ* is concomitantly expressed in the case of bacterial SOS response. Investigations of the mutagenicity of $K_2[RhCl_5]$ and $(NH_4)_3[RhCl_6]$ in the SOS chromotest have shown that their activity is lower than that of cisplatin.[114,115] Mutagenicity strongly depends on the structure and properties of the complexes. The complexes with 2,2′-biquinoline (biq) and 2-(2′-pyridyl)quinoline (pq), *cis*-$[RhCl_2(biq)_2]Cl$ and *cis*-$[RhBr_2(pq)_2]Br$ are highly mutagenic, *cis*-$[RhCl_2(pq)_2]Cl$ and *cis*-$[RhBr_2(biq)_2]Br$ are moderately mutagenic, while *fac*- and *mer*-$[RhCl_3(pq)(H_2O)]$, as well as *cis*-$[RhCl_2(pq)_2][RhCl_2(CO)_2]\cdot H_2O$, are slightly mutagenic.[116] Concentration of rhodium and other platinum group elements (PGE) in the earth crust is very low. However, the increasing uses of PGE in industry (mainly in catalytic processes), in vehicle exhaust catalysts, in jewelry and in medicine cause their anthropogenic emission and increase of their concentration in the environment.[117] The metallic form of the PGE is generally considered to be inert as regards biochemical processes. However, the platinum metals in the environment are partly transformed into soluble compounds. In water at pH values between 5 and 8 the transformation in the model samples prepared from a milled catalytic vehicle converter is low and nearly constant: 0.01–0.025% for Pt and 0.05% for Rh. However, it was demonstrated that even 2–10% of the total PGE present in traffic dust was soluble. The bioaccumulation of PGE is low. The transfer coefficient (TC) (the ratio of the metal concentration in the plant to that in the soil) decreases in the order Pd > Pt ≥ Rh. The TC values of Pt for cabbage were found to be in the range of 0.004–0.008. However, the PGE can be accumulated by some plants and animals very efficiently. In the case of the hydroponically grown cucumbers the TC values for roots ranged from 1700 to 2100. Water hyacinth is capable of recovering PGE even from diluted solutions (5 ppm) according to the order: Pt(II) > Pd(II) > Os(IV) ≈ Ru(III) > Ir(III) ≈ Rh(III). The toxicity at the 10 ppm level decreases in the same order. The Pt(II) compounds at this concentration are toxic, in contrast to the $Na_3[RhCl_6]$, causing increase of the biomass of hyacinths some 6.7%. The exposure of the freshwater isopod *Asellus aquaticus* to PGE standard solutions for 24 h yielded the TC values

equal to 150, 85 and 7 for Pd, Pt and Rh, respectively. The PGE are also biologically available for earthworms, rats, zebra mussels, barbels and European eels.

References

1. E. Wong, C.M. Giandomenico, *Chem. Rev.* **99**, 2451–66 (1999).
2. M.J. Clarke, F. Zhu, D.R. Frasca, *Chem. Rev.* **99**, 2511–33 (1999).
3. B. Keppler, ed., *Metal Complexes in Cancer Chemotherapy*, VCH, Weinheim, New York, 1993.
4. M. Gielen, *Appl. Organomet. Chem.* **16**, 481–94 (2002) and references therein.
5. M. Gielen, M. Biesemans, D. de Vos, R. Willem, *J. Inorg. Biochem.* **79**, 139–45 (2000).
6. I. Haiduc, C. Silvestru, *Coord. Chem. Rev.* **99**, 253–96 (1990); I. Haiduc, C. Silvestru, *In Vivo* **3**, 285–93 (1992).
7. N. Katsaros, A. Anagnostopoulou, *Crit. Rev. Oncol. Hematol.* **42**, 297–308 (2002).
8. G. Sava, T. Giraldi, G. Mestroni, G. Zassinovich, *Chem. Biol. Interact.* **45**, 1–6 (1983).
9. T. Giraldi, G. Sava, G. Mestroni *et al.*, *Chem. Biol. Interact.* **22**, 231–8 (1978).
10. T. Giraldi, G. Sava, G. Bertoli *et al.*, *Cancer Res.* **37**, 2662–6 (1977).
11. G. Sava, S. Zorzet, G. Mestroni, G. Zassinovich, *Anticancer Res.* **5**, 249–52 (1985).
12. G. Sava, S. Zorzet, S. Pacor *et al.*, *Cancer Chemother. Pharmacol.* **24**, 302–6 (1989).
13. G. Sava, S. Pacor, V. Ceschia *et al.*, *Anticancer Res.* **9**, 787–90 (1989).
14. G. Sava, S. Zorzet, L. Perrisin, *Inorg. Chim. Acta* **137**, 69–71 (1989).
15. G. Craciunescu, V. Scarcia, A. Furlani *et al.*, *Anticancer Res.* **9**, 781–5 (1989).
16. D.G. Craciunescu, V. Scarcia, A. Furlani *et al.*, *In Vivo* **5**, 329–32 (1991).
17. D.G. Craciunescu, C. Molina, M.P. Alonso *et al.*, *An. R. Acad. Farm.* **56**, 469–86 (1990).
18. D.G. Craciunescu, E. Parrondo Iglesias, M.T. Gutierrez-Rios *et al.*, *An. R. Acad. Farm.* **59**, 73–97 (1993).
19. R. Sartori, G. Rencoret, A. Mora *et al.*, *Anticancer Drugs* **7**, 87–92 (1996).
20. F.A. Cotton, R.A. Walton, *Multiple Bonds Between Metal Atoms*, Clarendon Press, Oxford, 1993.
21. F.P. Pruchnik, *Pure Appl. Chem.* **61**, 795–804 (1989).
22. R.G. Hughes, J.L. Bear, A.P. Kimball, *Proc. Am. Assoc. Cancer Res.* **13**, 120 (1972).
23. A. Erck, L. Rainen, J. Whileyman *et al.*, *Proc. Soc. Exp. Biol. Med.* **145**, 1278–83 (1974).
24. J.L. Bear, R.A. Howard, A.M. Dennis, *Curr. Chemother.* 1321 (1978).
25. P.N. Rao, M.L. Smith, S. Pathak *et al.*, *J. Natl. Cancer Inst.* **64**, 905–12 (1980).
26. L. Rainen, R.A. Howard, A.P. Kimball, J.L. Bear, *Inorg. Chem.* **14**, 2752–4 (1975).
27. J.L. Bear, H.B. Gray, L. Rainen *et al.*, *Cancer Chemother. Rep.* **59**, 611–20 (1975).
28. R.A. Howard, T.G. Spring, J.L. Bear, *Cancer Res.* **36**, 4402–5 (1976).
29. R.A. Howard, A.P. Kimball, J.L. Bear, *Cancer Res.* **39**, 2568–73 (1979).
30. R.A. Howard, E. Sherwood, A. Erck *et al.*, *J. Med. Chem.* **20**, 943–6 (1977).
31. A.R. Souza, R. Najjar, S. Glikmanas, S.B. Zyngier, *J. Inorg. Biochem.* **64**, 1–5 (1996).

32. B.P. Esposito, E. Oliveira, S.B. Zyngier, R. Najjar, *J. Braz. Chem. Soc.* **11**, 447–52 (2000).
33. B.P. Esposito, S.B. Zyngier, A.R. Souza, R. Najjar, *Metal-Based Drugs* **4**, 333–8 (1997).
34. B.P. Esposito, S.B. Zyngier, R. Najjar *et al.*, *Metal-Based Drugs* **6**, 17–8 (1999).
35. E.M. Reibscheid, S. Zyngier, D.A. Maria *et al.*, *Braz. J. Med. Biol. Res.* **27**, 91–4 (1994).
36. S. Zyngier, E. Kimura, R. Najjar, *Braz. J. Med. Biol. Res.* **22**, 397–401 (1989).
37. A.E. Burgos, J.C. Belchior, R.D. Sinisterra, *Biomaterials* **23**, 2519–26 (2002).
38. E.I. Ferreira, S.B. Zyngier, R. Najjar, *Metal-Based Drugs* **6**, 19–24 (1999).
39. V. Fimiani, T. Ainis, T. Cavallaro, P. Piraino, *J. Chemother.* **2**, 319–26 (1990).
40. K.V. Catalan, J.S. Hess, M.M. Maloney *et al.*, *Inorg. Chem.* **38**, 3904–13 (1999).
41. F.P. Pruchnik, *Pure Appl. Chem.* **61**, 795–804 (1989).
42. F.P. Pruchnik, D. Duś, *J. Inorg. Biochem.* **61**, 55–61 (1996).
43. M. Bień, F.P. Pruchnik, A. Seniuk *et al.*, *J. Inorg. Biochem.* **73**, 49–55 (1999).
44. F.P. Pruchnik, M. Bień, T. Lachowicz, *Metal-Based Drugs* **3**, 185–95 (1996).
45. M. Bień, T.M. Lachowicz, A. Rybka *et al.*, *Metal-Based Drugs* **4**, 81–8 (1997).
46. F.P. Pruchnik, G. Kluczewska, A.Wilczok *et al.*, *J. Inorg. Biochem.* **65**, 25–34 (1997).
47. L. Trynda-Lemiesz, F.P. Pruchnik, *J. Inorg. Biochem.* **66**, 187–92 (1997).
48. F.P. Pruchnik, R. Starosta, Z. Ciunik *et al.*, *Can. J. Chem.* **79**, 868–77 (2001).
49. F.P. Pruchnik, R. Starosta, M.W. Kowalska *et al.*, *J. Organometal. Chem.* **597**, 20–8 (2000).
50. M. Koralewicz, F.P. Pruchnik, A. Szymaszek *et al.*, *Transition Met. Chem.* **23**, 523–5 (1998).
51. H.T. Chifotides, J.S. Hess, A.M. Sorasaenee, K.R. Dunbar, *J. Chem. Soc., Dalton Trans.* 4426–30 (2003).
52. J.M. Asara, J.S. Hess, E. Lozada *et al.*, *J. Am. Chem. Soc.* **122**, 8–13 (2000).
53. K. Sorasaenee, J.R. Galan-Mascaros, K.R. Dunbar, *Inorg. Chem.* **41**, 433–6 (2002).
54. K. Sorasaenee, J.R. Galan-Mascaros, K.R. Dunbar, *Inorg. Chem.* **42**, 661–3 (2003).
55. P. Jakimowicz, Ł. Ostropolska, F.P. Pruchnik, *Metal-Based Drugs* **7**, 201–9 (2000).
56. F.P. Pruchnik, P. Jakimowicz, Z. Ciunik *et al.*, *Inorg. Chem. Commun.* **4**, 19–22 (2001).
57. F.P. Pruchnik, P. Jakimowicz, Z. Ciunik, *Inorg. Chem. Commun.* **4**, 726–9 (2001).
58. F.P. Pruchnik, A. Jutarska, Z. Ciunik, M. Pruchnik, *Inorg. Chim. Acta* **350**, 609–16 (2003).
59. F.P. Pruchnik, A. Jutarska, Z. Ciunik, M. Pruchnik, *Inorg. Chim. Acta* **357**, 3019–26 (2004).
60. K. Matsumoto, M. Ochiai, *Coord. Chem. Rev.* **231**, 229–38 (2002).
61. Y. Dohta, C.S. Browning, P. Rekonen *et al.*, *Inorg. Chim. Acta* **263**, 69–79 (1997).
62. K. Sorasaenee, P.K.-L. Fu, A.M. Angelez-Boza *et al.*, *Inorg. Chem.* **42**, 1267–71 (2003).
63. L. Trynda, F.P. Pruchnik, *J. Inorg. Biochem.* **58**, 69–77 (1995).
64. B.P. Esposito, A. Faljoni-Alario, J.F.S. Menezes *et al.*, *J. Inorg. Biochem.* **75**, 55–61 (1999).
65. B.P. Esposito, R. Najjar, *Coord. Chem. Rev.* **232**, 137–49 (2002).
66. M.J. Cleare, P.C. Hydes, in H. Sigel, ed., *Metal Ions Biol. Syst.* **11**, 1–62 (1980).
67. G. Mestroni, G. Zassinovich, E. Alessio, A. Bontempi, *Inorg. Chim. Acta* **137**, 63–7 (1987).

68. G. Mestroni, E. Alessio, A. Sessanta o Santi *et al.*, *Inorg. Chim. Acta* **273**, 62–71 (1998).
69. K.S. McCully, M.P. Vezeridis, *Cancer Invest.* **5**, 25–30 (1987).
70. C. Cascales, P. Martin-Sanz, R.A. Pittner *et al.*, *Biochem. Pharmacol.* **35**, 2655–61 (1986).
71. M. Cascales, P. Martin-Sanz, D.G. Craciunescu *et al.*, *Carcinogenesis* **12**, 233–40 (1991).
72. F.P. Pruchnik, P. Jakimowicz, Z. Ciunik *et al.*, *Inorg. Chim. Acta* **334**, 59–66 (2002).
73. K.E. Erkkila, D.T. Odom, J.K. Barton, *Chem. Rev.* **99**, 2777–95 (1999).
74. C. Metcalfe, J.A. Thomas, *Chem. Soc. Rev.* **32**, 215–24 (2003).
75. D.T. Odom, C.S. Parker, J.K. Barton, *Biochemistry* **38**, 5155–63 (1999).
76. K.D. Copeland, M.P. Fitzsimmons, R.P. Houser, J.K. Barton, *Biochemistry* **41**, 343–56 (2002).
77. B.A. Jackson, V.Y. Alekseyev, J.K. Barton, *Biochemistry* **38**, 4655–62 (1999).
78. E.S. Gil, S.H.P. Serrano, E.I. Ferreira, L.T. Kubota, *J. Pharm. Biomed. Anal.* **29**, 579–84 (2002).
79. V. Sharma, D. Piwnica-Worms, *Chem. Rev.* **99**, 2545–60 (1999).
80. R. Sanchez-Delgado, M. Navarro, H. Perez, J. Urbina, *J. Med. Chem.* **39**, 1095–9 (1996).
81. C. Mesa-Valle, V. Moraleda-Lindez, D. Craciunescu *et al.*, *Arzneim.-Forsch./Drug Res.* **43**, 1010–13 (1993).
82. S.L. Croft, R.A. Neal, D.G. Craciunescu, G. Certad-Fombona, *Trop. Med. Parasitol.* **43**, 24–8 (1992).
83. M.N. Rodriguez-Cabezas, C.M. Mesa-Valle, S. Azzouz *et al.*, *Pharmacology* **63**, 112–19 (2001).
84. D.G. Craciunescu, C. Molina, E. Parrondo-Iglesias *et al.*, *An. R. Acad. Farm.* **57**, 15–35 (1991).
85. N.P. Farrell, J. Williamson, D.J. McLaren, *Biochem. Pharmacol.* **33**, 961–71 (1984).
86. L.M. Ruiz-Perez, A. Osuna, S. Castanys *et al.*, *Arzneimittelforschung* **36**, 13–16 (1986); L.M. Ruiz-Perez, A. Osuna, M.C. Lopez *et al.*, *Arzneimittelforschung* **38**, 312–14 (1988).
87. R.A. Sanchez-Delgado, M. Navarro, K. Lazardi *et al.*, *Inorg. Chim. Acta* **275–276**, 528–40 (1988).
88. P.M. Loiseau, J.J. Jaffe, D.G. Craciunescu, *Int. J. Parasit.* **28**, 1279–82 (1998).
89. R.J. Bromfield, R.H. Dainty, R.D. Gillard, B.T. Heaton, *Nature* **223**, 735–6 (1969); R.D. Gillard, *Kem. Kozlemenyek* **48**, 107–18 (1977).
90. M.J. Heeg, S.S. Jurisson, *Acc. Chem. Res.* **32**, 1053–60 (1999).
91. N.A. Powell, A.M. Hill, W. Levason, M. Webster, *J. Chem. Soc., Dalton Trans.* 467–71 (1996).
92. M. Venkatesh, N. Goswami, W.A. Volkert *et al.*, *Nucl. Med. Biol.* **23**, 33–40 (1996).
93. N. Goswami, C. Higginbotham, W. Volkert *et al.*, *Nucl. Med. Biol.* **26**, 951–7 (1999).
94. R.C. Brooks, P. Carnochan, J.F. Vollano *et al.*, *Nucl. Med. Biol.* **26**, 421–30 (1999).
95. N. Li, M. Struttman, C. Higginbotham *et al.*, *Nucl. Med. Biol.* **24**, 85–92 (1997).
96. P. Bernhardt, E. Forsell-Aronsson, L. Jacobsson, G. Skarnemark, *Acta Oncol.* **40**, 602–8 (2001).
97. D. Hesk, J.R. Jones, W.J. Lockley, *J. Pharm. Sci.* **80**, 887–90 (1991).
98. P.K. Lommatzsch, C. Werschnik, E. Schuster, *Graefes Arch. Clin. Exp. Ophthalmol.* **238**, 129–37 (2000).

99. L.W. Brady, J.C. Hernandez, *Strahlenter Oncol.* **168**, 61–5 (1992).
100. A.C. Lim, J.K. Barton, *Bioorg. Med. Chem.* **5**, 1131–6 (1997).
101. H. Junicke, J.R. Hart, J. Kisko *et al.*, *Proc. Natl. Acad. Sci. USA* **100**, 3737–42 (2003).
102. H.L. Harmon, H. Morrison, *Inorg. Chem.* **34**, 4937–8 (1995).
103. T. Mohammad, C. Chen, P. Guo, H. Morrison, *Bioorg. Med. Chem. Lett.* **9**, 1703–8 (1999).
104. T. Mohammad, *Toxicol. In Vitro* **18**, 45–53 (2004).
105. R.C. Richmond, N.P. Farrell, H. K. Mahtani, *Radiat. Res.* **120**, 403–15 (1989).
106. R.C. Richmond, H.K. Mahtani, *Radiat. Res.* **127**, 36–44 (1991).
107. R. Chibber, I.J. Stratford, P. O'Neill *et al.*, *Int. J. Radiat. Biol. Relat. Stud. Phys. Chem. Med.* **48**, 513–24 (1985).
108. C.W. Bradford, B.J. Chase, in *Handbook on Toxicity of Inorganic Compounds*, eds H.G. Seiler, H. Sigel, A. Sigel; Marcel Dekker, New York, 1988, pp. 561–5.
109. K.-H. König, M. Schuster, in *Handbook on Metals in Clinical and Analytical Chemistry*, eds H.G. Seiler, A. Sigel, H. Sigel; Marcel Dekker, New York, 1994, pp. 521–30.
110. B. Venugopal, T.D. Luckey, *Metal Toxicity in Mammals. 2. Chemical Toxicity of Metals and Metalloids*, Plenum Press, New York, 1978, pp. 298–300.
111. H. Renner, G. Schmuckler, in *Metals and Their Compounds in the Environment*, ed. E. Merian, VCH, Weinheim, 1991, pp. 1135–51.
112. B. Santucci, C. Valenzano, M. de Rocco, A. Cristaudo, *Contact Dermatitis* **43**, 333–8 (2000).
113. N. Kanematsu, H. Nakamine, Y. Fukuta *et al.*, *Gifu Shika Gakkai Zasshi* **17**, 575–81 (1990).
114. H. Lantzsch, T. Gebel, *Mutat. Res.* **389**, 191–7 (1997).
115. T. Gebel, A. Koenig, *Mutat. Res.* **444**, 405–11 (1999).
116. M.F. Sadiq, M.H. Zaghal, *Polyhedron* **16**, 1483–6 (1997).
117. K. Ravindra, L. Bencs, R. Van Grieken, *Sci. Total Environ.* **318**, 1–43 (2004).

21 $_{46}$Pd The Use of Palladium Complexes in Medicine

Achilleas Garoufis, Sotiris K. Hadjikakou and Nick Hadjiliadis

21.1 Introduction

Palladium with atomic number 46 is a member of the Platinum-group metals lying above platinum and below nickel and belongs to the 2nd transition metal series. It has a very similar chemistry to that of platinum. The most stable oxidation states for both metals are the +2 and +4 states. In their most common oxidation state (+2), both metals have a d^8 configuration and constantly form square planar diamagnetic complexes with various ligands of dsp^2 hybridization. More rarely, five coordinate complexes of trigonal bipyramidal structures are also formed. Both metals are characterized as soft acids, forming stronger bonds with nitrogen or sulfur donors than oxygen donors. Pd(II) complexes are generally less stable than Pt(II) both thermodynamically and kinetically.

The use of palladium and its complexes in medicine is limited, compared to other metals. ^{103}Pd has been used as a radioactive isotope in the treatment of rapidly growing, high-grade prostate cancer.[1,2] The discovery of the antitumor properties of *cis*-platin (*cis*-$Pt(NH_3)_2Cl_2$) and related complexes raised

Metallotherapeutic Drugs and Metal-Based Diagnostic Agents: The Use of Metals in Medicine
Edited by Gielen and Tiekink

interest in the discovery of other efficient complexes with other metals and ligands. Palladium(II) was among the first to be used for clinical trials against tumors, due to the very similar aqueous chemistry of its complexes to those of Pt(II). Thus, the low anti-tumor activity of *cis*-$Pd(NH_3)_2Cl_2$ and *cis*-$Pd(DACH)_2Cl_2$ and their high toxicity were attributed to the faster aquation of the Pd(II) complexes as compared to the corresponding Pt(II) complexes.[3]

In more recent years, however, an attempt was made to prepare Pd(II) complexes with inert ligands (e.g. sulfur donors). It was suggested by Das and Livingstone[4] that Pd(II) chelates of sulfur-containing ligands may be more effective anti-tumor agents than those of other metals, for example Ni(II), Cu(II), Zn(II), etc. N, S chelates do not have sufficient thermodynamic stability, Pt(II) chelates are very inert kinetically, while Pd(II) chelates possess the proper lability to bring the metal to its target (DNA) and allow it to interact with it.

The N, S donors used to prepare anti-tumor Pd(II) complexes included mainly thiosemicarbazones, known to possess anti-viral, -malarial, -fungal, -microbial and -tumor activity.[5] Their mechanism of action is believed to be due to the inhibition of ribonucleotide reductase which converts ribonucleotides to deoxyribonucleotides.[6]

Thiosemicarbazones and dithiocarbazates were also used to prepare anti-microbial complexes, which is another possible use of Pd(II) complexes in medicine.

Ligands used as anti-microbial or anti-tumor drugs have also been used to prepare Pd(II) complexes with enhanced activities. The present chapter is a brief review of the use of Pd(II) complexes as anti-tumor and anti-microbial agents discovered in recent years, with these ligands and others. The results are promising and encouraging for their future use.

21.2 Anti-viral, Anti-fungal and Anti-microbial Activity of Pd(II) Complexes

21.2.1 Sulfur-donor ligands

Thiosemicarbazones

Thiosemicarbazones have anti-viral, -malarial, -fungal, -microbial and -tumor activity.[5] Their Pd(II) complexes are expected to enhance the properties of these ligands. The ligands used are included in Figure 21.1.

Thus, Pd(II) complexes of general formula $[PdLCl_2]$ (L is a thiosemicarbazone derivative) (Figure 21.1a) were prepared and screened *in vitro* against HK-9 strain of *Entamoeba histolytica* and found active. All complexes were more

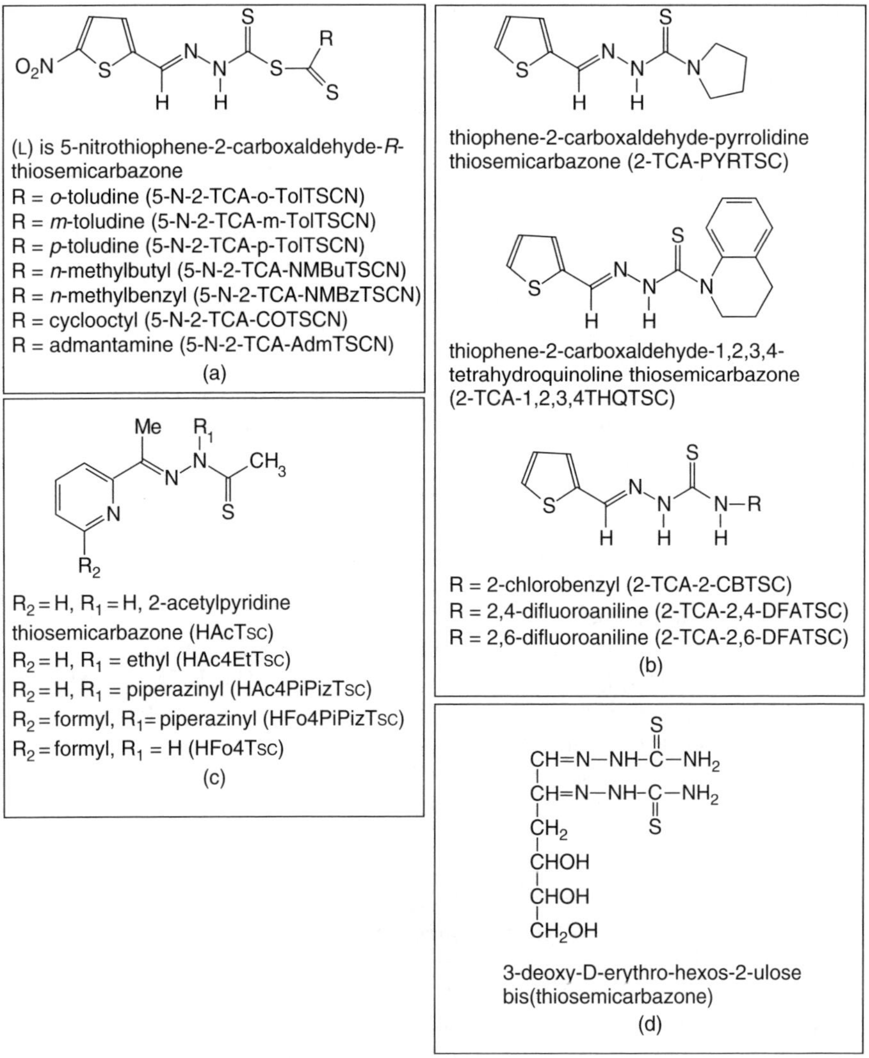

Figure 21.1

active than the free ligands themselves. The complexes [Pd(5-N-2-TCA-COTSCN)Cl_2] and [Pd(5-N-2-TCA-AdmTSCDN)Cl_2] displayed activities higher than that of the drug metronidazole (2-methyl-5-nitro-1*H*-imidazole-1-ethanol) with IC_{50} values of 0.81 ± 0.22 and $0.73 \pm 0.18\,\mu M$, $2.05 \pm 0.33\,\mu M$, respectively.[7]

More thiosemicarbazone derivative complexes of Pd(II) (Figure 21.1b) of related formulae were also screened for anti-amoebic activity against strain of

E. histolytica (HM-1:1 MSS).[8] The best results were found with [Pd(2-TCA-1,2,3,4-THQTSCD)Cl_2] ($IC_{50} = 1.15 \pm 0.34\,\mu M$).

Complexes of formulae [Pd(AcTSC)Cl], [Pd$(HAcTSCD)_2$]Cl_2 and [Pd$(AcTsc)_2$] (HAcTsc is 2-acetylpyridine thiosemicarbazone, Figure 21.1c) showed a completely lethal effect on Gram(+) bacteria (*Staphylococcus aureus*, *Bacillus cereus*) but no effect against Gram(−) bacteria (*E. coli*, *P. aeruginoza*).[9]

Another series of Pd(II) complexes with 2-acetylpyridine *N*(4)-ethyl-thiosemicarbazones (HAc4EtTSCD), 2-acetylpyridine *N*(4)-1-(2-pyridyl)-piperazinyl thiosemicarbazones (HAc4PiPizTSCD) and 2-formylpyridine *N*(4)-1-(2-pyridyl)-piperazinyl thiosemicarbazone (HFo4PiPizTSCD) of various formulae were tested for anti-bacterial activity against Gram(−) *E. coli* and *B. subtilis*, and Gram(+) *B. cereus* and *S. aureus*. Complexes Pd(Ac4EtTSCD), Pd$(HAc4PiPizTSCD)_2Cl_2$, Pd$(Ac4EtTSCD)_2$ and Pd$(Fo4PiPizTSCD)_2$ were more effective, with maximum inhibitory concentration (MIC) values of 6–12 μg/ml but not one affected *E. coli* growth, at a concentration of 100 μg/ml.[10]

3-Deoxy-D-erythro-hexos-2-ulose bis(thiosemicarbazone) (Figure 21.1d) acts as a tetradentate N_2S_2-donor ligand. Its complex of Pd(II) has shown anti-viral activity against poliovirus type I.[11]

The ligand pyridine-2-carbaldehyde thiosemicarbazone (HFo4TSCD) and its three complexes with Pd(II) were studied for their activity against herpes simplex virus 1 (HSV-1) infection in cultured cells.[12] [Pd$(HFo4TSCD)_2$]Cl_2, [Pd$(Fo4TSCD)_2$] and [Pd(Fo4TSCD)Cl] complexes were cytotoxic, with a maximum non-toxic concentration (MNC) of 1–10 μmol/l and a 50% cytotoxic concentration (CC_{50}) of 60–100 μmol/l after 72 h, for all.

Dithiocarbazates

The ligands *S*-acetyl N^{β}-acetyldithiocarbazate (aadt) (Figure 21.2a), *S*-methyl-N^{β}-[5-nitrothiophene-2-yl) methylene]dithiocarbazate (mndt) (Figure 21.2b) and *S*-benzyl N^{β}-[5-nitrothiophene-2-yl) methylene]dithiocarbazate (bntdt) (Figure 21.2c) afforded complexes of formulae [Pd(L)Cl_2][13] (L are ligands Figure 21.2a–c) with Pd(II). Ligand (Figure 21.2a) coordinates with the thioxo *S*-atom and the carbazate *N*(β)-atom while ligands (Figure 21.2b) and (Figure 21.2c) use the thioxo *S*-atom and the azomethine *N*-atom.[13]

Screening for anti-amebic activity of these compounds against the HK-9 strain showed all to be more active than their respective free ligands and comparable to the activity of metronidazole, a highly effective drug against the acute disease, with an IC_{50} value of 0.33 μg/ml, compared with those for the complexes [Pd(aadt)Cl_2], [Pd(mntdt)Cl_2] and [Pd(bntdt)Cl_2] of 0.53 ± 0.09, 0.32 ± 0.08 and 0.28 ± 0.07, respectively. These results indicate that complexation to Pd(II) increases the activities of the free ligands and modifies them from amebostatic to amebicidal.[13]

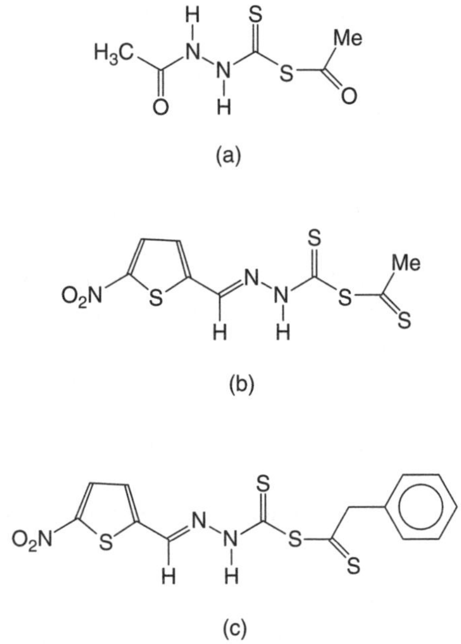

Figure 21.2

Complexes of formulae $[Pd(L)_2]Cl_2$ and $[Pd(L)_2]$[14] with L = *o*-FC_6H_4C-(H)NNC(SH)SCH_2Ph or *o*-FC_6H_4C(Me)NNC(SH)SCH_2Ph (Figure 21.3) were synthesized and tested against selected fungi and bacteria. The results have shown that the Pd(II) complexes exhibit higher inhibitory effects than the free ligands.

The synthesis, characterization and biochemical behavior of Pd(II) complexes of benzylidenehydrazinecarboxamide and -carbothiamide have been described.[15] Complexes correspond to the formulae $[Pd(RN{-}XH)_2]Cl_2$ and $[Pd(RN{-}X)_2]$ (X is O or S). Against selected pathogenic fungi and bacteria, the complexes showed higher inhibitory effects than the free ligands. This observation was attributed to the inherent properties of metal ions to denaturate proteins.[15]

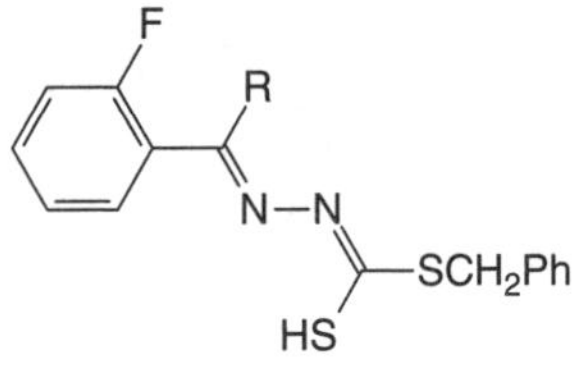

Figure 21.3

It is noteworthy that the complex [Pd(H_2Cu)Cl·Py] (H_2Cu is 5-carboxyuracil (isoorotic acid)), prepared amongst a series of Ag(I), Pt(II) and Pt(IV) complexes of isoorotic and thioisoorotic acid, was found active against *Pseudomonas* sp. and *Microcus* sp. at MIC values lower than 50 μg/cm^3.[16]

The mixed ligand complexes involving methionine (meth), 2-mercaptopyrimidine (2-Spym) and 2-aminopyrimidine (2-Spym) of formulae [Pd(meth)-Cl_2], [Pd(meth)(2-pymerpy)Cl]Cl and [Pd(meth)(2ampy)Cl]Cl, and those of ethionine (ethio), cytosine (cyt) and guanosine (guo) of formulae [Pd(ethio)Cl_2], [Pd(ethio)(cyt)Cl]Cl and [Pd(ethio)(guo)Cl]Cl were tested for anti-microbial activity against various bacteria (*Klebsiella pneumoniae*, *Vibrio cholerae*, etc.) and showed a broad spectrum of activity. It was shown that [Pd(meth)(2merpy)Cl]Cl eliminated plasmids from various human microbial pathogens with a 100% frequency.[17] The above complexes were also screened *in vitro* for anti-tumor activity against Hela (*Epidermoid Carcinoma Cervix*) and CHO (Chinese hamster ovary) cell lines. An excellent correlation between the anti-tumor activity of the complexes and their ability to cure plasmids was found.

21.2.2 Metal complexes of drugs used as ligands

Sadler *et al.*[18] used bis-tetraazamacrocyclo-xylyl-bicyclan, shown in Figure 21.4, a potent anti-HIV (human immunodeficiency virus) agent, to prepare a series of metal complexes with Zn(II) and Pd(II), and tried to achieve recognition by the co-receptor CXCR4, a G-protein coupled receptor used by the HIV for membrane fusion and cell entry. While Zn(II)$_2$-xylyl-bicyclam complexes give indications of binding to CXCR4, involving the carboxylate groups Asp262 (Zn(II) coordination), the corresponding square planar Pd(II) complexes interact only weakly with axial ligands and are thus inactive.

Complexes of Pd(II) and Pt(II) with metronidazole, a highly effective drug of formula [M(mnz)$_2$Cl$_2$] (against the acute disease caused by *E. histolytica*), were found more active than the free drug against the same disease (IC_{50} 0.103 μM for the Pd(II) analogue vs 1.50 μM).[19]

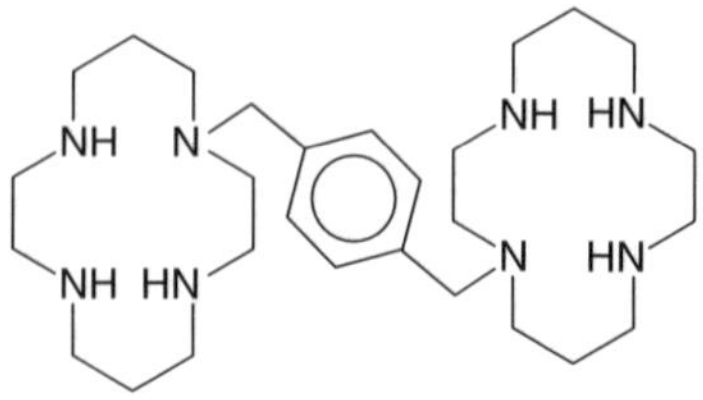

Figure 21.4

Two complexes of Pd(II) with diphenylpyraline (DPH) and isothipendyl (IPH), showing anti-serotonin, -histaminic, -convulsant and -fungal activity, of formulae $[Pd(DPH)_2Cl_2]$ and $[Pd(IPH)Cl_2]$, were synthesized and studied for their anti-bacterial and -fungal activities against *Alternaria alternata*, *A. tenius*, *Aspergillus flavus*, *A. niger*, *Staphylococcus aureus* and *E. coli*. Their anti-microbial activity was higher than that of the free ligands.[20]

The anti-viral activity of Pd(II) complexes with the anti-herpic drug penciclovir (pen) of formulae *cis*-$[(pen)_2PdCl_2]$ and *cis*-$[(nucl)_2Pd(pen)_2]Cl_2$ (nucl = guanosine, inosine, cytidine or penciclovir) were tested. All prepared complexes were markedly active against HIV-1 and HSV-2 (herpes simplex virus) strains but not against the thymidine kinase-deficient HSV-1 strain in E_6SM diploid fibroblastic cell cultures.[21]

21.2.3 Palladium(II) complexes with Schiff base ligands

Schiff base complexes of Pd(II), derived by the template condensation of 1,10-phenanthroline-2,9-dicarboxaldehyde, 2,3-diamino-1,4-naphthoquinone and 1,2-dibromoethane in ethanol (L) (Figure 21.5), of formula $[Pd(L)Br_2]\cdot 2H_2O$,[22] together with the corresponding Mn(II) complex were synthesized and tested for anti-bacterial activity, against *B. cereus* and other G(+) and G(−) strains.[22] The complexes were more active than sodium penicillin against *B. cereus*. The Pd(II) complex showed good activity against G(−) strains.

Azole (thiazolo and triazolo)-2,4-pentanedione (L_1 and L_2) were condensed with *o*-phenylenediamine in the presence of $NiCl_2$, $CuCl_2$ and $PdCl_2$, and the resulting trinuclear/polynuclear complexes were characterized and tested against the crop-damaging fungi, such as *Alternaria brasicae* and *Fusarium lycopersici*.[23] The fungi-toxicities of the compounds were evaluated and the evaluations indicated that all metal complexes had a better anti-fungal activity than their parent ligands, with the more active complex being the Pd(II) complex[23] (Figure 21.6).

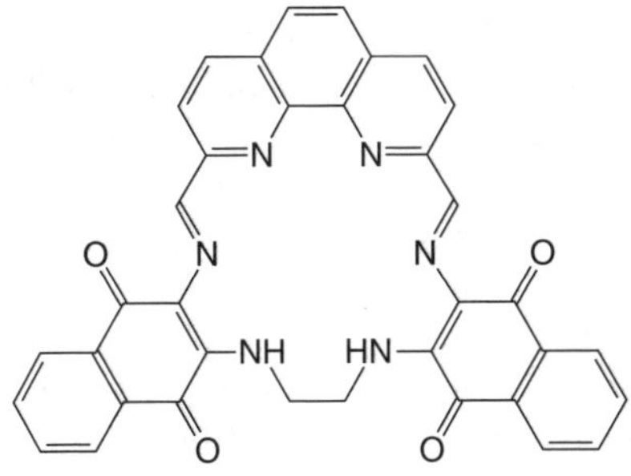

Figure 21.5

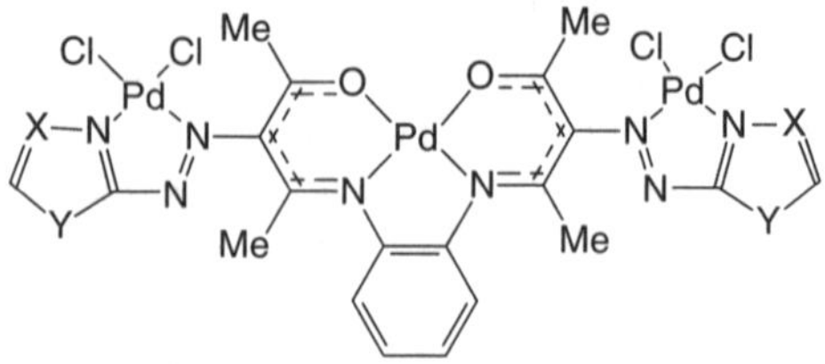

L_1: X = CH, Y = S; L_2: X = N, Y = NH

Figure 21.6

New Pd(II) complexes of biologically active ketimines containing NOH groups of formulae $Pd(L)_2$ and $Pd(L)_2Cl_2$ were synthesized, characterized and tested for anti-microbial activity[24] together with their ligands (L), i.e. 2-acetylfuran semicarbazone ($HSCZ_1$), 2-acetylthiophene semicarbazone ($HSCZ_2$), 2-acetylpyridine semicarbazone ($HSCZ_3$) and 2-acetylnaphthalene semicarbazone ($HSCZ_4$). The metal chelates showed greater inhibitory effects than the parent ligands.

Complexes of Pd(II) with Schiff bases derived from the condensation of 2-acetylpyridine and *S*-methyldithiocarbazate (SMDT), *S*-benzyldithiocarbazate (SBDT) and thiosemicarbazide (TSCD) of formula $[Pd(L)Cl_2]$ were synthesized and tested against *E. histolytica* (strain HK-9).[25] The first two showed greater activity and the third one, activity similar to that of metronidazole of IC_{50} 0.33 μg/mL.

21.2.4 Miscellaneous

Sharma *et al.*[26] reported the synthesis and characterization of heterobimetallic complexes of formula $[M(C_6H_6N_2)_2(M')_2(R)_4]Cl_2$ or $[M(C_6H_6N_2)_2 (M'')_2 (R)_4] Cl_2$ by the direct reaction of $[M(C_6H_8N_2)_2]Cl_2$ with the group 4 or 14 organometallic dichlorides $Ph_2M'Cl_2$, $Me_2M'Cl_2$ or $Cp_2M''Cl_2$. (M is Pd(II) or Pt(II), M′ is Si or Sn, M″ is Ti or Zr, $C_6H_8N_2$ is 1,2-phenylene diamine, R is Ph or Me.) All complexes were screened for anti-fungal and -bacterial activity against a number of bacteria and were found to be active. No postulation of structure–activity relationship was made. Thus, the results showed that the tin compounds were more active than the silicon analogues. Except in the case of Ph_2SiCl_2 and Me_2SnCl_2, silicon compounds were more active than tin, due to the presence of Ph groups in the former.

Finally, the new ligands *N*,*N*′-bis[2-hydroxy-3-methoxy-*N*-(pyridyl)-benzylamine]-2,6-diacetylidenepyridine and *N*,*N*′-bis[2-hydroxy-3-methoxy-*N*-(pyridyl)-benzylamine]-1,2-phthaldialdimine and their Pd(II) complexes were described and they showed anti-microbial activity against bacteria and yeasts.[27]

21.3 Anti-tumor Activity of Pd(II) Complexes

21.3.1 Sulfur-donor ligands

Thiosemicarbazones

Thiosemicarbazones and other sulfur-containing ligands have been used to prepare Pd(II) complexes with possible anti-tumor properties. Such ligands are included in Figure 21.7a–h.

Thiosemicarbazones have anti-tumor properties in their own right, and this activity was found to increase in their complexes with Pd(II).

Thus, the reaction of *p*-isopropylbenzaldehyde thiosemicarbazone (*p*-isoTSCN) with Pd(II) acetate yielded two tetrameric orthopalladate isomeric complexes of formula [Pd(*p*-isoTSCN)$_4$] with a CNS_2 donor environment around the metal.[6] The complexes were found active in several types of tumor cell lines (for the most active isomer values of $IC_{50} \pm SD$ in μM/vs *cis*-DDP: Jurkat (human T-cell leukemic cell line) $7 \pm 0.3/22 \pm 3$, Hela $4 \pm 0.1/7 \pm 0.5$, 3T3 $3 \pm 0.2/35 \pm 2$ and PAM (normal murine Keratinocytes cell line) $8 \pm 0.3/164 \pm 8$). With the Pam-Ras (normal murine Keratinocytes cell line transformed with the H-ras ongogen) tumor cells resistant to *cis*-DDP, this Pd(II) isomer was 33 times more active than *cis*-DDP (the $IC_{50} \pm SD$ in μM/vs *cis*-DDP: Pam-Ras $5 \pm 0.3/165 \pm 5$).[6] All three complexes exhibited higher cytotoxic activity than the drugs etoposide and adriamycin in the *cis*-DDP resistant cell lines.[6]

The binuclear complex [Pd(*p*-isoTSCN)(μ-Cl)]$_2$ shows IC_{50} values of the same range as *cis*-DDP, and it also displays cytotoxic activity against tumor cell lines resistant to this drug.[28]

Cyclometallated Pd(II) complexes of substituted thiosemicarbazones such as [Pd(*p*-isoTSCN-NHMe)] and [Pd(*p*-isoTSCN-NHex)]$_4$ show specific cytotoxic properties (apoptosis induction) in tumor cells transformed by *cis* oncogenes (Pam 212-ras) resistant to *cis*-DDP.[29] Similarly, the complexes Pd(phenylacetaldehyde thiosemicarbazone)Cl_2 and its Pt(II) analogue exhibit IC_{50} values higher than *cis*-DDP, etoposide and adriamycin against several tumor lines sensitive to *cis*-DDP [(HL-60 (human myeloid leukemia cell line), U-937 (human monoblastic cell line), Jurkat, Hela)] and resistant to *cis*-DDP (Pam–Ras).[30]

The polymeric complex [Pd(ESDT)Cl]$_n$, where ESDT is ethyl ethylaminoacetate dithiocarboxylate, with the ligand bound to Pd(II) using two sulfur atoms, one of which coordinates to Pd(II) of a neighboring complex, in a ClS_3 donor environment, demonstrated markedly greater cytotoxicity than *cis*-DDP in resistant tumor cell lines, such as human ovarian carcinoma cells and C13 adenocarcinoma cells. However, this cytotoxicity is not useful owing to the high nephrotoxicity induced by the complex.[31,32] In addition, the mixed ligand complex [Pd(ESDT)($PrNH_2$)Cl], with $ClNS_2$ donor atoms, showed a similar

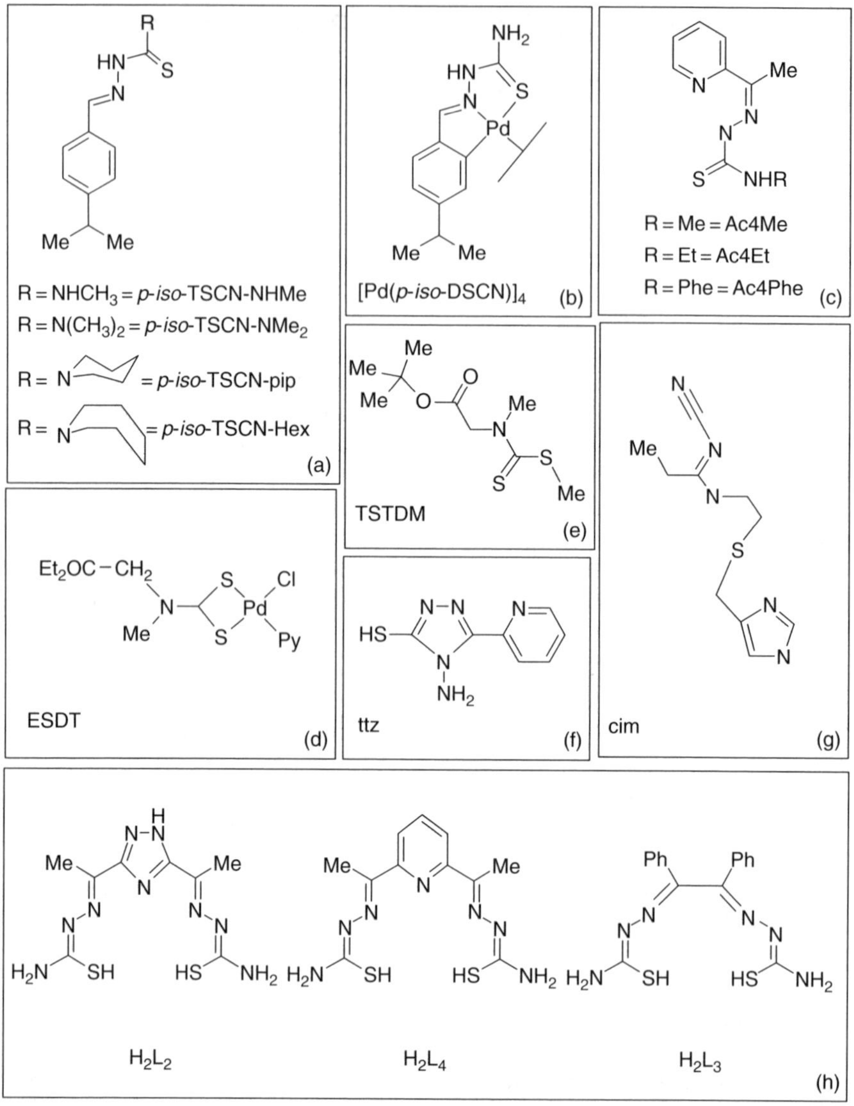

Figure 21.7

activity to that of $[Pd(ESDT)Cl]_n$,[32], while the complex [Pd(ESDT)(py)Cl] was characterized by IC_{50} values of 3.11 ± 0.13 for HL-60 and 5.61 ± 1.01 for Hela cell lines, lower than those of *cis*-DDP of 3.40 ± 0.11 and 6.33 ± 0.11, respectively.[32]

Of the two complexes of formulae $[Pd(TSDTM)X_2]$ (X is Cl or Br) and $[Pd(TSDT)X]_4$ (TSDTM is *tert*-butylsarcosine (*S*-methyl)dithiocarbamate and

TSDT is *tert*-butylsarcosinedithiocarbamate) tested against HL-60 and adenocarcinoma Hela cells, the former were completely inactive and only the complex $[Pd(TSDT)Br]_n$ showed an IC_{50} value comparable to that of *cis*-DDP (IC_{50} (μM) ± SD 5.20 ± 0.99 for HL-60 and 11.01 ± 0.43 for Hela cells).[33]

The Pd(II) complexes of formulae $[Pd(AcN(4)alkyl)X_2]$ and $[Pd(AcN(4)alkyl)_2]X_2$ (AcN(4)alkyl is 2-acetylpyridine N(4) alkyl thiosemicarbazone and X is Cl or Br) inhibited DNA synthesis in P388 lymphoblastic leukemia and L1210 lymphocytic leukemia cell cultures. More particularly (i) the two complexes differ in their activity from the free ligands, (ii) the chlorides were more cytotoxic than the bromides, and (iii) the less toxic and most effective against P388-bearing BDF_1 mice was the complex $[Pd(AcN(4)Et)_2]$.[34]

The two complexes [Pd(pyTSC)Cl] and $[Pd(pyTSC)_2]$ (pyTSC is pyridine-2-carbaldehyde thiosemicarbazone) with an N_2S_2 environment around Pd(II) were tested *in vivo* for anti-tumor activity. $[Pd(pyTSC)_2]$ showed eight times higher activity than the free ligand.

Thiosemicarbazone derivatives (Figure 21.7h) of H_2L_2, H_2L_3 and H_2L_4 of formulae $[PdCl_2(H_2L_2)]$, $[PdCl_2(H_2L_3)]$ and $[Pd(L_4)]$ were screened against several human, monkey and marine cell lines (Hela, Vero (cells derived from the kidney of *Cercopithicus aethiops*), and Pam 212) and resistant (Pam-ras) to *cis*-DDP. The $[Pd(L_4)]$ complex showed IC_{50} values similar to those of *cis*-DDP and exhibited notable cytotoxic activity in Pam-ras cells resistant to the drug.[35]

Other sulfur-containing ligands

Of the first Pd(II) complexes found to be active against 9KB, a human epidermoid carcinoma of the nasopharynx, was the one with the anionic form of the acetone Schiff base of *S*-methyldithiocarbazate (asme) of formula $[Pd(asme)_2]$.[4]

Both complexes $[Pd(asme)_2]$ and $[Pd(asbz)_2]$ (asbz is the anionic form of the acetone Schiff base of *S*-benzyldithiocarbazate) were active against T-lymphoplastic leukemia cells, with IC_{50} values of 2.5 and 2.9 $\mu g/cm^3$, respectively. It should be mentioned, however, that the corresponding $[Pt(asme)_2]$ and $[Pt(asbz)_2]$ complexes exhibited only a very weak activity against the same leukemic cells.[4]

The mixed ligand complex of Pd(II) with diethyldithiocarbamate (ddtc) and bidendate aromatic donors of formulae [Pd(ddtc)L] (L is 1,10-phenanthroline or 2,2′-bipyridine) were active against leukemia cells.[36] Similarly, Pd(II) complexes of dithiocarbamate, thiomorpholine, piperidine and morpholine, their methyl esters and the corresponding thiuramdisulfides showed noticeable cytostatic activity against KB, L1210 and P388 cells.[37,38]

Carrara *et al.*[39] tested the complex $[Pd(MPH)_3X]X$ (X is Cl or Br and MPH is 6-mercaptopyridine), isostructural to the corresponding Pt(II) complexes, against the tumor cell lines F10 (hybridoma), Fohn (renal cell carcinoma), LoVo (human colonic carcinoma), Hela and 3T3 (aneuploid murine cell line). The most active was $[Pd(MPH)_3Br]Br$ especially against LoVo cells.

The mixed ligand complexes of Pd(II) with *cis*-dichloromethionine, 2-mercaptopyrimidine and 2-aminopyrimidine were characterized and screened *in vitro* for anti-tumor activity against Hela and CHO cell lines, and were found active.[17] A correlation between the anti-tumor activity of Pd(II) complexes and their ability to cure plasmids exists.[17]

Finally, the complexes Pd(L-L)(SR)Cl: (L-L = $Ph_2PCH_2CH_2PPh_2$ (dppe) or $Ph_2AsCH_2CH_2PPh_2$ (dadpe); RSH glutathione, L-cysteine or *N*-acetyl-L-cysteine) have been prepared and the glutathione complexes were tested against the cell lines L1210, ADJ/PC6 (murine plasmocytoma) and CH1 (human ovarian carcinoma). The cytotoxicities toward L1210 cells were comparable to those of the parent dichlorocomplexes.[40] They could be useful in combination therapy with established drugs, such as *cis*-DDP.

21.3.2 Nitrogen and other donor atoms

Organopalladates

The cyclopalladate complexes of imidazoline (Imd) derivatives $[PdCl_2(C_6H_5CH_2C_3H_5N_2)_2]$ and $[PdCl(SEt_2)(C_6H_4\text{-}C_3H_5\text{-}N_2)_2]$ exhibited important cytotoxic activity against HL-60, with IC_{50} values, lower than 10 μg/ml, close to those of *cis*-DDP and considerably higher than the free ligand. Thus, the cytotoxic activity of free Imd increased upon coordination with Pd(II).[41]

Two diastereomeric cyclopalladated complexes arising from $[Pd_2(mpba)_2(\mu\text{-}OAc)_2]$ (mpba is 4-methoxy-benzoylbenzylideneimine, Figure 21.8b) were isolated, characterized, and tested against colon CX-1 and lung LX-1 human tumor cells, with compound II more active than I. The two complexes differ in the orientation of their benzoyl groups and this may also explain the higher activity of II over I.[42]

The orthopalladated complexes $[Pd(C^2\text{-dmba})(N_3)(dppe)]$ and $[Pd(C^2\text{-N-dmba})(cis\text{-dppet})](N_3)$ (dmba is *N*,*N*′-dimethylbenzylamine) were found to be the most active against a panel of three human cancer cell lines (Hela, Hep-2 and C_6) among a series of synthesized complexes containing the azido group (N_3) and diphosphines. The IC_{50} values were 1.0, 1.8 and <0.5 μM and 1.75, 2.20 and 0.85 μM, respectively.[43] Also, the complexes $[Pd(S(-)(C^2,N\text{-dmba})(dppe)]Cl_2$, $[Pd(S(-)(C^2,N\text{-dmba})_2(\mu\text{-dppe})]Cl_2$ and $[Pd_2(R(-)(C^2,N\text{-dmba})_2(\mu\text{-dppe})]Cl_2$ showed activity against B16-Nex2 cells in a dose, for 100% cell death *in vitro* with IC_{50} <1.25 μM[44] (Figure 21.8d).

The dimers $[Pd(cpba)X]_2$[45] and $[Pd(bpba)X]_2$[46]) (cpba is *N*-(4-chlorophenyl)-α-benzoylbenzylideneamine, bpba is *N*-(4-methoxyphenyl)-α-benzoylbenzylideneamine and X is Cl or OAc) were tested against the MDA-MB 468 (breast carcinoma) and HL-60 cell lines. The μ-Cl bridged complex with cpba was active with IC_{50} values close to that of *cis*-DDP.

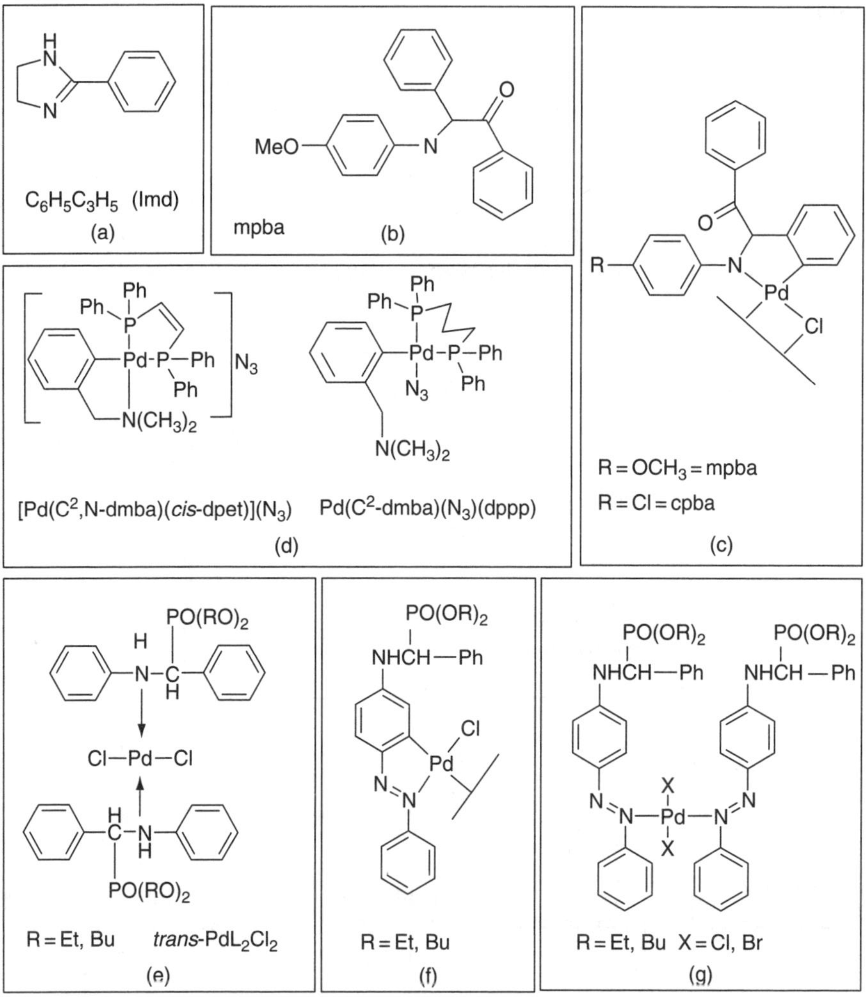

Figure 21.8

Palladium(II) halide complexes containing the diethyl and di-*n*-butyl esters of (α-anilino-*N*-benzyl)phosphonic acid and [α-(4-benzeneazoanilino)-*N*-benzyl] phosphonic acid of formula *trans*-PdL_2X_2 (L is these ligands and X is Cl, Br) are mononuclear (Figure 21.8g). By contrast, ligands containing azobenzene form binuclear organopalladium complexes with chloro-bridges, $[Pd(L\text{-}H^+)Cl]_2$. All complexes were found to be active against KB and L1210 cell lines. The most active was Pd(diethyl-α-benzeneazoaniline-*N*-benzyl-phosphonate)$_2$Cl$_2$ with IC_{50} values near 0.70 μM, comparable to those of *cis*-DDP.[47]

Amine complexes

The dimeric complexes of Pd(II) with putrescine (put), $[Pd_2(put)_2Cl_4]$ and $[putH_2][PdCl_4]$, and spermine (sperm), $[Pd_2(sperm)Cl_4]$ and $[PdCl_2(spermH)_2][PdCl_4]$ (Figure 21.9e) were assayed for *in vitro* anti-proliferative activity against MDA-MB468 and HL-60 human cancer cells. The putrescine complexes had IC_{50} values twofold lower than that of *cis*-DDP, but not those containing spermine.[48] However, the spermidine compounds of formulae $[Pd(sperH)_2][PdCl_4]$ and $[(PdCl_2)_3(sper)_2]$ were very active against MDA-

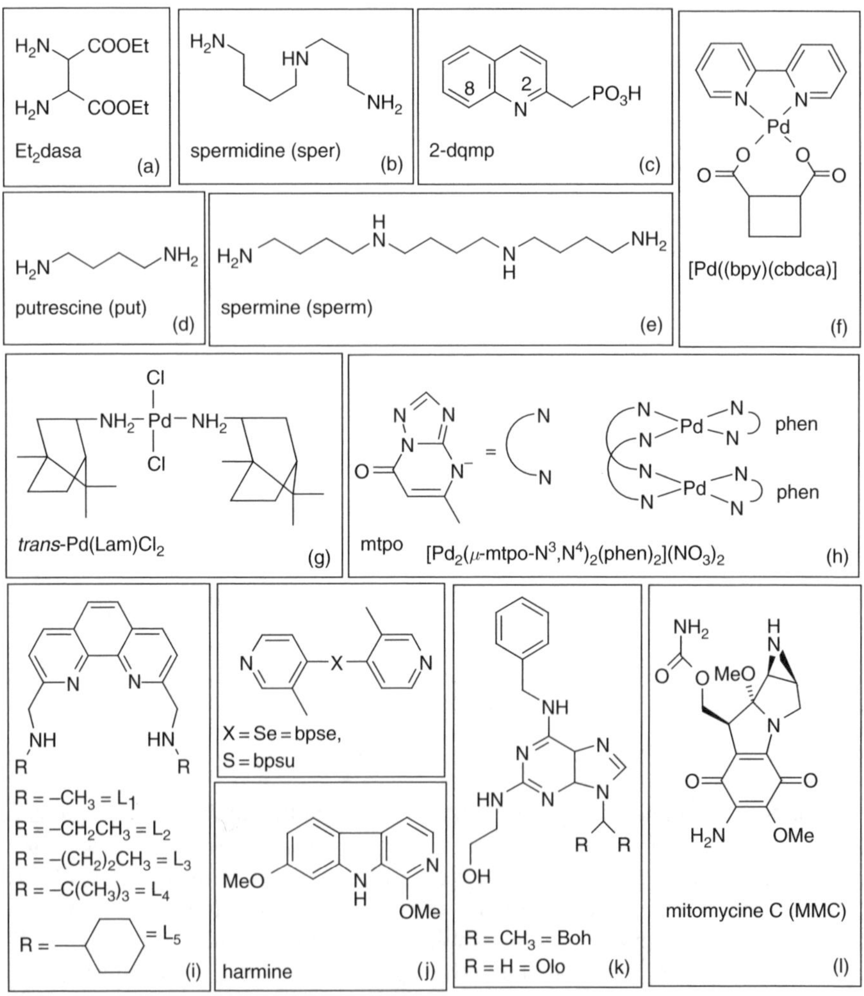

Figure 21.9

MB468 (human breast) cancer cell line with ID_{50} values, similar or even lower to those of *cis*-DDP (0.76 μg/ml, 0.56 μg/ml, and 0.80 μg/ml respectively[49]).

The anti-tumor properties against L929 (murine fibroblastoma), K562 (human myelogenous leukemia) and Hela cell lines of the enantiomerically pure *trans*-$PdCl_2$[(R)(−)bornylamino]$_2$ complex showed that its activity against Hela was comparable to those of cisplatin, carboplatin and oxalilplatin.[50]

Complexes of general formula [Pd(N∩N)(XO_3)] (N∩N is 1,10-phenanthroline, 2,2′-dipyridylamine, ethylenediamine or (+/−)-*trans*-1,2-diaminocyclohexane and XO_3 is $[SeO_3]^{2-}$ or $[TeO_3]^{2-}$) were tested against P388 cells. While the selenides showed ID_{50} values lower than cisplatin, the tellurides usually had higher values.[51]

The mixed ligand complexes of formulae [Pd(dach)(3-methylorot)] and [Pd(dach)(5-fluoro-orot)] (dach is 1,2-diaminocyclohexane and orot is orotic acid) were found to be active against sarcoma 180 cell lines with $T/C = 270\%$ and 267%, respectively, compared with 277% for cisplatin (T is the median survival time of treated mice, C is the median survival time of the control). It is worthwhile mentioning that the corresponding Pt(II) complexes were inactive and this result was attributed to the slow hydrolysis of the Pd(II) and the non-hydrolysis of the Pt(II) complexes.[3]

The 2-quinolylmethylphosphonate (2-dqmp) complexes of Pd(II) of formulae [(2-Hdqmp)][$PdCl_4$] and [Pd(2-dqmp)X_2] (X is Cl, Br) were found to be active against KB and L1210 cell lines, with the dimeric [Pd(2-dqmp)Br_2]$_2$ complex being the most active.[52]

The dipalladium complex [Pd_2(μ-meso-N^3, N^4)$_2$(phen)$_2$](NO_3)$_2$ (phen is 1,10-phenanthroline, Hmptpo is 5,7-dihydro-7-oxo-5-methyl[1,2,4]triazolopyrimidine) showed high activity against TG ovarian carcinoma and notable activity against BT-20 breast carcinoma.[53] It was more active than cisplatin and carboplatin.

Complexes of formula [Pd(L)]Cl_2 [L is *N*,*N*′-dimethyl-1,10-phenanthroline-2,9-dimethanamine (L_1) , *N*,*N*′-diethyl-1,10-phenanthroline-2,9-dimethanamine (L_2), *N*,*N*′-dipropyl-1,10-phenanthroline-2,9-dimethanamine (L_3), *N*,*N*′-di-*tert*-butyl-1,10-phenanthroline-2,9-dimethanamine (L_4), or *N*,*N*′-dicyclohexyl-1, 10-phenanthroline-2,9-dimethanamine (L_5)] were tested against mouse L1210 and Bel-7402 (liver carcinoma) tumor cells and were found active, with their activity dependent on the nature of L. The ID_{50} values of the complexes with ligands L_3, L_4 and L_5 against L1210 were 11.59, 10.26 and 8.20 μM, respectively (cisplatin, 15.33 μM) and against Bel-7402 were 10.62 μM, 9.53 μM and 8.20 μM, respectively (cisplatin, 23 μM).[54]

Only one of the four complexes of formulae [Pd(ethylenediamine)LCl](NO_3) (L is pyridine (py), 4-methylpyridine, 4-hydroxypyridine or 4-aminopyridine), namely [Pd(en)(py)Cl](NO_3), was found active against HL-60 cell lines with an ID_{50} value of 1.17 μM comparable to that exhibited by cisplatin (1.07 μM).[55]

The amino acid derivatives of formula [Pd(L)(amac)]Cl (L is 2,2-dipyridylamine or 1,10-phenanthroline, amac is the anion of the amino acids Gly,

L-Ala, L-Leu, L-Phenala, L-Tyr, L-Trp, L-Val, L-Pro, and L-Ser) were screened for cytotoxicity against P388 cells. The complex [Pd(dipy)(L-Ala)]Cl was found more active than *cis*-DDP, while [Pd(phen)(Gly)]Cl and [Pd(phen)(L-Val)]Cl showed comparable activity.[56,57] However, the complexes [Pd(phen)(L-Trp)]Cl·5H_2O and [Pd(5-NO_2-phen)(L-Trp)]Cl·5H_2O showed anti-cancer activity against the lung cancer cell AZGY-83a with IC_{50} values of 0.25 and 0.44 mM, respectively.[58]

The synthesis, characterization and cytotoxic properties of complex *trans*-[Pd(harmine)(DMSO)Cl_2] (harmine is an alkaloid) were tested against the P388, L1210 and K562 cell lines. The IC_{50} values for the complex, *cis*-DDP, carboplatin and 5-FU in μM were (a) against P388: 0.385, 0.500, >27, 1.15; (b) against L1210: 0.385, 0.833, >27, 1.15; and (c) against K562: 0.364, 20.0, 27.0, not tested.[59]

The cytokinin-derived compounds bohemine (Boh) (6-(benzylamino)-2-[(3-(hydroxypropyl)amino]-9-isopropylpurine) and olomoucine (Olo) (6-(benzylamino)-2-[(2-(hydroxyethyl)amino]-9-methylpurine) were used as ligands in Pd(II) complexes to screen against G3G1 (human malignant melanoma), HOS (human astrogenic sarcoma), K562 and MCF-7 (human breast adenocarcinoma). (Figure 21.9k). The complexes [Pd(BohH(+))Cl_3]·H_2O, [Pd(Boh)Cl_2(H_2O)], [Pd(Boh-H^+)Cl(H_2O)$_2$]·EtOH were found active.[60] For the G3G1 cell line, the most active complex was [Pd(Boh-H^+)Cl(H_2O)$_2$]·EtOH but the IC_{50} value was high at 26 μM compared with 3 μM for *cis*-DDP. For MCF-7, the IC_{50} value of the same complex was close to that exhibited by *cis*-DDP.

The complexes of Pd(II) with mitomycine (MMC, an anti-tumor agent) with formula [Pd(MMC)Cl_2] maintained an anti-tumor activity against K562 leukemia cells as compared to the free ligand ($IC_{50} = 15.8 \pm 5.3$ μM for the complex and 1.0 ± 0.3 μM for the free ligand).[61] Similarly, the complex of Pd(II) with the anti-tumor antibiotic Altromycin B was the most active against K562 leukemia and GLC4 lung tumor cell lines (resistant to deoxorubicin-RRD).[62] The IC_{50} value for the Pd(II) complex was 6.9 ± 1.5 nm and of the free ligand 3.8 ± 0.3 nm in K562 leukemia-resistant cell lines. For GLC4 lung tumor resistance lines, the values were 1.4 ± 0.1 nm for the complex and 1.1 ± 0.2 nm for the free ligand.

The complexes [Pd_2(bipy)$_2$(dsa)]Cl_2 and [Pd_2(bipy)$_2$(daa)]Cl_2 (dsa is the dianion of meso-α-α'-diaminosuberic acid and daa is the dianion of meso-α-α'-diaminoadipic acid) were tested against P388 cells. They showed ID_{50} values comparable or lower than *cis*-DDP.[63]

21.4 Concluding Remarks

The square planar Pd(II) complexes undergoing aquation and ligand exchange reaction 10^5 times faster than their corresponding Pt(II) complexes has was explained by the high toxicity and low anti-cancer activity of the former complexes compared to the latter (*cis*-DDP). This property of Pd(II) complexes is preventing them from reaching their pharmacological targets (DNA), since they react priorly with other biological molecules of the body, which is the cause of their toxicity. However, the results are different when S- and N-donor ligands are used.

Thus the suggestion of Das and Livingstone[4] that Pd(II) complexes with S-, N-chelated ligands might exhibit anti-microbial and anti-tumor activities seems to be substantiated by the results summarized in this review.

In certain cases molecules possessing anti-microbial and/or anti-tumor properties seem to have a synergistic effect with the Pd(II) center, in that the complex had greater cytotoxicity/anti-tumor activity compared with the uncomplexed molecules.

The lipophilicity of the ligands that allows the permeability of the Pd(II) complexes through cell membranes has also been claimed to improve their anti-microbial activity. However, a structure–activity relationship cannot be established at present and more work is required towards this goal, in order to allow definitive conclusions to be drawn.

References

1. N.N. Stone, P.G. Stock, 'Complications following permanent prostate brachytherapy', *Eur. Urology*, **41**, 427–433 (2002).
2. L. Potters, Y. Cao, E. Calugaru *et al.*, 'A comprehensive review of CT-based dosimetry parameters and biochemical control in patients treated with permanent prostate brachytherapy', *Int. J. Radiat. Oncol. Biol. Phys.*, **50**, 605–614 (2001).
3. J.L. Butour, S. Wimmer, F. Wimmer, P. Castan, 'Pd(II) compounds with potential antitumor properties and their Pt(II) analogues: A comparative study of the reaction of some orotiuc acid derivatives with DNA in vitro', *Chem. Biol. Inter.*, **104**, 165–178 (1997).
4. M. Das, S.E. Livingstone, 'Cytotoxic action of some transition-metal chelate of schiff-bases derived from S-methyldithiocarbazate', *Br. J. Cancer*, **37**, 466–469 (1978).
5. E. Bermejo, R. Carballa, A. Castineiras *et al.*, 'Synthesis, structural characteristics and biological activities of complexes of Zn(II), Cd(II), Hg(II), Pd(II) and Pt(II) with 2-acetylpyridine 4-methylthiosemicarbazone', *Eur. J. Inorg. Chem.*, 965–973 (1999).
6. A.G. Quiroga, J.M. Perez, I. Lopez-Solera *et al.*, 'Novel tetranuclear orthometalated complexes of Pd(II) and Pt(II) derived from *p*-isopropylbenzaldehyde thiosemicarbazone with cytotoxic activity in *cis*-DDP resistant tumor cell lines: Interactions of these compounds with DNA', *J. Med. Chem.*, **41**, 1399–1408 (1998).
7. N. Bharti, Shailendra, S. Sharma *et al.*, 'New palladium(II) complexes of 5-nitrothiophene-2-carboxaldehyde thiosemicarbazones: Synthesis, spectral studies and in vitro anti-amoebic activity', *Bioorg. Med. Chem.*, **11**(13), 2923–2929 (2003).
8. N. Bharti, Shailendra, F. Naqvi, A. Azam, 'Synthesis, spectral studies and screening for amoebicidal activity of new palladium(II) complexes derived from thiophene-2-carboxaldehyde thiosemicarbazones', *Bioorg. Med. Chem. Lett.*, **13**, 689–692 (2003).

9. D. Kovala-Demertzi, M.A. Demertzis, J.R. Miller *et al.*, 'Platinum(II) complexes with 2-acetyl pyridine thiosemicarbazone: Synthesis, crystal structure, spectral properties, antimicrobial and antitumour activity', *J. Inorg. Biochem.*, **86**, 555–563 (2001).
10. D. Kovala-Demertzi, M.A. Demertzis, E. Filiou *et al.*, 'Platinum(II) and Palladium(II) complexes with 2-acetyl pyridine 4N-ethyl thiosemicarbazone, able to overcome the cisplatin resistance: Structure, antibacterial activity and DNA strand breakage', *Biometals*, **16**, 411–418 (2003).
11. D. Harton, O. Varela, 'Cu, Pt, and Pd complexes of 3-deoxy-1,2-bis(thiosemicarbazone) derived from D-glucose', *Carbohydr. Res.*, **328**, 425–429 (2000).
12. T. Varadinova, D. Kovala-Demertzi, M. Rupelieva *et al.*, 'Antiviral activity of platinum(II) and palladium(II) complexes of pyridine-2-carbaldehyde thiosemicarbazone', *Acta Virol.*, **45**, 87–94 (2001).
13. N. Bharti, Shailendra, F. Naqvi, A. Azam, 'Synthesis, characterization and screening for amoebicidal activity of new palladium(II), platinum(II) and rutherenium(II) complexes with NS-donor ligands', *Helvetica Chim. Acta*, **85**, 2713–2720 (2002).
14. R. Singh, K. Sharma, N. Fahmi, 'Divalent manganese and palladium complexes of the phenylmethyl ester of hydrazinecarbodithioic acid', *Trans. Met. Chem.*, **24**, 562–565 (1999).
15. N. Fahmi, C. Saxena, R.V. Singh, 'Spectroscopic characterization and biological potential of palladium(II) complexes of benzylidenehydrazinecarboxamide or -carbothioamide', *Bull. Chem. Soc. Japan*, **69**, 963–969 (1996).
16. F. Hueso-Urena, M. Moreno-Carretero, J.M. Salas-Peregrin, G.A. de Cienfuegos-Lopez 'Silver(I), palladium(II), platinum(II) and platinum(IV) complexes with isoorotate and 2-thioisoorotate ligands: Synthesis, IR and NMR-spectra, thermal-behavior and antimicrobial activity', *Trans. Met. Chem.*, **20**, 262–269 (1995).
17. B. Khan, J. Bhatt, K. Najmuddin, 'Synthesis, antimicrobial and antitumor-activity of a series of palladium(II) mixed-ligand complexes', *J. Inorg. Biochem.*, **44**, 55–63 (1991).
18. Xi. Liang, J.A. Parkinson, M. Weishaupl *et al.*, 'Structure and dynamic of metallomacrocycles: Recognition of zinc Xylyl-Bicyclam by a HIV coreceptor', *J. Am. Chem. Soc.*, **124**, 9105–9112 (2002).
19. N. Bharti, Shailendra, S.J. Coles *et al.*, 'Synthesis, crystal structure and enhancement of the efficiency of metromidazole against entamoeba histolytica by complexation with palladium(II), platinum(II) or copper(II)', *Helvetica Chim. Acta*, **85**, 2704–2712 (2002).
20. P.G. Ramappa, K.G. Somasekharappa, 'Synthesis, characterization and antimicrobial activity of palladium(II) complexes of diphenylpyraline and isothipendyl', *Indian J. Chem.*, **33A**, 66–68 (1994).
21. A. Garoufis, K. Karidi, N. Hadjiliadis *et al.*, 'Synthesis, characterization and antiviral properties of Pd(II) complexes with penciclovir', *Metal-Based Drugs*, **8**, 57–63 (2001).
22. Juan R. Anacona, Erasto Bastardo, 'Manganese(II) and palladium(II) complexes containing a new macrocyclic schiff base ligand: Antibacterial properties', *Trans. Met. Chem.*, **24**, 478–480 (1999).
23. Lallan Mishra, Anjali Jha, Ajay K. Yadaw, 'Synthesis, spectroscopic and antifungal studies of transition metal trinuclear/polynuclear complexes with azolo-2,4-pentanedione part III (3)', *Trans. Met. Chem.*, **22**, 406–410 (1997).

24. Nighat Fahmi, R.V. Singh, 'Palladium(II) complexes of ketimines', *Trans. Met. Chem.*, **20**, 185–187 (1995).
25. Neelan Bharti, Mannar R. Manrya, Fehmida Naqvi *et al.*, 'Palladium(II) complexes of NS donor ligands derived from S-methyldithiocarbazate, S-benzyldithiocarbazate and thiosemicarbazides as antiamoebic agents', *Eur. J. Med. Chem.*, **35**, 481–486 (2000).
26. Kripa Sharma, Nighat Fahmi, Ranvir Singh, 'Synthesis, characterization and toxicity of new heterobimetallic complexes of platinum(II) and palladium(II)', *App. Organomet. Chem.*, **15**, 221–228 (2001).
27. H. Koksal, M. Dolaz, M. Tumer, S. Serin, 'Copper(II), cobalt(III), nickel(II), palladium(II), and zinc(II) complexes of the Schiff base ligands derived from 2,6-diacetylpyridine and phthaldialdehyde', *Synth. React. Inorg. Metal-Organ. Chem.*, **31**, 1141–1162 (2001).
28. A.G. Quiroga, J.M. Perez, I. Lopez-Solera *et al.*, 'Binuclear chloro-bridged palladated and platinated complexes derived from *p*-isopropylbenzaldehyde thiosemicarbazone with cytotoxicity against cisplatin resistant tumor cell lines', *J. Inorg. Biochem.*, **69**, 275–281 (1998).
29. A.G. Quiroga, J.M. Perez, E.I. Montero *et al.*, 'Synthesis and characterization of Pd(II) and Pt(II) complexes of *p*-isopropylbenzaldehyde N-protected thiosemicarbazones: Cytotoxic activity against *ras*-transformed cells', *J. Inorg. Biochem.*, **75**, 293–301 (1999).
30. A.G. Quiroga, J.M. Perez, I. Lopez-Solera *et al.*, 'Palladated and platinated complexes derived from phenylacetaldehyde thiosemicarbazone with cytotoxic activity in *cis*-DDP resistant tumor cells: Formation of DNA interstrand cross-links by these complexes', *J. Inorg. Biochem.*, **70**, 117–123 (1998).
31. A. Trevisan, C. Marzano, P. Cristofori *et al.*, 'Synthesis of a palladium-dithiocarbamate complex: Biological assay and nephrotoxicity in rats', *Arch. Toxicol.*, **76**, 262–268 (2002).
32. G. Faraglia, D. Fregona, S. Sitran *et al.*, 'Platinum(II) and palladium(II) complexes with dithiocarbamates and amines: Synthesis, characterization and cell assay', *J. Inorg. Biochem.*, **83**, 31–41 (2001).
33. D. Fregona, L. Givagnini, L. Ronconi *et al.*, 'Pt(II) and Pd(II) derivatives of terbutylsarcosinedithiocarbamate: Synthesis, chemical and biological characterization and in vitro nephrotoxicity', *J. Inorg. Biochem.*, **93**, 181–189 (2003).
34. D. Kovala-Demertzi, A. Domopoulou, M.A. Demertzis *et al.*, 'Palladium(II) complexes of 2-acetylpyridine N(4)-methyl, N(4)-ethyl and N(4)-phenyl-thiosemicarbazones: Crystal structure of chloro(2-acetylpyridine N(4)-methylthiosemicarbazonato) palladium(II); Synthesis, spectral studies, in vitro and in vivo antitumor activity', *J. Inorg. Biochem.*, **68**, 147–155 (1997).
35. A.I. Matesanz, J.M. Perez, P. Navarro *et al.*, 'Synthesis and characterization of novel palladium(II) complexes of bis(thiosemicarbazone): Structure, cytotoxic activity and DNA binding of Pd(II)benzyl bis(thiosemicarbazonate)', *J. Inorg. Biochem.*, **76**, 29–37 (1999).
36. R. Mital, N. Jain, T.S. Srivastava, 'Synthesis, characterization and cytotoxic studies of diamine and diamine palladium complexes of diethyldithiocarbamate and binding of these and analogous platinum(II) complexes with DNA', *Inorg. Chim. Acta*, **166**, 135–140 (1989).

37. L. Marcheselli, C. Preti, M. Tagliazucchi *et al.*, 'Synthesis characterization and evaluation of biological activity of palladium(II) and platinum(II) complexes with dithiocarbamic acids and their derivatives as ligands', *Eur. J. Med. Chem.*, **28**, 347–352 (1993).
38. V. Scarcia, G. Faraglia, L. Sindellari, L. Trincia, 'Synthesis and in vitro cytostatic effect of palladium(II) halide complexes with dithiocarbamic esters', *Eur. J. Med. Chem.*, **21**, 261–265 (1986).
39. M. Carrara, T. Berti, S. Dancona *et al.*, 'In vitro effect of Pt and Pd mercaptopyridine complexes', *Anticancer Res.*, **17**, 975–980 (1997).
40. O.A. Ni Dhubhghaill, P.J. Sadler, E.C. Fernandez, 'Cytotoxic phosphinnoarsino and diphosphino Pd(II) complexes of thiolate amino acids and glutathione', *Metal-Based Drugs*, **2**, 19–35 (1995).
41. F. Zamora, V.M. Gonzalez, J.M. Perez *et al.*, 'Pd(II) and Pt(II) complexes of 2-phenyl- and 2-benzyl-imidazoline: Synthesis, structural characterization, DNA modification and in vitro antileukaemic activity', *Appl. Organometal. Chem.*, **11**, 659–666 (1997).
42. C. Navarro-Ranninger, I. Lopez-Solera, J.M. Perez *et al.*, 'In vitro antitumor activity of 2 isomeric cyclopalladiated compounds derived from benzoylbenzylidenimines', *Appl. Organometal. Chem.*, **7**, 57–61 (1993).
43. A.C.F. Caires, E.T. Almeida, A.E. Mauro *et al.*, 'Synthesis and cytotoxicity of some cyclometallated palladium(II) complexes containing coordinated azide and diphosphines', *Quim. Nova*, **22**, 329–334 (1999).
44. E.G. Rodrigues, L.S. Silva, D.M. Fausto *et al.*, 'Cyclopalladated compounds a chemotherapeutic agents: Antitumor activity against a murine melanoma cell line', *Int. J. Cancer*, **107**, 498–504 (2003).
45. C. Navarro-Ranninger, I. Lopez-Solera, V.M. Gonzalez *et al.*, 'Cyclometalated complexes of platinum and palladium with N-(4-chlorophenyl)-alpha-benzoylbenzylideneamine: *In vitro* cytostatic activity, DNA modification, and interstrand cross-link studies', *Inorg. Chem.*, **35**, 5181–5187 (1996).
46. C. Navarro-Ranninger, I. Lopez-Solera, J. Rodriguez *et al.*, 'Analysis of 2 cycloplatinated compounds derived from N-(4-methoxy-phenyl)-α-benzoylbenylidenamine: Comparison of the activity of these compounds with other isostructural cyclopalladated compounds', *J. Med. Chem.*, **36**, 3795–3801 (1993).
47. M. Curic, L. Tusek-Bozic, D. Vikic-Topic *et al.*, 'Palladium(II) complexes of dialkyl α-anilinobenzylphosphonates: Synthesis, characterization and cytostatic activity', *J. Inorg. Biochem.*, **63**, 125–142 (1996).
48. C. Navarro-Ranninger, J.M. Pérez, F. Zamora *et al.*, 'Palladium(II) compounds of putrescine and spermine: Synthesis, characterization, and DNA-binding and antitumor properties', *J. Inorg. Biochem.*, **52**, 37–49 (1993).
49. C. Navarro-Ranninger, F. Zamora, I. López-Solera *et al.*, 'Palladium(II) salt and complexes of spermidine with a six-member chelate ring: Synthesis, characterization, and initial DNA-binding and antitumor studies', *J. Inorg. Biochem.*, **46**, 267–279 (1992).
50. A.S. Abu-Surrah, T.A.K. Al-Allaf, L.J. Rashan *et al.*, 'Synthesis, crystal structure and initial biological evaluation of the new enantiomerically pure chiral palladium(II) complex *trans*-bis {endo(1R)-1,7,7-trimethylbicyclo [2.2.1]-heptan-2-amino} palladium(II)dichloride', *Eur. J. Inorg. Chem.*, **37**, 919–922 (2002).

51. H. Mansuri-Torshizi, R. Mital, T.S. Srivastava *et al.*, 'Synthesis, characterization, and cytotoxic studies of -diimine/1,2-diamine platinum(II) and palladium(II) complexes of selenite and tellurite and binding of some of these complexes to DNA', *J. Inorg. Biochem.*, **44**, 239–247 (1991).
52. L. Tusek-Bozic, I. Matijasic, G. Bocelli *et al.*, 'Preparation, characterization and activity of Pd(II) halide-complexes with 2-quinolylmethylphosphonate (2-dqmp): X-ray crystal structures of *trans*-$Pd(2\text{-dqmp})_2X_2$ (X = Cl or Br)', *J. Chem. Soc., Dalton Trans.*, 195–201 (1991).
53. K. Akdi, R.A. Vilaplana, S. Kamah *et al.*, 'Study of the biological effects and DNA damage exerted by a new dipalladium-Hmtpo complex on human cancer cells', *J. Inorg. Biochem.*, **90**, 51–60 (2002).
54. G.H. Zhao, H.W. Sun, H.K. Lin *et al.*, 'Palladium(II) complexes with *N,N'*-Dialkyl-1,10-phenanthroline-2,9-dimathanamine: Synthesis, characterization and cytotoxic activity', *J. Inorg. Biochem.*, **72**, 173–177 (1998).
55. G.H. Zhao, H.K. Lin, P. Yu *et al.*, 'Ethylenediamine palladium(II) complexes with pyridine and its derivatives: Synthesis, molecular structure and initial antitumor studies', *J. Inorg. Biochem.*, **73**, 145–149 (1999).
56. A.K. Paul, H. Mansuri-Torshizi, T.S. Srivastava *et al.*, 'Some potential antitumor 2,2'-dipyridylamine Pt(II)/Pd(II) complexes with amino acids: Their synthesis, spectroscopy, DNA binding, and cytotoxic studies', *J. Inorg. Biochem.*, **50**, 9–20 (1993).
57. R. Mital, T.S. Srivastava, H.K. Parekh, M.P. Chitnis, 'Synthesis, characterization, DNA binding and cytotoxic studies of some mixed ligand Pd(II) and Pt(II) complexes of α-diimine and amino acids', *J. Inorg. Biochem.*, **41**, 93–103 (1991).
58. E.J. Gao, Q.T. Liu, 'Study on the synthesis, anticancer activity and interactions with DNA of compounds [Pd(L)(trp)]Cl $5H_2O$ (L = phen, 5-NO_2phen)', *Acta Chim. Sin.*, **60**, 674–680 (2002).
59. T.A.K. Al-Allaf, L.J. Rashan, 'Synthesis and cytotoxic evaluation of the first *trans*-palladium(II) complex with naturally occurring alkaloid harmine', *Eur. J. Inorg. Chem.*, **33**, 817–820 (1998).
60. Z. Travnicek, M. Malon, M. Zatloukal *et al.*, 'Mixed ligand complexes of platinum(II) and palladium(II) with cytokinin-derived compounds bohemine and olomoucine: X-ray structure of $[Pt(BohH^+\text{-N7})Cl_3]$ $9/5H_2O$ {Boh = 6-(benzylamino)-2-[(3-(hydroxypropyl)-amino]-9-isopropylpurine, Bohemine}', *J. Inorg. Biochem.*, **94**, 307–316 (2003).
61. M.M.L. Fiallo, E. Deydier, M. Bracci *et al.*, 'Mitomycin antitumor compounds: Interaction of transition metal ions with mitomycin C; Solution structure and biological activity of a Pd^{2+}-MMC complex', *J. Med. Chem.*, **46**, 1683–1689 (2003).
62. N. Nikolis, C. Methenitis, G. Pneumatikakis, M.M.L. Fiallo, 'Interactions of the anticancer antibiotic altromycine B with copper(II), palladium(II) and platinum(II) ions and invitro activity of the formed complexes', *J. Inorg. Biochem.*, **89**, 131–141 (2002).
63. H. Mansuritorshizi, T.S. Srivastava, H.K. Parekh, M.P. Chitnis, 'Synthesis, spectroscopic, cytotoxic and DNA-binding studies of binuclear 2,2'-bipyridine Pt(II) and 2,2'-bipyridine Pd(II) complexes of *meso*-α-α'-diaminoadipic acid and *meso*-α-α'-diaminosuberic acid', *J. Inorg. Biochem.*, **45**, 135–148 (1992).

22 $_{50}$Sn Tin Compounds and Their Therapeutic Potential

Marcel Gielen and Edward R.T. Tiekink

22.1 Introduction

Tin is an element that has a very long association with mankind with its first use probably being to strengthen copper by the formation of a copper–tin alloy, an event that marked the beginning of the Bronze Age, about 3500 BC. Since then, tin and its compounds have been exploited in a wide variety of applications such as agriculture, art, catalysis and medicine, to name but a few. The purpose of this chapter is to highlight current applications and potential of tin compounds in therapeutic medicine.

After the anti-tumour activity of cisplatin, *cis*-$Cl_2Pt(NH_3)_2$, was discovered, several research groups started to investigate the possible therapeutic applications of other metal-based, often organometallic, compounds, and at the forefront of these were organotin compounds.

The organotin compounds that were first tested were those that were available or easily synthesized, like tri- or di-organotin halides: these compounds

Metallotherapeutic Drugs and Metal-Based Diagnostic Agents: The Use of Metals in Medicine
Edited by Gielen and Tiekink

can be routinely prepared from tetraorganotin compounds, R_4Sn, dihalogens, X_2, or HX[1] in accord with:

$$R_4Sn + Y—X \rightarrow R_3SnX + R—Y \text{ and } R_3SnX + Y—X \rightarrow R_2SnX_2 + R—Y$$
$$Y = X \text{ or } H$$

Diorganotin halides can also be synthesized by direct synthesis in the presence of a suitable catalyst:[2]

$$2R—X + Sn \rightarrow R_2SnX_2$$

A number of early reviews recording advances in the screening for anti-tumour potential of organotins are available and the interested reader is referred to these.[3–8] The focus of the present review is upon results obtained in the past decade or so, as well as upon other therapeutic applications of tin compounds. Before summarizing the more recent developments in this field, it is appropriate to describe some aspects of the most useful spectroscopic method to determine the structural chemistry of organotin compounds in solution.

22.2 Structures of Organotin Compounds in Solution

The characterization of organotin compounds by multinuclear NMR provides important information about the structure of these compounds in solution. Tin has seven major isotopes of which two have spin 1/2, ^{117}Sn and ^{119}Sn; present in reasonable proportions to be usable for NMR experiments. Proton NMR spectra can reveal $^2J(^1H—C—^{117/119}Sn)$ coupling constants that provide information about the magnitude of the C—Sn—C angle in solution; $^3J(^1H—C—C—^{117/119}Sn)$ coupling constants can also be detected in favourable circumstances. In carbon-13 NMR spectra, the $^1J(^{13}C—^{117/119}Sn)$ coupling constants give similar angular information. $^2J(^{13}C—C—^{117/119}Sn)$ and $^3J(^{13}C—C—C—^{117/119}Sn)$ coupling constants are also easily measured providing equally useful structural information. The tin-119 (or tin-117) chemical shifts provide information about the coordination number of tin in solution.

22.3 Anti-tumour Pre-screening

In vivo testing of tetraorganotin compounds showed that they are inactive, whereas organotin halides and their complexes with amines and other ligands exhibit borderline activities against P388 and L1210 leukaemia.[9–14] The *in vivo* pre-screenings against these two cancers used initially by the National Cancer Institute (NCI), USA, were later replaced by *in vitro* pre-screenings against

Table 22.1 ID_{50} values (ng/ml) of oncology drugs: doxorubicin (DOX), cisplatin (CPT), 5-fluorouracil (5-FU), methotrexate (MTX) and etoposide (ETO) against seven cancer cell lines of human origin, MCF-7, EVSA-T (mammary cancers), WIDR (colon cancer), IGROV (ovarian cancer), M19 MEL (melanoma), A498 (renal cancer) and H226 (non-small-cell lung cancer)

	Drug				
Cell line	DOX	CPT	5-FU	MTX	ETO
MCF-7	10	699	750	18	2594
EVSA-T	8	422	475	5	317
WIDR	11	967	225	<3	150
IGROV	60	169	297	7	580
M19 MEL	16	558	442	23	505
A498	90	2253	143	37	1314
H226	199	3269	340	2287	3934

a panel of human tumour cell lines.[15–27] This was also the procedure used when organotin compounds were tested by the Rotterdam Cancer Institute. The seven human tumour cell lines comprising the panel were: MCF-7 and EVSA-T (two mammary cancers), WIDR (a colon cancer), IGROV (an ovarian cancer), M19 MEL (a melanoma), A498 (a renal cancer) and H226 (a non-small-cell lung cancer).

In Table 22.1, ID_{50} (the mass of material required to inhibit the growth of a particular cell line by 50%) values of some well-known oncology drugs are presented; ID_{50} values sometimes show some variation due to the circumstances of the test but these provide a reliable guide of cytotoxicity and a benchmark for comparison with test compounds. Normally, compounds were dissolved in ethanol solution containing 1 mg of compound in 200 μl and diluted to appropriate concentrations in the screen. Trial compounds that were found to be insoluble in ethanol were dissolved in dimethyl sulphoxide (DMSO).

22.4 Early *In Vitro* Cytotoxicity Testing of Diorganotin Carboxylates

When organotin halides are dissolved in water, the pH of the solution dramatically decreases because they are converted into organotin hydroxides that are afterwards possibly converted into insoluble bis(triorganotin) oxides or diorganotin oxides, thereby liberating protons. Di- or tri-organotin carboxylates do not suffer from this disadvantage as such derivatives, when dissolved in water, generally remain intact for long periods, i.e. days.

Reviews describing the anti-tumour properties of organotin compounds have been published recently[28–30] and, therefore, only salient aspects of their anti-tumour potential will be discussed here.

The influence of the R and R′ groups of $R_2Sn(OCOR')_2$ or $\{[R_2Sn(OCOR')]_2O\}_2$ on cytotoxicity has been determined, allowing for the development of a structure–activity relationship.[31–33] The di-*n*-butyltin compounds were among the more potent compounds examined. Di-*n*-butyltin derivatives have found several industrial applications[2] and, therefore, di-*n*-butyltin oxide, a useful starting material to synthesize di-*n*-butyltin derivatives, is commercially available and quite inexpensive.

The conversion of polymeric and insoluble $(R_2SnO)_n$ into $R_2Sn(OCOR')_2$ or $\{[R_2Sn(OCOR')]_2O\}_2$ is facile: $(R_2SnO)_n$ and R′COOH are mixed in a mixture of ethanol and toluene, and heated. Very quickly, the heterogeneous reaction mixture becomes a clear solution. The water/toluene/ethanol azeotrope is then distilled off and, when the concentrated remaining solution is cooled down, sometimes, the reaction product precipitates directly from solution.

If the $1/n(R_2SnO)/R'COOH$ molar ratio is 1:1, generally, dimeric distannoxanes $\{[R_2Sn(OCOR')]_2O\}_2$ of type **1** are obtained (Figure 22.1):[34–38]

$$(R_2SnO)_n + nR'COOH = n\{[R_2Sn(OCOR')]_2O\}_2 + H_2O$$

A 1:2 molar ratio generally yields diorganotin dihalides, $R_2Sn(OCOR')_2$, of type **2** (see Figure 22.1).[39–53]

$$(R_2SnO)_n + 2nR'COOH = nR_2Sn(OCOR')_2 + nH_2O$$

Tin NMR spectroscopy can be used to easily distinguish between diorganotin derivatives of carboxylic acids of types **1** and **2**: the tetraorganodicarboxylatodistannoxane dimers of type **1** are characterized by two tin resonances because they contain two heterotopic tin atoms, those forming the Sn_2O_2 four-membered

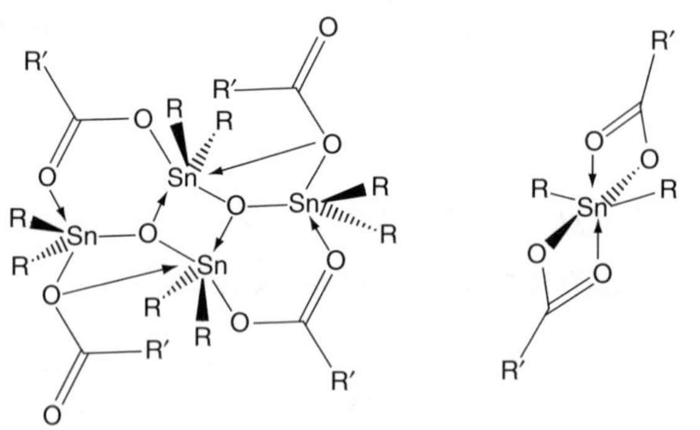

1 **2**

Figure 22.1 Chemical structures of tetraorganodicarboxylatodistannoxane dimers $\{[R_2Sn(OCOR')]_2O\}_2$ (type **1**) and diorganotin dicarboxylates $R_2Sn(OCOR')_2$ (type **2**)

Table 22.2 Typical ID_{50} values (ng/ml) of di-*n*-butyltin dicarboxylates: di-*n*-butyl digibberellate and di-*n*-butyltin diterebate. The ID_{50} values of cisplatin and doxorubicin are given for comparison

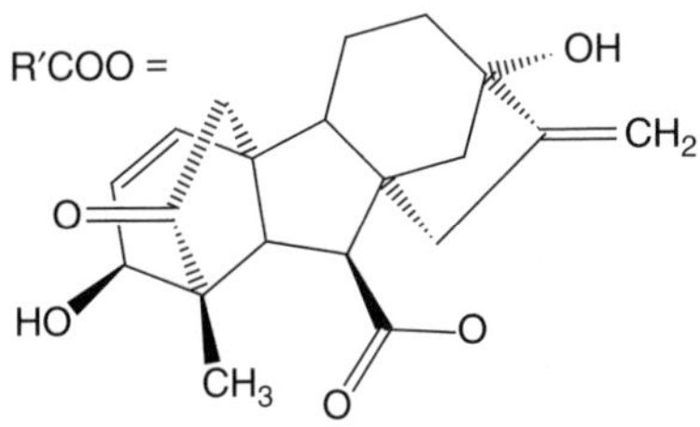

Gibberellate

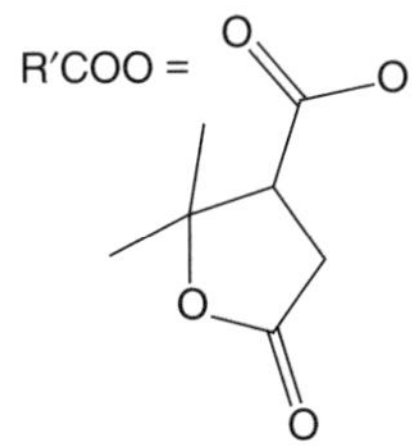

Terebate

Compound	MCF-7	EVSA-T	WIDR	IGROV	M19 MEL	A498	H226
2, R = Bu, R′COO = gibberellate	262	244	401	245	247	327	306
2, R = Bu, R′COO = terebate	27	25	134	18	61	61	104
CPT	699	422	967	169	558	2253	3269
DOX	10	8	11	60	16	90	199

ring and those that are substituents of that ring, whereas a single tin resonance characterizes the diorganotin dicarboxylates of type **2**.

The ID_{50} values observed for typical diorganotin derivatives of carboxylic acids are given in Table 22.2. They show that di-*n*-butyltin digibberellate is a little more potent *in vitro* than cisplatin, but less active than doxorubicin, whereas di-*n*-butyltin diterebate is much more potent than cisplatin *in vitro* against all cell lines. Its potency is similar to that of the best clinical drug doxorubicin.

Several other series of organotin derivatives of carboxylic and dicarboxylic acids were routinely synthesized and tested (see for instance Ref. 54–56).

22.5 The Di-*n*-Butyltin Analogue of Carboplatin

The di-*n*-butyltin analogue of carboplatin (Figure 22.2) was synthesized and screened against MCF-7 and WIDR because this was an early result that was obtained when the panel comprised only these two cell lines.[57,58]

The diorganotin compound was characterized by ID_{50} values of 63 and 121 ng/ml, respectively, whereas the ID_{50} values of cisplatin were 699 and 967 ng/ml.

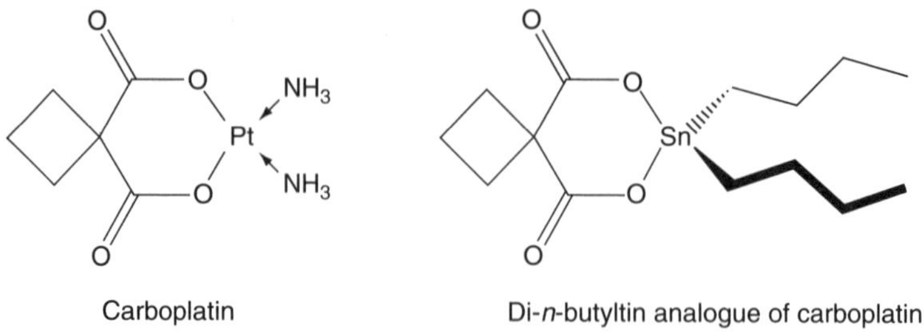

Figure 22.2 Carboplatin and the di-*n*-butyltin analogue of carboplatin

22.6 Anti-tumour Potential of Triorganotin Carboxylates

Triorganotin compounds, which probably adopt four-coordinate distorted tetrahedral structures in non-coordinating solvents as deduced from NMR studies, are quite well known as bactericides and fungicides[59,60] (Figure 22.3).

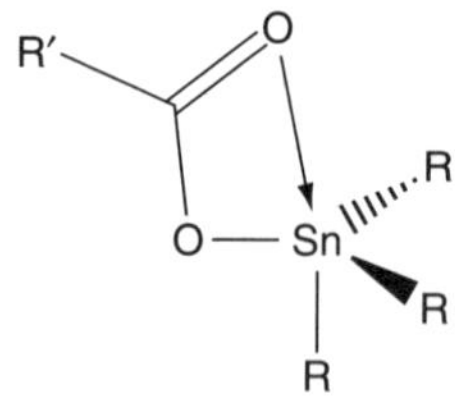

3

Figure 22.3 Likely chemical structure of a triorganotin carboxylate (type **3**) in solution

Table 22.3 Typical ID_{50} values (ng/ml) of triorganotin carboxylates: tri-*n*-butyl- and triphenyl-terebates (see Table 22.2 for the structure of terebate)

Compound	MCF-7	EVSA-T	WIDR	IGROV	M19 MEL	A498	H226
3, R = Bu, R′COO = terebate	3	<3	11	4	11	15	8
3, R = Ph, R′COO = terebate	17	<3	17	19	42	42	39
CPT	699	422	967	169	558	2253	3269
DOX	10	8	11	60	16	90	199

Such compounds were prepared for that purpose and consequently screened for their cytotoxicity and several of these were found to be quite potent.[61–63] Typical ID_{50} values of such compounds are given in Table 22.3.

The cytotoxicities of the tin compounds were, again, found to be comparable to those of doxorubicin and far superior to that of cisplatin.

22.7 Anti-tumour Screening of Organotin Derivatives of Biologically Relevant Substrates

A strategy for developing improved anti-cancer drugs is one where biologically active species are coordinated to the tin centre so that, for example, the metabolic pathway of the original drug is altered resulting in greater therapeutic benefit. In this context, diorganotin (of type **1**) and triorganotin steroidcarboxylates (of type **3**) were studied (Table 22.4).[64–66]

Triorganotin steroidcarboxylates are more potent than dimeric distannoxanes and possess high *in vitro* cytotoxicities, but their water solubility still remains a drawback,[67] which could possibly affect their *in vivo* properties owing to low bioavailability.

Table 22.4 Typical ID_{50} values (ng/ml) of three selected organotin steroidcarboxylates

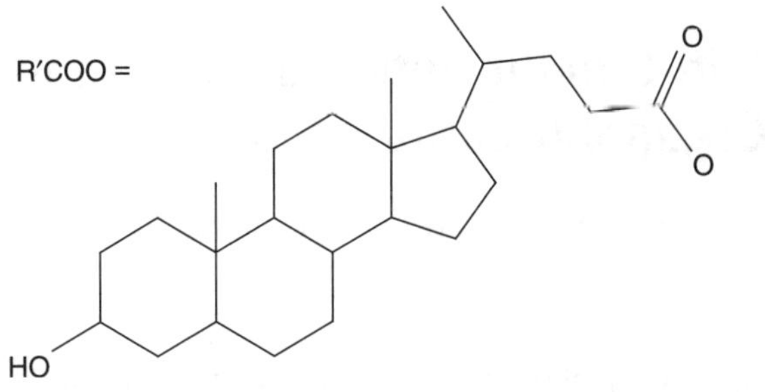

	Cell line						
Compound	MCF-7	EVSA-T	WIDR	IGROV	M19 MEL	A498	H226
1, R = *n*-Bu	160	60	390	160	120	220	420
3, R = Ph	16	<3	19	18	51	65	61
3, R = *n*-Bu	16	<3	15	<3	51	138	76
CPT	699	422	967	169	558	2253	3269
DOX	10	8	11	60	16	90	199

22.8 Anti-tumour Potential of Fluorine-Substituted Organotin Carboxylates

Several attempts were then made to synthesize organotin compounds with enhanced water solubility. The lipophilic/hydrophilic characteristics of the compounds are very important; their lipophilic properties being essential to allow for the crossing of the cell membrane and their hydrophilic character for being accepted by the water-rich cell.[67]

First, the hydrogen atoms of phenyl rings of benzoato ligands were replaced by hydroxyl groups.[68] This did not significantly affect the water solubility or the cytotoxicity of the resultant organotin compounds. A more successful strategy to improve water solubility was to incorporate fluorine in the molecules.

As early as 1984, such fluoride-substituted organotin compounds were synthesized to determine whether the replacement of hydrogen by fluorine improved cytotoxicity.[69] Indeed, a peculiar property of the fluorine-substituted compounds is that they are more soluble in water than their hydrogen analogues and yet still retain their solubility in non-polar solvents. Unfortunately, the fluorine-containing organotin compounds were not significantly more potent than their unfluorinated analogues as may be seen from the data in Table 22.5.[70–74]

Organotin salts, for instance stannates, were also synthesized by adding an anionic ligand to diorganotin derivatives.[75–79] Their cytotoxicities were not better than those of the starting compounds and this line of investigation was abandoned.

22.9 Anti-tumour Potential of Boron-Containing Organotin Carboxylates

A number of carboranyl-based organotin compounds have been screened for anti-cancer activity.[80–82]

Table 22.5 Typical ID_{50} values (ng/ml) of two selected fluorine-substituted di-*n*-butyltin carboxylates of types **1** and **2**

	Cell line					
Compound	MCF-7	EVSA-T	WIDr	IGROV	M19 MEL	A498
1, R = Bu, R′COO = $C_6F_5CH_2COO$	55	43	275	60	114	105
2, R = Bu, R′COO = $C_6F_5CH_2COO$	10	19	145	20	36	50
CPT	699	422	967	169	558	2253
DOX	10	8	11	60	16	90

A carborane moiety, $C_2B_{10}H_{11}$, is of about the same size as a phenyl ring but is spherical instead of planar. The replacement of an aromatic phenyl ring of an organic molecule by a hyper-aromatic carboranyl moiety is routinely used in the field of boron-based metallotherapeutic agents. In some of the compounds synthesized, the tin atom was linked directly to one of the boron or carbon atoms of the carboranyl moiety. In other compounds, the carboranyl moieties were bound to a CO_2 or to a CH_2CO_2 group that served as a linker to tin atoms of a distannoxane structure (type **1**). Their cytotoxicities are comparable to those of organotin carboxylates without the carborane moiety (Table 22.6).

An attempt to synthesize a carborane-based organotin compound with a polyoxaalkyl chain linked to the carboranyl moiety, in order to increase the water solubility (discussed later), produced a triphenylstannate[83] instead of a derivative of type **3**, which was expected to be formed (Figure 22.4).

The water solubility of this triphenylstannate is increased dramatically by the presence of the polyoxa substituent and probably primarily by the fact that the compound is a salt. Like other organotin salts, it exhibits a cytotoxicity similar to those obtained for carboranyltin compounds of type **3**.

Table 22.6 Typical ID_{50} values (ng/ml) of two selected carborane-based organotin compounds of type **1**: $\{[Bu_2(R'COO)Sn]_2O\}_2$

R′COO = 2-Ph-m-$C_2B_{10}H_{11}$-1-COO

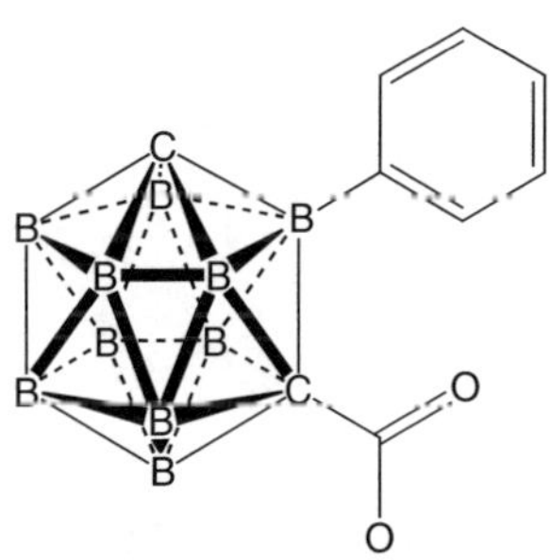

R′COO = 1,2-$C_2B_{10}H_{11}$-9 -COO

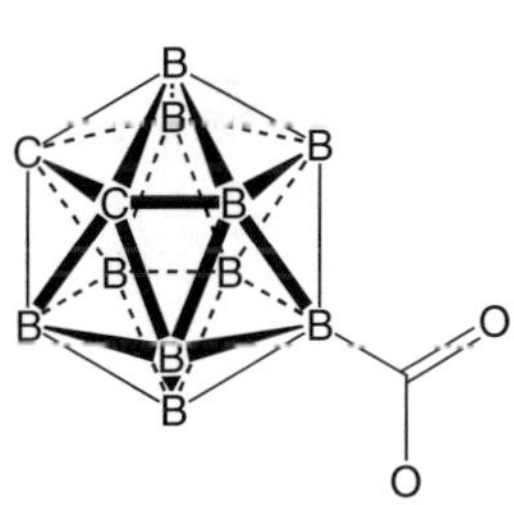

Compound	Cell line						
	MCF-7	EVSA-T	WIDR	IGROV	M19 MEL	A498	H226
2-Ph-m-$C_2B_{10}H_{11}$-1-COOH	56527	45168	42426	58292	>60000	55032	11747
1, R = Bu, R′COO = 2-Ph-m-$C_2B_{10}H_{11}$-1-COO	138	164	514	169	220	301	388
2, R′COO = 1,2-$C_2B_{10}H_{11}$-9-COO	146	142	439	139	174	195	291
CPT	699	422	967	169	558	2253	3269
DOX	10	8	11	60	16	90	199

22.10 Anti-tumour Potential of Organotin Carboxylates-Containing Polyoxaalkyl Moieties

Arguably, the most promising development in the search for anti-tumour organotin compounds has been achieved by the synthesis and testing of organotin compounds that contain a polyoxaalkyl moiety linked to tin either by a carbon–tin or by a carbon—oxygen bond.[84–86]

The polyoxaalkyl moiety can be either a linear one or cyclic such as a crown-ether. A number of these compounds, some of which are very soluble in water, exhibit exceptionally high cytotoxicities against the seven human cell lines studied as detailed in Table 22.7.

Of all the polyoxaalkyltin compounds tested, two distannoxanes of type **1** and two triorganotin derivatives of type **3** indeed exhibit very pronounced cytotoxicities as reported in Table 22.7.

Table 22.7 ID_{50} values (ng/ml) of four selected potent polyoxaalkyltin compounds tested against the seven human tumour cell lines

benzocrown—COO =

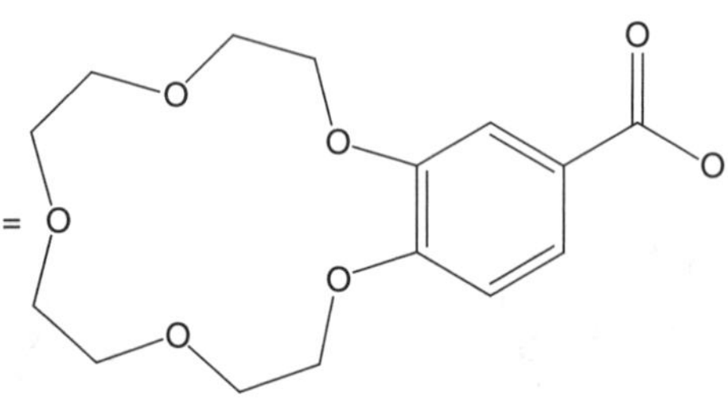

Compound	MCF-7	EVSA-T	WIDR	IGROV	M19	A498	H226
1, R = Bu, R′COO = $CH_3O(CH_2CH_2O)CH_2COO$	<1	<1	3.9	<1	<1	<1	3.3
1, R = Bu, R′COO = $CH_3O(CH_2CH_2O)_2CH_2COO$	<1	<1	<1.8	<1	<1	<1	<1
3, R = Ph, R′COO = benzocrown—COO	2.9	<2	<2	<2	<2	<2	<2
3, R = Bu, R′COO = benzocrown—COO	3.3	<2	<2	<2	<2	<2	<2
CPT	699	422	967	169	558	2253	3269
DOX	10	8	11	60	16	90	199

22.11 Mode of Action of Cytotoxic Organotin Compounds

Given the interest in generating anti-tumour organotin compounds, it is not surprising that significant effort has been devoted to elucidating the possible mode(s) of action of these compounds. Platinum anti-tumour drugs have been intensively studied and their probable mode of action elucidated: their interaction with DNA is responsible for their property to inhibit cell division; see Chapter 25 in this book for more details.

The interaction of diorganotin halides with DNA or DNA fragments was reported[87–95] but, at around pH 7, no effective interaction was detected.[89] In the same way, water-soluble cytotoxic organotin carboxylates were found not to interact significantly with DNA.[96]

22.12 Other Therapeutic Potential of Organotin Compounds

From the foregoing, it is plain that considerable effort has been devoted to investigating the anti-tumour potential of organotin compounds. However, by no means is the study of therapeutic potential of organotin compounds restricted to cancer diseases. In the following paragraphs, a brief overview of recent studies investigating the potential activity of organotin compounds against other diseases is given.

Following the screening for anti-tumour potential, many studies have coupled various carboxylate ligands to organotin centres and subjected the products to anti-bacterial trials.[97–100] In addition, studies on other species such as barbiturates[101] and thiones[102] have also been conducted. The principle of complexing pharmaceutically active substrates to organotin and investigating the resultant activity is one motivation for these trials. Thus, a series of tri-*n*-butyltin carboxylates, where the ligands have anti-inflammatory and analgesic properties, were prepared and of these, the smallest carboxylate of the series, i.e. derived from 3-furan-2-yl-acrylic acid anion, Figure 22.5(a), was the most active and had higher activity than the reference drugs Ampicillin® and Cephalexin® in most of the bacteria investigated.[97] A similar strategy was applied recently where tricyclohexyltin carboxylates derived from triorganogermyl propanoic acids (see Chapter 15 on therapeutic germanium compounds) were investigated.[98] Generally, the new compounds had little anti-bacterial activity. One of the compounds, i.e. $(C_6H_{11})_3SnO_2CH_2CH(Ph)Ge(o\text{-tolyl})_3$, displayed significant activity against a *Trichophyton* species, a human pathogen that is responsible for hair, nail and skin infections and, importantly, greater activity than the reference drug Imipenem®.

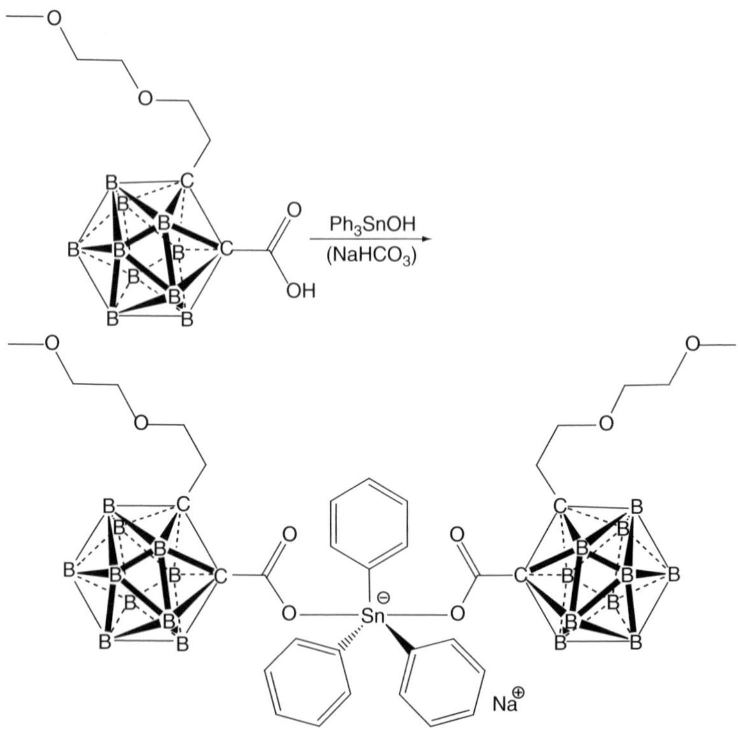

Figure 22.4 A boron-based triphenylstannate, sodium bis[2-(3′,6′,9′-trioxadecyl)-1,2-dicarba-closo-dodecaborane-1-carboxylato]triphenylstannate, $Na[(CH_3OCH_2CH_2OCH_2CH_2OCH_2CH_2)\text{-}1,2\text{-}C_2B_{10}H_{10}\text{-}9\text{-}COO)_2SnPh_3]$, formed unexpectedly during the synthesis of the type **3** analogue[83]

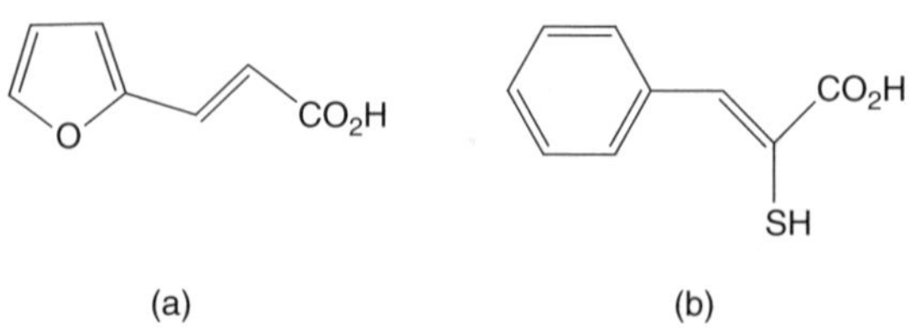

Figure 22.5 Chemical structures of (a) 3-furan-2-yl-acrylic acid and (b) 2-mercapto-3-phenyl-acrylic acid

The anti-microbial activity of a series of triphenyltin sulfanylpropenoates has been reported recently[99] (see Figure 22.5b for a chemical structure of a representative ligand). The essential role that tin plays in therapeutic potential was demonstrated in this study, in that the free ligands displayed no anti-bacterial activity but their anionic triphenyltin complexes did. Also noteworthy is the observation that the organotin compounds were more potent against Gram-positive bacteria.[99] This was correlated with the difficulty of the organotin species to penetrate the outer membrane of Gram-negative bacteria.

Organotin compounds have also been investigated for anti-trypanocidal activity.[103] Human sleeping sickness is a major health problem in parts of Africa and new drugs are required to overcome drug resistance developed by trypanosome strains. A series of water-soluble diorganotin thiolates, of the general formula $(CH_3OCH_2CH_2OCH_2CH_2CH_2)_2Sn(SR)_2$, i.e. containing a tin-bound water-solubilizing group were investigated for anti-trypanocidal activity, both *in vitro* and *in vivo*.[103] The organotin compounds investigated did not possess significant promise in this context although it was noted that the organotin compounds were well tolerated by the animals tested.[103]

22.13 Therapeutic Potential of Tin Compounds Without Tin-to-Carbon Bonds

While the emphasis of this review thus far focussed upon organotin compounds, owing to the large number of ongoing investigations seeking therapeutic potential of this class of compound, non-organotin compounds also play a role in medicine. In the same way, only Sn(IV) compounds have been mentioned but in the realm of 'inorganic' tin, both Sn(II) and Sn(IV) feature. Indeed, SnF_2 has been utilized in dental hygiene to inhibit dental plaque formation owing to its apparent prophylactic and therapeutic character.[7,8] Local destruction on a tooth arises from decalcification of the tooth enamel and may, after enzymatic lysis, lead to the formation of a cavity. SnF_2 is incorporated in less than 1% fraction in tooth paste marketed as, for example, Crest® and Gel Tin®.

Early work showed that Sn(IV) porphyrin complexes may be used clinically in the treatment of jaundice (icterus), a physiological condition found in new-born babies.[8] Neo-natal jaundice, hyperbilirubinaemia, occurs when the liver is not able to detoxify bilirubin, a bile pigment, and results in the yellow staining of tissue. The complex dichlorotin(IV) protoporphyrin IX (Figure 22.6a) is an inhibitor of haemoxidase, an enzyme responsible for the degradation, in the presence of O_2 and reduced Nicotinamide Adenine Dinucleotide Phosphate (NADPH), of haemoglobin to Fe(II) ions, CO and biliverdin; biliverdin is reduced to bilirubin by biliversin reductase. This control of haeme metabolism has an implication for human immunodeficiency virus (HIV) treatment. The observation that porphyrins can bind to the V3 loop of the envelope glycoprotein on HIV type 1 (HIV-1) and inhibit infection of cells by HIV-1 prompted a study of the anti-HIV-1 activity of dichlorotin(IV) protoporphyrin IX.[104] The conclusion of the study, which showed that the Sn(IV) complex binds and distorts the native conformation of the HIV-1 envelope and inhibits infection,

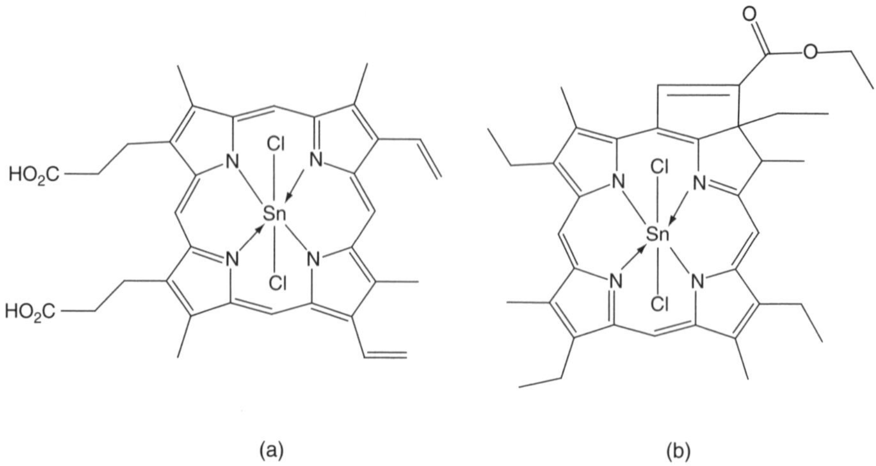

Figure 22.6 Chemical structures of (a) dichlorotin(IV) protoporphyrin IX, and (b) tin ethyl etiopurpurin

was that dichlorotin(IV) protoporphyrin IX could also be used as a prophylactic and therapeutic against HIV-1.[104] In a related study, using the principles of PhotoDynamic Therapy (PDT), dichlorotin(IV) protoporphyrin IX was evaluated for the treatment of Kaposi's sarcoma, a highly visible tumour that is frequently seen in patients with HIV.[105] The above led to two vitally important events in therapeutic tin pharmacology. Firstly, a patent has recently appeared with the title '*Methods for Preventing HIV-1 or HIV-2 Infection*' in which it is claimed that dichlorotin(IV) protoporphyrin IX, amongst others, is pharmaceutically effective against HIV-1 and HIV-2.[106] Secondly, other tin-based porphyrins were investigated for utility in PDT.

PhotoDynamic Theory is a relatively new procedure that can be employed in the treatment of skin cancers which comprise approximately one-third of all cancers suffered by humans. Examples of skin cancers are (i) the aforementioned Kaposi's sarcoma, (ii) solar keratoses, afflicting the face and scalp and (iii) basal cell carcinomas, one of the most common forms of skin cancer and one correlated with chronic exposure to sunlight. The basic principle of PDT is that photosensitizing agents, light and oxygen are employed in a chemical reaction localized at the cancer and designed to selectively kill cancer cells once light of a specific wavelength interacts with the photosensitizing agent. Tin ethyl etiopurpurin (Figure 22.6b) has attracted significant attention in this regard.[107–110] Tin ethyl etiopurpurin has also been investigated for other therapeutic potential.

Very recently, Miravant Medical Technologies,[111] a company specializing in 'photoselective medicine', has announced that the US Food and Drug Administration has accepted a New Drug Application submitted by Miravant. Tin ethyl etiopurpurin is being marketed for the treatment of wet age-related macular degeneration (AMD). AMD is associated with a loss of central vision owing to the leaking of various fluids, including blood, at the back of the eye by abnormal blood vessels; indeed, AMD is a leading cause of blindness in older adults. The drug is administered using the principles outlined above for PDT.

22.14 Conclusions

It is evident from this chapter that some very promising organotin compounds have been developed that exhibit very high *in vitro* cytotoxicities and that the study of other series of organotin derivatives might lead to even more active compounds. The next step is the screening of promising new derivatives in human tumour xenografts implanted in nude mice, i.e. the conduct of time-consuming, but essential, *in vivo* studies. *In vivo* testing has often been limited by the poor water solubility of the organotin compounds. Indeed, this is one of the most important factors emerging from the interpretation of the *in vivo* results and the evaluation of newly generated water-soluble organotin carboxylates may bring about a solution to this problem. In addition to the biological evaluation for therapeutic potential, further chemical and pharmacological studies are necessary to unravel a structure–activity relationship for organotin derivatives that can guide subsequent design of new drug candidates. While therapeutics-based organotin compounds continue to attract the greatest attention, 'inorganic' tin drugs are in or are finding their way into the clinic, being exploited now in PDT for the treatment of jaundice, cancers and wet AMD.

Acknowledgements

The Fonds voor Wetenschappelijk Onderzoek-Vlaanderen (contract G.0074.00), INTAS (grant 99-00806) and the National University of Singapore (R-143-000-186-112) are thanked for support.

References

1. M. Gielen, *J. Braz. Chem. Soc.*, **14**, 870–877 (2003).
2. I. Omae, Organotin chemistry, *J. Organomet. Chem. Libr.*, **21**, Elsevier, Amsterdam (1989).
3. F. Huber and R. Barbieri, in: *Tin as a Vital Nutrient: Implications in Cancer Prohylaxis and Other Physiological Processes*, N.F. Cardarelli (Ed.), CRC Press, Boca Raton, 175–187 (1986).

4. A.J. Crowe, *Drugs of the Future*, **12**, 255–275 (1987).
5. L.R. Sherman and F. Huber, *Appl. Organomet. Chem.*, **2**, 65–72 (1988).
6. A.K. Saxena and F. Huber, *Coord. Chem. Rev.*, **95**, 109–123 (1989).
7. A.J. Crowe, in: *Tin Compounds and their Potential as Pharmaceutical Agents*, NATO ASI Series – Vol. H37, M. Gielen (Ed.), Springer-Verlag, Berlin, Heidelberg, 69–114 (1990).
8. J.M. Tsangaris and D.R. Williams, *Appl. Organomet. Chem.*, **6**, 3–18 (1992).
9. I. Haiduc, C. Silvestru and M. Gielen, *Bull. Soc. Chim. Belg.*, **91**, 187–189 (1982).
10. M. Gielen, K. Jurkschat and G. Atassi, *Bull. Soc. Chim. Belg.*, **93**, 153–155 (1984).
11. A.H. Penninks, M. Bol-Schoenmakers, M. Gielen and W. Seinen, *Main Group Met. Chem.*, **12**, 1–15 (1989).
12. Z. Zhang, H. Pan, C. Hu *et al.*, *Appl. Organomet. Chem.*, **5**, 183–190 (1991).
13. C. Silvestru, I. Haiduc, E.R.T. Tiekink *et al.*, *Inorg. Chim. Acta*, **239**, 177–183 (1995).
14. M. Gielen, D. de Vos, A. Meriem *et al.*, *In Vivo*, **7**, 171–174 (1993).
15. M. Gielen and R. Willem, *Anticancer Res.*, **12**, 257–268 (1992).
16. M. Gielen and R. Willem, *Anticancer Res.*, **12**, 269–272 (1992).
17. M. Gielen and R. Willem, *Anticancer Res.*, **12**, 1323–1326 (1992).
18. M. Gielen, D. de Vos, H. Pan and R. Willem, *Main Group Met. Chem.*, **15**, 242–245 (1992).
19. M. Gielen and R. Willem, in: *Metal Ions in Biology and Medicine*, J. Anastassopoulou, P. Collery, J.C. Etienne and T. Theophanides (Eds), John Libbey Eurotext, Paris, Vol. II, 186–187 (1992).
20. M. Gielen and R. Willem, in: *Metal Ions in Biology and Medicine*, J. Anastassopoulou, P. Collery, J.C. Etienne and T. Theophanides (Eds), John Libbey Eurotext, Paris, Vol. II, 188–189 (1992).
21. M. Gielen, R. Willem, M. Bouâlam *et al.*, in: *Metal Ions in Biology and Medicine*, J. Anastassopoulou, P. Collery, J.C. Etienne and T. Theophanides (Eds), John Libbey Eurotext, Paris, Vol. II, 190–191 (1992).
22. M. Gielen, *Metal-Based Drugs*, **1**, 213–219 (1994).
23. M. Gielen, in: *Metal Ions in Biology and Medicine*, J. Anastassopoulou, P. Collery, J.C. Etienne and T. Theophanides (Eds), John Libbey Eurotext, Paris, Vol. III, 309–314 (1994).
24. M. Gielen and R. Willem, *Main Group Met. Chem.*, **17**, 1–8 (1994).
25. M. Gielen and R. Willem, in: *Main Group Elements and Their Compounds: Perspectives in Materials Science, Chemistry and Biology*, V.G. Kumar Das (Ed.), Springer-Verlag/Narosa Publishing House, 446–452 (1996).
26. M. Gielen and R. Willem, in: *Cytotoxic, Mutagenic and Carcinogenic Potential of Heavy Metals Related to Human Environment*, NATO ASI Series 2: Environment – Vol. 26, N.D. Hadjiliadis (Ed.), Kluwer Academic Publishers, Dordrecht/Boston/London, 445–455 (1997).
27. D. de Vos, M. Gielen, R. Willem *et al.*, *Metal-Based Drugs*, **5**, 179–188 (1998).
28. M. Gielen, *Coord. Chem. Rev.*, **151**, 41–51 (1996).
29. D. de Vos, R. Willem, M. Gielen *et al.*, *Metal-Based Drugs*, **5**, 179–188 (1998).
30. M. Gielen, *Appl. Organomet. Chem.*, **16**, 481–494 (2002).
31. M. Gielen, L. De Clercq, R. Willem and E. Joosen, in: *Tin and Malignant Cell Growth*, J.J. Zuckerman (Ed.), CRC Press, Boca Raton, Florida, 39–46 (1988).
32. M. Bouâlam, R. Willem, M. Biesemans and M. Gielen, *Appl. Organomet. Chem.*, **5**, 497–506 (1991).

33. M. Gielen, A. El Khloufi, M. Biesemans *et al.*, *Bull. Soc. Chim. Belg.*, **101**, 243–248 (1992).
34. M. Bouâlam, R. Willem, J. Gelan *et al.*, *Appl. Organomet. Chem.*, **4**, 335–343 (1990).
35. A. Meriem, M. Biesemans, R. Willem *et al.*, *Bull. Soc. Chim. Belg.*, **100**, 367–374 (1991).
36. M. Bouâlam, R. Willem, M. Biesemans *et al.*, *Main Group Met. Chem.*, **14**, 41–56 (1991).
37. M. Gielen, J. Meunier-Piret, M. Biesemans *et al.*, *Appl. Organomet. Chem.*, **6**, 59–67 (1992).
38. M. Gielen, J. Meunier-Piret, M. Biesemans *et al.*, *Polyhedron*, **11**, 1861–1868 (1992).
39. M. Gielen, E. Joosen, T. Mancilla *et al.*, *Main Group Met. Chem.*, **10**, 147–167 (1987).
40. M. Gielen, E. Joosen, T. Mancilla *et al.*, in: *Platinum and Other Metal Coordination Compounds in Cancer Chemotherapy*, M. Nicolini (Ed.), Martinus Nijhoff Publ, Boston/Dordrecht/Lancaster, 662–665 (1988).
41. A. Meriem, M. Gielen and R. Willem, *J. Organomet. Chem.*, **365**, 91–101 (1989).
42. M. Gielen, R. Willem, A. Delmotte *et al.*, *Main Group Met. Chem.*, **12**, 55–72 (1989).
43. A. Meriem, R. Willem, J. Meunier-Piret and M. Gielen, *Main Group Met. Chem.*, **12**, 187–198 (1989).
44. A. Meriem, R. Willem, J. Meunier-Piret *et al.*, *Main Group Met. Chem.*, **13**, 167–180 (1990).
45. A. Meriem, R. Willem, M. Biesemans *et al.*, *Appl. Organomet. Chem.*, **5**, 195–201 (1991).
46. M. Gielen, M. Acheddad, B. Mahieu and R. Willem, *Main Group Met. Chem.*, **14**, 73–79 (1991).
47. M. Gielen, M. Bouâlam, M. Biesemans *et al.*, *Main Group Met. Chem.*, **14**, 271–281 (1991).
48. M. Gielen, R. Willem, M. Biesemans *et al.*, *Appl. Organomet. Chem.*, **6**, 287–291 (1992).
49. M. Gielen, M. Bouâlam, A. Meriem *et al.*, *Heteroatom Chem.*, **3**, 449–452 (1992).
50. M. Gielen, A. Bouhdid, F. Kayser *et al.*, *Appl. Organomet. Chem.*, **9**, 251–257 (1995).
51. M. Gielen, R. Willem, A. Bouhdid *et al.*, *In Vivo*, **9**, 59–64 (1995).
52. M. Gielen, R. Willem, A. Bouhdid *et al.*, *Oncol. Rep.*, **3**, 583–587 (1996).
53. M. Gielen, R. Willem, H. Dalil *et al.*, *Metal-Based Drugs*, **5**, 83–90 (1998).
54. M. Gielen, E. Joosen, T. Mancilla *et al.*, *Main Group Met. Chem.*, **10**, 147–167 (1987).
55. A. Meriem, M. Gielen and R. Willem, *J. Organomet. Chem.*, **365**, 91–101 (1989).
56. M. Gielen, P. Lelieveld, D. de Vos and R. Willem, in: *Metal-Based Antitumour Drugs*, Vol. 2, M. Gielen (Ed.), Freund Publishing House, Tel Aviv, 29–54 (1992).
57. M. Gielen, M. Mélotte, G. Atassi and R. Willem, *Tetrahedron*, **45**, 1219–1229 (1989).
58. H. Preut, F. Huber and M. Gielen, *Acta Crystallogr.*, **C46**, 2071–2073 (1990).
59. S.J. Blunden and C.J. Evans, in: *Anthropogenic Compounds*, O. Hutziger (Ed.) Springer-Verlag, Heidelberg, 1–44 (1990).
60. C.J. Evans, in: *Chemistry of Tin*, P.J. Smith (Ed.) Blackie Academic & Professional, London, 442–479 (1998).

61. R. Willem, A. Bouhdid, B. Mahieu *et al.*, *J. Organomet. Chem.*, **531**, 151–158 (1997).
62. M. Gielen, A. El Khloufi, M. Biesemans *et al.*, *Metal-Based Drugs*, **1**, 305–309 (1994).
63. M. Bouâlam, M. Gielen, A. El Khloufi *et al.*, Pharmachemie B.V., *Eur. Pat.*, Publ. 538 517, Appl. 91/202,746, 22.10.91, *Chem. Abstr.*, **119**, 117548b (1993).
64. M. Gielen, P. Lelieveld, D. de Vos *et al.*, *Inorg. Chim. Acta*, **196**, 115–117 (1992).
65. R. Willem, H. Dalil, P. Broekaert *et al.*, *Main Group Met. Chem.*, **20**, 535–542 (1997).
66. M. Gielen, H. Pan, R. Willem and D. de Vos, Pharmachemie B.V., *Eur. Pat.*, Publ. 484 596, Appl. 90/202,936.2, 06/11/90, *Chem. Abstr.*, **117**, 111 807x (1992).
67. G. Atassi, *Rev. Si, Ge, Sn & Pb Cpds*, **8**, 219–235 (1985).
68. M. Gielen, R. Willem, A. Bouhdid and D. de Vos, Pharmachemie B.V., *Eur. Pat.*, Publ. 677 528 A1, Appl. 94/201012.5, 12/04/94, *Chem. Abstr.*, **124**, 87376c (1996).
69. L. De Clercq, R. Willem and M. Gielen, *Bull. Soc. Chim. Belg.*, **93**, 1089–1097 (1984).
70. M. Gielen, A. El Khloufi, M. Biesemans and R. Willem, *Appl. Organomet. Chem.*, **7**, 119–125 (1993).
71. M. Gielen, A. El Khloufi, M. Biesemans *et al.*, *Appl. Organomet. Chem.*, **7**, 201–206 (1993).
72. M. Gielen, M. Biesemans, A. El Khloufi *et al.*, *J. Fluorine Chem.*, **64**, 279–291 (1993).
73. M. Gielen, A. El Khloufi, D. de Vos *et al.*, *Bull. Soc. Chim. Belg.*, **102**, 761–764 (1993).
74. M. Gielen, E.R.T. Tiekink, A. Bouhdid *et al.*, *Appl. Organomet. Chem.*, **9**, 639–648 (1995).
75. S.W. Ng, V.G.K. Das, M. Gielen and E.R.T. Tiekink, *Appl. Organomet. Chem.*, **6**, 19–25 (1992).
76. R. Willem, M. Biesemans, F. Kayser *et al.*, *Inorg. Chim. Acta*, **197**, 25–30 (1992).
77. S.W. Ng, V.G.K. Das and M. Gielen, *Appl. Organomet. Chem.*, **6**, 489–492 (1992).
78. R. Willem, M. Biesemans, M. Bouâlam *et al.*, *Appl. Organomet. Chem.*, **7**, 311–317 (1993).
79. S.W. Ng, V.G.K. Das, J. Holecek *et al.*, *Appl. Organomet. Chem.*, **11**, 39–45 (1997).
80. M. Gielen, F. Kayser, O.B. Zhidkova *et al.*, *Metal-Based Drugs*, **2**, 37–42 (1995).
81. E.R.T. Tiekink, M. Gielen, A. Bouhdid *et al.*, *Metal-Based Drugs*, **4**, 75–80 (1997).
82. V.I. Bregadze, S.A. Glazun, P.V. Petrovskii *et al.*, *Appl. Organomet. Chem.*, **17**, 453–457 (2003).
83. V.I. Bregadze, S.A. Glazun, P.V. Petrovskii *et al.*, *Appl. Organomet. Chem.*, **18**, 191–194 (2003).
84. M. Kemmer, M. Gielen, M. Biesemans *et al.*, *Metal-Based Drugs*, **5**, 189–196 (1998).
85. M. Kemmer, L. Ghys, M. Gielen *et al.*, *J. Organomet. Chem.*, **582**, 195–203 (1999).
86. M. Gielen, R. Willem, M. Biesemans *et al.*, Pharmachemie B.V., *World Pat.*, Publ. C07F 7/22, A61K 31/32 Nr. WO 00/06583, Appl. PCT/NL98/00429, 28/7/98, Publ. 10/02/2000 (1998).
87. Q. Li, P. Yang, H. Wang and M. Guo, *J. Inorg. Biochem.*, **64**, 181–195 (1996).
88. Z. Yang, T. Bakas, A. Sanchez-Diaz *et al.*, *J. Inorg. Biochem.*, **72**, 133–140 (1998).
89. A. Jancso, L. Nagy, E. Moldrheim and E. Sletten, *J. Chem. Soc., Dalton Trans.*, 1587–1594 (1999).
90. R. Barbieri and A. Silvestri, *J. Inorg. Biochem.*, **41**, 31–35 (1991).
91. V. Piro, F. Di Simone, G. Madonia *et al.*, *Appl. Organomet. Chem.*, **6**, 537–542 (1992).

92. R. Barbieri, A. Silvestri, A.M. Giuliani *et al.*, *J. Chem. Soc., Dalton Trans.*, 585–590 (1992).
93. G. Barone, M. Cotta Ramusino, R. Barbieri and G. La Manna, *J. Mol. Struct.*, **469**, 143–149 (1999).
94. R. Barbieri, F. Huber, A. Silvestri *et al.*, *Appl. Organomet. Chem.*, **13**, 595–603 (1999).
95. L. Ghys, M. Biesemans, M. Gielen *et al.*, *Eur. J. Inorg. Chem.*, 513–522 (2000).
96. A. Casini, L. Messori, P. Orioli *et al.*, *J. Inorg. Biochem.*, **85**, 297–300 (2001).
97. S. Shahzadi, M.H. Bhatti, K. Shahid *et al.*, *Mont. Chem.*, **133**, 1089–1096 (2002).
98. Imtiaz-ud-Din, K.C. Molloy, M. Mazhar *et al.*, *Appl. Organomet. Chem.*, **17**, 781–787 (2003).
99. J.S. Casas, E. García-Martínez, M.L. Jorge *et al.*, *Appl. Organomet. Chem.*, **15**, 204–212 (2001).
100. F. Ahmad, S. Ali, M. Parvez *et al.*, *Hetereoatom Chem.*, **13**, 638–649 (2002).
101. M. Kidwai, B. Dave, P. Misra *et al.*, *Inorg. Chem. Commun.*, **3**, 465–468 (2000).
102. J.S. Casas, A. Castiñeiras, M.D. Couce *et al.*, *Appl. Organomet. Chem.*, **14**, 421–431 (2001).
103. J. Susperregui, M. Bayle, G. Lain *et al.*, *Eur. J. Med. Chem.*, **34**, 617–623 (1999).
104. A.R. Neurath, N. Strick, K. Lin *et al.*, *Antivir. Chem. Chemother.*, **5**; 322–330 (1994).
105. R.R. Allison, T.S. Mang, B.D. Wilson and V. Vongtama, *Curr. Therap. Res.*, **59**, 23–27 (1998).
106. A.R. Neurath, S. Jiang and A.K. Debnath, Patent Application US 94-342745 19941021 (2004).
107. L. Polo, E. Reddi, G.M. Garbo *et al.*, *Can. Lett.*, **66**, 217–223 (1992).
108. S.L. Rakestraw, W.E. Ford, R.G. Tompkins *et al.*, *Biotechnol. Prog.*, **8**, 30–39 (1992).
109. P. Sekher and G.M. Garbo, *J. Photochem. Photobiol. B: Biol.*, **20**, 117–125 (1993).
110. S.H. Selman, D. Albrecht, R.W. Keck *et al.*, *J. Urology*, **165**, 1795–1801 (2001).
111. http://www.miravant.com/Index3.htm.

23 $_{51}$Sb Antimony in Medicine

Siucheong Yan, Lan Jin and Hongzhe Sun

23.1 Introduction

Antimony has been used in medicinal and healthy applications for several centuries. Its earliest-recorded use was documented by the ancient Egyptians, who used antimony particularly for cosmetic purposes, due to the beautiful colors of the compounds.[1] During that period, antimony was taken as a medicine for the treatment of fevers and skin irritations. In the early twentieth century, antimony was found to be extremely useful in the therapy of tropical diseases, particularly leishmaniasis, and has been used as the first-line drug for the treatment of this disease.[2] In this chapter we review the medicinal application and biological chemistry of antimony.

23.2 The Chemistry of Antimony

Antimony is in group 15 of the periodic table, together with nitrogen, phosphorus, arsenic and bismuth. It has two stable naturally occurring isotopes: ^{121}Sb (57.25%) and ^{123}Sb (42.75%).[3] Antimony is mostly found in ore form as either stibnite (Sb_2S_3) or ullmanite (NiSbS). Similar to arsenic, antimony is

Metallotherapeutic Drugs and Metal-Based Diagnostic Agents: The Use of Metals in Medicine
Edited by Gielen and Tiekink

generally classified as a semi-metal. The most common and stable oxidation states of the element are trivalent (Sb^{III}) and pentavalent (Sb^{V}), with ionic radii of 0.76 and 0.62 Å, respectively.[3,4]

The simple oxyanion forms of both antimonite (Sb^{III}) and antimonate (Sb^{V}) have not been adequately characterized. Although it is generally believed that H_3SbO_3 and Sb_2O_5 are the correct formulae, examples in earlier literature suggested that hydrated antimonate has the formula Sb_2O_5 but these were later proved to contain $[Sb(OH)_6]^-$.[1] Both Sb_2O_3 and Sb_2O_5 form complexes with organic acids such as tartaric acid and gluconic acid. The resulting compounds are useful as therapeutic agents. The basic properties of antimony are summarized in Table 23.1.

Trivalent antimony forms many complexes by means of Lewis acid–base reactions. The valence orbital from the metal center accommodates the incoming electron pair. In the case of haloantimony(III) complexes, monomeric haloantimony(III) anions $[SbX_4]^-$, $[SbX_5]^{2-}$ and $[SbX_6]^{3-}$ are formed by the addition of X^- (halide) ligands to SbX_3.

Complexes of Sb^{III} with group 15 elements have been investigated and their structures have been determined. For example, the crystal structure of an ionic complex [H(py)][SbI_4(dmpe)] (py = pyridine, dmpe = 1,2-bis (dimethylphosphino) ethane) showed that Sb^{III} is coordinated to four iodine and two phosphorus atoms from 1,2-bis(dimethylphosphino) ethane.[5]

Sb^{III} is also coordinated to various oxygen- and sulfur-containing ligands and its coordination number can vary. For example, in the antimony complex with citric acid, various building units were observed. In Li[Sb(citrate)$_2$(H_2O)]·$2H_2O$ and Na[Sb(citrate)$_2$(H_2O)$_2$]·H_2O, the Sb^{III} center is coordinated to four oxygens from two citrates with each in a bidentate mode. Antimony adopts a distorted pseudo-trigonal bipyramidal geometry with an equatorial lone pair (Figure 23.1).[6–8] The basic building unit of the clinically used antiparasitic agent potassium antimony(III) tartrate (PAT) is the dimeric anion. Each Sb^{III} is also coordinated to four oxygens from two tartrates, with each in a bidentate mode. The coordination geometry of Sb^{III} is distorted pseudo-trigonal

Table 23.1 Basic properties of antimony

Atomic number	51
Earth occurrence	0.2 ppm
Relative atomic mass	121.75
Ground state electronic configuration	$[Kr]4d^{10}5s^25p^3$
Ionic radius (Å)	Sb^{III}(0.76), Sb^{V}(0.62)
Isotopic distribution (%)	^{121}Sb(57.25), ^{123}Sb(42.75)
Standard reduction potential E^0 (H^+)	$Sb^{III} + 3e^- \rightarrow Sb^0$ 0.204 V $Sb^{V} + 2e^- \rightarrow Sb^{III}$ 0.605 V
Major ores	Stibnite (Sb_2S_3), ullmanite (NiSbS)
Radioactive isotope	^{125}Sb, β^- emission, $t_{1/2} = 2.76$ years

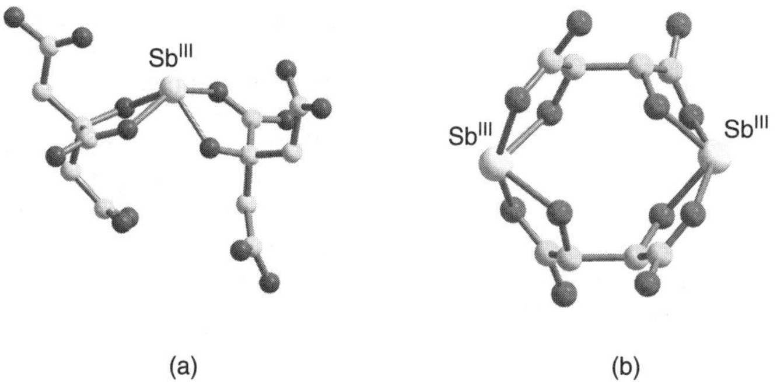

Figure 23.1 X-ray crystal structures of antimony(III) citrate (a) and potassium antimony(III) tartrate (PAT, $K_2Sb_2(C_4H_2O_6)_2$) (b) (see Plate 3)

bipyramidal (Figure 23.1).[9] Sb^{III} has a high affinity for thiolate ligands and the X-ray crystal structures of several Sb^{III}–thiolate complexes have been determined. In these complexes, Sb^{III} exhibits coordination numbers of either 3 or 4. The metal has trigonal pyramidal or bipyramidal geometry with Sb—S bond distances of ~2.4–2.6 Å.[10] Weak interactions to other thiolate sulfurs (either intra- or inter-molecular) are observed in some of the complexes.

Like the Sb^{III} complexes, Sb^{V} also forms many complexes. Pentahaloantimony compounds such as $SbCl_5$ are powerful halide acceptors, usually with the formation of the $[SbX_6]^-$ mono-ligand coordinated adduct. The sixth coordination site can also be occupied by a neutral donor ligand. One example is the SbF_5(NC—CN) dinitrile complex with Sb—N bond lengths of 2.23 Å.[11] Antimony pentachloride can act as a powerful halide Lewis acid in the presence of $AsCl_3$ and chloride.[12]

Complexes containing Sb^{V} and group 15 donor ligands have been investigated. The structures of these complexes are normally simple octahedral or distorted octahedral due to the absence of the lone pair electrons in the coordination environment, in contrast to the Sb^{III} complexes. The X-ray crystal structure of an Sb^{V} porphyrin complex, $[Sb(TTP)(OCH(CH_3)_2)_2]Cl$ (TTP = *meso*-tetra-*p*-tolylporphyrin), showed that Sb^{V} is inserted into the porphyrin ring and octahedrally coordinated by four nitrogens from the porphyrin and two oxygens from 2-propanol molecules at the *z*-direction.[13] The coordination chemistry of antimony has been extensively studied.[14]

23.3 Antimony in Medicine

23.3.1 Historic view of antimony drugs

Arsenic, antimony and bismuth in group 15 have been used for medicinal purposes for several centuries.[15–17] Anti-parasitic effects of antimony compounds

Table 23.2 Use of antimony compounds for the treatment of various diseases[18]

Antimony compounds	Medical treatment
Ancient antimony compounds	Inflammation of the upper respiratory tract and vagina, fever, arthritic conditions lupus, syphilis, internal tumors
Antimony cups (antimony wine)	'The Universal Medicine'
Potassium antimony tartrate (Tartar emetic®)	Tropical disease, fever Pneumonia, experimental anti-cancer drugs
Sodium stibogluconate (Pentosam®)	Cutaneous and visceral leishmaniasis
Meglumine antimonate (Glucantime®)	Cutaneous and visceral leishmaniasis

were known for a long time before they came into common use for the treatment of several parasitic diseases including leishmaniasis.[2] In the sixteenth century, antimony was first introduced into medicinal treatment and won widespread popularity due to its emetic effect. Antimony compounds were also used for the treatment of ulcers. Wine allowed to stand in antimony goblets acquired emetic properties, as the tartaric acid in wine interacts with the metal to form tartar emetic.[18]

Although antimony was a popular drug in the sixteenth century, its use was later abandoned due to its toxicity and lack of specific medicinal properties. However, interest in the use of antimony in medicine revived in the early twentieth century after it was discovered to have specific antiparasitic effects against trypanosomiasis. The use of antimony compounds for the treatment of various diseases is summarized in Table 23.2.

23.3.2 Applications of antimony drugs as anti-leishmanial agents

Antimony-based drugs have remained the major therapeutic treatment in the clinic for decades.[19] The use of antimony chemotherapy increases the survival rate of those patients with *Leishmania* infection to over 90%.[20] Although its main action is concentrated only on the visceral leishmaniasis, other forms of the disease are also responsive. The major disadvantages are a long treatment period, requiring hospitalization, and the need in many cases for subsequent course of treatment for an effective clinical cure in certain areas. The first anti-leishmanial drug, PAT (Tartar emetic®) was developed around 1910.[18,21,22] In spite of increasing the survival rate, the agent was highly toxic to humans. Several other antimony compounds including antimony bis(4,5-dihydroxybenzene-3,5-disulphonate)

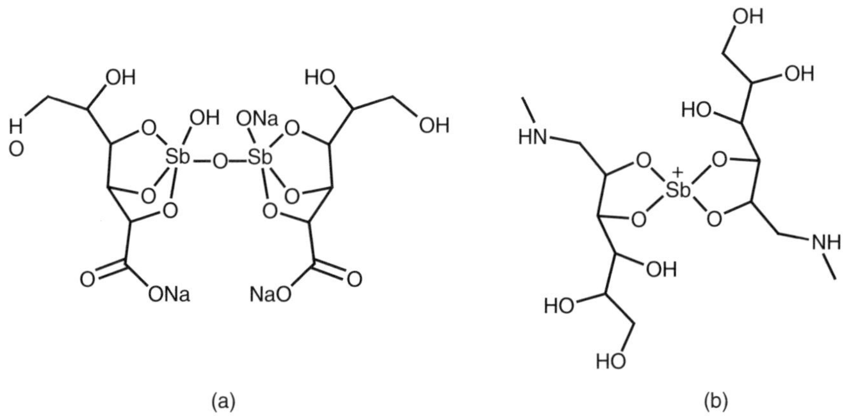

Figure 23.2 Proposed structures of clinically used Sb^V drugs: (a) Pentosam and (b) Glucantime

(Stibophen®), meglumine antimonate (Glucantime®), sodium antimony(III) gluconate (Triostam®) and sodium stibogluconate (Pentosam®), have also been used clinically for the treatment of different forms of *leishmaniasis*. Since pentavalent antimonials have a more efficient therapeutic index and are 10 times less toxic to humans than their trivalent analogues, they have become the first choices of chemotherapy.[23–25] The carbohydrates in the drugs increase the solubility of antimony and are responsible for the delivery of the antimony to the macrophages, where the protozoa that cause *leishmaniasis* undergo division. Sodium stibogluconate is an Sb^V complex of a carbohydrate (gluconic acid) and was first used for the treatment of *Leishmania* during the mid-1930s[26] and is still predominantly used in English-speaking countries.[20] The structure of the complex remains unknown. It is commonly believed that the agent contains only one 746 Da species, probably a dimer, with two Sb^{III} linked by an oxygen (Figure 23.2).[27] However, polymeric species with molecular weight up to 4 kDa may also be present in solution.[28]

Meglunmine antimonate has also been used as the first-choice drug predominantly in French- and Spanish-speaking countries.[20] A series of oligomers of antimony and *N*-methyl-D-glucamine (NMG) has been observed with the general formula of $Sb_n(NMG)_{n+1}$ or $Sb_n(NMG)_n$.[29] The major component of meglunmine antimonate is $Sb(NMG)_2$, with a formula mass of 507 Da. The proposed basic structures of the two drugs are shown in Figure 23.2.

23.3.3 Other medicinal applications

Anti-cancer activities of antimony agents have also been reported. Antimony tartrate has been found to be cytotoxic *in vitro* to various lung cancer cell lines with IC_{50} ranging from 4.2 to 322 μg ml^{-1}. The complex was found to be as effective as those clinically used anti-cancer drugs such as cisplatin and doxorubicin.[30] Other antimony compounds (including organometallic complexes) have been tested in both cell cultures and animal models, and have displayed moderate activities.[31]

The anti-cancer activity of sodium stibogluconate that synergizes with interferon-α (IFN-α) to eradicate IFN-α-resistant human cancer cells both *in vitro* and in mouse models has been reported.[32] It was shown that the complex functions as a protein tyrosine phosphatase (PTPase) inhibitor in cancer cells and augments IFN-α signaling.[33] The activity of Src homology PTPase1 (SPH-1) was almost completely inhibited by sodium stibogluconate at 10 μg ml^{-1}, which is comparable to the serum concentration of *Leishmania* treatment (~10 μg ml^{-1}). The inhibition is likely to target the catalytic domain of PTPase. Interference with the intracellular tyrosine phosphorylation resulted in the disruption of the cell proliferation, differentiation and signaling activities, with consequent anti-tumor activity.[34]

Since sodium stibogluconate is a polymeric complex, the inhibitory activities of different fractions from HPLC toward PTPase have been determined. It has become evident that a small portion of the fractions in sodium stibogluconate is mainly responsible for the PTPase inhibitory activity and that interestingly it is not solely defined by antimony contents. It might be feasible to develop more specific and effective inhibitors for phosphatase-targeted anti-cancer therapeutics through the screening of sodium stibogluconate-related compounds, comprised of antimony conjugated to different organic moieties. Sodium stibogluconate appears to be a better inhibitor than suramin, a compound known for its anti-neoplastic activity against several types of cancers,[35–38] and new therapeutic applications of sodium stibogluconate targeting the phosphatase have then been explored.

Hepatitis C virus (HCV) is a blood-borne virus, previously the most common causative agent of post-transfusion non-A and non-B hepatitis. Recently, it has been reported that sodium stibogluconate is effective in suppressing HCV replication in a HCV-permissive cell line as well as in fresh human liver slices obtained from HCV-infected patients.[39] The anti-viral effect of the drug was subsequently verified using a cell line (293EBNA-Sip-L) and was proved to be permissive for HCV infection/replication. A nearly complete suppression effect was achieved in four out of six human liver slices at the concentration of 100 μg/ml, lower than that required for the treatment of leishmaniasis (recommended dose: 20 mg kg^{-1} day^{-1}).

In the mid-1980s, antimoniotungstate ($[(NH_4)_{17}Na(NaSb_9W_{21}O_{86})\cdot 14H_2O]$, HPA-23) was reported to be active in reducing HIV levels in a patient with AIDS.[40] This observation stimulated the search for other polyoxometalates as potential anti-HIV agents. The median effective concentration (EC_{50}) and inhibitory concentration (IC_{50}) values of HPA-23 are 0.39 and 35 μM, respectively in human peripheral blood mononuclear (PBM) cells for its anti-HIV activity and toxicity.[41] Other polyoxometalates were also found to be active in this cell line. A structural study of HPA-23 showed that it has C_{3h} molecular symmetry with six possible sites where additional metal ions could bind.[42,43]

23.4 Pharmacology, Toxicology and Metabolism of Antimony Drugs

23.4.1 Pharmacology and human toxicology of antimony drugs

No effective vaccine for *Leishmania* has yet been developed; chemotherapy remains the only way at present of treating the disease. Dosage of antimony drugs is based on the amount of metal (antimony) administered per body weight. The early trial of treatment of visceral leishmaniasis by sodium stibogluconate was recommended to be $10\,mg\,kg^{-1}\,day^{-1}$ for the period of 6–10 days. Despite a high cure rate, increasing resistance and relapses have been reported.[44–46] Currently, the optimum overall dosage of the drug is roughly $20\,mg\,kg^{-1}\,day^{-1}$ with maximum 850 mg of antimony per day for 20–30 days.[47,48]

Antimony compounds have been known for their much lower toxicity in humans compared with arsenic. Generally, Sb^{III} is more toxic than its pentavalent analogue (Sb^{V}),[49] with 50% of a lethal dose (LD_{50}) ranging from 172 to $4000\,mg\,kg^{-1}$ in mice and rats (for Sb^{III}), in contrast to the highly toxic arsenic compound (e.g. Me_2AsCl $\sim 5\,mg\,kg^{-1}$).[15] Nevertheless, side-effects of Sb^{III} such as nausea, vomiting, weakness and myalgia, abdominal colic, diarrhea and skin rashes, together with the most important cardiotoxicity, have been observed.[50]

It was also reported that when 10 times of the daily recommended sodium stibogluconate was accidentally administered in a case of cutaneous leishmaniasis treatment (8500 mg per day), no serious toxicity was observed and the overdose was tolerated.[51] This suggests that although not recommended, a single high dose of Sb^{V} has little side-effect on patients.

Genotoxicity has been shown in $SbCl_3$ and Sb_2O_3 by inducing sister chromatid exchange (SCE) in V79 cells and human lymphocytes.[52–54] In contrast, no significant SCE activity was found in their pentavalent analogues.[52] In addition to elevating the SCE activity, $SbCl_3$ also triggers DNA damage and induces apoptosis in Chinese hamster and human cells after short-term exposure ($>50\,\mu M$). This again suggests the limited toxicity of Sb^{V} is preferable to the trivalent form. Sb^{III} compounds have also been found to inhibit the repair of DNA damage. The inhibition of repair of radiation-induced DNA double-strand breaks was observed to be dose dependent when $SbCl_3$ or PAT was incubated with Chinese hamster ovary cells.[55]

Antimony-induced oxidative stress and cardiotoxicity have been reported previously. PAT induces cell death and decrease of spontaneous beat rate;

reduced intracellular glutathione (GSH) levels and increased lipid pre-oxidation were also observed in cardiac myocytes.[56] Detailed studies on the mechanisms of PAT on cardiac myocytes revealed that depletion of cellular ATP levels due to inhibition of pyruvate dehydrogenase and disruption of cellular thiol homeostasis due to inhibition of glutathione peroxidase to cardiac myocytes conferred antimony cardiotoxicity.[57]

In spite of the relatively low toxicity of antimony, Sb^{V} and/or Sb^{III} administered for treatment dosage still places a heavy load on the parasite-weakened body. One study on the meglumine antimonate treatment of visceral leishmaniasis patients co-infected with human immunodeficiency virus (HIV) type-1 showed that severe side-effects and even fatality could occur at the dosing schedule recommended by the centers for disease control and prevention (CDCP) (Atlanta, GA).[58]

Since the mid-1980s, visceral leishmaniasis has been increasingly reported as a complication of AIDS. In 1999, more than 1400 cases of HIV–*Leishmania* co-infection were reported. Therefore pentavalent antimonials have been frequently used in co-infected patients. The treatment with 20 mg of antimonial compounds kg^{-1} day^{-1}, with a maximum daily dose of 850 mg in adults, seemed to be ineffective for most HIV-infected patients. Higher antimonial doses (>20 mg of Sb^{V} kg^{-1} day^{-1}), with no upper-limited daily dose, have been recommended for immunocompetent patients, and some reports have shown that this treatment could be more effective in HIV-related visceral leishmaniasis.[59]

23.4.2 Pharmacokinetics studies and clinical aspect of antimony drugs

After administration of an Sb^{V} compound (e.g. stibogluconate or meglumine antimonate), the peak concentration of antimony reaches to 9–12 μg ml^{-1} in blood 2 h after initial dose, followed by a rapid drop within 24 h.[60,69] About 80–95% of the drug appears in the urine within 6 h, indicating that the drug is not biotransformed appreciably.[61] About 12% of the drug accumulates in an extravascular compartment, primarily in liver, spleen or kidney, which becomes saturated after 5 days of therapy and from which antimony is slowly released.[62–64] The kinetics of Sb^{V} agents is best described by a two-compartment, three-term pharmacokinetics model. The initial absorption phase has a mean half-life of 0.85 h, followed by a rapid elimination phase with a half-life of 2.02 h and a slow elimination phase with a half-life of 76 h.[60]

The period of *Leishmania* treatment (30 days) by Sb^{V} is related to the slow accumulation of antimony in the human body. The minimum antimony concentration in blood increased from 0.04–0.08 μg ml^{-1} at the beginning of the treatment to 0.19–0.33 μg ml^{-1} at day 30 of the treatment.[60] The slow accumulation phase was believed to be due to the reduction of Sb^{V} to Sb^{III} *in vivo*.

On the other hand, modifications of the stibogluconate formulation improve the pharmacokinetic parameters of antimony agents for a better treatment of *L. donovani*.[65,66] The encapsulated stibogluconate favors the organ distribution and increases the tissue resistance time, both of which enhance the efficiency of the treatment.

23.5 Possible Mechanism of Action of Antimony Drugs: Interaction of Antimony with Biologically Relevant Molecules

23.5.1 General hypothesis on the mechanisms of action of antimony anti-leishmanial compounds

The actual mechanisms responsible for the anti-leishmanial activity are still poorly understood.[20,67,68] Although both Sb^V and Sb^{III} drugs show some anti-leishmanial activities, it has been suggested that the relatively non-toxic Sb^V acts only as a prodrug and is reduced into the more toxic trivalent form at or near the site of action. The actual anti-leishmanial activity is due to the trivalent ion (Sb^{III}). In fact, in the bloodstream, there is a mildly reducing environment, and the anti-leishmanial action is probably related to the interaction between the Sb^{III} and Cys-containing peptides, proteins and enzymes.

23.5.2 *In vitro* and *in vivo* reduction of antimony(V) drugs

The hypothesis that Sb^V agents act only as prodrugs, and that the reduction of Sb^V to Sb^{III} species *in vivo* is responsible for the actual mechanism of action was first proposed by Goodwin and Page after they found that a reduction of Sb^V occurred in the host organism.[69,70] This hypothesis was supported by the observation that the Sb^{III} contents in serum and urine samples of patients were 5–25% and 50%, respectively, after treatment with Sb^V drugs,[71,72] and significant amounts of Sb^{III} were detected in tissues and serum in *L. garnhami*-infected hamsters treated with meglumine antimonate.[73] The reduction/activation hypothesis was further supported by the fact that IC_{50} of antimony chemotherapy is 60- to over 600-fold higher for promastigotes (the parasites culture at pH 7.4 and 298 K to resume the extracellular stage) than for amastigotes (the parasites culture at pH 5.0 and 310 K to resume the intracellular form) despite the intracellular antimony concentrations being similar at the respective IC_{50}.[74]

Trivalent antimonials were generally more toxic than the pentavalent antimonials for both promastigotes and intramacrophage amastigotes.[75–78] It has been demonstrated that the extracellular amastigotes of *L. infantum* were most susceptible to Sb^{III}.[78] Macrophages can also retain both Sb^{III} and Sb^{V} from the medium up to 3 days during a 4 h exposure, and metabolic conversion of Sb^{V} to Sb^{III} was found inside the macrophage.[79] In addition, axenic amastigotes of *L. infantum* were poorly susceptible to meglumine antimonate (Sb^{V}), in contrast to those growing inside human leukemia monocyte cell lines (THP-1 cells).[79] Consistent with an observation that toxicity of Sb^{V} towards the parasitic protozoa indeed depends on the level of Sb^{III}, a strong cross-resistance to meglumine antimonate was noticed in Sb^{III}-resistant axenic *L. infantum* amastigotes growing in THP-1 cells.[79] These results again suggest that fairly toxic Sb^{V} is converted *in vivo* into highly active Sb^{III} compounds and the Sb^{V} reduction could take place at the host cells. However, the intracellular reduction of pentavalent antimony inside parasites was also possible since intracellular Sb^{V} reduction was apparent in amastigotes at a stage-specific manner in *L. donovani*.[80]

Recently, an *in vitro* experiment showed that up to 30% of the Sb^{V} was reduced to Sb^{III} in amastigotes (310 K), while little reduction of Sb^{V} to Sb^{III} was observed in promastigotes (299 K) after 12 h incubation.[80] In spite of similar standard redox potentials between trypanothione ($T(SH)_2$) and the tripeptide GSH (γ-L-Glu-L-Cys-Gly, Figure 23.3) ($E_0 = -0.242$ and -0.230 V, respectively),[81] trypanothione can reduce sodium stibogluconate (Sb^{V}) within several hours, in contrast to its analogue, GSH (days). The reduction process is both pH- and temperature-dependent and proceeds more effectively at higher temperature and acidic pH, indicating that rapid reduction of Sb^{V} by $T(SH)_2$ is favored in amastigotes.[82–84] This suggests that trypanothione may play an important role in antimonial activation. $T(SH)_2$ could therefore be important in the activation of Sb^{V} inside parasites, although the presence of an enzyme to reduce Sb^{V} cannot be excluded.

23.5.3 Interaction of antimony(III) compounds with thiolate molecules

Sb^{III} is classified as a borderline metal ion and has a high affinity towards nitrogen- and thiol-containing ligands.[85,86] The anti-leishmanial mechanisms of Sb^{III} drugs are probably related to their interaction with those thiol-containing polypeptides, proteins and enzymes. One of the possible targets specific to the parasitic protozoa is the trypanothione reductase (TR), the analogoue of glutathione reductase (GR) in mammalian systems. Although TR shares structural and mechanistic similarity with its analogy GR, differences in the disulfide binding site between TR and GR draw the interest for selective inhibition.[87] Since trypanothione is responsible for the survival of the protozoa

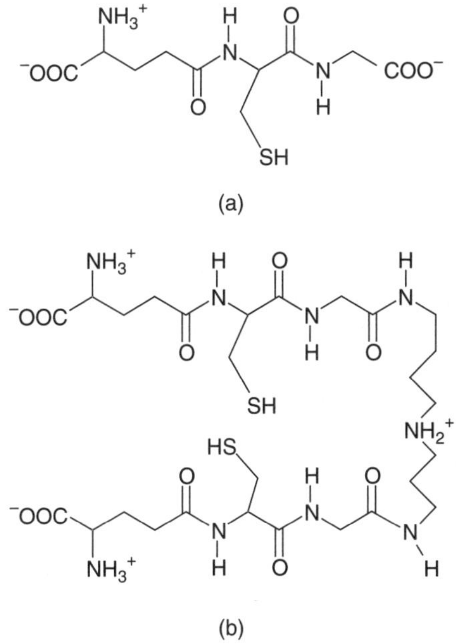

Figure 23.3 The structures of glutathione, GSH (γ-L-Glu-L-Cys-Gly) (a) and trypanothione, $T(SH)_2$ (b)

inside the macrophage of the host, alternation of the metabolism of trypanothione by the inhibition of TR resulted in a disruption of the reducing environment of the parasite, which is essential for the survival of parasites.

The active site of TR consists of a disulfide-binding site. In the crystal structure of TR from *Trypanosoma cruzi*, the two components of the GSH in trypanothione bind to the active site of TR through 25 amino acid residues.[88] The mutually exclusive substrate specificity between the trypanothione–TR system and its human analogy GSSG–GR system is revealed which is probably related to the structure differences in the orientation of the domains as well as the steric and electrostatic factors of the two enzymes.

Reversible inhibition of TR by trivalent antimonials has been observed in a time-dependent manner with a pseudo-first-order rate related to antimony concentration and the inhibition occurs within the disulfide binding site, provided that the redox-active disulfide binding site is first reduced by NADPH (reduced nicotinamide adenine dinucleotide phosphate).[87] This, in turn, suggested that the thiol is likely to react with the antimonial first. The interaction of the free antimonial with thiol generates a transient monothio-antimony adduct followed by a rapid intramolecular rearrangement and binding of the charge-transfer thiol from the enzyme to form a stable antimony–thiolate complex.[87]

Like As^{III} and Bi^{III}, Sb^{III} has been demonstrated to be able to form a stable complex with the tripeptide GSH with a stoichiometry $Sb(GS)_3$. The rate of uptake of the complex into red blood cells and complexation with intracellular GSH is rapid, a matter of minutes.[89] Strong binding of Sb^{III} to trypanothione ($T(SH)_2$) has been chemically characterized by both electrospray ionization mass spectrometry (ESI-MS) and NMR spectroscopy (pM = 24.5 and 22.1 for $Sb(TS_2)$ and $Sb(GS)_3$, respectively, where pM = −log[M], and [M] is the concentration of unchelated unhydrolyzed metal ion at equilibrium in a pH 7.4 solution of 1 μM metal and 10 μM ligand). In contrast to the GSH complex, each Sb^{III} coordinates to only one $T(SH)_2$ with two sulfurs provided by Cys residues and probably an oxygen from a water molecule (Figure 23.4).[90] $T(SH)_2$ has been known to be the most abundant low molecular mass thiol-containing ligand (>80%) in *Leishmania* parasites.[91,92] Together with TR, it is crucial for the growth and survival of the parasites.[93,94] The resistance to antimonials in *Leishmania* is related to the overproduction of $T(SH)_2$.[95] For this reason, trypanothione metabolism in parasites is considered as a valid target for drug design.

Recently, the crystal structure of Sb^{III}-bound ArsA ATPase complex has shown that there is an Sb^{III} cluster, and that Sb^{III} acts as a soft metal and coordinates to three donating atoms, with two of them from the protein residues and one from a non-protein ligand (Cl^-) (Figure 23.5).[96] The coordination of As^{III} to the protein at the metal cluster is essentially identical.[97] Among the coordinated ligands, two are from residues of domains A1 and A2 and a third is from a non-protein ligand chloride (Cl^-), and the binding of Sb^{III}/As^{III} brings together the two domains of the enzyme, triggering the ATP-dependent activation. Both $Sb^{III}(TS_2)$ and Sb^{III}-bound ArsA ATPase complexes can probably be regarded as intermediates, and the bound water or chloride (Cl^-) can readily be replaced by a nitrogen (e.g. histidine) or thiolate sulfur (e.g. cysteine) from another molecule (e.g. protein or enzyme). Indeed,

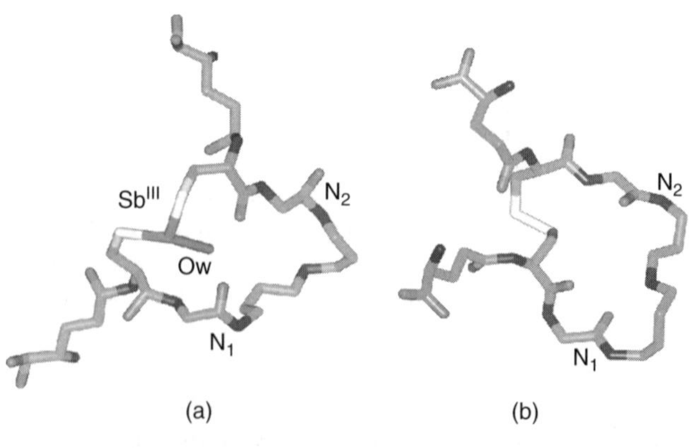

Figure 23.4 Structures of $Sb(T(S)_2)$ complex together with the oxidized form of trypanothione (adapted with permission from Ref. 90 copyright Wiley-VCH) (see Plate 4)

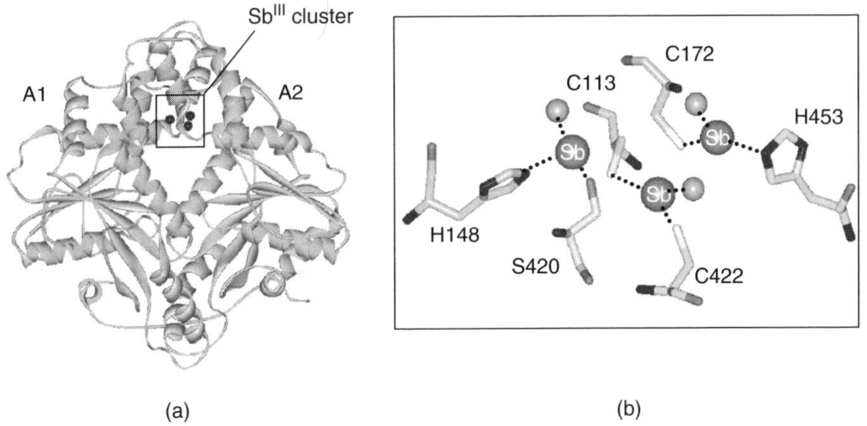

Figure 23.5 Overall structure of the ArsA ATPase (a) and coordination of Sb^{III} at the metal binding site (b) (coordinates from PDBID: 1F48). Residues His^{148}, Cys^{113} and Cys^{172} are from domain A1 while Ser^{420}, Cys^{422} and His^{453} are from domain A2. The purple balls represent the bound chlorides (Cl^-) (see Plate 5)

a ternary complex can readily form between Sb^{III}-trypanothione and other monothiol ligands such as GSH and cysteine.[83] Although they are thermodynamically stable, both the $Sb^{III}(TS_2)$ and the ternary complexes are kinetically labile, and the free and bound forms of either trypanothione or cysteins/GSH exchange on the ^1H NMR time scale. This exchange is also pH-dependent, and is relatively slow at low pH (~4), but considerably faster at pH 7. Such a facile exchange may be important for the transport and delivery of Sb^{III} *in vivo*.

23.5.4 Inhibition of topoisomerase I from *Leishmania donovani* and formation of DNA-protein complexes by Sb^V

Sodium stibogluconate has been shown to specifically inhibit type I DNA topoisomerase from *L. donovani* through the inhibition of the unwinding and cleavage of the supercoiled plasmid pBR322, and to stabilize topoisomerase and DNA covalent complexes but not calf-thymus topoisomerase I and *Escherichia coli* DNA gyrase.[98] The *in vivo* sensitivity and resistance of *Leishmania* is correlated with the effect of such a complex.[99] A similar mechanism was also found in the anti-tumor activity of the corresponding chemotherapies.[100] This finding indicated that topoisomerase is one of the targets for Sb^V.

23.5.5 Antimony-mediated DNA fragmentation, apoptosis and Ca^{2+}-related programmed cell death in *Leishmania*

While arsenite-mediated apoptosis in mammalian cells has been extensively studied,[101–105] little is known about the mechanisms of apoptosis mediated by antimonite.[106,107] The DNA-fragmentation-associated cell death in *L. infantum* amastigotes is induced by PAT at low concentrations (10 μg ml^{-1}), which is highly correlated with the toxicity of Sb^{III} species against axenic amastigotes.[108] Induced DNA fragmentation indicates the appearance of late events of programmed cell death (apoptosis).

Further studies on the mechanisms of the apoptotic cell death pathway in *Leishmania* indicated that intracellular Ca^{2+} plays a central role in intracellular parasite clearance,[109] a phenomenon previously documented in the oxidative-stress-induced apoptosis-like death in *L. donovani* promastigotes.[110]

23.5.6 Interference of the exclusive purine scavenge pathway of *Leishmania*

Unlike mammalian cells, all protozoan parasites studied to date are unable to synthesize the purine ring *de novo*.[111,112] Instead, a unique purine salvage strategy is evolved by the parasites, which is critical for its survival. The first step of the salvage process of *Leishmania donovani* is the translocation of preformed purines, mediated by two high-affinity nucleoside transporters, LdNT1 and LdNT2, located on the parasites' surface membrane.[113,114]

The uptake of a trivalent arsenical drug (an analogue of antimony) is regulated by a P2 nucleoside transporter in *Trypanosoma brucei* and resistance to the melarsen drugs (organoarsenic drugs for the treatment of sleeping sickness – melarsen oxide, *p*-(4,6-diamino-*s*-triazin-2-yl)aminophenylarsenoxide) is related to alterations in P2 transporter.[115] As other anti-leishmanial agents such as the purine analogues allopurinol riboside were also manipulated by this specific transporter,[116] a selective interference with this novel salvage pathway, which is the fundamental biochemical difference between parasite and host, may be one of the mechanisms of action of Sb^V drugs. Direct interactions of Sb^V with nucleosides under acidic conditions (pH ~5) have been reported[117] and the uptake of the drug may be mediated by nucleosides.

Guanosine diphosphate-mannose is the key monosaccharide for protein and lipid glycoconjugation in mannose-activated pathway. A large amount of unusual mannose-rich cell surface associated and secreted glycoconjugates is produced in *Leishmania*, including lipophosphoglycan (LPG), proteophosphoglycans (PPGs) and glucoinositolphospholipids (GIPLs), depending on the availability of GDP-mannose. GDP-mannose pyrophosphorylase (GDPMP)

Table 23.3 Biological features and therapeutic aspects of As, Sb and Bi[15–17,54]

	Arsenic	Antimony	Bismuth
Treatment for	Sleeping sickness, Leukemia	Parasite infection (*Leishmania*)	Ulcer, *Helicobacter pylori* infection
Human absorption	As^{V} (60%), As^{III} (~80–100%)	Sb^{V} (5%), Sb^{III} (5%)	RBC (0.5%)
Metabolism	$As^{V} \rightarrow As^{III} \rightarrow$ MMA	$Sb^{V} \rightarrow Sb^{III} \rightarrow$ thiol conjugate	$Bi^{III} \rightarrow$ thiol conjugate
Excretion	Urine	80–95% urine with within 6 h	95% via urine
Biological reductant	Arsenate reductase (ArsC)	$T(SH)_2$? Arsc?	/
Extent of reduction	> 60%	~5–10%	/
Cellular uptake	As^{V}, phosphate inorganic transporter (PiT) As^{III}, GlpF	Sb^{V}? Sb^{III}, GlpF	Transferrin, citrate
Cellular excretion	As^{III}, ArsAB ATPase pump, MT?	Sb^{III}, ArsAB ATPase pump, MT?	MT?
Thiol affinity	As^{V}, low As^{III}, medium	Sb^{V}, low Sb^{III}, medium high	Bi^{III}, High
Toxicity	As^{V} (high) As^{III} (high)	Sb^{V} (low) Sb^{III} (medium)	Bi^{III} (low)

is the central enzyme in the formation of GDP-mannose.[118] Disruption of the mannose-activated glycoconjugate synthesis by deletion of the corresponding GDPMP gene in *L. mexicana* resulted in the loss of the ability to infect macrophages.[119] The virulence character demonstrated that targeting the GDP-mannose biosynthesis pathway might be an effective way to control *Leishmania* infection. The possible mechanisms and resistance of antimony agents in *Leishmania* are summarized in Table 23.3 and Figure 23.6.

23.5.7 Modulation and increase of cellular signaling of the host macrophage as a means of intracellular *Leishmania* elimination

Beside their direct specific drug susceptibility on the protozoa, additional anti-leishmanial activity of antimony compounds is also related to the intracellular

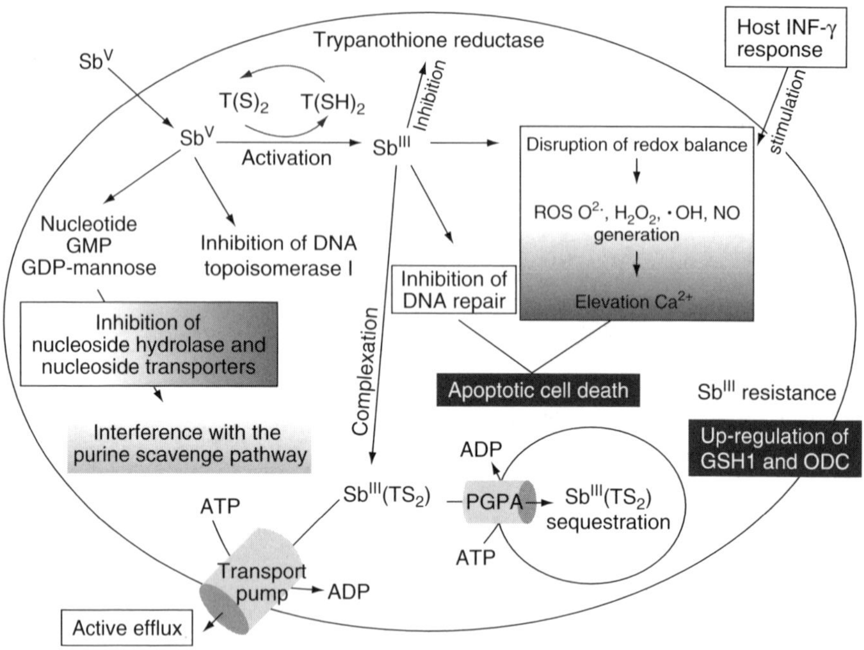

Figure 23.6 Possible mechanisms of action and resistance of Sb^V in *Leishmania*

clearance of the parasite which is medicated by interferon gamma (INF-γ) activity via the regulation of the chemotherapy response of host cytokines. INF-γ plays an important role in eliminating intracellular *Leishmania* and the impaired INF-γ signaling was detected in *Leishmania*-infected macrophage suggested that anti-leishmanial activity of Sb^V may derive from an increase of cell signaling.[120,121]

Massive generation of superoxide anion ($O_2^{-\bullet}$) and other reactive oxygen species (ROS) such as hydroxyl radicals ($OH^\bullet$), hydrogen peroxide (H_2O_2) and singlet oxygen ($^1O_2$) stimulated by particles or soluble mediators are known as respiratory burst. Respiratory burst activity plays an important role in the destruction of infected pathogens by phagocytes.[122] Activated macrophages kill intracellular parasites by generation of ROS and nitric oxide in *Leishmania* infection,[123–126] and the down-regulation of this immune defense response by *Leishmania* species is related to the resistance of the parasites to oxygen-dependent toxicity.[127,128] Treatment of Sb^V (Pentostam®) promotes the generation of ROS *in vivo*, an observation known as 'priming', when phagocytes' respiratory burst is enhanced by the treatment of low concentration of drugs.[129] An accelerated and enhanced overall efficiency is observed when INF-γ is synergistically injected with antimony.[130] Thus the effect of antimony on host cell reactivity may contribute to the mechanisms of action of antimony drugs.

23.6 Concluding Remarks

The variety of medicinal uses for antimony compounds suggests a diverse biological and medicinal relevance for the element, even though little research has been done on its possibilities in this respect. The biochemistry of antimony is poorly understood and the mechanism of action of the metal remains unclear, although it is commonly believed that the major targets for antimony appear to be proteins and enzymes. A systematic investigation of chemistry and biochemistry under physiologically relevant conditions will improve our understanding of its mechanism of action, and subsequently allow for a better drug design. Recent work on the uptake of antimony by microorganisms with the aid of molecular biology, structural biology and proteomics have established key directions for future research.[131–133] The evaluation of new antimony complexes for other purposes (e.g. anti-cancer and anti-virus) should also be emphasized.

Abbreviations

Amastigote, the parasites culture at pH 5.0 and 310 K to resume the intracellular form; citrate, tetra-anionic citrate ($C_6H_4O_7^{4-}$), similarly, Hcit, tri-anionic citrate ($^-OOCCH_2CCOO^-(OH)CH_2COO^-$); GR, glutathione reductase; GSH, glutathione (γ-L-Glu-L-Cys-Gly); GSSG, oxidized form of glutathione; Melarsen oxide, *p*-(4,6-diamino-*s*-triazin-2-yl)aminophenylarsenoxide; NMG, *N*-methyl-D-glucamine; PAT, potassium antimony tartrate; promastigotes, the parasites culture at pH 7.4 and 298 K to resume the extracellular stage; PTPase, protein tyrosine phosphatase; ROS, reactive oxygen species; $T(SH)_2$, trypanothione; TR, trypanothione reductase.

Acknowledgement

We are grateful for generous support from the Hong Kong Research Grants Council (RGC, HKU7110/02P), NSFC/RGC (N_HKU720/02), the Area of Excellence Scheme of the University Grants Committee (Hong Kong) and the University of Hong Kong. We acknowledge the use of the Protein Data Bank (http://www.rcsb.org/pdb/) and the Cambridge Structural Database (Cambridge, UK) for access to some of the structural data shown in this chapter.

References

1. M. Grayson, in *Kirk-Othmer Encyclopedia of Chemical Technology*, 3rd edn, John Wiley, New York, 1983.
2. R.I. MaCallum, *Antimony in Medical History*, Pentland Press, Edinburgh, 1999.

3. J. Emsley, *The Element*, 2nd edn, Clarendon Press, Oxford, 1991.
4. D.R. Lide, *Handbook of Chemistry*, 73rd edn, CRC Press, Boca Raton, FL, 1992–1993.
5. W. Clegg, M.R.J. Elsegood and V. Graham, *J. Chem. Soc., Dalton Trans.* 997–998 (1993).
6. D.W. Hartlry, G. Smith, D.S. Sagatys and C.H.L. Kennard, *J. Chem. Soc., Dalton Trans.* 2735–2739 (1991).
7. G. Smith, D.S. Sagatys, R.C. Bott *et al.*, *Polyhedron* **11**, 631–634 (1992).
8. G. Smith, D.S. Sagatys, R.C. Bott and D.E. Lynch, *Polyhedron* **12**, 1491–1497 (1993).
9. M.E. Gress and R.A. Jacobson, *Inorg. Chim. Acta* **8**, 209–217 (1974).
10. J. Wegener, K. Kirschbaum and D.M. Giolando, *J. Chem. Soc., Dalton Trans.* 1213–1218 (1994).
11. I.C. Tornieporth-Oetting, T.M. Klapotke and T.S. Cameron, *J. Chem. Soc., Dalton Trans.* 537–543 (1992).
12. H. Preiss, *Z. Anorg. Allg. Chem.* **380**, 71–75 (1971).
13. T. Barbour, W.J. Belcher, P.J. Brothers *et al.*, *Inorg. Chem.* **31**, 746–754 (1992).
14. W. Levason and G. Reid, in *Comprehensive Coordination Chemistry II*, Volume 3 (G.F.R. Parkin), Elsevier, Oxford, UK, 2004.
15. J. Reglinski, Environmental and medicinal chemistry of arsenic, antimony, and bismuth, in *Chemistry of Arsenic, Antimony and Bismuth*, N.C. Norman (Ed.), Blavkie Academic & Professional, London, 1998.
16. H. Sun, L. Zhang and K.Y. Szeto, *Metal Ions Biol. Syst.* **41**, 333–378 (2004).
17. J. Zhu, Z. Chen, V. Lallemand-Breitenbach and H. de Thé, *Nat. Rev. Cancer* **2**, 705–714 (2002).
18. H. Schmidt and F.M. Peter, *Advances in the Therapeutics of Antimony*, Thieme Verlag, Leipzig, 1938.
19. B.L. Herwaldt, *Lancet* **354**, 1191–1199 (1999).
20. J.D. Berman, *Rev. Infect. Dis.* **10**, 560–586 (1988).
21. (a) G. Vianna, *Arch. Bras. Méd.* **2**, 426 (1912). (b) G. Vianna, *Arch. Bras. Méd.* **3**, 1171 (1913).
22. (a) J.B. Christopherson, *Lancet* **II**, 325–327 (1918). (b) J.E.R. McDonagh, *Lancet* **II**, 371 (1918). (c) J.B. Christopherson, *Brit. Med. J.* **2**, 652 (1918).
23. T.D. Luckey and B. Venugopal, *Metal Toxicity in Mammals*, Volume 2, Plenum Press, New York, 1997.
24. P. Aggarwal, R. Handa, S. Singh and J.P. Wali, *Ind. J. Pediatr.* **66**, 63–71 (1999).
25. P.L. Olliaro and A.D.M. Bryceson, *Parasitol. Today* **9**, 323–328 (1993).
26. R. Kirk and M.H. Stai, *Ann. Trop. Med. Parasitol.* **41**, 14–21 (1947).
27. I.M. Rollo, *The Pharmacological Basis of Therapeutics*, 3rd edn, Macmillan, New York, 1965.
28. J.D. Berman and M. Grogl, *Exp. Parasitol.* **67**, 96–103 (1988).
29. W.L. Roberts, W.J. McMurray and P.M. Rainey, *Antimicrob. Agents Chemother.* **42**, 1076–1082 (1998).
30. J. Duffin and B.G. Campling, *J. Hist. Med.* **57**, 61–78 (2002).
31. E.R.T. Tiekink, *Crit. Rev. Onol. Hematol.* **42**, 217–224 (2002).
32. T.L. Yi, M.K. Pathak, D.J. Lindner *et al.*, *J. Immunol.* **169**, 5978–5985 (2002).
33. M.K. Pathak and T. Yi, *J. Immunol.* **167**, 3391–3397 (2001).
34. M.E. O'Dwyer and B.J. Druker, *Curr. Opin. Oncol.* **12**, 594–597 (2000).

35. C.A. Stein, R.V. LaRocca, R. Thomas *et al.*, *J. Clin. Oncol.* **7**, 499–508 (1989).
36. C.A. Stein, *Cancer Res.* **53**, 2239–2248 (1993).
37. D. Church, Y. Zhang, R. Rago and G. Wilding, *Cancer Chemother. Pharmacol.* **43**, 198–204 (1999).
38. Y.L. Zhang, Y.F. Keng, Y. Zhao *et al.*, *J. Biol. Chem.* **273**, 12281–12287 (1998).
39. C.T. Yeh, D.R. Hwang, H.Y. Lai and J.T.A. Hsu, *Biochem. Biophys. Res. Commun.* **310**, 537–541 (2003).
40. W. Rozenbaum, D. Dormont, B. Spire *et al.*, *Lancet* **325**, 450–451 (1985).
41. L.H. Craig, S.W. Mark and F.S. Raymond, *J. Med. Chem.* **33**, 2767–2772 (1990).
42. J. Fischer, L. Ricard and R. Wiess, *J. Am. Chem. Soc.* **98**, 3050–3052 (1976).
43. M. Michelon, G. Herve and M. Leyrie, *J. Inorg. Nucl. Chem.* **42**, 1583–1586 (1980).
44. J.D. Berman, J.D Chulay, L.D. Hendricks and C.N. Oster, *Am. J. Trop. Med. Hyg.* **31**, 459–465 (1982).
45. F. Faraut-Gambarelli, R. Piarroux, M. Deniau *et al.*, *Antimicrob. Agents Chemother.* **41**, 827–830 (1997).
46. M. Grogl, T.N. Thomason and E.D. Franke, *Am. J. Trop. Med. Hyg.* **47**, 117–126 (1992).
47. The leishmaniasis. Report of a WHO expert committee. *WHO Tech. Rep. Ser.* **701**, 1–140 (1984).
48. B.L. Herwaldt and J.D. Berman, *Am. J. Trop. Med. Hyg.* **46**, 296–306 (1992).
49. K. Beneze, *Handbook on Metals in Clinical and Analytical Chemistry*, Marcel Dekker, New York, 1994.
50. K.A. Winship, *Adv. Drug React. Acute Pois. Rev.* **6**, 67–90 (1987).
51. B.L. Herwaldt, E.T. Kaye, T.J. Lepore *et al.*, *Infect. Dis.* **165**, 968–971 (1992).
52. K. Kuroda, G. Endo, A. Okamoto *et al.*, *Mutat. Res.* **264**, 163–170 (1991).
53. T. Gebel, S. Christensen and H. Dunkelberg, *Anticancer Res.* **17**, 2603–2607 (1997).
54. T. Gebel, *Chem. Biol. Interact.* **107**, 131–144 (1997).
55. S. Takahashi, H. Sato, Y. Kubota *et al.*, *Toxicology* **180**, 249–256 (2002).
56. M.A. Tirmenstein, P.I. Plews, C.V. Walker *et al.*, *Toxicol. Appl. Pharmacol.* **130**, 41–47 (1995).
57. M.A. Tirmenstein, P.I. Mathias, J.E. Snawder *et al.*, *Toxicology* **119**, 203–211 (1997).
58. J. Delgado, J. Macias, J.A. Pineda *et al.*, *Am. J. Trop. Med. Hyg.* **61**, 766–769 (1999).
59. V. Pintado and R. Lopez-Velez, *Clin. Microbio. Infect.* **7**, 291–300 (2001).
60. J.D. Chulay, L. Fleckenstein and D.H. Smith, *Trans. R. Soc. Trop. Med. Hyg.* **82**, 69–72 (1988).
61. P.H. Rees, M.I. Keating, P.A. Kager and W.T. Hockmeyer, *Lancet* **2**, 226–229 (1980).
62. K.P. Chang and R.S. Bray, *Human Parasitic Disease*, Volume 1, *Leishmaniasis*, Elsevier, Netherlands, 1985.
63. M.P. Dieter, C.W. Jameson, M.R. Elwell *et al.*, *J. Toxicol. Environ. Health* **34**, 51–82 (1991).
64. S.A. Felicetti, R.G. Thomas and R.O. McClellan, *Am. Ind. Hyg. Assoc. J.* **35**, 292–300 (1974).
65. K.C. Carter, A.B. Mullen, S. Sundar and R.T. Kenney, *Antimicrob. Agents Chemother.* **45**, 3555–3559 (2001).
66. J. Nieto, J. Alvar, A.B. Mullen *et al.*, *Antimicrob. Agents Chemother.* **47**, 2781–2787 (2003).

67. J.D. Berman, D. Waddell and B.D. Hanson, *Antimicrob. Agents Chemother.* **27**, 916–920 (1985).
68. J.D. Berman, J.V. Gallalee and J.M. Best, *Biochem. Pharmacol.* **36**, 197–201 (1987).
69. L.G. Goodwin and J.E. Page, *Biochem. J.* **37**, 198–209 (1943).
70. L.C. Goodwin, *Trans. R. Soc. Trop. Med. Hyg.* **89**, 339–341 (1995).
71. J.L. Burguera, M. Burguera, Y. Petit de Pena *et al.*, *Trace Elem. Med.* **10**, 66–70 (1993).
72. Y. Petit de Pena, M. Gallignani, M. Burguera *et al.*, *J. Braz. Chem. Soc.* **1**, 72–75 (1990).
73. A. Lugo de Yarbuh, N. Anez, Y. Petit de Pena *et al.*, *Ann. Trop. Med. Parastiol.* **88**, 37–41 (1994).
74. W.L. Roberts, J.D. Berman and P.M. Rainey, *Antimicrob. Agents Chemother.* **39**, 1234–1239 (1995).
75. J.C. Mottram and G.H. Coombs, *Exp. Parasitol.* **59**, 151–160 (1985).
76. P. Borst and M. Ouellette, *Annu. Rev. Microbiol.* **49**, 427–460 (1995).
77. S. Detke, K. Katakura and K.P. Chang, *Exp. Cell Res.* **180**, 161–170 (1989).
78. D. Sereno and J.L. Lemesre, *Antimicrob. Agents Chemother.* **41**, 972–976 (1997).
79. D. Sereno, M. Cavaleyra, K. Zemzoumi *et al.*, *Antimicrob. Agents Chemother.* **42**, 3097–3102 (1998).
80. P. Shaked-Mishan, N. Ulrich, M. Ephros and D. Zilberstein, *J. Biol. Chem.* **276**, 3971–3976 (2001).
81. A.H. Fairlamb and G.B. Henderson, Metabolism of trypanothione and glutathionylspermidine in trypanosomatids, in *Host–Parasite Cellular and Molecular Interactions in Protozoal Infections*, K.-P. Chang and D. Snary (Eds), Springer-Verlag/NATO AST Series (Volume H11), Berlin, 1987.
82. S.C. Yan, I.L. Wong, L.M.C. Chow and H. Sun, *Chem. Commun.* 266–267 (2003).
83. S.C. Yan, F. Li, K.Y. Ding and H. Sun, *J. Biol. Inorg. Chem.* **8**, 689–697 (2003).
84. C.S. Ferreira, P.S. Martins, C. Demicheli *et al.*, *Biometals* **16**, 441–446 (2003).
85. K.L. Stemmer, *Pharmacol. Ther. A.* **1**, 157–160 (1976).
86. M.A. Basinger and M.M. Jones, *Res. Commun. Chem. Pathol. Pharmacol.* **32**, 355–363 (1981).
87. M.L. Cunningham and A.H. Fairlamb, *Eur. J. Biochem.* **230**, 460–468 (1995).
88. C.S. Bond, Y. Zhang, M. Berriman *et al.*, *Structure* **7**, 81–89 (1999).
89. H. Sun, S.C. Yan and W.S. Cheng, *Eur. J. Biochem.* **267**, 5450–5457 (2000).
90. S.C. Yan, K. Ding, L. Zhang and H. Sun, *Angew. Chem. Int. Ed.* **39**, 4260–4262 (2000).
91. A.H. Fairlamb and A. Cerami, *Annu. Rev. Microbiol.* **46**, 695–729 (1992).
92. A.H. Fairlamb, *Adv. Exp. Med. Biol.* **250**, 667–674 (1988).
93. C. Dumas, M. Ouellette, J. Tovar *et al.*, *EMBO J.* **16**, 2590–2598 (1997).
94. J. Tovar, M.L. Cunningham, A.C. Smith *et al.*, *Proc. Natl. Acad. Sci. USA* **95**, 5311–5316 (1998).
95. R. Mukhopadhyay, S. Dey, N. Xu *et al.*, *Proc. Natl. Acad. Sci. USA* **93**, 10383–10387 (1996).
96. T. Zhou, S. Radaev, B.P. Rosen and D.L. Gatti, *EMBO J.* **19**, 4838–4845 (2000).
97. T. Zhou, S. Radaev, B.P. Rosen and D.L. Gatti, *J. Biol. Chem.* **276**, 30414–30422 (2001).
98. A.K. Chakraborty and H.K. Majumder, *Biochem. Biophys. Res. Commun.* **152**, 605–611 (1988).
99. A. Lucumi, S. Robledo, V. Gama and N.G. Saravia, *Antimicrob. Agents Chemother.* **42**, 1990–1995 (1998).

100. M. Palumbo, M. Mabilia and A. Pozzan, *J. Mol. Recognit.* **7**, 227–231 (1994).
101. T.S. Wang, C.F. Kuo, K.Y. Jan and H. Huang, *J. Cell. Physiol.* **169**, 256–268 (1996).
102. D.C. Ma, Y.H. Sun, K.Z. Chang *et al.*, *Eur. J. Haematol.* **61**, 27–35 (1998).
103. Y.C. Chen, S.Y. Lin-Shiau and J.K. Lin, *J. Cell. Physiol.* **177**, 324–333 (1998).
104. R.W. Watson, H.P. Redmond, J.H. Wang and D. Bouchler-Hayes, *J. Leukoc. Biol.* **60**, 625–632 (1996).
105. N. Larochette, D. Decaudin, E. Jacobot *et al.*, *Exp. Cell Res.* **249**, 413–421 (1999).
106. H. Huang, S.C. Shu, J.H. Shih *et al.*, *Toxicology* **129**, 113–123 (1998).
107. V. Lecureur, D. Lagadic-Gossmann and O. Fardel, *Int. J. Oncol.* **20**, 1071–1076 (2002).
108. D. Sereno, P. Holzmuller, I. Mangot *et al.*, *Antimicrob. Agents Chemother.* **45**, 2064–2069 (2001).
109. G. Sudhandiran and C. Shaha, *J. Biol. Chem.* **278**, 25120–25132 (2003).
110. S.B. Mukherjee, M. Das, G. Sudhandiran and C. Shaha, *J. Biol. Chem.* **277**, 24717–24727 (2002).
111. R.L. Berens, E.C. Krug and J.J. Marr, Purine and pyrimidine metabolism, in *Biochemistry and Molecular Biology of Parasites*, J.J. Marr and M. Müller (Eds), Academic, New York, 1995.
112. D.J. Hammond and W.E. Gutteridge, *Mol. Biochem. Parasitol.* **13**, 243–261 (1984).
113. G. Vasudevan, N.S. Carter, M.E. Drew *et al.*, *Proc. Natl. Acad. Sci. USA* **95**, 9873–9878 (1998).
114. N.S. Carter, M.E. Drew, M.A. Sanchez *et al.*, *J. Biol. Chem.* **275**, 20935–20941 (2000).
115. N.S. Carter and A.H. Fairlamb, *Nature* **361**, 173–176 (1993).
116. J.J. Marr and R.L. Berens, *Mol. Biochem. Parasitol.* **7**, 339–356 (1983).
117. C. Demicheli, F. Frézard, M. Lecouvey and A. Garnier-Suillerot (2002) *Biochim. Biophys. Acta* **1570**, 192–198 (2002).
118. H. Ferrze and M. Aebi, *Biochim. Biophys. Acta* **1455**, 167–178 (1999).
119. A. Garami and T. Ilg, *EMBO J.* **20**, 3657–3666 (2001).
120. M. Kemp, J.A. Kurtzhals, A. Kaharazmi and T.G. Theander, *Immunol. Cell Biol.* **71**, 583–587 (1993).
121. J. Blanchette, N. Racette, R. Faure *et al.*, *Eur. J. Immunol.* **29**, 3737–3744 (1999).
122. B.M. Babior, *N. Engl. J. Med.* **298**, 721–725 (1978).
123. P. Scott, *Exp. Parasitol.* **68**, 369–372 (1989).
124. F.Y. Liew, S. Millott, C. Parkinson *et al.*, *J. Immunol.* **144**, 4794–4797 (1990).
125. I. Vouldoukis, V. Riveros-Moreno, B. Dugas *et al.*, *Proc. Natl. Acad. Sci. USA* **92**, 7804–7808 (1995).
126. H.W. Murray and C.F. Nathan, *J. Exp. Med.* **189**, 741–746 (1999).
127. M. Ho, D.K. Koech, D.W. Iha and A.D. Bryceson, *Clin. Exp. Immunol.* **51**, 207–214 (1983).
128. E.M. Carvalho, O. Bacellar, A. Barral *et al.*, *J. Clin. Investig.* **83**, 860–864 (1989).
129. M.B. Hallett and D. Lloyds, *Immunol. Today* **16**, 264–268 (1995).
130. H.W. Murray, *Int. J. Infect. Dis.* **4**, 158 (2000).
131. C. Guimond, N. Trudel, C. Brochu *et al.*, *Nucleic Acids Res.* **31**, 5886–5896 (2003).
132. J. Drummelsmith, V. Brochu, I. Girard *et al.*, *Mol. Cell. Proteomics* **2**, 146–155 (2003).
133. Y.L. Meng, Z. Liu and B.P. Rosen, *J. Biol. Chem.* **279**, 18334–18341 (2004).

24 $_{75}$Re Therapeutic Rhenium Radiopharmaceuticals

Jonathan R. Dilworth and Paul S. Donnelly

24.1 Introduction

Radiopharmaceutials are radiolabelled molecules that are used to achieve non-invasive imaging or to deliver a therapeutic dose of sterilizing radiation to specific disease sites. Following systemic administration, these molecular agents localize in specific areas as a consequence of a number of factors. In the case of the diagnostic imaging agents the goal is to use a radioisotope that emits radiation that can be readily detected as it leaves the body but has little effect on surrounding tissue. This is achieved with γ- or positron-emitting isotopes, the most commonly used isotope being ^{99m}Tc, which is used in over 85% of all diagnostic scans currently performed in hospital nuclear medicine departments.[1]

The treatment of cancers with conventional beam radiation is well established and plays an important role in modern oncology but it is not particularly appropriate for the treatment of metastatic sites or widely disseminated disease. Treatment of secondary tumours with a systemically administered radiopharmaceutical could be a useful supplement to conventional beam radiation treatments and surgery. Recent developments in targeting vectors such as receptor

Metallotherapeutic Drugs and Metal-Based Diagnostic Agents: The Use of Metals in Medicine
Edited by Gielen and Tiekink

avid molecules, monoclonal antibodies, fragment antibodies and site-specific peptides have led to the possibility of selective *in situ* delivery of radioisotopes that can administer a sterilizing dose of ionizing radiation to cancer sites. This is best achieved with α- and β-emitting (and Auger-electron emitting) isotopes. If the nuclide also emits γ-photons, they can be used to simultaneously image the distribution of the therapeutic agent, although such emissions should be of low abundance so they do not increase the dose to non-target tissue.[2] The effects of ionizing radiation on cells stem from modification of cellular DNA and the prime target is therefore the cell nucleus. The eventual death of the cell via apoptosis is a result of a series of complex steps which follow the original interactions of the radiation with the cellular environment. Chemical reactions initiated by the radiation result in the formation of reactive species such as radicals which can alter certain cell components.[3] Since the principal target is the cell nucleus, very low energy β-emitters and Auger-emitters, which have a short path length in tissue, must be delivered directly into the nucleus. When higher energy β-emitters which have a longer range in tissue are used, it is not necessary to deliver the radionuclide to the nucleus but the energy of the β-emissions dictates the size of the tumour that can be treated. It is also essential to minimize irradiation of non-target tissue although 'cross-fire' can be of use in non-uniform large tumour masses.[4]

Ideally, therapeutic radiopharmaceuticals should locate specifically in the target area whilst causing minimum damage to normal tissues. If specific targeting molecules are used, it is essential that the radionuclide is attached in a way that the bioconjugate is sufficiently stable *in vivo*, the binding ability of the receptor molecule is not compromised and that the therapeutic agent localizes in the target tissue within a time frame that is appropriate for the half-life of the radioisotope. The development of effective radiotherapeutic agents is a complex problem which cannot be achieved by just adding a radionuclide to a targeting vector in an indiscriminate fashion. The choice of radionuclide depends on a number of factors including the nature of the emissions, half-life, ease of production, availability, the nature of the decay products, the size of the tumour and cost. Rhenium has two β-emitting isotopes, ^{186}Re and ^{188}Re, that offer potential to be used as therapeutic radionuclides (Table 24.1). The superficial chemical similarity of rhenium to its group VII congener

Table 24.1 Radioactive isotopes of rhenium

Isotope	Half-life (h)	Max. β-energy (MeV)	Range in tissue (mm)	γ-Energy (MeV)	Method of production
^{186}Re	90	1.07 (71%)	5	137 (9%)	^{185}Re + n (reactor)
^{188}Re	17	2.1 (100%)	11	155 (15%)	^{188}W decay generator

technetium means that at least some of the research and development into technetium imaging agents can be extrapolated to rhenium-based therapeutic agents.

Rhenium-186 can be produced by neutron irradiation of natural rhenium in a nuclear reactor. Natural rhenium consists of 37.4% ^{185}Re and 62.6% ^{187}Re, which leads to a considerable impurity of ^{188}Re which can be removed by allowing it to decay but this also results in considerable loss in activity of the desired isotope. Higher specific activities can be obtained by using isotopically enriched ^{185}Re which minimizes both ^{188}Re contamination and the amount of residual non-radioactive 'carrier' rhenium. ^{186}Re has medium energy β-emissions (max. 1.08 MeV) with a range of about 5 mm in tissue which means it is suitable for use with small tumours. It has a long half-life of 3.7 days which makes it particularly useful in the labelling of large biomolecules that are not rapidly cleared from the blood stream.[1,4]

Rhenium-188 can be obtained as a solution of $^{188}ReO_4$ in high specific activity from a ^{188}W/^{188}Re generator similar in design to the commonly used ^{99m}Tc generator.[5] The generator design involves utilizing the ^{188}Re which forms as a decay product from ^{188}W, which has a half-life of 60 days. The generator has an alumina column to which ^{188}W-tungstate has been chemisorbed and the decay product ^{188}Re is then eluted with a saline solution as Na[$^{188}ReO_4$]. A single generator with 0.5 Ci of ^{188}W can last between 2 and 6 months and in this time it has the potential to provide sufficient radioactive rhenium for the treatment of several hundred patients. ^{188}Re is a high-energy β-emitter (max. 2.1 MeV) with a range in tissue of about 11 mm and a half-life of 17 h. This makes it potentially useful for the treatment of large tumour masses but its relatively short half-life dictates that it must be used in conjunction with agents that are rapidly cleared from the blood into the target tissue.[1,4]

Both nuclides also emit γ-photons of energy in a diagnostically useful range in relatively low abundance which means that simultaneous therapy and imaging is a possibility without too much irradiation of non-target tissue. In addition both isotopes give rise to secondary electron emissions that add to the radiation dose delivered within a zone much less than one cell diameter.

As therapeutic radionuclides, both ^{188}Re and ^{186}Re have advantages over other β-emitting nuclides such as ^{131}I and ^{90}Y. The isotope ^{131}I, a β-emitter, was the first isotope to be investigated for therapeutic applications. However, the success was limited due to relative ease of dehalogenation *in vivo* and subsequent accumulation in the thyroid and the abundance of high-energy γ-photons that can irradiate non-target tissue. Several other studies have focussed on the β-emitter, ^{90}Y; however, preliminary therapy trials in humans were plagued by accumulation of ^{90}Y in bone following premature dissociation of the metal ion from the chelate. In contrast, the thermodynamically favoured fate of rhenium *in vivo* is $[ReO_4]^-$, which is rapidly excreted via the kidneys, and does not accumulate in the thyroid or in bone.[1]

24.2 Synthetic Aspects of Rhenium Radiopharmaceuticals

This section deals with the coordination chemistry of rhenium complexes relevant to radiopharmacy and, following some general remarks, is organized according to the core involved. The emphasis of the latter sections is on complexes which have been tested biologically.

Rhenium is located in group VII of the transition elements immediately below technetium. Its position in one of the central groups confers the ability to exist in a range of oxidation states from +7 to −1, with well-defined complexes existing for all eight of these possibilities. The chemistry of the higher (5–7) oxidation states is dominated by strongly π-donating ligands such as oxo-, imido-(RN^{2-}) and nitride (e.g. $[Re^{VII}O_4]^-$, $[Re^{VII}(NBu^t)_4]^-$, $[Re^{VII}NCl_4]^-$) which are able to compensate for the electron deficiency of the metal centre. Such ligands exert a strong *trans* effect and a square pyramidal geometry is frequently found for Re(V) mono–oxo complexes with no ligand *trans* to the oxo group such as $[Re^VOCl_4]^-$. Otherwise hard donor ligands with nitrogen or oxygen donors are commonly found as co-ligands in the higher oxidation state compounds as in $[ReO_2(pyridine)_4]^+$. In view of the fact that $[ReO_4]^-$ is the obligatory starting point for the synthesis of rhenium-based radiopharmaceuticals, the higher oxidation states tend to dominate among the known pharmaceutically relevant complexes, and Re(V) oxo-complexes are the most prevalent.

As expected, in the lower oxidation states strong π-acceptor ligands such as CO and tertiary phosphines prevail, exemplified by complexes such as $[Re_2(CO)_{10}]$ and $[Re(CNBu^t)_6]^+$. Somewhat surprisingly such complexes are in fact accessible from $[ReO_4]^-$ by selection of the appropriate combination of reductant and co-ligand, a comparatively recent example being the synthesis and exploitation of complexes of the $[Re(CO)_3]^+$ core. Intermediate oxidation states as Re(IV), Re(III) and Re(II) are less well represented and the co-ligands are usually a combination of π-acceptors and σ-donors, e.g. $[ReCl_3(PMe_2Ph)_3]$.

The chemistry of rhenium is formally analogous to that of technetium, as might be expected from their location in the same group, and many of their complexes are isostructural. However, the fact that rhenium is a third-row element introduces some significant differences in its chemical behaviour which impacts strongly on its potential radiopharmaceutical use. The decrease in ionization energy to attain the +7 oxidation state encountered on moving from technetium to rhenium has its origin in the different penetration of 4*d* as opposed to 5*d* orbitals and results in higher stability for Re(VII) compared to Tc(VII). This is manifested in the values of the standard electrode potentials for the M(VI)/M(VII) couples (in acid solution M = Re, $E^o = +0.77$ V, M = Tc, $E^o = +0.70$ V). The higher thermodynamic stability of Re(VII) means that $[ReO_4]^-$ is harder to reduce than $[TcO_4]^-$. Coupled with the fact that rhenium complexes generally have higher kinetic stability due in part to the increased bond strengths descending the group, the production of radiopharmaceuticals from perrhenate is frequently not as straightforward as for technetium. If the ligands used are not reducing (e.g. thiolate or tertiary phosphine) then a reductant has to be employed. Tin(II) has frequently been used but does not always produce high yields; however, the use of oxalate in combination with stannous chloride results in quicker labelling with enhanced yields.[6] A recent paper has also reported good results for the synthesis of Re(V) complexes from perrhenate using acetylhydrazine as reductant.[7]

24.2.1 Complexes of bifunctional ligands

The use of β-emitting radionuclides for therapy imposes stringent requirements on the nature of the complex involved. The complex must to be resistant to oxidation or reduction which could modify the biodistribution, and have high kinetic stability to ensure that the highly toxic radioisotope is not lost en route to the target. Also, it has to have high specificity for the target, which is generally cancerous cells or a tumour. The so-called first-generation complexes depend on their intrinsic physical properties to achieve targeting and, with one or two exceptions, this does not achieve the level of selectivity required. The whole thrust of rhenium chemistry directed towards therapeutic applications therefore involves the use of bifunctional ligands which are presented schematically in Figure 24.1.

The chelate ligand is polydentate to confer kinetic stability and is conjugated to the biologically active molecule (BAM), usually via an amide bond. A diverse range of molecules can be used as the BAM and examples include monoclonal antibodies or peptides and both these types have been used with rhenium. Specific examples are discussed in the following sections.

24.3 The Radiopharmaceutical Chemistry of Rhenium

24.3.1 Complexes with oxo-cores

These are the most prevalent type of complex among potential radiopharmaceuticals using rhenium, and these invariably involve Re(V). The ReO^{3+} core

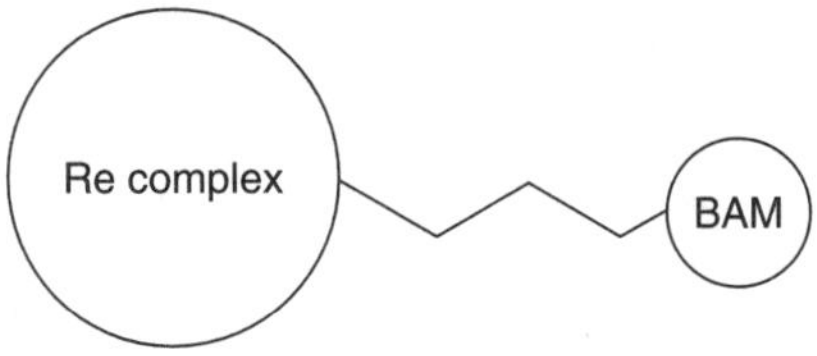

Figure 24.1

dominates and many examples of both square pyramidal and octahedral geometries have been characterized, by including X-ray crystallography. The ReO_2^+ core can be stabilized by amine or tertiary phosphine co-ligands but has been less explored in a radiopharmaceutical context.

MAG_3 type ligands

Mercaptoacetyltriglycine, MAG_3, has been used extensively for the generation of ^{99m}Tc radiopharmaceuticals and selectivity is generated by bioconjugation to an appropriate bioactive molecule. The synthetic strategy generally adopted is summarized in Figure 24.2, which shows labelling via pre-conjugate formation. The rhenium is present as a mono-oxo species in the +V oxidation state with N_3S coordination providing the characteristic square planar geometry. The X-ray crystal structure of the simple $ReOMAG_3$ complex has been reported.[8]

Studies of both the pre- and post-conjugation labelling of peptides such as the RC-160 somastatin analogue with ^{188}Re suggest that the pre-conjugation approach is preferable due to interaction with an S—S bond in the post-conjugation technique.[9] Biodistribution studies in mice showed high tumour uptake comparable with values obtained for the direct labelling of RC-160.[10]

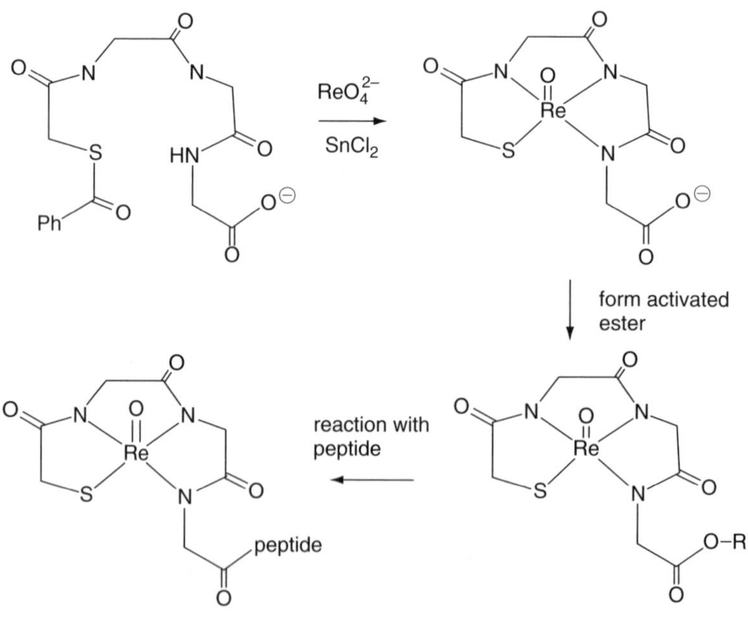

Figure 24.2

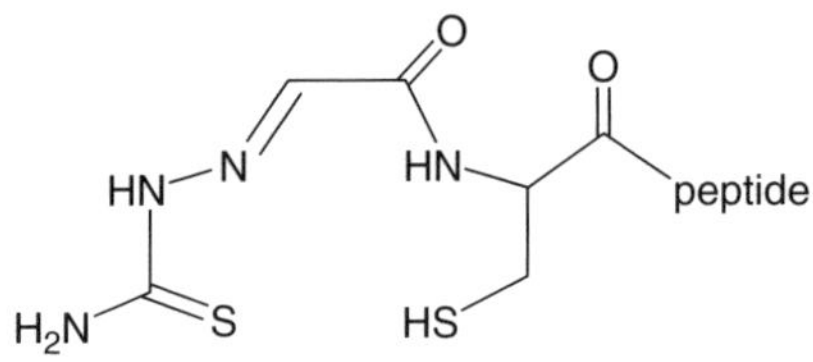

Figure 24.3

An analogous approach has been used for the ^{188}Re labelling of monoclonal antibodies using MAG_3.[11] Comparative studies of the radiolabelling of MORFs (synthetic oligomeric DNA analogues) with ^{90}Y-DOTA (DOTA = 1,4,7,10-tetraazacyclododecane-1,4,7,10-tetraacetic acid) and ^{188}Re-MAG_3 conjugates showed some disadvantages for both radionuclides. Self-radiolysis of the ^{90}Y species led to loss of the radionuclide and the rhenium species underwent slow decomposition to perrhenate, but both were nevertheless essentially intact after 48 h.[12]

A modification of the MAG_3-protein labelling procedure has been reported in which the cysteinyl group is replaced by a thiosemicarbazone, as shown in Figure 24.3. The tethering group was prepared by reaction of thiosemicarbazide with glyoxylic acid and was subsequently linked to a bispecific antibody (bsMAB) via peptides which bind specifically to one arm of the antibody.[13] The structure of the rhenium species formed on reaction with ^{188}Re-perrhenate was not established, but work with rhenium thiosemicarbazone complexes suggests it may be a Re(V) mono-oxo species.[14] Studies of biodistribution of the bsMAB–^{188}Re conjugate in mice showed promising tumour to blood, kidney and liver ratios, suggesting good potential for the radioimmunotherapy of colorectal cancer.[13]

Aminethiolate complexes

The coordination chemistry of rhenium with N_2S_2 ligands has been well explored and the known types are summarized in Figure 24.4. The aminethiolate type have neutral charge whereas the charge on the amide thiolate complexes is dependent on the reaction conditions used, the monoanionic complex being obtained at higher values of pH. Despite the extensive chemistry known with non-radioactive rhenium, and in marked contrast to technetium, there appear to have been few recent examples of the use of amine or amide thiolate rhenium oxo-complexes for bioconjugate coupling to BAMs. An indirect approach is illustrated by the use of lipidol, which is an iodinated esterified lipid from poppy seed oil, that has been used a contrast agent for the detection of liver cancer. The synthesis of N_2S_2 amine thiolate rhenium complexes with long chain substituents (Figure 24.5a) confers solubility in lipidol which acts as

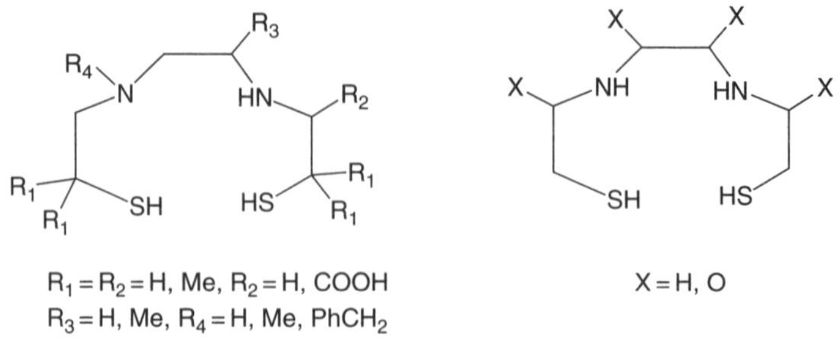

Figure 24.4

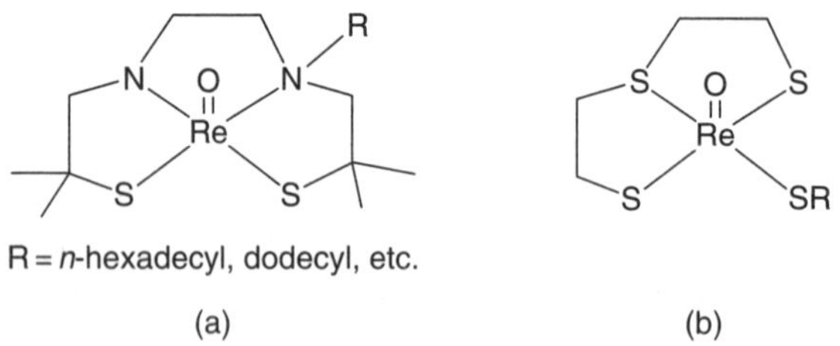

Figure 24.5

the delivery agent for the radiopharmaceutical. The biodistribution of such solutions was investigated in mice and high initial uptake in liver cancer was observed, but this was followed by washout and accumulation in the lung, but it was concluded that this approach was still of potential value for the treatment of liver cancer.[15]

The 'three plus one' and 'three plus two' approaches

This mixed ligand approach to targeting rhenium radioisotopes is exemplified by the complex shown in Figure 24.5b, prepared by reaction of an Re(V) precursor such as $[ReO(gluconate)_2]^-$ simultaneously with the tridentate ligand and the thiol RSH. Attachment of an appropriate functional group to the thiol R group then permits the introduction of the specific biological targeting group. This tactic has been studied extensively with conjugation to BAMs such as tropane derivatives for targeting dopamine transporter sites in the brain.[16] The disadvantage in terms of clinical applications is that the monodentate thiolate ligand is prone to exchange with other thiols in biological media such as glutathione.

Dioxo complexes

A reproducible protocol for the synthesis of ^{186}Re- or ^{188}Re-labelled polyamine complexes of the type $[ReO_2N_4]^+$ has been developed. Although acceptable yields were obtained for linear polyamines in water, the macrocyclic cyclam required the use of a two-phase water–dichloromethane solvent system with intermediate generation of $[ReOCl_3(PPh_3)_2]$. Yields of 64% and radiochemical purity (RCP) of 96% were obtained with the macrocyclic cylam after chromatography. The complexes were shown to have high stability in serum.[17]

The dioxo core has also been used with P_2S_2 ligands (Figure 24.6) to produce bioconjugates with BAMs such as bombesin analogues. Bombesin, a 14 amino acid peptide, and structurally modified analogues show an enhanced response in a variety of tumour tissues. The major thrust of this work has been with ^{99m}Tc, but studies with ^{188}Re have also been reported.

The presence of the reducing phosphine groups facilitates the perrhenate reduction and the dioxo-complex can be generated in good yield. Protein conjugation was carried out after binding of the radionuclide and the conjugate was formed in high yields. Biodistribution studies in mice with the analogous ^{99m}Tc species showed significant pancreatic uptake, suggesting that the complexes were being effectively targeted to gastrin releasing peptides (GRP) receptors.[18]

Direct labelling

This method has been widely used for radiolabelling of monoclonal antibodies with ^{186}Re or ^{188}Re. The antibody is treated with a reagent such as mercaptoethanol to reduce disulphide groups to thiol and then reacted with perrhenate in the presence of stannous chloride. The precise identity of the rhenium species present is not known, but a Re(V) mono-oxo species is likely. The direct labelling of peptides requires the peptide to have an amino acid sequence suitable for binding to the metal. Sequences such as Cys-Gly-Cys and Lys-Gly-Cys are incorporated in the required peptide at a location unlikely to

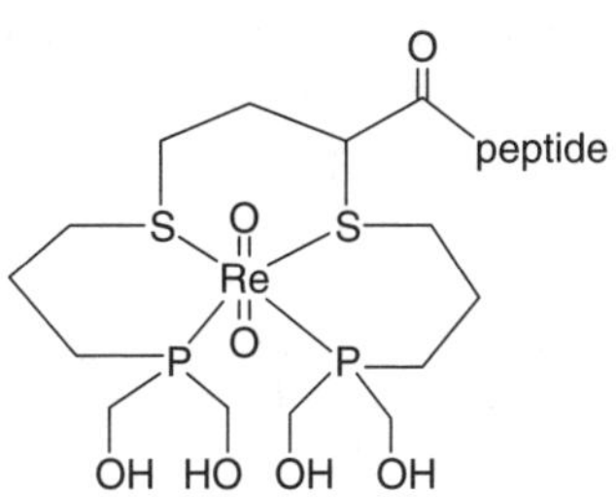

Figure 24.6

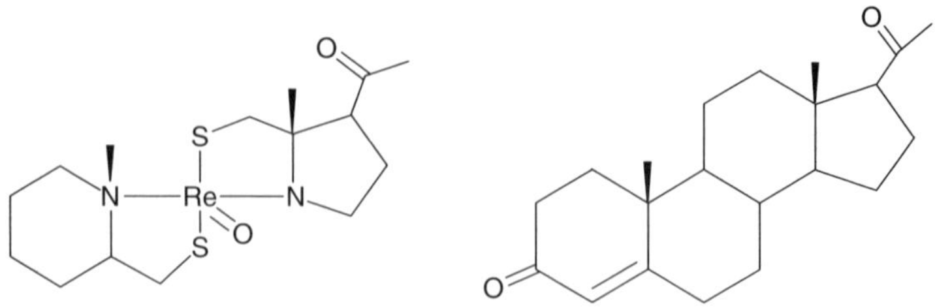

Figure 24.7

impair receptor binding and form mono-oxo chelates analogous to the MAG_3 species discussed above.[19]

Steroid modelling

An interesting alternative to conjugation of a steroid to a rhenium chelate has been proposed whereby the peripheral groups on a mixed ligand complex are selected so that the whole complex simulates the shape and receptor-binding groups of the steroid.[4] This is illustrated in Figure 24.7.

The mixed ligand Re(V) oxo-complex is formed preferentially in good yield using a 1:1 mixture of the bidentate ligands in good yield and the complex shows some affinity for steroid-binding receptors. It remains to be seen if this strategy can be extended to the targeting of other types of receptor as the synthetic challenges are considerable.

24.3.2 Complexes with cores containing metal–nitrogen multiple bonds

Nitrido-complexes

The first example of a rhenium complex of this core was reported in the 1960s when $[ReNCl_2(PPh_3)_2]$ was prepared from $[ReOCl_3(PPh_3)_2]$ and a hydrazine salt[20] and it was subsequently also prepared using azide as the nitride source.[21] However, it has only been comparatively recently that this chemistry has been extended to the radioactive isotopes using a procedure developed for technetium. Reaction of $[^{188}ReO_4]^-$ with a dithiocarbazate $[H_2NN(Me)CSSMe]$ generated the $Re(V)N^{2+}$ core which was then stabilized by the addition of a dithiocarbamate salt to give the Re(V) species $[ReN(S_2CNR_2)_2]$ (R = Me, Et) in 95% yield.[22] The biodistribution of the radiolabelled analogue, as expected, paralleled closely that of the known ^{99m}Tc analogue.

A potentially interesting approach to the targeting of rhenium nitride complexes is via Mixed PXP (where X = N or S) tridentate ligands combined with a bidentate ligand bearing a potential conjugation site (Figure 24.8).[23]

This chemistry has been thoroughly established for cold rhenium and for the ^{99m}Tc analogues, but there are no reports as yet of work with ^{188}Re or biological studies of the labelled rhenium derivatives.

Diazenide and isodiazene derivatives

The ability of rhenium to form strong metal–nitrogen multiple bonds allied with the oxidizing ability of the higher oxidation states permits the synthesis of a range of rhenium and technetium complexes containing diazenide (NNR) or isodiazene (NNR_2) ligands. Typical examples of rhenium complexes are shown in Figure 24.9.[24,25] These ligands are electronically flexible and can bind with linear or bent M—N—N systems depending on the requirements of the metal centre. The linear forms predominate for simple terminally bound ligands, and virtually all the radiopharmaceutically relevant complexes discussed exhibit this geometry.

The chemistry of the bis(diazenido) and bis(isodiazene) complexes is complicated by the presence of the two N—N ligands and a recent advance has seen the synthesis of the monodiazenide complexes $[ReCl_2(NNAr)(MeCN)(PPh_3)_2]$ directly from perrhenate, hydrazine and triphenylphosphine.[26] As well as

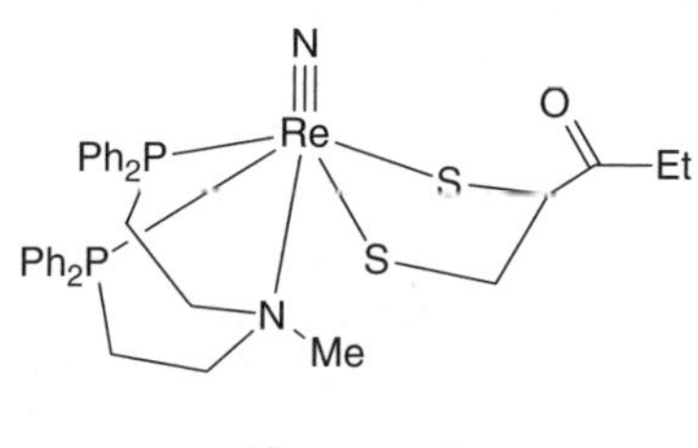

Figure 24.8

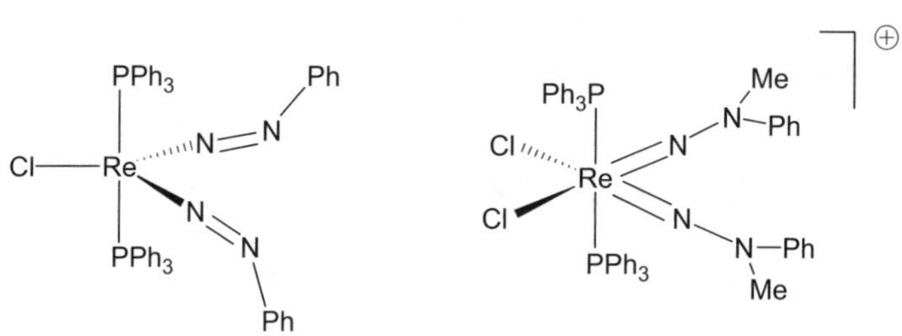

Figure 24.9

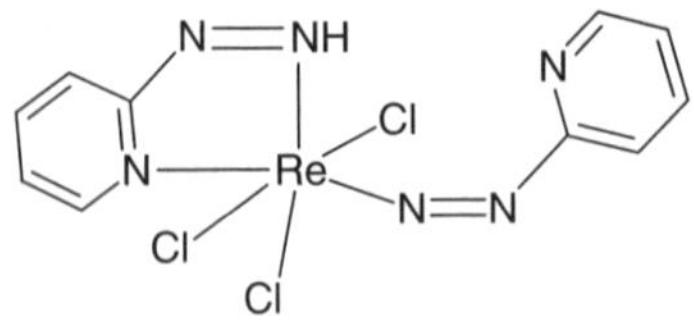

Figure 24.10

providing a derivatisable core, an ester group in the 4-position of the aryl group of the diazenide ligand provides a potential conjugation site.[26] The chloride and phosphine ligands can be substituted by a range of polydentate ligands,[27] but this chemistry has yet to be extended to using rhenium radioisotopes.

The coordination chemistry of pyridylhydrazines has been well developed with both technetium and rhenium and the use of carboxyl substituents to attach targeting groups has been widely exploited in the HYNIC system (HYNIC = *N*-oxysuccinimidylhydrazinonicotinamide). This involves conjugation via an amide linkage by reaction of the carboxyl group on the pyridyl ring with an amino group of a protein such as chemotactic peptide via an activated ester. This procedure has been widely used for the development of specific imaging agents with ^{99m}Tc, but addition of a further co-ligand is necessary. The exact molecular structure of the technetium complex is not certain, but is believed to involve diazenide- or isodiazene-type coordination of the C_5H_4NNN fragment.

Reaction of perrhenate with pyridylhydrazine dihydrochloride gives the complex shown in Figure 24.10 in good yield.[28] One of the pyridylNN units is chelated and protonated at the nitrogen adjacent to the metal and the other is bound analogously to the phenyl derivatives discussed above. The chloride ligands can be displaced by N-, S-donor ligands, giving access to a range of diazenide/diazene complexes. Although extensive use has been made of the HYNIC strategy for ^{99m}Tc, there are as yet few examples of its extension to radioactive rhenium isotopes. A comparison of ^{186}Re radiolabelling of a murine A7 antibody with a MAG_3 derivative and HYNIC suggested that the HYNIC conjugate, although easier to prepare, was less stable.[29] Also, an investigation of the Re-188-HYNIC labelling of octreotide reported that there were difficulties with the co-formation of insoluble ReO_2.[30]

24.3.3 Complexes of the Re(I) tricarbonyl core

There has been a lot of interest engendered by the relatively recent development of methods that allow the synthesis of technetium or rhenium compounds of the type *fac*-$[M(CO)_3(H_2O)_3]^+$ (where M = Tc(I) or Re(I)) under conditions amenable to the radiopharmaceutical applications. The use of what has become

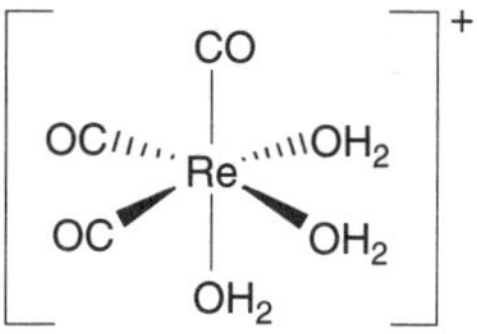

Figure 24.11

known as the 'carbonyl core' approach exploits the stability of the metal tricarbonyl core whilst manipulating the relatively labile water ligands to attach pertinent biomolecules (Figure 24.11). A variety of mono-, bi- and tri-dentate ligands react with the tricarbonyl core, displacing the substitutionally labile aqua ligands. The small size of the core should allow tethering of radiometals to biomolecules without deleterious effects on their activity and specificity.

These organometallic derivatives provide a novel approach to the design of radiopharmaceuticals. One of the most significant features of the chemistry has been the development of methods that allow synthesis of the carbonyl compounds under relatively mild conditions from pertechnetate or perrhenate, which are the starting materials for radiopharmaceutical preparations.[31]

One synthetic procedure involves reacting a tetrahydrofuran (THF) solution of $[NBu_4][MO_4]$ (M = Re(VII) or Tc(VII)) with a carbon dioxide purged solution of tetrabutylammonium chloride using borane in THF as the reducing agent. $[NEt_4]_2[\textit{fac}\text{-}MBr_3(CO)_3]$ can be isolated following acidification and addition of $[NEt_4]Cl$. The synthesis of the rhenium compound requires longer reaction times (about 4 h) than does the synthesis of the technetium analogue (about 2 h). In aqueous solutions this compound rapidly exchanges the halides for water ligands to give the useful precursor $[\textit{fac}\text{-}M(CO)_3(OH_2)_3]^+$ (M = Re(I) or Tc(I)).[32]

Further modifications have been directed towards developing a procedure that eliminates the requirement to use gaseous carbon monoxide, which is unsuitable for use in commercial radiopharmaceutical 'kits'. The technetium carbonyl compound $\textit{fac}\text{-}[Tc(CO)_3(OH_2)_3]^+$ can be prepared by using the innovative approach of using a boron-based carbonylating agent, boranocarbonate $K_2[H_3BCO_2]$, which acts as a source of carbon monoxide and as a reducing agent.[33] This development has allowed the introduction of the 'Isolink kit®' for the synthesis of technetium carbonyl compounds. The kit is commercially available from Mallinckrodt Medical B.[34] The same technology can be adapted to rhenium although this is not straightforward. Problems arise because perrhenate is harder to reduce than pertechnetate and any Re(III) or Re(V) complexes formed during the reductive carbonylation are unstable at basic pH. These problems can be circumvented by using aminoborane as the reducing agent and careful control of the pH (Figure 24.12).[35]

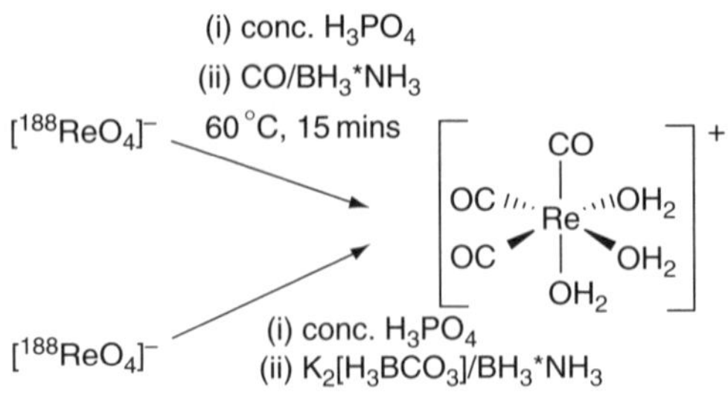

Figure 24.12

The *fac*-$[Re(CO)_3(H_2O)_3]^+$ cation has been referred to as a 'semi aquo ion' with three tightly held CO ligands and three labile coordination sites. Carbonyl ligands are known to stabilize low oxidation states via backbonding mechanisms and the *trans* effect increases the lability of the *trans* ligands allowing them to be readily substituted. This allows the attachment of a wide variety of ligands to the carbonyl core. The Re(I) complexes have d^6 low spin octahedral configurations, which are known to form stable complexes.

The ligand exchange reactions in these systems occur via a dissociative mechanism and for radiopharmaceutical applications, rapid exchange to form stable complexes is essential. It has been shown that aliphatic amines, carboxylates and other anions only coordinate weakly to the *fac*-$[Re(OH_2)_3(CO)_3]^+$ core. In contrast, thioether ligands were found to coordinate very slowly but the resulting complexes exhibited high stability. It has been suggested that the best ligands for the *fac*-$[M(CO)_3]^+$ core are aromatic amines as they exhibit fast complexation kinetics to form reasonably stable complexes and that any anchoring of biomolecules to the *fac*-$[Re(CO)_3]^+$ core should be done via a unit that contains at least one aromatic amine. Furthermore, the combination of an aromatic amine with an aliphatic amine or carboxylate group gave complexes that formed quickly whilst having high stability in aqueous solutions.[31]

A particularly good ligand is histidine, which binds facially as a tridentate ligand. This offers the potential of anchoring biomolecules to the $[Re(CO)_3]^+$ core via a histidine residue (Figure 24.13).[31]

If the carboxyl group of histidine is used in the formation of an amide bond as part of a protein sequence then the histidine residue binds as a bidentate ligand. Radiolabelling studies have been carried out with an N-terminal His-peptide by incubating $[Tc(CO)_3(H_2O)_3]^+$ with His-Gly-Gly-Ala-Ala-Leu for 90 min at 37 °C. This has been extended to the radiolabelling single-chain antibody fragments carrying a C-terminal His-tag. These conjugates displayed neither *trans* chelation nor loss of binding affinity.[36] It is anticipated that this labelling technology could be readily extended to $^{186/188}Re$ to provide radiotherapeutic analogues of the technetium imaging agents.

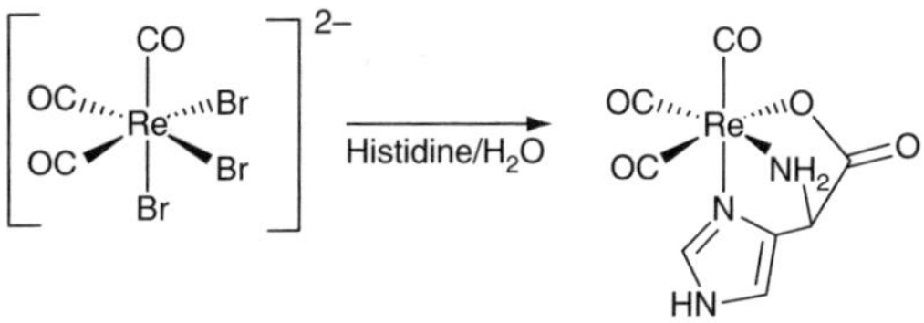

Figure 24.13

The [^{188}Re(CO)$_3$] core has been labelled by tridentate bifunctional ligands based on either bis(imidazol-2-yl)methylamine or iminodiacetic acid ligands with alkyl spacers of different lengths and a pendant primary amino or carboxylic acid functionality (Figure 24.14). The optimum radiolabelling conditions were shown to be incubating [^{188}Re(CO)$_3$(H$_2$O)$_2$] with the ligand in a phosphate buffer at 60 °C for 60 min. Shorter incubation at higher temperatures (100 °C) or longer incubation at lower temperatures (37 °C) resulted in poor radiolabelling yields due to the oxidation of the rhenium carbonyl precursor to perrhenate. The bis(imidazol-2-yl)methylamine complexes are more stable than the iminodiacetic acid complexes, the former showing about 20% degradation in human serum over 24 h, the major decomposition product being perrhenate.[35]

A wide variety of other ligands that coordinate to the *fac*-[Re(CO)$_3$] core and have the potential to act as the basis of new radiopharmaceuticals have been reported. Some selected examples are discussed in the following paragraphs.

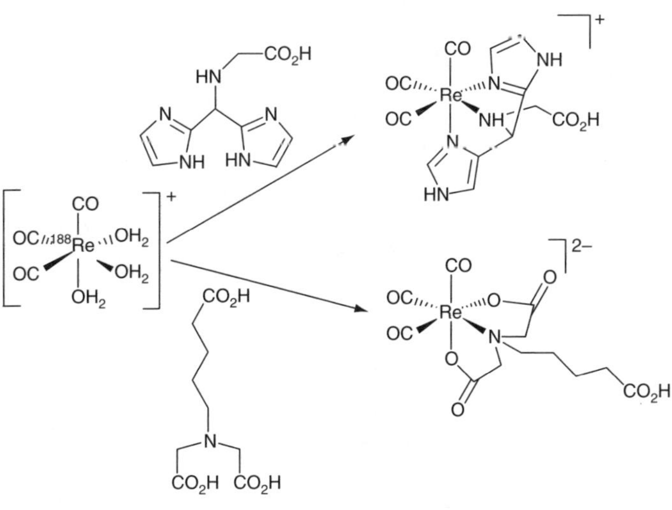

Figure 24.14

Tridentate ligands bearing two pyridyl groups derived from amino acids or amino acid analogues react with $[Et_4N]_2[ReBr_3(CO)_3]$ cleanly and in high yield to give complexes of the type $[Re(CO)_3L]^+$. The $[Re(CO)_3L]^+$ complex with Fmoc-Lysine (Fmoc = 9-fluorenylmethoxycarbonyl)-derived ligand of this type (Figure 24.15) has been used to incorporate the rhenium carbonyl core into a short peptide sequence through automated solid phase methods. This flexible approach should allow the rhenium chelate to be incorporated at any position in a peptide sequence.[37]

Poly(mercaptoimidazolyl)borate ligands form complexes of the type $[Re(L)(CO)_3]$, where the ligands act as a facially coordinating tridentate via the two sulphur donor and an unusual agostic B—H···Re bond (Figure 24.16).[38] These ligands have been further functionalized to include central nervous system receptor–avid molecules.[39]

$[NEt_4]_2[Re(CO)_3Br_3]$ has been shown to react with 2-acetylpyridine phenyl-thiosemicarbazone to form air-stable, neutral Re(I) complexes of the composition $[Re(CO)_3(L)]$. The thiosemicarbazone ligand deprotonates and binds to the metal in a facial manner as a tridentate N, N, S donor (Figure 24.17). To accommodate the facial coordination geometry the ligand considerably distorts from planarity, which is unusual for the extensively conjugated thiosemicarbazone ligand.[40]

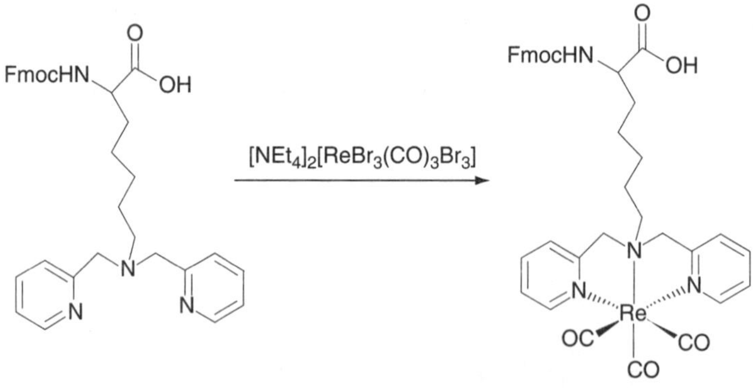

Figure 24.15

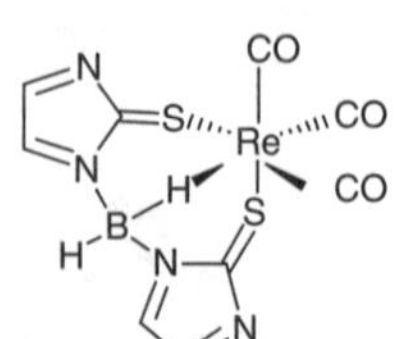

Figure 24.16

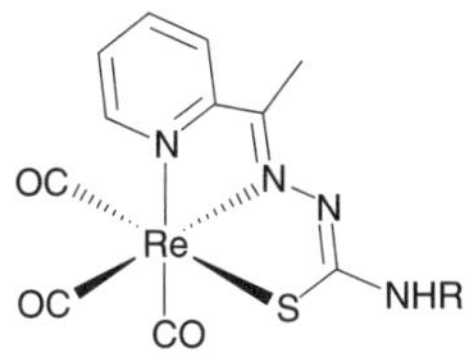

Figure 24.17

The examples listed here clearly show that the 'tricarbonyl core approach' offers considerable potential in the development of rhenium radiopharmaceuticals but the commercial viability of the analogous approach to technetium imaging radiopharmaceuticals has been questioned. Problems in the commercial development of imaging agents based on the $[Tc(CO)_3]^+$ core include intellectual property issues and regulatory hurdles.[34,41] The same issues may also hinder the development of rhenium tricarbonyl-based therapeutics and more work is required on a synthesis of $[^{188}Re(CO)_3(H_2O)_2]^+$ that can be performed on a routine basis. However, this versatile technology could lead to clinically viable therapeutic agents which could be developed in concert with the analogous technetium-based imaging agents.

24.3.4 Organometallic approaches

Another approach which utilizes rhenium in a low oxidation state and mainstream organometallic chemistry is the use of the $[CpRe(CO)_3]$ core (Cp = cyclopentadienyl). In this system the π-accepting properties of the cyclopentadienyl and carbonyl ligands are well suited to stabilizing the electron-rich Re(I) centre. Cyclopentadiene can be readily functionalized with various biologically active vectors which can then be used to make rhenium tricarbonyl cyclopentadienyl conjugates, and the small size and low molecular weight of the cyclopentadienyl ligand minimizes any potential steric interactions with the receptor-binding part of the biomolecule. A $[CpRe(CO)_3]$ derivative of Tamoxifen®, a drug widely used in the treatment of breast cancer, retains considerable binding affinity to the oestrogen receptor and an anti-proliferative activity in MCF-7-type cell lines comparable to Tamoxifen® (Figure 24.18).[42]

These compounds show some promise as non-radioactive anti-cancer agents but combining this activity with β-emitting ^{188}Re represents a considerable challenge. Cyclopentadienyl derivatives are often insoluble and unstable in water, which can prohibit the synthesis of these types of compounds in a radiopharmaceutical context. An elegant approach to circumvent these problems involves using a cyclopentadienyl derivative that is deprotonated to a reasonable extent at physiological pH. This can be achieved by introducing electron-withdrawing substituents such as an α-carbonyl group. The 1-(2-methoxy-phenyl)piperazine

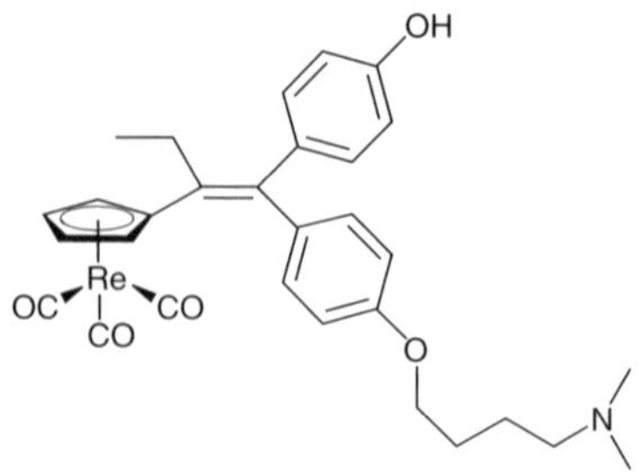

Figure 24.18

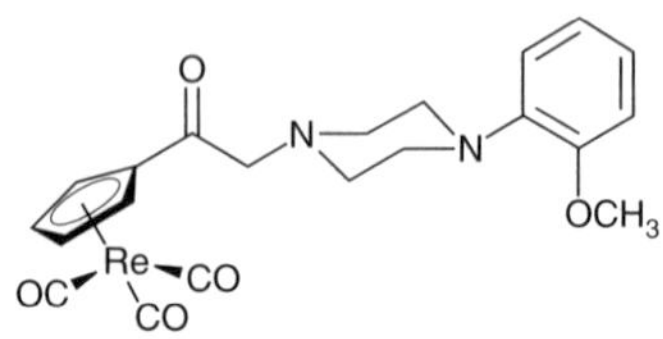

Figure 24.19

cyclopentadienyl derivative (Figure 24.19) reacts with $[Re(CO)_3(H_2O)_3]^+$ in a phosphate buffer (pH 7.4) and heating at 85 °C. The interest in aryl piperazines is due to their being receptor molecules for the 5-HT_{1A} subclass of serotonergic receptors.[43,44]

The synthesis of cyclopentadienyltricarbonylrhenium derivatives has recently been reported with ^{188}Re. A cyclopentadienylcarboxylic acid rhenium tricarbonyl and its glycine conjugate were synthesized by utilizing a double ligand transfer reaction. For example, 1,1′-bis(methoxycarbonyl)ferrocene, chromium hexacarbonyl and tin(II) chloride were added to $[^{188}ReO_4]^-$ in methanol and then sealed in a pressure tube and heated at 180 °C for 45 min. Purification by column chromatography followed by hydrolysis of the ester and further purification by SepPak cartridge and reverse phase HPLC gave the product. This procedure is not amenable to routine radiopharmaceutical synthesis but the compounds were shown to be stable in murine plasma and their *in vivo* metabolism was studied in mice.[45]

24.3.5 Miscellaneous approaches

Various porphyrin derivatives are known to be tumour-avid agents and a rhenium labelled porphyrin derivative has been evaluated as a potential therapeutic modality. A water-soluble porphyrin-bearing carboxylate group (Figure 24.20) was labelled with radioactive rhenium, a mixture of ^{186}Re and ^{188}Re, prepared by irradiation of natural rhenium.

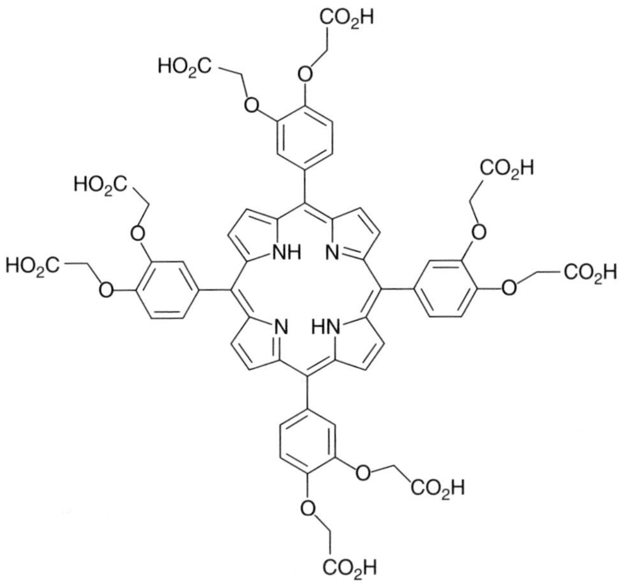

Figure 24.20

The two isotopes that were produced with this method were not separated. $[^{186/188}ReO_4]^-$ was added to a solution of the buffer dissolved in a bicarbonate buffer system and stannous tartrate was added as a reductant. The actual chemical nature of the complex is not known and the radiochemical purity decreased from 98 to 35% when the complex was stored at room temperature. Biodistribution studies were carried out in fibrosarcoma-bearing Swiss mice. An uptake of more than 3.5%/g of tumour was observed 30 min after the injection. The low stability of the complex was implicated in the fast clearance of the agent from non-target tissue and 24 h after the injection 89% of the activity showed renal clearance presumably as perrhenate.[46] In this case the rhenium probably coordinates to the peripheral carboxylate groups and a more stable rhenium–porphyrin conjugate could be synthesised by incorporating donor groups that are better suited to rhenium in mid–low oxidation states.

Another ligand system that offers potential in the development of rhenium-based radiopharmaceuticals is that based on tridentate formylpyridyl thiosemicarbazones (Figure 24.21). A series of Re(III) complexes, $[ReL_2]^+$, where LH = a 2-formylpyridine thiosemicarbazone, have been synthesized. Thiosemicarbazones are of considerable pharmacological interest since a number of derivatives have shown a broad spectrum of chemotherapeutic properties and their metal complexes often show enhanced biological activities. These thiosemicarbazone ligands can be readily functionalized with biomolecules to provide *in vivo* specificity or to control lipophilicity and biodistribution of any metal

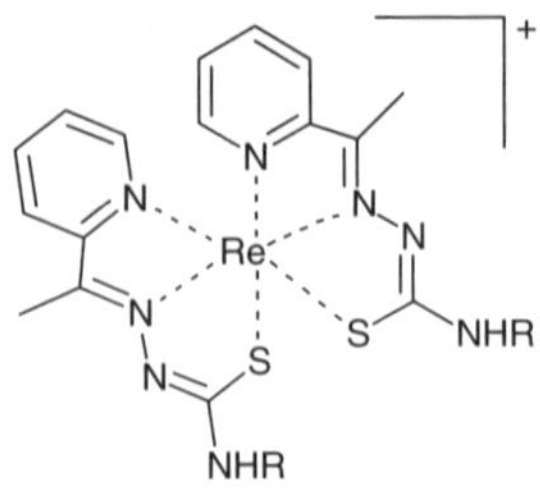

Figure 24.21

complexes. In these complexes the rhenium atom is in a distorted octahedral environment with two tridentate deprotonated thiosemicarbazones binding as monoanionic ligands through the sulphur, pyridyl nitrogen and azamethinic nitrogen in a *mer* (azomethine nitrogen atoms *trans*) configuration.[47] The compounds can be synthesized directly from perrhenate. Although radiolabelling of these compounds is yet to be published, these complexes are of particular interest in the development of radiopharmaceuticals, as they offer unique homoleptic complexes and the use of two tridentate chelating ligands binding in a *mer* fashion to form octahedral rhenium complexes. This does not result in *syn*/*anti*-isomeric mixtures which complicate the radiolabelling and pharmacokinetic properties of the well-known oxorhenium complexes formed with N_2S_2 ligands; also the complexes are stable and of a low molecular weight.

24.4 'First-Generation' Rhenium Agents

As mentioned earlier first-generation complexes are comparatively rare among rhenium complexes for therapy, and some examples are given in the following paragraphs.

24.4.1 Bone-targeting phosphonate complexes

Some of the most-studied rhenium radiopharmaceuticals are bone-targeting rhenium diphosphonate complexes which have been shown to be useful as relieving pain associated with bone metastases. Neither the cause of the pain nor the mechanism of pain relief by irradiation is well understood.[4] The ^{186}Re complex of hydroxyethylidenediphosphonate (HEDP) (Figure 24.22) is commercially available and is an approved radiopharmaceutical in Europe.[2]

A typical preparation of this therapeutic agent involves incubating $[^{186}ReO_4]^-$ with HEDP at 100 °C for 10 min in the presence of $SnCl_2$ which acts as a reductant, with the pH maintained between 3 and 3.5 and ascorbic

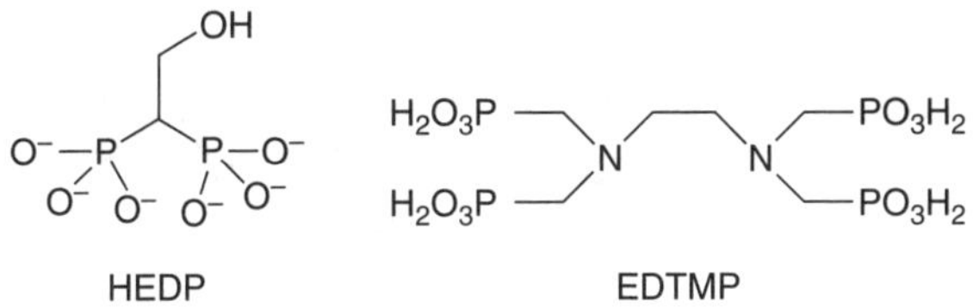

Figure 24.22

acid added as an anti-oxidant.[4] The exact nature of the structure of these compounds is not completely understood but has been studied by extended X-ray absorption fine structure (EXAFS) experiments which show that the structure is dependent on the method of preparation and that the material is often oligomeric mixture with rhenium–oxygen interactions as well as Re—Re bonds.[48] The accumulation in bone is believed to be similar to the technetium analogue which is thought to occur as a consequence of adsorption to the bone mineral (hydroxyapatite) through coordination of phosphoryl oxygen atoms or Re=O atoms to calcium ions on the hydroxyapatite surface. It is thought that increased blood flow and a higher fraction of amorphous calcium phosphate in diseased bone account for the selective uptake at metastatic sites.[1,4] The ^{188}Re analogue is a safe radiopharmaceutical for prostate cancer patients with osseous metastases. The biodistribution and radiation dosimetry characteristics appear similar to the ^{186}Re HEDP agent and have similar benefits and toxicities in patients with skeletal metastases.[49] The ^{188}Re complex of ethylenediamine-*N*,*N*,*N*′,*N*′-terakis(methylene phosphonic acid) (EDTMP, Figure 24.22) has also been investigated as a potential therapeutic bone agent. The biodistribution of the ^{188}Re–EDTMP complex studied in rats showed high bone uptake and clearance from other organs.[50]

24.4.2 [Re(V)-oxo(DMSA)]

The Re(V)-oxo complex of *meso*-2,3-dimercaptosuccinic acid (DMSA) has been investigated as a therapeutic analogue of technetium radiopharmaceutical, which is known as 'pentavalent technetium-99^{m}-DMSA' and which is used as an imaging agent in medullary thyroid carcinoma. The rhenium complex is a mixture of three isomers each with square pyramidal Re(V) (Figure 24.23).[4]

The biological properties of [^{188}Re(O)DMSA] have been investigated in humans with medullary thyroid carcinoma. The investigation showed selective uptake in tumour tissue and the compound is selectively taken up in bone metastases. In this respect it offers an alternative to ^{188}Re-HEDP in the palliation of painful bone metastases. A potential advantage of [^{188}Re(O)DMSA] is that it exhibits a lower uptake in normal bone leading to a lower radiation dose to bone marrow.[51,52] It is thought that selectivity for bone metastases arises

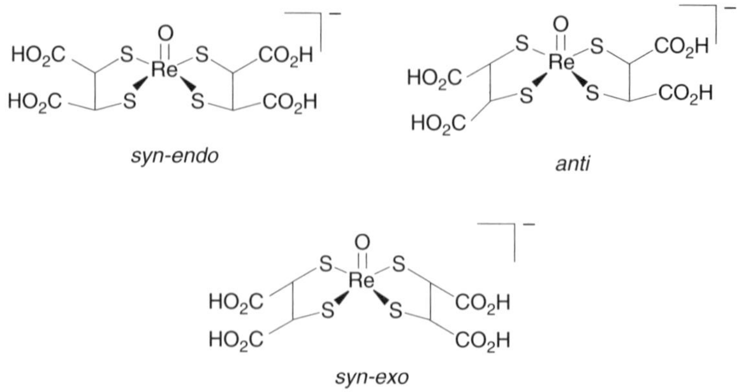

Figure 24.23

from interaction of the carboxylate and oxo groups with calcium-rich areas in the bone mineral. The γ-photons (15%) have been used to obtain good quality images that are useful in visualizing whole body distribution of the pharmaceutical.[52] A new efficient synthesis of [ReO(DMSA)] has been published, utilizing acetic hydrazide as a reductant which offers the advantage of the isolation of a single isomer.[7] New bioconjugates have been prepared using the oxobis(dithiolato)rhenate(V) core by dehydration of two carboxylic acid groups to form a cyclic dianhydride which was then used to tether salmon calcitonin analogue, a 32 amino acid peptide to the rhenium core.[7]

24.4.3 Rhenium agents for endovascular radiation therapy

Angioplasty is an invasive procedure performed to reduce or eliminate blockages in coronary arteries. The goal is to restore blood flow to blood-deprived heart tissue, reduce the need for medication and eliminate or reduce the number of episodes of angina (chest pain). One of the problems with angioplasty is that the arterial blockage often returns due to regrowth of plaque or restenosis as a result of the body's immune response to the injury of the angioplasty.

A method that has been investigated for the prevention of restenosis involves the use of radiation delivered to the coronary vessel wall. The delivery of the radiation can be achieved via catheter-based systems, radioactive stents or using unsealed radioactivity with a liquid-filled balloon filled with a β-emitter. The availability and specific properties of ^{188}Re mean that it is an attractive nuclide for such endovascular radiation therapy. It is important to consider the possibility and consequence of balloon rupture. For the safety of the patient it is important that any released radioactive substance should exhibit rapid excretion via the urinary bladder and not accumulate in any non-target tissue. The rhenium compounds that have been investigated for this purpose include Na[$^{188}ReO_4$], [^{188}Re-MAG$_3$], [^{188}Re(DTPA)] and ^{188}Re-ethylene dicysteine.[53,54] The biodistribution of the ^{188}Re complex of ethylene dicysteine was studied in Istar rats and showed fast blood clearance and low retention of activity in vital organs with a high renal excretion (91%).[55,56]

24.4.4 Rhenium agents for radiation synovectomy

Various radionuclides including ^{188}Re are being evaluated for treatment of rheumatoid arthritis in an alternative procedure known as surgical synovectomy which involves the removal of inflamed joint lining.[57] ^{188}Re microspheres have been developed as radiation synovectomy agents that are injected into the diseased joint. The exact nature of the microspheres was not reported but they were synthesized from anion exchange resin with a particle size of 2 μm, which were treated with a source of ^{188}Re. The levels of extra-articular radiation with this reagent are negligible.[58] An alternative radiation synovectomy agent is ^{188}Re–tin colloid. The colloid was prepared by reacting [$^{188}ReO_4$]$^-$ with $SnCl_2{\cdot}2H_2O$ in 0.5 mol l^{-1} HCl for 2 h at 100 °C. The pH was adjusted to the physiological range with buffer and then 1.0% of lidocaine was added. The ^{188}Re–tin colloid was used to perform radiation synovectomy in 22 knees of 21 patients with rheumatoid arthritis in the first clinical trial using a ^{188}Re-labelled agent for synovectomy. The treatment resulted in the improvement of arthritic symptoms with no clinical side effects.[59]

24.5 Summary

Although the use of metal-based imaging agents has become commonplace in modern hospitals, the development of metal-based therapeutic radiopharmaceuticals has been somewhat slower with only a limited number of agents in current clinical use for the treatment of cancers. Despite this, it has been predicted that there will be a huge growth in the therapeutic application of metal complexes in the next 10–20 years and biotechnology companies are beginning to place particular focus on medicinal inorganic chemistry. To utilize fully the potential of therapeutic metal complexes, more research is needed into the fundamental chemistry of the radionuclides with bifunctional chelators with the ability to deliver the radiometal selectively to the tumour site. Hopefully, this brief review will serve to summarize the current status of the rapidly developing field and provide some pointers to potentially productive future areas of research.

References

1. J.R. Dilworth and S.J. Parrott, *J. Chem. Soc. Rev.* 43–55 (1999).
2. W.A. Volkert and T.J. Hoffman, *Chem. Rev.* **99**, 2269–2292 (1999).
3. J.-P. Pouget and S.J. Mather, *Eur. J. Nucl. Med.* **28**, 541–561 (2001).
4. P.J. Blower and S. Prakash, *Perspect. Bioinorg. Chem.* **4**, 91–143 (1999).
5. F.F. Knapp Jr, S. Mirzadeh, A.L. Beets *et al.*, *Appl. Radiat. Isot.* **49**, 309–315 (1998).
6. C. Bolzati, A. Boschi, L. Uccelli *et al.*, *Nucl. Med. Biol.* **27**, 309–314 (2000).
7. U. Choudhry, W. Greenland, W.A. Goddard *et al.*, *Dalton Trans.*, 311–317 (2003).
8. L. Hansen, R. Cini, A.J. Taylor and L.G. Marzilli, *Inorg. Chem.* **31**, 2801–2808 (1992).
9. S. Guhlke, A. Schaffland, P.O. Zamora *et al.*, *Nucl. Med. Biol.* **25**, 621–631 (1998).
10. P.O. Zamora, S. Gulhke, H. Bender *et al.*, *Int. J. Cancer* **65**, 214–220 (1996).
11. G.W. Visser, M. Gerretsen, J.D. Herscheid *et al.*, *J. Nucl. Med.* **34**, 1953–1963 (1993).
12. C.-B. Liu, G.-Z. Liu, N. Liu *et al.*, *Nucl. Med. Biol.* **30**, 207–214 (2003).
13. H. Karacay, W.J. McBride, G.L. Griffiths *et al.*, *Bioconjug. Chem.* **11**, 842–854 (2000).
14. J.R. Dilworth, P.S. Donnelly and C. McGibbon, unpublished results.
15. Y.-S. Lee, J.M. Jeong, Y.J. Kim *et al.*, *Nucl. Med. Commun.* **23**, 237–242 (2002).
16. S. Meegalla, K. Ploessl, M.-P. Kung *et al.*, *J. Am. Chem. Soc.* **117**, 11037–11038 (1995).
17. S. Prakash, M.J. Went and P.J. Blower, *Nucl. Med. Biol.* **23**, 543–549 (1996).
18. S.R. Carra, R. Schibli, H. Gali *et al.*, *Bioconjug. Chem.* **10**, 254–260 (1999).
19. S.S. Jurisson, 'Tc and Re peptide chemistry: Direct labelling', in *Technetium, Rhenium and Other Metals in Chemistry and Nuclear Medicine 6*, M. Nicolini and U. Mazzi, eds (SGE Ditoriali, Padova, 2002), pp. 15–21.
20. J. Chatt, J.D. Garforth, N.P. Johnson and G.A. Rowe, *J. Chem. Soc.* 1012–1020 (1964).
21. J. Chatt, C.D. Falk, G.J. Leigh and R.J. Paske, *J. Chem. Soc.* (A), 2288–2293 (1969).
22. A. Boschi, C. Bolzati, L. Uccelli and A. Duatti, *Nucl. Med. Biol.* **30**, 381–387 (2003).
23. A. Boschi, C. Bolzati, E. Benini *et al.*, *Bioconjug. Chem.* 1035–1042 (2001).
24. J.R. Dilworth, S.A. Harrison, D.R.M. Walton and E. Schweda, *Inorg. Chem.* **24**, 2594–2595 (1985).
25. J.R. Dilworth, P. Jobanputra, S.J. Parrott *et al.*, *Polyhedron* **11**, 147–155 (1992).
26. A.R. Cowley, J.R. Dilworth and P.S. Donnelly, *Inorg. Chem.* **42**, 929–931 (2003).
27. A.R. Cowley, J.R. Dilworth, P.S. Donnelly and S.J. Ross, 'Towards rhenium radiopharmaceuticals with diazenide and dithiocarbamate ligands', in *Technetium, Rhenium and Other Metals in Chemistry and Nuclear Medicine 6*, M. Nicolini and U. Mazzi, eds (SGE Ditoriali, Padova, 2002), pp. 223–225.
28. D.J. Rose, K.P. Maresca, T. Nicholson *et al.*, *Inorg. Chem.* **37**, 2701–2716 (1998).
29. K. Yokoyama, H. Tega, S. Kinuya *et al.*, *Eur. J. Nucl. Med.* **26**, P567 (1999).
30. S. Guhlke, M. Behe, M. Bangard *et al.*, *Eur. J. Nucl. Med.* **26**, OS249 (1999).
31. R. Alberto, R. Schibli, R. Waibel *et al.*, *Coord. Chem. Rev.* **190–192**, 901–919 (1999).

32. R. Alberto, R. Schibli, P.A. Schubiger *et al.*, *Polyhedron* **15**, 1079–1089 (1996).
33. R. Alberto, K. Ortner, N. Wheatley *et al.*, *J. Am. Chem. Soc.* **123**, 3135–3136 (2001).
34. R. Alberto, *Eur. J. Nucl. Med. Mol. Imaging* **30**, 1299–1302 (2003).
35. R. Schibli, R. Schwarzbach, R. Alberto *et al.*, *Bioconjug. Chem.* **12**, 750–756 (2002).
36. R. Waibel, R. Alberto, J. Willuda *et al.*, *Nat. Biotechnol.* **17**, 897–901 (1999).
37. K.A. Stephenson, J. Zubieta, S.R. Banerjee *et al.*, *Bioconjug. Chem.* **15**, 128–136 (2004).
38. R. Garcia, A. Paulo, A. Domingos *et al.*, *J. Am. Chem. Soc.* **122**, 11240–11241 (2000).
39. R. Garcia, Y.-H. Xing, A. Paulo *et al.*, *J. Chem. Soc., Dalton Trans.*, 4236–4241 (2002).
40. I.G. Santos, U. Abram, R. Alberto *et al.*, *Inorg. Chem.* **43**, 1834–1836 (2004).
41. M.J. Welch and J.S. Lewis, *Eur. J. Nucl. Med. Mol. Imaging* **30**, 1302–1304 (2003).
42. F.L. Bideau, M. Salmain, S. Top and G. Jaouen, *Chem. Eur. J.* **7**, 2289–2294 (2001).
43. J. Wald, R. Alberto, K. Ortner and L. Candreia, *Angew. Chem. Int. Ed.* **40**, 3062–3066 (2001).
44. J. Bernard, K. Ortner, B. Spingler *et al.*, *Inorg. Chem.* **42**, 1014–1022 (2003).
45. T. Uehara, M. Koike, H. Nakata *et al.*, *Nucl. Med. Biol.* **30**, 327–334 (2003).
46. S. Banerjee, T. Das, G. Samuel *et al.*, *Nucl. Med. Commun.* **22**, 1101–1107 (2001).
47. A.R. Cowley, J.R. Dilworth, P.S. Donnelly and J. Woolard-Shore, *Dalton Trans.*, 748–754 (2003).
48. R.C. Elder, J. Yuan, B. Helmer *et al.*, *Inorg. Chem.* **36**, 3055–3063 (1997).
49. H. Palmedo, S. Guhlke, H. Bender *et al.*, *Eur. J. Nucl. Med.* **27**, 123–130 (2000).
50. S.J. Oh, K.S. Won, D.H. Moon *et al.*, *Nucl. Med. Commun.* **23**, 75–81 (2002).
51. P.J. Blower, A.S.K. Lam, M.J. O'Doherty *et al.*, *Eur. J. Nucl. Med.* **25**, 613–621 (1998).
52. P.J. Blower, A.G. Kettle, M.J. O'Doherty *et al.*, *Eur. J. Nucl. Med.* **27**, 1405–1409 (2000).
53. J. Kotzerke, H. Hanke and M. Hoher, *Eur. J. Nucl. Med.* **27**, 223–236 (2000).
54. J. Lee, D.S. Lee, K.M. Kim *et al.*, *Eur. J. Nucl. Med.* **27**, 76–82 (2000).
55. T. Das, S. Banerjee, G. Samuel *et al.*, *Nucl. Med. Biol.* **27**, 189–197 (2000).
56. T. Das, S. Banerjee, G. Samuel *et al.*, *Nucl. Med. Commun.* **21**, 939–945 (2000).
57. M. Frier, *Mini-Rev. Med. Chem.* **4**, 61–68 (2004).
58. S.-J. Wang, W.-Y. Lin, M.-N. Chen *et al.*, *Nucl. Med. Biol.* **28**, 727–732 (2001).
59. E.B. Lee, K.C. Shin, Y.J. Lee *et al.*, *Nucl. Med. Commun.* **24**, 689–696 (2003).

25 $_{78}$Pt Platinum-Based Drugs

Viktor Brabec and Jana Kasparkova

25.1 Introduction, Anti-tumor Activity of Cisplatin

Platinum compounds constitute a discrete class of chemotherapeutics, which are now widely used in the clinic as anti-tumor and anti-viral agents. The interest in platinum chemotherapeutics has its origin in the 1960s, with the serendipitous discovery by Rosenberg of the inhibition of division of bacterial cells by platinum complexes.[1] The first platinum anti-cancer drug introduced in the clinic is cisplatin [*cis*-diamminedichloroplatinum(II)]. This very simple and purely inorganic molecule (Figure 25.1) was first synthesized in 1844 and at that time was called Peyrone's chloride.

The ammonia ligands in this platinum(II) molecule represent non-leaving groups whereas the chlorides can be replaced by other ligands (nucleophiles). Hence, cisplatin belongs to the class of bifunctional reactants. In spite of its simplicity it is one of the most potent drugs available for anti-cancer chemotherapy.[2,3] Cisplatin is highly effective for the treatment of testicular and ovarian cancer and is used in combination regimens for a variety of other carcinomas, including bladder, small lung tumors and those of the head and neck. However, cisplatin toxicity in tumor cells is coupled with several drawbacks.[2]

Metallotherapeutic Drugs and Metal-Based Diagnostic Agents: The Use of Metals in Medicine
Edited by Gielen and Tiekink

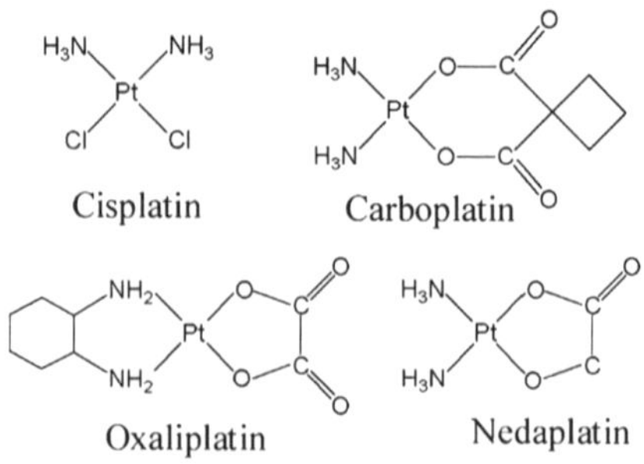

Figure 25.1 Platinum complexes used clinically

Its applicability is still limited to a relatively narrow range of tumors. Some tumors have natural resistance to cisplatin, while others develop resistance after initial treatment. Cisplatin also has limited solubility in aqueous solution and is administered intravenously. There are also significant problems in terms of inducing severe side-effects[2] (especially kidney damage, vomiting/nausea, neurotoxicity and emetogenesis).

The limitations associated with the clinical use of cisplatin prompted a search for more effective and less toxic alternative platinum anti-tumor agents. This effort led to the synthesis of thousands of platinum complexes which were evaluated as anti-tumor agents and tested in cancer cells, but only a few reached clinical trials and most of them have already been rejected. Those which have achieved routine clinical use and have been approved for clinical administration, together with cisplatin, are its direct analogues (Figure 25.1), namely carboplatin [*cis*-diamminecyclobutanedicarboxylatoplatinum(II)], oxaliplatin {[(*1R*,*2R*-diamminocyclohexane)oxalatoplatinum(II)] (1,2-diaminocyclohexane = DACH)} and nedaplatin [*cis*-diammineglycolatoplatinum(II)].

25.2 Anti-tumor Activity of Carboplatin, Oxaliplatin and Nedaplatin

The leaving group of carboplatin is a less labile cyclobutanedicarboxylato ligand which undergoes a slower rate of aquation than the chlorides in cisplatin. Unfortunately, carboplatin exhibits cross-resistance with cisplatin so that although less toxic and more easily administered than cisplatin, it has not demonstrated any substantial advantages over cisplatin.[4,5] The therapeutic index defined by the efficacy versus toxicity profile even favors cisplatin in most tumor systems (in testis, bladder and head and neck), particularly if cisplatin is administered in a multi-fractionated drug schedule. Cisplatin may be considered the superior analogue compared to carboplatin also in terms of the toxicity profile, in particular, the absence of hematologic toxicity. The dose-limiting toxicity of carboplatin is myelosuppression.[6] Therefore, carboplatin has replaced cisplatin in the treatment of several but not all platinum-sensitive tumors.

Another cisplatin analogue which has already received approval for use in Asia, Europe and America is oxaliplatin.[5,7] Importantly, oxaliplatin has no cross-resistance with cisplatin and is active in combination with 5-fluorouracil and leucovorin for the treatment of colorectal cancer, a disease in which cisplatin and carboplatin have no significant clinical activity.[8] Administration of oxaliplatin as a single agent has not been associated with nephrotoxicity, ototoxicity or myelosuppression. The dose-limiting toxicity of oxaliplatin is neurotoxicity.[6]

Nedaplatin is registered for use in head and neck, testicular, lung, esophageal, ovarian and cervical cancers in Japan. A randomized clinical trial has revealed that nedaplatin is cross-resistant with cisplatin, but less toxic.[9] It has shown no marked advantage over cisplatin in objective response and overall survival. More thrombocytopenia, but less leucopenia, nephrotoxicity and gastrointestinal toxicity were observed.

Thus, since the discovery of the anti-tumor effects of cisplatin, the continuing efforts directed toward finding alternative platinum compounds for clinical use have resulted in the development of only three new agents registered for use in the clinic, which do not represent a fundamental breakthrough in cancer treatment, however. Therefore, the search for new platinum anti-tumor drugs and their clinical testing continues to seek more significant improvements in cancer treatment with platinum complexes than during the last three decades.

25.3 Mechanism of Action of Cisplatin

Broadening the chemotherapeutic arsenal depends on understanding existing agents with a view toward developing new modes of attack. In the 35 years since the discovery of the anti-tumor activity of cisplatin, laboratory studies have provided considerable information as to how the drug exerts its anti-tumor effects and how some tumors are, or become, resistant to this drug. Although a number of features of the mechanism underlying the anti-tumor effects of cisplatin have been thoroughly described, some still remain to be disclosed. There is a large body of experimental evidence that the success of platinum complexes in killing tumor cells mainly results from their ability to form various types of adducts with DNA.[10] Hence, intensive research has focused on DNA modifications by platinum anti-tumor drugs and how these modifications modulate the 'downstream effects' such as specific protein recognition by DNA-processing enzymes and other cellular components (for a review, see for instance Ref. 11).

Cisplatin reacts with DNA in the cell nucleus, where the concentration of chloride is markedly lower than in extracellular fluids. The drug loses its chloride ligands in media containing low concentrations of chloride to form positively charged mono- and di-aqua species (Figure 25.2). It has been shown that only these aquated forms bind to DNA. Bifunctional cisplatin binds to DNA in a two-step process first forming mono-functional adducts preferentially at the guanine residues, which subsequently close to major intrastrand cross-links between adjacent purine residues (1,2-GG or 1,2-AG intrastrand cross-links). Other minor adducts are 1,3-GXG intrastrand cross-links (X = A, C, T), interstrand cross-links and mono-functional lesions. In all adducts, cisplatin is coordinated to the N7 atom of purine residues. The percentage of the 1,2-intrastrand adducts formed by cisplatin is larger than statistically expected so that this cross-link has generally been assumed to be the important adduct correlated with anti-cancer activity and was therefore most extensively investigated.

The adducts formed by cisplatin in DNA affect the DNA's secondary structure.[11] The formation of major intrastrand cross-links of cisplatin leads to marked conformational alterations in DNA. These adducts induce a roll between the platinated purine residues, displacement of the platinum atom from the planes of the purine rings, a directional and rigid bend of the helix axis toward the major groove and a local unwinding. In addition, severe

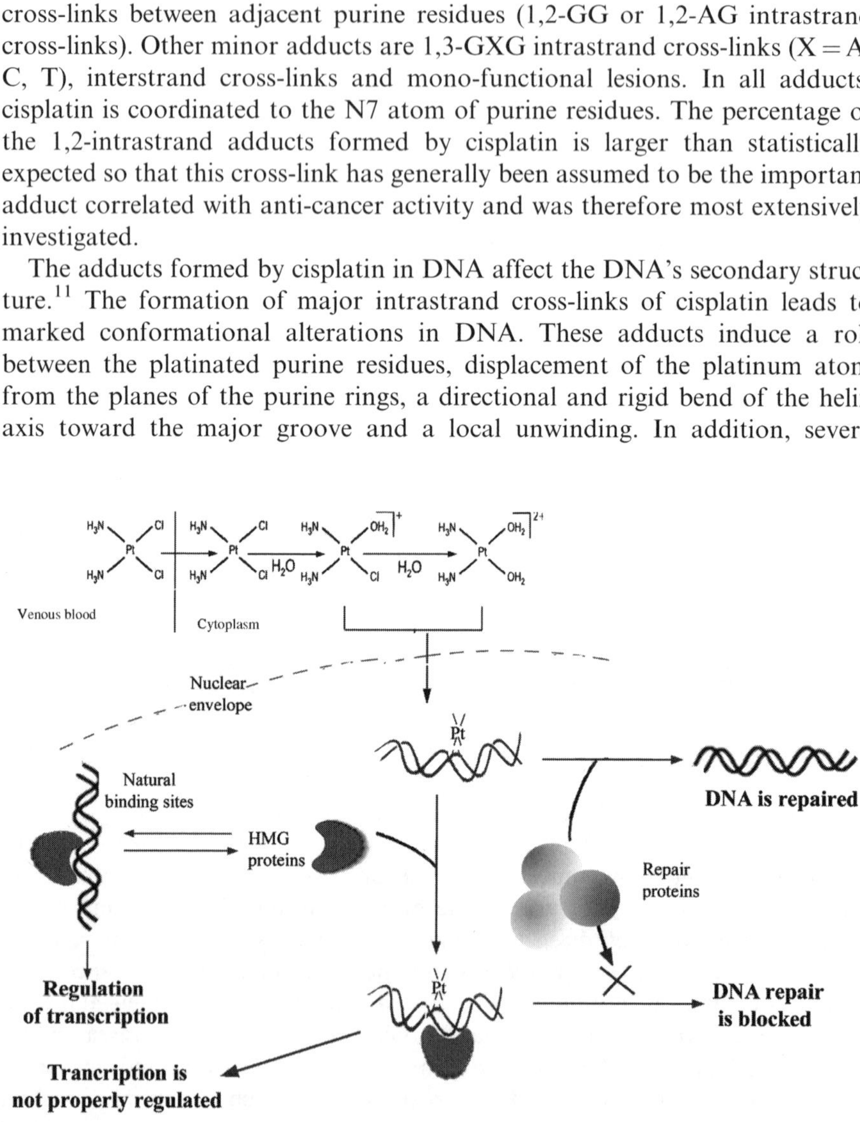

Figure 25.2 Molecular mechanism of anti-tumor activity of cisplatin

perturbation of hydrogen bonding within the 5′-coordinated GC base pair, widening and flattening of the minor groove opposite the cisplatin adduct, creation of a hydrophobic notch, global distortion extending over 4–5 base pairs and additional helical parameters characteristic of the A-form of DNA have also been reported. Other minor adducts of cisplatin also induce several irregularities in DNA, the details of which have been reviewed elsewhere.[11]

Cisplatin inhibits DNA and RNA synthesis by prokaryotic and eukaryotic DNA and RNA polymerases, respectively both *in vitro* and *in vivo*.[12,13] Inhibition of DNA or RNA synthesis occurs mainly at GG sites, consistent with the fact that these sites are preferential binding sites of cisplatin.

The limitations of using cisplatin in cancer chemotherapy are also associated with intrinsic and acquired resistance of tumor cells to this drug.[14,15] Resistance to cisplatin is multi-factorial and, in general, may consist of mechanisms either limiting the formation of DNA adducts and/or operating downstream of the interaction of cisplatin with DNA to promote cell survival. The formation of DNA adducts by cisplatin can be limited by reduced accumulation of the drug, enhanced drug efflux and cisplatin inactivation by coordination to sulfur-containing biomolecules including metallothioneins whose production may be increased as a consequence of cisplatin treatment. The second group of mechanisms includes enhanced repair of DNA adducts of cisplatin and increased tolerance of the resulting DNA damage.

In human cells, cisplatin intrastrand adducts are removed from DNA mainly by the nucleotide excision repair (NER) system.[11] It has been found using cell-free extracts or a reconstituted NER system that intrastrand cross-links of cisplatin are efficiently repaired. Importantly, this repair of 1,2-, but not 1,3-intrastrand cross-links is blocked upon addition of an HMG-domain protein (HMG = high mobility group) (discussed later). There is also evidence that other cellular repair mechanisms, such as recombination or mismatch repair, can affect anti-tumor efficiency of cisplatin.[11] Recent observations[16] support the view that mismatch repair mediates the cytotoxicity of cisplatin in tumor cells and that dysfunction of this type of DNA repair may result in the resistance of tumor cells to cisplatin or in drug tolerance.

It is generally believed that anti-tumor activity of cisplatin is mediated by the recognition of its DNA adducts by cellular proteins.[11,17] Several classes of these proteins have been identified and mechanisms have been proposed to explain how they mediate anti-tumor effects of cisplatin (discussed later). The greatest attention has been paid to the studies of recognition of platinated DNA by HMGB1 and HMGB2 proteins which belong to architectural chromatin proteins and play some kind of structural role in the formation of functional higher order protein–DNA or protein–protein complexes or as signaling molecules in genetically regulated repair pathways. These structure-specific proteins bind selectively to the 1,2-GG or 1,2-AG adducts of cisplatin, but not to its 1,3-intrastrand cross-links. Extensive reviews addressing these questions have been recently published.[11,17]

It is generally believed that the key intracellular pharmacological target for platinum compounds is DNA.[10] The adducts formed by cisplatin distort the DNA conformation. DNA adducts of cisplatin inhibit replication and transcription, but they are also bypassed by DNA or RNA polymerases. In addition, cisplatin adducts are removed from DNA mainly by NER. They are, however, also recognized by a number of proteins which could block DNA adducts of cisplatin from damage recognition needed for repair (Figure 25.2). In this way these adducts can persist for sufficient time, which potentiates the anti-tumor effect of the drug. The other hypothesis based on the observation that a number of various proteins recognize cisplatin-modified DNA is that cisplatin–DNA adducts hijack proteins away from their normal binding sites, thereby disrupting fundamental cellular processes (Figure 25.2). Experimental support for these hypothetical aspects of the mechanism underlying anti-tumor activity of cisplatin or resistance of some tumors to this drug has been thoroughly reviewed recently.[11,17] Initially, inhibition of DNA replication by cisplatin adducts was considered to be a process very likely relevant to its anti-tumor efficiency.[18] However, other studies have indicated that cisplatin inhibits tumor-cell growth at doses which are considerably lower than those needed to inhibit DNA synthesis.[19] Subsequent observations have revealed that cisplatin can trigger G2 cell-cycle arrest and programmed cell death (apoptosis),[20] exposing another mechanism of cytotoxicity of this drug. However, since apoptosis is a very complex process, a number of possible pathways still have to be explored for a complete understanding of the mechanism by which cisplatin triggers apoptosis.[21]

25.4 Mechanism of Action of Carboplatin, Oxaliplatin and Nedaplatin

Carboplatin forms a similar spectrum of DNA adducts as cisplatin with slightly different sequence preferences.[22] Concentrations of carboplatin approximately two orders of magnitude higher are needed to obtain DNA platination levels equivalent to cisplatin. Conformational alterations induced in DNA by carboplatin in cell-free media have been characterized[23] and the changes of non-denaturational character similar to those observed for cisplatin at the same level of DNA platination have been found.

In order for the reaction of oxaliplatin with DNA to occur, the parent compound must become aquated. The hydrolysis of oxaliplatin to form reactive species, 1,2-DACH diaqua platinum(II) ($[Pt(R,R\text{-DACH})(H_2O)_2]^{2+}$), is a slower process than the hydrolysis of cisplatin. The sites in DNA of oxaliplatin adducts and their spectra are nearly identical to the situation when DNA is modified by cisplatin.[24] On the other hand, oxaliplatin is inherently less able than cisplatin to form DNA adducts.[25] Oxaliplatin adducts are repaired with kinetics similar as cisplatin adducts.[26] However, oxaliplatin is more efficient than cisplatin per equal number of DNA adducts in inhibiting DNA chain elongation. Despite lower DNA reactivity, oxaliplatin exhibits similar or greater cytotoxicity in several human tumor cell lines (discussed earlier). Thus, oxaliplatin requires fewer DNA lesions than does cisplatin to achieve cell growth inhibition.[25] The overall conformational alterations induced in DNA by the 1,2-GG intrastrand cross-link of cisplatin and oxaliplatin are similar, but also differ in several details.[27] The bulky DACH ring of the oxaliplatin adduct fills much of the DNA major groove, making it narrower and less polar at the site of the cross-link. The subtle differences in overall conformation of 1,2-GG intrastrand cross-links of cisplatin and oxaliplatin have been suggested[28] to influence further processing of the oxaliplatin cross-link in the cell.

Data on the molecular mechanism of anti-tumor effects of nedaplatin,[29] which is cross-resistant with cisplatin as for carboplatin, indicate that this drug reacts with guanine residues, but it requires a reaction time about 20 times that required by cisplatin. Nedaplatin is a drug structurally similar to carboplatin so that it is reasonable to expect that nedaplatin forms the same types of DNA adducts as cisplatin and carboplatin. Interestingly, survival of the cultured cells of human ovarian cancer correlated with interstrand cross-linking efficiency of this platinum drug.[29]

Taken together, the adducts produced on DNA by direct analogues of cisplatin, such as carboplatin, oxaliplatin and nedaplatin, are similar to those produced by the parent drug, though different in their relative rates of formation. Hence, from a mechanistic DNA-binding point of view, it is not too surprising that the introduction of the three new platinum anti-tumor drugs in the clinic does not represent a fundamental breakthrough in the treatment of cancer with platinum agents. This conclusion prompted a formulation of the hypothesis that development of platinum compounds structurally dissimilar to cisplatin may lead to agents with a spectrum of clinical activity complementary to the parent drug.[30] Moreover, as DNA is considered the major pharmacological target of platinum compounds, the latter hypothesis also implies that platinum agents which bind to DNA in a manner fundamentally different to that of cisplatin may have altered pharmacological properties. This concept has already led to the synthesis of several new unconventional platinum anti-tumor compounds that violate the original structure–activity relationships.

25.5 Activation of *trans* Geometry

Structure–activity studies often employ inactive compounds such as transplatin [*trans*-diamminedichloroplatinum(II), *trans*-isomer of cisplatin] (Figure 25.3).[31] Transplatin–DNA adducts are mainly interstrand cross-links preferentially formed between guanine and complementary cytosine residues[32] and a relatively

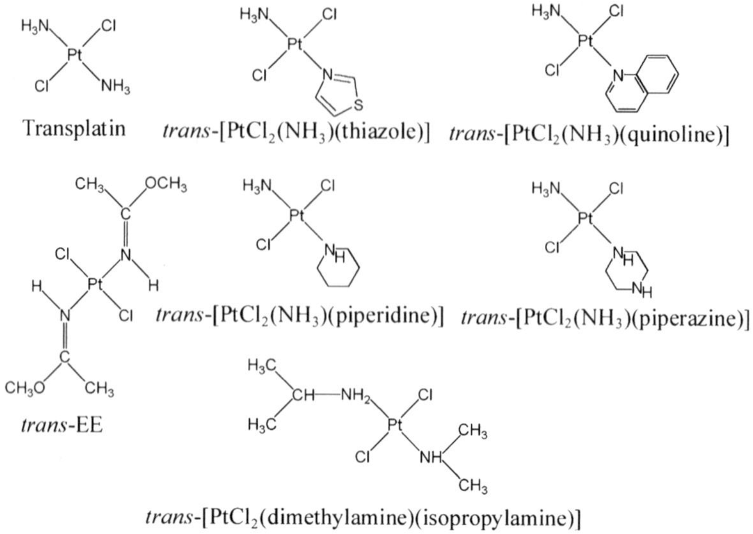

Figure 25.3 Transplatin and its anti-tumor analogues

large portion of adducts remains monofunctional. Intrastrand cross-links between non-adjacent guanine residues or between guanine and either adenine or cytosine residues have also been reported. However, while these adducts are stable in single-stranded DNA, they isomerize in double-helical DNA in a number of sequences into interstrand cross-links.[33] It has been also suggested[33] that in cells transplatin forms only a small amount of cross-links because of the slow closure of the monofunctional adducts coupled to their trapping by intracellular sulfur nucleophiles. Importantly, the adducts of transplatin also affect DNA conformation although less severely than the cross-links of cisplatin.

The clinical inactivity of transplatin is considered a paradigm for the classical structure–activity relationships of platinum drugs,[31] but, to this end, several new analogues of transplatin which exhibit a different spectrum of cytostatic activity including activity in tumor cells resistant to cisplatin have been identified. Examples are analogues containing iminoether groups of general formula *trans*-$[PtCl_2(E\text{-iminoether})_2]$ (imino ether = $HN{=}C(OCH_3)CH_3$) or analogues containing heterocyclic amine ligand of general structure *trans*-$[PtCl_2(NH_3)(L)]$, where L = planar amine such as quinoline or thiazole or non-planar amine such as piperidine or piperazine (Figure 25.3).

The replacement of both NH_3 groups in transplatin by an iminoether ligand results in a marked enhancement of its cytotoxicity so that it is more cytotoxic than its *cis* congener and exhibits significant anti-tumor activity including activity in cisplatin-resistant tumor cells.[34] The new *trans*- compound (*trans*-$[PtCl_2(E\text{-iminoether})_2]$) forms mainly stable monofunctional adducts at guanine residues when reacted with DNA.[35] These adducts locally distort DNA, are not recognized by HMG-domain proteins and are readily removed from DNA by the NER system. Importantly, the monofunctional adducts of *trans*-[$PtCl_2$(*E*-iminoether)$_2$] readily cross-link proteins, which markedly enhances the efficiency of this adduct to terminate DNA polymerization by DNA polymerases *in vitro* and to inhibit removal of this adduct from DNA by the NER system.[36] Hence, DNA–protein ternary cross-links produced by *trans*-[$PtCl_2$(*E*-iminoether)$_2$] can persist considerably longer than the non-cross-linked monofunctional adducts, which potentiates toxicity of this anti-tumor platinum compound toward tumor cells sensitive to this drug. Thus, *trans*-[$PtCl_2$(*E*-iminoether)$_2$] represents a quite new class of platinum anti-tumor drugs in which activation of the *trans* geometry is associated with an increased efficiency to form DNA–protein ternary cross-links thereby acting by a different mechanism than 'classical' cisplatin and its analogues.

The molecular mechanism of the anti-tumor activity of *trans*-platinum compounds containing heterocyclic ligands has also been extensively investigated.[37] Here, we will discuss two distinct series of this class of agents having the general structure *trans*-[$PtCl_2NH_3$(L)]: (i) L = planar ligand, such as quinoline or thiazole; (ii) L = non-planar ligand, such as piperidine or piperazine. There are two major chemical differences between NH_3 and a heterocyclic ligand. Steric hindrance by the appropriately positioned H atoms of the ring system reduces chemical reactivity, and surface and stacking interactions become more pronounced over the hydrogen bonding associated with NH_3 groups. Importantly, this class of new *trans*- compounds forms considerably more interstrand cross-links and, even more importantly, they also form 1,3-GXG intrastrand cross-links which are stable in double-helical DNA.

An interesting feature of DNA modification by *trans*-[$PtCl_2(NH_3)$(thiazole)] is that its monofunctional adduct creates a local conformational distortion, which is very similar to that produced in DNA by major 1,2-GG intrastrand cross-link of anti-tumor cisplatin.[38] In addition, these monofunctional adducts are recognized by HMG-domain proteins with an affinity similar to the 1,2-GG intrastrand cross-link of cisplatin. The effective removal of the monofunctional adducts of *trans*-[$PtCl_2(NH_3)$(thiazole)] from DNA by NER has also been observed, but it is inhibited if the NER system is supplemented by HMG-domain proteins. Additional work[39] has shown that *trans*-[$PtCl_2(NH_3)$(quinoline)] and *trans*-[$PtCl_2(NH_3)$(thiazole)] preferentially form DNA interstrand cross-links between guanine residues at the 5′-GC/5′-GC sites. It implies that these cross-links are formally equivalent to those formed by anti-tumor cisplatin, but different from those formed by clinically ineffective transplatin (which preferentially forms these adducts between complementary guanine and cytosine residues[32]). These results have shown for the first time that a simple chemical modification of the structure of an inactive platinum compound alters its DNA-binding mode into that of an active drug and that the processing of the monofunctional DNA adducts of these *trans*-platinum analogues in tumor cells may be similar to that of the major bifunctional adducts of conventional cisplatin.

The anti-tumor analogues of transplatin-containing non-planar heterocyclic ligands, such as piperidine or piperazine, form 1,3-GXG intrastrand cross-links which are considerably more stable in double-helical DNA than the same adducts of the parent transplatin, distort the DNA conformation and are not readily removed from DNA by the NER system.[40,41] Hence, the intrastrand cross-links of these analogues can also persist for sufficient time to potentiate toxicity toward tumor cells. These analogues also form DNA interstrand cross-links which are similar to those of ineffective transplatin. So, these adducts appear less likely candidates for genotoxic lesions responsible for anti-tumor effects of the analogues containing non-planar heterocyclic ligands. Hence, the role of structurally unique intrastrand cross-links in the anti-tumor effects of transplatin analogues in which one ammine group is replaced by a heterocyclic non-planar ligand may predominate.

Recently, the DNA-binding modes of new cytotoxic *trans*-platinum compounds containing different aliphatic amines (e.g. *trans*-[$PtCl_2$(dimethylamine)(isopropylamine)]) have been evaluated in cell-free media.[42] The complex *trans*-[$PtCl_2$(dimethylamine)(isopropylamine)] readily forms DNA interstrand cross-links. The amount of these cross-links is considerably higher than the amount of the lesions formed under the same conditions by transplatin or cisplatin. In addition, the complex exhibits a preferential binding affinity to alternating purine–pyrimidine sequences in DNA, forms the adduct that inhibits the B–Z transition in DNA and blocks DNA synthesis *in vitro* more efficiently than the adducts of cisplatin. These particular DNA-binding properties have been proposed to be related to the efficiency of *trans*-[$PtCl_2$(dimethylamine)(isopropylamine)] in inducing apoptosis and some selective killing in a *H-ras* over-expressing cell line.[43]

25.6 Polynuclear Platinum Compounds

Polynuclear platinum compounds comprise a unique class of anti-cancer platinum agents with distinct chemical and biological properties different from mononuclear platinum drugs.[44] The lead compound [{*trans*-$PtCl(NH_3)_2$}$_2\mu$-*trans*-$Pt(NH_3)_2\{H_2N(CH_2)_6NH_2\}_2]^{4+}$ (BBR3464, Figure 25.4) is a trinuclear,

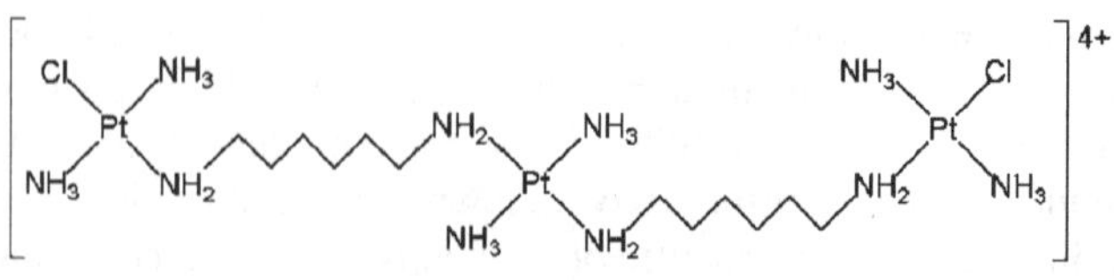

BBR3464

Figure 25.4 Anti-tumor trinuclear platinum(II) complex BBR3464

bifunctional DNA-binding agent with an overall 4+ charge. Phase I trials demonstrated a clear pattern of responses in cancers not normally treatable with cisplatin including responses in melanoma, pancreatic cancer and lung cancer. Objective responses in Phase II were verified in relapsed ovarian cancer and non-small-cell lung cancer. Pre-clinical studies indicated activity in p53-mutant tumors and a minimal induction of p53 following BBR3464 treatment. The interactions of anti-tumor polynuclear platinum compounds with target DNA (for reviews, see for instance Refs 11, 44) are distinct from the mono-nuclear-based cisplatin family and, indeed, unlike those of any DNA-damaging agent in clinical use. The ability of BBR3464 to induce DNA adducts such as long-range delocalized intra- and inter-strand cross-links[45] which are not produced by conventional mononuclear platinum compounds suggests that BBR3464 may avoid, at least in part, the classical mechanism of cisplatin resistance related to DNA damage recognition and repair. The cross-links of BBR3464 distort the DNA conformation, but these distortions are not recognized by HMG-domain proteins.[46,47] On the other hand, while intrastrand adducts of BBR3464 are readily removed from DNA by the NER system, the interstrand cross-links are not. It has been suggested[46,47] that interstrand and not intrastrand cross-links of BBR3464 could persist for a sufficiently long time in cells, which would potentiate toxicity of the interstrand lesions toward tumors sensitive to this polynuclear platinum drug.

25.7 Platinum(IV) Compounds

The mode of action of platinum agents can also be considerably affected by means other than changing of chemical structure. For instance, originally inactive compounds can be activated only after they are delivered to the target tumor cells or to their intracellular components. Platinum(IV) complexes represent one example of this class of anti-tumor agents. Such compounds are frequently designated as pro-drugs that have to be first, after their administration, activated by a reaction with reducing agents present in body liquids. In addition, being inert to substitution, Pt(IV) complexes theoretically have the advantage of demonstrating fewer side effects than their Pt(II) counterparts. One of the anti-tumor Pt(IV) drugs, which is an analogue of cisplatin, is *bis*-acetatoamminedichloro(cyclohexylamine)platinum(IV) (JM216, Figure 25.5) the first orally administrable platinum complex that has entered the clinic.[48] The DNA-binding properties of JM216 on naked DNA or within tumor cells are similar to those of cisplatin, although there are some differences in the nature of the adducts.[49] The search for a clinically useful orally available anti-tumor platinum drug also led to the design of several Pt(IV) compounds with *trans*-leaving ligands.[50] An example of one of these complexes is *trans*-amine(cyclohexylamine)dichlorodihydroxoplatinum(IV) (JM335, Figure 25.5).

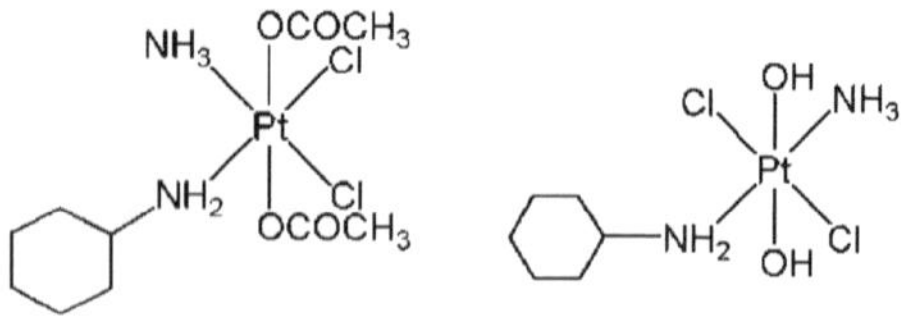

JM216 JM335

Figure 25.5 Tetravalent complexes

Mechanistic studies with JM335 in carcinoma cell lines[49] have revealed DNA-binding properties different from those of cisplatin. JM335 induces DNA single-strand breaks (formation of this DNA lesion is cell-line dependent) and interstrand cross-links. Interestingly, DNA extracted from cells treated with JM335 is not recognized by an antibody elicited against DNA modified by cisplatin. In addition, the kinetics of apoptosis is more rapid for JM335 than for its *cis*-isomer. However, JM335 is generally less cytotoxic than cisplatin probably because of its inactivation by thiols; so, it has not been selected for clinical trials. Nevertheless, it remains one of the more interesting and active *trans*-platinum complexes.

Another attractive design strategy might involve a relatively non-toxic precursor, a pro drug, which could be selectively photoactivated at the site of the tumor thereby minimizing side effects. Photoactive Pt(IV) azide complexes were recently synthesized[51] which are stable in human blood plasma and which, on irradiation with visible light, can undergo reductive elimination reactions in aqueous solution and liberate dinitrogen. Whereas no binding to DNA occurred in the dark, upon irradiation with blue light (457.9 nm) DNA was platinated at the same sites as by treatment with cisplatin.[52] Hence, the complexes can be viewed as 'pro-drugs' for cisplatin-type complexes although photoactivation may lead to novel reactive intermediates. Work is now needed to investigate potential photoinduced cytotoxicity toward cancer cells, and optimization of the wavelength of photoactivation via changes in the ligands in these octahedral *cis*-diazido Pt(IV) complexes.

25.8 Targeted Analogues

One way to affect the biological activity of cisplatin is also to target this drug to DNA differently by the attachment of the platinum moiety to a suitable carrier.[53,54] Compared with untargeted analogues, such compounds may exhibit a different DNA-binding mode, including sequence selectivity. Importantly, this change of the DNA-binding selectivity has been shown already to result in good *in vitro* and *in vivo* activity in tumors resistant to conventional agents. One class of platinum complexes targeted to DNA involves cisplatin linked to an

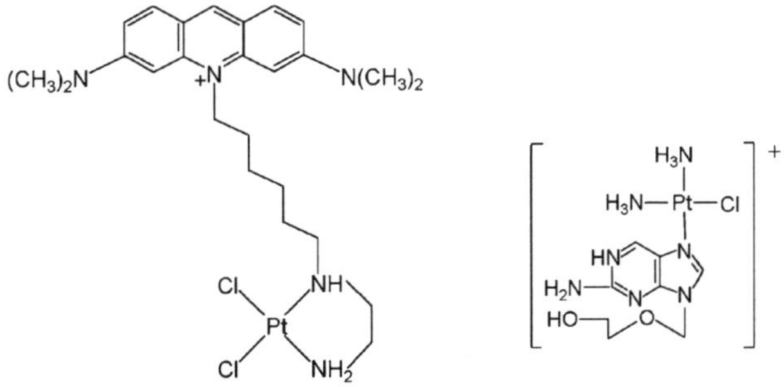

(1,2-diaminoethane)dichloroplatinum(II) linked to acridine orange by hexamethylene chain

cis-$[PtCl(NH_3)_2(N7\text{-}ACV)]^+$

Figure 25.6 Targeted and anti-viral analogues of cisplatin

intercalator. An example is (1,2-diaminoethane)dichloroplatinum(II) linked to acridine orange by trimethylene or hexamethylene chains (Figure 25.6).[55]

In the reaction with DNA, the platinum residues cross-link two base residues, while acridine orange is intercalated between the base pairs at a distance of one to two base pairs from the platinum-binding site. Other DNA-targeted platinum compounds that show anti-tumor activity include analogues of cisplatin containing as a carrier DNA minor groove non-intercalative binders netropsin and distamycin.[56–58] Polymer-based delivery of platinum to tumors has also been proposed to minimize the toxicity of the cytotoxic agent.[59–61] However, to date, there have been no clinically significant advances that have developed from this design approach.

Tumor targeting has also been the focus of much work in liposomal delivery of platinum drugs.[62,63] Other mechanisms of action, such as release from drug depot formulations, heat-triggered local drug release and transfection of genetic materials, may prove to be useful in human patients. Liposomal formulations of cisplatin and its analogues have shown promise *in vitro* and in animal models.[64,65] A novel method has been described[66] allowing the efficient encapsulation of cisplatin in a lipid formulation. The method is unique in that it generates nanocapsules which have an unprecedented drug-to-lipid ratio and an *in vitro* cytotoxicity up to 1000-fold higher than the free drug. Analysis of the mechanism of nanocapsule formation suggests that the method may be generalized to other drugs showing low water solubility and lipophilicity.

25.9 Anti-viral Activity

A number of platinum coordination compounds also exhibit anti-viral activities.[67] In the search for new, therapeutically more effective platinum drugs, platinum(II) compounds containing, as a part of the coordination sphere, certain selected anti-viral nucleosides were also synthesized.[67,68] Several compounds of this type exhibit similar or enhanced anti-viral activities *in vitro* and in many instances are less toxic to normal cells than either component.[67] An example of this class of anti-viral compounds is the *cis*-$[PtCl(NH_3)_2(N7\text{-ACV})]^+$ complex (Figure 25.6). This compound is based on anti-tumor cisplatin, but contains in its coordination sphere anti-viral acyclovir [ACV, the deoxyriboguanosine analogue 9-(2-hydroxyethoxymethyl)guanine].

25.10 Combinations of Platinum Complexes with Other Agents

Platinum anti-tumor drugs have been combined successfully with other anticancer drugs, e.g. doxorubicine, etoposide, fluorouracil, gemcitabine, ifosfamide, mytomycine and paclitaxel. For instance, combinations of cisplatin or carboplatin with paclitaxel appear highly active in advanced ovarian cancer[69] or urothelial[70] and endometrial[71] cancers respectively. Combinations of anti-tumor platinum compounds with DNA repair inhibitors such as cytarabine, hydroxyurea, gemcitabine and arabinosyl-2-fluoroadenine have also shown synergy in several experimental models.[72,73] In addition, platinum-based chemotherapy combined with radiotherapy represents a significant advance in the treatment of locally advanced cervical cancer compared with chemotherapy alone.[74]

25.11 Conclusions

Changing the chemical structure of platinum compounds may substantially modulate their DNA-binding mode, subsequent processing of DNA damage and consequently the mechanism of biological efficacy of these compounds. These structural modifications may also affect the spectrum of biological activity of the platinum agents, the development of drug resistance, and also their toxicity profile. Hence, a further understanding of how new platinum compounds modify DNA and how these modifications are further processed in cells may lead to further insight of 'downstream' effects, initiated through differential protein recognition and repair, that may produce unique biological effects. On the other hand, there is still a gap between these molecular events involving DNA interactions and an understanding as to why platinum anticancer drugs are more poisonous to cancer cells than to normal cells. The studies so far performed in this area have implicated multiple systems including several classes of DNA repair, replication, transcription, cell cycle and cell death responses involved in the processes associated with cellular sensitivity to the platinum drugs. It is also likely that many other determinants remain to be identified. Complete knowledge of how modifications of DNA by anti-tumor platinum compounds and other metal-based drugs affect the components of these pathways should provide a basis for understanding the mechanism of action of platinum drugs and, thereby, a rational basis for the design of new metal-based drugs and to identify optimal treatment strategies.

Acknowledgements

The support from the Grant Agency of the Academy of Sciences of the Czech Republic (Grants 5004101, B5004301 and S5004009) is gratefully acknowledged. JK is international research scholar at Howard Hughes Medical Institute. JK and VB acknowledge that their participation in the EC COST Chemistry Action D20 enabled them to exchange regularly the most recent ideas in the field of platinum anti-cancer drugs with several European colleagues. Note that the reference list is not an exhaustive review of all literature associated with the subject of this article. In many cases, the most recent papers or reviews are cited to enable the reader to trace back to earlier contributions.

References

1. B. Rosenberg, L. Van Camp, J.E. Trosko and V.H. Mansour, *Nature* **222**, 385–386 (1969).
2. E. Wong and C.M. Giandomenico, *Chem. Rev.* **99**, 2451–2466 (1999).
3. Y.P. Ho, S.C.F. AuYeung and K.K.W. To, *Med. Res. Rev.* **23**, 633–655 (2003).
4. M.S. Highley and A.H. Calvert, in: *Platinum-Based Drugs in Cancer Therapy*, L.R. Kelland and N.P. Farrell (Eds), Humana Press Inc., Totowa/NJ, pp. 171–194 (2000).
5. J. Lokich, *Cancer Invest.* **19**, 756–760 (2001).
6. P.J. O'Dwyer, J.P. Stevenson and S.W. Johnson, *Drugs* **59**, 19–27 (2000).
7. L.R. Wiseman, J.C. Adkins, G.L. Plosker and K.L. Goa, *Drugs Aging* **14**, 459–475 (1999).
8. F. Levi, G. Metzger, C. Massari and G. Milano, *Clin. Pharmacokinetics* **38**, 1–21 (2000).
9. B. Desoize and C. Madoulet, *Crit. Rev. Oncol. Hematol.* **42**, 317–325 (2002).
10. N.P. Johnson, J.-L. Butour, G. Villani *et al.*, *Prog. Clin. Biochem. Med.* **10**, 1–24 (1989).
11. V. Brabec, *Prog. Nucleic Acid Res. Mol. Biol.* **71**, 1–68 (2002).
12. G. Villani, N.T. Le Gac and J.-S. Hoffmann, in: *Cisplatin: Chemistry and Biochemistry of a Leading Anticancer Drug*, B. Lippert (Ed.), VHCA, Wiley-CH, Zürich, Weinheim, pp. 135–157 (1999).
13. Y. Jungand and S.J. Lippard, *J. Biol. Chem.* **278**, 52084–52092 (2003).

14. M. Kartalou and J.M. Essigmann, *Mutat. Res.* **478**, 23–43 (2001).
15. V. Brabec and J. Kasparkova, *Drug Resist. Updates* **5**, 147–161 (2002).
16. S.W. Johnson, K.V. Ferry and T.C. Hamilton, *Drug Resist. Updates* **1**, 243–254 (1998).
17. M. Kartalou and J.M. Essigmann, *Mutat. Res.* **478**, 1–21 (2001).
18. A.L. Pinto and S.J. Lippard, *Proc. Natl. Acad. Sci. USA* **82**, 4616–4620 (1985).
19. B. Salles, J.L. Butour, C. Lesca and J.P. Macquet, *Biochem. Biophys. Res. Commun.* **112**, 555–563 (1983).
20. M.A. Fuertes, J. Castilla, C. Alonso and J.M. Perez, *Curr. Med. Chem.* **10**, 257–266 (2003).
21. V.M. Gonzalez, M.A. Fuertes, C. Alonso and J.M. Perez, *Mol. Pharmacol.* **59**, 657–663 (2001).
22. F.A. Blommaert, H.C.M. van Dijk-Knijnenburg, F.J. Dijt *et al.*, *Biochemistry* **34**, 8474–8480 (1995).
23. O. Vrana, V. Brabec and V. Kleinwächter, *Anticancer Drug Des.* **1**, 95–109 (1986).
24. J.M. Woynarowski, W.G. Chapman, C. Napier *et al.*, *Mol. Pharmacol.* **54**, 770–777 (1998).
25. J.M. Woynarowski, S. Faivre, M.C.S. Herzig *et al.*, *Mol. Pharmacol.* **58**, 920–927 (2000).
26. J.T. Reardon, A. Vaisman, S.G. Chaney and A. Sancar, *Cancer Res.* **59**, 3968–3971 (1999).
27. B. Spingler, D.A. Whittington and S.J. Lippard, *Inorg. Chem.* **40**, 5596–5602 (2001).
28. A. Vaisman, S.E. Lim, S.M. Patrick *et al.*, *Biochemistry* **38**, 11026–11039 (1999).
29. K. Kawanishi, Y. Miyagi, J. Yamamoto *et al.*, *Cancer Chemother. Pharmacol.* **47**, 303–308 (2001).
30. N. Farrell, L.R. Kelland, J.D. Roberts and M. Van Beusichem, *Cancer Res.* **52**, 5065–5072 (1992).
31. V. Brabec, V. Kleinwächter, J.L. Butour and N.P. Johnson, *Biophys. Chem.* **35**, 129–141 (1990).
32. V. Brabec and M. Leng, *Proc. Natl. Acad. Sci. USA* **90**, 5345–5349 (1993).
33. M. Leng, A. Schwartz and M.J. Giraud-Panis, in: *Platinum-Based Drugs in Cancer Therapy*, L.R. Kelland and N.P. Farrell (Eds), Humana Press Inc., Totowa/NJ, pp. 63–85 (2000).
34. M. Coluccia, F. Nassii, F. Loseto *et al.*, *J. Med. Chem.* **36**, 510–512 (1993).
35. V. Brabec, O. Vrana, O. Novakova *et al.*, *Nucleic Acids Res.* **24**, 336–341 (1996).
36. O. Novakova, J. Kasparkova, J. Malina *et al.*, *Nucleic Acids Res.* **31**, 6450–6460 (2003).
37. G. Natile and M. Coluccia, *Coord. Chem.* **216**, 383–410 (2001).
38. J. Kasparkova, O. Novakova, N. Farrell and V. Brabec, *Biochemistry* **42**, 792–800 (2003).
39. V. Brabec, K. Neplechova, J. Kasparkova and N. Farrell, *J. Biol. Inorg. Chem.* **5**, 364–368 (2000).
40. J. Kasparkova, V. Marini, Y. Najajreh *et al.*, *Biochemistry* **42**, 6321–6332 (2003).
41. J. Kasparkova, O. Novakova, V. Marini *et al.*, *J. Biol. Chem.* **278**, 47516–47525 (2003).

42. J.M. Perez, E.I. Montero, A.M. Gonzalez *et al.*, *J. Med. Chem.* **43**, 2411–2418 (2000).
43. E.I. Montero, S. Diaz, A.M. Gonzalez-Vadillo *et al.*, *J. Med. Chem.* **42**, 4264–4268 (1999).
44. N. Farrell, in: *Platinum-Based Drugs in Cancer Therapy*, L.R. Kelland and N.P. Farrell (Eds), Humana Press Inc., Totowa/NJ, pp. 321–338 (2000).
45. V. Brabec, J. Kasparkova, O. Vrana *et al.*, *Biochemistry* **38**, 6781–6790 (1999).
46. J. Kasparkova, J. Zehnulova, N. Farrell and V. Brabec, *J. Biol. Chem.* **277**, 48076–48086 (2002).
47. J. Zehnulova, J. Kasparkova, N. Farrell and V. Brabec, *J. Biol. Chem.* **276**, 22191–22199 (2001).
48. E. Fokkema, H.J.M. Groen, M.N. Helder *et al.*, *Biochem. Pharmacol.* **63**, 1989–1996 (2002).
49. K.J. Mellish, C.F.J. Barnard, B.A. Murrer and L.R. Kelland, *Int. J. Cancer* **62**, 717–723 (1995).
50. L.R. Kelland, in: *Platinum-Based Drugs in Cancer Therapy*, L.R. Kelland and N.P. Farrell (Eds), Humana Press Inc., Totowa/NJ, pp. 299–319 (2000).
51. P. Muller, B. Schroder, J.A. Parkinson *et al.*, *Angew. Chem. Int. Ed.* **42**, 335–339 (2003).
52. J. Kasparkova, F.S. Mackay, V. Brabec and P.J. Sadler, *J. Biol. Inorg. Chem.* **8**, 741–745 (2003).
53. W.I. Sundquist and S.J. Lippard, *Coord. Chem. Rev.* **100**, 293–322 (1990).
54. M. Leng and V. Brabec, in: *DNA Adducts: Identification and Biological Significance*, K. Hemminki, A. Dipple, D.E.G. Shuker, F.F. Kadlubar, D. Segerbäck and H. Bartsch (Eds), International Agency for Research on Cancer, Lyon, pp. 339–348 (1994).
55. B.E. Bowler, L.S. Hollis and S.J. Lippard, *J. Am. Chem. Soc.* **106**, 6102–6104 (1984).
56. M. Lee, J.E. Simpson, A.J. Burns *et al.*, *Med. Chem. Res.* **6**, 365–371 (1996).
57. H. Loskotova and V. Brabec, *Eur. J. Biochem.* **266**, 392–402 (1999).
58. H. Kostrhunova and V. Brabec, *Biochemistry* **39**, 12639–12649 (2000).
59. A.A. Bogdanov, C. Martin, A.V. Bogdanova *et al.*, *Bioconjug. Chem.* 7, 144–149 (1996).
60. Anon., *Expert Opin. Ther. Patents* **9**, 661–662 (May 1999).
61. M. Bouma, B. Nuijen, D.R. Stewart *et al.*, *Anticancer Drugs* **13**, 915–924 (2002).
62. I. Han, A.R. Khokhar and R. PerezSoler, *Cancer Chemother. Pharmacol.* **39**, 17–24 (1996).
63. A. Mori, S.P. Wu, I. Han *et al.*, *Cancer Chemother. Pharmacol.* **37**, 435–444 (1996).
64. M.S. Newman, G.T. Colbern, P.K. Working *et al.*, *Cancer Chemother. Pharmacol.* **43**, 1–7 (1999).
65. T. Boulikas and M. Vougiouka, *Oncol. Rep.* **10**, 1663–1682 (2003).
66. K.N.J. Burger, R.W.H.M. Staffhorst, H.C. deVijlder *et al.*, *Nat. Med.* **8**, 81–84 (2002).
67. R.C. Taylor and S.G. Ward, in: *Lectures in Bioinorganic Chemistry*, M. Nicolini and L. Sindellari (Eds), Raven Press, New York, pp. 63–90 (1991).
68. M. Coluccia, A. Boccarelli, C. Cermelli *et al.*, *Metal-Based Drugs* **2**, 249–256 (1995).
69. N.A. Jones, J. Turner, A.J. McIlwrath *et al.*, *Mol. Pharmacol.* **53**, 819–826 (1998).
70. M. Hussain, U. Vaishampayan, W. Du *et al.*, *J. Clin. Oncol.* **19**, 2527–2533 (2001).

71. A. Vasuratna, A.P. Kudelka, C.L. Edwards *et al.*, *Anticancer Drugs* **9**, 283–285 (1998).
72. R.P. Perez, *Eur. J. Cancer* **34**, 1535–1542 (1998).
73. G. Giaccone, *Drugs* **59**, U4–U5 (2000).
74. M. Fujita, T. Fujita, T. Kodama *et al.*, *Int. J. Oncol.* **17**, 393–397 (2000).

26 $_{79}$Au Gold-Based Metallotherapeutics: Use and Potential

Soo Yei Ho and Edward R.T. Tiekink

26.1 Preamble

Since ancient times, gold has been regarded as the most beautiful of all the elements that have been known and treasured by man. Gold was amongst the first metals used to make tools, weapons and jewellery, and these rapidly extended into trade and money. All of these uses made gold inseparable from human civilization. In the same way, the rarity and beauty of gold has invoked greed and triggered wars. The purpose of this chapter is to overview the more salient physical and chemical properties of gold, to describe the use of gold in therapeutic medicine, and then, to outline future prospects for this element with a particular emphasis on the development of anti-tumour agents.

26.2 The Physical and Chemical Properties of Gold

26.2.1 Physical properties of the element

Gold has the highest ductility and malleability of any element, appearing as a soft, lustrous and yellow metal. For example, one gram of gold can be extended

Metallotherapeutic Drugs and Metal-Based Diagnostic Agents: The Use of Metals in Medicine
Edited by Gielen and Tiekink

into a string several kilometres long.[1] Gold has an atomic weight of 196.967 and a density of 19.30 g cm^{-3} at 293 K;[1] the pure metal melts at 1337 K and boils at 3129 K.[1] Being soft, gold is usually alloyed to give it more strength.[2] White gold, a substitute for platinum, is an alloy of gold with platinum, palladium, nickel, or nickel and zinc. Green gold, also used by jewellers, is usually an alloy of gold with silver. Alloys of gold with copper are reddish-yellow and are used for coinage and jewellery.

Gold crystallizes in a face-centred cubic lattice and the closest gold–gold distance in this structure is 289 pm. In the gas phase, gold exists as a mixture of atoms and diatomic molecules. The Au—Au bond length in Au_2 is 250 pm and the dissociation energy of Au_2 is 221 ± 2 kJ mol^{-1}.[3,4] The first ionization energy of gold is 890 kJ mol^{-1},[1] the electron affinity is 222.8 kJ mol^{-1},[1] and the electronegativity (Pauling scale) is 2.54, comparable to that of iodide at 2.78.[1] In the solid state, many gold complexes are characterized by aurophilic interactions so that Au···Au separations in the range 250–330 pm are observed. These distances are comparable to the closest gold–gold distance in elemental gold and in diatomic gold (gas phase), and are shorter than the sum of the van der Waals radius of gold, being 360 pm.[5,6] The energy of stabilization of such aurophilic interactions, i.e. 7–11 kcal mol^{-1}, is of the same order of magnitude as hydrogen-bonding interactions.[7] The strength of the Au—Au bond in gas phase Au_2, coupled with the high values of the above-mentioned thermodynamic quantities and the presence of aurophilic interactions, appears to be the result of relativistic effects.[8–11] One consequence of Einstein's 'Special Theory of Relativity' is that the speed limit for any object is the speed of light, $c = 3 \times 10^8$ m s^{-1}. The 1*s* electron of gold must travel at 1.6×10^8 m s^{-1}, i.e. comparable to the speed of light, in order to avoid falling into the gold atom's nucleus ($Z = 79$). Since the mass of an object depends on its speed, the mass of the 1*s* electron in gold is increased to $1.2 \times m_e$. This causes substantial stabilization (contraction) of the 6*s* orbital, a small stabilization of the 6*p* orbitals and a destabilization of the 5*d* orbitals, resulting in a greater 6*s*–6*p* and a smaller 6*s*–5*d* energy separation.[9,10]

26.2.2 Chemical reactivity of gold

Gold is the most noble of the metals, being the only metal inert to moisture, oxygen and sulphur, even at elevated temperatures. However, gold can be attacked by halogens, and dissolves readily in *aqua regia* to give tetrachloroauric(III) acid, $H[AuCl_4]$. Likewise, gold dissolves in the presence of the oxidants hypochlorite or Fe(III).[2]

26.2.3 Oxidation states and stereochemistries of gold complexes

Gold belongs to group 11 of the periodic table, along with copper and silver, and has the electronic configuration $[Xe]4f^{14}5d^{10}6s^1$. In gold complexes, the oxidation states I, II and III are now well established and a brief overview of each is given in the following paragraphs. Au(I) and Au(III) species are present in gold drugs and gold complexes undergoing biological evaluation. While there are no Au(II) species used in medicinal chemistry, it may be that Au(II) is an intermediate between the Au(I) and Au(III) species.

Oxidation state I

Gold(I) complexes disproportionate in aqueous solution unless they are either of very low solubility, e.g. Au(I), or in presence of a strong ligand, e.g. $[Au(CN)_2]^-$, according to the equation

$$3Au(I)(aq) \rightarrow 2Au(s) + Au(III)(aq) \quad E_0 = +0.47\,V$$

Since Au(I), the aurous ion, has a $5d^{10}$ electronic configuration, Au(I) complexes are diamagnetic. They usually adopt linear structures. As mentioned above, the relativistic effects operating for gold stabilize the 6*s* orbital, which decreases the energy gap between this and the 5*d* orbitals. As a result, the *s*, *p* and *d* orbitals have relatively small energy differences, making *sp* hybridization possible, and thus explaining the high tendency of Au(I) to form two-coordinate complexes, contrary to the isoelectronic Pt(0) and even Ag(1) centres.[12] The two-coordinate complexes are linear, in accord with the valence-shell electron repulsion theory (VSEPR), and may be cationic, e.g. $[Au(PPh_3)_2]^+$, neutral, e.g. $[Ph_3PAuCl]$ or anionic, e.g. $[AuCN_2]^-$. Being a soft metal centre, the most favoured ligands for gold are 'soft', i.e. those ligands that contain sulphur (thiolates, thioethers), selenium (selenates) and phosphorus (phosphines) donor atoms. Nitrogen (cyanide, amines) donors and alkyl (methyl, phenyl) groups are also possible but less common.

While two-coordinate linear geometries dominate Au(I) structures, higher coordination numbers are also known. Trigonal planar complexes with coordination number three such as $[Au(PPh_3)_3]^+$ and $[(Ph_3P)_2AuCl]$, as well as tetrahedral complexes with coordination number four such as $[Au(PMe_3)_4]^+$ are known.[13] Nevertheless, these complexes are rather unstable and readily lose phosphine ligands to revert to the more favoured linear geometry.[14,15]

Oxidation state II

Compared to the common oxidation states for gold, Au(I) and Au(III), Au(II) complexes are relatively scarce and have no known biological function. Gold(II) species may be formed as transient intermediates in redox reactions between the stable oxidation states I and III.[16] The lack of stability of Au(II) arises as these species readily disproportionate to more stable Au(I) and Au(III) species.[17] Complexes formed with Au(II) are often organoAu(II) species and are expected to be paramagnetic, owing to the outer $5d^9$ electronic configuration, unless a gold—gold bond is present.[18] Gold(II) complexes are also formed with certain unsaturated sulphur-donor ligands such as the maleonitriledithiolate ion leading to square planar Au(II) centres.[18]

Oxidation state III

Gold(III), the auric ion, has a $5d^8$ electronic configuration and, with the isoelectronic Pt(II) and Rh(I) ions, is diamagnetic and adopts, almost exclusively, four-coordinate, square-planar complexes. The range of ligands able to complex Au(III) is numerous and so complexes of Au(III) can have various chemical and physical properties. For example, they can carry charges from +3 (with four neutral ligands, e.g. NH_3) to −1 (with four negative ligands, e.g. Cl^-). Five-coordinate Au(III) complexes with rigid bidentate nitrogen-donor ligands are square pyramidal or distorted square pyramidal, e.g. [$AuCl_3$(2,2′-biquinolyl)][19] and [$AuCl_3$(2,9-dimethyl-1,10-phenanthroline)].[20] Trigonal bipyramidal Au(III) complexes are generally formed as intermediates during ligand-substitution reactions in square-planar Au(III) complexes.[20]

26.3 Gold in Medicine

26.3.1 Historical perspective

The Arabians, Chinese and Indians, from as early as 2500 BC, are known to have exploited gold for medicinal purposes.[21,22] In the first century, Pliny of Greece prescribed gold to cure warts. In medieval Europe, alchemists had numerous recipes for an elixir known as *aurum potabile*. This mixture was, in general, an alcoholic solution that contained various herb extracts and essential oils, along with gold, as a powder or as small flakes. In the seventeenth century, Nicholas Culpepper used gold for the treatment of ailments caused by a decrease in the 'vital spirits', such as melancholy, fainting and fever. Later in the nineteenth century, Chrestien prescribed the double chloride of gold and sodium, Na[$AuCl_4$], for the treatment of syphilis, and at the end of the century, Na[$AuCl_4$] was prescribed to cure chronic alcoholism.[23] It was in 1890 that the German bacteriologist Robert Koch discovered the bactericidal properties of gold cyanide.[24] It might be argued that this event marked the beginning of medicinal uses of gold based on scientific observation and data. Gold cyanide was shown to be lethal towards tubercle bacillus *in vitro*, the bacteria responsible for tuberculosis. This led to the use of gold therapy for tuberculosis in the 1920s.[25,26] Gold therapy was then extended to the treatment of other diseases such as rheumatoid arthritis (RA), on the mistaken belief that RA was in fact a bacterial infection. Other treatments include against systemic lupus erythematosus (a condition where the body's immune system attacks itself), ulcerative colitis (inflammation and ulcers occurring in the rectum and lower part of the colon) and other autoimmune conditions.[27] 'Chrysotherapy' is a term now used to designate the treatment of RA with Au(I) salts and is derived from the Greek name *Chryseis*, who, in Greek mythology, was the golden-haired daughter of *Chryses*, a priest of Apollo. Chrysotherapy continues to enjoy some acceptance for the treatment of RA even after over 80 years of use. However, in contemporary treatment, it is probably fair to state that Chrysotherapy is often used only as a last-line defence against RA. An interesting and thorough historical account of the development of gold and its complexes in medicine is available.[28]

26.3.2 Gold therapy for rheumatoid arthritis

Rheumatoid arthritis

It is generally agreed that RA is an inflammatory disease, of unknown causation, and one that is manifested, sometimes chronically, particularly in the peripheral joints of the body. The well-recognized symptoms of the disease, arising from inflammation, are the swellings of the joints and the incumbent physical disability and pain. Inflammation occurs first in the synovial membrane surrounding the joints and soon spreads to the synovium with progression of the disease. This has the result of eroding the articular cartilage on the bone joints that once damaged or removed means the bones directly rub against each other with concomitant bone erosion. Approximately 1–2% of the overall population of industrialized countries suffers from RA, although perhaps 10 times that number suffers from one of the approximately 150 known arthritic conditions. RA normally afflicts adults but not exclusively so. There is no pattern of inheritance with the disease, and females suffer RA more than males, in a ratio of 4:1. The pathogenesis of RA is linked to a breakdown of the body's immune system but the actual cause of RA is not known. However, a patient is diagnosed with RA if they have a combination of three factors, namely erosive joints, inflammation and the presence of the so-called rheumatoid factor (associated with the production of immunoglobulin).

Anti-arthritic Drugs

The first-generation gold compounds used for the treatment of RA are Au(I) thiolates such as aurothioglucose (**1**; marketed variously as Solganol®, Aureotan®, Auromyose®), disodium aurothiomalate (**2**; Myocrisin®, Myochrisin®, Mychrisis®, Miocrin®, Shiosol®, Tauredon®), tri-sodium bis(thiosulphato)gold (**3**; Sanocrysin®, Sanochrisin®), aurothiopropanol sulphonate (**4**; Allochrysin®, Allochrysine®, Limière®) and sodium gold 4-amino-2-mercaptobenzoic acid (**5**; Krysolgan®). Of these, drugs **1** and **2** are probably the most important. The chemical compositions of these drugs are shown in Figure 26.1. It should be noted that except for **3**,[29] the precise structures of these molecules are not known. In 1985, a new compound, a second-generation drug (2,3,4,6-tetra-*O*-acetyl-1-thio-*β*-D-glucopyranosyl) (triethylphosphine)gold(I) (**6**; Auranofin®, Aktil®, Crisinor®, Crisofin®, Ridaura®), was introduced as an orally active gold drug for the treatment of RA.

The first-generation gold complexes used to treat RA feature linear, two-coordinate Au(I) thiolates and, with the exception of **3**, are polymeric. Crystallographic evidence for this is found in the structure of $[Na_2Cs(H)][\{Au(SCH)(CO_2)CH_2CO_2\}_2]$, an analogue of **2**.[30] The +I oxidation state of gold in these complexes is stabilized against disproportionation to Au(0) and Au(III) by the thiolate ligand. Importantly, the open geometry allows for exchange reactions with biological ligands, via three-coordinate intermediates. A common feature of drugs **1**–**5** is that they are water soluble, owing to the presence of hydrophilic

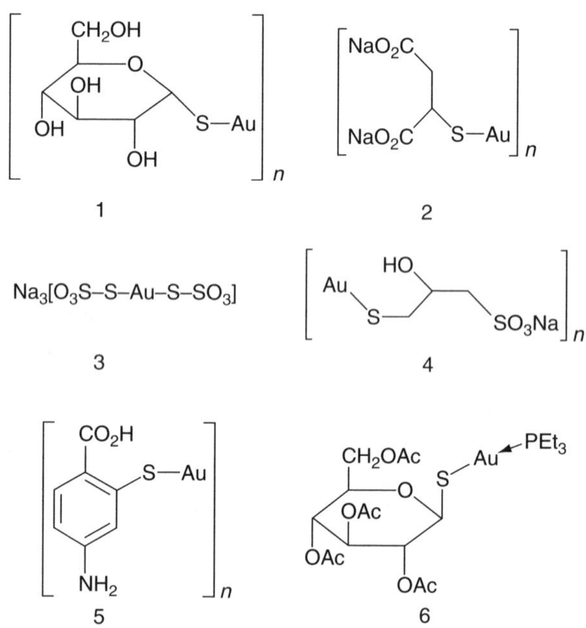

Figure 26.1 Chemical structures of aurothioglucose (**1**), disodium aurothiomalate (**2**), tri-sodium of bis(thiosulphato)gold (**3**), aurothiopropanol sulphonate (**4**), gold(I) 4-amino-2-mercaptobenzoate (**5**) and [tetra-*O*-acetyl-*β*-D-(glucopyranosyl)thio]-triethylphosphine)gold(I) (**6**)

groups and/or charge, and this property directs their mode of administration to patients, i.e. via intramuscular injection.[31,32] A disadvantage of the first-generation gold drugs is that they require frequent visits to the clinic as a doctor must administer the drug. Typically, doses are of the order of about 50 mg per week, at least initially, and usually decrease after an induction period. One potential problem with these gold drugs is that gold is accumulated in organs such as the kidney resulting in nephrotoxicity, which leads to a leakage of protein and blood into the urine; in severe cases, gold therapy must be stopped. Also, the beneficial effects of therapy may not be apparent for 6 months after the commencement of treatment and many patients suffer from other side effects of varying severity (see below) that may continue for a considerable time after Chrysotherapy has been discontinued.[33] Over and above this, not all patients respond to Chrysotherapy.[34]

A relatively new gold-based drug that finds wide use is Auranofin (**6**).[35] It has an effective linear P–Au–S geometry and has been structurally characterized by X-ray crystallography[36] in contrast with many of the other gold-based drugs. Being lipophilic, Auranofin can be administered orally, in doses of 3–6 mg per day, so that this 'second-generation' drug has a considerable advantage in this respect over the aforementioned drugs that require regular visits to a doctor. The likely role of the phosphine ligand in Auranofin is to enhance the lipid solubility of the drug to facilitate its absorption in the gut. The different solubility profile of Auranofin results in a different biodistribution of gold throughout the body compared with that of the conventional water-soluble drugs, which are injected immediately into the blood stream. A possible advantage of this is that the concentration of gold in the kidneys, with harmful effects as alluded to earlier, is considerably less for Auranofin. This advantage notwithstanding, there are, common with the water-soluble gold-based drugs, a number of deleterious side effects associated with Chrysotherapy.

Patients undergoing Chrysotherapy may suffer from a wide variety of unpleasant and harmful side effects. These may include a metallic taste in the mouth, discolouration of the skin, dermatitis, diarrhoea, nausea, flushing, vomiting, bone marrow damage, etc. Such a range of difficulties begs the question, why use gold-based drugs? The answer to this question is quite straightforward: there is no cure for RA. Even though there are various classes of drugs available for the treatment of RA, it should be noted that treatment is designed to alleviate the symptoms of the disease, to reduce the swelling and pain, or even halt the progression of the disease, but there is no cure for RA. Amongst the classes of drugs employed in the treatment of RA are (i) Non-steroidal anti-inflammatory drugs (NSAIDS), which function by blocking the production of (bad) prostaglandins that promote inflammation and pain. Examples of NSAIDS include Advil® (ibuprofen), Feldene® (piroxicam) and Naprosyn® (naproxen). The major problem with NSAIDS is that they do not differentiate between the so-called good (that afford protection to the stomach lining and kidneys) and bad prostaglandins; (ii) COX-II enzyme inhibitors that target inflammatory arthritis. Examples include Celebrex® (celecoxib) and

Mobic® (meloxicam); (iii) Salicylates, such as Aspirin®, are well known to relieve the symptoms associated with RA; and (iv) Disease-Modifying Anti-Rheumatic DrugS (DMARDS) which function to impede the progression of the disease as the name suggests. Examples of DMARDS include Cytoxan® (cyclophosphamide), Novo-chloroquine® (chloroquine) and Rheumatrex® (methotrexate). Gold drugs used in Chrysotherapy are classed as DMARDS. Thus, there are a large variety of therapies available for managing RA and the gold drugs are but one of these. Given the long use of gold drugs, it is not surprising that some considerable effort has been devoted to elucidating the possible mechanism(s) of action of gold drugs as summarized in the next section.

26.4 Metabolism of Gold Drugs

26.4.1 Biological ligand exchange

Research on humans and laboratory animals has shown that the clinically used gold drugs should be considered as prodrugs, which undergo rapid metabolism *in vivo* to form active pharmacological species.[37,38] The mechanism of action of these drugs is not known precisely but most probably involves a facile thiol exchange process, via a three-coordinate intermediate (see above), i.e. the substitution of the thiolate ligand with a biological thiol such as albumin or, less likely, glutathione. Albumin is the major protein in serum and being sulphur-rich, is the major source of biological thiols outside of the cells in blood. There are 35 cysteine residues in albumin and of these, cysteine-34 is likely to be deprotonated at physiological pH. Thus, this is most likely the target for gold drugs in blood. Subsequent reaction with albumin (or another thiol) ultimately results in gold being embedded within a ball of protein that is delivered to the site of inflammation, via white blood cells, where presumably the gold is extracted to allow for pharmacological benefit. In the case of Auranofin, the liberated triethylphosphine (Et_3P) is oxidized, with concomitant reduction of a disulphide link in albumin, to the harmless triethylphosphine oxide ($Et_3P{=}O$), which is excreted via the kidneys.[39]

26.4.2 Gold in cells: The thiol-shuttle model

First-generation gold drugs, such as aurothioglucose (**1**) and disodium aurothiomalate (**2**), are injected intramuscularly into patients and from there can interact with cells. It turns out that these usually water soluble and charged species do not enter cells but, rather, bind to the cell membranes via cell surface thiols. In this way, they are able to affect the cell metabolism by interfering with normal cell signalling pathways or starving cells by the inhibition of nutrient uptake.[40,41] By contrast, Auranofin (**6**) (or more precisely its metabolites) is taken up rapidly and extensively by various cells, and Mirabelli *et al.* postulated a model, known as the thiol-shuttle (or sulphydryl-shuttle), to explain the mechanism of Auranofin transport in and out of cultured macrophage cells.[42] Auranofin reaches the cell in its complete form where the drugs encounter sulphydryl-dependent membrane transport proteins (MSH), which populate the cell membrane. Before entering the cell, the thiolate ligand of Auranofin is displaced and remains outside of the cell, whereas the Et_3PAu^+ moiety binds to the MSH, which transports the cation through the membrane. Once inside the cell, the cation is transferred to proteins and low-molecular-weight thiols, just as in blood, and the phosphine can be oxidized to $Et_3P{=}O$ as described earlier. Whether or not bound to the phosphine, the gold can be taken out of the cell by effectively the same way as it entered it, possibly still bound to an intracellular thiol such as albumin and glutathione (R′SH). Once outside the cell, the gold species again binds to an extracellular thiol, such as albumin, resulting in the formation of an AlbS–Au–SR′ complex. This may indeed be the major circulating metabolite after displacement of the original carrier ligands of the drug. The question remains. How is the gold extracted from the above-described 'ball of protein' in order to effect therapeutic use? A proposal gaining acceptance is one that involves the extraction of gold with cyanide at the sites of inflammation.[43–46]

26.4.3 Cyanide metabolites

Cyanide, an exceptionally strong ligand for gold, as seen in the log *K* value of approximately 36 for $[Au(CN)_2]^-$, may be selectively generated at sites of inflammation. With such a driving force as indicated in the magnitude of the log *K* value, it is hardly surprising that $[Au(CN)_2]^-$ is well established as a metabolite of Chrysotherapy. It is known that $[Au(CN)_2]^-$ is generated by stimulated polymorphonuclear leukocytes and a possible substrate is thiocyanate, SCN^-, that is acquired naturally from food. At the sites of inflammation, the locally activated leukocytes can produce hydrogen peroxide (H_2O_2) and at the same time may produce myeloperoxidase (MPO), a peroxide-consuming enzyme. Thus, the potential to form CN^- from the reaction between SCN^- and H_2O_2/MPO exists.[44] Similar reactions involving iminoacids, products of D-aminoacid oxidase, can also produce CN^-.[46] Without MPO, reactions between glycine and OCl^- (arising from the reaction between Cl^- and H_2O_2) can lead to the production of CN^-.[45] So, it might be that it is the formation of $[Au(CN)_2]^-$ that is the key step of the metabolism of gold drugs as this species is taken up into cells where they limit the oxidative burst of active inflammatory leukocytes and thereby limit the production of the reactive and damaging oxygen-containing species $OH^\bullet$ and O_2^-.[37,47]

26.5 Pharmacological Potential of Gold Complexes

26.5.1 Motivation for drug development

At present, the major clinical use of gold complexes relates to RA (see later). Despite this, there seems to be a dearth of activity, at least as judged from open literature, devoted to generating new gold drugs for Chrysotherapy with improved efficacy and reduced toxic side effects.[48–50] Most recent efforts in pharmaceutical gold chemistry are dedicated to developing anti-tumour agents of both Au(I) and Au(III). Basically, cancer develops when cells in a particular part or parts of the body begin to proliferate out of control as a result of the uncontrolled growth of abnormal cells with damaged DNA that are not repaired by the normal repair mechanisms of the body. Tumour cells can spread through the bloodstream or lymph vessels, via a process known as metastasis, to other parts of the body where they begin to grow and replace normal tissue. Motivations for investigating the anti-cancer activity of gold complexes are varied, but there is increasing evidence of a link between inflammatory disorders and cancer.

Screening for anti-tumour activity of anti-arthritic gold drugs could have been motivated by long-term studies showing that patients undergoing Chrysotherapy had reduced risk from also suffering from cancer.[51] However, the first mention of a link between inflammation and cancer dates back to 1863 when Virchow observed leukocytes in neoplastic tissues as discussed in a review by Balkwill and Mantovani[52] that goes on to discuss possible links between the two conditions. Further, drugs such as cyclophosphamide (**7**), 6-mercaptopurine (**8**) and methotrexate (**9**) are known to have both anti-cancer and anti-inflammatory characteristics;[52–54] Figure 26.2 for chemical structures. Other reasons for developing gold complexes as anti-tumour agents relate to chemical similarities as well as to drug delivery strategies.

In terms of a chemical motivation for examining Au(III) species, Au(III) is isoelectronic with cisplatin, containing the Pt(II) centre and which is, the most widely used anti-cancer drug, and adopts similar square-planar structures, indicating perhaps that Au(III) complexes may also target DNA as does cisplatin. In terms of a drug delivery strategy, given that Au(I) drugs are tolerated, to a greater or lesser extent, by patients, and what is known about the metabolism of Au(I) drugs, i.e. they are prodrugs, the use of Au(I) as a platform to deliver biologically active molecules for cancer treatment is another motivation for their study. Amongst the first gold complexes to be investigated for cytotoxicity and anti-tumour activity were the gold drugs used in Chrysotherapy. Generally, drugs such as disodium aurothiomalate (**2**) were inactive, most likely due to their inability to enter cells. By contrast, Auranofin (**6**)

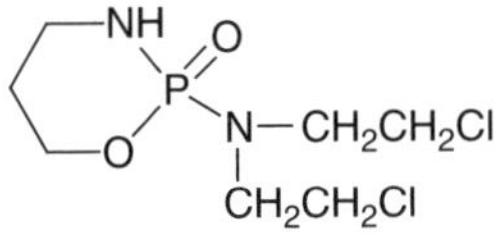

7

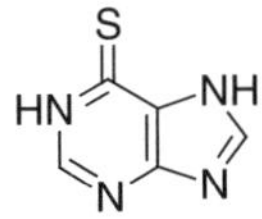

8

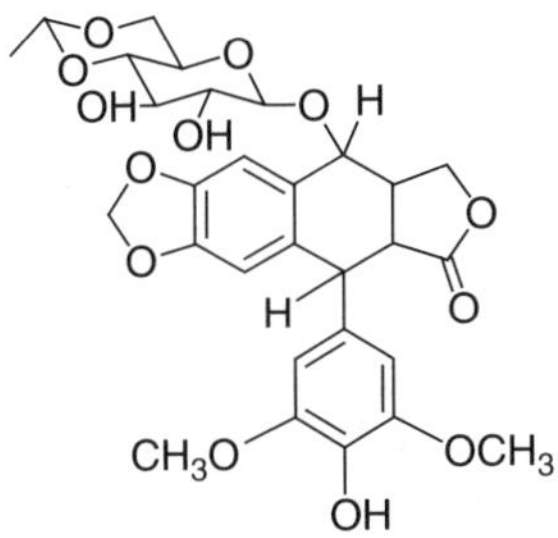

9

Figure 26.2 Chemical structures of cyclophosphamide (**7**), 6-mercaptopurine (**8**) and methotrexate (**9**)

demonstrated significant potential in this regard, sparking considerable interest in the study of the anti-cancer potential of phosphinegold(I) thiolates.[54,55]

26.5.2 Anti-tumour properties of gold drugs

The potential anti-tumour activity of gold complexes has been demonstrated in a number of experimental models but, to date, no gold complexes have entered clinical trials.[56] Auranofin (**6**) was shown to be cytotoxic towards Hela cells.[57] It was also active against P388 leukaemia *in vivo*, a well-documented primary screen for anti-tumour activity,[58] whereas the oligomeric Au(I) thiolates were inactive.[59,60] However, **6** was inactive in subsequent tests using solid tumour models.[61] A series of tertiary phosphinegold(I) complexes with thiosugar ligands was found to be active against both the P388 leukaemia and B16 melanoma, *in vitro*, and P388, *in vivo*.[55] In general, phosphinegold(I) thiolates were more cytotoxic than gold thiolates as well as their phosphinegold(I) chloride analogues, indicating the importance of the phosphine ligand and the thiolate for activity, and the investigation of Auranofin analogues continues today (see later). Besides demonstrating the potential of Au(I) complexes as anti-tumour agents, these early studies showed that Au(I) complexes that were inactive in *in vitro* trials were not active *in vivo*, thereby indicating a reliable protocol for screening.

26.5.3 Tetrahedral gold(I) complexes

As an extension of the above studies on mononuclear phosphinegold(I) thiolates, complexes with bidentate phosphines were investigated.[62] An interesting phenomenon was recognized for Au(I) complexes of bis(diphenylphosphino) ethane (dppe) in that the dinuclear species rearranged in solution to give the unusual bis-chelated, tetrahedrally coordinated Au(I) species, $[Au(dppe)_2]^+$ (**10**),[63–65] represented schematically in Figure 26.3. These, too, along with their Cu(I) and Ag(I) analogues, proved to present exciting cytotoxicity profiles. Another important result was the observation that $[Au(dppe)_2]^+$ when co-administered with cisplatin was more effective than either species administered individually, indicating some sort of synergic effect.[63] A series of bis-chelated phosphine analogues with the general formula $[Au(R_2PYPR_2)]X$ (where R and Y is alkyl or aryl; X is the counter ion such as halide, etc.) were also screened for anti-tumour potential in a subsequent study.[64]

From the data obtained in this study it was apparent that the nature of the X counter ion in $[Au(R_2PYPR_2)]X$ had little influence upon their potency against B16 melanoma.[64] The most active complexes had phosphorus-bound phenyl groups, i.e. $[Au(dppe)_2]^+$ (**10**). It was suggested that the reduced potency of complexes with phosphorus-bound ethyl substituents could be correlated to their enhanced reducing ability,[64] i.e. ability to be oxidized to phosphineoxides.

Despite the encouraging *in vitro* and *in vivo* results, clinical trials for $[Au(dppe)_2]^+$ could not proceed as a result of acute toxicity associated with this species in organs such as the heart, liver and lungs.[66]

Subsequent work in this area, conducted primarily by Berners-Price, McKeage *et al.* has focussed on moderating the lipophilicity of the tetrahedral phosphine gold(I) complexes so as to enhance their anti-mitochondrial activity,[67] as there is increasing evidence to suggest that these complexes target mitochondria in tumour cells.[68,69]

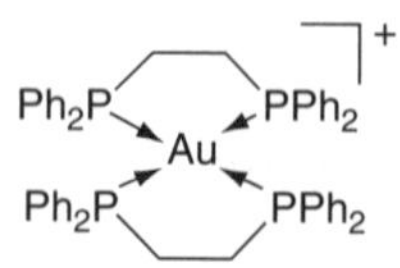

10

Figure 26.3 Chemical structure of cationic $[Au(dppe)_2]^+$ (**10**)

26.5.4 Auranofin analogues

As mentioned above, Auranofin (**6**) is known to possess some anti-tumour activity and this fact, coupled with the conclusion that this drug is in fact a prodrug, led to studies based on the following premise. If the phosphinegold(I) thiolate is metabolized upon administration to patients to liberate the thiolate, it makes sense to couple the phosphinegold(I) entity with thiols that possess anti-tumour activity. Thus, a particular focus of the study of Auranofin analogues has been to couple biologically active thiols with phosphinegold(I) entities in the hope that upon administration of the prodrug, both the phosphinegold(I) entity and thiol would provide therapeutic benefit.[70] An obvious example of a biologically active thiol, already mentioned, is 6-mercaptopurine (**8**), shown in Figure 26.2. That this approach may work can be ascertained from the following data. The IC_{50} values (i.e. the concentration of material required to inhibit the growth of a particular cell line by 50%), in this case against cisplatin-resistant L1210/DDP leukaemia cells, are 3.3, 0.14 and 0.05 μM for 6-mercaptopurine (**8**; 6-MPH), Ph_3PAuCl and Ph_3PAu(6-MP) (**11**; Figure 26.4), respectively.[71] These results demonstrate the obvious advantage of coupling the phosphinegold(I) and 6-mercaptopurine entities. Further, *in vivo* studies conducted on Balb/C mice inoculated with PC6 tumours (cisplatin-sensitive plastocytoma) show that while neither 6-mercaptopurine (**8**) nor Ph_3PAuCl had any measurable effect on the growth inhibition of the tumours, Ph_3PAu(6-MP) inhibited tumour growth by 60%.[71,72] Work in this area continues and a review summarizing this aspect of gold-drug development has appeared recently.[70] Notable aspects of this research include the observation that while no hard and fast structure–activity relationship exists, for the 6-mercaptopurinate series, potency (i.e. based on *in vitro* studies) varies in the order $Cy_3P > Ph_3P > Et_3P$.[73] However, exceptions to this rule exist. Thus, for the isomeric series of compounds, $R_3PAu(SC_6H_4CO_2H\text{-}n)$, $n = 2$, 3 and 4 (**12**) (Figure 26.4), the $n = 2$ complexes were the most potent and of these, the R = Et Et species was the most promising, in particular against the A498 (renal cancer) and H226 (non-small-cell lung cancer) cell lines.[74,75] Similarly, for the related dithiocarbamate complexes, it is the $Et_3PAu(S_2CNEt_2)$ complex (**13**) (Figure 26.4)

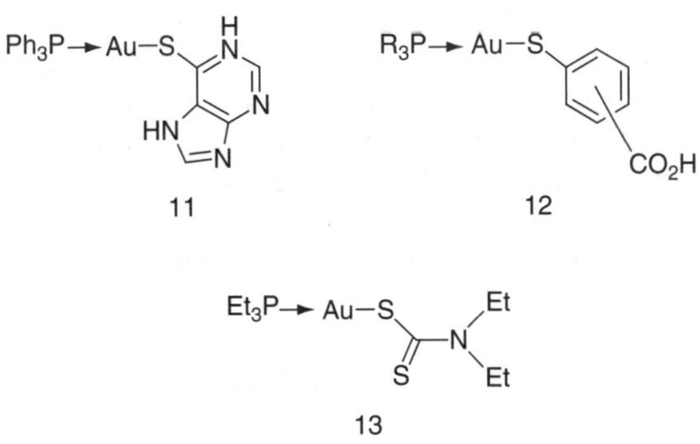

Figure 26.4 Chemical structures of Ph_3PAu(6-MP) (**11**), $R_3PAu(SC_6H_4CO_2H\text{-}n)$ (**12**) and $Et_3PAu(S_2CNEt_2)$ (**13**)

that is the most potent, at least of the derivatives investigated.[76] More encouraging have been the recent observations that the promising *in vitro* potency is translated into animal studies for these and related complexes.[77]

Despite the earlier observations indicating it was the analogues of Auranofin (**6**) that gave the greatest promise for the development of anti-tumour agents, new developments suggest that other types of gold complexes should also be investigated. For example, a recent report highlights the possible utility of Au(I) carbene complexes as anti-tumour agents.[78] Also very important in gold-drug development are complexes containing the Au(III) centre.

26.5.5 Gold(III) complexes

Gold(III) is harder than gold(I) and generally prefers hard donor atoms such as nitrogen, oxygen and carbon. Four-coordinate Au(III) compounds exist in square-planar geometries and in this regard, resemble cisplatin. Gold(III) is an oxidizing species and so might be considered unstable in the reducing mammalian environment. However, despite this, there is ample evidence that Au(III) complexes are cytotoxic; indeed, cytotoxic screening for Au(III) complexes can be traced back to the mid-1970s.[79]

In vivo *redox chemistry of gold*

In vivo, 'gold' exists as Au(I) species, usually metabolites of the administered drugs, as demonstrated by model studies.[37] This is also true when Au(III) is administered to rats.[80] The body is replete with reducing species with an affinity for gold, such as cysteine and methionine embedded in proteins. It has been known for over 20 years that disulphide bonds can reduce Au(III) to Au(I).[81] Thus, it is most likely that any administered Au(III) complex will be metabolized to an Au(I) species. The above notwithstanding, oxidation of Au(I) to Au(III), *in vivo*, has been observed.[82,83] Myeloperoxidase, already mentioned above in section 26.4.3 generates the oxidizing agent hypochlorous acid (HOCl) during the oxidative burst associated with inflammation, and can oxidize disodium aurothiomalate to Au(III).[84] A redox cycle can thus be established where Au(I) is oxidized to Au(III) and Au(III), in turn, can be reduced, by the above mentioned sulphur-containing species, to Au(I).

Gold(III) complexes with anti-tumour activity

As with some of the Au(I) complexes investigated for anti-tumour potential, the combination of Au(III) with biologically active molecules has been the motivation for some studies. For example, the combination of Au(III) with

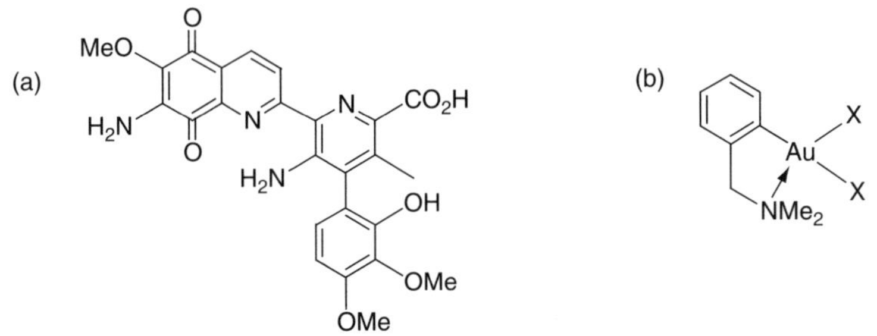

14 15

Figure 26.5 Chemical structures of streptonigrin (**14**) and generic 2-[(dimethylamino) methyl]phenylgold(III)X_2 (**15**)

streptonigrin (**14**) (Figure 26.5) to give a square-planar complex, for which the precise chemical structure is not known, is a cytotoxic species but one that has a comparable profile of potency to the uncomplexed streptonigrin.[85] Organogold(III) complexes have also been studied and arguably one of the most promising of these is based on 2-[(dimethylamino)methyl]phenylgold(III)X_2 (**15**), where X = halide, carboxylate, etc.;[86] the generic structure is shown in Figure 26.5. These complexes have a similar profile of cytotoxicity to cisplatin against the chosen panel of cell lines. Further, they demonstrate moderate antitumour activity in laboratory animals.[86] The question whether these and related square-planar Au(III) complexes interact with DNA, as do metabolites of cisplatin, is still controversial.[87]

A recent comprehensive review of the types of Au(III) complexes screened and their cytotoxicity profiles is available.[56]

26.6 Other Potential Clinical Applications of Au(I) Complexes

26.6.1 Anti-microbial activity

The bis-chelated tetrahedral Au(I) phosphine complexes described in section 26.5.3 have also been evaluated for anti-microbial activity and shown to have moderate activity but not as good as compared to their Ag(I) analogues.[88] Gold(I) complexes containing a P—Au—N core such as Ph_3PAu(imidazolate) (**16**) (Figure 26.6) show selective and significant anti-microbial activity against two Gram-positive bacteria, namely *Bacillus subtilis* and *Staphylococcus aureus*, moderate activity against a yeast (*Candida albicans*) and no activity against the Gram-negative bacteria *Escherichia coli* and *Pseudomonas aeruginosa*, and

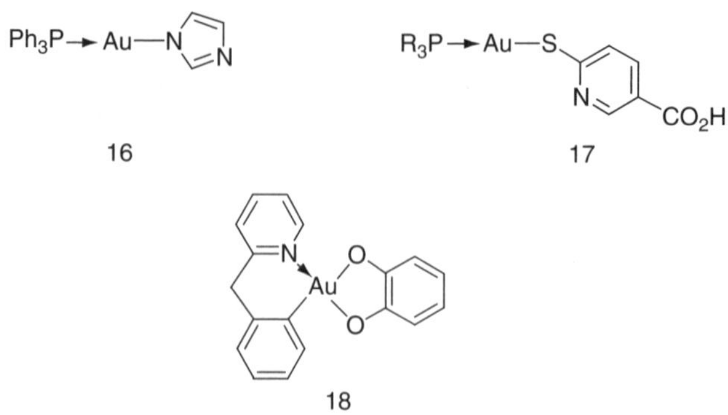

Figure 26.6 Chemical structures of Ph_3PAu(imidazolate) (**16**), Ph_3PAu(6-mercaptonicotinicate) (**17**) and 2-benyzlpyridylgold(III) catecholate (**18**)

the molds *Aspergillis niger* and *Penicillium citrinum*.[89] Analogues to Auranofin (**6**) can also show selective activity against Gram-positive bacteria, e.g. Ph_3PAu(6-mercaptonicotinate) (**17**), Figure 26.6. Nomiya *et al.* have suggested that the specific profiles of anti-microbial activity of the linear Au(I) complexes can be correlated with ease of ligand substitution, i.e. complexes that were stable to substitution to biologically relevant thiols were not active.[90] Gold(III) complexes have also been examined for anti-microbial activity.[91] Selectively in activity was determined for some complexes, such as the catecholate complex (**18**), shown in Figure 26.6, against *B. subtilis* and *P. aeruginosa*, i.e. Gram-positive and -negative bacteria, respectively.[91]

26.6.2 Anti-malaria activity

Shaw states in a recent review that a number of anti-malarial drugs, which can also be used for the treatment of RA, actually exhibit pharmacological profiles akin to that of gold drugs, an observation indicating that common features exist between the mode of action of these anti-malarial and anti-arthritic drugs.[37] Despite this, relatively little work has been devoted to exploring the possible application of gold complexes as anti-malarial agents. A series of metal complexes of the anti-malarial drugs Amodiaquine® (**19**) and Primaquine® (**20**) (Figure 26.7) were screened, *in vitro*, for activity against the parasitic protozoan *Plasmodium falciparum*, the causative agent for malaria.[92] The complexation of these drugs to a metal centre, including Au(III), did not appear to influence greatly the minimum inhibitory values.[92] More recently, a cationic phosphine-gold(I) chloroquine complex, $[Ph_3PAu(chloroquine)]PF_6$ (**21**) (Figure 26.7) was demonstrated to be active against the *P. falciparum*, *in vitro*, and *P. berghei*, both *in vitro* and *in vivo*.[93] It is noteworthy that both of these strains are

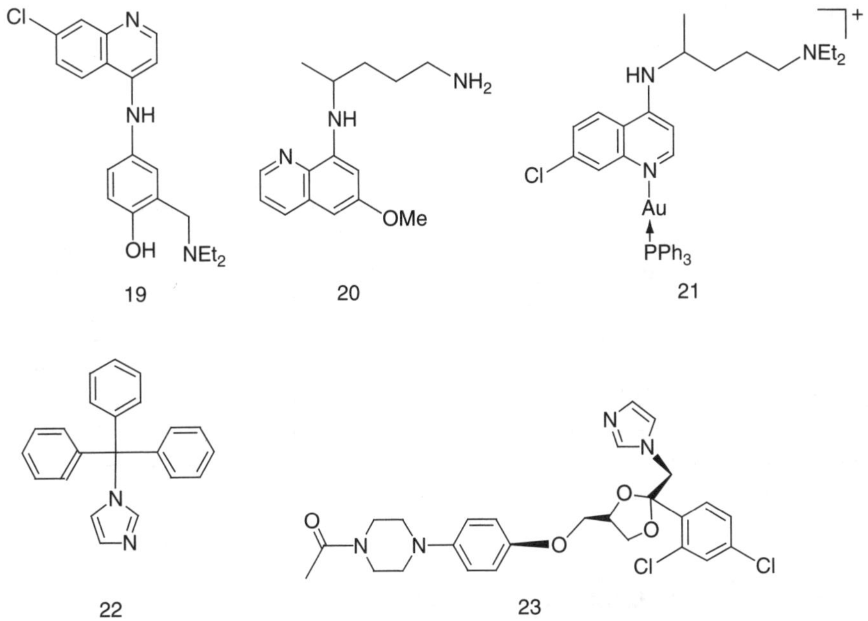

Figure 26.7 Chemical structures of Amodiaquine® (**19**), Primaquine® (**20**), [Ph_3PAu (chloroquine)]PF_6 (**21**), Clotrimazole® (**22**) and Ketoconazole® (**23**)

resistant to Chloroquine®. In addition and importantly, it was shown that the presence of Au(I) gave rise to enhanced activity compared with other metal chloroquine complexes as well as Chloroquine Diphosphate®.[93] Related Ph_3PAu(I) complexes of Clotrimazolc® (**22**) and Kctoconazolc® (**23**), shown in Figure 26.7, have been evaluated for activity against *Trypanosoma cruzi*, the causative agent of Chagas disease, a major parasitic disease afflicting people in Latin America. As found in the study of the anti-malarial activity of the Au(I) complexes discussed earlier, the presence of the phosphinegold(I) moiety resulted in an enhancement of growth inhibitory activity.[94]

26.6.3 Treatment of bronchial asthma

Studies in Japan suggest that gold drugs used in RA, particularly gold thioglucose (**3**), may also be used effectively for the treatment for bronchial asthma, thereby exploiting the immunosuppressant properties of the gold drugs.[95–97] Clinical trials have indicated that some patients with bronchial asthma have very good responses to Chrysotherapy, while others, akin to experience with arthritic patients, must drop the therapy altogether.[37]

26.6.4 Anti-HIV activity and AIDS

Interest in the use of gold complexes in this context was initiated by a clinical observation that indicated that an AIDS patient, who was being treated with Auranofin for psoriatic arthritis, experienced increasing CD4+ T lymphocyte counts. This is a significant observation, as patients with AIDS usually suffer an irreversible decline of these cells.[98] Consequently, the testing of a variety of gold compounds against human immunodeficiency virus (HIV) was undertaken. While the literature on advances made with gold complexes is sparse, it is known that $[Au(CN)_2]^-$, a known metabolite of Chrysotherapy, can inhibit the proliferation of HIV in cultured T-9 cells (a strain of CD4+ T-cells) at concentrations as low as 20 nm.[99] Other Au(I) species, including gold drugs, have also been studied. Gold thioglucose (**3**) does not readily enter cells (see earlier) and therefore is not effective against HIV in cultured cells.[37,100] Phosphinegold(I) cyanide species are too cytotoxic, killing cells at concentrations lower than those required to display anti-viral activity.[37]

26.7 Conclusions and Outlook

From the foregoing discussion, it is clear that gold complexes, containing both Au(I) and Au(III) centres, may play a vital role in therapeutic medicine. Thus, gold complexes are used in the treatment of arthritic and asthmatic conditions, and are actively being explored for potential use against diseases such as cancer and, more recently, tropical diseases. These advances indicate that the continued investigation of gold complexes, and indeed other metal complexes, is quite plainly warranted, as novel and beneficial therapeutic agents are likely to be generated.

Acknowledgements

The authors are pleased to acknowledge the National University of Singapore (R-143-000-139-112 & R-143-000-213-112) and the Biomedical Research Council (Singapore) for research funding.

References

1. http://www.webelements.com/webelements/text/Au/index.html.
2. R.J. Puddephatt, *The Chemistry of Gold*, Elsevier Scientific Publishing Company, New York, 1980.
3. J. Kordis, K.A. Gingerich and R.J. Seyse, *J. Chem. Phy.*, **61**, 5114–5121 (1974).

4. J. Donohue, *The Structures of the Elements*, Wiley, New York, 1974.
5. H. Schmidbaur, *Interdis. Sci. Rev.*, **17**, 213–220 (1992).
6. H. Schmidbaur, *Gold Bull.*, **23**, 11–21 (1990).
7. H. Schmidbaur, *Gold Bull.*, **33**, 3–10 (2000).
8. H. Schmidbaur, *Nature* (London), **413**, 31–33 (2001).
9. P. Schwerdtfeger, P.D.W. Boyd, A.K. Burrell *et al.*, *Inorg. Chem.*, **29**, 3593–3607 (1990).
10. P. Pyykkö, *Chem. Rev.*, **97**, 597–636 (1997).
11. P. Pyykkö, *Angew. Chem. Int. Ed.*, **41**, 3573–3578 (2002).
12. S.A. Cotton, *Chemistry of Precious Metals*, Blackie Academic & Professional, London (1997).
13. H. Schmidbaur and R. Franke, *Chem. Ber.*, **105**, 2985–2997 (1972).
14. P.L. Bellon, M. Manassero and M. Sansoni, *Ric. Sci.*, **39**, 173–175 (1969).
15. F. Klanberg, E.L. Muetterties and L.J. Guggenberger, *Inorg. Chem.*, **7**, 2272–2278 (1968).
16. R.L. Rich and H. Taube, *J. Phys. Chem.*, **58**, 6–11 (1954).
17. H. Schmidbauer, J.R. Mandl, A. Frank and G. Huttner, *Chem. Ber.*, **109**, 466–472 (1976).
18. J.G.M. Van Rens, M.P.A. Viegers and E. De Boer, *Chem. Phys. Lett.*, **28**, 104–108 (1974).
19. R.J. Charlton, C.M. Harris, H. Patil and N.C. Stephenson, *Inorg. Nucl. Chem. Lett.*, **2**, 409–414 (1966).
20. W.T. Robinson and E. Sinn, *J. Chem. Soc., Dalton Trans.*, 726–731 (1975).
21. G.J. Higby, *Gold Bull.*, **15**, 130–140 (1982).
22. E.R.T. Tiekink and M.W. Whitehouse, *Chem. Aust.*, **57**, 346–348 (1990).
23. S.P. Fricker, Medicinal chemistry of organogold compounds, in *The Chemistry of Organic Derivatives of Gold & Silver*, S. Patai and Z. Rappoport (Eds), John Wiley & Sons Ltd, New York, 1999.
24. R. Koch, *Dtsch. Med. Wochenschr.*, **16**, 756–757 (1890).
25. W. Unverzagt, *Dtsch. Med. Wochenschr.*, **53**, 1985–1986 (1927).
26. A. Macdowell, *Am. Rev. Tuberculosis*, **25**, 252–284 (1932).
27. B. Merchant, *Biologicals*, **26**, 49–59 (1998).
28. W.F. Kean, F. Forestier, Y. Kassam *et al.*, *Sem. Arth. Rheum.*, **14**, 180–186 (1985).
29. (a) R.F. Baggio and S. Baggio, *J. Inorg. Nucl. Chem.*, **35**, 3191–3200 (1973); (b) H. Reuben, A. Zalkin, M.O. Faltenns and D.H. Templeton, *Inorg. Chem.*, **13**, 1836–1838 (1974).
30. R. Bau, *J. Am. Chem. Soc.*, **120**, 9380–9381 (1998).
31. R.V. Parish and S.M. Cottrill, *Gold Bull.*, **20**, 3–12 (1987).
32. D.T. Walz, M.J. Di Martino and B.M. Sutton, *Med. Chem., Ser. Monogr.*, **13**, 209–244 (1974).
33. M.J. Lefevre, *Rev. Belge Homoeopath.*, **26**, 75–84 (1974).
34. M.J. Borigini and H.E. Paulus, in *Treatment of the Rheumatic Disease: Companion to the Textbook of Rheumatology*, M.H. Weisman and M.E. Weinblatt (Eds) W.B. Saunders, Philadelphia, 1995.
35. D.T. Walz, M.J. DiMartino and D.E. Griswold, *Scand. J. Rheumatol.*, **51** (Suppl.), 16–25 (1984).
36. D.T. Hill and B.M. Sutton, *Cryst. Struct. Commun.*, **9**, 679–686 (1980).

37. C.F. Shaw, III, *Chem. Rev.*, **99**, 2589–2600 (1999).
38. R. Eisler, *Inflamm. Res.*, **52**, 487–501 (2003).
39. C.F. Shaw III, A.A. Isab, J.D. Hoeschele *et al.*, *J. Am. Chem. Soc.*, **116**, 2254–2260 (1994).
40. W.E. Smith and J. Reglinski, *Persp. Bioinorg. Chem.*, **1**, 183–208 (1991).
41. W.E. Smith, J. Reglinski, S. Hoey *et al.*, *Inorg. Chem.*, **29**, 5190–5196 (1990).
42. R.M. Snyder, C. Mirabelli and S.J. Crooke, *Biochem. Pharmacol.*, **35**, 923–932 (1986).
43. R. Rudkowski, G.G. Graham, G.D. Champion and J.B. Ziegler, *Biochem. Pharmacol.*, **11**, 1687–1695 (1990)
44. G.G. Graham and M.M. Dale, *Biochem. Pharmacol.*, **11**, 1697–1702 (1990).
45. G.G. Graham, G.D. Champion and J.B. Ziegler, *Metal-Based Drugs*, **1**, 395–404 (1994).
46. E.R.T. Tiekink and M.W. Whitehouse, in *Handbook of Metal-Ligand Interactions in Biological Fluids*, Vol. 2, Marcel Dekker, New York, 1266–1273 (1995).
47. D.J. Porter and H.J. Bright, *J. Biol. Chem.*, **262**, 9154–9159 (1987).
48. P.D. Cookson, E.R.T. Tiekink and M.W. Whitehouse, *Aust. J. Chem.*, **47**, 577–586 (1994).
49. M.W. Whitehouse, P.D. Cookson, G. Siasios and E.R.T. Tiekink, *Metal-Based Drugs*, **5**, 245–249 (1998).
50. E.R.T. Tiekink, *Gold Bull.*, **36**, 117–124 (2003).
51. J.F. Fries, D. Bloch, P. Spitz and D.M. Mitchell, *Am. J. Med.*, **78** (Suppl. A), 56–59 (1985).
52. F. Balkwill and A. Mantovani, *Lancet*, **357**, 539–545 (2001).
53. I. Lorenzen, *Ann. Clin. Res.*, **7**, 195–201 (1975).
54. J.R. Ward, *Am. J. Med.*, **85** (Suppl. A), 39–44 (1988).
55. C.K. Mirabelli, R.K. Johnson, D.T. Hill *et al.*, *J. Med. Chem.*, **26**, 218–223 (1986).
56. E.R.T. Tiekink, *Crit. Rev. Oncol. Hematol.*, **42**, 225–248 (2002).
57. T.M. Simon, D.H. Kunishima, G.J. Vibert and A. Lorber, *Cancer*, **44**, 1865–1975 (1979).
58. T.M. Simon, D.H. Kunishima, G.J. Vibert and A. Lorber, *Cancer Res.*, **41**, 94–97 (1981).
59. T. Giraldi, G. Sava, L. Perissin and S. Zorzet, *Anticancer Res.*, **4**, 221–224 (1984).
60. T. Koide, T. Kojima and H. Kamei, *Cancer Biother. Radiopharm.*, **13**, 189–192 (1998).
61. C.K. Mirabelli, R.K. Johnson, C.-M. Sung *et al.*, *Cancer Res.*, **45**, 32–39 (1985).
62. S.J. Berners-Price and P.J. Sadler, *Struct. Bonding*, **70**, 27–102 (1988).
63. S.J. Berners-Price, C.K. Mirabelli, R.K. Johnson *et al.*, *Cancer Res.*, **46**, 5486–5493 (1986).
64. S.J. Berners-Price, G.R. Girard, D.T. Hill *et al.*, *J. Med. Chem.*, **33**, 1386–1392 (1990).
65. S.J. Berners-Price, M.A. Mazid and P.J. Sadler, *J. Chem. Soc., Dalton Trans.*, 969–974 (1984).
66. G.D. Hoke, R.A. Macia, P.C. Meunier *et al.*, *Toxicol. Appl. Pharmacol.*, **100**, 293–306 (1989).
67. M.J. McKeage, L. Maharaj and S.J. Berners-Price, *Coord. Chem. Rev.*, **232**, 127–135 (2002).
68. M.J. McKeage, *Brit. J. Pharmacol.*, **136**, 1081–1082 (2002).

69. M.P. Rigobello, G. Scutari, R. Boscolo and A. Bindoli, *Brit. J. Pharmacol.*, **136**, 1162–1168 (2002).
70. E.R.T. Tiekink, *Bioinorg. Chem. Appl.*, **1**, 53–67 (2003).
71. E.R.T. Tiekink, P.D. Cookson, B.M. Linahan and L.K. Webster, *Metal-Based Drugs*, **1**, 299–304 (1994).
72. L.K. Webster, S. Rainone, E. Horn and E.R.T. Tiekink, *Metal-Based Drugs*, **3**, 63–66 (1996).
73. D. Crump, G. Siasios and E.R.T. Tiekink, *Metal-Based Drugs*, **6**, 361–368 (1999).
74. D. de Vos, P. Clements, S.M. Pyke *et al.*, *Metal-Based Drugs*, **6**, 31–40 (1999).
75. D. de Vos, D.R. Smyth and E.R.T. Tiekink, *Metal-Based Drugs*, **8**, 303–306 (2001).
76. D. de Vos, S.Y. Ho and E.R.T. Tiekink, *Bioinorg. Chem. Appl.*, **2**, 141–154 (2004).
77. S.Y. Ho, Ph.D. Dissertation, National University of Singapore (2004).
78. P.J. Barnard, M.V. Baker, S.J. Berners-Price *et al.*, *Dalton Trans.*, 1038–1047 (2004).
79. R.V. Parish, *Metal-Based Drugs*, **6**, 271–276 (1999).
80. R.C. Elder and M.K. Eidsness, *Chem. Rev.*, **87**, 1027–1046 (1987).
81. C.F. Shaw III, M.P. Cancro, P.L. Witkiewicz and J.E. Eldridge, *Inorg. Chem.*, **19**, 3198–3201 (1980).
82. D. Schuhmann, M. Muranyi, J. Mirtcheva *et al.*, *J. Immunol.*, **145**, 2132–2139 (1990).
83. K. Takahashi, P. Griem, C. Goebel *et al.*, *Arthritis Rheum.*, **35**, 483–496 (1994).
84. C.F. Shaw III, S. Schraa, E. Gleichmann *et al.*, *Metal-Based Drugs*, **1**, 351–362 (1994).
85. A. Moustatih and A. Garnier-Suillerot, *J. Med. Chem.*, **32**, 1426–1431 (1989).
86. R.G. Buckley, A.M. Elsome, S.P. Fricker *et al.*, *J. Med. Chem.*, **39**, 5208–5214 (1996).
87. G. Marcon, S. Carotti, M. Coronnello *et al.*, *J. Med. Chem.*, **45**, 1672–1677 (2002).
88. S.J. Berners-Price, R.K. Johnson, A.J. Giovenella *et al.*, *J. Inorg. Biochem.*, **33**, 285–295 (1988).
89. K. Nomiya, R. Noguchi, K. Ohsawa *et al.*, *J. Inorg. Biochem.*, **78**, 363–370 (2000).
90. K. Nomiya, S. Yamamoto, R. Noguchi *et al.*, *J. Inorg. Biochem.*, **95**, 208–220 (2003).
91. C.H.A. Goss, W. Henderson, A.L. Wilkins and C. Evans, *J. Organomet. Chem.*, **679**, 194–201 (2003).
92. N. Wasi, H.B. Singh, A. Gajanana and A.N. Raichowdhary, *Inorg. Chim. Acta*, **135**, 133–137 (1987).
93. M. Navarro, H. Pérez and R.A. Sánchez-Delgado, *J. Med. Chem.*, **40**, 1937–1939 (1997).
94. M. Navarro, E.J. Cisneros-Fajardo, T. Lehmann *et al.*, *Inorg. Chem.*, **40**, 6879–6884 (2001).
95. M. Muranaka, T. Miyamoto, T. Shida *et al.*, *Ann. Allegy*, **40**, 132–137 (1978).
96. M. Adachi, H. Sato and T. Takahashi, *Ther. Res.*, **5**, 796–803 (1986).
97. A.C. Falcao, M.J. Rocha, A.M. Almeida and M.M. Caramona, *J. Clin. Pharma. Ther.*, **25**, 191–195 (2000).
98. D.L. Shapiro and J.R. Masci, *J. Rheumatol.*, **23**, 1818–1820 (1996).
99. K. Tepperman, Y. Shinyu, P.W. Roy *et al.*, *Metal-Based Drugs*, **1**, 433–444 (1994).
100. T. Okada, B.K. Patterson, S.-Q. Ye and M.E. Gurney, *Virology*, **192**, 631–642 (1993).

27 $_{83}$Bi Bismuth-Based Pharmaceuticals

Neil Burford and Melanie D. Eelman

27.1 Introduction

Most metal ions have the ability to promote an array of responses in biological systems that represent the recognition of deficiency or toxicity.[1] Heavy metals and metalloids are notorious for their toxic effects (e.g. toxic intake levels for a 70 kg human: Hg = 0.4 mg; Pb = 1 mg).[2] By contrast, the heaviest stable element, bismuth, is considered to be non-toxic and as much as 15 g can be tolerated by an individual. Bismuth is renowned for widespread medicinal applications. Some bismuth compounds exhibit anti-microbial activity at low concentrations, relatively low elemental human cell cytotoxicity and ill-defined gastric cytoprotective properties, highlighting their potential as therapeutic agents.[3–7] Consequently, developments in the chemistry of bismuth are driven by the potential for the discovery of new pharmaceutical agents.[8] However, mechanisms of bioactivity for bismuth are ill defined as complexes involving biological molecules have not been confirmed.[6]

27.2 Bismuth-Based Therapeutics

Bismuth subnitrate ('magisterium bismuti') was first used in the seventeenth century[9] as a treatment for syphilis and other bacterial infections. Applications

Metallotherapeutic Drugs and Metal-Based Diagnostic Agents: The Use of Metals in Medicine
Edited by Gielen and Tiekink

of bismuth compounds as remedies for gastrointestinal ailments began in the nineteenth century and some compounds have been approved for human use. The preparations Pepto-Bismol® (containing bismuth subsalicylate, BSS) and De-Nol® (containing colloidal bismuth subcitrate, CBS) are distributed over the counter to treat gastrointestinal disorders, including travelers' diarrhea, non-ulcer dyspepsia and non-steroidal anti-inflammatory drug damage. They are also prescribed for gastric and duodenal ulcer therapy[10] as well as for *Helicobacter pylori* infections (*Hp*, a bacterium associated with the pathogenesis of ulcers).[11] Anti-*Hp* clinical trials have been extensive, employing either CBS or BSS concurrently with an antibiotic (e.g. metronidazole) and an anti(acid)secretory (e.g. omeprazole). This 'triple-therapy' regimen has an efficacy of up to 95% eradication of *Hp* and is postulated to inhibit the adhesion of bacteria to the epithelial cells in the stomach lining. In addition, pharmacological studies suggest that the treatment and prevention of ulcers by CBS involves the fortification of gastric mucus and the stimulation of cytoprotective processes, such as the synthesis of endogenous prostaglandins and the secretion of mucosal bicarbonate.[12] Ranitidine (Figure 27.1) bismuth citrate (RBC)[13] has also been used as a therapy for peptic ulcer disease, often in combination with other medicines, such as clarithromycin. It is obtained from the reaction of ranitidine hydrochloride with bismuth citrate, and is speculated to protect epithelial cells through penetration of the mucus layer, where ranitidine serves to inhibit the secretion of stomach acid.

27.3 Fundamental Aspects of Bismuth Chemistry and Methods of Characterization

Bismuth is mined as either bismuth oxide (Bi_2O_3, 'bismite') or bismuth sulfide (Bi_2S_3, 'bismuthinite'). Bismuth(III) carbonate, $Bi_2(CO_3)_3$ is formed from the treatment of Bi_2O_3 with excess CO_2, or from $Bi(NO_3)_3$ with excess Na_2CO_3 or K_2CO_3. The composition of $Bi_2(CO_3)_3$ is sensitive to reaction conditions, which lead to the 'light' and 'heavy' terminology for 'bismuth subcarbonate' in the early pharmaceutical and medical literature.[14] $BiCl_3$ and $Bi(NO_3)_3$ are prepared from Bi_2O_3 by reaction with the corresponding mineral acids. In weakly acidic to basic conditions, these bismuth salts are 'hydrolyzed' to the oxides BiOCl and $BiONO_3$, respectively, which have lower solubility. They contain the bismuthyl ion (BiO^+) and are commonly referred to as 'basic', 'oxy-' or 'sub-' salts.

The inherently low solubility of bismuth compounds in both aqueous and organic solvents precludes the application of many characterization techniques. Most bismuth compounds have been identified by isolation of crystalline

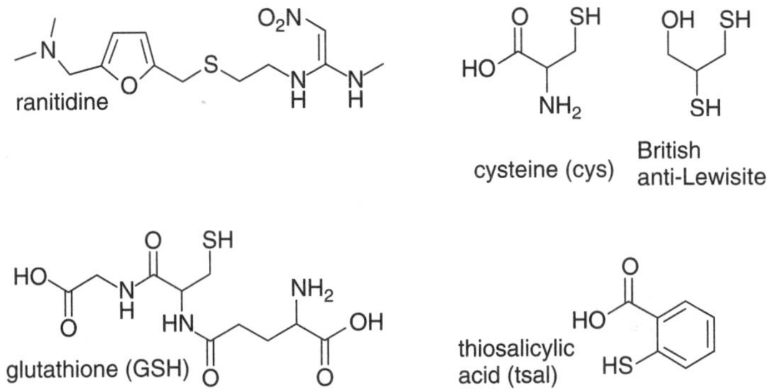

Figure 27.1 Molecular drawings and common abbreviations for compounds whose interaction with bismuth is recognized to be biologically relevant

material and determination of the solid-state structure by X-ray crystallography. As a result, numerous complexes of bismuth are unique examples involving a particular ligand, and series of related compounds are rare. Moreover, the isolated samples represent the most readily crystallized compounds and are not necessarily broadly indicative of bismuth chemistry. Therefore, few general synthetic procedures have been established to obtain series of compounds for systematic assessment of physical and chemical properties. Although compositions of solutions and stability constants have been assessed for some systems using bismuth-ion-selective electrodes, potentiometry, polarography or oscillopolarography, and ligands on bismuth may be evaluated by NMR spectroscopy (^{1}H and ^{13}C), these data do not provide definitive formulae or allow for structural assignments. Electrospray ionization mass spectrometry (ESI-MS) has been employed to investigate reaction mixtures containing complexes of bismuth,[15] and offers a sensitive and versatile technique to determine the formulae of ionic species in the gas-phase.

27.4 Hydroxycarboxylate Complexes of Bismuth

Bismuth compounds containing hydroxycarboxylate ligands have been most extensively studied in the context of medicinal applications. Despite the 'sub' terminology in BSS (Pepto-Bismol®) and CBS (De-Nol®), experimental data are inconclusive for the presence of the bismuthyl (BiO^+) cation in examples of these compounds. Hydroxycarboxylic acids that are observed to form complexes with bismuth are shown in Figure 27.2. Most complexes involve conjugate bases, which are abbreviated according to the degree of deprotonation of the acid; for example, (sal-H)$^-$ is a monoanionic conjugate of salicylic acid and

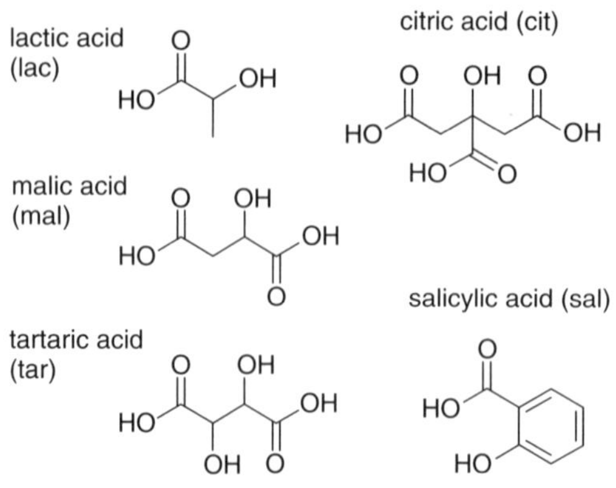

Figure 27.2 Drawings and abbreviations for hydroxycarboxylic acids that are observed to form complexes with bismuth

$(mal\text{-}3H)^{3-}$ is a trianionic conjugate of malic acid. Some formulae assignments for salicylate and citrate complexes of bismuth are based on elemental analysis data, and conclusions are often impeded by variable water content in solid samples.

The only structural reports of bismuth–salicylate complexes involve 2,2′-bipyridine (bipy) or 1,10-phenanthroline (phen) as auxiliary ligands,[16] which are believed to provide a stabilizing influence or disruption of the coordination polymeric lattice that is responsible for the low solubility of BSS. The solid-state structures of $[Bi(sal\text{-}H)_3(bipy)]_2$ and $[Bi(sal\text{-}H)(sal\text{-}2H)(phen)]_2$ are illustrated in Figure 27.3. Both involve a hydroxycarboxylate ligand interacting with bismuth through the carboxylate functionality, as either a terminal four-membered chelate or a bridging carbonyl. In the latter, a dianionic salicylate $(sal\text{-}2H)^{2-}$ behaves as an alkoxycarboxylate-chelating bridge. The potential structural complexity for these types of compounds is demonstrated by the octanuclear cluster observed for $Bi_8(tsal\text{-}2H)_{12}(DMF)_{12}$, which involves dianionic thiosalicylate (Figure 27.1) ligands engaging bismuth with thiolate and carboxylate donors (DMF = *N*,*N*-dimethylformamide).[17]

Bismuth citrate is essentially insoluble in water, but a dramatic increase in solubility is observed with increasing pH, yielding the 'bismuth citrate' anion. Formulation of these solutions is complicated and various examples have been determined by X-ray crystallography and classified in the literature as CBS.[18–26] In this context, the Merck index uses the non-stoichiometric name 'tripotassium dicitrato bismuthate'.[27] The reported structures of CBS are closely related and are generally constructed from the monoanionic complex of Bi^{3+} with a tetra-anionic (tricarboxylate/alkoxide) citrate ligand $(cit\text{-}4H)^{4-}$. As shown in Figure 27.4, two carboxylate oxygen centers and the alkoxide effect a tridentate interaction with bismuth. The third carboxylate is pendant due to the required tetrahedral geometry of the central carbon center in the citrate framework, and is

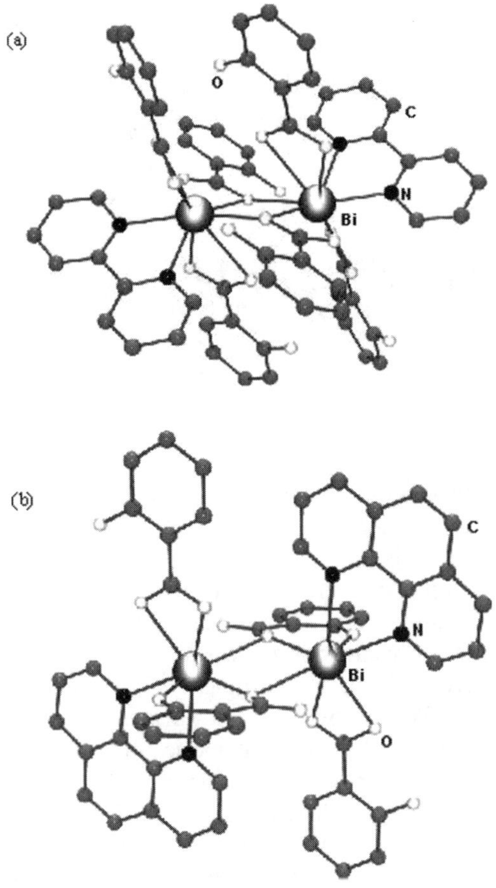

Figure 27.3 Solid-state structures of (a) [Bi(sal-H)$_3$(bipy)]$_2$ and (b) [Bi(sal-H)(sal-2H)(phen)]$_2$

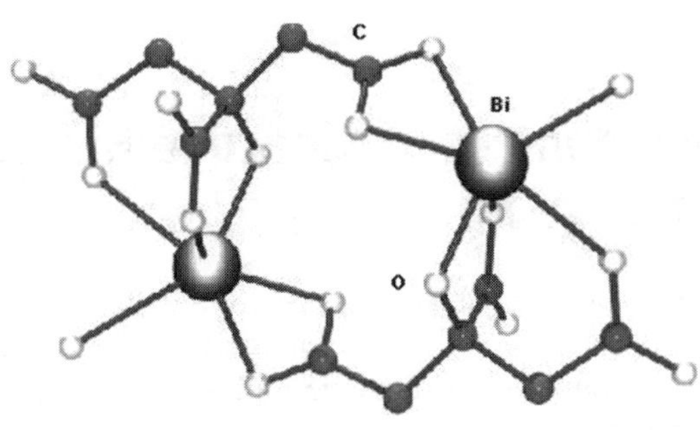

Figure 27.4 A dianionic dimeric unit observed in the solid-state structures of 'colloidal bismuth subcitrate' compounds

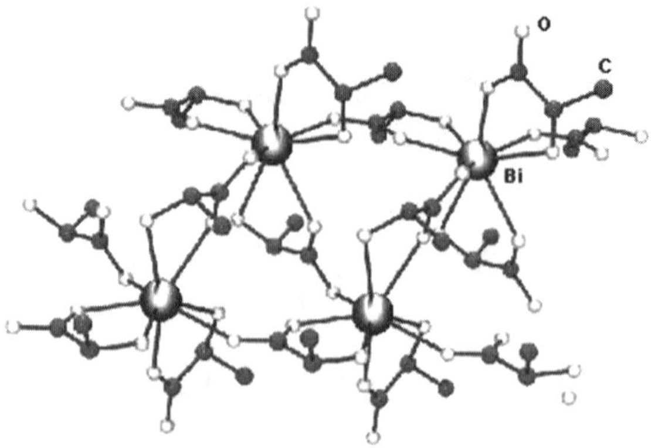

Figure 27.5 A section of the polymeric solid-state structure of $Bi(lac\text{-}H)_3$

available for interaction with a second bismuth center that involves a bidentate chelation. As a result, well-defined dimeric arrangements exist in most structures and the formulations are distinguished by differences in the interactions between the dimeric units and the degree of hydration.

The structural features of these salicylate and citrate complexes can be compared with those for bismuth complexes of other familiar hydroxycarboxylates. The multifunctional nature of conjugates from lactic acid, malic acid and tartaric acid (Figure 27.2) provide for a variety of structures that include intermolecular Bi—O coordination polymeric interactions. Tartrate engages bismuth as both mono- and di-anionic ligands in $Bi(tar\text{-}H)(tar\text{-}2H)\cdot H_2O$[28] and $[NH_4][Bi(tar\text{-}2H)_2\cdot H_2O]$.[29] Bismuth is chelated by carboxylate and hydroxy oxygen centers in the lactate salt $Bi(lac\text{-}H)_3$, as illustrated in Figure 27.5.[30] Three additional oxygen centers from carboxylate functionalities of neighboring complexes impose a coordination number of nine on bismuth. The multifunctional trianionic malate ligand also imposes nonacoordination on bismuth in Bi(mal-3H)[28] by intermolecular interactions and interaction with a water molecule (Figure 27.6).

27.5 Development of Bioactive Bismuth Compounds

The apparent chemical complexity of the commercially available bismuth-containing pharmaceutical agents has prompted the search for and development of new bioactive compounds. Therapeutic and anti-bacterial activity has been suggested for many 'model' compounds, with some having been assessed by *in vitro* and/or *in vivo* studies. Systematic synthetic studies coupled with bioactivity assessments confirm the biological significance of bismuth, provide comprehensive

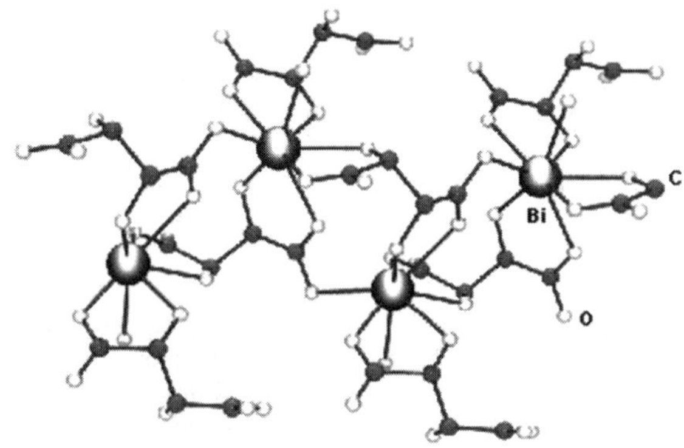

Figure 27.6 A section of the polymeric solid-state structure of Bi(mal-3H)·H_2O

characterization data for the compounds and reveal important trends. Tropolone complexes of bismuth(III) (Figure 27.7a) have been extensively derivatized, as well as less common examples of bismuth(V) complexes, and their anti-*Hp* activity is defined by minimum inhibitory concentrations.[31] Comparisons have been made with derivatives of bismuth(III) thiosemicarbazones and dithiocarbazonic acid methylester (Figure 27.7b),[32] implicating relationships between specific structural features and bioactivity.

Thiosemicarbazone complexes are representatives of a wide range of thiobismuth compounds, made possible by the relatively high thiophilicity of bismuth (thermodynamically favorable Bi—S interaction). Indeed, sulfur compounds represent the most extensive series of bismuth complexes. Systematic series of bismuth-dithiolate compounds (Figure 27.7c) have been isolated and comprehensively characterized.[33] The thiolate ligand serves to stabilize bismuth complexes with respect to hydrolysis and behaves as an anchor on bismuth for hetero-bifunctional thiolate ligands that promote or assist interaction of tethered weak Lewis donors with bismuth.[15,34–36] Otherwise, complexes of bismuth with weakly donating functionalities can only be isolated in the absence of moisture and many conventional types of ligands have yet to be observed on bismuth. Hetero-bifunctional thiolate ligands provide access to series of complexes (Figure 27.7d, 27.7e),[15,37] and offer a platform for systematic studies on models for medicinally relevant complexes. For example, dimeric structures are observed for tris(esterthiolate)bismuth and tris(methylthiosalicylate)bismuth complexes (Figure 27.8), in which two ligands on each bismuth center adopt a thiolatecarbonyl chelate interaction and the third behaves as a thiolatecarbonyl bridge. Both structures are reminiscent of the dimeric arrangement observed for 'colloidal bismuth subcitrate' compounds (Figure 27.4).

Minimum inhibitory concentrations (in the context of cytotoxicity assessments) for a range of compounds (Figure 27.7c) against *Clostridium difficile*, *Hp*,[36] *Escherichia coli*, *Pseudomonas aeruginosa* and *Proteus mirabilis*[38] reveal

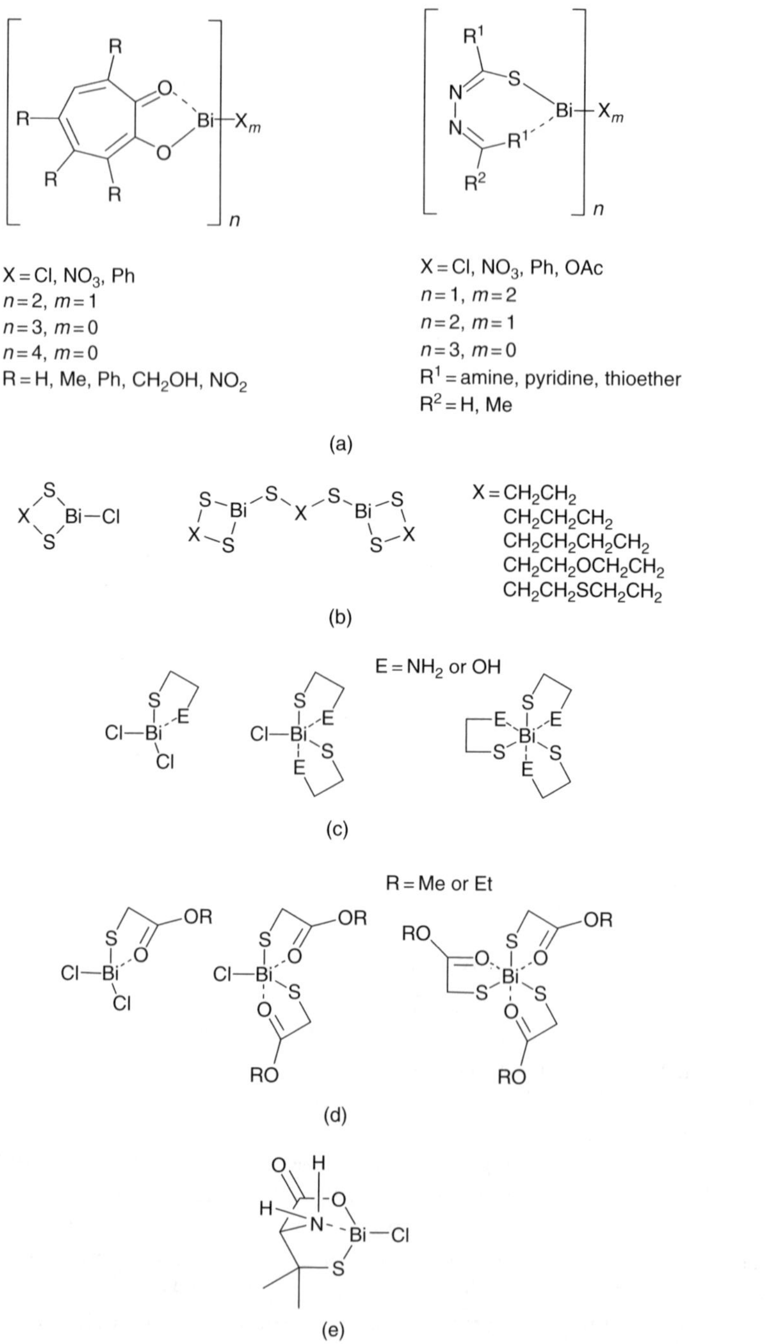

Figure 27.7 Complexes of bismuth with tropolone and biologically relevant thiolate ligands

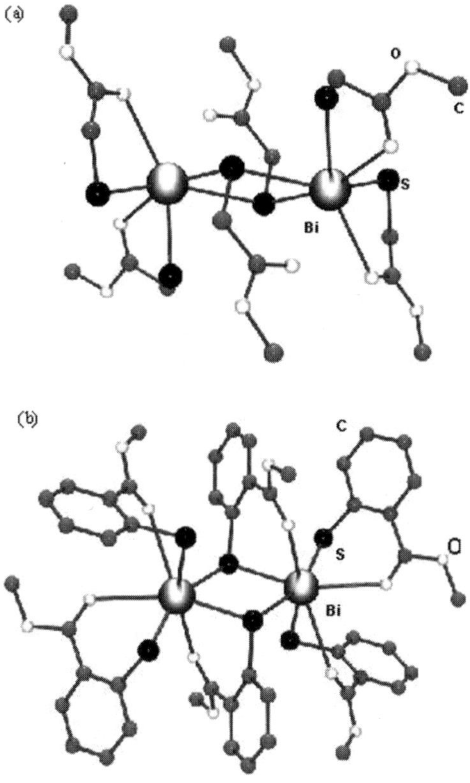

Figure 27.8 The dimeric units observed in the solid-state structures of (a) tris(esterthiolate)bismuth and (b) tris(methylthiosalicylate)bismuth

significant differences within a series suggesting a structure–activity relationship for the bismuth environment. In addition, studies using a rat model of gastric ulceration reveal distinct differences in ulcer-healing efficacy.[39] More general biological activity, including anti-bacterial (*Bacillus subtilis*, *E. coli*, *Candida tropicalis*, *Penicillium camembertii*),[40] fungicidal (yeast and moulds)[40] and anti-tumor behavior,[41] has been reported for a series of spectroscopically characterized bis(thiolate)bismuth compounds. The stability of bismuth complexes involving British anti-Lewisite (Figure 27.1) has highlighted them for anti-microbial assessment *in vitro* and *in vivo*.[42]

27.6 Interactions of Bismuth Compounds with Biological Molecules

The biochemical fate of bismuth is speculated to include BiOCl in the stomach, $Bi_2O_2CO_3$ in the small intestine and BiS_3 in the large intestine.[43] Bismuth is known to concentrate in the kidneys and the liver, with trace amounts in the skeleton and other organs. The binding of bismuth with proteins of exposed ulcer tissue and the formation of a protective coating is a likely mode of action for the anti-ulcer behavior exhibited by some bismuth compounds. In this context, the chemistry of bismuth complexes involving biological molecules as ligands represents an important component in understanding aspects of bioactivity. Interactions of bismuth with methionine,[44] cysteine (Figure 27.1),[45,46] glutathione (Figure 27.1),[46–48] metallothionein,[49] human serum transferin,[50–53] lactoferin[54] and other serum proteins[55] have been studied using IR, UV, fluorescence, ^{1}H NMR and ^{13}C NMR spectroscopy, inductively coupled plasma mass spectroscopy (ICP-MS), ESI-MS and extended X-ray absorption fine structure spectroscopy. The pronounced thiophilicity of bismuth highlights conjugates of thiol-carboxylic acids as obvious sites for interaction of bismuth with biological molecules. In this context, *in situ* ESI-MS (Figure 27.9) of solutions containing $Bi(NO_3)_3$ and glutathione (Figure 27.1) confirm the existence of the complex ions $[Bi(GSH)_2\text{-}2H]^+$ and $[Bi(GSH)\text{-}2H]^+$,[46] and X-ray crystallographic data confirm the interaction of a dianionic penicillaminate (a derivative of cysteinate) with bismuth (Figure 27.7f).[56]

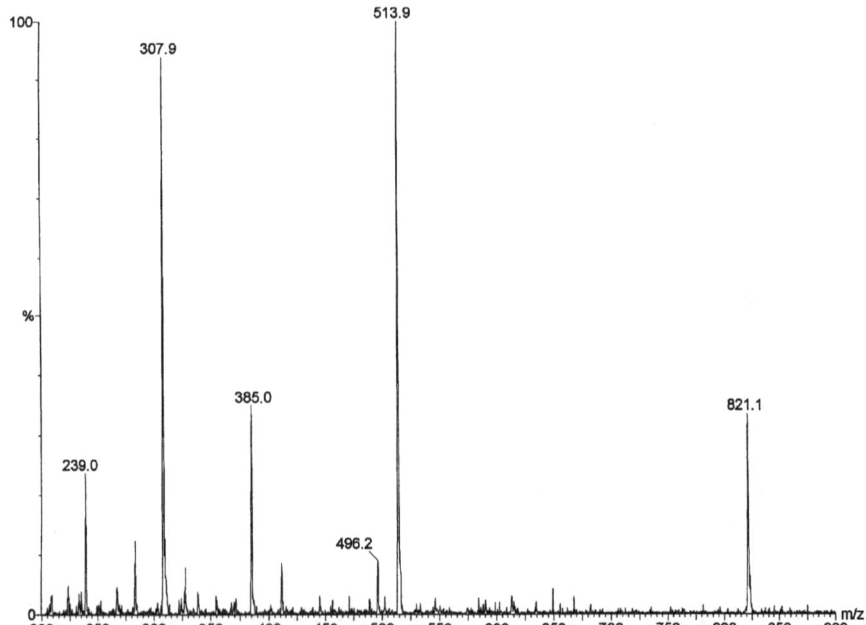

Figure 27.9 ESI-MS of a reaction mixture containing $Bi(NO_3)_3$ and glutathione in aqueous solution. Assignments for *m/z* peaks: 821.1 $[Bi(GSH)_2\text{-}2H]^+$; 513.9 $[Bi(GSH)\text{-}2H]^+$

27.7 Concluding Remarks

An array of bismuth compounds has been investigated for potential bioactivity, but few definitive interactions of bismuth with biological molecules have been reported. Fundamental developments in the chemistry of bismuth are important for the discovery and development of new pharmaceutical agents.

References

1. K.H. Thompson and C. Orvig, *Science*, **300**, 936–939 (2003).
2. J. Emsley, *The Elements*, Clarendon Press, Oxford (1990).
3. G.F. Baxter, *Pharm. J.*, **243**, 805–810 (1989).
4. G.F. Baxter, *Chem. Br.*, 445–448 (1992).
5. G.G. Briand and N. Burford, *Chem. Rev.*, **99**, 2601–2657 (1999).
6. P.J. Sadler, H. Li and H. Sun, *Coord. Chem. Rev.*, **185–186**, 689–709 (1999).
7. P.J. Sadler and Z. Guo, *Pure Appl. Chem.*, **70**, 863–871 (1998).
8. J. Reglinski, *Chemistry of Arsenic, Antimony, and Bismuth*, Blackie Academic & Professional, London, 403–440 (1998).
9. F. Lazarini, *Cryst. Struct. Commun.*, **8**, 69–74 (1979).
10. M.J. Abrams and B.A. Murrer, *Science*, **261**, 725–730 (1993).
11. B.J. Marshall, *Am. J. Gastroenterol.*, **86**, 16–25 (1991).
12. S.P. Lee, *Scand. J. Gastroenterol.*, **26** (suppl. 185), 1–6 (1991).
13. P.J. Sadler and H. Sun, *Dalton Trans.*, 1395–1401 (1995).
14. B.F. Howard and O. Chick, *Pharm. J.*, **115**, 661–662 (1925).
15. G.G. Briand, N. Burford, M.D. Eelman *et al.*, *Inorg. Chem.*, **42**, 3136–3141 (2003).
16. J.H. Thurston, E.M. Marlier and K.H. Whitmire, *Chem. Commun.*, 2834–2835 (2002).
17. E. Asato, K. Katsura, M. Arakaki *et al.*, *Chem. Lett.*, 2123–2126 (1994).
18. E. Asato, K. Katsura, M. Mikuriya *et al.*, *Inorg. Chem.*, **32**, 5322–5329 (1993).
19. E. Asato, K. Katsura, M. Mikuriya *et al.*, *Inorg. Chem.*, **34**, 2447–2454 (1995).
20. E. Asato, W.L. Driessen, R.A.G. de Graaff *et al.*, *Inorg. Chem.*, **30**, 4210–4218 (1991).
21. E. Asato, K. Katsura, M. Mikuriya *et al.*, *Chem. Lett.*, 1967–1970 (1992).
22. E. Asato, C.M. Hol, F.B. Hulsbergen *et al.*, *Inorg. Chim. Acta*, **214**, 159–167 (1993).
23. P.J. Barrie, M.I. Djuran, M.A. Mazid *et al.*, *Dalton Trans.*, 2417–2422 (1996).
24. W.A. Herrmann, E. Herdtweck and L. Pajdla, *Inorg. Chem.*, **30**, 2579–2581 (1991).
25. W.A. Herrmann, E. Herdtweck and L. Pajdla, *Z. Kristallogr.*, **198**, 257–264 (1992).
26. W. Li, L. Jin, N. Zhu *et al.*, *J. Am. Chem. Soc.*, **125**, 12408–12409 (2003).
27. *Merck Index*, 421 (1996).
28. W.A. Herrmann, E. Herdtweck, W. Scherer *et al.*, *Chem. Ber.*, **126**, 51–56 (1993).
29. D.S. Sagatys, E.J. O'Reilly, S. Patel *et al.*, *Aust. J. Chem.*, **45**, 1027–1034 (1992).
30. P. Kiprof, W. Scherer, L. Pajdla *et al.*, *Chem. Ber.*, **125**, 43–46 (1992).
31. U. Dittes, E. Vogel and B.K. Keppler, *Coord. Chem. Rev.*, **163**, 345–364 (1997).
32. R. Diemer, U. Dittes, B. Nuber *et al.*, *Metal-Based Drugs*, **2**, 271–292 (1995).

33. L. Agocs, N. Burford, T.S. Cameron *et al.*, *J. Am. Chem. Soc.*, **126**, 895–898 (1996).
34. L. Agocs, G.G. Briand, N. Burford *et al.*, *Inorg. Chem.*, **36**, 2855–2860 (1997).
35. G.G. Briand, N. Burford, T.S. Cameron and W. Kwiatkowski, *J. Am. Chem. Soc.*, **120**, 11374–11379 (1998).
36. E. Asato, K. Kamamuta, Y. Akamine *et al.*, *Bull. Chem. Soc. Jpn*, **70**, 639–648 (1997).
37. N. Burford, M.D. Eelman and T.S. Cameron, *Chem. Commun.*, 1402–1403 (2002).
38. D.E. Mahony, S. Lim-Morrison, L. Bryden *et al.*, *Antimicrob. Agents Chemother.*, **43**, 582–588 (1999).
39. G.S. Sandha, R. LeBlanc, S.J.O.V. van Zanten *et al.*, *Dig. Dis. Sci.*, **43**, 2727–2732 (1999).
40. T. Klapotke, *J. Organomet. Chem.*, **331**, 299–307 (1987).
41. P. Köpf-Maier and T. Klapotke, *Inorg. Chim. Acta*, **152**, 49–52 (1988).
42. P. Domenico, R.J. Salo, S.G. Novick *et al.*, *Antimicrob. Agents Chemother.*, **41**, 1697–1703 (1997).
43. D.Ws. Bierer, *Rev. Infect. Dis.*, **12** (suppl. 1), S3–S8 (1990).
44. C.A. McAuliffe, J.V. Quagliano and L.M. Vallarino, *Inorg. Chem.*, **5**, 1996–2003 (1966).
45. G. Alonzo, N. Bertazzi and M. Consiglio, *Inorg. Chim. Acta*, **85**, L35–L37 (1984).
46. N. Burford, M.D. Eelman, D. Mahony and M. Morash, *Chem. Commun.*, 146–147 (2003).
47. P.J. Sadler, H. Sun and H. Li, *Chem. Eur. J.*, **2**, 701–708 (1996).
48. H. Sun, H. Li and P.J. Sadler, *J. Inorg. Biochem.*, **59**, 190 (1995).
49. H. Sun, H. Li, I. Harvey and P.J. Sadler, *J. Biol. Chem.*, **274**, 29094–29101 (1999).
50. H. Li, P.J. Sadler and H. Sun, *J. Biol. Chem.*, **271**, 9483–9489 (1996).
51. H. Sun, H. Li, A.B. Mason *et al.*, *Biochem. J.*, **337**, 105–111 (1999).
52. H. Sun, H. Li, A.B. Mason *et al.*, *J. Bio. Chem.*, **276**, 8829–8835 (2001).
53. N. Rao and S. Feldman, *Pharm. Res.*, **7**, 188–191 (1990).
54. L. Zhang, K.Y. Szeto, W.B. Wong *et al.*, *Biochemistry*, **40**, 13281–13287 (2001).
55. H. Sun and K.Y. Szeto, *J. Inorg. Biochem.*, **94**, 114–120 (2003).
56. W.A. Herrmann, E. Herdtweck and L. Pajdla, *Chem. Ber.*, **126**, 895–898 (1993).

28 Paramagnetic Metal Complexes as Contrast Agents for Magnetic Resonance Imaging

Silvio Aime, Alessandro Barge, Eliana Gianolio, Simonetta Geninatti Crich, Walter Dastrù and Fulvio Uggeri

28.1 Generalities on Magnetic Resonance Imaging (MRI) and Paramagnetic Contrast Agents

Biological tissues are transparent to radiation at very short wavelengths (e.g. X-rays) but they are opaque at intermediate wavelengths such as ultraviolet, infrared and microwave. Surprisingly, the body is permeable to longer wavelengths such as radiowaves, the radiation used in magnetic resonance (MR).

The contrast in an MR image is the result of a complex interplay of numerous factors, including the relative T_1 and T_2 relaxation times, proton density of the imaged tissues and instrumental parameters. The excellent soft tissue discrimination has made MRI the modality of choice in medical diagnosis.[1,2]

Metallotherapeutic Drugs and Metal-Based Diagnostic Agents: The Use of Metals in Medicine
Edited by Gielen and Tiekink

The MR image contrast can be further enhanced by administration of suitable MRI contrast agents (CAs). The presence of a CA causes a dramatic variation of the water proton relaxation rates and then allows to add physiological information to the impressive anatomical resolution commonly obtained in the uncontrasted images. Thus the administration of CAs has entered into the pool of diagnostic protocols and is particularly useful to assess organ perfusion and any abnormalities in the blood–brain barrier or in kidney clearance.

Several other applications, primarily in the field of angiography and tumour targeting, are currently under intense scrutiny, with the promise of soon being available in clinical practice. Nowadays about 35% of MRI examinations make use of CAs, but this percentage is predicted to increase further following the development of more effective and specific contrast media than those currently available.

Unlike CAs used in X-ray-computed tomography and in nuclear medicine, MRI contrast agents are not directly visualized in the image. Only their effects are observed: contrast is affected by the variation that the CA causes on water proton relaxation times, and consequently on the intensity of the NMR signal.[3,4] Generally, the purpose is to reduce T_1 in order to obtain an intense signal in short times and a better signal-to-noise ratio with the acquisition of a higher number of measurements. CAs that reduce either T_1 or T_2 are called 'positive', whereas those that affect only T_2 are called 'negative'. Since unpaired electrons are able to remarkably reduce T_1 and T_2, the search for positive CAs is mainly oriented towards paramagnetic compounds, particularly towards paramagnetic metal complexes. The paramagnetic metal ions most extensively studied are both in the transition metals and in the lanthanide series.

As far as lanthanides are concerned,[5] the attention is essentially focused on Gd(III) both for its high paramagnetism (7 unpaired electrons) and for its favourable properties in terms of electronic relaxation.[6] This metal does not possess any physiological function in mammalians, and its administration as a free ion is strongly toxic even at low doses (10–20 μmol kg^{-1}). For this reason, it is necessary to use ligands that form very stable chelates with the lanthanide ion.[3,4] The high affinity shown by Gd(III) towards some polyaminocarboxylic acids, either cyclic or linear, has been exploited to form very stable complexes (up to log $K_{ML} > 20$). The first CA approved for clinical use was Gd-DTPA (Magnevist®, Schering AG, Germany), which, in more than 10 years of clinical experimentation, has been administered to more than twenty million patients. Other Gd(III)-based CAs similar to Magnevist® are now available: Gd-DOTA (Dotarem®, Guerbert SA, France), Gd-DTPA-BMA (Omniscan®, Nycomed Imaging AS, Norway) and Gd-HPDO3A (Prohance®, Bracco SpA, Italy) (Figure 28.1).[5] These CAs have very similar pharmacokinetic properties because they distribute in the extracellular fluid and are eliminated via glomerular filtration. They are particularly useful to delineate lesions in the blood–brain barrier. Similar to these systems are two Gd(III) complexes used in the imaging of the liver: Gd-EOB-DTPA[7] (Eovist®, Schering AG, Germany) and Gd-BOPTA[8] (Multihance®, Bracco SpA, Italy) (Figure 28.1). They are more lipophilic

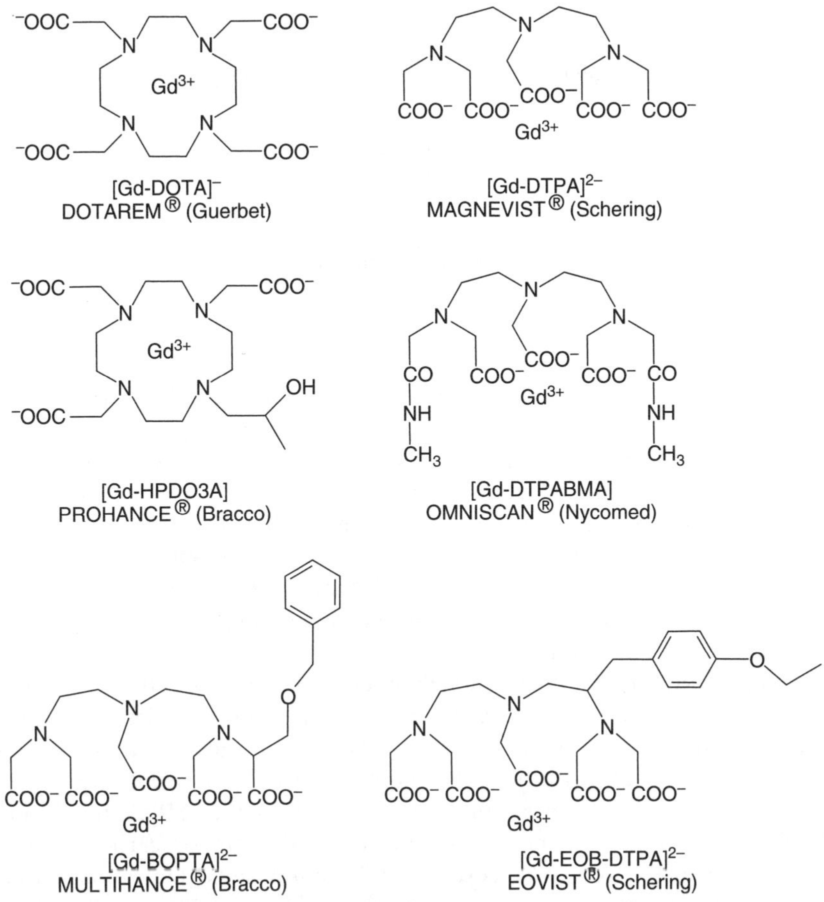

Figure 28.1 Gd(III)-based MRI contrast agents currently used in clinical practice

Gd-DTPA derivatives owing to the introduction of an aromatic substituent on the ligand surface.

Paramagnetic chelates of Mn(II) (five unpaired electrons) have also been considered. The main drawback appears to be related to the stability of these complexes. Mn(II) being an essential metal, the evolution has enabled biological structures to sequester Mn(II) ions with high efficiency. Thus, it has been difficult to design Mn(II) chelates that maintain their integrity when administered to living organisms. Actually, MnDPDP (Figure 28.2) has entered the clinical practice and is recommended as a hepatotropic agent.[9] It is the only agent that does its job by releasing metal ions to endogenous macromolecules. The huge proton relaxation enhancement brought about by the resulting Mn(II) protein adducts is responsible for the MRI visualization of hepatocytes even at low administered doses of MnDPDP.

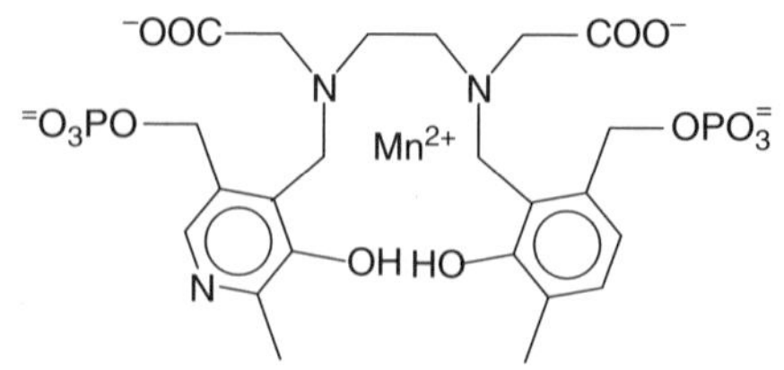

[Mn-DPDP]$^{4-}$
TESLASCAN® (Amersham)

Figure 28.2 A Mn(II)-based MRI contrast agent approved for diagnostic trials

Recently,[10] it has been shown that small focal injections of Mn^{2+} deeply within the mouse central nervous system combined with *in vivo* high resolution MRI mark neuronal tracts originating from the site of injection. Previous work[11] has shown that Mn^{2+} can be taken up through voltage-gated Ca^{2+} channels, transported along axons and across synapses. The combined features of transport along neurons and paramagnetism have been used to trace neuronal connections in mice, utilizing manganese enhanced MRI (MEMRI). This tract-tracing methodology will allow rapid mapping of neuronal connections in live animals from focal points of interest within the brain.

Although iron(III), having the same number of unpaired electrons as Mn(II), has entered the field of MRI CAs mainly as iron oxide particles, such water-insoluble systems yield very strong T_2 effects as the result of a dramatic long-range disturbance in the magnetic field homogeneity. These agents are made of a crystalline core of superparamagnetic iron(III) oxide (SPIO, maghemite, γ-Fe_2O_3) surrounded by coating materials like dextran or carboxydextran. The diameter of the iron oxide core is just 3–5 nm whereas the overall particle may be of 50–200 nm diameter. Two products are available for clinical use: Endorem® (Guerbet) and Resovist® (Schering).[12–13] These agents provide excellent (negative) contrast when administered at doses as low as 8–15 µmol/kg body weight. Once administered intravenously, as particles, these agents accumulate in the cells of the reticuloendothelial system. The pharmacodynamic properties of the iron oxide particles are affected by both the size and the overall electric charge. The smaller particles remain in the blood circuit for a time long enough to be considered as blood pool agents for angiographic assays.

28.2 Determinants of the Relaxivity of Paramagnetic Metal Complexes

The relaxivity represents the relaxation enhancement of water protons in solutions containing the paramagnetic agent at 1 mM concentration.[8] It usually refers to the value measured at the observation frequency of 20 MHz and the temperature of 25 °C or 39 °C. Much work has been done in the past 15 years or so to get an understanding of the relationships between structure and dynamics of a paramagnetic complex and the observed relaxivity.[14] The observed longitudinal relaxation rate (R_1^{obs}) of the water protons in an aqueous solution containing a paramagnetic complex is the sum of three contributions: (i) a diamagnetic one, whose value corresponds to the relaxation rate that would have been measured in the presence of a corresponding diamagnetic complex (R_1^0); (ii) a paramagnetic one (R_{1p}^{IS}), arising from the exchange of water molecules from the inner coordination sphere of the metal ion with the bulk water; and (iii) a paramagnetic one relative to the contribution of water molecules that diffuse in the outer coordination sphere of the paramagnetic centre (R_{1p}^{OS}):

$$R_{1p}^{obs} = R_1^o + R_{1p}^{IS} + R_{1p}^{OS} \tag{28.1}$$

Sometimes also a fourth paramagnetic contribution is taken into account that is due to the presence of mobile protons or water molecules (tightly interacting with the surface of the chelate) in the second coordination sphere of the metal ion.[15]

The inner-sphere contribution R_{1p}^{IS} is given by:

$$R_{1p}^{IS} = \frac{q[\mathrm{C}]}{55.5(T_{1M} + \tau_m)} \tag{28.2}$$

where [C] is the concentration of the paramagnetic agent, q is the number of water molecules (generally 1 or 2) coordinated to the paramagnetic complex, τ_m is their exchange lifetime and T_{1M} is the longitudinal relaxation time of their protons.

Commonly, T_{1M} is evaluated on the basis of the expression developed earlier for simple aqua-ions:[16]

$$\frac{1}{T_{1M}} = \frac{2}{15}\left(\frac{\mu_o}{4\pi}\right)^2 \frac{\gamma_I^2 g_e^2 \mu_B^2 S(S+1)}{r_H^6}\left(\frac{7\tau_c}{1+\omega_S^2\tau_c^2} + \frac{3\tau_c}{1+\omega_I^2\tau_c^2}\right) \tag{28.3}$$

where S is the electron spin quantum number, γ_I the proton nuclear magnetogyric ratio, g_e and μ_B the electronic g factor and Bohr magneton, respectively. ω_I and ω_S the proton and electron Larmor frequencies, respectively, and τ_c is the correlation time. τ_c is given by the sum:

$$\tau_c^{-1} = \tau_R^{-1} + \tau_m^{-1} + \tau_s^{-1} \tag{28.4}$$

Thus the shortest of the three correlation times (τ_R = reorientational; τ_s = electronic; τ_m = exchange) determines τ_c and in turn T_{1M} and the overall relaxivity.

The commercial CAs shown in Figure 28.1 are monohydrated ($q = 1$) systems with a molecular weight of about 600–800 Da. that corresponds to rotational correlation times τ_R, about 60–80 ps. For this class of polyaminocarboxylate complexes, the exchange lifetime τ_M is typically found to be in the range 50–500 ns and $T_{1e} \approx 1$ ns at 0.5 T; thus the inner sphere relaxivity, R_{1p}^{IS}, assumes a value of ca. 2.5–3.5 $mM^{-1} s^{-1}$, at 25 °C. Therefore, as it was recognized earlier, it is evident that at 0.5 T the overall correlation time τ_{ci} is largely dominated by the rotational correlation time, whereas the contribution of both the exchange lifetime and the electronic relaxation plays an almost negligible role.

An important structural parameter that influences the 'inner sphere' relaxivity is the hydration number q. This represents a scaling factor in Equation 28.2 and therefore a higher number of coordinated water molecules ($q > 1$) provides a clear advantage in terms of relaxivity. The use of hepta- or hexa-dentate ligands would in principle result in Gd(III) complexes with two- and three-coordinated water molecules, respectively, but the decrease of the oleuticity of the ligand is likely to be accompanied by a decrease of their thermodynamic stability and an increase of their toxicity. Furthermore, systems with $q = 2$ may suffer a 'quenching' effect upon interacting with proteins, as donor atoms from Asp or Glu residues may replace the coordinated water molecules.[17] However, some stable Gd(III) chelates containing two 'inner sphere' water molecules have been identified and are under intense scrutiny.

Among them, an interesting class is represented by Gd–HOPO complexes developed by Raymond and co-workers. HOPO ligands (Figure 28.3) are based on 4-carboxyamido-3,2-hydroxypyridinone chelating units and act as hepta-dentate ligands towards Gd(III), thus leaving two water molecules in the inner coordination sphere. However, the peculiar coordinating geometry of Gd–HOPO complexes does not allow an easy replacement of the two water molecules by other ligands. Moreover, the exchange rate of the coordinated water molecules is in the range of the optimal values and the electronic relaxation appears to be slow enough.[18]

Another system that looks very interesting in this regard is represented by the Gd(III) complex of PCP2A (Figure 28.3), a ligand based on a pyridine-containing macrocycle bearing two acetic and one methylenephosphonic arms.[19] Its relaxivity is about two times higher than the values reported for CAs currently used in clinical practice. This is the result of the presence of two water molecules in the inner coordination sphere and a significant contribution from water protons bonded to the phosphonate group.

A novel Gd(III) chelate with the heptadentate AAZTA ligand (AAZTA: 6-amino-6-methylperhydro-1,4-diazepinetetraacetic acid; Figure 28.3) has been recently characterized. AAZTA is readily obtained in high yields and its Gd(III) complex displays excellent relaxation enhancement properties so as to be considered the prototype of a new class of enhanced MRI agents. It is characterized by a quite high relaxivity value (7.1 $mM^{-1} s^{-1}$, 20 MHz and 298 K), a fast exchange rate of the coordinated water molecules, a high thermodynamic stability in aqueous solution and a nearly complete inertness towards the influence of bidentate endogenous anions.

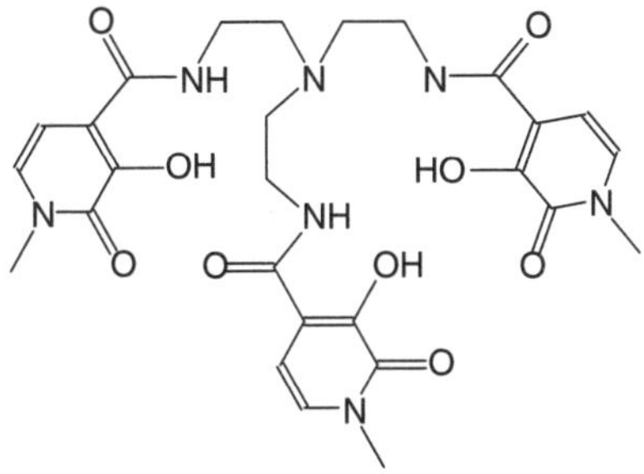

HOPO

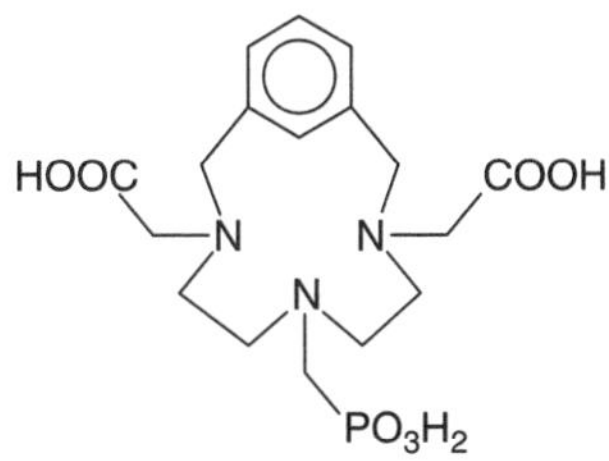

PCP2A

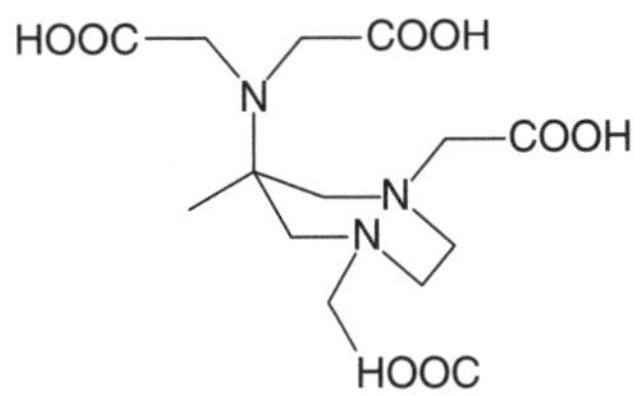

AAZTA

Figure 28.3 Schematic representations of the three heptadentate ligands (HOPO, PCP2A and AAZTA), whose Gd(III) complexes display two water molecules in the inner coordination sphere

It was earlier recognized that in the case of polyaminocarboxylate complexes of Gd(III), high relaxivities at the imaging field (0.5–1.5 T) can be obtained with long τ_R values. Therefore, slowly moving systems have been developed in order to reach high relaxivities. Basically, two routes have been explored to provide the Gd(III) chelates with long molecular reorientational times: (i) by forming a covalent linkage between the complex and a macromolecular substrate or (ii) by forming a non-covalent adduct between the complex and a slowly tumbling system.

The latter approach has been widely investigated by using the human serum albumin (HSA) as the interacting substrate. Research activities have been addressed along this direction to design Gd(III) chelates bearing on their surface suitable functionalities that promote the reversible binding of HSA.[20,21] Representative examples of such Gd(III) complexes are shown in Figure 28.4.

Besides the attainment of high relaxivities (see $R_{1p}{}^{bound}$ values in Figure 28.4), a high binding affinity to HSA enables the Gd(III) chelate to have a long intravascular retention time which is the property required for a good blood pool agent for MR angiography. For the reorientational time of HSA adducts (ca. 30 ns), the theory of paramagnetic relaxation foresees the attainment of relaxivity values much higher than the values actually obtained. It has been shown that the primary reason for the quenching of the relaxation enhancement

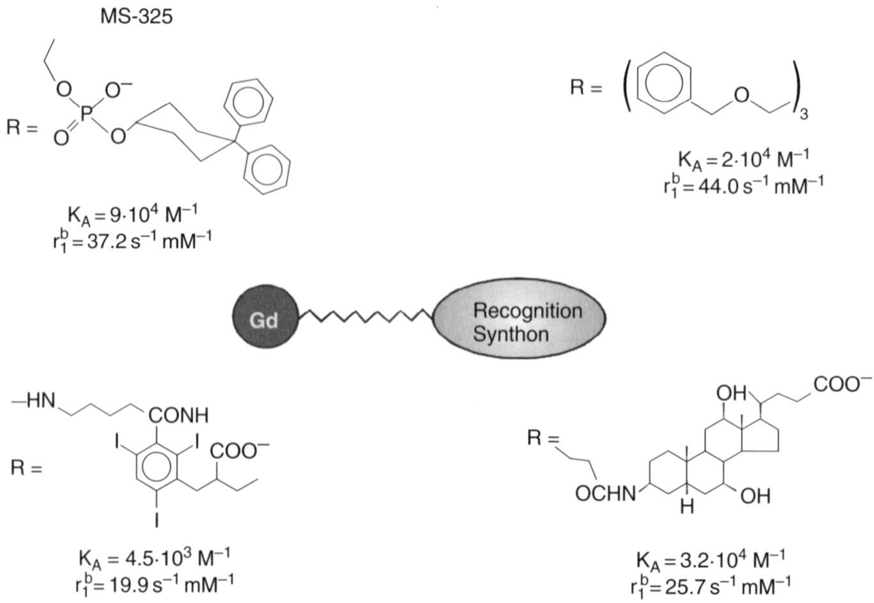

Figure 28.4 Gd(III) complexes with high binding affinity to HSA

is often associated with the occurrence of a relatively long exchange lifetime, τ_m, of the coordinated water.[22]

Slow rates of the coordinated water appear to be primarily characteristic of the complex rather than a consequence of the binding to the protein. Thus, for the attainment of high relaxivities one has to avoid Gd(III) chelates displaying slow exchange rates.

28.3 Responsive Agents

The term 'responsive' (sometimes also indicated as 'smart') refers to diagnostic agents whose contrasting properties are sensitive to a given physico-chemical variable that characterizes the micro-environment in which the probe distributes. Typical parameters of primary diagnostic relevance include pH, temperature, pO_2, enzymatic activity, redox potential and concentration of specific ions and low-weight metabolites.

Unfortunately, their clinical use is still rather far-off mainly because an accurate measurement of one of the above-mentioned parameters with a given relaxing probe requires a precise knowledge of the local concentration of the contrast medium in the region of interest. Only by this way, the observed change in the relaxation rate of water protons can be safely attributed to a change of the parameter to be measured. Recently, a novel class of MRI paramagnetic contrast media, the so-called PARACEST agents, has entered the field with the aim of acting as concentration-independent responsive agents.[23] These probes do not generate contrast by altering the relaxation rates of water protons, but through a different mechanism based on a transfer of saturated magnetization between exchangeable protons of the probe and water protons. Such a process yields a decrease of the NMR signal of bulk water protons, thus creating a negative contrast in the image. The saturation transfer is induced by the selective irradiation of the mobile protons of the probe with a proper radiofrequency field. When different pools of exchangeable protons are available, this modality makes possible the set-up of ratiometric methods in which a responsive PARACEST agent provides an MRI response no longer dependent on its absolute concentration.[24,25] Although the potential of PARACEST agents is very high, in the following paragraphs we will report some selected examples based on Gd(III) complexes whose responsiveness to given physico-chemical parameters is a function of the change in relaxivity.

28.3.1 pH sensitive agents

The design of a Gd(III)-based complex whose relaxivity is pH-dependent requires that at least one of the structural or dynamic parameters determining its relaxivity is rendered pH-dependent. In most of the examples so far reported, the pH dependence of the relaxivity reflects changes in the hydration of the metal complex.

For instance, Lowe *et al.* showed that the relaxivity of a series of macrocyclic Gd(III) complexes bearing β-arylsulfonamide groups is markedly pH-dependent (Figure 28.5) on passing from about $8\,s^{-1}mM^{-1}$ at pH <4 to about $2.2\,s^{-1}mM^{-1}$ at pH >8.[26]

It has been demonstrated that the observed decrease (about 4-fold) of r_1 is the result of a switch in the number of water molecules coordinated to the Gd(III) ion from 2 (at low pH values) to 0 (at basic pH values). This corresponds to a change in the coordination ability of the β-arylsulfonamide arm, that binds the metal ion only when it is in the deprotonated form.

In some cases the pH dependence of the relaxivity is associated with changes in the structure of the second hydration shell. Two such systems have been reported by Sherry's group. The first case deals with a macrocyclic tetramide derivative of DOTA (DOTA-4AmP, Figure 28.6) that possesses an unusual r_1 versus pH dependence.[27] In fact, the relaxivity of this complex increases from pH 4 to pH 6, decreases up to pH 8.5, remains constant up to pH 10.5 and, then, increases again. The authors suggested that this behaviour is related to the formation/disruption of the hydrogen bond network between the pendant phosphonate groups and the water bound to the Gd(III) ion. The deprotonation of phosphonate occurring at pH >4 promotes the formation of the hydrogen bond network that slows down the exchange of the metal-bound water protons.

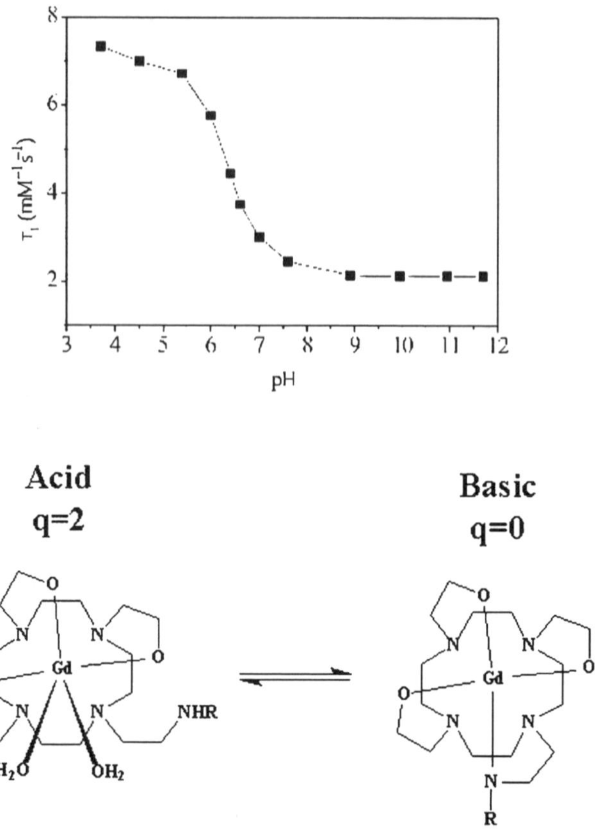

Figure 28.5 pH dependence of the relaxivity (20 MHz, 25 °C) of a Gd(III) complex bearing an arylsulfonamide group

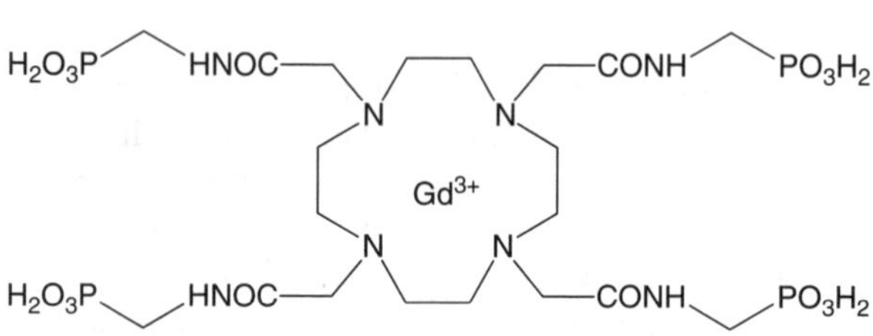

DOTA4-AmI

Figure 28.6 Schematic representation of the ligand DOTA-4AmP, whose Gd(III) complex has been proposed as pH-responsive probe

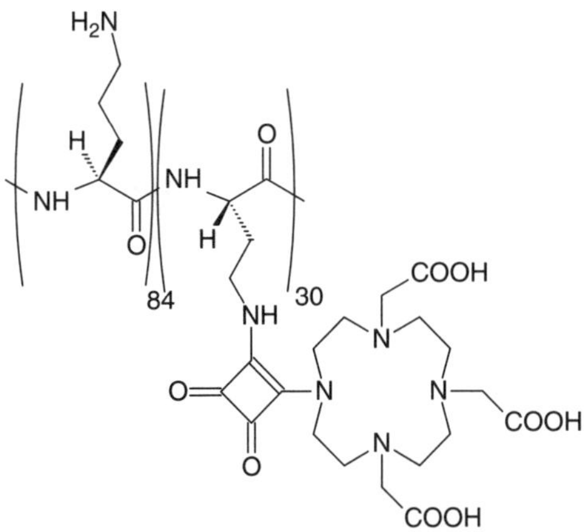

Figure 28.7 Macromolecular construct formed by 30 Gd(III) units covalently linked, by a squaric acid moiety, to a poly-ornithine

On the contrary, the behaviour observed at pH >10.5 was accounted for in terms of a shortening of τ_M catalyzed by OH^- ions. Recently, it has been demonstrated that this complex can be successfully used '*in vivo*' for mapping renal and systemic pH.[28] pH-dependent probes can also be obtained when the proton concentration is able to affect the rotational tumbling of a slowly moving Gd(III)-based system. An interesting example is represented by a macromolecular Gd(III) construction formed by 30 Gd(III) units covalently linked, by a squaric acid moiety, to a poly-ornithine (114 residues, Figure 28.7).[29]

At acidic pH the unreacted amino groups of the polymer are protonated and, therefore, tend to be localized as far apart as possible, whereas at basic pH the progressive deprotonation of the NH_3^+ groups determines an overall rigidification of the polymer structure owing to the formation of intramolecular hydrogen bonds between adjacent peptidic linkages. As expected, the reduced rotational mobility of the polymeric backbone upon increasing pH enhances the relaxivity of the system (Figure 28.8). Nevertheless, even if the enhancement is not particularly remarkable (ca. 40% in the 3–8 pH interval), it is worthy to note that the relaxivity of this system is considerably higher than the previous examples over the entire pH range, thus allowing, in principle, the detection of pH changes at lower concentration of the responsive probe.

28.3.2 Agents sensitive to the redox potential

A diagnostic MRI probe sensitive to the *in vivo* redox potential would be very useful for detecting regions with a reduced oxygen partial pressure (P_{O_2}), a typical symptom of several pathologies including strokes.

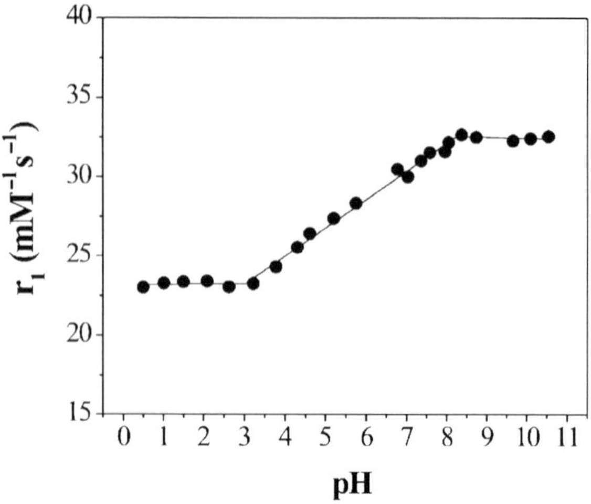

Figure 28.8 pH dependence of the relaxivity (20 MHz, 25 °C) for the paramagnetic polymeric system reported in Figure 28.7

Very few Gd(III) chelates sensitive to the tissue oxygenation have been reported so far. Our group has investigated the potential ability of GdDOTP (DOTP = 1,4,7,10-tetraazacyclododecane-*N*,*N*′,*N*″,*N*‴-tetrakis-methylenephosphonic acid) to act as an allosteric effector of haemoglobin.[30] In fact, it has been observed that this chelate binds specifically to the T-form of the protein, which is characterized by a lower affinity towards oxygen. The interaction is driven by electrostatic forces and leads to a significant relaxivity enhancement (ca. 5-fold) owing to the restricted molecular tumbling of the paramagnetic complex once it is bound to the protein. Although haemoglobin can be considered as an excellent indirect target for detecting P_{O_2}, the practical applicability of the method suffers from the inability of GdDOTP to enter red blood cells. Another approach deals with the use of liposomes built up with an amphiphilic Gd(III) complex containing a radical-sensitive disulphide bridge.[31] The relaxivity (at 20 MHz and 25 °C) of the liposomal paramagnetic agent is $13.6\,s^{-1}\,mM^{-1}$, i.e. 2-fold higher than the free Gd(III) complex (r_1 of $6.5\,s^{-1}\,mM^{-1}$). Possibly, the limited relaxivity enhancement is due to the rotational flexibility of the complex bound to the liposome. This system was tested '*in vitro*' by inducing the cleavage of the S—S bond chemically (by dithiothreitol) or physically (by γ-rays). In both cases the relaxivity decreased from the value of the liposomal agents to that of the free Gd(III) complex.

28.3.3 Enzyme responsive agents

One possible route to design enzyme-responsive agents is to synthesize paramagnetic inhibitors, whose binding to the active site of the protein can be signalled by the consequent relaxivity enhancement. An example of this approach has been provided by Anelli *et al.* who synthesized a linear Gd(III) complex bearing an arylsulfonamide moiety (Figure 28.9), which is a well-known inhibitor of carbonic anhydrase.[32]

In vitro experiments demonstrated that this complex binds quite strongly (K_A of about 1.5×10^4) to the enzyme and its relaxivity in the bound form is about 5-fold higher than the free complex ($27\,s^{-1}\,mM^{-1}$ vs $5\,s^{-1}\,mM^{-1}$ at 20 MHz). Unfortunately, an attempt to test the validity of this '*in vivo*' approach failed, possibly owing to the small amount of enzyme circulating in the blood.

An alternative approach is to design Gd(III) complexes acting as substrate for a specific enzyme. Along this direction, an example has been provided by Lauffer *et al.*, who prepared a Gd(III) chelate containing a phosphoric ester sensitive to the attack of the serum alkaline phosphatase (Figure 28.10).[33] The hydrolysis yields the exposure of a hydrophobic moiety well suited to bind to HSA. Upon binding, there is an increase of the relaxivity as a consequence of the lengthening of the molecular reorientational time. This approach was used by the same research group for designing Gd(III) complexes sensitive to TAFI

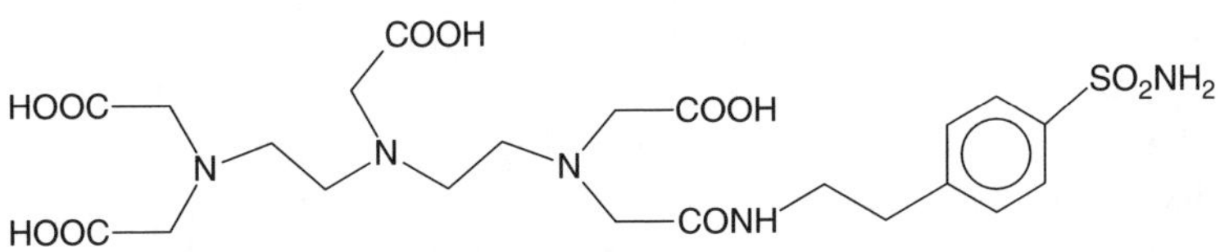

Figure 28.9 Schematic representation of a ligand bearing an arylsulfonamide moiety whose Gd(III) complex acts as an inhibitor of carbonic anhydrase

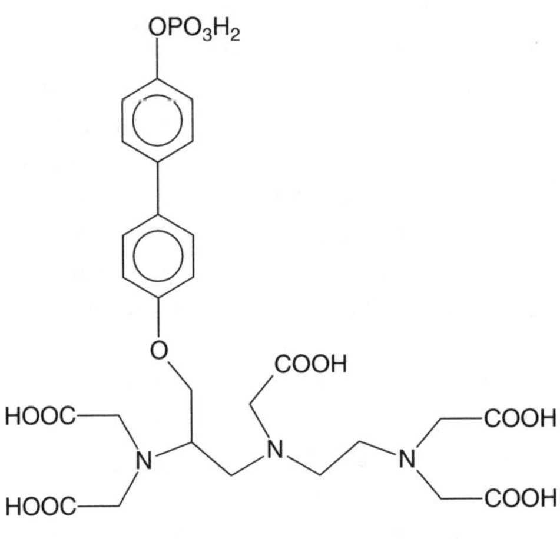

Figure 28.10 The Gd(III) complex of this ligand has been proposed as an agent responsive to serum alkaline phosphatase

(thrombin-activatable fibrinolysis inhibitor), a carboxypeptidase B involved in the clot degradation.[34]

Also particles have been considered as enzymatic responsive agents. To this purpose, insoluble Gd(III) chelates have been synthesized by introducing, on the ligand surface, long aliphatic chains via an ester or a peptidic bond. The particles can be internalized into cells having a phagocytic activity and then degraded by the action of the proper enzyme which cleaves the bond between the Gd(III) chelate and the insolubilizing moiety. Thus, the increase of intracellular relaxivity becomes a function of the activity of the enzyme of choice. In principle, such an approach provides a representative example of 'responsive particles'. Immediately after the internalization, the particles act as negative T_2 agents because they affect the bulk magnetic susceptibility. Then, their dissolution yields soluble Gd(III) chelates which act as positive T_1 agents.[35]

28.4 Targeting Cells with Gd(III) Chelates

Currently, an important task in the field consists of providing the CAs with targeting capabilities. After the implementation of extracellular and blood pool systems, the new frontier in MRI is represented by the possibility of imaging molecular targets that act as signatures of a given pathology. This will allow early diagnosis of diseases as the detection of an altered biochemical process largely anticipates the anatomical changes that are at the basis of current diagnostic modalities.

Several modalities have contributed to the early stages of this innovative approach, namely positron emission tomography (PET), single photon emission computed tomography (SPECT), MRI and Optical Imaging. MRI-Gd(III)-based agents are much less sensitive than radionuclear and optical imaging probes. Therefore, molecular imaging based on MRI, invariantly involves the need for accumulating a high number of contrast enhancing units at the site of interest. The basis for the design of a Gd-containing imaging probe is first dictated from the concentration and localization (vascular, extracellular matrix, on the cellular membrane, intracellular) of the target molecule.[36] Of course, the most accessible targets are those present on the surface of endothelial vessels. In principle they can be visualized by a number of macromolecular conjugates containing many Gd(III) complexes functionalized with the proper vector recognizing the given target. A nice example of targeting an endothelial site has been reported by Sipkins *et al.*[37] in the targeting of a specific angiogenesis marker, the endothelial integrin $\alpha_v\beta_3$, whose presence has been shown to correlate with the tumour growth. The imaging probe used in this work is a Gd-containing polymerized liposome. The target is first bound by a biotinilated antibody against $\alpha_v\beta_3$, which is successfully recognized by an avidin moiety on the surface of the liposome. Each liposome has a mean diameter of 300–350 nm,

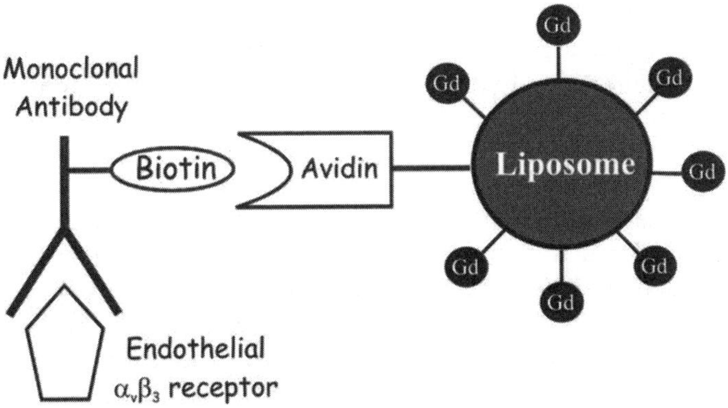

Figure 28.11 Targeting of the endothelial integrin $\alpha_v\beta_3$ as a specific angiogenesis marker

which appears suitable to avoid the uptake by the reticulo-endothelial system (Figure 28.11). This approach provided an enhanced and detailed detection of a rabbit carcinoma through the imaging of the angiogenetic vasculature.

Recently, the same $\alpha_v\beta_3$ target has been addressed with lipidic nanoparticles containing a huge number of Gd-chelated units (94,400 Gd/particle characterized by $r_1 = 19.1\ s^{-1}\ mM^{-1}$ (per Gd) $r_1 = 1{,}800{,}000$ per particle). One of the lipidic components, integrin peptidomimetic antagonist, is covalently bound to the $\alpha_v\beta_3$.[38]

The large molecular size of these systems limits their delivery to targets on the endothelial walls. To target receptors in solid tissues other routes have to be followed. Bhujwalla and co-workers[39] have recently developed and applied a two-component Gd-based avidin–biotin system for the visualization of HER 2/scan receptors. The latter is a member of the epidermal growth factor family and is amplified in multiple cancers. Their approach consisted of addressing the extracellular domain of the receptors by means of a biotinilated mAb. After clearance of the unbound mAb, Gd-labelled avidin is administered and binds, with a high affinity, to the biotinilated mAb. The expression level of the receptor was estimated at 7×10^5 receptors/cell and the average number of Gd-DTPA units per avidin molecule was 12.5. The method has been successfully applied in an experimental mouse model of breast carcinoma.

For labelling cells in culture, the most straightforward route is represented by incubating the cells in culture media added with the CA. In the absence of any specific interaction, the internalization of the agent may occur through pinocytosis, which is a process consisting of the invagination of the cell membrane to form small vescicles ($\leq$nm diameters) called endosomes. Therefore, incubation of cells, for a sufficiently long time, in a medium containing the imaging probe at relatively high concentration may lead to the internalization of the agent at amounts that may be sufficient for MRI visualization. Among a number of systems, the neutral, highly hydrophilic Gd-HPDO3A was considered as a good candidate for labelling stem cells by the pinocytotic route.[40] The *in vivo* MR visualization of labelled stem cells will allow their monitoring after transplantation. In a typical experiment of uptake via pinocytosis, a few millions of stem cells are incubated in a culture medium containing Gd-HPDO3A in the mM concentration range (10–50 mM) for a few hours. Upon incubation no saturation effect is observed and the amount of up taken Gd is linearly proportional to the concentration of the paramagnetic agent in the incubation medium. Once the cell has been internalized, the Gd-HPDO3A molecules get entrapped in endosomic vesicles as it can be checked out by observing the cells incubated with Eu-HPDO3A in the confocal microscope. In fact, Gd and Eu chelates with the same ligands display the same chemical/biological behaviour and the fluorescent response of Eu-HPDO3A acts as a histological reporter of the localization of Gd-HPDO3A in the cell. The potential of this approach has been proved by observing a mouse model of angiogenesis, on which blood-derived endothelial progenitor cells (EPCs) had been implanted subcutaneously, within a matrigel plug. After a few days, the histologic examination showed large capillary structures transposing the gel plugs. MR images parallel histologic findings as hyperintense spots corresponding to the labelled cells were clearly detected.

Phagocytosis is the process of internalization of particles by cells endowed with phagocytic activity. In such a case, this route appears highly efficient for a single-step internalization of a large amount of imaging probes. However, to be effective on MR images, Gd chelates must be water soluble. Therefore the particles must be biodegradable in order to release soluble Gd chelates once internalized into phagocytic cells. One may envisage several ways for the release of the Gd chelates. For instance, one may think of gel nano-particles of chitosan loaded with negatively charged Gd chelates. Such particles (200–400 nm diameter) are easily phagocyted and slowly degraded once internalized into the cells. Another approach to biodegradable Gd-containing particles has been pursued by designing particles whose insolubility is a property of the Gd chelates themselves. This goal is easily reached by introducing long aliphatic chains on the surface of the ligand. However, the fate of these systems can be controlled by means of the functionality used to link the insolubilizing moiety to the Gd chelate. In fact, by using ester or peptidic functionalities, the insolubilizing synthon can be displaced from the Gd chelates by the activity of the proper enzyme (Figure 28.12).

Another route for pursuing a high concentration of the imaging probe at the target of interest consists of its accumulation inside the cell. One way is offered by the 'cell internalization via receptors' route that is the procedure of choice in a number of nuclear medicine assays. For MRI, the design of the imaging probe requires the attachment of one or more Gd(III) chelates to the ligand molecule. Such structural modification may drastically affect the internalization process in respect to the mechanism occurring for the native ligand.

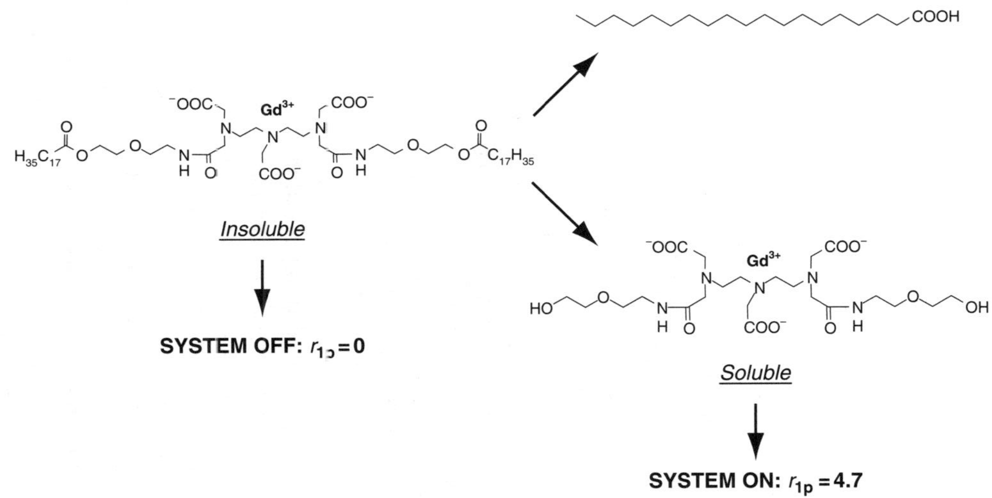

Figure 28.12 The insoluble system is represented by a Gd-DTPA-like complex functionalized with a long aliphatic chain which is bound to the chelate moiety through an ester bond; and the action of the esterase causes the cleavage of the bond between the insoluble moiety and the soluble Gd(III) chelate

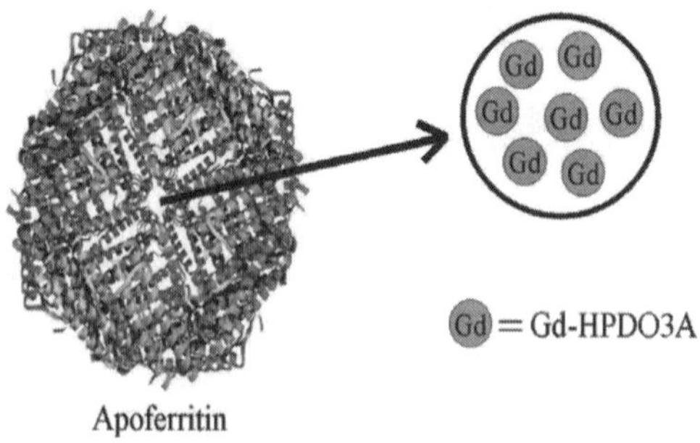

Figure 28.13 Schematic representation of Gd-HPDO3A entrapped into the Apoferritin cavity (see Plate 6)

In order to deal with a system whose structural characteristics were unaltered by the loading with Gd(III) chelates, Apoferritin was chosen because it allows the imaging probes to be entrapped inside its inner cavity.[41] (Figure 28.13) The exterior of such Gd(III)-loaded Apoferritin is exactly the same as in the parent Ferritin and then, once administered intravenously, it is quickly cleared up by the proper receptors on hepatocytes.[42]

In such a system, water can freely diffuse through the channels formed at the intersection of the protein subunits (10 channels), but the larger Gd-HPDO3A molecules cannot. The relaxivity shown by each Gd-HPDO3A entrapped in the Apoferritin cavity is very high (ca. $80\,s^{-1}\,mMGd^{-1}$ at 20 MHz and 25 °C). It has been possible to assess that the Gd-loaded Apoferritin maintains its integrity upon the cell internalization process as the relaxivity observed for the cytoplasmatic extract corresponds to that of the intact system. Finally, the amount of cell-internalized Gd-loaded Apoferritin is similar to that reported for the native Ferritin (6.5×10^6 molecules per cell in 6 h).

Probably, the most common route for the internalization of Gd(III) chelates is through a receptor-mediated endocytosis that is expected to occur any time a given ligand is modified by attaching one or more Gd-chelating units. Thus, the internalization process is no longer the one followed by the native ligand but the binding to the receptor stimulates an endocytotic process which starts with the folding of a portion of cellular membrane and ends up with the welding of its extremities. Through this process, a number of substances remain entrapped in endosomic vesicles, primarily the molecules bound to the membrane or in close proximity to the portion of membrane involved in the endosome formation.

Weiner *et al.*[43] showed that the uptake of a folate-conjugate dendrimer into tumour cells over expressing high-affinity Folate Receptor (hFR) occurs through this type of endocytotic pathway. Another example of receptor mediated endocytosis of a large dendrimer has been recently reported.[44] The macromolecular system comprising Avidin and a biotinilated dendrimer bearing 254 Gd-DTPA chelates (AV-G6Gd) has shown to accumulate '*in vitro*' into SHIN3 cells (a cell line obtained originally from human ovarian cancer) 50-fold greater than Gd-DTPA. The internalization process is driven by an Avidin molecule, a glycoprotein that binds to β-D-galactose receptors which are present on these tumour cells.

28.5 Concluding Remarks

The results obtained in the last two decades have shown the huge potential that paramagnetic agents may have in the field of MRI applications. Interestingly, the relaxivity is a property that can indicate the interactions the paramagnetic agent has in the microenvironment in which it distributes. This peculiar behaviour assigns a unique place to this class of agents among the imaging probes and deserves a bright future in the growing field of molecular imaging. We expect a further impact of these agents as chemists will design new systems endowed with improved relaxivity, responsiveness and targeting capabilities for the visualization of the outset of diseases and the follow-up of therapies as well.

References

1. Young, I.R., Ed. *Methods in Biomedical Magnetic Resonance Imaging and Spectroscopy*; John Wiley & Sons, Chichester, **2000**.
2. Rinck, P.A., *Magnetic Resonance in Medicine*; ABW Wissenschaftsverlag GmbH, Berlin, **2003**.
3. Tweedle, M.F., in *Lanthanide Probes in Life, Chemical and Earth Sciences: Theory and Practice* (J.-C.G. Bünzli and G.R. Choppin, Eds), Elsevier, Amsterdam, **1989**, pp. 127–179.
4. Brücher, E., Sherry, A.D., in *The Chemistry of Contrast Agents in Medical Magnetic Resonance Imaging* (A.E. Merbach and E. Tóth, Eds), Wiley, Chichester, **2001**, pp. 243–280.
5. Weinmann, H.-J., Mühler, A., Radüchel, B., in *Biomedical Magnetic Resonance Imaging and Spectroscopy* (I.R. Young, Ed.), John Wiley & Sons Ltd, Chichester, **2000**, pp. 705–712.
6. Caravan, P., Ellison, J.J., McMurry, T.J., Lauffer, R.B., *Chem. Rev.*, **1999**, *99*, pp. 2293–2352.
7. Schmitt-Willich, H., Brehm, M., Evers, C.L.J., *et al.*, *Inorg. Chem.*, **1999**, *38*, 1134–1144.
8. Uggeri, F., Aime, S., Anelli, P.L., *et al.*, *Inorg. Chem.*, **1995**, *34*, 633–642.
9. Rummeny, E.J., Marchal, G., *Acta Radiol.*, **1997**, *38*, 626–630.
10. Pautler, R.G., Mongeau, R., Jacobs, R.E., *Magn. Res. Med.*, **2003**, *50*, 33–39.
11. Narita, K., Kawasaki, F., Kita, H., *Brain Res.*, **1990**, *510*, 740–748.
12. Wang, Y.X.J., Hussain, S.M., Krestin, G.P., *Eur. Radiol.*, **2001**, *11*, 2319–2331.
13. Reimer, P., Balzer, T., *Eur. Radiol.*, **2003**, *13*, 1266–1276.

14. Koenig, S.H., Brown, R.D., *Prog. NMR Spectrosc.*, **1990**, *22*, 487–567.
15. Botta, M., *Eur. J. Inorg. Chem.*, **2000**, *3*, 399–407.
16. (a) Solomon, I., *Phys. Rev.*, **1995**, *99*, 559. (b) Bloembergen, N.J., *Chem. Phys.*, **1957**, *27*, 572. (c) Bloembergen, N., Morgan, L.O.J., *Chem. Phys.*, **1961**, *34*, 842–850.
17. Aime, S., Gianolio, E., Terreno, E., *et al.*, *J. Biol. Inorg. Chem.*, **2000**, *5*, 488–497.
18. Hajela, S., Botta, M., Giraudo, S., *et al.*, *JACS*, **2000**, *122*, 11228–11229.
19. Aime, S., Botta, M., Frullano, L., *et al.*, *J. Med. Chem.*, **2000**, *43*, 4017–4024.
20. Aime, S., Botta, M., Fasano, M., *et al.*, E. In Merbach, A.E., Tóth, E. (Eds) *The Chemistry of Contrast Agents in Medical Magnetic Resonance Imaging*. Chichester, John Wiley & Sons, LTD, **2001**, p. 193.
21. Lauffer, R.B., Parmalee, D.J., Dunham S.U., *et al.*, *Radiology*, **1998**, *207*, 529.
22. Aime, S., Chiaussa, M., Digilio, G., *et al.*, *J. Biol. Inorg. Chem.*, **1999**, *4*, 766–774.
23. Zhang, S.R., Merritt, M., Woessner, D.E., *et al.*, *Acc. Chem. Res.* **2003**, *36*, 783–790.
24. Aime, S., Barge, A., Castelli, D.D., *et al.*, *Magn. Res. Med.*, **2002**, *47*, 639–648.
25. Aime, S., Castelli, D.D., Terreno, E., *Angew. Chem. Int. Ed.*, **2002**, *41*, 4334–4336.
26. Lowe, M.P., Parker, D., Reany, O., *et al.*, *J. Am. Chem. Soc.*, **2001**, *123*, 7601–7609.
27. Zhang, S.R., Wu, K.C., Sherry, A.D., *Angew. Chem. Int. Ed.*, **1999**, *38*, 3192–3194.
28. Raghunand, N., Howison, C., Sherry, A.D., *et al.*, *Magn. Res. Med.*, **2003**, *49*, 249–257.
29. Aime, S., Botta, M., Geninatti, S., *et al.*, *Chem. Commun.*, **1999**, *16*, 1577–1578.
30. Aime, S., Ascenzi, P., Comoglio, E., *et al.*, *J. Am. Chem. Soc.*, **1995**, *117*, 9365–9366.
31. Glogard, C., Stensrud, G., Aime, S., *Magn. Res. Chem.*, **2003**, *41*, 585–588.
32. Anelli, P.L., Bertini, I., Fragai, M., *et al.*, *Eur. J. Inorg. Chem.*, **2000**, *4*, 625–630.
33. Lauffer, R.B., Mc Murry, T.J., Dunham, S.O., *et al.*, *PCT Int. Appl.*, WO9736619, **1997**.
34. Nivorozhkin, A.L., Kolodziej, A.F., Caravan, P., *et al.*, *Angew. Chem. Int. Ed.*, **2001**, *40*, 2903–2906.
35. Moats, R.A., Fraser, S.E., Meade, T.J., *Angew. Chem. Int. Ed.*, **1997**, *36*, 726–728.
36. Aime, S., Cabella, C., Colombatto, S., *et al.*, *JMRI*, **2002**, *16*, 394–406.
37. Sipkins, D.A., Cheresh, D.A., Kazemi, M.R., *et al.*, *Nat. Med.*, **1998**, *4*, 623–626.
38. Winter, P.M., Caruthers, S.D., Kassner A., *et al.*, *Cancer Res.*, **2003**, *63*, 5838–5843.
39. Bhujwalla, Z.M., Artemov, D., Mori, N., Ravi, R., *Cancer Res.*, **2003**, *63*, 2723–2727.
40. Aime, S., Crich, S.G., Camussi, G., *et al.*, *Magn. Reson. in Med.*, **2004**, *51*, 938–944.
41. Aime, S., Frullano, L., Crich, S.G., *Angew. Chem. Int. Ed.*, **2002**, *41*, 1017–1019.
42. Osterloh, K., Aisen, P., *Biochim. Biophys. Acta*, **1989**, *1011*, 40–45.
43. Wiener, E.C., Konda, S., Shadron, A., *et al.*, *Invest. Radiol.*, **1997**, *32*, 748–754.
44. Kobayashi, H., Brechbiel, M.W., *Mol. Imaging*, **2003**, *2*, 1–10.

Index

Note: Page numbers in *Italics* refer to figures; those in **bold** refer to tables

Index compiled by L.N. Derrick